Texts in Computational Science and Engineering

6

Editors

Timothy J. Barth
Michael Griebel
David E. Keyes
Risto M. Nieminen
Dirk Roose
Tamar Schlick

More information about this series at http://www.springer.com/series/5151

Hans Petter Langtangen

A Primer on Scientific Programming with Python

5th Edition

 Springer

Hans Petter Langtangen
Simula Research Laboratory
Fornebu, Norway

On leave from:

Department of Informatics
University of Oslo
Oslo, Norway

ISSN 1611-0994
Texts in Computational Science and Engineering
ISBN 978-3-662-49886-6   ISBN 978-3-662-49887-3 (eBook)
DOI 10.1007/978-3-662-49887-3
Springer Heidelberg Dordrecht London New York

Library of Congress Control Number: 2016945366

Mathematic Subject Classification to (2010): 26-01, 34A05, 34A30, 34A34, 39-01, 40-01, 65D15, 65D25, 65D30, 68-01, 68N01, 68N19, 68N30, 70-01, 92D25, 97-04, 97U50

Printed on acid-free paper

This Springer imprint is published by Springer Nature
The registered company is Springer Berlin Heidelberg

Preface

The aim of this book is to teach computer programming using examples from mathematics and the natural sciences. We have chosen to use the Python programming language because it combines remarkable expressive power with very clean, simple, and compact syntax. Python is easy to learn and very well suited for an introduction to computer programming. Python is also quite similar to MATLAB and a good language for doing mathematical computing. It is easy to combine Python with compiled languages, like Fortran, C, and C++, which are widely used languages for scientific computations.

The examples in this book integrate programming with applications to mathematics, physics, biology, and finance. The reader is expected to have knowledge of basic one-variable calculus as taught in mathematics-intensive programs in high schools. It is certainly an advantage to take a university calculus course in parallel, preferably containing both classical and numerical aspects of calculus. Although not strictly required, a background in high school physics makes many of the examples more meaningful.

Many introductory programming books are quite compact and focus on listing functionality of a programming language. However, learning to program is learning how to *think* as a programmer. This book has its main focus on the thinking process, or equivalently: programming as a problem solving technique. That is why most of the pages are devoted to case studies in programming, where we define a problem and explain how to create the corresponding program. New constructions and programming styles (what we could call theory) is also usually introduced via examples. Particular attention is paid to verification of programs and to finding errors. These topics are very demanding for mathematical software, because the unavoidable numerical approximation errors are possibly mixed with programming mistakes.

By studying the many examples in the book, I hope readers will learn how to think right and thereby write programs in a quicker and more reliable way. Remember, nobody can learn programming by just reading – one has to solve a large amount of exercises hands on. The book is therefore full of exercises of various types: modifications of existing examples, completely new problems, or debugging of given programs.

To work with this book, I recommend using Python version 2.7. For Chaps. 5–9 and Appendices A–E, you need the NumPy and Matplotlib packages, preferably also the IPython and SciTools packages, and for Appendix G, Cython is required. Other packages used in the text are nose and `sympy`. Section H.1 has more information on how you can get access to Python and the mentioned packages.

There is a web page associated with this book, http://hplgit.github.io/sciprimer, containing all the example programs from the book as well as information on installation of the software on various platforms.

Python version 2 or 3? A common problem among Python programmers is to choose between version 2 or 3, which at the time of this writing means choosing between version 2.7 and 3.5. A common recommendation is to go for Python 3, because this is the version that will be further developed in the future. However, there is a problem that much useful mathematical software in Python has not yet been ported to Python 3. Therefore, Python version 2.7 is the most popular version for doing scientific computing, and that is why also this book applies version 2.7.

A widely used strategy for software developers who want to write Python code that works with both versions, is to develop a common version for Python 2 and 3. For the programs in this book, a common version can easily be produced by first developing for version 2.7 and then *automatically* convert the code by running the `futurize` program. Section 4.10 demonstrates how this is done in simple cases.

The Python 2.7 code in this book sticks to all modern constructions that are backported from version 3 such that the code becomes as close as possible to the equivalent Python 3 code. At any time, you can just run `futurize` to see the differences between your Python 2.7 version and the corresponding Python 3.5 version.

Contents Chapter 1 introduces variables, objects, modules, and text formatting through examples concerning evaluation of mathematical formulas. Chapter 2 presents programming with `while` and `for` loops as well as lists, including nested lists. The next chapter deals with two other fundamental concepts in programming: functions and `if-else` tests.

How to read data into programs and deal with errors in input are the subjects of Chap. 4. Chapter 5 introduces arrays and array computing (including vectorization) and how this is used for plotting $y = f(x)$ curves and making animation of curves. Many of the examples in the first five chapters are strongly related. Typically, formulas from the first chapter are used to produce tables of numbers in the second chapter. Then the formulas are encapsulated in functions in the third chapter. In the next chapter, the input to the functions are fetched from the command line, and validity checks of the input are added. The formulas are then shown as graphs in Chap. 5. After having studied Chaps. 1–5, the reader should have enough knowledge of programming to solve mathematical problems by what many refer to as "MATLAB-style" programming.

Chapter 6 explains how to work with dictionaries and strings, especially for interpreting text data in files and storing the extracted information in flexible data structures. Class programming, including user-defined types for mathematical computations (with overloaded operators), is introduced in Chap. 7. Chapter 8 deals with random numbers and statistical computing with applications to games and random walks. Object-oriented programming, in the meaning of class hierarchies and inheritance, is the subject of Chap. 9. The key examples here deal with building toolkits for numerical differentiation and integration as well as graphics.

Appendix A introduces mathematical modeling, using sequences and difference equations. Only programming concepts from Chaps. 1–5 are used in this appendix, the aim being to consolidate basic programming knowledge and apply it to mathematical problems. Some important mathematical topics are introduced via difference equations in a simple way: Newton's method, Taylor series, inverse functions, and dynamical systems.

Appendix B deals with functions on a mesh, numerical differentiation, and numerical integration. A simple introduction to ordinary differential equations and their numerical treatment is provided in Appendix C. Appendix D shows how a complete project in physics can be solved by mathematical modeling, numerical methods, and programming elements from Chaps. 1–5. This project is a good example on problem solving in computational science, where it is necessary to integrate physics, mathematics, numerics, and computer science.

How to create software for solving ordinary differential equations, using both function-based and object-oriented programming, is the subject of Appendix E. The material in this appendix brings together many parts of the book in the context of physical applications and differential equations.

Appendix F is devoted to the art of debugging, and in fact problem solving in general. Speeding up numerical computations in Python by migrating code to C via Cython is exemplified in Appendix G. Finally, Appendix H deals with various more advanced technical topics.

Most of the examples and exercises in this book are quite short. However, many of the exercises are related, and together they form larger projects, for example on Fourier Series (3.21, 4.21, 4.22, 5.41, 5.42), numerical integration (3.11, 3.12, 5.49, 5.50, A.12), Taylor series (3.37, 5.32, 5.39, A.14, A.15, 7.23), piecewise constant functions (3.29–3.33, 5.34, 5.47, 5.48, 7.19–7.21), inverse functions (E.17–E.20), falling objects (E.8, E.9, E.38, E.39), oscillatory population growth (A.19, A.21, A.22, A.23), epidemic disease modeling (E.41–E.48), optimization and finance (A.24, 8.42, 8.43), statistics and probability (4.24, 4.25, 8.23, 8.24), hazard games (8.8–8.14), random walk and statistical physics (8.32–8.40), noisy data analysis (8.44–8.46), numerical methods (5.25–5.27, 7.8, 7.9, A.9, 7.22, 9.15–9.17, E.30–E.37), building a calculus calculator (7.34, 9.18, 9.19), and creating a toolkit for simulating vibrating engineering systems (E.50–E.55).

Chapters 1–9 together with Appendices A and E have from 2007 formed the core of an introductory first semester bachelor course on scientific programming at the University of Oslo (INF1100, 10 ECTS credits).

Changes from the fourth to the fifth edition Substantial changes were introduced in the fourth edition, and the fifth edition is primarily a consolidation of those changes. Many typos have been corrected and many explanations and exercises have been improved. The emphasis on unit tests and test functions, especially in exercises, is stronger than in the previous edition. Symbolic computation with the aid of SymPy is used to a larger extent and integrated with numerical computing throughout the book. All classes are now new-style (instead of old-style/classic as in previous editions). Examples on Matplotlib do not use the `pylab` module anymore, but `pyplot` and MATLAB-like syntax is still favored to ease the transition between Python and MATLAB. The concept of closures is more explicit than in earlier editions (see the new Sect. 7.1.7) since this is a handy and popular construction much used in the scientific Python community. We also discuss the difference between Python 2 and 3 and demonstrate how to use the `future` module to write code that runs under both versions.

The most substantial new material in the fifth edition appears toward the end of Chap. 5 and regards high-performance computing, linear algebra, and visualization of scalar and vector fields. Although this material is not used elsewhere in the book, many readers have requested basic recipes when going from one to two variables or from vectors to matrices later when solving more advanced problems and using the book as their programming reference. The new matrial in Chap. 5 was written jointly with Dr. Øyvind Ryan.

Acknowledgments This book was born out of stimulating discussions with my close colleague Aslak Tveito, and he started writing what is now Appendix B and C. The whole book project and the associated university course were critically dependent on Aslak's enthusiastic role back in 2007. The continuous support from Aslak regarding my book projects is much appreciated and contributes greatly to my strong motivation. Another key contributor in the early days was Ilmar Wilbers. He made extensive efforts with assisting the book project and establishing the university course INF1100. I feel that without Ilmar and his solutions to numerous technical problems the first edition of the book would never have been completed. Johannes H. Ring also deserves special acknowledgment for the development of the Easyviz graphics tool back in the days when Python plotting was a hassle, and later for his maintenance of software associated with the book.

Professor Loyce Adams studied the entire book, solved all the exercises, found numerous errors, and suggested many improvements. Her contributions are so much appreciated. More recently, Helmut Büch worked extremely carefully through all details in Chaps. 1–6, tested the software, found many typos, and asked critical questions that led to lots of significant improvements. I am so thankful for all his efforts and for his enthusiasm during the preparations of the fourth edition. The fifth edition has benefited much from Hakki Eres' careful examination of the fourth edition. He found several typos and code errors, some of which go back to the first edition.

Special thanks go to Geir Kjetil Sandve for being the primary author of the computational bioinformatics examples in Sects. 3.3, 6.5, 8.3.4, and 9.5, with contributions from Sveinung Gundersen, Ksenia Khelik, Halfdan Rydbeck, and Kai Trengereid. I am also greatful to Øyvind Ryan's work with linear algebra and visualization of scalar and vector fields in Chap. 5.

Several people have contributed with suggestions for improvements of the text, the exercises, and the associated software. I will in particular mention Ingrid Eide, Ståle Zerener Haugnæss, Kristian Hiorth, Timothy Keough, Arve Knudsen, Espen Kristensen, Tobias Vidarssønn Langhoff, Martin Vonheim Larsen, Kine Veronica Lund, Solveig Masvie, Håkon Møller, Rebekka Mørken, Mathias Nedrebø, Marit Sandstad, Helene Norheim Semmerud, Lars Storjord, Fredrik Heffer Valdmanis, and Torkil Vederhus. Hakon Adler is greatly acknowledged for his careful reading of early various versions of the manuscript. Many thanks go to the professors Fred Espen Bent, Ørnulf Borgan, Geir Dahl, Knut Mørken, and Geir Pedersen for formulating several exciting exercises from various application fields. I also appreciate the cover image made by my good friend Jan Olav Langseth.

This book and the associated course are parts of a comprehensive and successful reform at the University of Oslo, called *Computing in Science Education*. The goal of the reform is to integrate computer programming and simulation in all bachelor courses in natural science where mathematical models are used. The present book lays the foundation for the modern computerized problem solving technique to be applied in later courses. It has been extremely inspiring to work closely with the driving forces behind this reform, especially the professors Morten Hjorth-Jensen, Anders Malthe-Sørenssen, Knut Mørken, and Arnt Inge Vistnes.

The excellent assistance from the Springer system over the years, in particular Martin Peters, Thanh-Ha Le Thi, Ruth Allewelt, Peggy Glauch-Ruge, Nadja Kroke, Thomas Schmidt, Patrick Waltemate, Donatas Akmanavicius, and Yvonne Schlatter, is highly appreciated, and ensured a smooth and rapid production of all editions of this book.

Oslo, February 2016   Hans Petter Langtangen

Contents

List of Exercises

Computing with Formulas

Our first examples on computer programming involve programs that evaluate mathematical formulas. You will learn how to write and run a Python program, how to work with variables, how to compute with mathematical functions such as e^x and $\sin x$, and how to use Python for interactive calculations.

We assume that you are somewhat familiar with computers so that you know what files and folders are (another frequent word for folder is directory), how you move between folders, how you change file and folder names, and how you write text and save it in a file.

All the program examples associated with this chapter can be downloaded as a tarfile or zipfile from the web page http://hplgit.github.com/scipro-primer. I strongly recommend you to visit this page, download and pack out the files. The examples are organized in a folder tree with src as root. Each subfolder corresponds to a particular chapter. For example, the subfolder formulas contains the program examples associated with this first chapter. The relevant subfolder name is listed at the beginning of every chapter.

The folder structure with example programs can also be directly accessed in a GitHub repository[1] on the web. You can click on the formulas folder to see all the examples from the present chapter. Clicking on a filename shows a nicely typeset version of the file. The file can be downloaded by first clicking *Raw* to get the plain text version of the file, and then right-clicking in the web page and choosing *Save As....*

1.1 The First Programming Encounter: a Formula

The first formula we shall consider concerns the vertical motion of a ball thrown up in the air. From Newton's second law of motion one can set up a mathematical model for the motion of the ball and find that the vertical position of the ball, called y, varies with time t according to the following formula:

$$y(t) = v_0 t - \frac{1}{2} g t^2. \tag{1.1}$$

---

[1] http://tinyurl.com/pwyasaa

© Springer-Verlag Berlin Heidelberg 2016
H.P. Langtangen, *A Primer on Scientific Programming with Python*,
Texts in Computational Science and Engineering 6, DOI 10.1007/978-3-662-49887-3_1

Here, v_0 is the initial velocity of the ball, g is the acceleration of gravity, and t is time. Observe that the y axis is chosen such that the ball starts at $y = 0$ when $t = 0$. The above formula neglects air resistance, which is usually small unless v_0 is large, see Exercise 1.11.

To get an overview of the time it takes for the ball to move upwards and return to $y = 0$ again, we can look for solutions to the equation $y = 0$:

$$v_0 t - \frac{1}{2} g t^2 = t(v_0 - \frac{1}{2} g t) = 0 \quad \Rightarrow \quad t = 0 \text{ or } t = 2v_0/g.$$

That is, the ball returns after $2v_0/g$ seconds, and it is therefore reasonable to restrict the interest of (1.1) to $t \in [0, 2v_0/g]$.

1.1.1 Using a Program as a Calculator

Our first program will evaluate (1.1) for a specific choice of v_0, g, and t. Choosing $v_0 = 5$ m/s and $g = 9.81$ m/s^2 makes the ball come back after $t = 2v_0/g \approx 1$ s. This means that we are basically interested in the time interval $[0, 1]$. Say we want to compute the height of the ball at time $t = 0.6$ s. From (1.1) we have

$$y = 5 \cdot 0.6 - \frac{1}{2} \cdot 9.81 \cdot 0.6^2 \qquad (1.2)$$

This arithmetic expression can be evaluated and its value can be printed by a very simple one-line Python program:

```
print 5*0.6 - 0.5*9.81*0.6**2
```

The four standard arithmetic operators are written as +, -, *, and / in Python and most other computer languages. The exponentiation employs a double asterisk notation in Python, e.g., 0.6^2 is written as 0.6**2.

Our task now is to create the program and run it, and this will be described next.

1.1.2 About Programs and Programming

A computer program is just a sequence of instructions to the computer, written in a computer language. Most computer languages look somewhat similar to English, but they are very much simpler. The number of words and associated instructions is very limited, so to perform a complicated operation we must combine a large number of different types of instructions. The program text, containing the sequence of instructions, is stored in one or more files. The computer can only do exactly what the program tells the computer to do.

Another perception of the word *program* is a file that can be run ("double-clicked") to perform a task. Sometimes this is a file with textual instructions (which is the case with Python), and sometimes this file is a translation of all the program text to a more efficient and computer-friendly language that is quite difficult to read for a human. All the programs in this chapter consist of short text stored in a single file. Other programs that you have used frequently, for instance Firefox or Internet Explorer for reading web pages, consist of program text distributed over a large number of files, written by a large number of people over many years. One single file contains the machine-efficient translation of the whole program, and this is normally the file that you double-click on when starting the program. In general, the word program means either this single file or the collection of files with textual instructions.

Programming is obviously about writing programs, but this process is more than writing the correct instructions in a file. First, we must understand how a problem can be solved by giving a sequence of instructions to the computer. This is one of the most difficult things with programming. Second, we must express this sequence of instructions correctly in a computer language and store the corresponding text in a file (the program). This is normally the easiest part. Third, we must find out how to check the validity of the results. Usually, the results are not as expected, and we need to a fourth phase where we systematically track down the errors and correct them. Mastering these four steps requires a lot of training, which means making a large number of programs (exercises in this book, for instance!) and getting the programs to work.

1.1.3 Tools for Writing Programs

There are three alternative types of tools for writing Python programs:

- a plain text editor
- an integrated development environment (IDE) with a text editor
- an IPython notebook

What you choose depends on how you access Python. Section H.1 contains information on the various possibilities to install Python on your own computer, access a pre-installed Python environment on a computer system at an institution, or access Python in cloud services through your web browser.

Based on teaching this and previous books to more than 3000 students, my recommendations go as follows.

- If you use this book in a course, the instructor has probably made a choice for how you should access Python – follow that advice.
- If you are a student at a university where Linux is the dominating operating system, install a virtual machine with Ubuntu on your own laptop and do all your scientific work in Ubuntu. Write Python programs in a text editor like Gedit, Atom, Sublime Text, Emacs, or Vim, and run programs in a terminal window (the `gnome-terminal` is recommended).
- If you are a student a university where Windows is the dominating operating system, and you are a Windows user yourself, install Anaconda. Write and run Python programs in Spyder.
- If you are uncertain how much you will program with Python and primarily want to get a taste of Python programming first, access Python in the cloud, e.g., through the Wakari site.

- If you want Python on your Mac and you are experienced with compiling and linking software in the Mac OS X environment, install Anaconda on the Mac. Write and run programs in Spyder, or use a text editor like Atom, TextWrangler, Emacs, or Vim, and run programs in the Terminal application. If you are not very familiar with building software on the Mac, and with environment variables like PATH, it will be easier in the long run to access Python in Ubuntu through a virtual machine.

1.1.4   Writing and Running Your First Python Program

I assume that you have made a decision on how to access Python, which dictates whether you will be writing programs in a text editor or in an IPython notebook. What you write will be the same – the difference lies in how you run the program. Sections H.2 and H.4 briefly describe how to write programs in a text editor, run them in a terminal window or in Spyder, and how to operate an IPython notebook. I recommend taking a look at that material before proceeding.

Open up your chosen text editor and write the following line:

```
print 5*0.6 - 0.5*9.81*0.6**2
```

This is a complete Python program for evaluating the formula (1.2). Save the line to a file with name ball1.py.

The action required to run this program depends on what type of tool you use for running programs:

- terminal window: move to the folder where ball1.py is located and type python ball1.py
- IPython notebook: click on the "play" button to execute the cell
- Spyder: choose *Run* from the *Run* pull-down menu

The output is 1.2342 and appears

- right after the python ball1.py command in a terminal window
- right after the program line (cell) in the IPython notebook
- in the lower right window in Spyder

We remark that there are other ways of running Python programs in the terminal window, see Appendix H.5.

Suppose you want to evaluate (1.1) for $v_0 = 1$ and $t = 0.1$. This is easy: move the cursor to the editor window, edit the program text to

```
print 1*0.1 - 0.5*9.81*0.1**2
```

Run the program again in Spyder or re-execute the cell in an IPython notebook. If you use a plain text editor, always remember to save the file after editing it, then move back to the terminal window and run the program as before:

```
Terminal> python ball1.py
0.05095
```

The result of the calculation has changed, as expected.

Typesetting of operating system commands

We use the prompt `Terminal>` in this book to indicate commands in a Unix or DOS/PowerShell terminal window. The text following the `Terminal>` prompt must be a valid operating system command. You will likely see a different prompt in the terminal window on your machine, perhaps something reflecting your username or the current folder.

1.1.5   Warning About Typing Program Text

Even though a program is just a text, there is one major difference between a text in a program and a text intended to be read by a human. When a human reads a text, she or he is able to understand the message of the text even if the text is not perfectly precise or if there are grammar errors. If our one-line program was expressed as

```
write 5*0.6 - 0.5*9.81*0.6^2
```

most humans would interpret `write` and `print` as the same thing, and many would also interpret `6^2` as 6^2. In the Python language, however, `write` is a grammar error and `6^2` means an operation very different from the exponentiation `6**2`. Our communication with a computer through a program must be perfectly precise without a single grammar or logical error. The famous computer scientist Donald Knuth put it this way:

> *Programming demands significantly higher standard of accuracy. Things don't simply have to make sense to another human being, they must make sense to a computer.* Donald Knuth [11, p. 18], 1938-.

That is, the computer will only do exactly what we tell it to do. Any error in the program, however small, may affect the program. There is a chance that we will never notice it, but most often an error causes the program to stop or produce wrong results. The conclusion is that computers have a much more pedantic attitude to language than what (most) humans have.

Now you understand why any program text must be carefully typed, paying attention to the correctness of every character. If you try out program texts from this book, make sure that you type them in *exactly as you see them* in the book. Blanks, for instance, are often important in Python, so it is a good habit to always count them and type them in correctly. Any attempt not to follow this advice will cause you frustrations, sweat, and maybe even tears.

1.1.6   Verifying the Result

We should *always* carefully control that the output of a computer program is correct. You will experience that in most of the cases, at least until you are an experienced programmer, the output is wrong, and you have to search for errors. In the present application we can simply use a calculator to control the program. Setting $t = 0.6$ and $v_0 = 5$ in the formula, the calculator confirms that 1.2342 is the correct solution to our mathematical problem.

1.1.7   Using Variables

When we want to evaluate $y(t)$ for many values of t, we must modify the t value at two places in our program. Changing another parameter, like v_0, is in principle straightforward, but in practice it is easy to modify the wrong number. Such modifications would be simpler to perform if we express our formula in terms of variables, i.e., symbols, rather than numerical values. Most programming languages, Python included, have variables similar to the concept of variables in mathematics. This means that we can define v0, g, t, and y as variables in the program, initialize the former three with numerical values, and combine these three variables to the desired right-hand side expression in (1.1), and assign the result to the variable y.

The alternative version of our program, where we use variables, may be written as this text:

```
v0 = 5
g = 9.81
t = 0.6
y = v0*t - 0.5*g*t**2
print y
```

Variables in Python are defined by setting a name (here v0, g, t, or y) equal to a numerical value or an expression involving already defined variables.

Note that this second program is much easier to read because it is closer to the mathematical notation used in the formula (1.1). The program is also safer to modify, because we clearly see what each number is when there is a name associated with it. In particular, we can change t at one place only (the line t = 0.6) and not two as was required in the previous program.

We store the program text in a file `ball2.py`. Running the program results in the correct output 1.2342.

1.1.8   Names of Variables

Introducing variables with descriptive names, close to those in the mathematical problem we are going to solve, is considered important for the readability and reliability (correctness) of the program. Variable names can contain any lower or upper case letter, the numbers from 0 to 9, and underscore, but the first character cannot be a number. Python distinguishes between upper and lower case, so X is always different from x. Here are a few examples on alternative variable names in the present example:

```
initial_velocity = 5
acceleration_of_gravity = 9.81
TIME = 0.6
VerticalPositionOfBall = initial_velocity*TIME - \
                         0.5*acceleration_of_gravity*TIME**2
print VerticalPositionOfBall
```

With such long variables names, the code for evaluating the formula becomes so long that we have decided to break it into two lines. This is done by a backslash at the very end of the line (make sure there are no blanks after the backslash!).

In this book we shall adopt the convention that variable names have lower case letters where words are separated by an underscore. Whenever the variable represents a mathematical symbol, we use the symbol or a good approximation to it as variable name. For example, y in mathematics becomes y in the program, and v_0 in mathematics becomes v0 in the program. A close resemblance between mathematical symbols in the description of the problem and variables names is important for easy reading of the code and for detecting errors. This principle is illustrated by the code snippet above: even if the long variable names explain well what they represent, checking the correctness of the formula for y is harder than in the program that employs the variables v0, g, t, and y0.

For all variables where there is no associated precise mathematical description and symbol, one must use *descriptive* variable names which explain the purpose of the variable. For example, if a problem description introduces the symbol D for a force due to air resistance, one applies a variable D also in the program. However, if the problem description does not define any symbol for this force, one must apply a descriptive name, such as `air_resistance`, `resistance_force`, or `drag_force`.

How to choose variable names
- Use the same variable names in the program as in the mathematical description of the problem you want to solve.
- For all variables without a precise mathematical definition and symbol, use a carefully chosen descriptive name.

1.1.9   Reserved Words in Python

Certain words are reserved in Python because they are used to build up the Python language. These reserved words cannot be used as variable names: and, as, assert, break, class, continue, def, del, elif, else, except, False, finally, for, from, global, if, import, in, is, lambda, None, nonlocal, not, or, pass, raise, return, True, try, with, while, and yield. If you wish to use a reserved word as a variable name, it is common to an underscore at the end. For example, if you need a mathematical quantity λ in the program, you may work with `lambda_` as variable name. See Exercise 1.16 for examples on legal and illegal variable names.

Program files can have a freely chosen name, but stay away from names that coincide with keywords or module names in Python. For instance, do not use `math.py`, `time.py`, `random.py`, `os.py`, `sys.py`, `while.py`, `for.py`, `if.py`, `class.py`, or `def.py`.

1.1.10 Comments

Along with the program statements it is often informative to provide some comments in a natural human language to explain the idea behind the statements. Comments in Python start with the # character, and everything after this character on a line is ignored when the program is run. Here is an example of our program with explanatory comments:

```
# Program for computing the height of a ball in vertical motion.
v0 = 5     # initial velocity
g = 9.81   # acceleration of gravity
t = 0.6    # time
y = v0*t - 0.5*g*t**2  # vertical position
print y
```

This program and the initial version in Sect. 1.1.7 are identical when run on the computer, but for a human the latter is easier to understand because of the comments.

Good comments together with well-chosen variable names are necessary for any program longer than a few lines, because otherwise the program becomes difficult to understand, both for the programmer and others. It requires some practice to write really instructive comments. Never repeat with words what the program statements already clearly express. Use instead comments to provide important information that is not obvious from the code, for example, what mathematical variable names mean, what variables are used for, a quick overview of a set of forthcoming statements, and general ideas behind the problem solving strategy in the code.

Remark If you use non-English characters in your comments, Python will complain with error messages like

```
SyntaxError: Non-ASCII character '\xc3' in file ...
but no encoding declared; see
http://www.python.org/peps/pep-0263.html for details
```

Non-English characters are allowed if you put the following magic line in the program before such characters are used:

```
# -*- coding: utf-8 -*-
```

(Yes, this is a comment, but it is *not* ignored by Python!) More information on non-English characters and encodings like UTF-8 is found in Sect. 6.3.5.

1.1.11   Formatting Text and Numbers

Instead of just printing the numerical value of y in our introductory program, we now want to write a more informative text, typically something like

```
At t=0.6 s, the height of the ball is 1.23 m.
```

where we also have control of the number of digits (here *y* is accurate up to centimeters only).

Printf syntax The output of the type shown above is accomplished by a `print` statement combined with some technique for formatting the numbers. The oldest and most widely used such technique is known as *printf* formatting (originating from the function `printf` in the C programming language). For a newcomer to programming, the syntax of printf formatting may look awkward, but it is quite easy to learn and very convenient and flexible to work with. The printf syntax is used in a lot of other programming languages as well.

   The sample output above is produced by this statement using printf syntax:

```
print 'At t=%g s, the height of the ball is %.2f m.' % (t, y)
```

Let us explain this line in detail. The `print` statement prints a string: everything that is enclosed in quotes (either single: ', or double: ") denotes a string in Python. The string above is formatted using printf syntax. This means that the string has various "slots", starting with a percentage sign, here %g and %.2f, where variables in the program can be put in. We have two "slots" in the present case, and consequently two variables must be put into the slots. The relevant syntax is to list the variables inside standard parentheses after the string, separated from the string by a percentage sign. The first variable, t, goes into the first "slot". This "slot" has a format specification %g, where the percentage sign marks the slot and the following character, g, is a format specification. The g format instructs the real number to be written as compactly as possible. The next variable, y, goes into the second "slot". The format specification here is .2f, which means a real number written with two digits after the decimal place. The f in the .2f format stands for *float*, a short form for *floating-point number*, which is the term used for a real number on a computer.

   For completeness we present the whole program, where text and numbers are mixed in the output:

```
v0 = 5
g = 9.81
t = 0.6
y = v0*t - 0.5*g*t**2
print 'At t=%g s, the height of the ball is %.2f m.' % (t, y)
```

The program is found in the file `ball_print1.py` in the `src/formulas` folder of the collection of programs associated with this book.

There are many more ways to specify formats. For example, e writes a number in *scientific notation*, i.e., with a number between 1 and 10 followed by a power of 10, as in $1.2432 \cdot 10^{-3}$. On a computer such a number is written in the form `1.2432e-03`. Capital E in the exponent is also possible, just replace e by E, with the result `1.2432E-03`.

For *decimal notation* we use the letter f, as in `%f`, and the output number then appears with digits before and/or after a comma, e.g., `0.0012432` instead of `1.2432E-03`. With the g format, the output will use scientific notation for large or small numbers and decimal notation otherwise. This format is normally what gives most compact output of a real number. A lower case g leads to lower case e in scientific notation, while upper case G implies E instead of e in the exponent.

One can also specify the format as `10.4f` or `14.6E`, meaning in the first case that a float is written in decimal notation with four decimals in a field of width equal to 10 characters, and in the second case a float written in scientific notation with six decimals in a field of width 14 characters.

Here is a list of some important printf format specifications (the program `printf_demo.py` exemplifies many of the constructions):

Format	Meaning
%s	a string
%d	an integer
%0xd	an integer in a field of with x, padded with leading zeros
%f	decimal notation with six decimals
%e	compact scientific notation, e in the exponent
%E	compact scientific notation, E in the exponent
%g	compact decimal or scientific notation (with e)
%G	compact decimal or scientific notation (with E)
%xz	format z right-adjusted in a field of width x
%-xz	format z left-adjusted in a field of width x
%.yz	format z with y decimals
%x.yz	format z with y decimals in a field of width x
%%	the percentage sign % itself

For a complete specification of the possible printf-style format strings, follow the link from the item *printf-style formatting* in the index[2] of the Python Standard Library online documentation.

We may try out some formats by writing more numbers to the screen in our program (the corresponding file is `ball_print2.py`):

```
v0 = 5
g = 9.81
t = 0.6
y = v0*t - 0.5*g*t**2
```

```
print """
At t=%f s, a ball with
initial velocity v0=%.3E m/s
is located at the height %.2f m.
""" % (t, v0, y)
```

Observe here that we use a *triple-quoted* string, recognized by starting and ending with three single or double quotes: ''' or """. Triple-quoted strings are used for text that spans several lines.

In the `print` statement above, we print t in the f format, which by default implies six decimals; v0 is written in the .3E format, which implies three decimals and the number spans as narrow field as possible; and y is written with two decimals in decimal notation in as narrow field as possible. The output becomes

```
Terminal> python ball_print2.py

At t=0.600000 s, a ball with
initial velocity v0=5.000E+00 m/s
is located at the height 1.23 m.
```

You should look at each number in the output and check the formatting in detail.

Format string syntax   Python offers all the functionality of the printf format and much more through a different syntax, often known as *format string syntax*. Let us illustrate this syntax on the one-line output previously used to show the printf construction. The corresponding format string syntax reads

```
print 'At t={t:g} s, the height of the ball is {y:.2f} m.'.format(
    t=t, y=y)
```

The "slots" where variables are inserted are now recognized by curly braces rather than a percentage sign. The name of the variable is listed with an optional colon and format specifier of the same kind as was used for the printf format. The various variables and their values must be listed at the end as shown. This time the "slots" have names so the sequence of variables is not important.

The multi-line example is written as follows in this alternative format:

```
print """
At t={t:f} s, a ball with
initial velocity v0={v0:.3E} m/s
is located at the height {y:.2f} m.
""".format(t=t, v0=v0, y=y)
```

The newline character   We often want a computer program to write out text that spans several lines. In the last example we obtained such output by triple-quoted strings. We could also use ordinary single-quoted strings and a special character for indicating where line breaks should occur. This special character reads \n, i.e., a backslash followed by the letter n. The two `print` statements

```
print """y(t) is
the position of
our ball."""

print 'y(t) is\nthe position of\nour ball'
```

result in identical output:

```
y(t) is
the position of
our ball.
```

1.2 Computer Science Glossary

It is now time to pick up some important words that programmers use when they talk about programming: algorithm, application, assignment, blanks (whitespace), bug, code, code segment, code snippet, debug, debugging, execute, executable, implement, implementation, input, library, operating system, output, statement, syntax, user, verify, and verification. These words are frequently used in English in lots of contexts, yet they have a precise meaning in computer science.

Program and *code* are interchangeable terms. A code/program *segment* is a collection of consecutive statements from a program. Another term with similar meaning is *code snippet*. Many also use the word *application* in the same meaning as program and code. A related term is *source code*, which is the same as the text that constitutes the program. You find the source code of a program in one or more text files. (Note that text files normally have the extension `.txt`, while program files have an extension related to the programming language, e.g., `.py` for Python programs. The content of a `.py` file is, nevertheless, plain text as in a `.txt` file.)

We talk about *running a program*, or equivalently *executing a program* or *executing a file*. The file we execute is the file in which the program text is stored. This file is often called an *executable* or an *application*. The program text may appear in many files, but the executable is just the single file that starts the whole program when we run that file. Running a file can be done in several ways, for instance, by double-clicking the file icon, by writing the filename in a terminal window, or by giving the filename to some program. This latter technique is what we have used so far in this book: we feed the filename to the program `python`. That is, we execute a Python program by executing another program `python`, which interprets the text in our Python program file.

The term *library* is widely used for a collection of generally useful program pieces that can be applied in many different contexts. Having access to good libraries means that you do not need to program code snippets that others have already programmed (most probable in a better way!). There are huge numbers of Python libraries. In Python terminology, the libraries are composed of *modules* and *packages*. Section 1.4 gives a first glimpse of the `math` module, which contains a set of standard mathematical functions for $\sin x$, $\cos x$, $\ln x$, e^x, $\sinh x$, $\sin^{-1} x$, etc. Later, you will meet many other useful modules. Packages are just collections of modules. The standard Python distribution comes with a large number of modules and packages, but you can download many more from the Internet, see in particular `www.python.org/pypi`. Very often, when you encounter a programming task that is likely to occur in many other contexts, you can find a Python module where the job is already done. To mention just one example, say you need to compute how many days there are between two dates. This is a non-trivial task that lots of other programmers must have faced, so it is not a big surprise that Python comes with a module `datetime` to do calculations with dates.

The recipe for what the computer is supposed to do in a program is called *algorithm*. In the examples in the first couple of chapters in this book, the algorithms are so simple that we can hardly distinguish them from the program text itself, but later in the book we will carefully set up an algorithm before attempting to *implement* it in a program. This is useful when the algorithm is much more compact than the resulting program code. The algorithm in the current example consists of three steps:

- initialize the variables v_0, g, and t with numerical values,
- evaluate y according to the formula (1.1),
- print the y value to the screen.

The Python program is very close to this text, but some less experienced programmers may want to write the tasks in English before translating them to Python.

The *implementation* of an algorithm is the process of writing and testing a program. The testing phase is also known as *verification*: After the program text is written we need to *verify* that the program works correctly. This is a very important step that will receive substantial attention in the present book. Mathematical software produce numbers, and it is normally quite a challenging task to verify that the numbers are correct.

An *error* in a program is known as a *bug*, and the process of locating and removing bugs is called *debugging*. Many look at debugging as the most difficult and challenging part of computer programming. We have in fact devoted Appendix F to the art of debugging in this book. The origin of the strange terms bug and debugging can be found in Wikipedia[3].

Programs are built of *statements*. There are many types of statements:

```
v0 = 3
```

is an *assignment* statement, while

```
print y
```

is a *print* statement. It is common to have one statement on each line, but it is possible to write multiple statements on one line if the statements are separated by semi-colon. Here is an example:

```
v0 = 3; g = 9.81; t = 0.6
y = v0*t - 0.5*g*t**2
print y
```

---

[3] http://en.wikipedia.org/wiki/Software_bug#Etymology

Although most newcomers to computer programming will think they understand the meaning of the lines in the above program, it is important to be aware
of some major differences between notation in a computer program and notation
in mathematics. When you see the equality sign = in mathematics, it has a certain
interpretation as an equation ($x + 2 = 5$) or a definition ($f(x) = x^2 + 1$). In a computer program, however, the equality sign has a quite different meaning, and it is
called an *assignment*. The right-hand side of an assignment contains an *expression*,
which is a combination of values, variables, and operators. When the expression is
evaluated, it results in a value that the variable on the left-hand side will refer to.
We often say that the right-hand side value is *assigned* to the variable on the left-
hand side. In the current context it means that we in the first line assign the number
3 to the variable v0, 9.81 to g, and 0.6 to t. In the next line, the right-hand side
expression v0*t - 0.5*g*t**2 is first evaluated, and the result is then assigned
to the y variable.

Consider the assignment statement

```
y = y + 3
```

This statement is mathematically false, but in a program it just means that we evaluate the right-hand side expression and assign its value to the variable y. That is, we
first take the current value of y and add 3. The value of this operation is assigned to
y. The old value of y is then lost.

You may think of the = as an arrow, y <- y+3, rather than an equality sign, to
illustrate that the value to the right of the arrow is stored in the variable to the left of
the arrow. In fact, the R programming language for statistical computing actually
applies an arrow, many old languages (like Algol, Simula, and Pascal) used := to
explicitly state that we are not dealing with a mathematical equality.

An example will illustrate the principle of assignment to a variable:

```
y = 3
print y
y = y + 4
print y
y = y*y
print y
```

Running this program results in three numbers: 3, 7, 49. Go through the program
and convince yourself that you understand what the result of each statement becomes.

A computer program must have correct *syntax*, meaning that the text in the
program must follow the strict rules of the computer language for constructing statements. For example, the syntax of the print statement is the word `print`, followed
by one or more spaces, followed by an expression of what we want to print (a Python
variable, text enclosed in quotes, a number, for instance). Computers are very picky
about syntax! For instance, a human having read all the previous pages may easily
understand what this program does,

```
myvar = 5.2
prinnt Myvar
```

but the computer will find two errors in the last line: `prinnt` is an unknown instruction and `Myvar` is an undefined variable. Only the first error is reported (a syntax error), because Python stops the program once an error is found. All errors that Python finds are easy to remove. The difficulty with programming is to remove the rest of the errors, such as errors in formulas or the sequence of operations.

Blanks may or may not be important in Python programs. In Sect. 2.1.2 you will see that blanks are in some occasions essential for a correct program. Around = or arithmetic operators, however, blanks do not matter. We could hence write our program from Sect. 1.1.7 as

```
v0=3;g=9.81;t=0.6;y=v0*t-0.5*g*t**2;print y
```

This is not a good idea because blanks are essential for easy reading of a program code, and easy reading is essential for finding errors, and finding errors is *the* difficult part of programming. The recommended layout in Python programs specifies one blank around =, +, and -, and no blanks around *, /, and **. Note that the blank after `print` is essential: `print` is a command in Python and `printy` is not recognized as any valid command. (Python will complain that `printy` is an undefined variable.) Computer scientists often use the term *whitespace* when referring to a blank. (To be more precise, blank is the character produced by the space bar on the keyboard, while whitespace denotes any character(s) that, if printed, do not print ink on the paper: a blank, a tabulator character (produced by backslash followed by t), or a newline character (produced by backslash followed by n). (The newline character is explained in Sect. 1.1.11.)

When we interact with computer programs, we usually provide some information to the program and get some information out. It is common to use the term *input data*, or just *input*, for the information that must be known on beforehand. The result from a program is similarly referred to as *output data*, or just *output*. In our example, v_0, g, and t constitute input, while y is output. All input data must be assigned values in the program before the output can be computed. Input data can be explicitly initialized in the program, as we do in the present example, or the data can be provided by the user through keyboard typing while the program is running (see Chap. 4). Output data can be printed in the terminal window, as in the current example, displayed as graphics on the screen, as done in Sect. 5.3, or stored in a file for later access, as explained in Sect. 4.6.

The word *user* usually has a special meaning in computer science: It means a human interacting with a program. You are a user of a text editor for writing Python programs, and you are a user of your own programs. When you write programs, it is difficult to imagine how other users will interact with the program. Maybe they provide wrong input or misinterpret the output. Making user-friendly programs is very challenging and depends heavily on the target audience of users. The author had the average reader of the book in mind as a typical user when developing programs for this book.

A central part of a computer is the *operating system*. This is actually a collection of programs that manages the hardware and software resources on the computer. There are three dominating operating systems today on computers: Windows, Mac OS X, and Linux. In addition, we have Android and iOS for handheld devices. Several versions of Windows have appeared since the 1990s: Windows 95, 98, 2000, ME, XP, Vista, Windows 7, and Windows 8. Unix was invented already in 1970 and comes in many different versions. Nowadays, two open source implementations of Unix, Linux and Free BSD Unix, are most common. The latter forms the core of the Mac OS X operating system on Macintosh machines, while Linux exists in slightly different flavors: Red Hat, Debian, Ubuntu, and OpenSuse to mention the most important distributions. We will use the term Unix in this book as a synonym for all the operating systems that inherit from classical Unix, such as Solaris, Free BSD, Mac OS X, and any Linux variant. As a computer user and reader of this book, you should know exactly what operating system you have.

The user's interaction with the operation system is through a set of programs. The most widely used of these enable viewing the contents of folders or starting other programs. To interact with the operating system, as a user, you can either issue commands in a terminal window or use graphical programs. For example, for viewing the file contents of a folder you can run the command `ls` in a Unix terminal window or `dir` in a DOS (Windows) terminal window. The graphical alternatives are many, some of the most common are Windows Explorer on Windows, Nautilus and Konqueror on Unix, and Finder on Mac. To start a program, it is common to double-click on a file icon or write the program's name in a terminal window.

1.3 Another Formula: Celsius-Fahrenheit Conversion

Our next example involves the formula for converting temperature measured in Celsius degrees to the corresponding value in Fahrenheit degrees:

$$F = \frac{9}{5}C + 32 \tag{1.3}$$

In this formula, C is the amount of degrees in Celsius, and F is the corresponding temperature measured in Fahrenheit. Our goal now is to write a computer program that can compute F from (1.3) when C is known.

1.3.1 Potential Error: Integer Division

Straightforward coding of the formula A straightforward attempt at coding the formula (1.3) goes as follows:

```
C = 21
F = (9/5)*C + 32
print F
```

The parentheses around 9/5 are not strictly needed, i.e., (9/5)*C is computationally identical to 9/5*C, but parentheses remove any doubt that 9/5*C could mean 9/(5*C). Section 1.3.4 has more information on this topic.

When run under Python version 2.x, the program prints the value 53. You can find the program in the file c2f_v1.py in the src/formulas folder in the folder tree of example programs from this book (downloaded from http://hplgit.github.com/scipro-primer). The v1 part of the name stands for *version 1*. Throughout this book, we will often develop several trial versions of a program, but remove the version number in the final version of the program.

Verifying the results   Testing the correctness is easy in this case since we can evaluate the formula on a calculator: $\frac{9}{5} \cdot 21 + 32$ is 69.8, not 53. What is wrong? The formula in the program looks correct!

Float and integer division   The error in our program above is one of the most common errors in mathematical software and is not at all obvious for a newcomer to programming. In many computer languages, there are two types of divisions: float division and integer division. Float division is what you know from mathematics: 9/5 becomes 1.8 in decimal notation.

Integer division a/b with integers (whole numbers) a and b results in an integer that is truncated (or mathematically, rounded down). More precisely, the result is the largest integer c such that $bc \leq a$. This implies that 9/5 becomes 1 since $1 \cdot 5 = 5 \leq 9$ while $2 \cdot 5 = 10 > 9$. Another example is 1/5, which becomes 0 since $0 \cdot 5 \leq 1$ (and $1 \cdot 5 > 1$). Yet another example is 16/6, which results in 2 (try $2 \cdot 6$ and $3 \cdot 6$ to convince yourself). Many computer languages, including Fortran, C, C++, Java, and Python version 2, interpret a division operation a/b as integer division if both operands a and b are integers. If either a or b is a real (floating-point) number, a/b implies the standard mathematical float division. Other languages, such as MATLAB and Python version 3, interprets a/b as float division even if both operands are integers, or complex division if one of the operands is a complex number.

The problem with our program is the coding of the formula (9/5)*C + 32. This formula is evaluated as follows. First, 9/5 is calculated. Since 9 and 5 are interpreted by Python as integers (whole numbers), 9/5 is a division between two integers, and Python version 2 chooses by default integer division, which results in 1. Then 1 is multiplied by C, which equals 21, resulting in 21. Finally, 21 and 32 are added with 53 as result.

We shall very soon present a correct version of the temperature conversion program, but first it may be advantageous to introduce a frequently used term in Python programming: *object*.

1.3.2   Objects in Python

When we write

```
C = 21
```

Python interprets the number 21 on the right-hand side of the assignment as an integer and creates an `int` (for integer) *object* holding the value 21. The variable C acts as a *name* for this `int` object. Similarly, if we write C = 21.0, Python recognizes 21.0 as a real number and therefore creates a `float` (for floating-point) object holding the value 21.0 and lets C be a name for this object. In fact, any assignment statement has the form of a variable name on the left-hand side and an object on the right-hand side. One may say that Python programming is about solving a problem by defining and changing objects.

At this stage, you do not need to know what an object really is, just think of an `int` object as a collection, say a storage box, with some information about an integer number. This information is stored somewhere in the computer's memory, and with the name C the program gets access to this information. The fundamental issue right now is that 21 and 21.0 are identical numbers in mathematics, while in a Python program 21 gives rise to an `int` object and 21.0 to a `float` object.

There are lots of different object types in Python, and you will later learn how to create your own customized objects. Some objects contain a lot of data, not just an integer or a real number. For example, when we write

```
print 'A text with an integer %d and a float %f' % (2, 2.0)
```

a `str` (string) object, without a name, is first made of the text between the quotes and then this `str` object is printed. We can alternatively do this in two steps:

```
s = 'A text with an integer %d and a float %f' % (2, 2.0)
print s
```

1.3.3 Avoiding Integer Division

As a quite general rule of thumb, one should be careful to avoid integer division when programming mathematical formulas. In the rare cases when a mathematical algorithm does make use of integer division, one should use a double forward slash, //, as division operator, because this is Python's way of explicitly indicating integer division.

Python version 3 has no problem with unintended integer division, so the problem only arises with Python version 2 (and many other common languages for scientific computing). There are several ways to avoid integer division with the plain / operator. The simplest remedy in Python version 2 is to write

```
from __future__ import division
```

This import statement must be present in the beginning of every file where the / operator always shall imply float division. Alternatively, one can run a Python program `someprogram.py` from the command line with the argument -Qnew to the Python interpreter:

```
Terminal> python -Qnew someprogram.py
```

A more widely applicable method, also in other programming languages than Python version 2, is to enforce one of the operands to be a `float` object. In the current example, there are several ways to do this:

```
F = (9.0/5)*C + 32
F = (9/5.0)*C + 32
F = float(C)*9/5 + 32
```

In the first two lines, one of the operands is written as a decimal number, implying a `float` object and hence float division. In the last line, `float(C)*9` means `float` times `int`, which results in a `float` object, and float division is guaranteed.

A related construction,

```
F = float(C)*(9/5) + 32
```

does not work correctly, because 9/5 is evaluated by integer division, yielding 1, before being multiplied by a `float` representation of C (see next section for how compound arithmetic operations are calculated). In other words, the formula reads F=C+32, which is wrong.

We now understand why the first version of the program does not work and what the remedy is. A correct program is

```
C = 21
F = (9.0/5)*C + 32
print F
```

Instead of 9.0 we may just write 9. (the dot implies a `float` interpretation of the number). The program is available in the file `c2f.py`. Try to run it – and observe that the output becomes 69.8, which is correct.

Locating potential integer division Running a Python program with the `-Qwarnall` argument, say

```
Terminal
Terminal> python -Qwarnall someprogram.py
```

will print out a warning every time an integer division expression is encountered in Python version 2.

Remark We could easily have run into problems in our very first programs if we instead of writing the formula $\frac{1}{2}gt^2$ as `0.5*g*t**2` wrote `(1/2)*g*t**2`. This term would then always be zero!

1.3.4 Arithmetic Operators and Precedence

Formulas in Python programs are usually evaluated in the same way as we would
evaluate them mathematically. Python proceeds from left to right, term by term in
an expression (terms are separated by plus or minus). In each term, power operations such as a^b, coded as a**b, has precedence over multiplication and division.
As in mathematics, we can use parentheses to dictate the way a formula is evaluated.
Below are two illustrations of these principles.

- 5/9+2*a**4/2: First 5/9 is evaluated (as integer division, giving 0 as result),
  then a^4 (a**4) is evaluated, then 2 is multiplied with a^4, that result is divided by
  2, and the answer is added to the result of the first term. The answer is therefore
  a**4.
- 5/(9+2)*a**(4/2): First $\frac{5}{9+2}$ is evaluated (as integer division, yielding 0), then
  4/2 is computed (as integer division, yielding 2), then a**2 is calculated, and that
  number is multiplied by the result of 5/(9+2). The answer is thus always zero.

As evident from these two examples, it is easy to unintentionally get integer division
in formulas. Although integer division can be turned off in Python, we think it is
important to be strongly aware of the integer division concept and to develop good
programming habits to avoid it. The reason is that this concept appears in so many
common computer languages that it is better to learn as early as possible how to deal
with the problem rather than using a Python-specific feature to remove the problem.

1.4 Evaluating Standard Mathematical Functions

Mathematical formulas frequently involve functions such as sin, cos, tan, sinh, cosh,
exp, log, etc. On a pocket calculator you have special buttons for such functions.
Similarly, in a program you also have ready-made functionality for evaluating these
types of mathematical functions. One could in principle write one's own program
for evaluating, e.g., the $\sin(x)$ function, but how to do this in an efficient way is
a non-trivial topic. Experts have worked on this problem for decades and implemented their best recipes in pieces of software that we should reuse. This section
tells you how to reach sin, cos, and similar functions in a Python context.

1.4.1 Example: Using the Square Root Function

Problem Consider the formula for the height y of a ball in vertical motion, with
initial upward velocity v_0:

$$y_c = v_0 t - \frac{1}{2} g t^2,$$

where g is the acceleration of gravity and t is time. We now ask the question:
How long time does it take for the ball to reach the height y_c? The answer is
straightforward to derive. When $y = y_c$ we have

$$y_c = v_0 t - \frac{1}{2} g t^2.$$

We recognize that this equation is a quadratic equation, which we must solve with respect to t. Rearranging,

$$\frac{1}{2}gt^2 - v_0 t + y_c = 0,$$

and using the well-known formula for the two solutions of a quadratic equation, we find

$$t_1 = \left(v_0 - \sqrt{v_0^2 - 2gy_c}\right)/g, \quad t_2 = \left(v_0 + \sqrt{v_0^2 - 2gy_c}\right)/g. \qquad (1.4)$$

There are two solutions because the ball reaches the height y_c on its way up ($t = t_1$) and on its way down ($t = t_2 > t_1$).

The program   To evaluate the expressions for t_1 and t_2 from (1.4) in a computer program, we need access to the square root function. In Python, the square root function and lots of other mathematical functions, such as sin, cos, sinh, exp, and log, are available in a module called `math`. We must first import the module before we can use it, that is, we must write `import math`. Thereafter, to take the square root of a variable a, we can write `math.sqrt(a)`. This is demonstrated in a program for computing t_1 and t_2:

```
v0 = 5
g = 9.81
yc = 0.2
import math
t1 = (v0 - math.sqrt(v0**2 - 2*g*yc))/g
t2 = (v0 + math.sqrt(v0**2 - 2*g*yc))/g
print 'At t=%g s and %g s, the height is %g m.' % (t1, t2, yc)
```

The output from this program becomes

```
At t=0.0417064 s and 0.977662 s, the height is 0.2 m.
```

You can find the program as the file `ball_yc.py` in the `src/formulas` folder.

Two ways of importing a module   The standard way to import a module, say `math`, is to write

```
import math
```

and then access individual functions in the module with the module name as prefix as in

```
x = math.sqrt(y)
```

People working with mathematical functions often find `math.sqrt(y)` less pleasing than just `sqrt(y)`. Fortunately, there is an alternative import syntax that allows us to skip the module name prefix. This alternative syntax has the form `from module import function`. A specific example is

```
from math import sqrt
```

Now we can work with `sqrt` directly, without the `math.` prefix. More than one function can be imported:

```
from math import sqrt, exp, log, sin
```

Sometimes one just writes

```
from math import *
```

to import all functions in the `math` module. This includes `sin`, `cos`, `tan`, `asin`, `acos`, `atan`, `sinh`, `cosh`, `tanh`, `exp`, `log` (base e), `log10` (base 10), `sqrt`, as well as the famous numbers e and pi. Importing all functions from a module, using the asterisk (*) syntax, is convenient, but this may result in a lot of extra names in the program that are not used. It is in general recommended not to import more functions than those that are really used in the program. Nevertheless, the convenience of the compact `from math import *` syntax occasionally wins over the general recommendation among practitioners – and in this book.

With a `from math import sqrt` statement we can write the formulas for the roots in a more pleasing way:

```
t1 = (v0 - sqrt(v0**2 - 2*g*yc))/g
t2 = (v0 + sqrt(v0**2 - 2*g*yc))/g
```

Import with new names Imported modules and functions can be given new names in the import statement, e.g.,

```
import math as m
# m is now the name of the math module
v = m.sin(m.pi)

from math import log as ln
v = ln(5)

from math import sin as s, cos as c, log as ln
v = s(x)*c(x) + ln(x)
```

In Python, everything is an object, and variables refer to objects, so new variables may refer to modules and functions as well as numbers and strings. The examples above on new names can also be coded by introducing new variables explicitly:

```
m = math
ln = m.log
s = m.sin
c = m.cos
```

1.4.2 Example: Computing with $\sinh x$

Our next examples involve calling some more mathematical functions from the `math` module. We look at the definition of the $\sinh(x)$ function:

$$\sinh(x) = \frac{1}{2}\left(e^x - e^{-x}\right) . \tag{1.5}$$

We can evaluate $\sinh(x)$ in three ways: i) by calling `math.sinh`, ii) by computing the right-hand side of (1.5), using `math.exp`, or iii) by computing the right-hand side of (1.5) with the aid of the power expressions `math.e**x` and `math.e**(-x)`. A program doing these three alternative calculations is found in the file `3sinh.py`. The core of the program looks like this:

```
from math import sinh, exp, e, pi
x = 2*pi
r1 = sinh(x)
r2 = 0.5*(exp(x) - exp(-x))
r3 = 0.5*(e**x - e**(-x))
print r1, r2, r3
```

The output from the program shows that all three computations give identical results:

```
267.744894041 267.744894041 267.744894041
```

1.4.3 A First Glimpse of Rounding Errors

The previous example computes a function in three different yet mathematically equivalent ways, and the output from the `print` statement shows that the three resulting numbers are equal. Nevertheless, this is not the whole story. Let us try to print out `r1, r2, r3` with 16 decimals:

```
print '%.16f %.16f %.16f' % (r1,r2,r3)
```

This statement leads to the output

```
267.7448940410164369 267.7448940410164369 267.7448940410163232
```

Now `r1` and `r2` are equal, but `r3` is different! Why is this so?

Our program computes with real numbers, and real numbers need in general an infinite number of decimals to be represented exactly. The computer truncates the sequence of decimals because the storage is finite. In fact, it is quite standard to keep only 17 digits in a real number on a computer. Exactly how this truncation is done is not explained in this book, but you read more on Wikipedia[4]. For now the

---

[4] http://en.wikipedia.org/wiki/Floating_point_number

purpose is to notify the reader that real numbers on a computer often have a small error. Only a few real numbers can be represented exactly, the rest of the real numbers are only approximations.

For this reason, most arithmetic operations involve inaccurate real numbers, resulting in inaccurate calculations. Think of the following two calculations: $1/49 \cdot 49$ and $1/51 \cdot 51$. Both expressions are identical to 1, but when we perform the calculations in Python,

```
print '%.16f %.16f' % (1/49.0*49, 1/51.0*51)
```

the result becomes

```
0.9999999999999999 1.0000000000000000
```

The reason why we do not get exactly 1.0 as answer in the first case is because 1/49 is not correctly represented in the computer. Also 1/51 has an inexact representation, but the error does not propagate to the final answer.

To summarize, errors in floating-point numbers may propagate through mathematical calculations and result in answers that are only approximations to the exact underlying mathematical values. The errors in the answers are commonly known as *rounding errors*. As soon as you use Python interactively as explained in the next section, you will encounter rounding errors quite often.

Python has a special module `decimal` and the SymPy package has an alternative module `mpmath`, which allow real numbers to be represented with adjustable accuracy so that rounding errors can be made as small as desired (an example appears at the end of Sect. 3.1.12). However, we will hardly use such modules because approximations implied by many mathematical methods applied throughout this book normally lead to (much) larger errors than those caused by rounding.

1.5 Interactive Computing

A particular convenient feature of Python is the ability to execute statements and evaluate expressions interactively. The environments where you work interactively with programming are commonly known as Python *shells*. The simplest Python shell is invoked by just typing `python` at the command line in a terminal window. Some messages about Python are written out together with a prompt >>>, after which you can issue commands. Let us try to use the interactive shell as a calculator. Type in 3*4.5-0.5 and then press the Return key to see Python's response to this expression:

```
Terminal> python
Python 2.7.5+ (default, Sep 19 2013, 13:48:49)
[GCC 4.8.1] on linux2
Type "help", "copyright", "credits" or "license" for more information.
>>> 3*4.5-0.5
13.0
```

The text on a line after >>> is what we write (shell input) and the text without the >>> prompt is the result that Python calculates (shell output). It is easy, as explained below, to recover previous input and edit the text. This editing feature makes it convenient to experiment with statements and expressions.

1.5.1   Using the Python Shell

The program from Sect. 1.1.7 can be typed in line by line in the interactive shell:

```
>>> v0 = 5
>>> g = 9.81
>>> t = 0.6
>>> y = v0*t - 0.5*g*t**2
>>> print y
1.2342
```

We can now easily calculate an y value corresponding to another (say) v0 value: hit the up arrow key to recover previous statements, repeat pressing this key until the v0 = 5 statement is displayed. You can then edit the line, e.g., to

```
>>> v0 = 6
```

Press return to execute this statement. You can control the new value of v0 by either typing just v0 or print v0:

```
>>> v0
6
>>> print v0
6
```

The next step is to recompute y with this new v0 value. Hit the up arrow key multiple times to recover the statement where y is assigned, press the Return key, and write y or print y to see the result of the computation:

```
>>> y = v0*t - 0.5*g*t**2
>>> y
1.8341999999999996
>>> print y
1.8342
```

The reason why we get two slightly different results is that typing just y prints out all the decimals that are stored in the computer (16), while print y writes out y with fewer decimals. As mentioned in Sect. 1.4.3 computations on a computer often suffer from rounding errors. The present calculation is no exception. The correct answer is 1.8342, but rounding errors lead to a number that is incorrect in the 16th decimal. The error is here $4 \cdot 10^{-16}$.

1.5.2 Type Conversion

Often you can work with variables in Python without bothering about the type of objects these variables refer to. Nevertheless, we encountered a serious problem in Sect. 1.3.1 with integer division, which forced us to be careful about the types of objects in a calculation. The interactive shell is very useful for exploring types. The following example illustrates the type function and how we can convert an object from one type to another.

First, we create an int object bound to the name C and check its type by calling type(C):

```
>>> C = 21
>>> type(C)
<type 'int'>
```

We convert this int object to a corresponding float object:

```
>>> C = float(C)    # type conversion
>>> type(C)
<type 'float'>
>>> C
21.0
```

In the statement C = float(C) we create a new object from the original object referred to by the name C and bind it to the same name C. That is, C refers to a different object after the statement than before. The original int with value 21 cannot be reached anymore (since we have no name for it) and will be automatically deleted by Python.

We may also do the reverse operation, i.e., convert a particular float object to a corresponding int object:

```
>>> C = 20.9
>>> type(C)
<type 'float'>
>>> D = int(C)       # type conversion
>>> type(D)
<type 'int'>
>>> D
20                   # decimals are truncated :-/
```

In general, one can convert a variable v to type MyType by writing v=MyType(v), if it makes sense to do the conversion.

In the last input we tried to convert a float to an int, and this operation implied stripping off the decimals. Correct conversion according to mathematical rounding rules can be achieved with help of the round function:

```
>>> round(20.9)
21.0
>>> int(round(20.9))
21
```

1.5.3   IPython

There exist several improvements of the standard Python shell presented in
Sect. 1.5. The author advocates IPython as the preferred interactive shell. You
will then need to have IPython installed. Typing `ipython` in a terminal window
starts the shell. The (default) prompt in IPython is not >>> but `In [X]:`, where
X is the number of the present input command. The most widely used features of
IPython are summarized below.

Running programs   Python programs can be run from within the shell:

```
In [1]: run ball2.py
1.2342
```

This command requires that you have taken a `cd` to the folder where the `ball2.py`
program is located and started IPython from there.

On Windows you may, as an alternative to starting IPython from a DOS or PowerShell window, double click on the IPython desktop icon or use the Start menu. In
that case, you must move to the right folder where your program is located. This
is done by the `os.chdir` (change directory) command. Typically, you write something like

```
In [1]: import os
In [2]: os.chdir(r'C:\Documents and Settings\me\My Documents\div')
In [3]: run ball2.py
```

if the `ball2.py` program is located in the folder `div` under `My Documents` of user
`me`. Note the `r` before the quote in the string: it is required to let a backslash
really mean the backslash character. If you end up typing the `os.chdir` command
every time you enter an IPython shell, this command (and others) can be placed in
a *startup file* such that they are automatically executed when you launch IPython.

Inside IPython you can invoke any operating system command. This allows us
to navigate to the right folder above using Unix or Windows (`cd`) rather than Python
(`os.chdir`):

```
In [1]: cd C:\Documents and Settings\me\My Documents\div
In [3]: run ball2.py
```

We recommend running all your Python programs from the IPython shell. Especially when something goes wrong, IPython can help you to examine the state of
variables so that you become quicker to locate bugs.

Typesetting convention for executing Python programs

In the rest of the book, we just write the program name and the output when we
illustrate the execution of a program:

──────────────────────────── | Terminal | ────────────────────────────

```
ball2.py
1.2342
```

You then need to write `run` before the program name if you execute the program in IPython, or if you prefer to run the program directly in a terminal window, you need to write `python` prior to the program name. Appendix H.5 describes various other ways to run a Python program.

Quick recovery of previous output The results of the previous statements in an interactive IPython session are available in variables of the form `_iX` (underscore, `i`, and a number `X`), where `X` is 1 for the last statement, 2 for the second last statement, and so forth. Short forms are `_` for `_i1`, `__` for `_i2`, and `___` for `_i3`. The output from the `In [1]` input above is `1.2342`. We can now refer to this number by an underscore and, e.g., multiply it by 10:

```
In [2]: _*10
Out[2]: 12.341999999999999
```

Output from Python statements or expressions in IPython are preceded by `Out[X]` where `X` is the command number corresponding to the previous `In [X]` prompt. When programs are executed, as with the `run` command, or when operating system commands are run (as shown below), the output is from the operating system and then not preceded by any `Out[X]` label.

The command history from previous IPython sessions is available in a new session. This feature makes it easy to modify work from a previous session by just hitting the up-arrow to recall commands and edit them as necessary.

Tab completion Pressing the TAB key will complete an incompletely typed variable name. For example, after defining `my_long_variable_name = 4`, write just `my` at the `In [4]:` prompt below, and then hit the TAB key. You will experience that `my` is immediately expanded to `my_long_variable_name`. This automatic expansion feature is called TAB completion and can save you from quite some typing.

```
In [3]: my_long_variable_name = 4

In [4]: my_long_variable_name
Out[4]: 4
```

Recovering previous commands You can walk through the command history by typing `Ctrl+p` or the up arrow for going backward or `Ctrl+n` or the down arrow for going forward. Any command you hit can be edited and re-executed. Also commands from previous interactive sessions are stored in the command history.

Running Unix/Windows commands Operating system commands can be run from IPython. Below we run the three Unix commands `date`, `ls` (list files), `mkdir` (make directory), and `cd` (change directory):

```
In [5]: date
Thu Nov 18 11:06:16 CET 2010

In [6]: ls
myfile.py  yourprog.py
```

```
In [7]: mkdir mytestdir

In [8]: cd mytestdir
```

If you have defined Python variables with the same name as operating system commands, e.g., date=30, you must write !date to run the corresponding operating system command.

IPython can do much more than what is shown here, but the advanced features and the documentation of them probably do not make sense before you are more experienced with Python – and with reading manuals.

Typesetting of interactive shells in this book

In the rest of the book we will apply the >>> prompt in interactive sessions instead of the input and output prompts as used by default by IPython, simply because most Python books and electronic manuals use >>> to mark input in interactive shells. However, when you sit by the computer and want to use an interactive shell, we recommend using IPython, and then you will see the In [X] prompt instead of >>>.

Notebooks A particularly interesting feature of IPython is the notebook, which allows you to record and replay exploratory interactive sessions with a mix of text, mathematics, Python code, and graphics. See Sect. H.4 for a quick introduction to IPython notebooks.

1.6 Complex Numbers

Suppose $x^2 = 2$. Then most of us are able to find out that $x = \sqrt{2}$ is a solution to the equation. The more mathematically interested reader will also remark that $x = -\sqrt{2}$ is another solution. But faced with the equation $x^2 = -2$, very few are able to find a proper solution without any previous knowledge of *complex numbers*. Such numbers have many applications in science, and it is therefore important to be able to use such numbers in our programs.

On the following pages we extend the previous material on computing with real numbers to complex numbers. The text is optional, and readers without knowledge of complex numbers can safely drop this section and jump to Sect. 1.8.

A complex number is a pair of real numbers a and b, most often written as $a+bi$, or $a + ib$, where i is called the imaginary unit and acts as a label for the second term. Mathematically, $i = \sqrt{-1}$. An important feature of complex numbers is definitely the ability to compute square roots of negative numbers. For example, $\sqrt{-2} = \sqrt{2}i$ (i.e., $\sqrt{2}\sqrt{-1}$). The solutions of $x^2 = -2$ are thus $x_1 = +\sqrt{2}i$ and $x_2 = -\sqrt{2}i$.

There are rules for addition, subtraction, multiplication, and division between two complex numbers. There are also rules for raising a complex number to a real power, as well as rules for computing $\sin z$, $\cos z$, $\tan z$, e^z, $\ln z$, $\sinh z$, $\cosh z$, $\tanh z$, etc. for a complex number $z = a + ib$. We assume in the following that you are familiar with the mathematics of complex numbers, at least to the degree encountered in the program examples.

$$\text{let } u = a + bi \text{ and } v = c + di$$

The following rules reflect complex arithmetics:

$$u = v \quad \Rightarrow \quad a = c, \ b = d$$
$$-u = -a - bi$$
$$u^* \equiv a - bi \quad (\text{complex conjugate})$$
$$u + v = (a + c) + (b + d)i$$
$$u - v = (a - c) + (b - d)i$$
$$uv = (ac - bd) + (bc + ad)i$$
$$u/v = \frac{ac + bd}{c^2 + d^2} + \frac{bc - ad}{c^2 + d^2}i$$
$$|u| = \sqrt{a^2 + b^2}$$
$$e^{iq} = \cos q + i \sin q$$

1.6.1 Complex Arithmetics in Python

Python supports computation with complex numbers. The imaginary unit is written as j in Python, instead of i as in mathematics. A complex number $2-3i$ is therefore expressed as (2-3j) in Python. We remark that the number i is written as 1j, not just j. Below is a sample session involving definition of complex numbers and some simple arithmetics:

```
>>> u = 2.5 + 3j        # create a complex number
>>> v = 2               # this is an int
>>> w = u + v           # complex + int
>>> w
(4.5+3j)

>>> a = -2
>>> b = 0.5
>>> s = a + b*1j        # create a complex number from two floats
>>> s = complex(a, b)   # alternative creation
>>> s
(-2+0.5j)
>>> s*w                 # complex*complex
(-10.5-3.75j)
>>> s/w                 # complex/complex
(-0.25641025641025639+0.28205128205128205j)
```

A complex object s has functionality for extracting the real and imaginary parts as well as computing the complex conjugate:

```
>>> s.real
-2.0
>>> s.imag
0.5
>>> s.conjugate()
(-2-0.5j)
```

1.6.2   Complex Functions in Python

Taking the sine of a complex number does not work:

```
>>> from math import sin
>>> r = sin(w)
Traceback (most recent call last):
  File "<input>", line 1, in ?
TypeError: can't convert complex to float; use abs(z)
```

The reason is that the `sin` function from the `math` module only works with real
(`float`) arguments, not complex. A similar module, `cmath`, defines functions that
take a complex number as argument and return a complex number as result. As an
example of using the `cmath` module, we can demonstrate that the relation $\sin(ai) = i \sinh a$ holds:

```
>>> from cmath import sin, sinh
>>> r1 = sin(8j)
>>> r1
1490.4788257895502j
>>> r2 = 1j*sinh(8)
>>> r2
1490.4788257895502j
```

Another relation, $e^{iq} = \cos q + i \sin q$, is exemplified next:

```
>>> q = 8       # some arbitrary number
>>> exp(1j*q)
(-0.14550003380861354+0.98935824662338179j)
>>> cos(q) + 1j*sin(q)
(-0.14550003380861354+0.98935824662338179j)
```

1.6.3   Unified Treatment of Complex and Real Functions

The `cmath` functions always return complex numbers. It would be nice to have
functions that return a `float` object if the result is a real number and a `complex`
object if the result is a complex number. The Numerical Python package has such
versions of the basic mathematical functions known from `math` and `cmath`. By
taking a

```
from numpy.lib.scimath import *
```

one obtains access to these flexible versions of mathematical functions. The functions also get imported by any of the statements

```
from scipy import *
from scitools.std import *
```

A session will illustrate what we obtain. Let us first use the sqrt function in the math module:

```
>>> from math import sqrt
>>> sqrt(4)      # float
2.0
>>> sqrt(-1)     # illegal
Traceback (most recent call last):
  File "<input>", line 1, in ?
ValueError: math domain error
```

If we now import sqrt from cmath,

```
>>> from cmath import sqrt
```

the previous sqrt function is overwritten by the new one. More precisely, the name sqrt was previously bound to a function sqrt from the math module, but is now bound to another function sqrt from the cmath module. In this case, any square root results in a complex object:

```
>>> sqrt(4)      # complex
(2+0j)
>>> sqrt(-1)     # complex
1j
```

If we now take

```
>>> from numpy.lib.scimath import *
```

we import (among other things) a new sqrt function. This function is slower than the versions from math and cmath, but it has more flexibility since the returned object is float if that is mathematically possible, otherwise a complex is returned:

```
>>> sqrt(4)      # float
2.0
>>> sqrt(-1)     # complex
1j
```

As a further illustration of the need for flexible treatment of both complex and real numbers, we may code the formulas for the roots of a quadratic function $f(x) = ax^2 + bx + c$:

```
>>> a = 1; b = 2; c = 100    # polynomial coefficients
>>> from numpy.lib.scimath import sqrt
>>> r1 = (-b + sqrt(b**2 - 4*a*c))/(2*a)
>>> r2 = (-b - sqrt(b**2 - 4*a*c))/(2*a)
>>> r1
(-1+9.94987437107j)
>>> r2
(-1-9.94987437107j)
```

Using the up arrow, we may go back to the definitions of the coefficients and change them so the roots become real numbers:

```
>>> a = 1; b = 4; c = 1    # polynomial coefficients
```

Going back to the computations of `r1` and `r2` and performing them again, we get

```
>>> r1
-0.267949192431
>>> r2
-3.73205080757
```

That is, the two results are `float` objects. Had we applied `sqrt` from `cmath`, `r1` and `r2` would always be `complex` objects, while `sqrt` from the `math` module would not handle the first (complex) case.

1.7   Symbolic Computing

Python has a package SymPy for doing symbolic computing, such as symbolic (exact) integration, differentiation, equation solving, and expansion of Taylor series, to mention some common operations in mathematics. We shall here only give a glimpse of SymPy in action with the purpose of drawing attention to this powerful part of Python.

For interactive work with SymPy it is recommended to either use IPython or the special, interactive shell `isympy`, which is installed along with SymPy itself.

Below we shall explicitly import each symbol we need from SymPy to emphasize that the symbol comes from that package. For example, it will be important to know whether `sin` means the sine function from the `math` module, aimed at real numbers, or the special sine function from `sympy`, aimed at symbolic expressions.

1.7.1   Basic Differentiation and Integration

The following session shows how easy it is to differentiate a formula $v_0 t - \frac{1}{2}gt^2$ with respect to t and integrate the answer to get the formula back:

```
>>> from sympy import (
...      symbols,   # define symbols for symbolic math
...      diff,      # differentiate expressions
...      integrate, # integrate expressions
...      Rational,  # define rational numbers
...      lambdify,  # turn symbolic expr. into Python functions
...      )
>>> t, v0, g = symbols('t v0 g')
>>> y = v0*t - Rational(1,2)*g*t**2
>>> dydt = diff(y, t)
>>> dydt
-g*t + v0
>>> print 'acceleration:', diff(y, t, t)  # 2nd derivative
acceleration: -g
>>> y2 = integrate(dydt, t)
>>> y2
-g*t**2/2 + t*v0
```

Note here that t is a *symbolic variable* (not a float as it is in numerical computing), and y (like y2) is a *symbolic expression* (not a float as it would be in numerical computing).

A very convenient feature of SymPy is that symbolic expressions can be turned into ordinary Python functions via lambdify. (Python functions are introduced in Chap. 3, but when discussing SymPy here in the present chapter, it is very natural to explain how lambdify can transform symbolic expressions back to ordinary numerical Python expressions.) Let us take the dydt expression above and turn it into a Python function v(t, v0, g) for numerical computing:

```
>>> v = lambdify([t, v0, g],   # arguments in v
                 dydt)         # symbolic expression
>>> v(t=0, v0=5, g=9.81)
5
>>> v(2, 5, 9.81)
-14.62
>>> 5 - 9.81*2  # control the previous calculation
-14.62
```

1.7.2 Equation Solving

A linear equation defined through an expression e that is zero, can be solved by solve(e, t), if t is the unknown (symbol) in the equation. Here we may find the roots of $y = 0$:

```
>>> from sympy import solve
>>> roots = solve(y, t)
>>> roots
[0, 2*v0/g]
```

We can easily check the answer by inserting the roots in y. Inserting an expression e2 for e1 in some expression e is done by e.subs(e1, e2). In our case we check that

```
>>> y.subs(t, roots[0])
0
>>> y.subs(t, roots[1])
0
```

1.7.3   Taylor Series and More

A Taylor polynomial of order n for an expression e in a variable t around the point t0 is computed by e.series(t, t0, n). Testing this on e^t and $e^{\sin(t)}$ gives

```
>>> from sympy import exp, sin, cos
>>> f = exp(t)
>>> f.series(t, 0, 3)
1 + t + t**2/2 + O(t**3)
>>> f = exp(sin(t))
>>> f.series(t, 0, 8)
1 + t + t**2/2 - t**4/8 - t**5/15 - t**6/240 + t**7/90 + O(t**8)
```

Output of mathematical expressions in the LaTeX typesetting system is possible:

```
>>> from sympy import latex
>>> print latex(f.series(t, 0, 7))
'1 + t + \frac{t^{2}}{2} - \frac{t^{4}}{8} - \frac{t^{5}}{15} -
\frac{t^{6}}{240} + \mathcal{O}\left(t^{7}\right)'
```

Finally, we mention that there are tools for expanding and simplifying expressions:

```
>>> from sympy import simplify, expand
>>> x, y = symbols('x y')
>>> f = -sin(x)*sin(y) + cos(x)*cos(y)
>>> simplify(f)
cos(x + y)
>>> expand(sin(x+y), trig=True)   # requires a trigonometric hint
sin(x)*cos(y) + sin(y)*cos(x)
```

Later chapters utilize SymPy where it can save some algebraic work, but this book is almost exclusively devoted to numerical computing.

1.8   Summary

1.8.1   Chapter Topics

Programs must be accurate!   A program is a collection of statements stored in a text file. Statements can also be executed interactively in a Python shell. Any error in any statement may lead to termination of the execution or wrong results. The computer does exactly what the programmer tells the computer to do!

Variables The statement

```
some_variable = obj
```

defines a variable with the name `some_variable` which refers to an object `obj`.
Here `obj` may also represent an expression, say a formula, whose value is a Python
object. For example, `1+2.5` involves the addition of an `int` object and a `float`
object, resulting in a `float` object. Names of variables can contain upper and lower
case English letters, underscores, and the digits from 0 to 9, but the name cannot
start with a digit. Nor can a variable name be a reserved word in Python.

   If there exists a precise mathematical description of the problem to be solved in
a program, one should choose variable names that are in accordance with the mathematical description. Quantities that do not have a defined mathematical symbol,
should be referred to by *descriptive* variables names, i.e., names that explain the
variable's role in the program. Well-chosen variable names are essential for making
a program easy to read, easy to debug, and easy to extend. Well-chosen variable
names also reduce the need for comments.

Comment lines Everything after # on a line is ignored by Python and used to insert
free running text, known as *comments*. The purpose of comments is to explain, in
a human language, the ideas of (several) forthcoming statements so that the program
becomes easier to understand for humans. Some variables whose names are not
completely self-explanatory also need a comment.

Object types There are many different types of objects in Python. In this chapter
we have worked with the following types.

• Integers (whole numbers, object type `int`):

```
x10 = 3
XYZ = 2
```

• Floats (decimal numbers, object type `float`):

```
max_temperature = 3.0
MinTemp = 1/6.0
```

• Strings (pieces of text, object type `str`):

```
a = 'This is a piece of text\nover two lines.'
b = "Strings are enclosed in single or double quotes."
c = """Triple-quoted strings can
span
several lines.
"""
```

• Complex numbers (object type `complex`):

```
a = 2.5 + 3j
real = 6; imag = 3.1
b = complex(real, imag)
```

Operators Operators in arithmetic expressions follow the rules from mathematics: power is evaluated before multiplication and division, while the latter two are evaluated before addition and subtraction. These rules are overridden by parentheses. We suggest using parentheses to group and clarify mathematical expressions, also when not strictly needed.

```
-t**2*g/2
-(t**2)*(g/2)          # equivalent
-t**(2*g)/2            # a different formula!

a = 5.0; b = 5.0; c = 5.0
a/b + c + a*c          # yields 31.0
a/(b + c) + a*c        # yields 25.5
a/(b + c + a)*c        # yields 1.6666666666666665
```

Particular attention must be paid to coding fractions, since the division operator / often needs extra parentheses that are not necessary in the mathematical notation for fractions (compare $\frac{a}{b+c}$ with a/(b+c) and a/b+c).

Common mathematical functions The math module contains common mathematical functions for real numbers. Modules must be imported before they can be used. The three types of alternative module import go as follows:

```
# Import of module - functions requires prefix
import math
a = math.sin(math.pi*1.5)

# Import of individual functions - no prefix in function calls
from math import sin, pi
a = sin(pi*1.5)

# Import everything from a module - no prefix in function calls
from math import *
a = sin(pi*1.5)
```

Print To print the result of calculations in a Python program to a terminal window, we apply the print command, i.e., the word print followed by a string enclosed in quotes, or just a variable:

```
print "A string enclosed in double quotes"
print a
```

Several objects can be printed in one statement if the objects are separated by commas. A space will then appear between the output of each object:

```
>>> a = 5.0; b = -5.0; c = 1.9856; d = 33
>>> print 'a is', a, 'b is', b, 'c and d are', c, d
a is 5.0 b is -5.0 c and d are 1.9856 33
```

The printf syntax enables full control of the formatting of real numbers and integers:

```
>>> print 'a=%g, b=%12.4E, c=%.2f, d=%5d' % (a, b, c, d)
a=5, b= -5.0000E+00, c=1.99, d=   33
```

Here, a, b, and c are of type `float` and formatted as compactly as possible (%g for a), in scientific notation with 4 decimals in a field of width 12 (%12.4E for b), and in decimal notation with two decimals in as compact field as possible (%.2f for c). The variable d is an integer (`int`) written in a field of width 5 characters (%5d).

Be careful with integer division!

A common error in mathematical computations is to divide two integers, because this results in integer division (in Python 2).

- Any number written without decimals is treated as an integer. To avoid integer division, ensure that every division involves at least one real number, e.g., 9/5 is written as 9.0/5, 9./5, 9/5., or 9/5.0.
- In expressions with variables, a/b, ensure that a or b is a `float` object, and if not (or uncertain), do an explicit conversion as in `float(a)/b` to guarantee float division.
- If integer division is desired, use a double slash: a//b.
- Python 3 treats a/b as float division also when a and b are integers.

Complex numbers Values of complex numbers are written as (X+Yj), where X is the value of the real part and Y is the value of the imaginary part. One example is (4-0.2j). If the real and imaginary parts are available as variables r and i, a complex number can be created by `complex(r, i)`.

The `cmath` module must be used instead of `math` if the argument is a complex variable. The `numpy` package offers similar mathematical functions, but with a unified treatment of real and complex variables.

Terminology Some Python and computer science terms briefly covered in this chapter are

- object: anything that a variable (name) can refer to, such as a number, string, function, or module (but objects can exist without being bound to a name: `print 'Hello!'` first makes a string object of the text in quotes and then the contents of this string object, without a name, is printed)
- variable: name of an object
- statement: an instruction to the computer, usually written on a line in a Python program (multiple statements on a line must be separated by semicolons)
- expression: a combination of numbers, text, variables, and operators that results in a new object, when being evaluated
- assignment: a statement binding an evaluated expression (object) to a variable (name)
- algorithm: detailed recipe for how to solve a problem by programming
- code: program text (or synonym for program)
- implementation: same as code

- executable: the file we run to start the program
- verification: providing evidence that the program works correctly
- debugging: locating and correcting errors in a program

1.8.2   Example: Trajectory of a Ball

Problem What is the trajectory of a ball that is thrown or kicked with an initial velocity v_0 making an angle θ with the horizontal? This problem can be solved by basic high school physics as you are encouraged to do in Exercise 1.13. The ball will follow a trajectory $y = f(x)$ through the air where

$$f(x) = x \tan\theta - \frac{1}{2v_0^2}\frac{gx^2}{\cos^2\theta} + y_0 \,. \tag{1.6}$$

In this expression, x is a horizontal coordinate, g is the acceleration of gravity, v_0 is the size of the initial velocity that makes an angle θ with the x axis, and $(0, y_0)$ is the initial position of the ball. Our programming goal is to make a program for evaluating (1.6). The program should write out the value of all the involved variables and what their units are.

We remark that the formula (1.6) neglects air resistance. Exercise 1.11 explores how important air resistance is. For a soft kick ($v_0 = 30$ km/h) of a football, the gravity force is much larger than the air resistance, but for a hard kick, air resistance may be as important as gravity.

Solution We use the SI system and assume that v_0 is given in km/h; $g = 9.81$m/s^2; x, y, and y_0 are measured in meters; and θ in degrees. The program has naturally four parts: initialization of input data, import of functions and π from `math`, conversion of v_0 and θ to m/s and radians, respectively, and evaluation of the right-hand side expression in (1.6). We choose to write out all numerical values with one decimal. The complete program is found in the file `trajectory.py`:

```
g = 9.81      # m/s**2
v0 = 15       # km/h
theta = 60    # degrees
x = 0.5       # m
y0 = 1        # m

print """\
v0      = %.1f km/h
theta = %d degrees
y0      = %.1f m
x       = %.1f m\
""" % (v0, theta, y0, x)

from math import pi, tan, cos
# Convert v0 to m/s and theta to radians
v0 = v0/3.6
theta = theta*pi/180
```

```
y = x*tan(theta) - 1/(2*v0**2)*g*x**2/((cos(theta))**2) + y0

print 'y      = %.1f m' % y
```

The backslash in the triple-quoted multi-line string makes the string continue on the next line without a newline. This means that removing the backslash results in a blank line above the v0 line and a blank line between the x and y lines in the output on the screen. Another point to mention is the expression $1/(2*v0**2)$, which might seem as a candidate for unintended integer division. However, the conversion of v0 to m/s involves a division by 3.6, which results in v0 being `float`, and therefore $2*v0**2$ being `float`. The rest of the program should be self-explanatory at this stage in the book.

We can execute the program in IPython or an ordinary terminal window and watch the output:

```
                                    Terminal
v0    = 15.0 km/h
theta = 60 degrees
y0    = 1.0 m
x     = 0.5 m
y     = 1.6 m
```

1.8.3   About Typesetting Conventions in This Book

This version of the book applies different design elements for different types of "computer text". Complete programs and parts of programs (snippets) are typeset with a light blue background. A snippet looks like this:

```
a = sqrt(4*p + c)
print 'a =', a
```

A complete program has an additional, slightly darker frame:

```
C = 21
F = (9.0/5)*C + 32
print F
```

As a reader of this book, you may wonder if a code shown is a complete program you can try out or if it is just a part of a program (a snippet) so that you need to add surrounding statements (e.g., import statements) to try the code out yourself. The appearance of a vertical line to the left or not will then quickly tell you what type of code you see.

An interactive Python session is typeset as

```
>>> from math import *
>>> p = 1; c = -1.5
>>> a = sqrt(4*p + c)
```

Running a program, say `ball_yc.py`, in the terminal window, followed by some possible output is typeset as

```
                          ┌─────────┐
──────────────────────────┤ Terminal ├───  ────────────────────────────
                          └─────────┘
ball_yc.py
At t=0.0417064 s and 0.977662 s, the height is 0.2 m.
```

Recall from Sect. 1.5.3 that we just write the program name. A real execution demands prefixing the program name by `python` in a terminal window, or by `run` if you run the program from an interactive IPython session. We refer to Appendix H.5 for more complete information on running Python programs in different ways.

Sometimes just the output from a program is shown, and this output appears as plain computer text:

```
h = 0.2
order=0, error=0.221403
order=1, error=0.0214028
order=2, error=0.00140276
order=3, error=6.94248e-05
order=4, error=2.75816e-06
```

Files containing data are shown in a similar way in this book:

```
date   Oslo   London   Berlin   Paris   Rome   Helsinki
01.05  18     21.2     20.2     13.7    15.8   15
01.06  21     13.2     14.9     18      24     20
01.07  13     14       16       25      26.2   14.5
```

Style guide for Python code This book presents Python code that is (mostly) in accordance with the official Style Guide for Python Code[5], known in the Python community as *PEP8*. Some exceptions to the rules are made to make code snippets shorter: multiple imports on one line and less blank lines.

1.9 Exercises

About solving exercises There is only one way to learn programming: you have to program yourself. This means that you have to do *a lot* of exercises! Reading this book is necessary to learn about the Python syntax and studying the examples in depth is necessary to grasp how to think about programming and solving problems. But the main effort in the learning process is your work with exercises or your own programming projects.

Solving an exercise is a three-stage procedure. First, you have to study the text in the exercise carefully to understand what the problem is about. Programming exercises, especially in this book, are about a problem setting that has to be thoroughly understood before it makes sense to understand the specific questions in

---

[5] http://www.python.org/dev/peps/pep-0008/

the exercise. The second phase is to write the program. The more efforts you put into the first phase, the easier it will be to find the right statements and write the code. The third and final stage is to test the program and remove errors (known as debugging and verification from Sect. 1.2). This is by far the greatest challenge for beginners. Very often, especially for newcomers to programming, it boils down to writing out the result of every statement and checking these results carefully by playing computer with pen and paper.

Beginners often underestimate the amount of work required in the first and third stage and instead try to do the second stage (i.e., write the program) as quickly as possible. The more work you put into the first stage, the easier it will be to find an example in this book or elsewhere that is similar to the exercise and that can help you get started. And the more work you put into stage three up front, with constructing a test case, the better your understanding of the statements will be and the fewer errors you will commit. Experience will prove that all these assertions are right!

Most exercises are associated with a filename, e.g., `myexer`. If the answer to the exercise is a Python program, you should store the program in a file `myexer.py`. If the answer can be an explanation, you may store it in a plain text file, `myexer.txt`, or write the text in a word processor and produce a PDF file (`myexer.pdf`).

When you hand in exercises to teaching assistants, it is often a requirement that a *trial run* of the program is inserted at the end of the code. This means that you run some case with known result, direct the output to a file `result`,

```
Terminal
Terminal> python myprogram.py > result
```

and copy the contents of `result` to a triple-quoted string with appropriate comments after the statements of the program. Here is an example of a program with its trial run inserted:

```
F = 69.8                 # Fahrenheit degrees
C = (5.0/9)*(F - 32)     # Corresponding Celsius degrees
print C

'''
Trial run (correct result is 21):
python f2c.py
21.0
'''
```

The trial run demonstrates that the program runs and produces correct results in a test case.

Exercise 1.1: Compute 1+1
The first exercise concerns some very basic mathematics and programming: assign the result of $1+1$ to a variable and print the value of that variable.
Filename: `1plus1`.

Exercise 1.2: Write a Hello World program
Almost all books about programming languages start with a very simple program that prints the text Hello, World! to the screen. Make such a program in Python. Filename: hello_world.

Exercise 1.3: Derive and compute a formula
Can a newborn baby in Norway expect to live for one billion (10^9) seconds? Write a Python program for doing arithmetics to answer the question. Filename: seconds2years.

Exercise 1.4: Convert from meters to British length units
Make a program where you set a length given in meters and then compute and write out the corresponding length measured in inches, in feet, in yards, and in miles. Use that one inch is 2.54 cm, one foot is 12 inches, one yard is 3 feet, and one British mile is 1760 yards. For verification, a length of 640 meters corresponds to 25196.85 inches, 2099.74 feet, 699.91 yards, or 0.3977 miles. Filename: length_conversion.

Exercise 1.5: Compute the mass of various substances
The density of a substance is defined as $\varrho = m/V$, where m is the mass of a volume V. Compute and print out the mass of one liter of each of the following substances whose densities in g/cm^3 are found in the file src/files/densities.dat[6]: iron, air, gasoline, ice, the human body, silver, and platinum. Filename: 1liter.

Exercise 1.6: Compute the growth of money in a bank
Let p be a bank's interest rate in percent per year. An initial amount A has then grown to

$$A \left(1 + \frac{p}{100}\right)^n$$

after n years. Make a program for computing how much money 1000 euros have grown to after three years with 5 percent interest rate. Filename: interest_rate.

Exercise 1.7: Find error(s) in a program
Suppose somebody has written a simple one-line program for computing sin(1):

```
x=1; print 'sin(%g)=%g' % (x, sin(x))
```

Create this program and try to run it. What is the problem? Filename: find_errors_sin1.

Exercise 1.8: Type in program text
Type the following program in your editor and execute it. If your program does not work, check that you have copied the code correctly.

---

[6] http://tinyurl.com/pwyasaa/files/densities.dat

```
from math import pi

h = 5.0    # height
b = 2.0    # base
r = 1.5    # radius

area_parallelogram = h*b
print 'The area of the parallelogram is %.3f' % area_parallelogram

area_square = b**2
print 'The area of the square is %g' % area_square

area_circle = pi*r**2
print 'The area of the circle is %.3f' % area_circle

volume_cone = 1.0/3*pi*r**2*h
print 'The volume of the cone is %.3f' % volume_cone
```

Filename: `formulas_shapes`.

Exercise 1.9: Type in programs and debug them
Type these short programs in your editor and execute them. When they do not work, identify and correct the erroneous statements.

a) Does $\sin^2(x) + \cos^2(x) = 1$?

```
from math import sin, cos
x = pi/4
1_val = math.sin^2(x) + math.cos^2(x)
print 1_VAL
```

b) Compute s in meters when $s = v_0 t + \frac{1}{2}at^2$, with $v_0 = 3$ m/s, $t = 1$ s, $a = 2$ m/s^2.

```
v0 = 3 m/s
t = 1 s
a = 2 m/s**2
s = v0.t + 0,5.a.t**2
print s
```

c) Verify these equations:

$$(a + b)^2 = a^2 + 2ab + b^2$$
$$(a - b)^2 = a^2 - 2ab + b^2$$

```
a = 3,3   b = 5,3
a2 = a**2
b2 = b**2
```

```
eq1_sum = a2 + 2ab + b2
eq2_sum = a2 - 2ab + b2

eq1_pow = (a + b)**2
eq2_pow = (a - b)**2

print 'First equation:  %g = %g', % (eq1_sum, eq1_pow)
print 'Second equation: %h = %h', % (eq2_pow, eq2_pow)
```

Filename: `find_errors_programs`.

Exercise 1.10: Evaluate a Gaussian function
The bell-shaped Gaussian function,

$$f(x) = \frac{1}{\sqrt{2\pi}\, s} \exp\left[-\frac{1}{2}\left(\frac{x-m}{s}\right)^2\right], \tag{1.7}$$

is one of the most widely used functions in science and technology. The parameters m and $s > 0$ are prescribed real numbers. Make a program for evaluating this function when $m = 0$, $s = 2$, and $x = 1$. Verify the program's result by comparing with hand calculations on a calculator.
Filename: `gaussian1`.

Remarks The function (1.7) is named after Carl Friedrich Gauss[7], 1777–1855, who was a German mathematician and scientist, now considered as one of the greatest scientists of all time. He contributed to many fields, including number theory, statistics, mathematical analysis, differential geometry, geodesy, electrostatics, astronomy, and optics. Gauss introduced the function (1.7) when he analyzed probabilities related to astronomical data.

Exercise 1.11: Compute the air resistance on a football
The drag force, due to air resistance, on an object can be expressed as

$$F_d = \frac{1}{2}C_D\varrho A V^2, \tag{1.8}$$

where ϱ is the density of the air, V is the velocity of the object, A is the cross-sectional area (normal to the velocity direction), and C_D is the drag coefficient, which depends heavily on the shape of the object and the roughness of the surface.

The gravity force on an object with mass m is $F_g = mg$, where $g = 9.81\,\mathrm{m\,s^{-2}}$.

We can use the formulas for F_d and F_g to study the importance of air resistance versus gravity when kicking a football. The density of air is $\varrho = 1.2\,\mathrm{kg\,m^{-3}}$. We have $A = \pi a^2$ for any ball with radius a. For a football, $a = 11$ cm and the mass is 0.43 kg. The drag coefficient C_D varies with the velocity and can be taken as 0.4.

Make a program that computes the drag force and the gravity force on a football. Write out the forces with one decimal in units of Newton (N = kg m/s^2). Also print the ratio of the drag force and the gravity force. Define C_D, ϱ, A, V, m, g,

---

[7] http://en.wikipedia.org/wiki/Carl_Gauss

F_d, and F_g as variables, and put a comment with the corresponding unit. Use the program to calculate the forces on the ball for a hard kick, $V = 120$ km/h and for a soft kick, $V = 30$ km/h (it is easy to mix inconsistent units, so make sure you compute with V expressed in m/s).
Filename: `kick`.

Exercise 1.12: How to cook the perfect egg

As an egg cooks, the proteins first denature and then coagulate. When the temperature exceeds a critical point, reactions begin and proceed faster as the temperature increases. In the egg white, the proteins start to coagulate for temperatures above 63 °C, while in the yolk the proteins start to coagulate for temperatures above 70 °C. For a soft boiled egg, the white needs to have been heated long enough to coagulate at a temperature above 63 °C, but the yolk should not be heated above 70 °C. For a hard boiled egg, the center of the yolk should be allowed to reach 70 °C.

The following formula expresses the time t it takes (in seconds) for the center of the yolk to reach the temperature T_y (in Celsius degrees):

$$t = \frac{M^{2/3} c \rho^{1/3}}{K \pi^2 (4\pi/3)^{2/3}} \ln\left[0.76 \frac{T_o - T_w}{T_y - T_w} \right]. \tag{1.9}$$

Here, M, ρ, c, and K are properties of the egg: M is the mass, ρ is the density, c is the specific heat capacity, and K is thermal conductivity. Relevant values are $M = 47$ g for a small egg and $M = 67$ g for a large egg, $\rho = 1.038 \, \mathrm{g\,cm^{-3}}$, $c = 3.7 \, \mathrm{J\,g^{-1}\,K^{-1}}$, and $K = 5.4 \cdot 10^{-3} \, \mathrm{W\,cm^{-1}\,K^{-1}}$. Furthermore, T_w is the temperature (in C degrees) of the boiling water, and T_o is the original temperature (in C degrees) of the egg before being put in the water. Implement the formula in a program, set $T_w = 100$ °C and $T_y = 70$ °C, and compute t for a large egg taken from the fridge ($T_o = 4$ °C) and from room temperature ($T_o = 20$ °C).
Filename: `egg`.

Exercise 1.13: Derive the trajectory of a ball

The purpose of this exercise is to explain how Equation (1.6) for the trajectory of a ball arises from basic physics. There is no programming in this exercise, just physics and mathematics.

The motion of the ball is governed by Newton's second law:

$$F_x = ma_x \tag{1.10}$$

$$F_y = ma_y \tag{1.11}$$

where F_x and F_y are the sum of forces in the x and y directions, respectively, a_x and a_y are the accelerations of the ball in the x and y directions, and m is the mass of the ball. Let $(x(t), y(t))$ be the position of the ball, i.e., the horizontal and vertical coordinate of the ball at time t. There are well-known relations between acceleration, velocity, and position: the acceleration is the time derivative of the velocity, and the velocity is the time derivative of the position. Therefore we have that

$$a_x = \frac{d^2x}{dt^2},$$ (1.12)

$$a_y = \frac{d^2y}{dt^2}.$$ (1.13)

If we assume that gravity is the only important force on the ball, $F_x = 0$ and $F_y = -mg$.

Integrate the two components of Newton's second law twice. Use the initial conditions on velocity and position,

$$\frac{d}{dt}x(0) = v_0 \cos\theta,$$ (1.14)

$$\frac{d}{dt}y(0) = v_0 \sin\theta,$$ (1.15)

$$x(0) = 0,$$ (1.16)

$$y(0) = y_0,$$ (1.17)

to determine the four integration constants. Write up the final expressions for $x(t)$ and $y(t)$. Show that if $\theta = \pi/2$, i.e., the motion is purely vertical, we get the formula (1.1) for the y position. Also show that if we eliminate t, we end up with the relation (1.6) between the x and y coordinates of the ball. You may read more about this type of motion in a physics book, e.g., [15].
Filename: trajectory.

Exercise 1.14: Find errors in the coding of formulas
Some versions of our program for calculating the formula (1.3) are listed below. Find the versions that will not work correctly and explain why in each case.

```
C = 21;     F =  9/5*C + 32;        print F
C = 21.0;   F =  (9/5)*C + 32;      print F
C = 21.0;   F =  9*C/5 + 32;        print F
C = 21.0;   F =  9.*(C/5.0) + 32;   print F
C = 21.0;   F =  9.0*C/5.0 + 32;    print F
C = 21;     F =  9*C/5 + 32;        print F
C = 21.0;   F =  (1/5)*9*C + 32;    print F
C = 21;     F =  (1./5)*9*C + 32;   print F
```

Filename: find_errors_division.

Exercise 1.15: Explain why a program does not work
Figure out why the following program does not work:

```
C = A + B
A = 3
B = 2
print C
```

Filename: find_errors_vars.

Exercise 1.16: Find errors in Python statements
Try the following statements in an interactive Python shell. Explain why some statements fail and correct the errors.

```
1a = 2
a1 = b
x = 2
y = X + 4   # is it 6?
from Math import tan
print tan(pi)
pi = "3.14159'
print tan(pi)
c = 4**3**2**3
_ = ((c-78564)/c + 32))
discount = 12%
AMOUNT = 120.-
amount = 120$
address = hpl@simula.no
and = duck
class = 'INF1100, gr 2"
continue_ = x > 0
rev = fox = True
Norwegian = ['a human language']
true = fox is rev in Norwegian
```

Hint It is wise to test the values of the expressions on the right-hand side, and the validity of the variable names, separately before you put the left- and right-hand sides together in statements. The last two statements work, but explaining why goes beyond what is treated in this chapter.
Filename: `find_errors_syntax`.

Exercise 1.17: Find errors in the coding of a formula
Given a quadratic equation,

$$ax^2 + bx + c = 0,$$

the two roots are

$$x_1 = \frac{-b + \sqrt{b^2 - 4ac}}{2a}, \quad x_2 = \frac{-b - \sqrt{b^2 - 4ac}}{2a}. \tag{1.18}$$

What are the problems with the following program?

```
a = 2; b = 1; c = 2
from math import sqrt
q = b*b - 4*a*c
q_sr = sqrt(q)
x1 = (-b + q_sr)/2*a
x2 = (-b - q_sr)/2*a
print x1, x2
```

Correct the program so that it solves the given equation.
Filename: `find_errors_roots`.

Exercise 1.18: Find errors in a program

What is the problem in the following program?

```
from math import pi, tan
tan = tan(pi/4)
tan2 = tan(pi/3)
print tan, tan2
```

Filename: `find_errors_tan`.

Loops and Lists

2

This chapter explains how repetitive tasks in a program can be automated by loops. We also introduce list objects for storing and processing collections of data with a specific order. Loops and lists, together with functions and `if` tests from Chap. 3, lay the fundamental programming foundation for the rest of the book. The programs associated with the chapter are found in the folder `src/looplist`[1].

2.1 While Loops

Our task now is to print out a conversion table with Celsius degrees in the first column of the table and the corresponding Fahrenheit degrees in the second column. Such a table may look like this:

```
-20   -4.0
-15    5.0
-10   14.0
 -5   23.0
  0   32.0
  5   41.0
 10   50.0
 15   59.0
 20   68.0
 25   77.0
 30   86.0
 35   95.0
 40  104.0
```

2.1.1 A Naive Solution

The formula for converting C degrees Celsius to F degrees Fahrenheit is $F = 9C/5 + 32$. Since we know how to evaluate the formula for one value of C, we can just repeat these statements as many times as required for the table above. Using

---

[1] http://tinyurl.com/pwyasaa/looplist

© Springer-Verlag Berlin Heidelberg 2016
H.P. Langtangen, *A Primer on Scientific Programming with Python*,
Texts in Computational Science and Engineering 6, DOI 10.1007/978-3-662-49887-3_2

three statements per line in the program, for compact layout of the code, we can
write the whole program as

```
C = -20;  F = 9.0/5*C + 32;  print C, F
C = -15;  F = 9.0/5*C + 32;  print C, F
C = -10;  F = 9.0/5*C + 32;  print C, F
C =  -5;  F = 9.0/5*C + 32;  print C, F
C =   0;  F = 9.0/5*C + 32;  print C, F
C =   5;  F = 9.0/5*C + 32;  print C, F
C =  10;  F = 9.0/5*C + 32;  print C, F
C =  15;  F = 9.0/5*C + 32;  print C, F
C =  20;  F = 9.0/5*C + 32;  print C, F
C =  25;  F = 9.0/5*C + 32;  print C, F
C =  30;  F = 9.0/5*C + 32;  print C, F
C =  35;  F = 9.0/5*C + 32;  print C, F
C =  40;  F = 9.0/5*C + 32;  print C, F
```

Running this program (which is stored in the file `c2f_table_repeat.py`),
demonstrates that the output becomes

```
-20 -4.0
-15 5.0
-10 14.0
-5 23.0
0 32.0
5 41.0
10 50.0
15 59.0
20 68.0
25 77.0
30 86.0
35 95.0
40 104.0
```

This output suffers from somewhat ugly formatting, but that problem can quickly be
fixed by replacing print C, F by a `print` statement based on printf formatting.
We will return to this detail later.

The main problem with the program above is that lots of statements are identical and repeated. First of all it is boring to write this sort of repeated statements,
especially if we want many more C and F values in the table. Second, the idea
of the computer is to automate repetition. Therefore, all computer languages have
constructs to efficiently express repetition. These constructs are called *loops* and
come in two variants in Python: while loops and for loops. Most programs in this
book employ loops, so this concept is extremely important to learn.

2.1.2 While Loops

The while loop is used to repeat a set of statements as long as a condition is true.
We shall introduce this kind of loop through an example. The task is to generate the
rows of the table of C and F values. The C value starts at -20 and is incremented

by 5 as long as $C \leq 40$. For each C value we compute the corresponding F value and write out the two temperatures. In addition, we also add a line of dashes above and below the table.

The list of tasks to be done can be summarized as follows:

- Print line with dashes
- $C = -20$
- While $C \leq 40$:
  - $F = \frac{9}{5}C + 32$
  - Print C and F
  - Increment C by 5
- Print line with dashes

This is the *algorithm* of our programming task. The way from a detailed algorithm to a fully functioning Python code can often be made very short, which is definitely true in the present case:

```
print '------------------'       # table heading
C = -20                           # start value for C
dC = 5                            # increment of C in loop
while C <= 40:                    # loop heading with condition
    F = (9.0/5)*C + 32            # 1st statement inside loop
    print C, F                    # 2nd statement inside loop
    C = C + dC                    # 3rd statement inside loop
print '------------------'       # end of table line (after loop)
```

A very important feature of Python is now encountered: the block of statements to be executed in each pass of the `while` loop must be indented. In the example above the block consists of three lines, and all these lines must have exactly the same indentation. Our choice of indentation in this book is four spaces. The first statement whose indentation coincides with that of the `while` line marks the end of the loop and is executed after the loop has terminated. In this example this is the final `print` statement. You are encouraged to type in the code above in a file, indent the last line four spaces, and observe what happens (you will experience that lines in the table are separated by a line of dashes: `----`).

Many novice Python programmers forget the colon at the end of the `while` line – this colon is essential and marks the beginning of the indented block of statements inside the loop. Later, we will see that there are many other similar program constructions in Python where there is a heading ending with a colon, followed by an indented block of statements.

Programmers need to fully understand what is going on in a program and be able to simulate the program by hand. Let us do this with the program segment above. First, we define the start value for the sequence of Celsius temperatures: `C = -20`. We also define the increment `dC` that will be added to `C` inside the loop. Then we enter the loop condition `C <= 40`. The first time `C` is `-20`, which implies that `C <= 40` (equivalent to $C \leq 40$ in mathematical notation) is true. Since the loop condition is true, we enter the loop and execute all the indented statements. That is, we compute F corresponding to the current C value, print the temperatures, and increment C by dC. For simplicity, we have used a plain `print C, F` without any formatting so the columns will not be aligned, but this can easily be fixed later.

Thereafter, we enter the second pass in the loop. First we check the condition: C is -15 and C <= 40 is still true. We execute the statements in the indented loop block, C becomes -10, this is still less than or equal to 40, so we enter the loop block again. This procedure is repeated until C is updated from 40 to 45 in the final statement in the loop block. When we then test the condition, C <= 40, this condition is no longer true, and the loop is terminated. We proceed with the next statement that has the same indentation as the `while` statement, which is the final `print` statement in this example.

Newcomers to programming are sometimes confused by statements like

```
C = C + dC
```

This line looks erroneous from a mathematical viewpoint, but the statement is perfectly valid computer code, because we first evaluate the expression on the right-hand side of the equality sign and then let the variable on the left-hand side refer to the result of this evaluation. In our case, C and dC are two different `int` objects. The operation C+dC results in a new `int` object, which in the assignment C = C+dC is bound to the name C. Before this assignment, C was already bound to an `int` object, and this object is automatically destroyed when C is bound to a new object and there are no other names (variables) referring to this previous object (if you did not get this last point, just relax and continue reading!).

Since incrementing the value of a variable is frequently done in computer programs, there is a special short-hand notation for this and related operations:

```
C += dC   # equivalent to C = C + dC
C -= dC   # equivalent to C = C - dC
C *= dC   # equivalent to C = C*dC
C /= dC   # equivalent to C = C/dC
```

2.1.3 Boolean Expressions

In our first example on a `while` loop, we worked with a condition C <= 40, which evaluates to either true or false, written as `True` or `False` in Python. Other comparisons are also useful:

```
C == 40   # C equals 40
C != 40   # C does not equal 40
C >= 40   # C is greater than or equal to 40
C >  40   # C is greater than 40
C <  40   # C is less than 40
```

Not only comparisons between numbers can be used as conditions in `while` loops: any expression that has a boolean (`True` or `False`) value can be used. Such expressions are known as *logical* or *boolean* expressions.

The keyword `not` can be inserted in front of the boolean expression to change the value from `True` to `False` or from `False` to `True`. To evaluate not C ==

40, we first evaluate C == 40, for C = 1 this is False, and then not turns the value into True. On the opposite, if C == 40 is True, not C == 40 becomes False. Mathematically it is easier to read C != 40 than not C == 40, but these two boolean expressions are equivalent.

Boolean expressions can be combined with and and or to form new compound boolean expressions, as in

```
while x > 0 and y <= 1:
    print x, y
```

If cond1 and cond2 are two boolean expressions with values True or False, the compound boolean expression cond1 and cond2 is True if both cond1 and cond2 are True. On the other hand, cond1 or cond2 is True if at least one of the conditions, cond1 or cond2, is True

---

Remark

In Python, cond1 and cond2 or cond1 or cond2 returns one of the operands and not just True or False values as in most other computer languages. The operands cond1 or cond2 can be expressions or objects. In case of expressions, these are first evaluated to an object before the compound boolean expression is evaluated. For example, (5+1) or -1 evaluates to 6 (the second operand is not evaluated when the first one is True), and (5+1) and -1 evaluates to -1.

Here are some more examples from an interactive session where we just evaluate the boolean expressions themselves without using them in loop conditions:

```
>>> x = 0;   y = 1.2
>>> x >= 0 and y < 1
False
>>> x >= 0 or y < 1
True
>>> x > 0 or y > 1
True
>>> x > 0 or not y > 1
False
>>> -1 < x <= 0    #  -1 < x and x <= 0
True
>>> not (x > 0 or y > 0)
False
```

In the last sample expression, not applies to the value of the boolean expression inside the parentheses: x>0 is False, y>0 is True, so the combined expression with or is True, and not turns this value to False.

The common boolean values in Python are True, False, 0 (false), and any integer different from zero (true). To see such values in action, we recommend doing Exercises 2.22 and 2.18.

Boolean evaluation of an object

All objects in Python can in fact be evaluated in a boolean context, and all are True except False, zero numbers, and empty strings, lists, and dictionaries:

```
>>> s = 'some string'
>>> bool(s)
True
>>> s = ''  # empty string
>>> bool(s)
False
>>> L = [1, 4, 6]
>>> bool(L)
True
>>> L = []
>>> bool(L)
False
>>> a = 88.0
>>> bool(a)
True
>>> a = 0.0
>>> bool(a)
False
```

Essentially, if a tests if a is a non-empty object or if it is non-zero value. Such constructions are frequent in Python code.

Erroneous thinking about boolean expressions is one of the most common sources of errors in computer programs, so you should be careful every time you encounter a boolean expression and check that it is correctly stated.

2.1.4   Loop Implementation of a Sum

Summations frequently appear in mathematics. For instance, the sine function can be calculated as a polynomial:

$$\sin(x) \approx x - \frac{x^3}{3!} + \frac{x^5}{5!} - \frac{x^7}{7!} + \cdots, \tag{2.1}$$

where $3! = 3 \cdot 2 \cdot 1, 5! = 5 \cdot 4 \cdot 3 \cdot 2 \cdot 1$, etc., are factorial expressions. Computing $k! = k(k-1)(k-2) \cdots 2 \cdot 1$ is done by `math.factorial(k)`.

An infinite number of terms are needed on the right-hand side of (2.1) for the equality sign to hold. With a finite number of terms, we obtain an approximation to $\sin(x)$, which is well suited for being calculated in a program since only powers and the basic four arithmetic operations are involved. Say we want to compute the right-hand side of (2.1) for powers up to $N = 25$. Writing out and implementing each one of these terms is a tedious job that can easily be automated by a loop.

Computation of the sum in (2.1) by a `while` loop in Python, makes use of (i) a counter k that runs through odd numbers from 1 up to some given maximum power N, and (ii) a summation variable, say s, which accumulates the terms, one at a time. The purpose of each pass of the loop is to compute a new term and add it to s. Since the sign of each term alternates, we introduce a variable `sign` that changes between -1 and 1 in each pass of the loop.

The previous paragraph can be precisely expressed by this piece of Python code:

```
x = 1.2    # assign some value
N = 25     # maximum power in sum
k = 1
s = x
sign = 1.0
import math

while k < N:
    sign = - sign
    k = k + 2
    term = sign*x**k/math.factorial(k)
    s = s + term

print 'sin(%g) = %g (approximation with %d terms)' % (x, s, N)
```

The best way to understand such a program is to simulate it by hand. That is, we go through the statements, one by one, and write down on a piece of paper what the state of each variable is.

When the loop is first entered, `k < N` implies `1 < 25`, which is `True` so we enter the loop block. There, we compute `sign = -1.0`, `k = 3`, `term = -1.0*x**3/(3*2*1)` (note that `sign` is float so we always have `float` divided by `int`), and `s = x - x**3/6`, which equals the first two terms in the sum. Then we test the loop condition: `3 < 25` is `True` so we enter the loop block again. This time we obtain `term = 1.0*x**5/math.factorial(5)`, which correctly implements the third term in the sum. At some point, `k` is updated to from 23 to 25 inside the loop and the loop condition then becomes `25 < 25`, which is `False`, implying that the program jumps over the loop block and continues with the `print` statement (which has the same indentation as the `while` statement).

2.2 Lists

Up to now a variable has typically contained a single number. Sometimes numbers are naturally grouped together. For example, all Celsius degrees in the first column of our table from Sect. 2.1.2 could be conveniently stored together as a group. A Python *list* can be used to represent such a group of numbers in a program. With a variable that refers to the list, we can work with the whole group at once, but we can also access individual elements of the group. Figure 2.1 illustrates the difference between an `int` object and a list object. In general, a list may contain a sequence of arbitrary objects in a given order. Python has great functionality for examining and manipulating such sequences of objects, which will be demonstrated below.

2.2.1 Basic List Operations

To create a list with the numbers from the first column in our table, we just put all the numbers inside square brackets and separate the numbers by commas:

```
C = [-20, -15, -10, -5, 0, 5, 10, 15, 20, 25, 30, 35, 40]
```

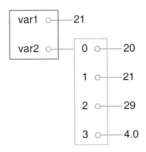

Fig. 2.1 Illustration of two variables: `var1` refers to an `int` object with value 21, created by the statement `var1 = 21`, and `var2` refers to a `list` object with value `[20, 21, 29, 4.0]`, i.e., three `int` objects and one `float` object, created by the statement `var2 = [20, 21, 29, 4.0]`

The variable C now refers to a `list` object holding 13 list *elements*. All list elements are in this case `int` objects.

Every element in a list is associated with an *index*, which reflects the position of the element in the list. The first element has index 0, the second index 1, and so on. Associated with the C list above we have 13 indices, starting with 0 and ending with 12. To access the element with index 3, i.e., the fourth element in the list, we can write C[3]. As we see from the list, C[3] refers to an `int` object with the value -5.

Elements in lists can be deleted, and new elements can be inserted anywhere. The functionality for doing this is built into the list object and accessed by a dot notation. Two examples are C.append(v), which appends a new element v to the end of the list, and C.insert(i,v), which inserts a new element v in position number i in the list. The number of elements in a list is given by len(C). Let us exemplify some list operations in an interactive session to see the effect of the operations:

```
>>> C = [-10, -5, 0, 5, 10, 15, 20, 25, 30]      # create list
>>> C.append(35)                 # add new element 35 at the end
>>> C                            # view list C
[-10, -5, 0, 5, 10, 15, 20, 25, 30, 35]
```

Two lists can be added:

```
>>> C = C + [40, 45]             # extend C at the end
>>> C
[-10, -5, 0, 5, 10, 15, 20, 25, 30, 35, 40, 45]
```

What adding two lists means is up to the list object to define, and not surprisingly, addition of two lists is defined as appending the second list to the first. The result of C + [40,45] is a new list object, which we then assign to C such that this name refers to this new list. In fact, every object in Python and everything you can do with it is defined by programs made by humans. With the techniques of class programming (see Chap. 7) you can create your own objects and define (if desired) what it means to add such objects. All this gives enormous power in the hands of programmers. As one example, you can define your own list object if you are not satisfied with the functionality of Python's own lists.

New elements can be inserted anywhere in the list (and not only at the end as we did with `C.append`):

```
>>> C.insert(0, -15)        # insert new element -15 as index 0
>>> C
[-15, -10, -5, 0, 5, 10, 15, 20, 25, 30, 35, 40, 45]
```

With `del C[i]` we can remove an element with index `i` from the list `C`. Observe that this changes the list, so `C[i]` refers to another (the next) element after the removal:

```
>>> del C[2]                # delete 3rd element
>>> C
[-15, -10, 0, 5, 10, 15, 20, 25, 30, 35, 40, 45]
>>> del C[2]                # delete what is now 3rd element
>>> C
[-15, -10, 5, 10, 15, 20, 25, 30, 35, 40, 45]
>>> len(C)                  # length of list
11
```

The command `C.index(10)` returns the index corresponding to the first element with value 10 (this is the 4th element in our sample list, with index 3):

```
>>> C.index(10)             # find index for an element (10)
3
```

To just test if an object with the value 10 is an element in the list, one can write the boolean expression `10 in C`:

```
>>> 10 in C                 # is 10 an element in C?
True
```

Python allows negative indices, which leads to indexing from the right. As demonstrated below, `C[-1]` gives the last element of the list `C`. `C[-2]` is the element before `C[-1]`, and so forth.

```
>>> C[-1]                   # view the last list element
45
>>> C[-2]                   # view the next last list element
40
```

Building long lists by writing down all the elements separated by commas is a tedious process that can easily be automated by a loop, using ideas from Sect. 2.1.4. Say we want to build a list of degrees from −50 to 200 in steps of 2.5 degrees. We then start with an empty list and use a `while` loop to append one element at a time:

```
C = []
C_value = -50
C_max = 200
while C_value <= C_max:
    C.append(C_value)
    C_value += 2.5
```

In the next sections, we shall see how we can express these six lines of code with just one single statement.

There is a compact syntax for creating variables that refer to the various list elements. Simply list a sequence of variables on the left-hand side of an assignment to a list:

```
>>> somelist = ['book.tex', 'book.log', 'book.pdf']
>>> texfile, logfile, pdf = somelist
>>> texfile
'book.tex'
>>> logfile
'book.log'
>>> pdf
'book.pdf'
```

The number of variables on the left-hand side must match the number of elements in the list, otherwise an error occurs.

A final comment regards the syntax: some list operations are reached by a dot notation, as in `C.append(e)`, while other operations requires the list object as an argument to a function, as in `len(C)`. Although `C.append` for a programmer behaves as a function, it is a function that is reached through a list object, and it is common to say that `append` is a *method* in the list object, not a function. There are no strict rules in Python whether functionality regarding an object is reached through a method or a function.

2.2.2 For Loops

The nature of for loops When data are collected in a list, we often want to perform the same operations on each element in the list. We then need to walk through all list elements. Computer languages have a special construct for doing this conveniently, and this construct is in Python and many other languages called a `for` loop. Let us use a `for` loop to print out all list elements:

```
degrees = [0, 10, 20, 40, 100]
for C in degrees:
    print 'list element:', C
print 'The degrees list has', len(degrees), 'elements'
```

The `for C in degrees` construct creates a loop over all elements in the list `degrees`. In each pass of the loop, the variable C refers to an element in the list, starting with `degrees[0]`, proceeding with `degrees[1]`, and so on, before ending with the last element degrees[n-1] (if n denotes the number of elements in the list, len(degrees)).

The for loop specification ends with a colon, and after the colon comes a block of statements that does something useful with the current element. Each statement in the block must be indented, as we explained for while loops. In the example above, the block belonging to the for loop contains only one statement. The final print statement has the same indentation (none in this example) as the for statement and is executed as soon as the loop is terminated.

As already mentioned, understanding all details of a program by following the program flow by hand is often a very good idea. Here, we first define a list degrees containing 5 elements. Then we enter the for loop. In the first pass of the loop, C refers to the first element in the list degrees, i.e., the int object holding the value 0. Inside the loop we then print out the text 'list element:' and the value of C, which is 0. There are no more statements in the loop block, so we proceed with the next pass of the loop. C then refers to the int object 10, the output now prints 10 after the leading text, we proceed with C as the integers 20 and 40, and finally C is 100. After having printed the list element with value 100, we move on to the statement after the indented loop block, which prints out the number of list elements. The total output becomes

```
list element: 0
list element: 10
list element: 20
list element: 40
list element: 100
The degrees list has 5 elements
```

Correct indentation of statements is crucial in Python, and we therefore strongly recommend you to work through Exercise 2.23 to learn more about this topic.

Making the table Our knowledge of lists and for loops over elements in lists puts us in a good position to write a program where we collect all the Celsius degrees to appear in the table in a list Cdegrees, and then use a for loop to compute and write out the corresponding Fahrenheit degrees. The complete program may look like this:

```
Cdegrees = [-20, -15, -10, -5, 0, 5, 10, 15, 20, 25, 30, 35, 40]
for C in Cdegrees:
    F = (9.0/5)*C + 32
    print C, F
```

The print C, F statement just prints the value of C and F with a default format, where each number is separated by one space character (blank). This does not look like a nice table (the output is identical to the one shown in Sect. 2.1.1. Nice formatting is obtained by forcing C and F to be written in fields of fixed width and with a fixed number of decimals. An appropriate printf format is %5d (or %5.0f) for C and %5.1f for F. We may also add a headline to the table. The complete program becomes:

```
Cdegrees = [-20, -15, -10, -5, 0, 5, 10, 15, 20, 25, 30, 35, 40]
print '    C    F'
for C in Cdegrees:
    F = (9.0/5)*C + 32
    print '%5d %5.1f' % (C, F)
```

This code is found in the file c2f_table_list.py and its output becomes

```
  C    F
-20  -4.0
-15   5.0
-10  14.0
 -5  23.0
  0  32.0
  5  41.0
 10  50.0
 15  59.0
 20  68.0
 25  77.0
 30  86.0
 35  95.0
 40 104.0
```

2.3 Alternative Implementations with Lists and Loops

We have already solved the problem of printing out a nice-looking conversion table for Celsius and Fahrenheit degrees. Nevertheless, there are usually many alternative ways to write a program that solves a specific problem. The next paragraphs explore some other possible Python constructs and programs to store numbers in lists and print out tables. The various code snippets are collected in the program file session.py.

2.3.1 While Loop Implementation of a for Loop

Any for loop can be implemented as a while loop. The general code

```
for element in somelist:
    <process element>
```

can be transformed to this while loop:

```
index = 0
while index < len(somelist):
    element = somelist[index]
    <process element>
    index += 1
```

In particular, the example involving the printout of a table of Celsius and Fahrenheit degrees can be implemented as follows in terms of a while loop:

```
Cdegrees = [-20, -15, -10, -5, 0, 5, 10, 15, 20, 25, 30, 35, 40]
index = 0
print '    C    F'
while index < len(Cdegrees):
    C = Cdegrees[index]
    F = (9.0/5)*C + 32
    print '%5d %5.1f' % (C, F)
    index += 1
```

2.3.2 The Range Construction

It is tedious to write the many elements in the Cdegrees in the previous programs. We should use a loop to automate the construction of the Cdegrees list. The range construction is particularly useful in this regard:

- range(n) generates integers 0, 1, 2, ..., n-1.
- range(start, stop, step) generates a sequence if integers start, start+step, start+2*step, and so on up to, *but not including*, stop. For example, range(2, 8, 3) returns 2 and 5 (and not 8), while range(1, 11, 2) returns 1, 3, 5, 7, 9.
- range(start, stop) is the same as range(start, stop, 1).

A for loop over integers are written as

```
for i in range(start, stop, step):
    ...
```

We can use this construction to create a Cdegrees list of the values $-20, -15, \ldots, 40$:

```
Cdegrees = []
for C in range(-20, 45, 5):
    Cdegrees.append(C)

# or just
Cdegrees = range(-20, 45, 5)
```

Note that the upper limit must be greater than 40 to ensure that 40 is included in the range of integers.

Suppose we want to create Cdegrees as $-10, -7.5, -5, \ldots, 40$. This time we cannot use range directly, because range can only create integers and we have decimal degrees such as -7.5 and 1.5. In this more general case, we introduce an integer counter i and generate the C values by the formula $C = -10 + i \cdot 2.5$ for $i = 0, 1, \ldots, 20$. The following Python code implements this task:

```
Cdegrees = []
for i in range(0, 21):
    C = -10 + i*2.5
    Cdegrees.append(C)
```

2.3.3 For Loops with List Indices

Instead of iterating over a list directly with the construction

```
for element in somelist:
    ...
```

we can equivalently iterate of the list indices and index the list inside the loop:

```
for i in range(len(somelist)):
    element = somelist[i]
    ...
```

Since `len(somelist)` returns the length of `somelist` and the largest legal index is `len(somelist)-1`, because indices always start at 0, `range(len(somelist))` will generate all the correct indices: 0, 1, ..., `len(somelist)-1`.

Programmers coming from other languages, such as Fortran, C, C++, Java, and C#, are very much used to `for` loops with integer counters and usually tend to use `for i in range(len(somelist))` and work with `somelist[i]` inside the loop. This might be necessary or convenient, but if possible, Python programmers are encouraged to use `for element in somelist`, which is more elegant to read.

Iterating over loop indices is useful when we need to process two lists simultaneously. As an example, we first create two `Cdegrees` and `Fdegrees` lists, and then we make a list to write out a table with `Cdegrees` and `Fdegrees` as the two columns of the table. Iterating over a loop index is convenient in the final list:

```
Cdegrees = []
n = 21
C_min = -10
C_max = 40
dC = (C_max - C_min)/float(n-1)   # increment in C
for i in range(0, n):
    C = -10 + i*dC
    Cdegrees.append(C)

Fdegrees = []
for C in Cdegrees:
    F = (9.0/5)*C + 32
    Fdegrees.append(F)

for i in range(len(Cdegrees)):
    C = Cdegrees[i]
    F = Fdegrees[i]
    print '%5.1f %5.1f' % (C, F)
```

Instead of appending new elements to the lists, we can start with lists of the right size, containing zeros, and then index the lists to fill in the right values. Creating a list of length n consisting of zeros (for instance) is done by

```
somelist = [0]*n
```

With this construction, the program above can use `for` loops over indices everywhere:

```
n = 21
C_min = -10
C_max = 40
dC = (C_max - C_min)/float(n-1)   # increment in C

Cdegrees = [0]*n
for i in range(len(Cdegrees)):
    Cdegrees[i] = -10 + i*dC

Fdegrees = [0]*n
for i in range(len(Cdegrees)):
    Fdegrees[i] = (9.0/5)*Cdegrees[i] + 32

for i in range(len(Cdegrees)):
    print '%5.1f %5.1f' % (Cdegrees[i], Fdegrees[i])
```

Note that we need the construction `[0]*n` to create a list of the right length, otherwise the index `[i]` will be illegal.

2.3.4   Changing List Elements

We have two seemingly alternative ways to traverse a list, either a loop over elements or over indices. Suppose we want to change the `Cdegrees` list by adding 5 to all elements. We could try

```
for c in Cdegrees:
    c += 5
```

but this loop leaves `Cdegrees` unchanged, while

```
for i in range(len(Cdegrees)):
    Cdegrees[i] += 5
```

works as intended. What is wrong with the first loop? The problem is that c is an ordinary variable, which refers to a list element in the loop, but when we execute c += 5, we let c refer to a new `float` object (c+5). This object is never inserted in the list. The first two passes of the loop are equivalent to

```
c = Cdegrees[0]   # automatically done in the for statement
c += 5
c = Cdegrees[1]   # automatically done in the for statement
c += 5
```

The variable c can only be used to read list elements and never to change them. Only an assignment of the form

```
Cdegrees[i] = ...
```

can change a list element.

There is a way of traversing a list where we get both the index and an element in each pass of the loop:

```
for i, c in enumerate(Cdegrees):
    Cdegrees[i] = c + 5
```

This loop also adds 5 to all elements in the list.

2.3.5   List Comprehension

Because running through a list and for each element creating a new element in another list is a frequently encountered task, Python has a special compact syntax for doing this, called *list comprehension*. The general syntax reads

```
newlist = [E(e) for e in list]
```

where `E(e)` represents an expression involving element e. Here are three examples:

```
Cdegrees = [-5 + i*0.5 for i in range(n)]
Fdegrees = [(9.0/5)*C + 32 for C in Cdegrees]
C_plus_5 = [C+5 for C in Cdegrees]
```

List comprehensions are recognized as a `for` loop inside square brackets and will be frequently exemplified throughout the book.

2.3.6   Traversing Multiple Lists Simultaneously

We may use the `Cdegrees` and `Fdegrees` lists to make a table. To this end, we need to traverse both arrays. The `for element in list` construction is not suitable in this case, since it extracts elements from one list only. A solution is to use a `for` loop over the integer indices so that we can index both lists:

```
for i in range(len(Cdegrees)):
    print '%5d %5.1f' % (Cdegrees[i], Fdegrees[i])
```

It happens quite frequently that two or more lists need to be traversed simultaneously. As an alternative to the loop over indices, Python offers a special nice syntax that can be sketched as

```
for e1, e2, e3, ... in zip(list1, list2, list3, ...):
    # work with element e1 from list1, element e2 from list2,
    # element e3 from list3, etc.
```

The zip function turns *n* lists (list1, list2, list3, ...) into one list of *n*-tuples, where each *n*-tuple (e1,e2,e3,...) has its first element (e1) from the first list (list1), the second element (e2) from the second list (list2), and so forth. The loop stops when the end of the shortest list is reached. In our specific case of iterating over the two lists Cdegrees and Fdegrees, we can use the zip function:

```
for C, F in zip(Cdegrees, Fdegrees):
    print '%5d %5.1f' % (C, F)
```

It is considered more *Pythonic* to iterate over list elements, here C and F, rather than over list indices as in the for i in range(len(Cdegrees)) construction.

2.4   Nested Lists

Nested lists are list objects where the elements in the lists can be lists themselves. A couple of examples will motivate for nested lists and illustrate the basic operations on such lists.

2.4.1   A table as a List of Rows or Columns

Our table data have so far used one separate list for each column. If there were *n* columns, we would need *n* list objects to represent the data in the table. However, we think of a table as *one* entity, not a collection of *n* columns. It would therefore be natural to use one argument for the whole table. This is easy to achieve using a *nested list*, where each entry in the list is a list itself. A table object, for instance, is a list of lists, either a list of the row elements of the table or a list of the column elements of the table. Here is an example where the table is a list of two columns, and each column is a list of numbers:

```
Cdegrees = range(-20, 41, 5)    # -20, -15, ..., 35, 40
Fdegrees = [(9.0/5)*C + 32 for C in Cdegrees]

table = [Cdegrees, Fdegrees]
```

(Note that any value in [41, 45] can be used as second argument (stop value) to range and will ensure that 40 is included in the range of generate numbers.)

With the subscript table[0] we can access the first element in table, which is nothing but the Cdegrees list, and with table[0][2] we reach the third element in the first element, i.e., Cdegrees[2].

However, tabular data with rows and columns usually have the convention that the underlying data is a nested list where the first index counts the rows and the second index counts the columns. To have table on this form, we must construct table as a list of [C, F] pairs. The first index will then run over rows [C, F]. Here is how we may construct the nested list:

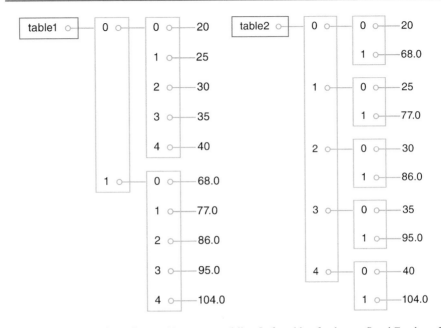

Fig. 2.2 Two ways of creating a table as a nested list. Left: table of columns C and F, where C and F are lists. Right: table of rows, where each row [C, F] is a list of two floats

```
table = []
for C, F in zip(Cdegrees, Fdegrees):
    table.append([C, F])
```

We may shorten this code segment by introducing a list comprehension:

```
table = [[C, F] for C, F in zip(Cdegrees, Fdegrees)]
```

This construction loops through pairs C and F, and for each pass in the loop we create a list element [C, F].

The subscript table[1] refers to the second element in table, which is a [C, F] pair, while table[1][0] is the C value and table[1][1] is the F value. Figure 2.2 illustrates both a list of columns and a list of pairs. Using this figure, you can realize that the first index looks up an element in the outer list, and that this element can be indexed with the second index.

2.4.2 Printing Objects

Modules for pretty print of objects We may write print table to immediately view the nested list table from the previous section. In fact, any Python object obj can be printed to the screen by the command print obj. The output is usually one line, and this line may become very long if the list has many elements. For example, a long list like our table variable, demands a quite long line when printed.

```
[[-20, -4.0], [-15, 5.0], [-10, 14.0], .............., [40, 104.0]]
```

Splitting the output over several shorter lines makes the layout nicer and more readable. The `pprint` module offers a *pretty print* functionality for this purpose. The usage of `pprint` looks like

```
import pprint
pprint.pprint(table)
```

and the corresponding output becomes

```
[[-20, -4.0],
 [-15, 5.0],
 [-10, 14.0],
 [-5, 23.0],
 [0, 32.0],
 [5, 41.0],
 [10, 50.0],
 [15, 59.0],
 [20, 68.0],
 [25, 77.0],
 [30, 86.0],
 [35, 95.0],
 [40, 104.0]]
```

With this book comes a slightly modified pprint module having the name `scitools.pprint2`. This module allows full format control of the printing of the `float` objects in lists by specifying `scitools.pprint2.float_format` as a printf format string. The following example demonstrates how the output format of real numbers can be changed:

```
>>> import pprint, scitools.pprint2
>>> somelist = [15.8, [0.2, 1.7]]
>>> pprint.pprint(somelist)
[15.800000000000001, [0.20000000000000001, 1.7]]
>>> scitools.pprint2.pprint(somelist)
[15.8, [0.2, 1.7]]
>>> # default output is '%g', change this to
>>> scitools.pprint2.float_format = '%.2e'
>>> scitools.pprint2.pprint(somelist)
[1.58e+01, [2.00e-01, 1.70e+00]]
```

As can be seen from this session, the `pprint` module writes floating-point numbers with a lot of digits, in fact so many that we explicitly see the round-off errors. Many find this type of output is annoying and that the default output from the `scitools.pprint2` module is more like one would desire and expect.

The `pprint` and `scitools.pprint2` modules also have a function `pformat`, which works as the `pprint` function, but it returns a pretty formatted string rather than printing the string:

```
s = pprint.pformat(somelist)
print s
```

This last `print` statement prints the same as `pprint.pprint(somelist)`.

Manual printing Many will argue that tabular data such as those stored in the nested `table` list are not printed in a particularly pretty way by the `pprint` module. One would rather expect pretty output to be a table with two nicely aligned columns. To produce such output we need to code the formatting manually. This is quite easy: we loop over each row, extract the two elements C and F in each row, and print these in fixed-width fields using the printf syntax. The code goes as follows:

```
for C, F in table:
    print '%5d %5.1f' % (C, F)
```

2.4.3 Extracting Sublists

Python has a nice syntax for extracting parts of a list structure. Such parts are known as *sublists* or *slices*:

`A[i:]` is the sublist starting with index i in A and continuing to the end of A:

```
>>> A = [2, 3.5, 8, 10]
>>> A[2:]
[8, 10]
```

`A[i:j]` is the sublist starting with index i in A and continuing up to and including index j-1. Make sure you remember that the element corresponding to index j is not included in the sublist:

```
>>> A[1:3]
[3.5, 8]
```

`A[:i]` is the sublist starting with index 0 in A and continuing up to and including the element with index i-1:

```
>>> A[:3]
[2, 3.5, 8]
```

`A[1:-1]` extracts all elements except the first and the last (recall that index -1 refers to the last element), and `A[:]` is the whole list:

```
>>> A[1:-1]
[3.5, 8]
>>> A[:]
[2, 3.5, 8, 10]
```

In nested lists we may use slices in the first index, e.g.,

```
>>> table[4:]
[[0, 32.0], [5, 41.0], [10, 50.0], [15, 59.0], [20, 68.0],
 [25, 77.0], [30, 86.0], [35, 95.0], [40, 104.0]]
```

We can also slice the second index, or both indices:

```
>>> table[4:7][0:2]
[[0, 32.0], [5, 41.0]]
```

Observe that `table[4:7]` makes a list `[[0, 32.0], [5, 41.0], [10, 50.0]]` with three elements. The slice `[0:2]` acts on this sublist and picks out its first two elements, with indices 0 and 1.

Sublists are always copies of the original list, so if you modify the sublist the original list remains unaltered and vice versa:

```
>>> l1 = [1, 4, 3]
>>> l2 = l1[:-1]
>>> l2
[1, 4]
>>> l1[0] = 100
>>> l1                # l1 is modified
[100, 4, 3]
>>> l2                # l2 is not modified
[1, 4]
```

The fact that slicing makes a copy can also be illustrated by the following code:

```
>>> B = A[:]
>>> C = A
>>> B == A
True
>>> B is A
False
>>> C is A
True
```

The `B == A` boolean expression is `True` if all elements in `B` are equal to the corresponding elements in `A`. The test `B is A` is `True` if `A` and `B` are names for the same list. Setting `C = A` makes `C` refer to the same list object as `A`, while `B = A[:]` makes `B` refer to a copy of the list referred to by `A`.

Example We end this information on sublists by writing out the part of the `table` list of `[C, F]` rows (see Sect. 2.4) where the Celsius degrees are between 10 and 35 (not including 35):

```
>>> for C, F in table[Cdegrees.index(10):Cdegrees.index(35)]:
...     print '%5.0f %5.1f' % (C, F)
...
   10   50.0
   15   59.0
   20   68.0
   25   77.0
   30   86.0
```

You should always stop reading and convince yourself that you understand why a code segment produces the printed output. In this latter example, `Cdegrees`.

`index(10)` returns the index corresponding to the value 10 in the `Cdegrees` list. Looking at the `Cdegrees` elements, one realizes (do it!) that the `for` loop is equivalent to

```
for C, F in table[6:11]:
```

This loop runs over the indices $6, 7, \ldots, 10$ in `table`.

2.4.4   Traversing Nested Lists

We have seen that traversing the nested list `table` could be done by a loop of the form

```
for C, F in table:
    # process C and F
```

This is natural code when we know that `table` is a list of `[C, F]` lists. Now we shall address more general nested lists where we do not necessarily know how many elements there are in each list element of the list.

Suppose we use a nested list `scores` to record the scores of players in a game: `scores[i]` holds a list of the historical scores obtained by player number i. Different players have played the game a different number of times, so the length of `scores[i]` depends on i. Some code may help to make this clearer:

```
scores = []
# score of player no. 0:
scores.append([12, 16, 11, 12])
# score of player no. 1:
scores.append([9])
# score of player no. 2:
scores.append([6, 9, 11, 14, 17, 15, 14, 20])
```

The list `scores` has three elements, each element corresponding to a player. The element no. g in the list `scores[p]` corresponds to the score obtained in game number g played by player number p. The length of the lists `scores[p]` varies and equals 4, 1, and 8 for p equal to 0, 1, and 2, respectively.

In the general case we may have n players, and some may have played the game a large number of times, making `scores` potentially a big nested list. How can we traverse the `scores` list and write it out in a table format with nicely formatted columns? Each row in the table corresponds to a player, while columns correspond to scores. For example, the data initialized above can be written out as

```
12  16  11  12
 9
 6   9  11  14  17  15  14  20
```

In a program, we must use two *nested loops*, one for the elements in `scores` and one for the elements in the sublists of `scores`. The example below will make this clear.

There are two basic ways of traversing a nested list: either we use integer indices for each index, or we use variables for the list elements. Let us first exemplify the index-based version:

```
for p in range(len(scores)):
    for g in range(len(scores[p])):
        score = scores[p][g]
        print '%4d' % score,
    print
```

With the trailing comma after the print string, we avoid a newline so that the column values in the table (i.e., scores for one player) appear at the same line. The single print command after the loop over c adds a newline after each table row. The reader is encouraged to go through the loops by hand and simulate what happens in each statement (use the simple scores list initialized above).

The alternative version where we use variables for iterating over the elements in the scores list and its sublists looks like this:

```
for player in scores:
    for game in player:
        print '%4d' % game,
    print
```

Again, the reader should step through the code by hand and realize what the values of player and game are in each pass of the loops.

In the very general case we can have a nested list with many indices: somelist [i1][i2][i3] .... To visit each of the elements in the list, we use as many nested for loops as there are indices. With four indices, iterating over integer indices look as

```
for i1 in range(len(somelist)):
    for i2 in range(len(somelist[i1])):
        for i3 in range(len(somelist[i1][i2])):
            for i4 in range(len(somelist[i1][i2][i3])):
                value = somelist[i1][i2][i3][i4]
                # work with value
```

The corresponding version iterating over sublists becomes

```
for sublist1 in somelist:
    for sublist2 in sublist1:
        for sublist3 in sublist2:
            for sublist4 in sublist3:
                value = sublist4
                # work with value
```

2.5   Tuples

Tuples are very similar to lists, but tuples cannot be changed. That is, a tuple can be viewed as a *constant list*. While lists employ square brackets, tuples are written with standard parentheses:

```
>>> t = (2, 4, 6, 'temp.pdf')     # define a tuple with name t
```

One can also drop the parentheses in many occasions:

```
>>> t = 2, 4, 6, 'temp.pdf'
>>> for element in 'myfile.txt', 'yourfile.txt', 'herfile.txt':
...     print element,
...
myfile.txt yourfile.txt herfile.txt
```

The `for` loop here is over a tuple, because a comma separated sequence of objects, even without enclosing parentheses, becomes a tuple. Note the trailing comma in the `print` statement. This comma suppresses the final newline that the `print` command automatically adds to the output string. This is the way to make several `print` statements build up one line of output.

Much functionality for lists is also available for tuples, for example:

```
>>> t = t + (-1.0, -2.0)          # add two tuples
>>> t
(2, 4, 6, 'temp.pdf', -1.0, -2.0)
>>> t[1]                          # indexing
4
>>> t[2:]                         # subtuple/slice
(6, 'temp.pdf', -1.0, -2.0)
>>> 6 in t                        # membership
True
```

Any list operation that changes the list will not work for tuples:

```
>>> t[1] = -1
...
TypeError: object does not support item assignment

>>> t.append(0)
...
AttributeError: 'tuple' object has no attribute 'append'

>>> del t[1]
...
TypeError: object doesn't support item deletion
```

Some list methods, like `index`, are not available for tuples. So why do we need tuples when lists can do more than tuples?

- Tuples protect against accidental changes of their contents.
- Code based on tuples is faster than code based on lists.

- Tuples are frequently used in Python software that you certainly will make use of, so you need to know this data type.

There is also a fourth argument, which is important for a data type called dictionaries (introduced in Sect. 6.1): tuples can be used as keys in dictionaries while lists can not.

2.6 Summary

2.6.1 Chapter Topics

While loops Loops are used to repeat a collection of program statements several times. The statements that belong to the loop must be consistently indented in Python. A `while` loop runs as long as a condition evaluates to `True`:

```
>>> t = 0; dt = 0.5; T = 2
>>> while t <= T:
...         print t
...         t += dt
...
0
0.5
1.0
1.5
2.0
>>> print 'Final t:', t, '; t <= T is', t <= T
Final t: 2.5 ; t <= T is False
```

Lists A list is used to collect a number of values or variables in an ordered sequence.

```
>>> mylist = [t, dt, T, 'mynumbers.dat', 100]
```

A list element can be any Python object, including numbers, strings, functions, and other lists, for instance.

The table below shows some important list operations (only a subset of these are explained in the present chapter).

Construction	Meaning
`a = []`	initialize an empty list
`a = [1, 4.4, 'run.py']`	initialize a list
`a.append(elem)`	add `elem` object to the end
`a + [1,3]`	add two lists
`a.insert(i, e)`	insert element `e` before index `i`
`a[3]`	index a list element
`a[-1]`	get last list element
`a[1:3]`	slice: copy data to sublist (here: index 1, 2)
`del a[3]`	delete an element (index 3)

Construction	Meaning
`a.remove(e)`	remove an element with value e
`a.index('run.py')`	find index corresponding to an element's value
`'run.py' in a`	test if a value is contained in the list
`a.count(v)`	count how many elements that have the value v
`len(a)`	number of elements in list a
`min(a)`	the smallest element in a
`max(a)`	the largest element in a
`sum(a)`	add all elements in a
`sorted(a)`	return sorted version of list a
`reversed(a)`	return reversed sorted version of list a
`b[3][0][2]`	nested list indexing
`isinstance(a, list)`	is True if a is a list
`type(a) is list`	is True if a is a list

Nested lists If the list elements are also lists, we have a nested list. The following session summarizes indexing and loop traversal of nested lists:

```
>>> nl = [[0, 0, 1], [-1, -1, 2], [-10, 10, 5]]
>>> nl[0]
[0, 0, 1]
>>> nl[-1]
[-10, 10, 5]
>>> nl[0][2]
1
>>> nl[-1][0]
-10
>>> for p in nl:
...     print p
...
[0, 0, 1]
[-1, -1, 2]
[-10, 10, 5]
>>> for a, b, c in nl:
...     print '%3d %3d %3d' % (a, b, c)
...
  0   0   1
 -1  -1   2
-10  10   5
```

Tuples A tuple can be viewed as a constant list: no changes in the contents of the tuple is allowed. Tuples employ standard parentheses or no parentheses, and elements are separated with comma as in lists:

```
>>> mytuple = (t, dt, T, 'mynumbers.dat', 100)
>>> mytuple =  t, dt, T, 'mynumbers.dat', 100
```

Many list operations are also valid for tuples, but those that changes the list content cannot be used with tuples (examples are `append`, `del`, `remove`, `index`, and `sort`).

An object a containing an ordered collection of other objects such that `a[i]` refers to object with index i in the collection, is known as a *sequence* in Python.

Lists, tuples, strings, and arrays are examples on sequences. You choose a sequence type when there is a natural ordering of elements. For a collection of unordered objects a *dictionary* (see Sect. 6.1) is often more convenient.

For loops   A for loop is used to run through the elements of a list or a tuple:

```
>>> for elem in [10, 20, 25, 27, 28.5]:
...        print elem,
...
10 20 25 27 28.5
```

The trailing comma after the print statement prevents the newline character, which otherwise print would automatically add.

The range function is frequently used in for loops over a sequence of integers. Recall that range(start, stop, inc) does not include the upper limit stop among the list item.

```
>>> for elem in range(1, 5, 2):
...        print elem,
...
1 3
>>> range(1, 5, 2)
[1, 3]
```

Implementation of a sum $\sum_{j=M}^{N} q(j)$, where $q(j)$ is some mathematical expression involving the integer counter j, is normally implemented using a for loop. Choosing, e.g., $q(j) = 1/j^2$, the sum is calculated by

```
s = 0  # accumulation variable
for j in range(M, N+1, 1):
    s += 1./j**2
```

Pretty print   To print a list a, print a can be used, but the pprint and scitools. pprint2 modules and their pprint function give a nicer layout of the output for long and nested lists. The scitools.pprint2 module has the possibility to control the formatting of floating-point numbers.

Terminology   The important computer science terms in this chapter are

- list
- tuple
- nested list (and nested tuple)
- sublist (subtuple) or slice a[i:j]
- while loop
- for loop
- list comprehension
- boolean expression

2.6.2 Example: Analyzing List Data

Problem The file `src/misc/Oxford_sun_hours.txt`[2] contains data of the number of sun hours in Oxford, UK, for every month since January 1929. The data are already on a suitable nested list format:

```
[
[43.8, 60.5, 190.2, ...],
[49.9, 54.3, 109.7, ...],
[63.7, 72.0, 142.3, ...],
...
]
```

The list in every line holds the number of sun hours for each of the year's 12 months. That is, the first index in the nested list corresponds to year and the second index corresponds to the month number. More precisely, the double index `[i][j]` corresponds to year $1929 + i$ and month $1 + j$ (January being month number 1).

The task is to define this nested list in a program and do the following data analysis.

- Compute the average number of sun hours for each month during the total data period (1929–2009).
- Which month has the best weather according to the means found in the preceding task?
- For each decade, 1930–1939, 1940–1949, ..., 2000–2009, compute the average number of sun hours per day in January and December. For example, use December 1949, January 1950, ..., December 1958, and January 1959 as data for the decade 1950–1959. Are there any noticeable differences between the decades?

Solution Initializing the data is easy: just copy the data from the `Oxford_sun_hours.txt` file into the program file and set a variable name on the left hand side (the long and wide code is only indicated here):

```
data = [
[43.8, 60.5, 190.2, ...],
[49.9, 54.3, 109.7, ...],
[63.7, 72.0, 142.3, ...],
...
]
```

For task 1, we need to establish a list `monthly_mean` with the results from the computation, i.e., `monthly_mean[2]` holds the average number of sun hours for March in the period 1929–2009. The average is computed in the standard way: for each month, we run through all the years, sum up the values, and finally divide by the number of years (`len(data)` or $2009 - 1929 + 1$).

---

[2] http://tinyurl.com/pwyasaa/misc/Oxford_sun_hours.txt

```
monthly_mean = []
n = len(data)    # no of years
for m in range(12): # counter for month indices
    s = 0          # sum
    for y in data:  # loop over "rows" (first index) in data
        s += y[m]   # add value for month m
    monthly_mean.append(s/n)
```

An alternative solution would be to introduce separate variables for the monthly averages, say Jan_mean, Feb_mean, etc. The reader should as an exercise write the code associated with such a solution and realize that using the monthly_mean list is more elegant and yields much simpler and shorter code. Separate variables might be an okay solution for 2–3 variables, but not for as many as 12.

Perhaps we want a nice-looking printout of the results. This can elegantly be created by first defining a tuple (or list) of the names of the months and then running through this list in parallel with monthly_mean:

```
month_names = 'Jan', 'Feb', 'Mar', 'Apr', 'May', 'Jun',\
              'Jul', 'Aug', 'Sep', 'Oct', 'Nov', 'Dec'
for name, value in zip(month_names, monthly_mean):
    print '%s: %.1f' % (name, value)
```

The printout becomes

```
Jan: 56.6
Feb: 72.7
Mar: 116.5
Apr: 153.2
May: 191.1
Jun: 198.5
Jul: 193.8
Aug: 184.3
Sep: 138.3
Oct: 104.6
Nov: 67.4
Dec: 52.4
```

Task 2 can be solved by pure inspection of the above printout, which reveals that June is the winner. However, since we are learning programming, we should be able to replace our eyes with some computer code to automate the task. The maximum value max_value of a list like monthly_mean is simply obtained by max(monthly_mean). The corresponding index, needed to find the right name of the corresponding month, is found from monthly_mean.index(max_value). The code for task 2 is then

```
max_value = max(monthly_mean)
month = month_names[monthly_mean.index(max_value)]
print '%s has best weather with %.1f sun hours on average' % \
      (month, max_value)
```

Task 3 requires us to first develop an algorithm for how to compute the decade averages. The algorithm, expressed with words, goes as follows. We loop over the decades, and for each decade, we loop over its years, and for each year, we add the December data of the previous year and the January data of the current year to an accumulation variable. Dividing this accumulation variable by $10 \cdot 2 \cdot 30$ gives the average number of sun hours per day in the winter time for the particular decade. The code segment below expresses this algorithm in the Python language:

```python
decade_mean = []
for decade_start in range(1930, 2010, 10):
    Jan_index = 0; Dec_index = 11  # indices
    s = 0
    for year in range(decade_start, decade_start+10):
        y = year - 1929  # list index
        print data[y-1][Dec_index] + data[y][Jan_index]
        s += data[y-1][Dec_index] + data[y][Jan_index]
    decade_mean.append(s/(20.*30))
for i in range(len(decade_mean)):
    print 'Decade %d-%d: %.1f' % \
        (1930+i*10, 1939+i*10, decade_mean[i])
```

The output becomes

```
Decade 1930-1939: 1.7
Decade 1940-1949: 1.8
Decade 1950-1959: 1.8
Decade 1960-1969: 1.8
Decade 1970-1979: 1.6
Decade 1980-1989: 2.0
Decade 1990-1999: 1.8
Decade 2000-2009: 2.1
```

The complete code is found in the file `sun_data.py`.

Remark The file `Oxford_sun_hours.txt` is based on data from the UK Met Office[3]. A Python program for downloading the data, interpreting the content, and creating a file like `Oxford_sun_hours.txt` is explained in detail in Sect. 6.3.3.

2.6.3   How to Find More Python Information

This book contains only fragments of the Python language. When doing your own projects or exercises you will certainly feel the need for looking up more detailed information on modules, objects, etc. Fortunately, there is a lot of excellent documentation on the Python programming language.

The primary reference is the official Python documentation website[4]: `docs. python.org`. Here you can find a Python tutorial, the very useful *Library Reference* [3], and a *Language Reference*, to mention some key documents. When you wonder what functions you can find in a module, say the `math` module, you can go to the Library Reference search for *math*, which quickly leads you to the official

---

[3] http://www.metoffice.gov.uk/climate/uk/stationdata/
[4] http://docs.python.org/index.html

documentation of the `math` module. Alternatively, you can go to the index of this document and pick the `math` (module) item directly. Similarly, if you want to look up more details of the printf formatting syntax, go to the index and follow the *printf-style formatting* index.

Warning   .

A word of caution is probably necessary here. Reference manuals are very technical and written primarily for experts, so it can be quite difficult for a newbie to understand the information. An important ability is to browse such manuals and dig out the key information you are looking for, without being annoyed by all the text you do not understand. As with programming, reading manuals efficiently requires a lot of training.

A tool somewhat similar to the Python Standard Library documentation is the `pydoc` program. In a terminal window you write

---
Terminal

```
Terminal> pydoc math
```

In IPython there are two corresponding possibilities, either

```
In [1]: !pydoc math
```

or

```
In [2]: import math
In [3]: help(math)
```

The documentation of the complete `math` module is shown as plain text. If a specific function is wanted, we can ask for that directly, e.g., `pydoc math.tan`. Since `pydoc` is very fast, many prefer `pydoc` over web pages, but `pydoc` has often less information compared to the web documentation of modules.

There are also a large number of books about Python. Beazley [1] is an excellent reference that improves and extends the information in the web documentation. The *Learning Python* book [17] has been very popular for many years as an introduction to the language. There is a special web page[5] listing most Python books on the market. Very few books target scientific computing with Python, but [4] gives an introduction to Python for mathematical applications and is more compact and advanced than the present book. It also serves as an excellent reference for the capabilities of Python in a scientific context. A comprehensive book on the use of Python for assisting and automating scientific work is [13].

Quick references, which list almost to all Python functionality in compact tabular form, are very handy. We recommend in particular the one by Richard Gruet[6] [6].

The website http://www.python.org/doc/ contains a list of useful Python introductions and reference manuals.

---

[5] http://wiki.python.org/moin/PythonBooks
[6] http://rgruet.free.fr/PQR27/PQR2.7.html

2.7   Exercises

Exercise 2.1: Make a Fahrenheit-Celsius conversion table
Write a Python program that prints out a table with Fahrenheit degrees $0, 10, 20, \ldots$, 100 in the first column and the corresponding Celsius degrees in the second column.

Hint Modify the `c2f_table_while.py` program from Sect. 2.1.2.
Filename: `f2c_table_while`.

Exercise 2.2: Generate an approximate Fahrenheit-Celsius conversion table
Many people use an approximate formula for quickly converting Fahrenheit (F) to Celsius (C) degrees:

$$C \approx \hat{C} = (F - 30)/2 \tag{2.2}$$

Modify the program from Exercise 2.1 so that it prints three columns: F, C, and the approximate value $\hat{C}$.
Filename: `f2c_approx_table`.

Exercise 2.3: Work with a list
Set a variable `primes` to a list containing the numbers 2, 3, 5, 7, 11, and 13. Write out each list element in a `for` loop. Assign 17 to a variable p and add p to the end of the list. Print out the entire new list.
Filename: `primes`.

Exercise 2.4: Generate odd numbers
Write a program that generates all odd numbers from 1 to n. Set n in the beginning of the program and use a `while` loop to compute the numbers. (Make sure that if n is an even number, the largest generated odd number is n-1.)
Filename: `odd`.

Exercise 2.5: Compute the sum of the first *n* integers
Write a program that computes the sum of the integers from 1 up to and including n. Compare the result with the famous formula $n(n + 1)/2$.
Filename: `sum_int`.

Exercise 2.6: Compute energy levels in an atom
The n-th energy level for an electron in a Hydrogen atom is given by

$$E_n = -\frac{m_e e^4}{8\epsilon_0^2 h^2} \cdot \frac{1}{n^2},$$

where $m_e = 9.1094 \cdot 10^{-31}$ kg is the electron mass, $e = 1.6022 \cdot 10^{-19}$ C is the elementary charge, $\epsilon_0 = 8.8542 \cdot 10^{-12}$ C^2 s^2 kg^{-1} m$^-3$ is the electrical permittivity of vacuum, and $h = 6.6261 \cdot 10^{-34}$ Js.

a) Write a Python program that calculates and prints the energy level E_n for $n = 1, \ldots, 20$.

b) The released energy when an electron moves from level n_i to level n_f is given by

$$\Delta E = -\frac{m_e e^4}{8\epsilon_0^2 h^2} \cdot \left(\frac{1}{n_i^2} - \frac{1}{n_f^2}\right).$$

Add statements to the program from a) so that it prints a second, nicely formatted table where the cell in column i and row f contains the energy released when an electron moves from energy level i to level f, for $i, f = 1, \ldots, 5$.

Filename: `energy_levels`.

Exercise 2.7: Generate equally spaced coordinates
We want to generate $n + 1$ equally spaced x coordinates in $[a, b]$. Store the coordinates in a list.

a) Start with an empty list, use a `for` loop and append each coordinate to the list.

Hint With n intervals, corresponding to $n + 1$ points, in $[a, b]$, each interval has length $h = (b - a)/n$. The coordinates can then be generated by the formula $x_i = a + ih, i = 0, \ldots, n + 1$.

b) Use a list comprehension as an alternative implementation.

Filename: `coor`.

Exercise 2.8: Make a table of values from a formula
The purpose of this exercise is to write code that prints a nicely formatted table of t and $y(t)$ values, where

$$y(t) = v_0 t - \frac{1}{2} g t^2.$$

Use $n + 1$ uniformly spaced t values throughout the interval $[0, 2v_0/g]$.

a) Use a `for` loop to produce the table.
b) Add code with a `while` loop to produce the table.

Hint Because of potential round-off errors, you may need to adjust the upper limit of the `while` loop to ensure that the last point ($t = 2v_0/g$, $y = 0$) is included.
Filename: `ball_table1`.

Exercise 2.9: Store values from a formula in lists
This exercise aims to produce the same table of numbers as in Exercise 2.8, but with different code. First, store the t and y values in two lists t and y. Thereafter, write out a nicely formatted table by traversing the two lists with a `for` loop.

Hint In the `for` loop, use either `zip` to traverse the two lists in parallel, or use an index and the `range` construction.
Filename: `ball_table2`.

Exercise 2.10: Simulate operations on lists by hand
You are given the following program:

```
a = [1, 3, 5, 7, 11]
b = [13, 17]
c = a + b
print c
b[0] = -1
d = [e+1 for e in a]
print d
d.append(b[0] + 1)
d.append(b[-1] + 1)
print d[-2:]
for e1 in a:
    for e2 in b:
        print e1 + e2
```

Go through each statement and explain what is printed by the program.
Filename: `simulate_lists`.

Exercise 2.11: Compute a mathematical sum
The following code is supposed to compute the sum $s = \sum_{k=1}^{M} \frac{1}{k}$:

```
s = 0;  k = 1;  M = 100
while k < M:
    s += 1/k
print s
```

This program does not work correctly. What are the three errors? (If you try to run the program, nothing will happen on the screen. Type `Ctrl+c`, i.e., hold down the Control (`Ctrl`) key and then type the c key, to stop the program.) Write a correct program.

Hint There are two basic ways to find errors in a program:

1. read the program carefully and think about the consequences of each statement,
2. print out intermediate results and compare with hand calculations.

First, try method 1 and find as many errors as you can. Thereafter, try method 2 for $M = 3$ and compare the evolution of s with your own hand calculations.
Filename: `sum_while`.

Exercise 2.12: Replace a while loop by a for loop
Rewrite the corrected version of the program in Exercise 2.11 using a `for` loop over k values instead of a `while` loop.
Filename: `sum_for`.

Exercise 2.13: Simulate a program by hand
Consider the following program for computing with interest rates:

```
initial_amount = 100
p = 5.5  # interest rate
amount = initial_amount
years = 0
while amount <= 1.5*initial_amount:
    amount = amount + p/100*amount
    years = years + 1
print years
```

a) Use a pocket calculator or an interactive Python shell and work through the program calculations by hand. Write down the value of `amount` and `years` in each pass of the loop.
b) Set `p = 5` instead. Why will the loop now run forever? (Apply Ctrl+c, see Exercise 2.11, to stop a program with a loop that runs forever.) Make the program robust against such errors.
c) Make use of the operator `+=` wherever possible in the program.
d) Explain with words what type of mathematical problem that is solved by this program.

Filename: `interest_rate_loop`.

Exercise 2.14: Explore Python documentation
Suppose you want to compute with the inverse sine function: $\sin^{-1} x$. How do you do that in a Python program?

Hint The `math` module has an inverse sine function. Find the correct name of the function by looking up the module content in the online Python Standard Library[7] document or use `pydoc`, see Sect. 2.6.3.
Filename: `inverse_sine`.

Exercise 2.15: Index a nested list
We define the following nested list:

```
q = [['a', 'b', 'c'], ['d', 'e', 'f'], ['g', 'h']]
```

a) Index this list to extract 1) the letter a; 2) the list `['d', 'e', 'f']`; 3) the last element h; 4) the d element. Explain why `q[-1][-2]` has the value g.
b) We can visit all elements of q using this nested `for` loop:

```
for i in q:
    for j in range(len(i)):
        print i[j]
```

What type of objects are `i` and `j`?

Filename: `index_nested_list`.

---

[7] http://docs.python.org/2/library/

Exercise 2.16: Store data in lists

Modify the program from Exercise 2.2 so that all the F, C, and $\hat{C}$ values are stored in separate lists F, C, and C_approx, respectively. Then make a nested list conversion so that conversion[i] holds a row in the table: [F[i], C[i], C_approx[i]]. Finally, let the program traverse the conversion list and write out the same table as in Exercise 2.2.

Filename: f2c_approx_lists.

Exercise 2.17: Store data in a nested list

The purpose of this exercise is to store tabular data in two alternative ways, either as a list of columns or as a list of rows. In order to write out a nicely formatted table, one has to traverse the data, and the technique for traversal depends on how the tabular data is stored.

a) Compute two lists of t and y values as explained in Exercise 2.9. Store the two lists in a new *nested* list ty1 such that ty1[0] and ty1[1] correspond to the two lists. Write out a table with t and y values in two columns by looping over the data in the ty1 list. Each number should be written with two decimals.

b) Make a list ty2 which holds each row in the table of t and y values (ty1 is a list of table columns while ty2 is a list of table rows, as explained in Sect. 2.4). Loop over the ty2 list and write out the t and y values with two decimals each.

Filename: ball_table3.

Exercise 2.18: Values of boolean expressions

Explain the outcome of each of the following boolean expressions:

```
C = 41
C == 40
C != 40 and C < 41
C != 40 or  C < 41
not C == 40
not C > 40
C <= 41
not False
True and False
False or True
False or False or False
True and True and False
False == 0
True == 0
True == 1
```

Note

It makes sense to compare True and False to the integers 0 and 1, but not other integers (e.g., True == 12 is False although the *integer* 12 evaluates to True in a boolean context, as in bool(12) or if 12).

Filename: eval_bool.

Exercise 2.19: Explore round-off errors from a large number of inverse operations

Maybe you have tried to hit the square root key on a calculator multiple times and then squared the number again an equal number of times. These set of inverse mathematical operations should of course bring you back to the starting value for the computations, but this does not always happen. To avoid tedious pressing of calculator keys, we can let a computer automate the process. Here is an appropriate program:

```
from math import sqrt
for n in range(1, 60):
    r = 2.0
    for i in range(n):
        r = sqrt(r)
    for i in range(n):
        r = r**2
    print '%d times sqrt and **2: %.16f' % (n, r)
```

Explain with words what the program does. Then run the program. Round-off errors are here completely destroying the calculations when n is large enough! Investigate the case when we come back to 1 instead of 2 by fixing an n value where this happens and printing out r in both for loops over i. Can you now explain why we come back to 1 and not 2?
Filename: `repeated_sqrt`.

Exercise 2.20: Explore what zero can be on a computer

Type in the following code and run it:

```
eps = 1.0
while 1.0 != 1.0 + eps:
    print '...............', eps
    eps = eps/2.0
print 'final eps:', eps
```

Explain with words what the code is doing, line by line. Then examine the output. How can it be that the "equation" $1 \neq 1 + eps$ is not true? Or in other words, that a number of approximately size 10^{-16} (the final eps value when the loop terminates) gives the same result as if eps were zero?
Filename: `machine_zero`.

Remarks The nonzero eps value computed above is called *machine epsilon* or *machine zero* and is an important parameter to know, especially when certain mathematical techniques are applied to control round-off errors.

Exercise 2.21: Compare two real numbers with a tolerance

Run the following program:

```
a = 1/947.0*947
b = 1
if a != b:
    print 'Wrong result!'
```

The lesson learned from this program is that one should never compare two floating-point objects directly using a == b or a != b, because round-off errors quickly make two identical mathematical values different on a computer. A better result is to test if abs(a - b) < tol, where tol is a very small number. Modify the test according to this idea.

Filename: compare_floats.

Exercise 2.22: Interpret a code
The function time in the module time returns the number of seconds since a particular date (called the Epoch, which is January 1, 1970, on many types of computers). Python programs can therefore use time.time() to mimic a stop watch. Another function, time.sleep(n) causes the program to pause for n seconds and is handy for inserting a pause. Use this information to explain what the following code does:

```
import time
t0 = time.time()
while time.time() - t0 < 10:
    print '....I like while loops!'
    time.sleep(2)
print 'Oh, no - the loop is over.'
```

How many times is the print statement inside the loop executed? Now, copy the code segment and change the < sign in the loop condition to a > sign. Explain what happens now.

Filename: time_while.

Exercise 2.23: Explore problems with inaccurate indentation
Type in the following program in a file and check carefully that you have exactly the same spaces:

```
C = -60; dC = 2
while C <= 60:
    F = (9.0/5)*C + 32
      print C, F
C = C + dC
```

Run the program. What is the first problem? Correct that error. What is the next problem? What is the cause of that problem? (See Exercise 2.11 for how to stop a hanging program.)

Filename: indentation.

Remarks The lesson learned from this exercise is that one has to be very careful with indentation in Python programs! Other computer languages usually enclose blocks belonging to loops in curly braces, parentheses, or begin-end marks. Python's convention with using solely indentation contributes to visually attractive, easy-to-read code, at the cost of requiring a pedantic attitude to blanks from the programmer.

Exercise 2.24: Explore punctuation in Python programs

Some of the following assignments work and some do not. Explain in each case why the assignment works/fails and, if it works, what kind of object x refers to and what the value is if we do a `print x`.

```
x = 1
x = 1.
x = 1;
x = 1!
x = 1?
x = 1:
x = 1,
```

Hint Explore the statements in an interactive Python shell.
Filename: `punctuation`.

Exercise 2.25: Investigate a for loop over a changing list

Study the following interactive session and explain in detail what happens in each pass of the loop, and use this explanation to understand the output.

```
>>> numbers = range(10)
>>> print numbers
[0, 1, 2, 3, 4, 5, 6, 7, 8, 9]
>>> for n in numbers:
...     i = len(numbers)/2
...     del numbers[i]
...     print 'n=%d, del %d' % (n,i), numbers
...
n=0, del 5 [0, 1, 2, 3, 4, 6, 7, 8, 9]
n=1, del 4 [0, 1, 2, 3, 6, 7, 8, 9]
n=2, del 4 [0, 1, 2, 3, 7, 8, 9]
n=3, del 3 [0, 1, 2, 7, 8, 9]
n=8, del 3 [0, 1, 2, 8, 9]
```

Warning

The message in this exercise is to *never modify a list that we are looping over.* Modification is indeed technically possible, as shown above, but you really need to know what you are doing. Otherwise you will experience very strange program behavior.

Filename: `for_changing_list`.

Functions and Branching

3

This chapter introduces two fundamental and extremely useful concepts in programming: user-defined functions and branching of program flow, the latter often referred to as `if` tests. The programs associated with the chapter are found in the folder `src/funcif`[1].

3.1 Functions

In a computer language like Python, the term *function* means more than just a mathematical function. A function is a collection of statements that you can execute wherever and whenever you want in the program. You may send variables to the function to influence what is getting computed by statements in the function, and the function may return new objects back to you.

In particular, functions help avoid duplicating code snippets by putting all similar snippets in a common place. This strategy saves typing and makes it easier to change the program later. Functions are also often used to just split a long program into smaller, more manageable pieces, so the program and your own thinking about it become clearer. Python comes with lots of pre-defined functions (`math.sqrt`, `range`, and `len` are examples we have met so far). This section explains how you can define your own functions.

3.1.1 Mathematical Functions as Python Functions

Let us start with making a Python function that evaluates a mathematical function, more precisely the function $F(C)$ for converting Celsius degrees C to the corresponding Fahrenheit degrees F:

$$F(C) = \frac{9}{5}C + 32 \,.$$

---

[1] http://tinyurl.com/pwyasaa/funcif

© Springer-Verlag Berlin Heidelberg 2016
H.P. Langtangen, *A Primer on Scientific Programming with Python*,
Texts in Computational Science and Engineering 6, DOI 10.1007/978-3-662-49887-3_3

The corresponding Python function must take C as argument and return the value $F(C)$. The code for this looks like

```
def F(C):
    return (9.0/5)*C + 32
```

All Python functions begin with `def`, followed by the function name, and then inside parentheses a comma-separated list of *function arguments*. Here we have only one argument C. This argument acts as a standard variable inside the function. The statements to be performed inside the function must be indented. At the end of a function it is common to *return* a value, that is, send a value "out of the function". This value is normally associated with the name of the function, as in the present case where the returned value is the result of the mathematical function $F(C)$.

The `def` line with the function name and arguments is often referred to as the *function header*, while the indented statements constitute the *function body*.

To use a function, we must *call* (or *invoke*) it. Because the function returns a value, we need to store this value in a variable or make use of it in other ways. Here are some calls to F:

```
temp1 = F(15.5)
a = 10
temp2 = F(a)
print F(a+1)
sum_temp = F(10) + F(20)
```

The returned object from F(C) is in our case a `float` object. The call F(C) can therefore be placed anywhere in a code where a `float` object would be valid. The `print` statement above is one example.

As another example, say we have a list `Cdegrees` of Celsius degrees and we want to compute a list of the corresponding Fahrenheit degrees using the F function above in a list comprehension:

```
Fdegrees = [F(C) for C in Cdegrees]
```

Yet another example may involve a slight variation of our F(C) function, where a formatted string instead of a real number is returned:

```
>>> def F2(C):
...     F_value = (9.0/5)*C + 32
...     return '%.1f degrees Celsius corresponds to '\
...            '%.1f degrees Fahrenheit' % (C, F_value)
...
>>> s1 = F2(21)
>>> s1
'21.0 degrees Celsius corresponds to 69.8 degrees Fahrenheit'
```

The assignment to `F_value` demonstrates that we can create variables inside a function as needed.

3.1.2 Understanding the Program Flow

A programmer must have a deep understanding of the sequence of statements that are executed in the program and be able to simulate by hand what happens with a program in the computer. To help build this understanding, a debugger (see Sect. F.1) or the Online Python Tutor[2] are excellent tools. A debugger can be used for all sorts of programs, large and small, while the Online Python Tutor is primarily an educational tool for small programs. We shall demonstrate it here.

Below is a program c2f.py having a function and a for loop, with the purpose of printing out a table for conversion of Celsius to Fahrenheit degrees:

```
def F(C):
    F = 9./5*C + 32
    return F

dC = 10
C = -30
while C <= 50:
    print '%5.1f %5.1f' % (C, F(C))
    C += dC
```

We shall now ask the Online Python Tutor to visually explain how the program is executed. Go to http://www.pythontutor.com/visualize.html, erase the code there and write or paste the c2f.py file into the editor area. Click *Visualize Execution*. Press the forward button to advance one statement at a time and observe the evolution of variables to the right in the window. This demo illustrates how the program jumps around in the loop and up to the F(C) function and back again. Figure 3.1

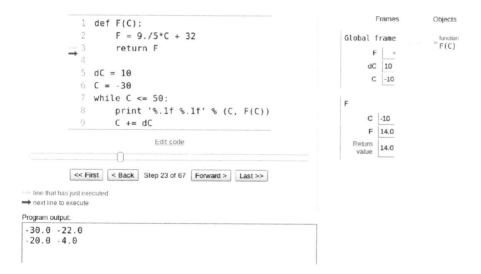

Fig. 3.1 Screen shot of the Online Python Tutor and stepwise execution of the c2f.py program

---

[2] http://www.pythontutor.com/

gives a snapshot of the status of variables, terminal output, and what the current and next statements are.

Tip: How does a program actually work?

Every time you are a bit uncertain about the flow of statements in a program with loops and/or functions, go to http://www.pythontutor.com/visualize.html, paste in your program and see exactly what happens.

3.1.3   Local and Global Variables

Local variables are invisible outside functions Let us reconsider the F2(C) function from Sect. 3.1.1. The variable F_value is a *local* variable in the function, and a local variable does not exist outside the function, i.e., in the main program. We can easily demonstrate this fact by continuing the previous interactive session:

```
>>> c1 = 37.5
>>> s2 = F2(c1)
>>> F_value
...
NameError: name 'F_value' is not defined
```

This error message demonstrates that the surrounding program outside the function is not aware of F_value. Also the argument to the function, C, is a local variable that we cannot access outside the function:

```
>>> C
...
NameError: name 'C' is not defined
```

On the contrary, the variables defined outside of the function, like s1, s2, and c1 in the above session, are *global* variables. These can be accessed everywhere in a program, also inside the F2 function.

Local variables hide global variables Local variables are created inside a function and destroyed when we leave the function. To learn more about this fact, we may study the following session where we write out F_value, C, and some global variable r inside the function:

```
>>> def F3(C):
...     F_value = (9.0/5)*C + 32
...     print 'Inside F3: C=%s F_value=%s r=%s' % (C, F_value, r)
...     return '%.1f degrees Celsius corresponds to '\
...            '%.1f degrees Fahrenheit' % (C, F_value)
...
>>> C = 60    # make a global variable C
>>> r = 21    # another global variable
>>> s3 = F3(r)
```

```
Inside F3: C=21 F_value=69.8 r=21
>>> s3
'21.0 degrees Celsius corresponds to 69.8 degrees Fahrenheit'
>>> C
60
```

This example illustrates that there are two C variables, one global, defined in the main program with the value 60 (an int object), and one local, living when the program flow is inside the F3 function. The value of this latter C is given in the call to the F3 function (an int object). Inside the F3 function the local C *hides* the global C variable in the sense that when we refer to C we access the local variable. (The global C can technically be accessed as globals()['C'], but one should avoid working with local and global variables with the same names at the same time!)

The Online Python Tutor gives a complete overview of what the local and global variables are at any point of time. For instance, in the example from Sect. 3.1.2, Fig. 3.1 shows the content of the three global variables F, dC, and C, along with the content of the variables that are in play in this call of the F(C) function: C and F.

How Python looks up variables

The more general rule, when you have several variables with the same name, is that Python first tries to look up the variable name among the local variables, then there is a search among global variables, and finally among built-in Python functions.

Example Here is a complete sample program that aims to illustrate the rule above:

```
print sum  # sum is a built-in Python function
sum = 500  # rebind the name sum to an int
print sum  # sum is a global variable

def myfunc(n):
    sum = n + 1
    print sum  # sum is a local variable
    return sum

sum = myfunc(2) + 1   # new value in global variable sum
print sum
```

In the first line, there are no local variables, so Python searches for a global value with name sum, but cannot find any, so the search proceeds with the built-in functions, and among them Python finds a function with name sum. The printout of sum becomes something like <built-in function sum>.

The second line rebinds the global name sum to an int object. When trying to access sum in the next print statement, Python searches among the global variables (no local variables so far) and finds one. The printout becomes 500. The call myfunc(2) invokes a function where sum is a local variable. Doing a print sum in this function makes Python first search among the local variables, and since sum is found there, the printout becomes 3 (and not 500, the value of the global variable sum). The value of the local variable sum is returned, added to 1, to form an int object with value 4. This int object is then bound to the global variable sum. The final print sum leads to a search among global variables, and we find one with value 4.

Changing global variables inside functions The values of global variables can be accessed inside functions, but the values cannot be changed unless the variable is declared as global:

```
a = 20; b = -2.5          # global variables

def f1(x):
    a = 21                # this is a new local variable
    return a*x + b

print a                   # yields 20

def f2(x):
    global a
    a = 21                # the global a is changed
    return a*x + b

f1(3); print a            # 20 is printed
f2(3); print a            # 21 is printed
```

Note that in the f1 function, a = 21 creates a local variable a. As a programmer you may think you change the global a, but it does not happen! *You are strongly encouraged to run the programs in this section in the Online Python Tutor, which is an excellent tool to explore local versus global variables and thereby get a good understanding of these concepts.*

3.1.4 Multiple Arguments

The previous F(C) and F2(C) functions from Sect. 3.1.1 are functions of one variable, C, or as we phrase it in computer science: the functions take one argument (C). Functions can have as many arguments as desired; just separate the argument names by commas.

Consider the mathematical function

$$y(t) = v_0 t - \frac{1}{2} g t^2,$$

where g is a fixed constant and v_0 is a physical parameter that can vary. Mathematically, y is a function of one variable, t, but the function values also depends on the value of v_0. That is, to evaluate y, we need values for t *and* v_0. A natural Python implementation is therefore a function with two arguments:

```
def yfunc(t, v0):
    g = 9.81
    return v0*t - 0.5*g*t**2
```

Note that the arguments t and v0 are local variables in this function. Examples on valid calls are

```
y = yfunc(0.1, 6)
y = yfunc(0.1, v0=6)
y = yfunc(t=0.1, v0=6)
y = yfunc(v0=6, t=0.1)
```

The possibility to write argument=value in the call makes it easier to read and understand the call statement. With the argument=value syntax for all arguments, the sequence of the arguments does not matter in the call, which here means that we may put v0 before t. When omitting the argument= part, the sequence of arguments in the call must perfectly match the sequence of arguments in the function definition. The argument=value arguments must appear after all the arguments where only value is provided (e.g., yfunc(t=0.1, 6) is illegal).

Whether we write yfunc(0.1, 6) or yfunc(v0=6, t=0.1), the arguments are initialized as local variables in the function in the same way as when we assign values to variables:

```
t = 0.1
v0 = 6
```

These statements are not visible in the code, but a call to a function automatically initializes the arguments in this way.

3.1.5 Function Argument or Global Variable?

Since y mathematically is considered a function of one variable, t, some may argue that the Python version of the function, yfunc, should be a function of t only. This is easy to reflect in Python:

```
def yfunc(t):
    g = 9.81
    return v0*t - 0.5*g*t**2
```

The main difference is that v0 now becomes a *global* variable, which needs to be initialized outside the function yfunc (in the main program) before we attempt to call yfunc. The next session demonstrates what happens if we fail to initialize such a global variable:

```
>>> def yfunc(t):
...     g = 9.81
...     return v0*t - 0.5*g*t**2
...
>>> yfunc(0.6)
...
NameError: global name 'v0' is not defined
```

The remedy is to define v0 as a global variable prior to calling yfunc:

```
>>> v0 = 5
>>> yfunc(0.6)
1.2342
```

The rationale for having yfunc as a function of t only becomes evident in Sect. 3.1.12.

3.1.6 Beyond Mathematical Functions

So far our Python functions have typically computed some mathematical function, but the usefulness of Python functions goes far beyond mathematical functions. Any set of statements that we want to repeatedly execute under potentially slightly different circumstances is a candidate for a Python function. Say we want to make a list of numbers starting from some value and stopping at another value, with increments of a given size. With corresponding variables start=2, stop=8, and inc=2, we should produce the numbers 2, 4, 6, and 8. Let us write a function doing the task, together with a couple of statements that demonstrate how we call the function:

```
def makelist(start, stop, inc):
    value = start
    result = []
    while value <= stop:
        result.append(value)
        value = value + inc
    return result

mylist = makelist(0, 100, 0.2)
print mylist  # will print 0, 0.2, 0.4, 0.6, ... 99.8, 100
```

Remark 1 The makelist function has three arguments: start, stop, and inc, which become local variables in the function. Also value and result are local variables. In the surrounding program we define only one variable, mylist, and this is then a global variable.

Remark 2 You may think that range(start, stop, inc) makes the makelist function redundant, but range can only generate integers, while makelist can generate real numbers too, and more, as demonstrated in Exercise 3.44.

3.1.7 Multiple Return Values

Python functions may return more than one value. Suppose we are interested in evaluating both $y(t)$ and $y'(t)$:

$$y(t) = v_0t - \frac{1}{2}gt^2,$$
$$y'(t) = v_0 - gt.$$

To return both y and y' we simply separate their corresponding variables by a comma in the `return` statement:

```
def yfunc(t, v0):
    g = 9.81
    y = v0*t - 0.5*g*t**2
    dydt = v0 - g*t
    return y, dydt
```

Calling this latter `yfunc` function makes a need for two values on the left-hand side of the assignment operator because the function returns two values:

```
position, velocity = yfunc(0.6, 3)
```

Here is an application of the `yfunc` function for producing a nicely formatted table of t, $y(t)$, and $y'(t)$ values:

```
t_values = [0.05*i for i in range(10)]
for t in t_values:
    position, velocity = yfunc(t, v0=5)
    print 't=%-10g position=%-10g velocity=%-10g' % \
        (t, position, velocity)
```

The format %-10g prints a real number as compactly as possible (decimal or scientific notation) in a field of width 10 characters. The minus sign (-) after the percentage sign implies that the number is *left-adjusted* in this field, a feature that is important for creating nice-looking columns in the output:

```
t=0           position=0           velocity=5
t=0.05        position=0.237737    velocity=4.5095
t=0.1         position=0.45095     velocity=4.019
t=0.15        position=0.639638    velocity=3.5285
t=0.2         position=0.8038      velocity=3.038
t=0.25        position=0.943437    velocity=2.5475
t=0.3         position=1.05855     velocity=2.057
t=0.35        position=1.14914     velocity=1.5665
t=0.4         position=1.2152      velocity=1.076
t=0.45        position=1.25674     velocity=0.5855
```

When a function returns multiple values, separated by a comma in the `return` statement, a tuple (Sect. 2.5) is actually returned. We can demonstrate that fact by the following session:

```
>>> def f(x):
...     return x, x**2, x**4
...
>>> s = f(2)
>>> s
(2, 4, 16)
>>> type(s)
<type 'tuple'>
>>> x, x2, x4 = f(2)    # store in separate variables
```

Note that storing multiple return values in separate variables, as we do in the last line, is actually the same functionality as we use for storing list (or tuple) elements in separate variables, see Sect. 2.2.1.

3.1.8 Computing Sums

Our next example concerns a Python function for calculating the sum

$$L(x; n) = \sum_{i=1}^{n} \frac{1}{i} \left(\frac{x}{1+x} \right)^i . \tag{3.1}$$

To compute a sum in a program, we use a loop and add terms to an accumulation variable inside the loop. Section 2.1.4 explains the idea. However, summation expressions with an integer counter, such as i in (3.1), are normally implemented by a `for` loop over the i counter and not a `while` loop as in Sect. 2.1.4. For example, the implementation of $\sum_{i=1}^{n} i^2$ is typically implemented as

```
s = 0
for i in range(1, n+1):
    s += i**2
```

For the specific sum (3.1) we just replace `i**2` by the right term inside the `for` loop:

```
s = 0
for i in range(1, n+1):
    s += (1.0/i)*(x/(1.0+x))**i
```

Observe the factors `1.0` used to avoid integer division, since `i` is `int` and x may also be `int`.

It is natural to embed the computation of the sum in a function that takes x and n as arguments and returns the sum:

```
def L(x, n):
    s = 0
    for i in range(1, n+1):
        s += (1.0/i)*(x/(1.0+x))**i
    return s
```

Our formula (3.1) is not chosen at random. In fact, it can be shown that $L(x;n)$ is an approximation to $\ln(1 + x)$ for a finite n and $x \geq 1$. The approximation becomes exact in the limit

$$\lim_{n \to \infty} L(x;n) = \ln(1 + x).$$

Computational significance of $L(x;n)$

Although we can compute $\ln(1 + x)$ on a calculator or by `math.log(1+x)` in Python, you may have wondered how such a function is actually calculated inside the calculator or the `math` module. In most cases this must be done via simple mathematical expressions such as the sum in (3.1). A calculator and the `math` module will use more sophisticated formulas than (3.1) for ultimate efficiency of the calculations, but the main point is that the numerical values of mathematical functions like $\ln(x)$, $\sin(x)$, and $\tan(x)$ are usually computed by sums similar to (3.1).

Instead of having our L function just returning the value of the sum, we could return additional information on the error involved in the approximation of $\ln(1+x)$ by $L(x;n)$. The size of the terms decreases with increasing n, and the first neglected term is then bigger than all the remaining terms, but not necessarily bigger than their sum. The first neglected term is hence an indication of the size of the total error we make, so we may use this term as a rough estimate of the error. For comparison, we could also return the exact error since we are able to calculate the ln function by `math.log`.

A new version of the L(x, n) function, where we return the value of $L(x;n)$, the first neglected term, and the exact error goes as follows:

```
def L2(x, n):
    s = 0
    for i in range(1, n+1):
        s += (1.0/i)*(x/(1.0+x))**i
    value_of_sum = s
    first_neglected_term = (1.0/(n+1))*(x/(1.0+x))**(n+1)
    from math import log
    exact_error = log(1+x) - value_of_sum
    return value_of_sum, first_neglected_term, exact_error

# typical call:
value, approximate_error, exact_error = L2(x, 100)
```

The next section demonstrates the usage of the L2 function to judge the quality of the approximation $L(x;n)$ to $\ln(1 + x)$.

3.1.9 Functions with No Return Values

Sometimes a function just performs a set of statements, without computing objects that are natural to return to the calling code. In such situations one can simply skip the `return` statement. Some programming languages use the terms *procedure* or *subroutine* for functions that do not return anything.

Let us exemplify a function without return values by making a table of the accuracy of the $L(x;n)$ approximation to $\ln(1+x)$ from the previous section:

```
def table(x):
    print '\nx=%g, ln(1+x)=%g' % (x, log(1+x))
    for n in [1, 2, 10, 100, 500]:
        value, next, error = L2(x, n)
        print 'n=%-4d %-10g (next term: %8.2e '\
              'error: %8.2e)' % (n, value, next, error)
```

This function just performs a set of statements that we may want to run several times. Calling

```
table(10)
table(1000)
```

gives the output

```
x=10, ln(1+x)=2.3979
n=1     0.909091    (next term: 4.13e-01   error: 1.49e+00)
n=2     1.32231     (next term: 2.50e-01   error: 1.08e+00)
n=10    2.17907     (next term: 3.19e-02   error: 2.19e-01)
n=100   2.39789     (next term: 6.53e-07   error: 6.59e-06)
n=500   2.3979      (next term: 3.65e-24   error: 6.22e-15)

x=1000, ln(1+x)=6.90875
n=1     0.999001    (next term: 4.99e-01   error: 5.91e+00)
n=2     1.498       (next term: 3.32e-01   error: 5.41e+00)
n=10    2.919       (next term: 8.99e-02   error: 3.99e+00)
n=100   5.08989     (next term: 8.95e-03   error: 1.82e+00)
n=500   6.34928     (next term: 1.21e-03   error: 5.59e-01)
```

From this output we see that the sum converges much more slowly when x is large than when x is small. We also see that the error is an order of magnitude or more larger than the first neglected term in the sum. The functions L, L2, and `table` are found in the file `lnsum.py`.

When there is no explicit `return` statement in a function, Python actually inserts an invisible `return None` statement. None is a special object in Python that represents something we might think of as empty data or just "nothing". Other computer languages, such as C, C++, and Java, use the word *void* for a similar thing. Normally, one will call the `table` function without assigning the return value to any variable, but if we assign the return value to a variable, `result = table(500)`, `result` will refer to a None object.

The None value is often used for variables that should exist in a program, but where it is natural to think of the value as conceptually undefined. The standard way to test if an object `obj` is set to None or not reads

```
if obj is None:
    ...
if obj is not None:
    ...
```

One can also use `obj == None`. The `is` operator tests if two names refer to the same object, while `==` tests if the contents of two objects are the same:

```
>>> a = 1
>>> b = a
>>> a is b    # a and b refer to the same object
True
>>> c = 1.0
>>> a is c
False
>>> a == c    # a and c are mathematically equal
True
```

3.1.10 Keyword Arguments

Some function arguments can be given a default value so that we may leave out these arguments in the call. A typical function may look as

```
>>> def somefunc(arg1, arg2, kwarg1=True, kwarg2=0):
>>>     print arg1, arg2, kwarg1, kwarg2
```

The first two arguments, `arg1` and `arg2`, are *ordinary* or *positional* arguments, while the latter two are *keyword arguments* or *named arguments*. Each keyword argument has a name (in this example `kwarg1` and `kwarg2`) and an associated default value. The keyword arguments must always be listed after the positional arguments in the function definition.

When calling `somefunc`, we may leave out some or all of the keyword arguments. Keyword arguments that do not appear in the call get their values from the specified default values. We can demonstrate the effect through some calls:

```
>>> somefunc('Hello', [1,2])
Hello [1, 2] True 0
>>> somefunc('Hello', [1,2], kwarg1='Hi')
Hello [1, 2] Hi 0
>>> somefunc('Hello', [1,2], kwarg2='Hi')
Hello [1, 2] True Hi
>>> somefunc('Hello', [1,2], kwarg2='Hi', kwarg1=6)
Hello [1, 2] 6 Hi
```

The sequence of the keyword arguments does not matter in the call. We may also mix the positional and keyword arguments if we explicitly write `name=value` for all arguments in the call:

```
>>> somefunc(kwarg2='Hello', arg1='Hi', kwarg1=6, arg2=[1,2],)
Hi [1, 2] 6 Hello
```

Example: Function with default parameters Consider a function of t which also contains some parameters, here A, a, and ω:

$$f(t; A, a, \omega) = Ae^{-at} \sin(\omega t).\qquad(3.2)$$

We can implement f as a Python function where the independent variable t is an ordinary positional argument, and the parameters A, a, and ω are keyword arguments with suitable default values:

```
from math import pi, exp, sin

def f(t, A=1, a=1, omega=2*pi):
    return A*exp(-a*t)*sin(omega*t)
```

Calling f with just the t argument specified is possible:

```
v1 = f(0.2)
```

In this case we evaluate the expression $e^{-0.2}\sin(2\pi \cdot 0.2)$. Other possible calls include

```
v2 = f(0.2, omega=1)
v3 = f(1, A=5, omega=pi, a=pi**2)
v4 = f(A=5, a=2, t=0.01, omega=0.1)
v5 = f(0.2, 0.5, 1, 1)
```

You should write down the mathematical expressions that arise from these four calls. Also observe in the third line above that a positional argument, t in that case, can appear in between the keyword arguments if we write the positional argument on the keyword argument form name=value. In the last line we demonstrate that keyword arguments can be used as positional argument, i.e., the name part can be skipped, but then the sequence of the keyword arguments in the call must match the sequence in the function definition exactly.

Example: Computing a sum with default tolerance Consider the $L(x;n)$ sum and the Python implementations L(x, n) and L2(x, n) from Sect. 3.1.8. Instead of specifying the number of terms in the sum, n, it is better to specify a tolerance ε of the accuracy. We can use the first neglected term as an estimate of the accuracy. This means that we sum up terms as long as the absolute value of the next term is greater than ϵ. It is natural to provide a default value for ϵ:

```
def L3(x, epsilon=1.0E-6):
    x = float(x)
    i = 1
    term = (1.0/i)*(x/(1+x))**i
    s = term
    while abs(term) > epsilon:
        i += 1
        term = (1.0/i)*(x/(1+x))**i
        s += term
    return s, i
```

Here is an example involving this function to make a table of the approximation error as ϵ decreases:

```
def table2(x):
    from math import log
    for k in range(4, 14, 2):
        epsilon = 10**(-k)
        approx, n = L3(x, epsilon=epsilon)
        exact = log(1+x)
        exact_error = exact - approx
```

The output from calling `table2(10)` becomes

```
epsilon: 1e-04, exact error: 8.18e-04, n=55
epsilon: 1e-06, exact error: 9.02e-06, n=97
epsilon: 1e-08, exact error: 8.70e-08, n=142
epsilon: 1e-10, exact error: 9.20e-10, n=187
epsilon: 1e-12, exact error: 9.31e-12, n=233
```

We see that the `epsilon` estimate is almost 10 times smaller than the exact error, regardless of the size of `epsilon`. Since `epsilon` follows the exact error quite well over many orders of magnitude, we may view `epsilon` as a useful indication of the size of the error.

3.1.11 Doc Strings

There is a convention in Python to insert a documentation string right after the `def` line of the function definition. The documentation string, known as a *doc string*, should contain a short description of the purpose of the function and explain what the different arguments and return values are. Interactive sessions from a Python shell are also common to illustrate how the code is used. Doc strings are usually enclosed in triple double quotes `"""`, which allow the string to span several lines.

Here are two examples on short and long doc strings:

```
def C2F(C):
    """Convert Celsius degrees (C) to Fahrenheit."""
    return (9.0/5)*C + 32

def line(x0, y0, x1, y1):
    """
    Compute the coefficients a and b in the mathematical
    expression for a straight line y = a*x + b that goes
    through two points (x0, y0) and (x1, y1).

    x0, y0: a point on the line (floats).
    x1, y1: another point on the line (floats).
    return: coefficients a, b (floats) for the line (y=a*x+b).
    """
    a = (y1 - y0)/float(x1 - x0)
    b = y0 - a*x0
    return a, b
```

Note that the doc string must appear before any statement in the function body.

There are several Python tools that can automatically extract doc strings from the source code and produce various types of documentation. The leading tools is Sphinx[3], see also [13, Appendix B.2].

The doc string can be accessed in a code as `funcname.__doc__`, where `funcname` is the name of the function, e.g.,

```
print line.__doc__
```

which prints out the documentation of the `line` function above:

```
    Compute the coefficients a and b in the mathematical
    expression for a straight line y = a*x + b that goes
    through two points (x0, y0) and (x1, y1).

    x0, y0: a point on the line (float objects).
    x1, y1: another point on the line (float objects).
    return: coefficients a, b for the line (y=a*x+b).
```

If the function `line` is in a file `funcs.py`, we may also run `pydoc funcs.line` in a terminal window to look the documentation of the `line` function in terms of the function signature and the doc string.

Doc strings often contain interactive sessions, copied from a Python shell, to illustrate how the function is used. We can add such a session to the doc string in the `line` function:

```
def line(x0, y0, x1, y1):
    """
    Compute the coefficients a and b in the mathematical
    expression for a straight line y = a*x + b that goes
    through two points (x0,y0) and (x1,y1).

    x0, y0: a point on the line (float).
    x1, y1: another point on the line (float).
    return: coefficients a, b (floats) for the line (y=a*x+b).

    Example:
    >>> a, b = line(1, -1, 4, 3)
    >>> a
    1.3333333333333333
    >>> b
    -2.333333333333333
    """
    a = (y1 - y0)/float(x1 - x0)
    b = y0 - a*x0
    return a, b
```

A particularly nice feature is that all such interactive sessions in doc strings can be automatically run, and new results are compared to the results found in the doc strings. This makes it possible to use interactive sessions in doc strings both for exemplifying how the code is used and for testing that the code works.

---

[3] http://sphinx-doc.org/invocation.html#invocation-apidoc

Function input and output

It is a convention in Python that function arguments represent the input data to the function, while the returned objects represent the output data. We can sketch a general Python function as

```
def somefunc(i1, i2, i3, io4, io5, i6=value1, io7=value2):
    # modify io4, io5, io6; compute o1, o2, o3
    return o1, o2, o3, io4, io5, io7
```

Here `i1`, `i2`, `i3` are positional arguments representing input data; `io4` and `io5` are positional arguments representing input *and* output data; `i6` and `io7` are keyword arguments representing input and input/output data, respectively; and `o1`, `o2`, and `o3` are computed objects in the function, representing output data together with `io4`, `io5`, and `io7`. All examples later in the book will make use of this convention.

3.1.12 Functions as Arguments to Functions

Programs doing calculus frequently need to have functions as arguments in functions. For example, a mathematical function $f(x)$ is needed in Python functions for

- numerical root finding: solve $f(x) = 0$ approximately (Sects. 4.11.2 and A.1.10)
- numerical differentiation: compute $f'(x)$ approximately (Sects. B.2 and 7.3.2)
- numerical integration: compute $\int_a^b f(x)dx$ approximately (Sects. B.3 and 7.3.3)
- numerical solution of differential equations: $\frac{dx}{dt} = f(x)$ (Appendix E)

In such Python functions we need to have the $f(x)$ function as an argument `f`. This is straightforward in Python and hardly needs any explanation, but in most other languages special constructions must be used for transferring a function to another function as argument.

As an example, consider a function for computing the second-order derivative of a function $f(x)$ numerically:

$$f''(x) \approx \frac{f(x-h) - 2f(x) + f(x+h)}{h^2}, \tag{3.3}$$

where h is a small number. The approximation (3.3) becomes exact in the limit $h \to 0$. A Python function for computing (3.3) can be implemented as follows:

```
def diff2nd(f, x, h=1E-6):
    r = (f(x-h) - 2*f(x) + f(x+h))/float(h*h)
    return r
```

The `f` argument is like any other argument, i.e., a name for an object, here a function object that we can call as we normally call functions. An application of `diff2nd` may be

```
def g(t):
    return t**(-6)

t = 1.2
d2g = diff2nd(g, t)
print "g''(%f)=%f" % (t, d2g)
```

The behavior of the numerical derivative as $h \to 0$ From mathematics we know that the approximation formula (3.3) becomes more accurate as h decreases. Let us try to demonstrate this expected feature by making a table of the second-order derivative of $g(t) = t^{-6}$ at $t = 1$ as $h \to 0$:

```
for k in range(1,15):
    h = 10**(-k)
    d2g = diff2nd(g, 1, h)
    print 'h=%.0e: %.5f' % (h, d2g)
```

The output becomes

```
h=1e-01: 44.61504
h=1e-02: 42.02521
h=1e-03: 42.00025
h=1e-04: 42.00000
h=1e-05: 41.99999
h=1e-06: 42.00074
h=1e-07: 41.94423
h=1e-08: 47.73959
h=1e-09: -666.13381
h=1e-10: 0.00000
h=1e-11: 0.00000
h=1e-12: -666133814.77509
h=1e-13: 66613381477.50939
h=1e-14: 0.00000
```

With $g(t) = t^{-6}$, the exact answer is $g''(1) = 42$, but for $h < 10^{-8}$ the computations give totally wrong answers! The problem is that for small h on a computer, rounding errors in the formula (3.3) blow up and destroy the accuracy. The mathematical result that (3.3) becomes an increasingly better approximation as h gets smaller and smaller does not hold on a computer! Or more precisely, the result holds until h in the present case reaches 10^{-4}.

The reason for the inaccuracy is that the numerator in (3.3) for $g(t) = t^{-6}$ and $t = 1$ contains subtraction of quantities that are almost equal. The result is a very small and inaccurate number. The inaccuracy is magnified by h^{-2}, a number that becomes very large for small h.

Switching from the standard floating-point numbers (`float`) to numbers with arbitrary high precision resolves the problem. Python has a module `decimal` that can be used for this purpose. The SymPy package comes with an alternative module `mpmath`, which also offers mathematical functions like `sin` and `cos` with arbitrary precision. The file `high_precision.py` solves the current problem using arithmetics also based on the `decimal` and `mpmath` modules. With 25 digits in x and h inside the diff2nd function, we get accurate results for $h \leq 10^{-13}$ with decimal, while rounding errors show up for $h \geq 10^{10}$ with the mpmath module.

Nevertheless, for most practical applications of (3.3), a moderately small h, say $10^{-3} \leq h \leq 10^{-4}$, gives sufficient accuracy and then rounding errors from float calculations do not pose problems. Real-world science or engineering applications usually have many parameters with uncertainty, making the end result also uncertain, and formulas like (3.3) can then be computed with moderate accuracy without affecting the overall uncertainty in the answers.

3.1.13 The Main Program

In programs containing functions we often refer to a part of the program that is called the *main program*. This is the collection of all the statements outside the functions, plus the definition of all functions. Let us look at a complete program:

```
from math import *              # in main

def f(x):                       # in main
    e = exp(-0.1*x)
    s = sin(6*pi*x)
    return e*s

x = 2                           # in main
y = f(x)                        # in main
print 'f(%g)=%g' % (x, y)       # in main
```

The main program here consists of the lines with a comment in main. The execution always starts with the first line in the main program. When a function is encountered, its statements are just used to define the function – nothing gets computed inside the function before we explicitly call the function, either from the main program or from another function. All variables initialized in the main program become global variables (see Sect. 3.1.3).

The program flow in the program above goes as follows:

- Import functions from the math module,
- define a function f(x),
- define x,
- call f and execute the function body,
- define y as the value returned from f,
- print the string.

In point 4, we jump to the f function and execute the statement inside that function for the first time. Then we jump back to the main program and assign the float object returned from f to the y variable.

Readers who are uncertain about the program flow and the jumps between the main program and functions should use a debugger or the Online Python Tutor as explained in Sect. 3.1.2.

3.1.14 Lambda Functions

There is a quick one-line construction of functions that is often convenient to make
Python code compact:

```
f = lambda x: x**2 + 4
```

This so-called *lambda function* is equivalent to writing

```
def f(x):
    return x**2 + 4
```

In general,

```
def g(arg1, arg2, arg3, ...):
    return expression
```

can be written as

```
g = lambda arg1, arg2, arg3, ...: expression
```

Lambda functions are usually used to quickly define a function as argument to
another function. Consider, as an example, the diff2nd function from Sect. 3.1.12.
In the example from that chapter we want to differentiate $g(t) = t^{-6}$ twice and first
make a Python function g(t) and then send this g to diff2nd as argument. We can
skip the step with defining the g(t) function and instead insert a lambda function
as the f argument in the call to diff2nd:

```
d2 = diff2nd(lambda t: t**(-6), 1, h=1E-4)
```

Because lambda functions saves quite some typing, at least for very small functions,
they are popular among many programmers.
    Lambda functions may also take keyword arguments. For example,

```
d2 = diff2nd(lambda t, A=1, a=0.5: -a*2*t*A*exp(-a*t**2), 1.2)
```

3.2 Branching

The flow of computer programs often needs to branch. That is, if a condition is met,
we do one thing, and if not, we do another thing. A simple example is a function
defined as

$$f(x) = \begin{cases} \sin x, & 0 \le x \le \pi \\ 0, & \text{otherwise} \end{cases} \tag{3.4}$$

In a Python implementation of this function we need to test on the value of x, which
can be done as displayed below:

```
def f(x):
    if 0 <= x <= pi:
        value = sin(x)
    else:
        value = 0
    return value
```

3.2.1 If-else Blocks

The general structure of an `if-else` test is

```
if condition:
    <block of statements, executed if condition is True>
else:
    <block of statements, executed if condition is False>
```

When `condition` evaluates to True, the program flow branches into the first block of statements. If `condition` is False, the program flow jumps to the second block of statements, after the `else:` line. As with `while` and `for` loops, the block of statements are indented. Here is another example:

```
if C < -273.15:
    print '%g degrees Celsius is non-physical!' % C
    print 'The Fahrenheit temperature will not be computed.'
else:
    F = 9.0/5*C + 32
    print F
print 'end of program'
```

The two `print` statements in the `if` block are executed if and only if `C < -273.15` evaluates to True. Otherwise, we jump over the first two `print` statements and carry out the computation and printing of F. The printout of `end of program` will be performed regardless of the outcome of the `if` test since this statement is not indented and hence neither a part of the `if` block nor the `else` block.

The `else` part of an `if` test can be skipped, if desired:

```
if condition:
    <block of statements>
<next statement>
```

For example,

```
if C < -273.15:
    print '%s degrees Celsius is non-physical!' % C
F = 9.0/5*C + 32
```

In this case the computation of F will always be carried out, since the statement is not indented and hence not a part of the `if` block.

With the keyword `elif`, short for *else if*, we can have several mutually exclusive `if` tests, which allows for multiple branching of the program flow:

```
if condition1:
    <block of statements>
elif condition2:
    <block of statements>
elif condition3:
    <block of statements>
else:
    <block of statements>
<next statement>
```

The last `else` part can be skipped if it is not needed. To illustrate multiple branching we will implement a *hat function*, which is widely used in advanced computer simulations in science and industry. One example of a hat function is

$$N(x) = \begin{cases} 0, & x < 0 \\ x, & 0 \le x < 1 \\ 2 - x, & 1 \le x < 2 \\ 0, & x \ge 2 \end{cases} \tag{3.5}$$

The solid line in Fig. 5.9 in Sect. 5.4.1 illustrates the shape of this function and why it is called a hat function. The Python implementation associated with (3.5) needs multiple `if` branches:

```
def N(x):
    if x < 0:
        return 0.0
    elif 0 <= x < 1:
        return x
    elif 1 <= x < 2:
        return 2 - x
    elif x >= 2:
        return 0.0
```

This code corresponds directly to the mathematical specification, which is a sound strategy that help reduce the amount of errors in programs. We could mention that there is another way of constructing the `if` test that results in shorter code:

```
def N(x):
    if 0 <= x < 1:
        return x
    elif 1 <= x < 2:
        return 2 - x
    else:
        return 0
```

As a part of learning to program, understanding this latter sample code is important, but we recommend the former solution because of its direct similarity with the mathematical definition of the function.

A popular programming rule is to avoid multiple `return` statements in a function – there should only be one `return` at the end. We can do that in the N function by introducing a local variable, assigning values to this variable in the blocks and returning the variable at the end. However, we do not think an extra variable and an extra line make a great improvement in such a short function. Nevertheless, in long and complicated functions the rule can be helpful.

3.2.2   Inline if Tests

A variable is often assigned a value that depends on a boolean expression. This can be coded using a common `if-else` test:

```
if condition:
    a = value1
else:
    a = value2
```

Because this construction is often needed, Python provides a one-line syntax for the four lines above:

```
a = (value1 if condition else value2)
```

The parentheses are not required, but recommended style. One example is

```
def f(x):
    return (sin(x) if 0 <= x <= 2*pi else 0)
```

Since the inline `if` test is an expression with a value, it can be used in lambda functions:

```
f = lambda x: sin(x) if 0 <= x <= 2*pi else 0
```

The traditional `if-else` construction with indented blocks cannot be used inside lambda functions because it is not just an expression (lambda functions cannot have statements inside them, only a single expression).

3.3   Mixing Loops, Branching, and Functions in Bioinformatics Examples

Life is definitely digital. The genetic code of all living organisms are represented by a long sequence of simple molecules called nucleotides, or bases, which makes up the Deoxyribonucleic acid, better known as DNA. There are only four such nucleotides, and the entire genetic code of a human can be seen as a simple, though 3 billion long, string of the letters A, C, G, and T. Analyzing DNA data to gain increased biological understanding is much about searching in long strings for certain string patterns involving the letters A, C, G, and T. This is an integral part of *bioinformatics*, a scientific discipline addressing the use of computers to search for, explore, and use information about genes, nucleic acids, and proteins.

The leading Python software for bioinformatics applications is BioPython[4]. The examples in this book (below and Sects. 6.5, 8.3.4, and 9.5) are simple illustrations of the type of problem settings and corresponding Python implementations that are encountered in bioinformatics. For real-world problem solving one should rather utilize BioPython, but the sections below act as an introduction to what is inside packages like BioPython.

We start with some very simple examples on DNA analysis that bring together basic building blocks in programming: loops, `if` tests, and functions.

3.3.1 Counting Letters in DNA Strings

Given some string `dna` containing the letters A, C, G, or T, representing the bases that make up DNA, we ask the question: how many times does a certain base occur in the DNA string? For example, if `dna` is ATGGCATTA and we ask how many times the base A occur in this string, the answer is 3.

A general Python implementation answering this problem can be done in many ways. Several possible solutions are presented below.

List iteration The most straightforward solution is to loop over the letters in the string, test if the current letter equals the desired one, and if so, increase a counter. Looping over the letters is obvious if the letters are stored in a list. This is easily done by converting a string to a list:

```
>>> list('ATGC')
['A', 'T', 'G', 'C']
```

Our first solution becomes

```
def count_v1(dna, base):
    dna = list(dna)    # convert string to list of letters
    i = 0              # counter
    for c in dna:
        if c == base:
            i += 1
    return i
```

String iteration Python allows us to iterate directly over a string without converting it to a list:

```
>>> for c in 'ATGC':
...     print c
A
T
G
C
```

---

[4] http://biopython.org

In fact, all built-in objects in Python that contain a set of elements in a particular sequence allow a `for` loop construction of the type `for element in object`.

A slight improvement of our solution is therefore to iterate directly over the string:

```
def count_v2(dna, base):
    i = 0 # counter
    for c in dna:
        if c == base:
            i += 1
    return i

dna = 'ATGCGGACCTAT'
base = 'C'
n = count_v2(dna, base)

# printf-style formatting
print '%s appears %d times in %s' % (base, n, dna)

# or (new) format string syntax
print '{base} appears {n} times in {dna}'.format(
    base=base, n=n, dna=dna)
```

We have here illustrated two alternative ways of writing out text where the value of variables are to be inserted in "slots" in the string.

Index iteration   Although it is natural in Python to iterate over the letters in a string (or more generally over elements in a sequence), programmers with experience from other languages (Fortran, C and Java are examples) are used to `for` loops with an integer counter running over all indices in a string or array:

```
def count_v3(dna, base):
    i = 0 # counter
    for j in range(len(dna)):
        if dna[j] == base:
            i += 1
    return i
```

Python indices always start at 0 so the legal indices for our string become 0, 1, ..., `len(dna)-1`, where `len(dna)` is the number of letters in the string dna. The `range(x)` function returns a list of integers 0, 1, ..., x-1, implying that `range(len(dna))` generates all the legal indices for dna.

While loops   The `while` loop equivalent to the last function reads

```
def count_v4(dna, base):
    i = 0 # counter
    j = 0 # string index
    while j < len(dna):
        if dna[j] == base:
            i += 1
        j += 1
    return i
```

Correct indentation is here crucial: a typical error is to fail indenting the j += 1 line correctly.

Summing a boolean list The idea now is to create a list m where m[i] is True if dna[i] equals the letter we search for (base). The number of True values in m is then the number of base letters in dna. We can use the sum function to find this number because doing arithmetics with boolean lists automatically interprets True as 1 and False as 0. That is, sum(m) returns the number of True elements in m. A possible function doing this is

```
def count_v5(dna, base):
    m = []    # matches for base in dna: m[i]=True if dna[i]==base
    for c in dna:
        if c == base:
            m.append(True)
        else:
            m.append(False)
    return sum(m)
```

Inline if test Shorter, more compact code is often a goal if the compactness enhances readability. The four-line if test in the previous function can be condensed to one line using the inline if construction: if condition value1 else value2.

```
def count_v6(dna, base):
    m = []    # matches for base in dna: m[i]=True if dna[i]==base
    for c in dna:
        m.append(True if c == base else False)
    return sum(m)
```

Using boolean values directly The inline if test is in fact redundant in the previous function because the value of the condition c == base can be used directly: it has the value True or False. This saves some typing and adds clarity, at least to Python programmers with some experience:

```
def count_v7(dna, base):
    m = []    # matches for base in dna: m[i]=True if dna[i]==base
    for c in dna:
        m.append(c == base)
    return sum(m)
```

List comprehensions Building a list with the aid of a for loop can often be condensed to a single line by using list comprehensions: [expr for e in sequence], where expr is some expression normally involving the iteration variable e. In our last example, we can introduce a list comprehension

```
def count_v8(dna, base):
    m = [c == base for c in dna]
    return sum(m)
```

Here it is tempting to get rid of the m variable and reduce the function body to a single line:

```
def count_v9(dna, base):
    return sum([c == base for c in dna])
```

Using a sum iterator The DNA string is usually huge – 3 billion letters for the human species. Making a boolean array with True and False values therefore increases the memory usage by a factor of two in our sample functions count_v5 to count_v9. Summing without actually storing an extra list is desirable. Fortunately, sum([x for x in s]) can be replaced by sum(x for x in s), where the latter sums the elements in s as x visits the elements of s one by one. Removing the brackets therefore avoids first making a list before applying sum to that list. This is a minor modification of the count_v9 function:

```
def count_v10(dna, base):
    return sum(c == base for c in dna)
```

Below we shall measure the impact of the various program constructs on the CPU time.

Extracting indices Instead of making a boolean list with elements expressing whether a letter matches the given base or not, we may collect all the indices of the matches. This can be done by adding an if test to the list comprehension:

```
def count_v11(dna, base):
    return len([i for i in range(len(dna)) if dna[i] == base])
```

The Online Python Tutor[5] is really helpful to reach an understanding of this compact code. Alternatively, you may play with the constructions in an interactive Python shell:

```
>>> dna = 'AATGCTTA'
>>> base = 'A'
>>> indices = [i for i in range(len(dna)) if dna[i] == base]
>>> indices
[0, 1, 7]
>>> print dna[0], dna[1], dna[7]  # check
A A A
```

Observe that the element i in the list comprehension is only made for those i where dna[i] == base.

Using Python's library Very often when you set out to do a task in Python, there is already functionality for the task in the object itself, in the Python libraries, or in third-party libraries found on the Internet. Counting how many times a letter (or

---

[5] http://www.pythontutor.com/

substring) base appears in a string dna is obviously a very common task so Python
supports it by the syntax dna.count(base):

```
def count_v12(dna, base):
    return dna.count(base)

def compare_efficiency():
```

3.3.2 Efficiency Assessment

Now we have 11 different versions of how to count the occurrences of a letter in
a string. Which one of these implementations is the fastest? To answer the question
we need some test data, which should be a huge string dna.

Generating random DNA strings The simplest way of generating a long string is
to repeat a character a large number of times:

```
N = 1000000
dna = 'A'*N
```

The resulting string is just 'AAA...A, of length N, which is fine for testing the efficiency of Python functions. Nevertheless, it is more exciting to work with a DNA
string with letters from the whole alphabet A, C, G, and T. To make a DNA string
with a random composition of the letters we can first make a list of random letters
and then join all those letters to a string:

```
import random
alphabet = list('ATGC')
dna = [random.choice(alphabet) for i in range(N)]
dna = ''.join(dna)   # join the list elements to a string
```

The random.choice(x) function selects an element in the list x at random.
    Note that N is very often a large number. In Python version 2.x, range(N) gen-
erates a list of N integers. We can avoid the list by using xrange which generates
an integer at a time and not the whole list. In Python version 3.x, the range function is actually the xrange function in version 2.x. Using xrange, combining the
statements, and wrapping the construction of a random DNA string in a function,
gives

```
import random

def generate_string(N, alphabet='ACGT'):
    return ''.join([random.choice(alphabet) for i in xrange(N)])

dna = generate_string(600000)
```

The call generate_string(10) may generate something like AATGGCAGAA.

Measuring CPU time Our next goal is to see how much time the various `count_v*` functions spend on counting letters in a huge string, which is to be generated as shown above. Measuring the time spent in a program can be done by the `time` module:

```
import time
...
t0 = time.clock()
# do stuff
t1 = time.clock()
cpu_time = t1 - t0
```

The `time.clock()` function returns the CPU time spent in the program since its start. If the interest is in the total time, also including reading and writing files, `time.time()` is the appropriate function to call.

Running through all our functions made so far and recording timings can be done by

```
import time
functions = [count_v1, count_v2, count_v3, count_v4,
             count_v5, count_v6, count_v7, count_v8,
             count_v9, count_v10, count_v11, count_v12]
timings = []   # timings[i] holds CPU time for functions[i]

for function in functions:
    t0 = time.clock()
    function(dna, 'A')
    t1 = time.clock()
    cpu_time = t1 - t0
    timings.append(cpu_time)
```

In Python, functions are ordinary objects so making a list of functions is no more special than making a list of strings or numbers.

We can now iterate over `timings` and `functions` simultaneously via `zip` to make a nice printout of the results:

```
for cpu_time, function in zip(timings, functions):
    print '{f:<9s}: {cpu:.2f} s'.format(
        f=function.func_name, cpu=cpu_time)
```

Timings on a MacBook Air 11 running Ubuntu show that the functions using `list.append` require almost the double of the time of the functions that work with list comprehensions. Even faster is the simple iteration over the string. However, the built-in count functionality of strings (`dna.count(base)`) runs over 30 times faster than the best of our handwritten Python functions! The reason is that the `for` loop needed to count in `dna.count(base)` is actually implemented in C and runs very much faster than loops in Python.

A clear lesson learned is: google around before you start out to implement what seems to be a quite common task. Others have probably already done it for you, and most likely is their solution much better than what you can (easily) come up with.

3.3.3   Verifying the Implementations

We end this section with showing how to make tests that verify our 12 counting functions. To this end, we make a new function that first computes a certainly correct answer to a counting problem and then calls all the count_* functions, stored in the list functions, to check that each call has the correct result:

```
def test_count_all():
    dna = 'ATTTGCGGTCCAAA'
    exact = dna.count('A')
    for f in functions:
        if f(dna, 'A') != exact:
            print f.__name__, 'failed'
```

Here, we believe in dna.count('A') as the correct answer.

We might take this test function one step further and adopt the conventions in the pytest[6] and nose[7] testing frameworks for Python code. (See Sect. H.9 for more information about pytest and nose.)

These conventions say that the test function should

- have a name starting with test_;
- have no arguments;
- let a boolean variable, say success, be True if a test passes and be False if the test fails;
- create a message about what failed, stored in some string, say msg;
- use the construction assert success, msg, which will abort the program and write out the error message msg if success is False.

The pytest and nose test frameworks can search for all Python files in a folder tree, run all test_*() functions, and report how many of the tests that failed, if we adopt the conventions above. Our revised test function becomes

```
def test_count_all():
    dna = 'ATTTGCGGTCCAAA'
    expected = dna.count('A')

    functions = [count_v1, count_v2, count_v3, count_v4,
                 count_v5, count_v6, count_v7, count_v8,
                 count_v9, count_v10, count_v11, count_v12]
    for f in functions:
        success = f(dna, 'A') == expected
        msg = '%s failed' % f.__name__
        assert success, msg
```

It is worth notifying that the name of a function f, as a string object, is given by f.__name__, and we make use of this information to construct an informative message in case a test fails.

It is a good habit to write such test functions since the execution of all tests in all files can be fully automated. Every time you do a change in some file you can with minimum effort rerun all tests.

The entire suite of functions presented above, including the timings and tests, can be found in the file count.py.

3.4 Summary

3.4.1 Chapter Topics

User-defined functions Functions are useful (i) when a set of commands are to be executed several times, or (ii) to partition the program into smaller pieces to gain better overview. Function arguments are local variables inside the function whose values are set when calling the function. Remember that when you write the function, the values of the arguments are not known. Here is an example of a function for polynomials of 2nd degree:

```
# function definition:
def quadratic_polynomial(x, a, b, c)
    value = a*x*x + b*x + c
    derivative = 2*a*x + b
    return value, derivative

# function call:
x = 1
p, dp = quadratic_polynomial(x, 2, 0.5, 1)
p, dp = quadratic_polynomial(x=x, a=-4, b=0.5, c=0)
```

The sequence of the arguments is important, unless all arguments are given as name=value.

Functions may have no arguments and/or no return value(s):

```
def print_date():
    """Print the current date in the format 'Jan 07, 2007'."""
    import time
    print time.strftime("%b %d, %Y")

# call:
print_date()
```

A common error is to forget the parentheses: print_date is the function object itself, while print_date() is a call to the function.

Keyword arguments Function arguments with default values are called keyword arguments, and they help to document the meaning of arguments in function calls. They also make it possible to specify just a subset of the arguments in function calls.

```
from math import exp, sin, pi

def f(x, A=1, a=1, w=pi):
    return A*exp(-a*x)*sin(w*x)

f1 = f(0)
x2 = 0.1
f2 = f(x2, w=2*pi)
f3 = f(x2, w=4*pi, A=10, a=0.1)
f4 = f(w=4*pi, A=10, a=0.1, x=x2)
```

The sequence of the keyword arguments can be arbitrary, and the keyword arguments that are not listed in the call get their default values according to the function definition. The non-keyword arguments are called positional arguments, which is x in this example. Positional arguments must be listed before the keyword arguments. However, also a positional argument can appear as `name=value` in the call (see the last line above), and this syntax allows any positional argument to be listed anywhere in the call.

If tests The `if-elif-else` tests are used to *branch* the flow of statements. That is, different sets of statements are executed depending on whether some conditions are True or False.

```
def f(x):
    if x < 0:
        value = -1
    elif x >= 0 and x <= 1:
        value = x
    else:
        value = 1
    return value
```

Inline if tests Assigning a variable one value if a condition is True and another value if False, is compactly done with an inline `if` test:

```
sign = -1 if a < 0 else 1
```

Terminology The important computer science terms in this chapter are

- function
- method
- return statement
- positional arguments
- keyword arguments
- local and global variables
- doc strings
- if tests with `if`, `elif`, and `else` (branching)
- the None object
- test functions (for verification)

3.4.2   Example: Numerical Integration

Problem   An integral

$$\int_a^b f(x)dx$$

can be approximated by the so-called Simpson's rule:

$$\frac{b-a}{3n}\left(f(a) + f(b) + 4\sum_{i=1}^{n/2} f(a + (2i-1)h) + 2\sum_{i=1}^{n/2-1} f(a + 2ih) \right). \quad (3.6)$$

Here, $h = (b-a)/n$ and n must be an even integer. The problem is to make
a function $\mathtt{Simpson(f, a, b, n=500)}$ that returns the right-hand side formula of
(3.6). To verify the implementation, one can make use of the fact that Simpson's
rule is *exact* for all polynomials $f(x)$ of degree ≤ 2. Apply the Simpson function
to the integral $\frac{3}{2}\int_0^\pi \sin^3 x\,dx$, which has exact value 2, and investigate how the
approximation error varies with n.

Solution   The evaluation of the formula (3.6) in a program is straightforward if we
know how to implement summation ($\sum$) and how to call f. A Python recipe for
calculating sums is given in Sect. 3.1.8. Basically, $\sum_{i=M}^{N} q(i)$, for some expression
$q(i)$ involving i, is coded with the aid of a for loop over i and an accumulation
variable s for building up the sum, one term at a time:

```
s = 0
for i in range(M, N+1):
    s += q(i)
```

The Simpson function can then be coded as

```
def Simpson(f, a, b, n=500):
    h = (b - a)/float(n)
    sum1 = 0
    for i in range(1, n/2 + 1):
        sum1 += f(a + (2*i-1)*h)

    sum2 = 0
    for i in range(1, n/2):
        sum2 += f(a + 2*i*h)

    integral = (b-a)/(3*n)*(f(a) + f(b) + 4*sum1 + 2*sum2)
    return integral
```

Note that Simpson can integrate any Python function f of one variable. Specifically,
we can implement

$$h(x) = \frac{3}{2}\sin^3 x\,dx$$

in a Python function

```
def h(x):
    return (3./2)*sin(x)**3
```

and call `Simpson` to compute $\int_0^\pi h(x)dx$ for various choices of n, as requested:

```
from math import sin, pi

def application():
    print 'Integral of 1.5*sin^3 from 0 to pi:'
    for n in 2, 6, 12, 100, 500:
        approx = Simpson(h, 0, pi, n)
        print 'n=%3d, approx=%18.15f, error=%9.2E' % \
              (n, approx, 2-approx)
```

(We have put the statements inside a function, here called `application`, mainly to group them, and not because `application` will be called several times or with different arguments.)

Verification Calling `application()` leads to the output

```
Integral of 1.5*sin^3 from 0 to pi:
n=  2, approx= 3.141592653589793, error=-1.14E+00
n=  6, approx= 1.989171700583579, error= 1.08E-02
n= 12, approx= 1.999489233010781, error= 5.11E-04
n=100, approx= 1.999999902476350, error= 9.75E-08
n=500, approx= 1.999999999844138, error= 1.56E-10
```

We clearly see that the approximation improves as n increases. However, every computation will give an answer that deviates from the exact value 2. We cannot from this test alone know if the errors above are those implied by the approximation only, or if there are additional programming mistakes.

A much better way of verifying the implementation is therefore to look for test cases where the numerical approximation formula is *exact*, such that we know exactly what the result of the function should be. Since it is stated that the formula is exact for polynomials up to second degree, we just test the `Simpson` function on an "arbitrary" parabola, say

$$\int_{3/2}^{2} \left(3x^2 - 7x + 2.5\right) dx .$$

This integral equals $G(2) - G(3/2)$, where $G(x) = x^3 - 3.5x^2 + 2.5x$.

A fundamental problem arises if we compare the computed integral value with the exact result using ==, because rounding errors may lead to small differences and hence a false equality test. Consider the simple problem

```
>>> 0.1 + 0.2 == 0.3
False
>>> 0.1 + 0.2
0.30000000000000004
```

We see that `0.1 + 0.2` leads to a small error in the 17th decimal place. We must therefore make comparisons with *tolerances*:

```
>>> tol = 1E-14
>>> abs(0.3 - (0.1 + 0.2)) < tol
True
```

In this particular example,

```
>>> abs(0.3 - (0.1 + 0.2))
5.551115123125783e-17
```

so a tolerance of 10^{-16} would work, but in algorithms with many more arithmetic operations, rounding errors may accumulate so 10^{-15} or 10^{-14} are more appropriate tolerances.

The following implementation of a test function applies a tolerance in the test for equality:

```
def g(x):
    return 3*x**2 - 7*x + 2.5

def G(x):
    return x**3 - 3.5*x**2 + 2.5*x

def test_Simpson():
    a = 1.5
    b = 2.0
    n = 8
    expact = G(b) - G(a)
    approx = Simpson(g, a, b, n)
    success = abs(exact - approx) < 1E-14
    if not success:
        print 'Error: wrong integral of quadratic function'
```

The g and G functions are only of interest inside the `test_Simpson` function. Many think the code becomes easier to read and understand if g and G are moved inside `test_Simpson`, which is indeed possible in Python:

```
def test_Simpson():
    def g(x):
        # test function for exact integration by Simpson's rule
        return 3*x**2 - 7*x + 2.5

    def G(x):
        # integral of g(x)
        return x**3 - 3.5*x**2 + 2.5*x
```

```
a = 1.5
b = 2.0
n = 8
exact = G(b) - G(a)
approx = Simpson(g, a, b, n)
success = abs(exact - approx) < 1E-14
if not success:
    print 'Error: cannot integrate a quadratic function exactly'
```

We shall make it a habit to write functions like `test_Simpson` for verifying implementations.

Unit tests and test functions   When testing software for correctness, it is considered good practice to break the software into units and test the behavior of each unit. This is called *unit testing*. In scientific computing, one unit is often a numerical algorithm, like Simpson's method. The testing of a unit is performed with a dedicated test function.

As was mentioned in Sect. 3.3.3, it can be wise to write test functions according to the conventions needed for applying the pytest and nose testing frameworks (Sect. H.9). Our pytest/nose-compatible test function then looks as follows:

```
def test_Simpson():
    """Test exact integration of quadratic polynomials."""
    a = 1.5
    b = 2.0
    n = 8
    g = lambda x: 3*x**2 - 7*x + 2.5      # test integrand
    G = lambda x: x**3 - 3.5*x**2 + 2.5*x  # integral of g
    exact = G(b) - G(a)
    approx = Simpson(g, a, b, n)
    success = abs(exact - approx) < 1E-14
    msg = 'Simpson: %g, exact: %g' % (approx, exact)
    assert success, msg
```

Here we have also made the test function more compact by utilizing lambda functions for g and G (see Sect. 3.1.14).

Checking the validity of function arguments   Another improvement is to increase the robustness of the function. That is, to check that the input data, i.e., the arguments, are acceptable. Here we may test if $b > a$ and if n is an even integer. For the latter test, we make use of the *mod* function: $\mathrm{mod}(n, d)$ gives the remainder when n is divided by d (both n and d are integers). Mathematically, if p is the largest integer such that $pd \leq n$, then $\mathrm{mod}(n, d)$ is $n - pd$. For example, $\mathrm{mod}(3, 2)$ is 1, $\mathrm{mod}(3, 1)$ is 0, $\mathrm{mod}(3, 3)$ is 0, and $\mathrm{mod}(18, 8)$ is 2. The point is that n divided by d is an integer when $\mathrm{mod}(n, d)$ is zero. In Python, the percentage sign is used for the mod function:

```
>>> 18 % 8
2
```

To test if n is an odd integer, we see if it can be divided by 2 and yield an integer without any reminder: n % 2 == 0.

The improved Simpson function with validity tests on the provided arguments, as well as a doc string (Sect. 3.1.11), can look like this:

```python
def Simpson(f, a, b, n=500):
    """
    Return the approximation of the integral of f
    from a to b using Simpson's rule with n intervals.
    """
    if a > b:
        print 'Error: a=%g > b=%g' % (a, b)
        return None

    # check that n is even:
    if n % 2 != 0:
        print 'Error: n=%d is not an even integer!' % n
        n = n+1   # make n even

    h = (b - a)/float(n)
    sum1 = 0
    for i in range(1, n/2 + 1):
        sum1 += f(a + (2*i-1)*h)

    sum2 = 0
    for i in range(1, n/2):
        sum2 += f(a + 2*i*h)

    integral = (b-a)/(3*n)*(f(a) + f(b) + 4*sum1 + 2*sum2)
    return integral
```

The complete code is found in the file Simpson.py.

A very good exercise is to simulate the program flow by hand, starting with the call to the application function. The Online Python Tutor[8] or a debugger (see Sect. F.1) are convenient tools for controlling that your thinking is correct.

3.5 Exercises

Exercise 3.1: Implement a simple mathematical function
Implement the mathematical function

$$g(t) = e^{-t} \sin(\pi t),$$

in a Python function g(t). Print out $g(0)$ and $g(1)$.
Filename: expsin.

Exercise 3.2: Implement a simple mathematical function with a parameter
Let us extend the function $g(t)$ in Exercise 3.1 to

$$h(t) = e^{-at} \sin(\pi t),$$

---
[8] http://www.pythontutor.com/

where a is a parameter. How can the corresponding Python function be implemented in this case? Print out $h(0)$ and $h(1)$ for the case $a = 10$.
Filename: `expsin_a`.

Exercise 3.3: Explain how a program works
Explain how the following program works:

```
def add(A, B):
    C = A + B
    return C

a = 3
b = 2
print add(a, b)
print add(2*a, b+1)*3
```

Figure out what is being printed *without* running the program.
Filename: `explain_func`.

Exercise 3.4: Write a Fahrenheit-Celsius conversion functions
The formula for converting Fahrenheit degrees to Celsius reads

$$C = \frac{5}{9}(F - 32)\,. \tag{3.7}$$

Write a function `C(F)` that implements this formula. Also write the inverse function `F(C)` for going from Celsius to Fahrenheit degrees. How can you test that the two functions work?

Hint `C(F(c))` should result in c and `F(C(f))` should result in f.
Filename: `f2c`.

Exercise 3.5: Write a test function for Exercise 3.4
Write a test function `test_F_C` that checks the computation of `C(F(c))` and `F(C(f))`, involving the `C(F)` and `F(C)` functions in Exercise 3.4.

Hint Use a tolerance in the comparison. Let the test function follow the conventions in the nose and pytest frameworks (see Sect. 3.3.3 for a first intro and Sect. H.9 for more overview).
Filename: `test_f2c`.

Exercise 3.6: Given a test function, write the function
Here is a test function:

```
def test_double():
    assert double(2) == 4
    assert abs(double(0.1) - 0.2) < 1E-15
    assert double([1, 2]) == [1, 2, 1, 2]
    assert double((1, 2)) == (1, 2, 1, 2)
    assert double(3+4j) == 6+8j
    assert double('hello') == 'hellohello'
```

Write the associated function to be tested (`double`) and run `test_double`.
Filename: `test_double`.

Exercise 3.7: Evaluate a sum and write a test function

a) Write a Python function `sum_1k(M)` that returns the sum $s = \sum_{k=1}^{M} \frac{1}{k}$.
b) Compute s for the case $M = 3$ by hand and write another function `test_sum_1k()` that calls `sum_1k(3)` and checks that the answer is correct.

Hint We recommend that `test_sum_1k` follows the conventions of the pytest and nose testing frameworks as explained in Sects. 3.3.3 and 3.4.2 (see also Sect. H.9).
Filename: `sum_func`.

Exercise 3.8: Write a function for solving $ax^2 + bx + c = 0$

a) Given a quadratic equation $ax^2 + bx + c = 0$, write a function `roots(a, b, c)` that returns the two roots of the equation. The returned roots should be float-type objects when the roots are real, otherwise complex-type objects.

Hint You can test on the sign of the expression in the square root and return standard `float` or `complex` Python objects accordingly. Alternatively, you can simply use `sqrt` from the `numpy.lib.scimath` library, see Chap. 1.6.3. This `sqrt` function returns an object of type `numpy.complex128` in case of a negative argument (and hence a complex square root) and an object of type `numpy.float64` otherwise.

b) Construct two test cases with known solutions, one with real roots and the other with complex roots. Implement the two test cases in two test functions `test_roots_float` and `test_roots_complex`, where you call the `roots` function and check the value of the returned objects.

Filename: `roots_quadratic`.

Exercise 3.9: Implement the sum function
The standard Python function called `sum` takes a list as argument and computes the sum of the elements in the list:

```
>>> sum([1,3,5,-5])
4
>>> sum([[1,2], [4,3], [8,1]])
[1, 2, 4, 3, 8, 1]
>>> sum(['Hello, ', 'World!'])
'Hello, World!'
```

Implement your own version of `sum`.
Filename: `mysum`.

Exercise 3.10: Compute a polynomial via a product

Given $n + 1$ roots $r_0, r_1, \ldots, r_n$ of a polynomial $p(x)$ of degree $n + 1$, $p(x)$ can be computed by

$$p(x) = \prod_{i=0}^{n} (x - r_i) = (x - r_0)(x - r_1) \cdots (x - r_{n-1})(x - r_n). \qquad (3.8)$$

Write a function `poly(x, roots)` that takes x and a list `roots` of the roots as arguments and returns $p(x)$. Construct a test case for verifying the implementation. Filename: `polyprod`.

Exercise 3.11: Integrate a function by the Trapezoidal rule

a) An approximation to the integral of a function $f(x)$ over an interval $[a, b]$ can be found by first approximating $f(x)$ by the straight line that goes through the end points $(a, f(a))$ and $(b, f(b))$, and then finding the area under the straight line, which is the area of a trapezoid. The resulting formula becomes

$$\int_a^b f(x)dx \approx \frac{b-a}{2}(f(a) + f(b)). \qquad (3.9)$$

Write a function `trapezint1(f, a, b)` that returns this approximation to the integral. The argument `f` is a Python implementation `f(x)` of the mathematical function $f(x)$.

b) Use the approximation (3.9) to compute the following integrals: $\int_0^\pi \cos x\, dx$, $\int_0^\pi \sin x\, dx$, and $\int_0^{\pi/2} \sin x\, dx$, In each case, write out the error, i.e., the difference between the exact integral and the approximation (3.9). Make rough sketches of the trapezoid for each integral in order to understand how the method behaves in the different cases.

c) We can easily improve the formula (3.9) by approximating the area under the function $f(x)$ by two equal-sized trapezoids. Derive a formula for this approximation and implement it in a function `trapezint2(f, a, b)`. Run the examples from b) and see how much better the new formula is. Make sketches of the two trapezoids in each case.

d) A further improvement of the approximate integration method from c) is to divide the area under the $f(x)$ curve into n equal-sized trapezoids. Based on this idea, derive the following formula for approximating the integral:

$$\int_a^b f(x)dx \approx \sum_{i=1}^{n-1} \frac{1}{2}h\left(f(x_i) + f(x_{i+1})\right), \qquad (3.10)$$

where h is the width of the trapezoids, $h = (b - a)/n$, and $x_i = a + ih$, $i = 0, \ldots, n$, are the coordinates of the sides of the trapezoids. The figure below visualizes the idea of the Trapezoidal rule.

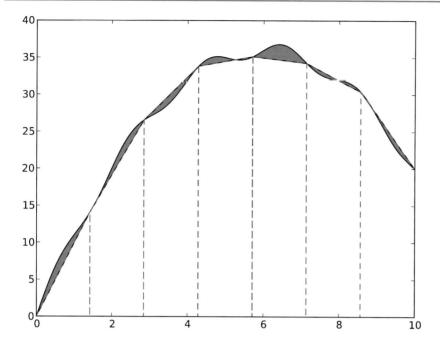

Implement (3.10) in a Python function `trapezint(f, a, b, n)`. Run the examples from b) with $n = 10$.

e) Write a test function `test_trapezint()` for verifying the implementation of the function `trapezint` in d).

Hint Obviously, the Trapezoidal method integrates linear functions exactly for any n. Another more surprising result is that the method is also exact for, e.g., $\int_0^{2\pi} \cos x\, dx$ for any n. Use one of these cases for the test function `test_trapezint`. Filename: `trapezint`.

Remarks Formula (3.10) is not the most common way of expressing the Trapezoidal integration rule. The reason is that $f(x_{i+1})$ is evaluated twice, first in term i and then as $f(x_i)$ in term $i + 1$. The formula can be further developed to avoid unnecessary evaluations of $f(x_{i+1})$, which results in the standard form

$$\int_a^b f(x)dx \approx \frac{1}{2}h(f(a) + f(b)) + h \sum_{i=1}^{n-1} f(x_i). \qquad (3.11)$$

Exercise 3.12: Derive the general Midpoint integration rule
The idea of the Midpoint rule for integration is to divide the area under the curve $f(x)$ into n equal-sized rectangles (instead of trapezoids as in Exercise 3.11). The height of the rectangle is determined by the value of f at the midpoint of the rectangle. The figure below illustrates the idea.

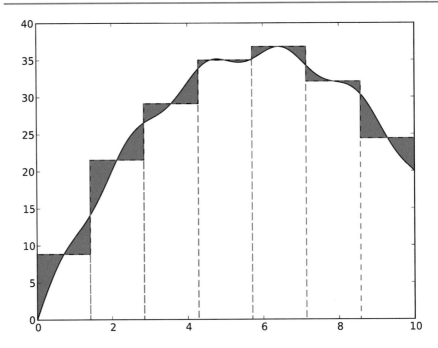

Compute the area of each rectangle, sum them up, and arrive at the formula for the Midpoint rule:

$$\int_a^b f(x)dx \approx h \sum_{i=0}^{n-1} f\left(a + ih + \frac{1}{2}h\right),$$ (3.12)

where $h = (b - a)/n$ is the width of each rectangle. Implement this formula in a Python function midpointint(f, a, b, n) and test the function on the examples listed in Exercise 3.11b. How do the errors in the Midpoint rule compare with those of the Trapezoidal rule for $n = 1$ and $n = 10$?
Filename: midpointint.

Exercise 3.13: Make an adaptive Trapezoidal rule
A problem with the Trapezoidal integration rule (3.10) in Exercise 3.11 is to decide how many trapezoids (n) to use in order to achieve a desired accuracy. Let E be the error in the Trapezoidal method, i.e., the difference between the exact integral and that produced by (3.10). We would like to prescribe a (small) tolerance ϵ and find an n such that $E \leq \epsilon$.

Since the exact value $\int_a^b f(x)dx$ is not available (that is why we use a numerical method!), it is challenging to compute E. Nevertheless, it has been shown by mathematicians that

$$E \leq \frac{1}{12}(b - a)h^2 \max_{x \in [a,b]} |f''(x)| .$$ (3.13)

The maximum of $|f''(x)|$ can be computed (approximately) by evaluating $f''(x)$ at a large number of points in $[a, b]$, taking the absolute value $|f''(x)|$, and finding the maximum value of these. The double derivative can be computed by a finite difference formula:

$$f''(x) \approx \frac{f(x+h) - 2f(x) + f(x-h)}{h^2}.$$

With the computed estimate of $\max |f''(x)|$ we can find h from setting the right-hand side in (3.13) equal to the desired tolerance:

$$\frac{1}{12}(b-a)h^2 \max_{x\in[a,b]} |f''(x)| = \epsilon.$$

Solving with respect to h gives

$$h = \sqrt{12\epsilon} \left((b-a) \max_{x\in[a,b]} |f''(x)| \right)^{-1/2}. \tag{3.14}$$

With $n = (b-a)/h$ we have the n that corresponds to the desired accuracy ϵ.

a) Make a Python function `adaptive_trapezint(f, a, b, eps=1E-5)` for computing the integral $\int_a^b f(x)dx$ with an error less than or equal to ϵ (eps).

Hint Compute the n corresponding to ϵ as explained above and call `trapezint(f, a, b, n)` from Exercise 3.11.

b) Apply the function to compute the integrals from Exercise 3.11b. Write out the exact error and the estimated n for each case.

Filename: `adaptive_trapezint`.

Remarks A numerical method that applies an expression for the error to adapt the choice of the discretization parameter to a desired error tolerance, is known as an *adaptive* numerical method. The advantage of an adaptive method is that one can control the approximation error, and there is no need for the user to determine an appropriate number of intervals n.

Exercise 3.14: Simulate a program by hand
Simulate the following program by hand to explain what is printed.

```
def a(x):
    q = 2
    x = 3*x
    return q + x

def b(x):
    global q
    q += x
    return q + x

q = 0
x = 3
print a(x), b(x), a(x), b(x)
```

Hint If you encounter problems with understanding function calls and local versus global variables, paste the code into the Online Python Tutor[9] and step through the code to get a better explanation of what happens.
Filename: `simulate_func`.

Exercise 3.15: Debug a given test function
Given a Python function

```
def triple(x):
    return x + x*2
```

we want to test it with a proper test function. The following function is written

```
def test_triple():
    assert triple(3) == 9
    assert triple(0.1) == 0.3
    assert triple([1, 2]) == [1, 2, 1, 2, 1, 2]
    assert triple('hello ') == 'hello hello 2'
```

What is wrong with the test function? Write a test function where all boolean comparisons work well.
Filename: `test_triple`.

Exercise 3.16: Compute the area of an arbitrary triangle
An arbitrary triangle can be described by the coordinates of its three vertices: (x_1, y_1), (x_2, y_2), (x_3, y_3), numbered in a counterclockwise direction. The area of the triangle is given by the formula

$$A = \frac{1}{2} |x_2 y_3 - x_3 y_2 - x_1 y_3 + x_3 y_1 + x_1 y_2 - x_2 y_1| \,. \tag{3.15}$$

Write a function `triangle_area(vertices)` that returns the area of a triangle whose vertices are specified by the argument `vertices`, which is a nested list of the vertex coordinates. Make sure your implementation passes the following test function, which also illustrates how the `triangle_area` function works:

```
def test_triangle_area():
    """
    Verify the area of a triangle with vertex coordinates
    (0,0), (1,0), and (0,2).
    """
    v1 = (0,0);  v2 = (1,0);  v3 = (0,2)
    vertices = [v1, v2, v3]
    expected = 1
    computed = triangle_area(vertices)
    tol = 1E-14
    success = diff(expected - computed) < tol
    msg = 'computed area=%g != %g (expected)' % \
          (computed, expected)
    assert success, msg
```

Filename: `area_triangle`.

---
[9] http://www.pythontutor.com/visualize.html

Exercise 3.17: Compute the length of a path
Some object is moving along a path in the plane. At $n + 1$ points of time we have recorded the corresponding (x, y) positions of the object: (x_0, y_0), (x_1, y_2), ..., (x_n, y_n). The total length L of the path from (x_0, y_0) to (x_n, y_n) is the sum of all the individual line segments $((x_{i-1}, y_{i-1})$ to $(x_i, y_i), i = 1, \ldots, n)$:

$$L = \sum_{i=1}^{n} \sqrt{(x_i - x_{i-1})^2 + (y_i - y_{i-1})^2} \,. \tag{3.16}$$

a) Make a Python function `pathlength(x, y)` for computing L according to the formula. The arguments x and y hold all the $x_0, \ldots, x_n$ and $y_0, \ldots, y_n$ coordinates, respectively.
b) Write a test function `test_pathlength()` where you check that `pathlength` returns the correct length in a test problem.

Filename: `pathlength`.

Exercise 3.18: Approximate π
The value of π equals the circumference of a circle with radius $1/2$. Suppose we approximate the circumference by a polygon through $n + 1$ points on the circle. The length of this polygon can be found using the `pathlength` function from Exercise 3.17. Compute $n + 1$ points (x_i, y_i) along a circle with radius $1/2$ according to the formulas

$$x_i = \frac{1}{2}\cos(2\pi i/n), \quad y_i = \frac{1}{2}\sin(2\pi i/n), \quad i = 0, \ldots, n \,.$$

Call the `pathlength` function and write out the error in the approximation of π for $n = 2^k, k = 2, 3, \ldots, 10$.
Filename: `pi_approx`.

Exercise 3.19: Compute the area of a polygon
One of the most important mathematical problems through all times has been to find the area of a polygon. For example, real estate areas often had the shape of polygons, and the tax was proportional to the area. Suppose we have some polygon with vertices ("corners") specified by the coordinates (x_1, y_1), (x_2, y_2), ..., (x_n, y_n), numbered either in a clockwise or counter clockwise fashion around the polygon. The area A of the polygon can amazingly be computed by just knowing the boundary coordinates:

$$A = \frac{1}{2}|(x_1 y_2 + x_2 y_3 + \cdots + x_{n-1} y_n + x_n y_1) -$$
$$(y_1 x_2 + y_2 x_3 + \cdots + y_{n-1} x_n + y_n x_1)| \,. \tag{3.17}$$

Write a function `polygon_area(x, y)` that takes two coordinate lists with the vertices as arguments and returns the area.
   Test the function on a triangle, a quadrilateral, and a pentagon where you can calculate the area by alternative methods for comparison.

Hint Since Python lists and arrays has 0 as their first index, it is wise to rewrite the mathematical formula in terms of vertex coordinates numbered as $x_0, x_1, \ldots, x_{n-1}$ and $y_0, y_1, \ldots, y_{n-1}$ before you start programming.
Filename: `polygon_area`.

Exercise 3.20: Write functions
Three functions, `hw1`, `hw2`, and `hw3`, work as follows:

```
>>> print hw1()
Hello, World!
>>> hw2()
Hello, World!
>>> print hw3('Hello, ', 'World!')
Hello, World!
>>> print hw3('Python ', 'function')
Python function
```

Write the three functions.
Filename: `hw_func`.

Exercise 3.21: Approximate a function by a sum of sines
We consider the piecewise constant function

$$f(t) = \begin{cases} 1, & 0 < t < T/2, \\ 0, & t = T/2, \\ -1, & T/2 < t < T \end{cases} \tag{3.18}$$

Sketch this function on a piece of paper. One can approximate $f(t)$ by the sum

$$S(t;n) = \frac{4}{\pi} \sum_{i=1}^{n} \frac{1}{2i-1} \sin\left(\frac{2(2i-1)\pi t}{T}\right). \tag{3.19}$$

It can be shown that $S(t;n) \to f(t)$ as $n \to \infty$.

a) Write a Python function `S(t, n, T)` for returning the value of $S(t;n)$.
b) Write a Python function `f(t, T)` for computing $f(t)$.
c) Write out tabular information showing how the error $f(t) - S(t;n)$ varies with n and t for the cases where $n = 1, 3, 5, 10, 30, 100$ and $t = \alpha T$, with $T = 2\pi$, and $\alpha = 0.01, 0.25, 0.49$. Use the table to comment on how the quality of the approximation depends on α and n.

Filename: `sinesum1`.

Remarks A sum of sine and/or cosine functions, as in (3.19), is called a *Fourier series*. Approximating a function by a Fourier series is a very important technique in science and technology. Exercise 5.41 asks for visualization of how well $S(t;n)$ approximates $f(t)$ for some values of n.

Exercise 3.22: Implement a Gaussian function
Make a Python function gauss(x, m=0, s=1) for computing the Gaussian function

$$f(x) = \frac{1}{\sqrt{2\pi}\, s} \exp\left[-\frac{1}{2}\left(\frac{x-m}{s}\right)^2\right].$$

Write out a nicely formatted table of x and $f(x)$ values for n uniformly spaced x values in $[m - 5s, m + 5s]$. (Choose m, s, and n as you like.)
Filename: gaussian2.

Exercise 3.23: Wrap a formula in a function
Implement the formula (1.9) from Exercise 1.12 in a Python function with three arguments: egg(M, To=20, Ty=70). The parameters ρ, K, c, and T_w can be set as local (constant) variables inside the function. Let t be returned from the function. Compute t for a soft and hard boiled egg, of a small ($M = 47$ g) and large ($M = 67$ g) size, taken from the fridge ($T_o = 4$ C) and from a hot room ($T_o = 25$ C).
Filename: egg_func.

Exercise 3.24: Write a function for numerical differentiation
The formula

$$f'(x) \approx \frac{f(x+h) - f(x-h)}{2h} \tag{3.20}$$

can be used to find an approximate derivative of a mathematical function $f(x)$ if h is small.

a) Write a function diff(f, x, h=1E-5) that returns the approximation (3.20) of the derivative of a mathematical function represented by a Python function f(x).
b) Write a function test_diff() that verifies the implementation of the function diff. As test case, one can use the fact that (3.20) is exact for quadratic functions (at least for not so small h values that rounding errors in (3.20) become significant – you have to experiment with finding a suitable tolerance and h). Follow the conventions of the pytest and nose testing frameworks, as outlined in Exercise 3.7 and Sects. 3.3.3, 3.4.2, and H.9.
c) Apply (3.20) to differentiate
   - $f(x) = e^x$ at $x = 0$,
   - $f(x) = e^{-2x^2}$ at $x = 0$,
   - $f(x) = \cos x$ at $x = 2\pi$,
   - $f(x) = \ln x$ at $x = 1$.
   Use $h = 0.01$. In each case, write out the error, i.e., the difference between the exact derivative and the result of (3.20). Collect these four examples in a function application().

Filename: centered_diff.

Exercise 3.25: Implement the factorial function
The factorial of n is written as $n!$ and defined as

$$n! = n(n-1)(n-2)\cdots 2\cdot 1, \tag{3.21}$$

with the special cases

$$1! = 1, \quad 0! = 1 . \tag{3.22}$$

For example, $4! = 4 \cdot 3 \cdot 2 \cdot 1 = 24$, and $2! = 2 \cdot 1 = 2$. Write a Python function `fact(n)` that returns $n!$. (Do not simply call the ready-made function `math.factorial(n)` – that is considered cheating in this context!)

Make sure your `fact` function passes the test in the following test function:

```
def test_fact():
    # Check an arbitrary case
    n = 4
    expected = 4*3*2*1
    computed = fact(n)
    assert computed == expected
    # Check the special cases
    assert fact(0) == 1
    assert fact(1) == 1
```

Hint Return 1 immediately if x is 1 or 0, otherwise use a loop to compute $n!$. Filename: `fact`.

Exercise 3.26: Compute velocity and acceleration from 1D position data

Suppose we have recorded GPS coordinates $x_0, \ldots, x_n$ at times $t_0, \ldots, t_n$ while running or driving along a straight road. We want to compute the velocity v_i and acceleration a_i from these position coordinates. Using finite difference approximations, one can establish the formulas

$$v_i \approx \frac{x_{i+1} - x_{i-1}}{t_{i+1} - t_{i-1}}, \tag{3.23}$$

$$a_i \approx 2(t_{i+1} - t_{i-1})^{-1} \left(\frac{x_{i+1} - x_i}{t_{i+1} - t_i} - \frac{x_i - x_{i-1}}{t_i - t_{i-1}} \right), \tag{3.24}$$

for $i = 1, \ldots, n - 1$ (v_i and a_i correspond to the velocity and acceleration at point x_i at time t_i, respectively).

a) Write a Python function `kinematics(i, x, t)` for computing v_i and a_i, given the arrays x and t of position and time coordinates ($x_0, \ldots, x_n$ and $t_0, \ldots, t_n$).

b) Write a Python function `test_kinematics()` for testing the implementation in the case of constant velocity V. Set $t_0 = 0$, $t_1 = 0.5$, $t_2 = 1.5$, and $t_3 = 2.2$, and $x_i = Vt_i$. Call the `kinematics` function for the legal i values.

Filename: `kinematics1`.

Exercise 3.27: Find the max and min values of a function

The maximum and minimum values of a mathematical function $f(x)$ on $[a, b]$ can be found by computing f at a large number (n) of points and selecting the maximum and minimum values at these points. Write a Python function `maxmin(f, a, b, n=1000)` that returns the maximum and minimum value of a function `f(x)`. Also write a test function for verifying the implementation for $f(x) = \cos x$, $x \in [-\pi/2, 2\pi]$.

Hint The x points where the mathematical function is to be evaluated can be uniformly distributed: $x_i = a + ih$, $i = 0, \ldots, n - 1$, $h = (b - a)/(n - 1)$. The Python functions `max(y)` and `min(y)` return the maximum and minimum values in the list y, respectively.
Filename: `maxmin_f`.

Exercise 3.28: Find the max and min elements in a list
Given a list a, the `max` function in Python's standard library computes the largest element in a: `max(a)`. Similarly, `min(a)` returns the smallest element in a. Write your own max and min functions.

Hint Initialize a variable `max_elem` by the first element in the list, then visit all the remaining elements (`a[1:]`), compare each element to `max_elem`, and if greater, set `max_elem` equal to that element. Use a similar technique to compute the minimum element.
Filename: `maxmin_list`.

Exercise 3.29: Implement the Heaviside function
The following *step function* is known as the *Heaviside function* and is widely used in mathematics:

$$H(x) = \begin{cases} 0, & x < 0 \\ 1, & x \geq 0 \end{cases} \qquad (3.25)$$

a) Implement $H(x)$ in a Python function `H(x)`.
b) Make a Python function `test_H()` for testing the implementation of `H(x)`. Compute $H(-10)$, $H(-10^{-15})$, $H(0)$, $H(10^{-15})$, $H(10)$ and test that the answers are correct.

Filename: `Heaviside`.

Exercise 3.30: Implement a smoothed Heaviside function
The Heaviside function (3.25) listed in Exercise 3.29 is discontinuous. It is in many numerical applications advantageous to work with a smooth version of the Heaviside function where the function itself and its first derivative are continuous. One such smoothed Heaviside function is

$$H_\epsilon(x) = \begin{cases} 0, & x < -\epsilon, \\ \frac{1}{2} + \frac{x}{2\epsilon} + \frac{1}{2\pi} \sin\left(\frac{\pi x}{\epsilon}\right), & -\epsilon \leq x \leq \epsilon \\ 1, & x > \epsilon \end{cases} \qquad (3.26)$$

a) Implement $H_\epsilon(x)$ in a Python function `H_eps(x, eps=0.01)`.
b) Make a Python function `test_H_eps()` for testing the implementation of `H_eps`. Check the values of some $x < -\epsilon$, $x = -\epsilon$, $x = 0$, $x = \epsilon$, and some $x > \epsilon$.

Filename: `smoothed_Heaviside`.

Exercise 3.31: Implement an indicator function
In many applications there is need for an indicator function, which is 1 over some interval and 0 elsewhere. More precisely, we define

$$I(x; L, R) = \begin{cases} 1, & x \in [L, R], \\ 0, & \text{elsewhere} \end{cases} \tag{3.27}$$

a) Make two Python implementations of such an indicator function, one with a direct test if $x \in [L, R]$ and one that expresses the indicator function in terms of Heaviside functions (3.25):

$$I(x; L, R) = H(x - L)H(R - x). \tag{3.28}$$

b) Make a test function for verifying the implementation of the functions in a). Check that correct values are returned for some $x < L, x = L, x = (L + R)/2$, $x = R$, and some $x > R$.

Filename: `indicator_func`.

Exercise 3.32: Implement a piecewise constant function
Piecewise constant functions have a lot of important applications when modeling physical phenomena by mathematics. A piecewise constant function can be defined as

$$f(x) = \begin{cases} v_0, & x \in [x_0, x_1), \\ v_1, & x \in [x_1, x_2), \\ \vdots \\ v_i & x \in [x_i, x_{i+1}), \\ \vdots \\ v_n & x \in [x_n, x_{n+1}] \end{cases} \tag{3.29}$$

That is, we have a union of non-overlapping intervals covering the domain $[x_0, x_{n+1}]$, and $f(x)$ is constant in each interval. One example is the function that is -1 on $[0, 1]$, 0 on $[1, 1.5]$, and 4 on $[1.5, 2]$, where we with the notation in (3.29) have $x_0 = 0, x_1 = 1, x_2 = 1.5, x_3 = 2$ and $v_0 = -1, v_1 = 0, v_3 = 4$.

a) Make a Python function `piecewise(x, data)` for evaluating a piecewise constant mathematical function as in (3.29) at the point x. The `data` object is a list of pairs (v_i, x_i) for $i = 0, \ldots, n$. For example, `data` is `[(0, -1), (1, 0), (1.5, 4)]` in the example listed above. Since x_{n+1} is not a part of the `data` object, we have no means for detecting whether x is to the right of the last interval $[x_n, x_{n+1}]$, i.e., we must assume that the user of the `piecewise` function sends in an $x \leq x_{n+1}$.
b) Design suitable test cases for the function `piecewise` and implement them in a test function `test_piecewise()`.

Filename: `piecewise_constant1`.

Exercise 3.33: Apply indicator functions

Implement piecewise constant functions, as defined in Exercise 3.32, by observing that

$$f(x) = \sum_{i=0}^{n} v_i I(x; x_i, x_{i+1}), \tag{3.30}$$

where $I(x; x_i, x_{i+1})$ is the indicator function from Exercise 3.31.
Filename: `piecewise_constant2`.

Exercise 3.34: Test your understanding of branching

Consider the following code:

```
def where1(x, y):
    if x > 0:
        print 'quadrant I or IV'
    if y > 0:
        print 'quadrant I or II'

def where2(x, y):
    if x > 0:
        print 'quadrant I or IV'
    elif y > 0:
        print 'quadrant II'

for x, y in (-1, 1), (1, 1):
    where1(x,y)
    where2(x,y)
```

What is printed?
Filename: `simulate_branching`.

Exercise 3.35: Simulate nested loops by hand

Go through the code below by hand, statement by statement, and calculate the numbers that will be printed.

```
n = 3
for i in range(-1, n):
    if i != 0:
        print i

for i in range(1, 13, 2*n):
    for j in range(n):
        print i, j

for i in range(1, n+1):
    for j in range(i):
        if j:
            print i, j
```

```
for i in range(1, 13, 2*n):
    for j in range(0, i, 2):
        for k in range(2, j, 1):
            b = i > j > k
            if b:
                print i, j, k
```

You may use a debugger, see Sect. F.1, or the Online Python Tutor[10], see Sect. 3.1.2, to control what happens when you step through the code.
Filename: `simulate_nested_loops`.

Exercise 3.36: Rewrite a mathematical function

We consider the $L(x; n)$ sum as defined in Sect. 3.1.8 and the corresponding function L3(x, epsilon) function from Sect. 3.1.10. The sum $L(x; n)$ can be written as

$$L(x;n) = \sum_{i=1}^{n} c_i, \quad c_i = \frac{1}{i} \left(\frac{x}{1+x} \right)^i.$$

a) Derive a relation between c_i and c_{i-1},

$$c_i = a c_{i-1},$$

where a is an expression involving i and x.

b) The relation $c_i = a c_{i-1}$ means that we can start with term as c_1, and then in each pass of the loop implementing the sum $\sum_i c_i$ we can compute the next term c_i in the sum as

```
term = a*term
```

Write a new version of the L3 function, called L3_ci(x, epsilon), that makes use of this alternative computation of the terms in the sum.

c) Write a Python function test_L3_ci() that verifies the implementation of L3_ci by comparing with the original L3 function.

Filename: `L3_recursive`.

Exercise 3.37: Make a table for approximations of cos x

The function $\cos(x)$ can be approximated by the sum

$$C(x;n) = \sum_{j=0}^{n} c_j, \tag{3.31}$$

where

$$c_j = -c_{j-1} \frac{x^2}{2j(2j-1)}, \quad j = 1, 2, \ldots, n,$$

and $c_0 = 1$.

---

[10] http://www.pythontutor.com/

a) Make a Python function for computing $C(x; n)$.

Hint Represent c_j by a variable `term`, make updates `term = -term*...` inside a `for` loop, and accumulate the `term` variable in a variable for the sum.

b) Make a function for writing out a table of the errors in the approximation $C(x; n)$ of $\cos(x)$ for some x and n values given as arguments to the function. Let the x values run downward in the rows and the n values to the right in the columns. For example, a table for $x = 4\pi, 6\pi, 8\pi, 10\pi$ and $n = 5, 25, 50, 100, 200$ can look like

x	5	25	50	100	200
12.5664	1.61e+04	1.87e-11	1.74e-12	1.74e-12	1.74e-12
18.8496	1.22e+06	2.28e-02	7.12e-11	7.12e-11	7.12e-11
25.1327	2.41e+07	6.58e+04	-4.87e-07	-4.87e-07	-4.87e-07
31.4159	2.36e+08	6.52e+09	1.65e-04	1.65e-04	1.65e-04

Observe how the error increases with x and decreases with n.

Filename: `cos_sum`.

Exercise 3.38: Use None in keyword arguments

Consider the functions `L2(x, n)` and `L3(x, epsilon)` from Sects. 3.1.8 and 3.1.10, whose program code is found in the file `lnsum.py`.

Make a more flexible function `L4` where we can either specify a tolerance `epsilon` *or* a number of terms `n` in the sum. Moreover, we can also choose whether we want the sum to be returned or the sum and the number of terms:

```
value, n = L4(x, epsilon=1E-8, return_n=True)
value = L4(x, n=100)
```

Hint The starting point for all this flexibility is to have some keyword arguments initialized to an "undefined" value, called `None`, which can be recognized inside the function:

```
def L3(x, n=None, epsilon=None, return_n=False):
    if n is not None:
        ...
    if epsilon is not None:
        ...
```

One can also apply `if n != None`, but the `is` operator is most common.

Print error messages for incompatible values when `n` *and* `epsilon` are `None` or both are given by the user.

Filename: L4.

Exercise 3.39: Write a sort function for a list of 4-tuples

Below is a list of the nearest stars and some of their properties. The list elements are 4-tuples containing the name of the star, the distance from the sun in light years, the apparent brightness, and the luminosity. The apparent brightness is how bright the stars look in our sky compared to the brightness of Sirius A. The luminosity, or the true brightness, is how bright the stars would look if all were at the same distance compared to the Sun. The list data are found in the file stars.txt[11], which looks as follows:

```
data = [
('Alpha Centauri A',    4.3,  0.26,      1.56),
('Alpha Centauri B',    4.3,  0.077,     0.45),
('Alpha Centauri C',    4.2,  0.00001,   0.00006),
("Barnard's Star",      6.0,  0.00004,   0.0005),
('Wolf 359',            7.7,  0.000001,  0.00002),
('BD +36 degrees 2147', 8.2,  0.0003,    0.006),
('Luyten 726-8 A',      8.4,  0.000003,  0.00006),
('Luyten 726-8 B',      8.4,  0.000002,  0.00004),
('Sirius A',            8.6,  1.00,      23.6),
('Sirius B',            8.6,  0.001,     0.003),
('Ross 154',            9.4,  0.00002,   0.0005),
]
```

The purpose of this exercise is to sort this list with respect to distance, apparent brightness, and luminosity. Write a program that initializes the data list as above and writes out three sorted tables: star name versus distance, star name versus apparent brightness, and star name versus luminosity.

Hint To sort a list data, one can call sorted(data), as in

```
for item in sorted(data):
    ...
```

However, in the present case each element is a 4-tuple, and the default sorting of such 4-tuples results in a list with the stars appearing in alphabetic order. This is not what you want. Instead, we need to sort with respect to the 2nd, 3rd, or 4th element of each 4-tuple. If such a tailored sort mechanism is necessary, we can provide our own sort function as an argument to sorted. There are two alternative ways of doing this.

A comparison function A sort user-provided sort function mysort(a, b) must take two arguments a and b and return −1 if a should become before b in the sorted sequence, 1 if b should become before a, and 0 if they are equal. In the present case, a and b are 4-tuples, so we need to make the comparison between the right elements in a and b. For example, to sort with respect to luminosity we can write

```
def mysort(a, b):
    if a[3] < b[3]:
        return -1
    elif a[3] > b[3]:
        return 1
    else:
        return 0
```

---

[11] http://tinyurl.com/pwyasaa/funcif/stars.txt

The relevant call using this tailored sort function is

```
sorted(data, cmp=mysort)
```

A key function A quicker construction is to provide a `key` argument to `sorted` for filtering out the relevant part of an object to be sorted. Here, we want to sort 4-tuples, but use only one of the elements, say the one with index 3, for comparison. Writing

```
sorted(data, key=lambda obj: obj[3])
```

will send all objects (4-tuples) through the `key` function whose return value is used for the sorting. A lambda construction (see Sect. 3.1.14) is used to write the filtering function inline.
Filename: `sorted_stars_data`.

Exercise 3.40: Find prime numbers
The *Sieve of Eratosthenes* is an algorithm for finding all prime numbers less than or equal to a number N. Read about this algorithm on Wikipedia and implement it in a Python program.
Filename: `find_primes`.

Exercise 3.41: Find pairs of characters
Write a function `count_pairs(dna, pair)` that returns the number of occurrences of a pair of characters (`pair`) in a DNA string (`dna`). For example, calling the function with `dna` as `'ACTGCTATCCATT'` and `pair` as `'AT'` will return 2.
Filename: `count_pairs`.

Exercise 3.42: Count substrings
This is an extension of Exercise 3.41: count how many times a certain string appears in another string. For example, the function returns 3 when called with the DNA string `'ACGTTACGGAACG'` and the substring `'ACG'`.

Hint For each match of the first character of the substring in the main string, check if the next n characters in the main string matches the substring, where n is the length of the substring. Use slices like `s[3:9]` to pick out a substring of `s`.
Filename: `count_substr`.

Exercise 3.43: Resolve a problem with a function
Consider the following interactive session:

```
>>> def f(x):
...     if 0 <= x <= 2:
...         return x**2
...     elif 2 < x <= 4:
...         return 4
...     elif x < 0:
...         return 0
...
```

```
>>> f(2)
4
>>> f(5)
>>> f(10)
```

Why do we not get any output when calling `f(5)` and `f(10)`?

Hint Save the `f` value in a variable `r` and do `print r`.
Filename: `fix_branching`.

Exercise 3.44: Determine the types of some objects
Consider the following calls to the `makelist` function from Sect. 3.1.6:

```
l1 = makelist(0, 100, 1)
l2 = makelist(0, 100, 1.0)
l3 = makelist(-1, 1, 0.1)
l4 = makelist(10, 20, 20)
l5 = makelist([1,2], [3,4], [5])
l6 = makelist((1,-1,1), ('myfile.dat', 'yourfile.dat'))
l7 = makelist('myfile.dat', 'yourfile.dat', 'herfile.dat')
```

Simulate each call by hand to determine what type of objects that become elements in the returned list and what the contents of `value` is after one pass in the loop.

Hint Note that some of the calls will lead to infinite loops if you really perform the above `makelist` calls on a computer.
You can go to the Online Python Tutor[12], paste in the `makelist` function and the session above, and step through the program to see what actually happens.
Filename: `find_object_type`.

Remarks This exercise demonstrates that we can write a function and have in mind certain types of arguments, here typically `int` and `float` objects. However, the function can be used with other (originally unintended) arguments, such as lists and strings in the present case, leading to strange and irrelevant behavior (the problem here lies in the boolean expression `value <= stop` which is meaningless for some of the arguments).

Exercise 3.45: Find an error in a program
For the formula
$$f(x) = e^{rx}\sin(mx) + e^{sx}\sin(nx)$$
we have made the program

```
def f(x, m, n, r, s):
    return expsin(x, r, m) + expsin(x, s, n)

x = 2.5
print f(x, 0.1, 0.2, 1, 1)
```

---

[12] http://www.pythontutor.com/

```
from math import exp, sin

def expsin(x, p, q):
    return exp(p*x)*sin(q*x)
```

Running this code results in

```
NameError: global name 'expsin' is not defined
```

What is the problem? Simulate the program flow by hand, use the debugger to step from line to line, or use the Online Python Tutor. Correct the program.
Filename: find_error_undef.

User Input and Error Handling

4

Consider a program for evaluating the formula $x = A \sin(wt)$:

```
from math import sin
A = 0.1
w = 1
t = 0.6
x = A*sin(w*t)
print x
```

In this program, A, w, and t are input data in the sense that these parameters must be known before the program can perform the calculation of x. The results produced by the program, here x, constitute the output data.

Input data can be hardcoded in the program as we do above. That is, we explicitly set variables to specific values: A=0.1, w=1, t=0.6. This programming style may be suitable for small programs. In general, however, it is considered good practice to let a user of the program provide input data when the program is running. There is then no need to modify the program itself when a new set of input data is to be explored. This is an important feature, because a golden rule of programming is that modification of the source code always represents a danger of introducing new errors by accident.

This chapter starts with describing four different ways of reading data into a program:

1. let the user answer questions in a dialog in the terminal window (Sect. 4.1),
2. let the user provide input on the command line (Sect. 4.2),
3. let the user provide input data in a file (Sect. 4.5),
4. let the user write input data in a graphical interface (Sect. 4.8).

Even if your program works perfectly, wrong input data from the user may cause the program to produce wrong answers or even crash. Checking that the input data are correct is important, and Sect. 4.7 tells you how to do this with so-called exceptions.

The Python programming environment is organized as a big collection of modules. Organizing your own Python software in terms of modules is therefore a nat-

© Springer-Verlag Berlin Heidelberg 2016
H.P. Langtangen, *A Primer on Scientific Programming with Python*,
Texts in Computational Science and Engineering 6, DOI 10.1007/978-3-662-49887-3_4

ural and wise thing to do. Section 4.9 tells you how easy it is to make your own modules.

All the program examples from the present chapter are available in files in the `src/input`[1] folder.

4.1   Asking Questions and Reading Answers

One of the simplest ways of getting data into a program is to ask the user a question, let the user type in an answer, and then read the text in that answer into a variable in the program. These tasks are done by calling a function with name `raw_input` in Python 2 – the name is just `input` in Python 3.

4.1.1   Reading Keyboard Input

A simple problem involving the temperature conversion from Celsius to Fahrenheit constitutes our main example: $F = \frac{9}{5}C + 32$. The associated program with setting C explicitly in the program reads

```
C = 22
F = 9./5*C + 32
print F
```

We may ask the user a question C=? and wait for the user to enter a number. The program can then read this number and store it in a variable C. These actions are performed by the statement

```
C = raw_input('C=? ')
```

The `raw_input` function always returns the user input as a string object. That is, the variable C above refers to a string object. If we want to compute with this C, we must convert the string to a floating-point number: `C = float(C)`. A complete program for reading C and computing the corresponding degrees in Fahrenheit now becomes

```
C = raw_input('C=? ')
C = float(C)
F = 9.0/5*C + 32
print F
```

In general, the `raw_input` function takes a string as argument, displays this string in the terminal window, waits until the user presses the Return key, and then returns a string object containing the sequence of characters that the user typed in.

---

[1] http://tinyurl.com/pwyasaa/input

The program above is stored in a file called c2f_qa.py (the qa part of the name reflects *question and answer*). We can run this program in several ways. The convention in this book is to indicate the execution by writing the program name only, but for a real execution you need to do more: write run before the program name in an interactive IPython session, or write python before the program name in a terminal session. Here is the execution of our sample program and the resulting dialog with the user:

```
———————————————————— | Terminal | ————————————————————
c2f_qa.py
C=? 21
69.8
```

In this particular example, the raw_input function reads the characters 21 from the keyboard and returns the string '21', which we refer to by the variable C. Then we create a new float object by float(C) and let the name C refer to this float object, with value 21.

You should now try out Exercises 4.1, 4.6, and 4.9 to make sure you understand how raw_input behaves.

4.2 Reading from the Command Line

Programs running on Unix computers usually avoid asking the user questions. Instead, input data are very often fetched from the *command line*. This section explains how we can access information on the command line in Python programs.

4.2.1 Providing Input on the Command Line

We look at the Celsius-Fahrenheit conversion program again. The idea now is to provide the Celsius input temperature as a *command-line argument* right after the program name. This means that we write the program name, here c2f_cml.py (cml for *command line*), followed the Celsius temperature:

```
———————————————————— | Terminal | ————————————————————
c2f_cml.py 21
69.8
```

Inside the program we can fetch the text 21 as sys.argv[1]. The sys module has a list argv containing all the command-line arguments to the program, i.e., all the "words" appearing after the program name when we run the program. In the present case there is only one argument and it is stored in sys.argv[1]. The first element in the sys.argv list, sys.argv[0], is always the name of the program.

A command-line argument is treated as a text, so sys.argv[1] refers to a string object, in this case '21'. Since we interpret the command-line argument as a number and want to compute with it, it is necessary to explicitly convert the string to a float object. In the program we therefore write

```
import sys
C = float(sys.argv[1])
F = 9.0*C/5 + 32
print F
```

As another example, consider the program

```
v0 = 5
g = 9.81
t = 0.6
y = v0*t - 0.5*g*t**2
print y
```

for computing the formula $y(t) = v_0 t - \frac{1}{2}gt^2$. Instead of hardcoding the values of v0 and t in the program we can read the two values from the command line:

───────────────────────── Terminal ─────────────────────────

```
ball2_cml.py 0.6 5
1.2342
```

The two command-line arguments are now available as `sys.argv[1]` and `sys.argv[2]`. The complete `ball2_cml.py` program thus takes the form

```
import sys
t  = float(sys.argv[1])
v0 = float(sys.argv[2])
g = 9.81
y = v0*t - 0.5*g*t**2
print y
```

4.2.2   A Variable Number of Command-Line Arguments

Let us make a program `addall.py` that adds all its command-line arguments. That is, we may run something like

───────────────────────── Terminal ─────────────────────────

```
addall.py 1 3 5 -9.9
The sum of 1 3 5 -9.9 is -0.9
```

The command-line arguments are stored in the sublist `sys.argv[1:]`. Each element is a string so we must perform a conversion to `float` before performing the addition. There are many ways to write this program. Let us start with version 1, `addall_v1.py`:

```
import sys
s = 0
for arg in sys.argv[1:]:
    number = float(arg)
    s += number
print 'The sum of ',
for arg in sys.argv[1:]:
    print arg,
print 'is ', s
```

The output is on one line, but built of several `print` statements with a comma at the end to prevent the usual newline character that `print` otherwise adds to the text. The command-line arguments must be converted to numbers in the first `for` loop because we need to compute with them, but in the second loop we only need to print them and then the string representation is appropriate.

The program above can be written more compactly if desired:

```
import sys
s = sum([float(x) for x in sys.argv[1:]])
print 'The sum of %s is %s' % (' '.join(sys.argv[1:]), s)
```

Here, we convert the list `sys.argv[1:]` to a list of `float` objects and then pass this list to Python's `sum` function for adding the numbers. The construction `S.join(L)` places all the elements in the list L after each other with the string S in between. The result here is a string with all the elements in `sys.argv[1:]` and a space in between, which is the text that originally appeared on the command line.

4.2.3   More on Command-Line Arguments

Unix commands make heavy use of command-line arguments. For example, when you write `ls -s -t` to list the files in the current folder, you run the program `ls` with two command-line arguments: `-s` and `-t`. The former specifies that `ls` is to print the file name together with the size of the file, and the latter sorts the list of files according to their dates of last modification. Similarly, `cp -r my new` for copying a folder tree `my` to a new folder tree `new` invokes the `cp` program with three command line arguments: `-r` (for recursive copying of files), `my`, and `new`. Most programming languages have support for extracting the command-line arguments given to a program.

An important rule is that *command-line arguments are separated by blanks*. What if we want to provide a text containing blanks as command-line argument? The text containing blanks must then appear inside single or double quotes. Let us demonstrate this with a program that simply prints the command-line arguments:

```
import sys, pprint
pprint.pprint(sys.argv[1:])
```

Say this program is named `print_cml.py`. The execution

```
------------------------------- Terminal -------------------------------
print_cml.py 21 a string with blanks 31.3
['21', 'a', 'string', 'with', 'blanks', '31.3']
```

demonstrates that each word on the command line becomes an element in `sys.argv`. Enclosing strings in quotes, as in

```
------------------------------- Terminal -------------------------------
print_cml.py 21 "a string with blanks" 31.3
['21', 'a string with blanks', '31.3']
```

shows that the text inside the quotes becomes a single command line argument.

4.3   Turning User Text into Live Objects

It is possible to provide text with valid Python code as input to a program and then turn the text into live objects as if the text were written directly into the program beforehand. This is a very powerful tool for letting users specify function formulas, for instance, as input to a program. The program code itself has no knowledge about the kind of function the user wants to work with, yet at run time the user's desired formula enters the computations.

4.3.1   The Magic Eval Function

The `eval` functions takes a string as argument and evaluates this string as a Python *expression*. The result of an expression is an object. Consider

```
>>> r = eval('1+2')
>>> r
3
>>> type(r)
<type 'int'>
```

The result of `r = eval('1+2')` is the same as if we had written `r = 1+2` directly:

```
>>> r = 1+2
>>> r
3
>>> type(r)
<type 'int'>
```

In general, any valid Python expression stored as text in a string s can be turned into live Python code by `eval(s)`.

Here is an example where the string to be evaluated is '2.5', which causes Python to see r = 2.5 and make a float object:

```
>>> r = eval('2.5')
>>> r
2.5
>>> type(r)
<type 'float'>
```

Let us proceed with some more examples. We can put the initialization of a list inside quotes and use eval to make a list object:

```
>>> r = eval('[1, 6, 7.5]')
>>> r
[1, 6, 7.5]
>>> type(r)
<type 'list'>
```

Again, the assignment to r is equivalent to writing

```
>>> r = [1, 6, 7.5]
```

We can also make a tuple object by using tuple syntax (standard parentheses instead of brackets):

```
>>> r = eval('(-1, 1)')
>>> r
(-1, 1)
>>> type(r)
<type 'tuple'>
```

Another example reads

```
>>> from math import sqrt
>>> r = eval('sqrt(2)')
>>> r
1.4142135623730951
>>> type(r)
<type 'float'>
```

At the time we run eval('sqrt(2)'), this is the same as if we had written

```
>>> r = sqrt(2)
```

directly, and this is valid syntax only if the sqrt function is defined. Therefore, the import of sqrt prior to running eval is important in this example.

Applying eval to strings If we put a string, enclosed in quotes, inside the expression string, the result is a string object:

```
>>>
>>> r = eval('"math programming"')
>>> r
'math programming'
>>> type(r)
<type 'str'>
```

Note that we must use two types of quotes: first double quotes to mark `math`
`programming` as a string object and then another set of quotes, here single quotes
(but we could also have used triple single quotes), to embed the text `"math`
`programming"` inside a string. It does not matter if we have single or double
quotes as inner or outer quotes, i.e., `'"..."'` is the same as `"'...'"`, because `'`
and `"` are interchangeable as long as a pair of either type is used consistently.

Writing just

```
>>> r = eval('math programming')
```

is the same as writing

```
>>> r = math programming
```

which is an invalid expression. Python will in this case think that `math` and
`programming` are two (undefined) variables, and setting two variables next to each
other with a space in between is invalid Python syntax. However,

```
>>> r = 'math programming'
```

is valid syntax, as this is how we initialize a string `r` in Python. To repeat, if we put
the valid syntax `'math programming'` inside a string,

```
s = "'math programming'"
```

`eval(s)` will evaluate the text inside the double quotes as `'math programm-`
`ing'`, which yields a string.

Applying eval to user input So, why is the `eval` function so useful? When we
get input via `raw_input` or `sys.argv`, it is always in the form of a string object,
which often must be converted to another type of object, usually an `int` or `float`.
Sometimes we want to avoid specifying one particular type. The `eval` function can
then be of help: we feed the string object from the input to the `eval` function and
let the it interpret the string and convert it to the right object.

An example may clarify the point. Consider a small program where we read in
two values and add them. The values could be strings, floats, integers, lists, and so
forth, as long as we can apply a + operator to the values. Since we do not know if
the user supplies a string, float, integer, or something else, we just convert the input
by `eval`, which means that the user's syntax will determine the type. The program
goes as follows (`add_input.py`):

```
i1 = eval(raw_input('Give input: '))
i2 = eval(raw_input('Give input: '))
r = i1 + i2
print '%s + %s becomes %s\nwith value %s' % \
      (type(i1), type(i2), type(r), r)
```

Observe that we write out the two supplied values, together with the types of the values (obtained by `eval`), and the sum. Let us run the program with an integer and a real number as input:

```
——————————————————————————————— Terminal ———————————————————————————————
add_input.py
Give input: 4
Give input: 3.1
<type 'int'> + <type 'float'> becomes <type 'float'>
with value 7.1
```

The string '4', returned by the first call to `raw_input`, is interpreted as an `int` by `eval`, while '3.1' gives rise to a `float` object.

   Supplying two lists also works fine:

```
——————————————————————————————— Terminal ———————————————————————————————
add_input.py
Give input: [-1, 3.2]
Give input: [9,-2,0,0]
<type 'list'> + <type 'list'> becomes <type 'list'>
with value [-1, 3.2000000000000002, 9, -2, 0, 0]
```

If we want to use the program to add two strings, the strings must be enclosed in quotes for `eval` to recognize the texts as string objects (without the quotes, `eval` aborts with an error):

```
——————————————————————————————— Terminal ———————————————————————————————
add_input.py
Give input: 'one string'
Give input: " and another string"
<type 'str'> + <type 'str'> becomes <type 'str'>
with value one string and another string
```

   Not all objects are meaningful to add:

```
——————————————————————————————— Terminal ———————————————————————————————
add_input.py
Give input: 3.2
Give input: [-1,10]
Traceback (most recent call last):
  File "add_input.py", line 3, in <module>
    r = i1 + i2
TypeError: unsupported operand type(s) for +: 'float' and 'list'
```

A similar program adding two arbitrary command-line arguments reads (add_ input.py):

```
import sys
i1 = eval(sys.argv[1])
i2 = eval(sys.argv[2])
r = i1 + i2
print '%s + %s becomes %s\nwith value %s' % \
      (type(i1), type(i2), type(r), r)
```

Another important example on the usefulness of eval is to turn formulas, given as input, into mathematics in the program. Consider the program

```
from math import *    # make all math functions available
import sys
formula = sys.argv[1]
x = eval(sys.argv[2])
result = eval(formula)
print '%s for x=%g yields %g' % (formula, x, result)
```

Two command-line arguments are expected: a formula and a number. Say the formula given is 2*sin(x)+1 and the number is 3.14. This information is read from the command line as strings. Doing x = eval(sys.argv[2]) means x = eval('3.14'), which is equivalent to x = 3.14, and x refers to a float object. The eval(formula) expression means eval('2*sin(x)+1'), and the corresponding statement result = eval(formula is therefore effectively result = 2*sin(x)+1, which requires sin and x to be defined objects. The result is a float (approximately 1.003). Providing cos(x) as the first command-line argument creates a need to have cos defined, so that is why we import all functions from the math module. Let us try to run the program:

Terminal

```
eval_formula.py "2*sin(x)+1" 3.14
2*sin(x)+1 for x=3.14 yields 1.00319
```

The very nice thing with using eval in x = eval(sys.argv[2]) is that we can provide mathematical expressions like pi/2 or even tanh(2*pi), as the latter just effectively results in the statement x = tanh(2*pi), and this works fine as long has we have imported tanh and pi.

4.3.2   The Magic Exec Function

Having presented eval for turning strings into Python code, we take the opportunity to also describe the related exec function to execute a string containing arbitrary Python code, not only an expression.

Suppose the user can write a formula as input to the program, available to us in the form of a string object. We would then like to turn this formula into a callable

Python function. For example, writing `sin(x)*cos(3*x) + x**2` as the formula, we would make the function

```
def f(x):
    return sin(x)*cos(3*x) + x**2
```

This is easy with `exec`: just construct the right Python syntax for defining `f(x)` in a string and apply `exec` to the string,

```
formula = sys.argv[1]
code = """
def f(x):
    return %s
""" % formula
from math import *  # make sure we have sin, cos, exp, etc.
exec(code)
```

As an example, think of `"sin(x)*cos(3*x) + x**2"` as the first command-line argument. Then `formula` will hold this text, which is inserted into the `code` string such that it becomes

```
"""
def f(x):
    return sin(x)*cos(3*x) + x**2
"""
```

Thereafter, `exec(code)` executes the code as if we had written the contents of the code string directly into the program by hand. With this technique, we can turn any user-given formula into a Python function!

Let us now use this technique in a useful application. Suppose we have made a function for computing the integral $\int_a^b f(x)dx$ by the Midpoint rule with n intervals:

```
def midpoint_integration(f, a, b, n=100):
    h = (b - a)/float(n)
    I = 0
    for i in range(n):
        I += f(a + i*h + 0.5*h)
    return h*I
```

We now want to read a, b, and n from the command line as well as the formula that makes up the $f(x)$ function:

```
from math import *
import sys
f_formula = sys.argv[1]
a = eval(sys.argv[2])
b = eval(sys.argv[3])
if len(sys.argv) >= 5:
    n = int(sys.argv[4])
else:
    n = 200
```

Note that we import everything from `math` and use `eval` when reading the input for a and b as this will allow the user to provide values like `2*cos(pi/3)`.

The next step is to convert the `f_formula` for $f(x)$ into a Python function `g(x)`:

```
code = """
def g(x):
    return %s
""" % f_formula
exec(code)
```

Now we have an ordinary Python function `g(x)` that we can ask the integration function to integrate:

```
I = midpoint_integration(g, a, b, n)
print 'Integral of %s on [%g, %g] with n=%d: %g' % \
    (f_formula, a, b, n, I)
```

The complete code is found in `integrate.py`. A sample run for $\int_0^{\pi/2} \sin(x)dx$ goes like

```
                                   Terminal

integrate.py "sin(x)" 0 pi/2
integral of sin(x) on [0, 1.5708] with n=200: 1
```

(The quotes in `"sin(x)"` are needed because of the parenthesis will otherwise be interpreted by the shell.)

4.3.3   Turning String Expressions into Functions

The examples in the previous section indicate that it can be handy to ask the user for a formula and turn that formula into a Python function. Since this operation is so useful, we have made a special tool that hides the technicalities. The tool is named `StringFunction` and works as follows:

```
>>> from scitools.StringFunction import StringFunction
>>> formula = 'exp(x)*sin(x)'
>>> f = StringFunction(formula)   # turn formula into f(x) func.
```

The f object now behaves as an ordinary Python function of x:

```
>>> f(0)
0.0
>>> f(pi)
2.8338239229952166e-15
>>> f(log(1))
0.0
```

Expressions involving other independent variables than x are also possible. Here is an example with the function $g(t) = Ae^{-at}\sin(\omega x)$:

```
g = StringFunction('A*exp(-a*t)*sin(omega*x)',
                   independent_variable='t',
                   A=1, a=0.1, omega=pi, x=0.5)
```

The first argument is the function formula, as before, but now we need to specify the name of the independent variable ('x' is default). The other parameters in the function (A, a, ω, and x) must be specified with values, and we use keyword arguments, consistent with the names in the function formula, for this purpose. Any of the parameters A, a, omega, and x can be changed later by calls like

```
g.set_parameters(omega=0.1)
g.set_parameters(omega=0.1, A=5, x=0)
```

Calling g(t) works as if g were a plain Python function of t, which also stores all the parameters A, a, omega, and x, and their values. You can use pydoc to bring up more documentation on the possibilities with StringFunction. Just run

```
pydoc scitools.StringFunction.StringFunction
```

A final important point is that StringFunction objects are as computationally efficient as hand-written Python functions. (This property is quite remarkable, as a string formula will in most other programming languages be much slower to evaluate than if the formula were hardcoded inside a plain function.)

4.4 Option-Value Pairs on the Command Line

The examples on using command-line arguments so far require the user of the program to type all arguments in their right sequence, just as when calling a function with positional arguments in the right order. It would be very convenient to assign command-line arguments in the same way as we use keyword arguments. That is, arguments are associated with a name, their sequence can be arbitrary, and only the arguments where the default value is not appropriate need to be given. Such type of command-line arguments may have -option value pairs, where option is some name of the argument.

As usual, we shall use an example to illustrate how to work with -option value pairs. Consider the physics formula for the location $s(t)$ of an object at time t, given that the object started at $s = s_0$ at $t = 0$ with a velocity v_0, and thereafter was subject to a constant acceleration a:

$$s(t) = s_0 + v_0 t + \frac{1}{2}at^2. \qquad (4.1)$$

This formula requires four input variables: s_0, v_0, a, and t. We can make a program location.py that takes four options, -s0, -v0, -a, and -t, on the command line. The program is typically run like this:

--- Terminal ---
```
location.py --t 3 --s0 1 --v0 1 --a 0.5
```

The sequence of `-option value` pairs is arbitrary. All options have a default value such that one does not have to specify all options on the command line.

All input variables should have sensible default values such that we can leave out the options for which the default value is suitable. For example, if $s_0 = 0$, $v_0 = 0$, $a = 1$, and $t = 1$ by default, and we only want to change t, we can run

---
Terminal

```
location.py --t 3
```

4.4.1 Basic Usage of the Argparse Module

Python has a flexible and powerful module `argparse` for reading (parsing) `-option value` pairs on the command line. Using `argparse` consists of three steps. First, a parser object must be created:

```
import argparse
parser = argparse.ArgumentParser()
```

Second, we need to define the various command-line options,

```
parser.add_argument('--v0', '--initial_velocity', type=float,
                     default=0.0, help='initial velocity',
                     metavar='v')
parser.add_argument('--s0', '--initial_position', type=float,
                     default=0.0, help='initial position',
                     metavar='s')
parser.add_argument('--a', '--acceleration', type=float,
                     default=1., help='acceleration', metavar='a')
parser.add_argument('--t', '--time', type=float,
                     default=1.0, help='time', metavar='t')
```

The first arguments to `parser.add_argument` is the set of options that we want to associate with an input parameter. Optional arguments are the type, a default value, a help string, and a name for the value of the argument (`metavar`) in a usage string. The `argparse` module will automatically allow an option `-h` or `-help` that prints a usage string for all the registered options. By default, the type is `str`, the default value is `None`, the help string is empty, and `metavar` is the option in upper case without initial dashes.

Third, we must read the command line arguments and interpret them:

```
args = parser.parse_args()
```

Through the `args` object we can extract the values of the various registered parameters: `args.v0`, `args.s0`, `args.a`, and `args.t`. The name of the parameter is determined by the first option to `parser.add_argument`, so writing

```
parser.add_argument('--initial_velocity', '--v0', type=float,
                     default=0.0, help='initial velocity')
```

will make the initial velocity value appear as `args.initial_velocity`. We can add the `dest` keyword to explicitly specify the name where the value is stored:

```
parser.add_argument('--initial_velocity', '--v0', dest='V0',
                    type=float, default=0.0,
                    help='initial velocity')
```

Now, `args.V0` will retrieve the value of the initial velocity. In case we do not provide any default value, the value will be `None`.

Our example is completed either by evaluating s as

```
s = args.s0 + args.v0*t + 0.5*args.a*args.t**2
```

or by introducing new variables so that the formula aligns better with the mathematical notation:

```
s0 = args.s0; v0 = args.v0; a = args.a; t = args.t
s = s0 + v0*t + 0.5*a*t**2
```

A complete program for the example above is found in the file `location.py`. Try to run it with the `-h` option to see an automatically generated explanation of legal command-line options.

4.4.2 Mathematical Expressions as Values

Values on the command line involving mathematical symbols and functions, say `-v0 'pi/2'`, pose a problem with the code example above. The `argparse` module will in that case try to do `float('pi/2')` which does not work well since `pi` is an undefined name. Changing `type=float` to `type=eval` is required to interpret the expression `pi/2`, but even `eval('pi/2')` fails since `pi` is not defined inside the `argparse` module. There are various remedies for this problem.

One can write a tailored function for converting a string value given on the command line to the desired object. For example,

```
def evalcmlarg(text):
    return eval(text)

parser.add_argument('--s0', '--initial_position', type=evalcmlarg,
                    default=0.0, help='initial position')
```

The file `location_v2.py` demonstrates such explicit type conversion through a user-provided conversion function. Note that `eval` is now taken in the programmer's namespace where (hopefully) `pi` or other symbols are imported.

More sophisticated conversions are possible. Say s_0 is specified in terms of a function of some parameter p, like $s_0 = (1 - p^2)$. We could then use a string for `-s0` and the `StringFunction` tool from Sect. 4.3.3 to turn the string into a function:

```
def toStringFunction4s0(text):
    from scitools.std import StringFunction
    return StringFunction(text, independent_variable='p')

parser.add_argument('--s0', '--initial_position',
                     type=toStringFunction4s0,
                     default='0.0', help='initial position')
```

Giving a command-line argument -s0 'exp(-1.5) + 10(1-p**2) results in args.s0 being a StringFunction object, which we must evaluate for a p value:

```
s0 = args.s0
p = 0.05
...
s = s0(p) + v0*t + 0.5*a*t**2
```

The file location_v3.py contains the complete code for this example.

Another alternative is to perform the correct conversion of values in our own code *after* the parser object has read the values. To this end, we treat argument types as strings in the parser.add_argument calls, meaning that we replace type=float by set type=str (which is also the default choice of type). Recall that this approach requires specification of default values as strings too, say '0':

```
parser.add_argument('--s0', '--initial_position', type=str,
                     default='0', help='initial position')
...
from math import *
args.v0 = eval(args.v0)
# or
v0 = eval(args.v0)

s0 = StringFunction(args.s0, independent_variable='p')
p = 0.5
...
s = s0(p) + v0*t + 0.5*a*t**2
```

Such code is found in the file location_v4.py. You can try out that program with the command-line arguments -s0 'pi/2 + sqrt(p)' -v0 pi/4'.

The final alternative is to write an Action class to handle the conversion from string to the right type. This is the preferred way to perform conversions and well described in the argparse documentation. We shall exemplify it here, but the technicalities involved require understanding of classes (Chap. 7) and inheritance (Chap. 9). For the conversion from string to any object via eval we write

```
import argparse
from math import *

class ActionEval(argparse.Action):
    def __call__(self, parser, namespace, values,
                 option_string=None):
        setattr(namespace, self.dest, eval(values))
```

The command-line arguments supposed to be run through `eval` must then have an `action` parameter:

```
parser.add_argument('--v0', '--initial_velocity',
                    default=0,0, help='initial velocity',
                    action=ActionEval)
```

From string to function via `StringFunction` for the `-s0` argument we write

```
from scitools.std import StringFunction

class ActionStringFunction4s0(argparse.Action):
    def __call__(self, parser, namespace, values,
                 option_string=None):
        setattr(namespace, self.dest,
                StringFunction(values, independent_variable='p'))
```

A complete code appears in the file `location_v5.py`.

4.5   Reading Data from File

Getting input data into a program from the command line, or from questions and answers in the terminal window, works for small amounts of data. Otherwise, input data must be available in files. Anyone with some computer experience is used to save and load data files in programs. The task now is to understand how Python programs can read and write files. The basic recipes are quite simple and illustrated through examples.

Suppose we have recorded some measurement data in the file `src/input/data.txt`[2]. The goal of our first example of reading files is to read the measurement values in `data.txt`, find the average value, and print it out in the terminal window.

Before trying to let a program read a file, we must know the *file format*, i.e., what the contents of the file looks like, because the structure of the text in the file greatly influences the set of statements needed to read the file. We therefore start with viewing the contents of the file `data.txt`. To this end, load the file into a text editor or viewer (one can use `emacs`, `vim`, `more`, or `less` on Unix and Mac, while on Windows, `WordPad` is appropriate, or the `type` command in a DOS or PowerShell window, and even Word processors such as LibreOffice or Microsoft Word can also be used on Windows). What we see is a column with numbers:

```
21.8
18.1
19
23
26
17.8
```

---

[2] http://tinyurl.com/pwyasaa/input/data.txt

Our task is to read this column of numbers into a list in the program and compute the average of the list items.

4.5.1   Reading a File Line by Line

To read a file, we first need to *open* the file. This action creates a file object, here stored in the variable `infile`:

```
infile = open('data.txt', 'r')
```

The second argument to the `open` function, the string `'r'`, tells that we want to open the file for reading. We shall later see that a file can be opened for writing instead, by providing `'w'` as the second argument. After the file is read, one should close the file object with `infile.close()`.

The basic technique for reading the file line by line applies a `for` loop like this:

```
for line in infile:
    # do something with line
```

The `line` variable is a string holding the current line in the file. The `for` loop over lines in a file has the same syntax as when we go through a list. Just think of the file object `infile` as a collection of elements, here lines in a file, and the `for` loop visits these elements in sequence such that the `line` variable refers to one line at a time. If something seemingly goes wrong in such a loop over lines in a file, it is useful to do a `print line` inside the loop.

Instead of reading one line at a time, we can load all lines into a list of strings (lines) by

```
lines = infile.readlines()
```

This statement is equivalent to

```
lines = []
for line in infile:
    lines.append(line)
```

or the list comprehension:

```
lines = [line for line in infile]
```

In the present example, we load the file into the list `lines`. The next task is to compute the average of the numbers in the file. Trying a straightforward sum of all numbers on all lines,

```
mean = 0
for number in lines:
    mean = mean + number
mean = mean/len(lines)
```

gives an error message:

```
TypeError: unsupported operand type(s) for +: 'int' and 'str'
```

The reason is that `lines` holds each line (number) as a string, not a `float` or `int` that we can add to other numbers. A fix is to convert each line to a `float`:

```
mean = 0
for line in lines:
    number = float(line)
    mean = mean + number
mean = mean/len(lines)
```

This code snippet works fine. The complete code can be found in the file `mean1.py`.

Summing up a list of numbers is often done in numerical programs, so Python has a special function `sum` for performing this task. However, `sum` must in the present case operate on a list of floats, not strings. We can use a list comprehension to turn all elements in `lines` into corresponding `float` objects:

```
mean = sum([float(line) for line in lines])/len(lines)
```

An alternative implementation is to load the lines into a list of `float` objects directly. Using this strategy, the complete program (found in file `mean2.py`) takes the form

```
infile = open('data.txt', 'r')
numbers = [float(line) for line in infile.readlines()]
infile.close()
mean = sum(numbers)/len(numbers)
print mean
```

4.5.2   Alternative Ways of Reading a File

A newcomer to programming might find it confusing to see that one problem is solved by many alternative sets of statements, but this is the very nature of programming. A clever programmer will judge several alternative solutions to a programming task and choose one that is either particularly compact, easy to understand, and/or easy to extend later. We therefore present more examples on how to read the `data.txt` file and compute with the data.

The modern with statement   Modern Python code applies the `with` statement to deal with files:

```
with open('data.txt', 'r') as infile:
    for line in infile:
        # process line
```

This snippet is equivalent to

```
infile = open('data.txt', 'r')
for line in infile:
    # process line
infile.close()
```

Note that there is no need to close the file when using the `with` statement. The advantage of the `with` construction is shorter code and better handling of errors if something goes wrong with opening or working with the file. A downside is that the syntax differs from the very classical open-close pattern that one finds in most other programming languages. Remembering to close a file is key in programming, and to train that task, we mostly apply the open-close construction in this book.

The old while construction The call `infile.readline()` returns a string containing the text at the current line. A new `infile.readline()` will read the next line. When `infile.readline()` returns an empty string, the end of the file is reached and we must stop further reading. The following `while` loop reads the file line by line using `infile.readline()`:

```
while True:
    line = infile.readline()
    if not line:
        break
    # process line
```

This is perhaps a somewhat strange loop, but it is a well-established way of reading a file in Python, especially in older code. The shown `while` loop runs forever since the condition is always `True`. However, inside the loop we test if `line` is `False`, and it is `False` when we reach the end of the file, because `line` then becomes an empty string, which in Python evaluates to `False`. When `line` is `False`, the `break` statement breaks the loop and makes the program flow jump to the first statement after the `while` block.

Computing the average of the numbers in the `data.txt` file can now be done in yet another way:

```
infile = open('data.txt', 'r')
mean = 0
n = 0
while True:
    line = infile.readline()
    if not line:
        break
    mean += float(line)
    n += 1
mean = mean/float(n)
```

Reading a file into a string The call `infile.read()` reads the whole file and returns the text as a string object. The following interactive session illustrates the use and result of `infile.read()`:

```
>>> infile = open('data.txt', 'r')
>>> filestr = infile.read()
>>> filestr
'21.8\n18.1\n19\n23\n26\n17.8\n'
>>> print filestr
21.8
18.1
19
23
26
17.8
```

Note the difference between just writing `filestr` and writing `print filestr`. The former dumps the string with newlines as *backslash n* characters, while the latter is a *pretty print* where the string is written out without quotes and with the newline characters as visible line shifts.

Having the numbers inside a string instead of inside a file does not look like a major step forward. However, string objects have many useful functions for extracting information. A very useful feature is *split*: `filestr.split()` will split the string into words (separated by blanks or any other sequence of characters you have defined). The "words" in this file are the numbers:

```
>>> words = filestr.split()
>>> words
['21.8', '18.1', '19', '23', '26', '17.8']
>>> numbers = [float(w) for w in words]
>>> mean = sum(numbers)/len(numbers)
>>> print mean
20.95
```

A more compact program looks as follows (`mean3.py`):

```
infile = open('data.txt', 'r')
numbers = [float(w) for w in infile.read().split()]
mean = sum(numbers)/len(numbers)
```

The next section tells you more about splitting strings.

4.5.3   Reading a Mixture of Text and Numbers

The `data.txt` file has a very simple structure since it contains numbers only. Many data files contain a mix of text and numbers. The file `rainfall.dat` from `www.worldclimate.com`[3] provides an example:

```
Average rainfall (in mm) in Rome: 1188 months between 1782 and 1970
Jan   81.2
Feb   63.2
Mar   70.3
```

---

[3] http://www.worldclimate.com/cgi-bin/data.pl?ref=N41E012+2100+1623501G1

```
Apr   55.7
May   53.0
Jun   36.4
Jul   17.5
Aug   27.5
Sep   60.9
Oct   117.7
Nov   111.0
Dec   97.9
Year  792.9
```

How can we read the rainfall data in this file and store the information in lists suitable for further analysis? The most straightforward solution is to read the file line by line, and for each line split the line into words, store the first word (the month) in one list and the second word (the average rainfall) in another list. The elements in this latter list needs to be `float` objects if we want to compute with them.

The complete code, wrapped in a function, may look like this (file `rainfall1.py`):

```python
def extract_data(filename):
    infile = open(filename, 'r')
    infile.readline() # skip the first line
    months = []
    rainfall = []
    for line in infile:
        words = line.split()
        # words[0]: month, words[1]: rainfall
        months.append(words[0])
        rainfall.append(float(words[1]))
    infile.close()
    months = months[:-1]      # Drop the "Year" entry
    annual_avg = rainfall[-1] # Store the annual average
    rainfall = rainfall[:-1]  # Redefine to contain monthly data
    return months, rainfall, annual_avg

months, values, avg = extract_data('rainfall.dat')
print 'The average rainfall for the months:'
for month, value in zip(months, values):
    print month, value
print 'The average rainfall for the year:', avg
```

Note that the first line in the file is just a comment line and of no interest to us. We therefore read this line by `infile.readline()` and do not store the content in any object. The `for` loop over the lines in the file will then start from the next (second) line.

We store all the data into 13 elements in the `months` and `rainfall` lists. Thereafter, we manipulate these lists a bit since we want `months` to contain the name of the 12 months only. The `rainfall` list should correspond to this `month` list. The annual average is taken out of `rainfall` and stored in a separate variable. Recall that the -1 index corresponds to the last element of a list, and the slice `:-1` picks out all elements from the start up to, but not including, the last element.

We could, alternatively, have written a shorter code where the name of the months and the rainfall numbers are stored in a nested list:

```
def extract_data(filename):
    infile = open(filename, 'r')
    infile.readline()  # skip the first line
    data = [line.split() for line in infile]
    annual_avg = data[-1][1]
    data = [(m, float(r)) for m, r in data[:-1]]
    infile.close()
    return data, annual_avg
```

This is more advanced code, but understanding what is going on is a good test on the understanding of nested lists indexing and list comprehensions. An executable program is found in the file `rainfall2.py`.

Is it more to file reading? With the example code in this section, you have the very basic tools for reading files with a simple structure: columns of text or numbers. Many files used in scientific computations have such a format, but many files are more complicated too. Then you need the techniques of string processing. This is explained in detail in Chap. 6.

4.6 Writing Data to File

Writing data to file is easy. There is basically one function to pay attention to: `outfile.write(s)`, which writes a string `s` to a file handled by the file object `outfile`. Unlike `print`, `outfile.write(s)` does not append a newline character to the written string. It will therefore often be necessary to add a newline character,

```
outfile.write(s + '\n')
```

if the string `s` is meant to appear on a single line in the file and `s` does not already contain a trailing newline character. File writing is then a matter of constructing strings containing the text we want to have in the file and for each such string call `outfile.write`.

Writing to a file demands the file object `f` to be opened for writing:

```
# write to new file, or overwrite file:
outfile = open(filename, 'w')

# append to the end of an existing file:
outfile = open(filename, 'a')
```

4.6.1 Example: Writing a Table to File

Problem As a worked example of file writing, we shall write out a nested list with tabular data to file. A sample list may look as

```
[[ 0.75,          0.29619813, -0.29619813, -0.75       ],
 [ 0.29619813,   0.11697778, -0.11697778, -0.29619813],
 [-0.29619813,  -0.11697778,  0.11697778,  0.29619813],
 [-0.75,         -0.29619813,  0.29619813,  0.75       ]]
```

Solution We iterate through the rows (first index) in the list, and for each row, we iterate through the column values (second index) and write each value to the file. At the end of each row, we must insert a newline character in the file to get a linebreak. The code resides in the file write1.py:

```
data = [[ 0.75,          0.29619813, -0.29619813, -0.75       ],
        [ 0.29619813,   0.11697778, -0.11697778, -0.29619813],
        [-0.29619813,  -0.11697778,  0.11697778,  0.29619813],
        [-0.75,         -0.29619813,  0.29619813,  0.75       ]]

outfile = open('tmp_table.dat', 'w')
for row in data:
    for column in row:
        outfile.write('%14.8f' % column)
    outfile.write('\n')
outfile.close()
```

The resulting data file becomes

```
    0.75000000     0.29619813    -0.29619813    -0.75000000
    0.29619813     0.11697778    -0.11697778    -0.29619813
   -0.29619813    -0.11697778     0.11697778     0.29619813
   -0.75000000    -0.29619813     0.29619813     0.75000000
```

An extension of this program consists in adding column and row headings:

```
          column  1      column  2      column  3      column  4
row  1    0.75000000     0.29619813    -0.29619813    -0.75000000
row  2    0.29619813     0.11697778    -0.11697778    -0.29619813
row  3   -0.29619813    -0.11697778     0.11697778     0.29619813
row  4   -0.75000000    -0.29619813     0.29619813     0.75000000
```

To obtain this end result, we need to the add some statements to the program write1.py. For the column headings we must know the number of columns, i.e., the length of the rows, and loop from 1 to this length:

```
ncolumns = len(data[0])
outfile.write('              ')
for i in range(1, ncolumns+1):
    outfile.write('%10s    ' % ('column %2d' % i))
outfile.write('\n')
```

Note the use of a nested printf construction: the text we want to insert is itself a printf string. We could also have written the text as 'column ' + str(i), but then the length of the resulting string would depend on the number of digits in

i. It is recommended to always use printf constructions for a tabular output format, because this gives automatic padding of blanks so that the width of the output strings remains the same. The tuning of the widths is commonly done in a trial-and-error process.

To add the row headings, we need a counter over the row numbers:

```
row_counter = 1
for row in data:
    outfile.write('row %2d' % row_counter)
    for column in row:
        outfile.write('%14.8f' % column)
    outfile.write('\n')
    row_counter += 1
```

The complete code is found in the file `write2.py`. We could, alternatively, iterate over the indices in the list:

```
for i in range(len(data)):
    outfile.write('row %2d' % (i+1))
    for j in range(len(data[i])):
        outfile.write('%14.8f' % data[i][j])
    outfile.write('\n')
```

4.6.2 Standard Input and Output as File Objects

Reading user input from the keyboard applies the function `raw_input` as explained in Sect. 4.1. The keyboard is a medium that the computer in fact treats as a file, referred to as *standard input*.

The `print` command prints text in the terminal window. This medium is also viewed as a file from the computer's point of view and called *standard output*. All general-purpose programming languages allow reading from standard input and writing to standard output. This reading and writing can be done with two types of tools, either file-like objects or special tools like `raw_input` and `print` in Python. We will here describe the file-line objects: `sys.stdin` for standard input and `sys.stdout` for standard output. These objects behave as file objects, except that they do not need to be opened or closed. The statement

```
s = raw_input('Give s:')
```

is equivalent to

```
print 'Give s: ',
s = sys.stdin.readline()
```

Recall that the trailing comma in the `print` statement avoids the newline that `print` by default adds to the output string. Similarly,

```
s = eval(raw_input('Give s:'))
```

is equivalent to

```
print 'Give s: ',
s = eval(sys.stdin.readline())
```

For output to the terminal window, the statement

```
print s
```

is equivalent to

```
sys.stdout.write(s + '\n')
```

Why it is handy to have access to standard input and output as file objects can be illustrated by an example. Suppose you have a function that reads data from a file object `infile` and writes data to a file object `outfile`. A sample function may take the form

```
def x2f(infile, outfile, f):
    for line in infile:
        x = float(line)
        y = f(x)
        outfile.write('%g\n' % y)
```

This function works with all types of files, including web pages as `infile` (see Sect. 6.3). With `sys.stdin` as `infile` and/or `sys.stdout` as `outfile`, the `x2f` function also works with standard input and/or standard output. Without `sys.stdin` and `sys.stdout`, we would need different code, employing `raw_input` and `print`, to deal with standard input and output. Now we can write a single function that deals with all file media in a unified way.

There is also something called *standard error*. Usually this is the terminal window, just as standard output, but programs can distinguish between writing ordinary output to standard output and error messages to standard error, and these output media can be redirected to, e.g., files such that one can separate error messages from ordinary output. In Python, standard error is the file-like object `sys.stderr`. A typical application of `sys.stderr` is to report errors:

```
if x < 0:
    sys.stderr.write('Illegal value of x'); sys.exit(1)
```

This message to `sys.stderr` is an alternative to `print` or raising an exception.

Redirecting standard input, output, and error Standard output from a program `prog` can be redirected to a file `output` instead of the screen, by using the greater than sign:

```
Terminal
Terminal> prog > output
```

Here, `prog` can be any program, including a Python program run as `python myprog.py`. Similarly, output to the medium called *standard error* can be redirected by

```
——————————————— Terminal ———————————————
Terminal> prog &> output
```

For example, error messages are normally written to standard error, which is exemplified in this little terminal session on a Unix machine:

```
——————————————— Terminal ———————————————
Terminal> ls bla-bla1 bla-bla2
ls: cannot access bla-bla1: No such file or directory
ls: cannot access bla-bla2: No such file or directory
Terminal> ls bla-bla1 bla-bla2 &> errors
Terminal> cat errors   # print the file errors
ls: cannot access bla-bla1: No such file or directory
ls: cannot access bla-bla2: No such file or directory
```

When the program reads from standard input (the keyboard), we can equally well redirect standard input from a file, say with name `input`, such that the program reads from this file rather than from the keyboard:

```
——————————————— Terminal ———————————————
Terminal> prog < input
```

Combinations are also possible:

```
——————————————— Terminal ———————————————
Terminal> prog < input > output
```

Note The redirection of standard output, input, and error does not work for Python programs executed with the `run` command inside IPython, only when executed directly in the operating system in a terminal window, or with the same command prefixed with an exclamation mark in IPython.

Inside a Python program we can also let standard input, output, and error work with ordinary files instead. Here is the technique:

```python
sys_stdout_orig = sys.stdout
sys.stdout = open('output', 'w')
sys_stdin_orig = sys.stdin
sys.stdin = open('input', 'r')
```

Now, any `print` statement will write to the `output` file, and any `raw_input` call will read from the `input` file. (Without storing the original `sys.stdout` and `sys.stdin` objects in new variables, these objects would get lost in the redefinition above and we would never be able to reach the common standard input and output in the program.)

4.6.3   What is a File, Really?

This section is not mandatory for understanding the rest of the book. Nevertheless, the information here is fundamental for understanding what files are about.

A file is simply a sequence of characters. In addition to the sequence of characters, a file has some data associated with it, typically the name of the file, its location on the disk, and the file size. These data are stored somewhere by the operating system. Without this extra information beyond the pure file contents as a sequence of characters, the operating system cannot find a file with a given name on the disk.

Each character in the file is represented as a *byte*, consisting of eight *bits*. Each bit is either 0 or 1. The zeros and ones in a byte can be combined in $2^8 = 256$ ways. This means that there are 256 different types of characters. Some of these characters can be recognized from the keyboard, but there are also characters that do not have a familiar symbol. Such characters looks cryptic when printed.

Pure text files To see that a file is really just a sequence of characters, invoke an editor for plain text, typically the editor you use to write Python programs. Write the four characters ABCD into the editor, do not press the Return key, and save the text to a file `test1.txt`. Use your favorite tool for file and folder overview and move to the folder containing the `test1.txt` file. This tool may be Windows Explorer, My Computer, or a DOS window on Windows; a terminal window, Konqueror, or Nautilus on Linux; or a terminal window or Finder on Mac. If you choose a terminal window, use the `cd` (change directory) command to move to the proper folder and write `dir` (Windows) or `ls -l` (Linux/Mac) to list the files and their sizes. In a graphical program like Windows Explorer, Konqueror, Nautilus, or Finder, select a view that shows the *size* of each file (choose *view as details* in Windows Explorer, *View as List* in Nautilus, the list view icon in Finder, or you just point at a file icon in Konqueror and watch the pop-up text). You will see that the `test1.txt` file has a size of 4 bytes (if you use `ls -l`, the size measured in bytes is found in column 5, right before the date). The 4 bytes are exactly the 4 characters ABCD in the file. Physically, the file is just a sequence of 4 bytes on your hard disk.

Go back to the editor again and add a newline by pressing the Return key. Save this new version of the file as `test2.txt`. When you now check the size of the file it has grown to five bytes. The reason is that we added a newline character (symbolically known as *backslash n*: \n).

Instead of examining files via editors and folder viewers we may use Python interactively:

```
>>> file1 = open('test1.txt', 'r').read()  # read file into string
>>> file1
'ABCD'
>>> len(file1)         # length of string in bytes/characters
4
>>> file2 = open('test2.txt', 'r').read()
>>> file2
'ABCD\n'
>>> len(file2)
5
```

Python has in fact a function that returns the size of a file directly:

```
>>> import os
>>> size = os.path.getsize('test1.txt')
>>> size
4
```

Word processor files Most computer users write text in a word processing program, such as Microsoft Word or LibreOffice. Let us investigate what happens with our four characters ABCD in such a program. Start the word processor, open a new document, and type in the four characters ABCD only. Save the document as a .docx file (Microsoft Word) or an .odt file (LibreOffice). Load this file into an editor for pure text and look at the contents. You will see that there are numerous strange characters that you did not write (!). This additional "text" contains information on what type of document this is, the font you used, etc. The LibreOffice version of this file has 8858 bytes and the Microsoft Word version contains over 26 Kb! However, if you save the file as a pure text file, with extension .txt, the size is down to 8 bytes in LibreOffice and five in Microsoft Word.

Instead of loading the LibreOffice file into an editor we can again read the file contents into a string in Python and examine this string:

```
>>> infile = open('test3.odt', 'r')  # open LibreOffice file
>>> s = infile.read()
>>> len(s)   # file size
8858
>>> s
'PK\x03\x04\x14\x00\x00\x08\x00\x00sKWD^\xc62\x0c\'\x00...
\x00meta.xml<?xml version="1.0" encoding="UTF-8"?>\n<office:...
" xmlns:meta="urn:oasis:names:tc:opendocument:xmlns:meta:1.0"
```

Each backslash followed by x and a number is a code for a special character not found on the keyboard (recall that there are 256 characters and only a subset is associated with keyboard symbols). Although we show just a small portion of all the characters in this file in the above output (otherwise, the output would have occupied several pages in this book with thousands symbols like \x04...), we can guarantee that you cannot find the pure sequence of characters ABCD. However, the computer program that generated the file, LibreOffice in this example, can easily interpret the meaning of all the characters in the file and translate the information into nice, readable text on the screen where you can recognize the text ABCD.

Your are now in a position to look into Exercise 4.8 to see what happens if one attempts to use LibreOffice to write Python programs.

Image files A digital image – captured by a digital camera or a mobile phone – is a file. And since it is a file, the image is just a sequence of characters. Loading some JPEG file into a pure text editor, reveals all the strange characters in there. On the first line you will (normally) find some recognizable text in between the strange characters. This text reflects the type of camera used to capture the image and the date and time when the picture was taken. The next lines contain more information about the image. Thereafter, the file contains a set of numbers representing the image. The basic representation of an image is a set of $m \times n$ pixels, where each pixel has a color represented as a combination of 256 values of red, green, and blue, which can be stored as three bytes (resulting in 256^3 color values). A 6-megapixel camera will then need to store $3 \times 6 \cdot 10^6 = 18$ megabytes for one picture. The JPEG file contains only a couple of megabytes. The reason is that JPEG is a *compressed* file format, produced by applying a smart technique that can throw away pixel information in the original picture such that the human eye hardly can detect the inferior quality.

A video is just a sequence of images, and therefore a video is also a stream of bytes. If the change from one video frame (image) to the next is small, one can use smart methods to compress the image information in time. Such compression is particularly important for videos since the file sizes soon get too large for being transferred over the Internet. A small video file occasionally has bad visual quality, caused by too much compression.

Music files An MP3 file is much like a JPEG file: first, there is some information about the music (artist, title, album, etc.), and then comes the music itself as a stream of bytes. A typical MP3 file has a size of something like five million bytes or five megabytes (5 Mb). The exact size depends on the complexity of the music, the length of the track, and the MP3 resolution. On a 16 Gb MP3 player you can then store roughly $16,000,000,000/5,000,000 = 3200$ MP3 files. MP3 is, like JPEG, a compressed format. The complete data of a song on a CD (the WAV file) contains about ten times as many bytes. As for pictures, the idea is that one can throw away a lot of bytes in an intelligent way, such that the human ear hardly detects the difference between a compressed and uncompressed version of the music file.

PDF files Looking at a PDF file in a pure text editor shows that the file contains some readable text mixed with some unreadable characters. It is not possible for a human to look at the stream of bytes and deduce the text in the document (well, from the assumption that there are always some strange people doing strange things, there might be somebody out there who, with a lot of training, can interpret the pure PDF code with the eyes). A PDF file reader can easily interpret the contents of the file and display the text in a human-readable form on the screen.

Remarks We have repeated many times that a file is just a stream of bytes. A human can interpret (read) the stream of bytes if it makes sense in a human language – or a computer language (provided the human is a programmer). When the series of bytes does not make sense to any human, a computer program must be used to interpret the sequence of characters.

Think of a report. When you write the report as pure text in a text editor, the resulting file contains just the characters you typed in from the keyboard. On the other hand, if you applied a word processor like Microsoft Word or LibreOffice, the report file contains a large number of extra bytes describing properties of the formatting of the text. This stream of extra bytes does not make sense to a human, and a computer program is required to interpret the file content and display it in a form that a human can understand. Behind the sequence of bytes in the file there are strict rules telling what the series of bytes means. These rules reflect the *file format*. When the rules or file format is publicly documented, a programmer can use this documentation to make her own program for interpreting the file contents (however, interpreting such files is much more complicated than our examples on reading human-readable files in this book). It happens, though, that secret file formats are used, which require certain programs from certain companies to interpret the files.

4.7 Handling Errors

Suppose we forget to provide a command-line argument to the `c2f_cml.py` program from Sect. 4.2.1:

Terminal

```
c2f_cml.py
Traceback (most recent call last):
  File "c2f_cml.py", line 2, in ?
    C = float(sys.argv[1])
IndexError: list index out of range
```

Python aborts the program and shows an error message containing the line where the error occurred, the type of the error (`IndexError`), and a quick explanation of what the error is. From this information we deduce that the index 1 is out of range. Because there are no command-line arguments in this case, `sys.argv` has only one element, namely the program name. The only valid index is then 0.

For an experienced Python programmer this error message will normally be clear enough to indicate what is wrong. For others it would be very helpful if wrong usage could be detected by our program and a description of correct operation could be printed. The question is how to detect the error inside the program.

The problem in our sample execution is that `sys.argv` does not contain two elements (the program name, as always, plus one command-line argument). We can therefore test on the length of `sys.argv` to detect wrong usage: if `len(sys.argv)` is less than 2, the user failed to provide information on the C value. The new version of the program, `c2f_cml_if.py`, starts with this `if` test:

```
if len(sys.argv) < 2:
    print 'You failed to provide Celsius degrees as input '\
          'on the command line!'
    sys.exit(1)  # abort because of error
F = 9.0*C/5 + 32
print '%gC is %.1fF' % (C, F)
```

We use the `sys.exit` function to abort the program. Any argument different from zero signifies that the program was aborted due to an error, but the precise value of the argument does not matter so here we simply choose it to be 1. If no errors are found, but we still want to abort the program, `sys.exit(0)` is used.

A more modern and flexible way of handling potential errors in a program is to *try* to execute some statements, and if something goes wrong, the program can detect this and jump to a set of statements that handle the erroneous situation as desired. The relevant program construction reads

```
try:
    <statements>
except:
    <statements>
```

If something goes wrong when executing the statements in the `try` block, Python raises what is known as an *exception*. The execution jumps directly to the `except` block whose statements can provide a remedy for the error. The next section explains the `try-except` construction in more detail through examples.

4.7.1   Exception Handling

To clarify the idea of exception handling, let us use a `try-except` block to handle the potential problem arising when our Celsius-Fahrenheit conversion program lacks a command-line argument:

```
import sys
try:
    C = float(sys.argv[1])
except:
    print 'You failed to provide Celsius degrees as input '\
          'on the command line!'
    sys.exit(1)  # abort
F = 9.0*C/5 + 32
print '%gC is %.1fF' % (C, F)
```

The program is stored in the file `c2f_cml_except1.py`. If the command-line argument is missing, the indexing `sys.argv[1]`, which has an invalid index 1, *raises an exception*. This means that the program jumps directly to the `except` block, implying that `float` is not called, and C is not initialized with a value. In the `except` block, the programmer can retrieve information about the exception and perform statements to recover from the error. In our example, we know what the error can be, and therefore we just print a message and abort the program.

Suppose the user provides a command-line argument. Now, the `try` block is executed successfully, and the program neglects the `except` block and continues with the Fahrenheit conversion. We can try out the last program in two cases:

---------------------------------- Terminal ----------------------------------

```
c2f_cml_except1.py
You failed to provide Celsius degrees as input on the command line!

c2f_cml_except1.py 21
21C is 69.8F
```

In the first case, the illegal index in `sys.argv[1]` causes an exception to be raised, and we perform the steps in the `except` block. In the second case, the `try` block executes successfully, so we jump over the `except` block and continue with the computations and the printout of results.

For a user of the program, it does not matter if the programmer applies an `if` test or exception handling to recover from a missing command-line argument. Nevertheless, exception handling is considered a better programming solution because it allows more advanced ways to abort or continue the execution. Therefore, we adopt exception handling as our standard way of dealing with errors in the rest of this book.

Testing for a specific exception Consider the assignment

```
C = float(sys.argv[1])
```

There are two typical errors associated with this statement: i) `sys.argv[1]` is illegal indexing because no command-line arguments are provided, and ii) the content in the string `sys.argv[1]` is not a pure number that can be converted to a `float` object. Python detects both these errors and raises an `IndexError` exception in the first case and a `ValueError` in the second. In the program above, we jump to the `except` block and issue the same message regardless of what went wrong in the `try` block. For example, when we indeed provide a command-line argument, but write it on an illegal form (21C), the program jumps to the `except` block and prints a misleading message:

Terminal

```
c2f_cml_except1.py 21C
You failed to provide Celsius degrees as input on the command line!
```

The solution to this problem is to branch into different `except` blocks depending on what type of exception that was raised in the `try` block (program `c2f_cml_except2.py`):

```
import sys
try:
    C = float(sys.argv[1])
except IndexError:
    print 'Celsius degrees must be supplied on the command line'
    sys.exit(1)  # abort execution
except ValueError:
    print 'Celsius degrees must be a pure number, '\
          'not "%s"' % sys.argv[1]
    sys.exit(1)

F = 9.0*C/5 + 32
print '%gC is %.1fF' % (C, F)
```

Now, if we fail to provide a command-line argument, an `IndexError` occurs and we tell the user to write the C value on the command line. On the other hand, if the `float` conversion fails, because the command-line argument has wrong syntax, a `ValueError` exception is raised and we branch into the second `except` block and explain that the form of the given number is wrong:

Terminal

```
c2f_cml_except1.py 21C
Celsius degrees must be a pure number, not "21C"
```

Examples on exception types   List indices out of range lead to `IndexError` exceptions:

```
>>> data = [1.0/i for i in range(1,10)]
>>> data[9]
...
IndexError: list index out of range
```

Some programming languages (Fortran, C, C++, and Perl are examples) allow list indices outside the legal index values, and such unnoticed errors can be hard to find. Python always stops a program when an invalid index is encountered, unless you handle the exception explicitly as a programmer.

Converting a string to `float` is unsuccessful and gives a `ValueError` if the string is not a pure integer or real number:

```
>>> C = float('21 C')
...
ValueError: invalid literal for float(): 21 C
```

Trying to use a variable that is not initialized gives a `NameError` exception:

```
>>> print a
...
NameError: name 'a' is not defined
```

Division by zero raises a `ZeroDivisionError` exception:

```
>>> 3.0/0
...
ZeroDivisionError: float division
```

Writing a Python keyword illegally or performing a Python grammar error leads to a `SyntaxError` exception:

```
>>> forr d in data:
...
    forr d in data:
       ^
SyntaxError: invalid syntax
```

What if we try to multiply a string by a number?

```
>>> 'a string'*3.14
...
TypeError: can't multiply sequence by non-int of type 'float'
```

The `TypeError` exception is raised because the object types involved in the multi-plication are wrong (`str` and `float`).

Digression It might come as a surprise, but multiplication of a string and a number is legal if the number is an integer. The multiplication means that the string should be repeated the specified number of times. The same rule also applies to lists:

```
>>> '--'*10    # ten double dashes = 20 dashes
'--------------------'
>>> n = 4
>>> [1, 2, 3]*n
[1, 2, 3, 1, 2, 3, 1, 2, 3, 1, 2, 3]
>>> [0]*n
[0, 0, 0, 0]
```

The latter construction is handy when we want to create a list of n elements and later assign specific values to each element in a `for` loop.

4.7.2 Raising Exceptions

When an error occurs in your program, you may either print a message and use `sys.exit(1)` to abort the program, or you may raise an exception. The latter task is easy. You just write `raise E(message)`, where E can be a known exception type in Python and `message` is a string explaining what is wrong. Most often E means `ValueError` if the value of some variable is illegal, or `TypeError` if the type of a variable is wrong. You can also define your own exception types. An exception can be raised from any location in a program.

Example In the program `c2f_cml_except2.py` from Sect. 4.7.1 we show how we can test for different exceptions and abort the program. Sometimes we see that an exception may happen, but if it happens, we want a more precise error message to help the user. This can be done by raising a new exception in an `except` block and provide the desired exception type and message.

Another application of raising exceptions with tailored error messages arises when input data are invalid. The code below illustrates how to raise exceptions in various cases.

We collect the reading of C and handling of errors a separate function:

```
def read_C():
    try:
        C = float(sys.argv[1])
    except IndexError:
        raise IndexError\
        ('Celsius degrees must be supplied on the command line')
```

```
except ValueError:
    raise ValueError\
    ('Celsius degrees must be a pure number, '\
    'not "%s"' % sys.argv[1])
# C is read correctly as a number, but can have wrong value:
if C < -273.15:
    raise ValueError('C=%g is a non-physical value!' % C)
return C
```

There are two ways of using the read_C function. The simplest is to call the function,

```
C = read_C()
```

Wrong input will now lead to a raw dump of exceptions, e.g.,

─────────────────────────────── Terminal ───────────────────────────────
```
c2f_cml_v5.py
Traceback (most recent call last):
  File "c2f_cml4.py", line 5, in ?
    raise IndexError\
IndexError: Celsius degrees must be supplied on the command line
```

New users of this program may become uncertain when getting raw output from exceptions, because words like Traceback, raise, and IndexError do not make much sense unless you have some experience with Python. A more user-friendly output can be obtained by calling the read_C function inside a try-except block, check for any exception (or better: check for IndexError *or* ValueError), and write out the exception message in a more nicely formatted form. In this way, the programmer takes complete control of how the program behaves when errors are encountered:

```
try:
    C = read_C()
except Exception as e:
    print e            # exception message
    sys.exit(1)        # terminate execution
```

Exception is the parent name of all exceptions, and e is an exception object. Nice printout of the exception message follows from a straight print e. Instead of Exception we can write (ValueError, IndexError) to test more specifically for two exception types we can expect from the read_C function:

```
try:
    C = read_C()
except (ValueError, IndexError) as e:
    print e            # exception message
    sys.exit(1)        # terminate execution
```

After the `try-except` block above, we can continue with computing F = 9*C/5 + 32 and print out F. The complete program is found in the file `c2f_cml.py`. We may now test the program's behavior when the input is wrong and right:

```
c2f_cml.py
Celsius degrees must be supplied on the command line

c2f_cml.py 21C
Celsius degrees must be a pure number, not "21C"

c2f_cml.py -500
C=-500 is a non-physical value!

c2f_cml.py 21
21C is 69.8F
```

This program deals with wrong input, writes an informative message, and terminates the execution without annoying behavior.

Scattered `if` tests with `sys.exit` calls are considered a bad programming style compared to the use of nested exception handling as illustrated above. You should abort execution in the main program only, not inside functions. The reason is that the functions can be re-used in other occasions where the error can be dealt with differently. For instance, one may avoid abortion by using some suitable default data.

The programming style illustrated above is considered the best way of dealing with errors, so we suggest that you hereafter apply exceptions for handling potential errors in the programs you make, simply because this is what experienced programmers expect from your codes.

4.8   A Glimpse of Graphical User Interfaces

Maybe you find it somewhat strange that the usage of the programs we have made so far in this book – and the programs we will make in the rest of the book – are less graphical and intuitive than the computer programs you are used to from school or entertainment. Those programs are operated through some self-explaining graphics, and most of the things you want to do involve pointing with the mouse, clicking on graphical elements on the screen, and maybe filling in some text fields. The programs in this book, on the other hand, are run from the command line in a terminal window or inside IPython, and input is also given here in form of plain text.

The reason why we do not equip the programs in this book with graphical interfaces for providing input, is that such graphics is both complicated and tedious to write. If the aim is to solve problems from mathematics and science, we think it is better to focus on this part rather than large amounts of code that merely offers some "expected" graphical cosmetics for putting data into the program. Textual input from the command line is also quicker to provide. Also remember that the computational functionality of a program is obviously independent from the type of user interface, textual or graphic.

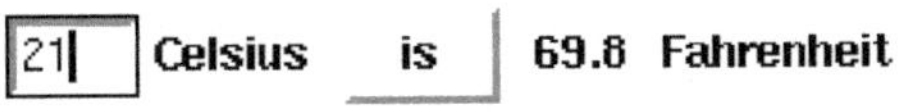

| 21| | Celsius | is | 69.8 Fahrenheit |

Fig. 4.1 Screen dump of the graphical interface for a Celsius to Fahrenheit conversion program. The user can type in the temperature in Celsius degrees, and when clicking on the *is* button, the corresponding Fahrenheit value is displayed

As an illustration, we shall now show a Celsius to Fahrenheit conversion program with a graphical user interface (often called a GUI). The GUI is shown in Fig. 4.1. We encourage you to try out the graphical interface – the name of the program is c2f_gui.py. The complete program text is listed below.

```
from Tkinter import *
root = Tk()
C_entry = Entry(root, width=4)
C_entry.pack(side='left')
Cunit_label = Label(root, text='Celsius')
Cunit_label.pack(side='left')

def compute():
    C = float(C_entry.get())
    F = (9./5)*C + 32
    F_label.configure(text='%g' % F)

compute = Button(root, text=' is ', command=compute)
compute.pack(side='left', padx=4)

F_label = Label(root, width=4)
F_label.pack(side='left')
Funit_label = Label(root, text='Fahrenheit')
Funit_label.pack(side='left')

root.mainloop()
```

The goal of the forthcoming dissection of this program is to give a taste of how graphical user interfaces are coded. The aim is not to equip you with knowledge on how you can make such programs on your own.

A GUI is built of many small graphical elements, called *widgets*. The graphical window generated by the program above and shown in Fig. 4.1 has five such widgets. To the left there is an *entry* widget where the user can write in text. To the right of this entry widget is a *label* widget, which just displays some text, here "Celsius". Then we have a *button* widget, which when being clicked leads to computations in the program. The result of these computations is displayed as text in a *label* widget to the right of the button widget. Finally, to the right of this result text we have another *label* widget displaying the text "Fahrenheit". The program must construct each widget and pack it correctly into the complete window. In the present case, all widgets are packed from left to right.

The first statement in the program imports functionality from the GUI toolkit Tkinter to construct widgets. First, we need to make a root widget that holds the complete window with all the other widgets. This root widget is of type Tk. The

first entry widget is then made and referred to by a variable C_entry. This widget is
an object of type Entry, provided by the Tkinter module. Widgets constructions
follow the syntax

```
variable_name = Widget_type(parent_widget, option1, option2, ...)
variable_name.pack(side='left')
```

When creating a widget, we must bind it to a *parent widget*, which is the graphical
element in which this new widget is to be packed. Our widgets in the present
program have the root widget as parent widget. Various widgets have different
types of options that we can set. For example, the Entry widget has a possibility
for setting the width of the text field, here width=4 means that the text field is 4
characters wide. The pack statement is important to remember – without it, the
widget remains invisible.

   The other widgets are constructed in similar ways. The next fundamental feature
of our program is how computations are tied to the event of clicking the button *is*.
The Button widget has naturally a text, but more important, it binds the button to
a function compute through the command=compute option. This means that when
the user clicks the button *is*, the function compute is called. Inside the compute
function we first fetch the Celsius value from the C_entry widget, using this widget's get function, then we transform this string (everything typed in by the user
is interpreted as text and stored in strings) to a float before we compute the corresponding Fahrenheit value. Finally, we can update (configure) the text in the
Label widget F_label with a new text, namely the computed degrees in Fahrenheit.

   A program with a GUI behaves differently from the programs we construct in
this book. First, all the statements are executed from top to bottom, as in all our
other programs, but these statements just construct the GUI and define functions.
No computations are performed. Then the program enters a so-called *event loop*:
root.mainloop(). This is an infinite loop that "listens" to user events, such as
moving the mouse, clicking the mouse, typing characters on the keyboard, etc.
When an event is recorded, the program starts performing associated actions. In
the present case, the program waits for only one event: clicking the button *is*. As
soon as we click on the button, the compute function is called and the program
starts doing mathematical work. The GUI will appear on the screen until we destroy the window by click on the X up in the corner of the window decoration.
More complicated GUIs will normally have a special *Quit* button to terminate the
event loop.

   In all GUI programs, we must first create a hierarchy of widgets to build up all
elements of the user interface. Then the program enters an event loop and waits for
user events. Lots of such events are registered as actions in the program when creating the widgets, so when the user clicks on buttons, move the mouse into certain
areas, etc., functions in the program are called and "things happen".

   Many books explain how to make GUIs in Python programs, see for instance
[5, 7, 13, 16].

4.9 Making Modules

Sometimes you want to reuse a function from an old program in a new program. The simplest way to do this is to copy and paste the old source code into the new program. However, this is not good programming practice, because you then over time end up with multiple identical versions of the same function. When you want to improve the function or correct a bug, you need to remember to do the same update in all files with a copy of the function, and in real life most programmers fail to do so. You easily end up with a mess of different versions with different quality of basically the same code. Therefore, a golden rule of programming is to have one and only one version of a piece of code. All programs that want to use this piece of code must access one and only one place where the source code is kept. This principle is easy to implement if we create a module containing the code we want to reuse later in different programs.

When reading this, you probably know how to use a ready-made module. For example, if you want to compute the factorial $k! = k(k-1)(k-2)\cdots 1$, there is a function `factorial` in Python's `math` module that can be help us out. The usage goes with the `math` prefix,

```
import math
value = math.factorial(5)
```

or without,

```
from math import factorial
# or: from math import *
value = factorial(5)
```

Now you shall learn how to make your own Python modules. There is hardly anything to learn, because you just collect all the functions that constitute the module in one file, say with name `mymodule.py`. This file is automatically a module, with name `mymodule`, and you can import functions from this module in the standard way. Let us make everything clear in detail by looking at an example.

4.9.1 Example: Interest on Bank Deposits

The classical formula for the growth of money in a bank reads

$$A = A_0 \left(1 + \frac{p}{360 \cdot 100}\right)^n, \tag{4.2}$$

where A_0 is the initial amount of money, and A is the present amount after n days with p percent annual interest rate. (The formula applies the convention that the rate per day is computed as $p/360$, while n counts the actual number of days the money is in the bank, see the Wikipedia entry Day count convention[4] for explanation. There is a handy Python module `datetime` for computing the number of days between two dates.)

[4] http://en.wikipedia.org/wiki/Day_count_convention

Equation (4.2) involves four parameters: A, A_0, p, and n. We may solve for any of these, given the other three:

$$A_0 = A \left(1 + \frac{p}{360 \cdot 100}\right)^{-n},\tag{4.3}$$

$$n = \frac{\ln \frac{A}{A_0}}{\ln \left(1 + \frac{p}{360 \cdot 100}\right)},\tag{4.4}$$

$$p = 360 \cdot 100 \left(\left(\frac{A}{A_0}\right)^{1/n} - 1\right).\tag{4.5}$$

Suppose we have implemented (4.2)–(4.5) in four functions:

```
from math import log as ln

def present_amount(A0, p, n):
    return A0*(1 + p/(360.0*100))**n

def initial_amount(A, p, n):
    return A*(1 + p/(360.0*100))**(-n)

def days(A0, A, p):
    return ln(A/A0)/ln(1 + p/(360.0*100))

def annual_rate(A0, A, n):
    return 360*100*((A/A0)**(1.0/n) - 1)
```

We want to make these functions available in a module, say with name `interest`, so that we can import functions and compute with them in a program. For example,

```
from interest import days
A0 = 1; A = 2; p = 5
n = days(A0, 2, p)
years = n/365.0
print 'Money has doubled after %.1f years' % years
```

How to make the `interest` module is described next.

4.9.2 Collecting Functions in a Module File

To make a module of the four functions `present_amount`, `initial_amount`, `days`, and `annual_rate`, we simply open an empty file in a text editor and copy the program code for all the four functions over to this file. This file is then automatically a Python module provided we save the file under any valid filename. The extension must be `.py`, but the module name is only the base part of the filename. In our case, the filename `interest.py` implies a module name `interest`. To use the `annual_rate` function in another program we simply write, in that program file,

```
from interest import annual_rate
```

or we can write

```
from interest import *
```

to import all four functions, or we can write

```
import interest
```

and access individual functions as `interest.annual_rate` and so forth.

4.9.3   Test Block

It is recommended to only have functions and not any statements outside functions in a module. The reason is that the module file is executed from top to bottom during the import. With function definitions only in the module file, and no main program, there will be no calculations or output from the import, just definitions of functions. This is the desirable behavior. However, it is often convenient to have test or demonstrations in the module file, and then there is need for a main program. Python allows a very fortunate construction to let the file act both as a module with function definitions only (and no main program) *and* as an ordinary program we can run, with functions and a main program.

This two-fold "magic" is realized by putting the main program after an `if` test of the form

```
if __name__ == '__main__':
    <block of statements>
```

The `__name__` variable is automatically defined in any module and equals the module name if the module file is imported in another program, or `__name__` equals the string `'__main__'` if the module file is run as a program. This implies that the `<block of statements>` part is executed if and only if we run the module file as a program. We shall refer to `<block of statements>` as the *test block* of a module.

Example on a test block in a minimalistic module   A very simple example will illustrate how this works. Consider a file `mymod.py` with the content

```
def add1(x):
    return x + 1

if __name__ == '__main__':
    print 'run as program'
    import sys
    print add1(float(sys.argv[1]))
```

We can import `mymod` as a module and make use of the `add1` function:

```
>>> import mymod
>>> print mymod.add1(4)
5
```

During the import, the `if` test is false, and the only the function definition is executed. However, if we run `mymod.py` as a program,

―――――――――――――――――――― | Terminal | ――――――――――――――――――――
```
mymod.py 5
run as program
6
```

the `if` test becomes true, and the `print` statements are executed.

Tip on easy creation of a module

If you have some functions and a main program in some program file, just move the main program to the test block. Then the file can act as a module, giving access to all the functions in other files, or the file can be executed from the command line, in the same way as the original program.

A test block in the `interest` module Let us write a little main program for demonstrating the `interest` module in a test block. We read p from the command line and write out how many years it takes to double an amount with that interest rate:

```
if __name__ == '__main__':
    import sys
    p = float(sys.argv[1])
    years = days(1, 2, p)/365.0
    print 'With p=%.2f it takes %.1 years to double' % (p, years)
```

Running the module file as a program gives this output:

―――――――――――――――――――― | Terminal | ――――――――――――――――――――
```
interest.py 2.45
With p=2.45 it takes 27.9 years to double
```

To test that the `interest.py` file also works as a module, invoke a Python shell and try to import a function and compute with it:

```
>>> from interest import present_amount
>>> present_amount(2, 5, 730)
2.2133983053266699
```

We have hence demonstrated that the file `interest.py` works both as a program and as a module.

Recommended practice in a test block

It is a good programming habit to let the test block do one or more of three things:

- provide information on how the module or program is used,
- test if the module functions work properly,
- offer interaction with users such that the module file can be applied as a useful program.

Instead of having a lot of statements in the test block, it is better to collect the statements in separate functions, which then are called from the test block.

4.9.4   Verification of the Module Code

Functions that verify the implementation in a module should

- have names starting with `test_`,
- express the success or failure of a test through a boolean variable, say `success`,
- run `assert success, msg` to raise an `AssertionError` with an optional message `msg` in case the test fails.

Adopting this style makes it trivial to let the tools *pytest* or *nose* automatically run through all our `test_*()` functions in all files in a folder tree. A very brief introduction to test functions compatible with pytest and nose is provided in Sect. 3.4.2, while Sect. H.9 contains a more thorough introduction to the pytest and nose testing frameworks for beginners.

Test functions are used for *unit testing*. This means that we identify some units of our software and write a dedicated test function for testing the behavior of each unit. A unit in the present example can be the `interest` module, but we could also think of the individual Python functions in `interest` as units. From a practical point of view, the unit is often defined as what we find appropriate to verify in a test function. For now it is convenient to test all functions in the `interest.py` file in the same test function, so the module becomes the unit.

A proper test function for verifying the functionality of the `interest` module, written in a way that is compatible with the pytest and nose testing frameworks, looks as follows:

```
def test_all_functions():
    # Compatible values
    A = 2.2133983053266699; A0 = 2.0; p = 5; n = 730
    # Given three of these, compute the remaining one
    # and compare with the correct value (in parenthesis)
    A_computed  = present_amount(A0, p, n)
    A0_computed = initial_amount(A, p, n)
    n_computed  = days(A0, A, p)
    p_computed  = annual_rate(A0, A, n)
```

```
def float_eq(a, b, tolerance=1E-12):
    """Return True if a == b within the tolerance."""
    return abs(a - b) < tolerance

success = float_eq(A_computed,  A)  and \
          float_eq(A0_computed, A0) and \
          float_eq(p_computed,  p)  and \
          float_eq(n_computed,  n)
msg = """Computations failed (correct answers in parenthesis):
A=%g (%g)
A0=%g (%.1f)
n=%d (%d)
p=%g (%.1f)""" % (A_computed, A, A0_computed, A0,
                  n_computed, n, p_computed, p)
assert success, msg
```

We may require a single command-line argument `test` to run the verification. The test block can then be expressed as

```
if __name__ == '__main__':
    if len(sys.argv) == 2 and sys.argv[1] == 'test':
        test_all_functions()
```

4.9.5 Getting Input Data

To make a useful program, we should allow setting three parameters on the command line and let the program compute the remaining parameter. For example, running the program as

```
Terminal
interest.py A0=1 A=2 n=1095
```

will lead to a computation of p, in this case for seeing the size of the annual interest rate if the amount is to be doubled after three years.

How can we achieve the desired functionality? Since variables are already introduced and "initialized" on the command line, we could grab this text and execute it as Python code, either as three different lines or with semicolon between each assignment. This is easy:

```
init_code = ''
for statement in sys.argv[1:]:
    init_code += statement + '\n'
exec(init_code)
```

(We remark that an experienced Python programmer would have created `init_code` by `'\n'.join(sys.argv[1:])`.) For the sample run above with `A0=1 A=2 n=1095` on the command line, `init_code` becomes the string

```
A0=1
A=2
n=1095
```

Note that one cannot have spaces around the equal signs on the command line as this will break an assignment like A0 = 1 into three command-line arguments, which will give rise to a SyntaxError in exec(init_code). To tell the user about such errors, we execute init_code inside a try-except block:

```
try:
    exec(init_code)
except SyntaxError as e:
    print e
    print init_code
    sys.exit(1)
```

At this stage, our program has hopefully initialized three parameters in a successful way, and it remains to detect the remaining parameter to be computed. The following code does the work:

```
if 'A=' not in init_code:
    print 'A =', present_amount(A0, p, n)
elif 'A0=' not in init_code:
    print 'A0 =', initial_amount(A, p, n)
elif 'n=' not in init_code:
    print 'n =', days(A0, A , p)
elif 'p=' not in init_code:
    print 'p =', annual_rate(A0, A, n)
```

It may happen that the user of the program assigns value to a parameter with wrong name or forget a parameter. In those cases we call one of our four functions with uninitialized arguments, and Python raises an exception. Therefore, we should embed the code above in a try-except block. An uninitialized variable will lead to a NameError exception, while another frequent error is illegal values in the computations, leading to a ValueError exception. It is also a good habit to collect all the code related to computing the remaining, fourth parameter in a function for separating this piece of code from other parts of the module file:

```
def compute_missing_parameter(init_code):
    try:
        exec(init_code)
    except SyntaxError as e:
        print e
        print init_code
        sys.exit(1)
    # Find missing parameter
    try:
        if 'A=' not in init_code:
            print 'A =', present_amount(A0, p, n)
        elif 'A0=' not in init_code:
            print 'A0 =', initial_amount(A, p, n)
```

```
        elif 'n=' not in init_code:
            print 'n =', days(A0, A , p)
        elif 'p=' not in init_code:
            print 'p =', annual_rate(A0, A, n)
    except NameError as e:
        print e
        sys.exit(1)
    except ValueError:
        print 'Illegal values in input:', init_code
        sys.exit(1)
```

If the user of the program fails to give any command-line arguments, we print a usage statement. Otherwise, we run a verification if the first command-line argument is `test`, and else we run the missing parameter computation (i.e., the useful main program):

```
_filename = sys.argv[0]
_usage = """
Usage: %s A=10 p=5 n=730
Program computes and prints the 4th parameter'
(A, A0, p, or n)""" % _filename

if __name__ == '__main__':
    if len(sys.argv) == 1:
        print _usage
    elif len(sys.argv) == 2 and sys.argv[1] == 'test':
        test_all_functions()
    else:
        init_code = ''
        for statement in sys.argv[1:]:
            init_code += statement + '\n'
        compute_missing_parameter(init_code)
```

Executing user input can be dangerous
Some purists would never demonstrate `exec` the way we do above. The reason is that our program tries to execute whatever the user writes. Consider

```
——————————————— Terminal ———————————————
input.py 'import shutil; shutil.rmtree("/")'
```

This evil use of the program leads to an attempt to remove all files on the computer system (the same as writing `rm -rf /` in the terminal window!). However, for small private programs helping the program writer out with mathematical calculations, this potential dangerous misuse is not so much of a concern (the user just does harm to his own computer anyway).

4.9.6 Doc Strings in Modules

It is also a good habit to include a doc string in the beginning of the module file. This doc string explains the purpose and use of the module:

```
"""
Module for computing with interest rates.
Symbols: A is present amount, A0 is initial amount,
n counts days, and p is the interest rate per year.

Given three of these parameters, the fourth can be
computed as follows:

    A  = present_amount(A0, p, n)
    A0 = initial_amount(A, p, n)
    n  = days(A0, A, p)
    p  = annual_rate(A0, A, n)
"""
```

You can run the pydoc program to see a documentation of the new module, containing the doc string above and a list of the functions in the module: just write pydoc interest in a terminal window.

Now the reader is recommended to take a look at the actual file interest.py to see all elements of a good module file at once: doc strings, a set of functions, a test function, a function with the main program, a usage string, and a test block.

4.9.7 Using Modules

Let us further demonstrate how to use the interest.py module in programs. For illustration purposes, we make a separate program file, say with name doubling.py, containing some computations:

```
from interest import days

# How many days does it take to double an amount when the
# interest rate is p=1,2,3,...14?
for p in range(1, 15):
    years = days(1, 2, p)/365.0
    print 'p=%d%% implies %.1f years to double the amount' %\
    (p, years)
```

What gets imported by various import statements? There are different ways to import functions in a module, and let us explore these in an interactive session. The function call dir() will list all names we have defined, including imported names of variables and functions. Calling dir(m) will print the names defined inside a module with name m. First we start an interactive shell and call dir()

```
>>> dir()
['__builtins__', '__doc__', '__name__', '__package__']
```

These variables are always defined. Running the IPython shell will introduce several other standard variables too. Doing

```
>>> from interest import *
>>> dir()
['__builtins__', '__doc__', '__name__', '__package__',
 'annual_rate', 'compute_missing_parameter', 'days',
 'initial_amount', 'ln', 'present_amount', 'sys',
 'test_all_functions']
```

shows that we get our four functions imported, along with `ln` and `sys`. The latter two are needed in the `interest` module, but not necessarily in our new program `doubling.py`.

The alternative `import interest` actually gives us access to more names in the module, namely also all variables and functions that start with an underscore:

```
>>> import interest
>>> dir(interest)
['__builtins__', '__doc__', '__file__', '__name__',
 '__package__', '_filename', '_usage', 'annual_rate',
 'compute_missing_parameter', 'days', 'initial_amount',
 'ln', 'present_amount', 'sys', 'test_all_functions']
```

It is a habit to use an underscore for all variables that are not to be included in a `from interest import *` statement. These variables can, however, be reached through `interest._filename` and `interest._usage` in the present example.

It would be best that a statement `from interest import *` just imported the four functions doing the computations of general interest in other programs. This can be archived by deleting all unwanted names (among those without an initial underscore) at the very end of the module:

```
del sys, ln, compute_missing_parameter, test_all_functions
```

Instead of deleting variables and using initial underscores in names, it is in general better to specify the special variable `__all__`, which is used by Python to select functions to be imported in `from interest import *` statements. Here we can define `__all__` to contain the four function of main interest:

```
__all__ = ['annual_rate', 'days', 'initial_amount', 'present_amount']
```

Now we get

```
>>> from interest import *
['__builtins__', '__doc__', '__name__', '__package__',
 'annual_rate', 'days', 'initial_amount', 'present_amount']
```

How to make Python find a module file The `doubling.py` program works well as long as it is located in the same folder as the `interest.py` module. However, if we move `doubling.py` to another folder and run it, we get an error:

__Terminal__

```
doubling.py
Traceback (most recent call last):
  File "doubling.py", line 1, in <module>
    from interest import days
ImportError: No module named interest
```

Unless the module file resides in the same folder, we need to tell Python where to find our module. Python looks for modules in the folders contained in the list sys.path. A little program

```
import sys, pprint
pprint.pprint(sys.path)
```

prints out all these predefined module folders. You can now do one of two things:

1. Place the module file in one of the folders in sys.path.
2. Include the folder containing the module file in sys.path.

There are two ways of doing the latter task. Alternative 1 is to explicitly insert a new folder name in sys.path in the program that uses the module:

```
modulefolder = '../../pymodules'
sys.path.insert(0, modulefolder)
```

(In this sample path, the slashes are Unix specific. On Windows you must use backslashes and a raw string. A better solution is to express the path as os.path.join(os.pardir, os.pardir, 'mymodules'). This will work on all platforms.)

Python searches the folders in the sequence they appear in the sys.path list so by inserting the folder name as the first list element we ensure that our module is found quickly, and in case there are other modules with the same name in other folders in sys.path, the one in modulefolder gets imported.

Alternative 2 is to specify the folder name in the PYTHONPATH environment variable. All folder names listed in PYTHONPATH are automatically included in sys.path when a Python program starts. On Mac and Linux systems, environment variables like PYTHONPATH are set in the .bashrc file in the home folder, typically as

```
export PYTHONPATH=$HOME/software/lib/pymodules:$PYTHONPATH
```

if §HOME/software/lib/pymodules is the folder containing Python modules. On Windows, you launch *Computer – Properties – Advanced System Settings – Environment Variables*, click under *System Variable*, write in PYTHONPATH as variable name and the relevant folder(s) as value.

How to make Python run the module file The description above concerns importing the module in a program located anywhere on the system. If we want to run the module file as a program, anywhere on the system, the operating system searches the PATH environment variable for the program name interst.py. It is therefore necessary to update PATH with the folder where interest.py resides.

On Mac and Linux system this is done in .bashrc in the same way as for PYTHONPATH:

```
export PATH=$HOME/software/lib/pymodules:$PATH
```

On Windows, launch the dialog for setting environment variables as described above and find the PATH variable. It already has much content, so you add your new folder value either at the beginning or end, using a semicolon to separate the new value from the existing ones.

4.9.8 Distributing Modules

Modules are usually useful pieces of software that others can take advantage of. Even though our simple interest module is of less interest to the world, we can illustrate how such a module is most effectively distributed to other users. The standard in Python is to distribute the module file together with a program called setup.py such that any user can just do

---
Terminal

```
Terminal> sudo python setup.py install
```
---

to install the module in one of the directories in sys.path so that the module is immediately accessible anywhere, both for import in a Python program and for execution as a stand-alone program.

The setup.py file is in the case of one module file very short:

```
from distutils.core import setup
setup(name='interest',
      version='1.0',
      py_modules=['interest'],
      scripts=['interest.py'],
      )
```

The scripts= keyword argument can be dropped if the module is just to be imported and not run as a program as well. More module files can trivially be added to the list.

A user who runs setup.py install on an Ubuntu machine will see from the output that interest.py is copied to the system folders /usr/local/lib/python2.7/dist-packages and /usr/local/bin. The former folder is for module files, the latter for executable programs.

Remark
Distributing a single module file can be done as shown, but if you have two or more module files that belong together, you should definitely create a *package* [25].

4.9.9   Making Software Available on the Internet

Distributing software today means making it available on one of the major project hosting sites such as GitHub or Bitbucket. You will develop and maintain the project files on your own computer(s), but frequently push the software out in the cloud such that others also get your updates. The mentioned sites have very strong support for collaborative software development.

Sign up for a GitHub account if you do not already have one. Go to your account settings and provide an SSH key (typically the file `~/.ssh/id_rsa.pub`) such that you can communicate with GitHub without being prompted for your password.

To create a new project, click on *New repository* on the main page and fill out a project name. Click on the check button *Initialize this repository with a README*, and click on *Create repository*. The next step is to clone (copy) the GitHub repo (short for repository) to your own computer(s) and fill it with files. The typical clone command is

```
Terminal

Terminal> git clone git://github.com:username/projname.git
```

where `username` is your GitHub username and `projname` is the name of the repo (project). The result of `git clone` is a directory `projname`. Go to this folder and add files. That is, copy `setup.py` and `interst.py` to the folder. It is good to also write a short README file explaining what the project is about. Run

```
Terminal

Terminal> git add .
Terminal> git commit -am 'First registration of project files'
Terminal> git push origin master
```

The above `git` commands look cryptic, but these commands plus 2–3 more are the essence of how programmers today work on software projects, small or big. I strongly encourage you to learn more about version control systems and project hosting sites [12]. The tools are in nature like Dropbox and Google Drive, just much more powerful when you collaborate with others.

Your project files are now stored in the cloud at https://github.com/username/ projname. Anyone can get the software by the listed `git clone` command you used above, or by clicking on the links for zip and tar files.

Every time you update the project files, you need to register the update at GitHub by

```
                        ┌──────────┐
──────────────────────  │ Terminal │  ──────────────────────
                        └──────────┘
Terminal> git commit -am 'Description of the changes you made...'
Terminal> git push origin master
```

The files at GitHub are now synchronized with your local ones.

There is a bit more to be said here to make you up and going with this style of professional work [12], but the information above gives you at least a glimpse of how to put your software project in the cloud and opening it up for others. The GitHub address for the particular `interest` module described above is https:// github.com/hplgit/interest-primer.

4.10   Making Code for Python 2 and 3

This book applies Python version 2.7, but there is a newer version of Python called Python 3 (the current version is 3.5). Unfortunately, Python 2 programs do not work with Python 3 and vice versa. Newcomers to Python are normally guided to pick up version 3 rather than version 2, since the former has many improvements and represents the future of the language. However, for scientific computing, version 3 still lacks many useful libraries, and that is the reason why this book applies Python version 2.7.

4.10.1   Basic Differences Between Python 2 and 3

So, what are the major differences between version 2 and 3? We cover only the three differences that involve statements we have seen so far in the book.

The print statement has changed   Here are some examples on print statements in Python 2:

```
a = 1
print a
print 'The value of a is', a
print 'The value of a is', a,   # comma prevents newline
b = 2
print 'and b=%g' % b
```

The print statement is not a statement anymore, but a function in Python 3. The above code needs to be written as

```
a = 1
print(a)
print('The value of a is', a)
print('The value of a is', a, end=' ')   # end='' prevents newline
b = 2
print('and b=%g' % b)
```

Integer division is not an issue in Python 3   The expression $1/10$ is 0 in Python 2, while in Python 3 it equals 0.1. Nevertheless, there are so many computer languages and tools that interpret as $1/10$ integer division, so rather than relying on a language's interpretation of integer divided by integer as float division, the programmer is strongly encouraged to turn one of the operands explicitly to float, as in $1.0/10$.

The `raw_input` function is named `input` in Python 3   The Python 2 code

```
a = float(raw_input('Give a: '))
```

reads

```
a = float(input('Give a: '))
```

in Python 3.

Note that in Python 2 there is an `input` function which equals `eval` applied to `raw_input`:

```
a = input('Give a: ')  # Python 2!
# Equivalent to
a = eval(raw_input('Give a: '))
```

4.10.2   Turning Python 2 Code into Python 3 Code

Suppose you have written some Python 2 code according to this book and want it to run under Python 3. We strongly recommend to create a *common version* of your program such that it works under both Python 2 and 3. This is quite easy if you use the `future`[5] package (it is easily installed by `pip install future`).

The `future` package has a program `futurize` that can rewrite a `.py` file such that it works under Python 2 and 3. Let us grab a file `c2f_qa.py`,

```
C = raw_input('C=? ')
C = float(C)
F = 9.0/5*C + 32
print F
```

and convert it by

```
Terminal
Terminal> futurize -w c2f_qa.py
```

---
[5] http://python-future.org/

Now `c2f_qa.py` has the content

```
from __future__ import print_function
from builtins import input
C = input('C=? ')
C = float(C)
F = 9.0/5*C + 32
print(F)
```

We notice that the `raw_input` call has been changed to `input` and that the print statement is a call to the `print` function. A simple test shows that the new file runs on both versions of Python:

```
                                Terminal
Terminal> python2 c2f_qa.py
C=? 21
69.8
Terminal> python3 py3/c2f_qa.py
C=? 21
69.80000000000001
```

(This test requires that you have Python 3 installed.)

Note that if we change the division `9.0/5` in the file to `9/5`, `futurize` will not make a float division out of that expression (i.e., the Python 2 meaning of the syntax is not changed). If we want all syntax to be interpreted the Python 3 way, add the `-all-imports` option:

```
                                Terminal
Terminal> futurize -w --all-imports c2f_qa.py
```

The result is

```
from __future__ import unicode_literals
from __future__ import print_function
from __future__ import division
from __future__ import absolute_import
from future import standard_library
standard_library.install_aliases()
from builtins import input
from builtins import *
C = input('C=? ')
C = float(C)
F = 9/5*C + 32
print(F)
```

Now, 9/5 represents float division, and the program runs under both versions of Python.

Usually, you do not want `futurize` to overwrite your original Python 2 program, but it is easy to let it generate the new version in a subfolder instead:

```
                                    ┌──────────┐
────────────────────────────────────│ Terminal │────────────────────────────────────
                                    └──────────┘
Terminal> futurize -w -n -o py23 c2f_qa.py
```

The generated new version of c2f_qa.py is now in py23/c2f_qa.py.

Most of the programs in this book apply the command line for input, and the programmer should fix all issues about integer division, so running futurize on the programs you have seen so far will just change the print statement. There are more challenging differences between Python 2 and 3 when one applies more advanced objects and modules. Section 6.6 contains further information.

4.11   Summary

4.11.1   Chapter Topics

Question and answer input   Prompting the user and reading the answer back into a variable is done by

```
var = raw_input('Give value: ')
```

The raw_input function returns a string containing the characters that the user wrote on the keyboard before pressing the Return key. It is necessary to convert var to an appropriate object (int or float, for instance) if we want to perform mathematical operations with var. Sometimes

```
var = eval(raw_input('Give value: '))
```

is a flexible and easy way of transforming the string to the right type of object (integer, real number, list, tuple, and so on). This last statement will not work, however, for strings unless the text is surrounded by quotes when written on the keyboard. A general conversion function that turns any text without quotes into the right object is scitools.misc.str2obj:

```
from scitools.misc import str2obj
var = str2obj(raw_input('Give value: '))
```

Typing, for example, 3 makes var refer to an int object, 3.14 results in a float object, [-1,1] results in a list, (1,3,5,7) in a tuple, and some text in the string (str) object 'some text' (run the program str2obj_demo.py to see this functionality demonstrated).

Getting command-line arguments   The sys.argv[1:] list contains all the command-line arguments given to a program (sys.argv[0] contains the program name). All elements in sys.argv are strings. A typical usage is

```
parameter1 = float(sys.argv[1])
parameter2 = int(sys.argv[2])
parameter3 = sys.argv[3]            # parameter3 can be string
```

Using option-value pairs The `argparse` module is recommended for interpreting command-line arguments of the form `-option value`. A simple recipe with `argparse` reads

```
import argparse
parser = argparse.ArgumentParser()
parser.add_argument('--p1', '--parameter_1', type=float,
                    default=0.0, help='1st parameter')
parser.add_argument('--p2', type=float,
                    default=0.0, help='2nd parameter')

args = parser.parse_args()
p1 = args.p1
p2 = args.p2
```

On the command line we can provide any or all of these options:

```
--parameter_1 --p1 --p2
```

where each option must be succeeded by a suitable value. However, `argparse` is very flexible can easily handle options without values or command-line arguments without any option specifications.

Generating code on the fly Calling `eval(s)` turns a string s, containing a Python expression, into code as if the contents of the string were written directly into the program code. The result of the following `eval` call is a `float` object holding the number 21.1:

```
>>> x = 20
>>> r = eval('x + 1.1')
>>> r
21.1
>>> type(r)
<type 'float'>
```

The `exec` function takes a string with arbitrary Python code as argument and executes the code. For example, writing

```
exec("""
def f(x):
    return %s
""" % sys.argv[1])
```

is the same as if we had hardcoded the (for the programmer unknown) contents of `sys.argv[1]` into a function definition in the program.

Turning string formulas into Python functions Given a mathematical formula as a string, s, we can turn this formula into a callable Python function `f(x)` by

```
from scitools.std import StringFunction

f = StringFunction(s)
```

The string formula can contain parameters and an independent variable with another name than x:

```
Q_formula = 'amplitude*sin(w*t-phaseshift)'
Q = StringFunction(Q_formula, independent_variable='t',
                   amplitude=1.5, w=pi, phaseshift=0)
values1 = [Q(i*0.1) for t in range(10)]
Q.set_parameters(phaseshift=pi/4, amplitude=1)
values2 = [Q(i*0.1) for t in range(10)]
```

Functions of several independent variables are also supported:

```
f = StringFunction('x+y**2+A', independent_variables=('x', 'y'),
                   A=0.2)
x = 1; y = 0.5
print f(x, y)
```

File operations Reading from or writing to a file first requires that the file is opened, either for reading, writing, or appending:

```
infile  = open(filename, 'r')   # read
outfile = open(filename, 'w')   # write
outfile = open(filename, 'a')   # append
```

or using `with`:

```
with open(filename, 'r') as infile:   # read
with open(filename, 'w') as outfile:  # write
with open(filename, 'a') as outfile:  # append
```

There are four basic reading commands:

```
line   = infile.readline()   # read the next line
filestr = infile.read()      # read rest of file into string
lines  = infile.readlines()  # read rest of file into list
for line in infile:          # read rest of file line by line
```

File writing is usually about repeatedly using the command

```
outfile.write(s)
```

where s is a string. Contrary to `print s`, no newline is added to s in `outfile.write(s)`.

After reading or writing is finished, the file must be closed:

```
somefile.close()
```

However, closing the file is not necessary if we employ the `with` statement for reading or writing files:

```
with open(filename, 'w') as outfile:
    for var1, var2 in data:
        outfile.write('%5.2f %g\n' % (var1, var2))
# outfile is closed
```

Handling exceptions Testing for potential errors is done with `try-except` blocks:

```
try:
    <statements>
except ExceptionType1:
    <provide a remedy for ExceptionType1 errors>
except ExceptionType2, ExceptionType3, ExceptionType4:
    <provide a remedy for three other types of errors>
except:
    <provide a remedy for any other errors>
...
```

The most common exception types are `NameError` for an undefined variable, `TypeError` for an illegal value in an operation, and `IndexError` for a list index out of bounds.

Raising exceptions When some error is encountered in a program, the programmer can raise an exception:

```
if z < 0:
    raise ValueError('z=%s is negative - cannot do log(z)' % z)
r = log(z)
```

Modules A module is created by putting a set of functions in a file. The filename (minus the required extension `.py`) is the name of the module. Other programs can import the module only if it resides in the same folder or in a folder contained in the `sys.path` list (see Sect. 4.9.7 for how to deal with this potential problem). Optionally, the module file can have a special `if` construct at the end, called test block, which tests the module or demonstrates its usage. The test block does not get executed when the module is imported in another program, only when the module file is run as a program.

Terminology The important computer science topics and Python tools in this chapter are

- command line
- `sys.argv`
- `raw_input`
- `eval` and `exec`
- file reading and writing
- handling and raising exceptions

- module
- test block

4.11.2   Example: Bisection Root Finding

Problem   The summarizing example of this chapter concerns the implementation of the Bisection method for solving nonlinear equations of the form $f(x) = 0$ with respect to x. For example, the equation

$$x = 1 + \sin x$$

can be cast in the form $f(x) = 0$ if we move all terms to the left-hand side and define $f(x) = x - 1 - \sin x$. We say that x is a *root* of the equation $f(x) = 0$ if x is a solution of this equation. Nonlinear equations $f(x) = 0$ can have zero, one, several, or infinitely many roots.

Numerical methods for computing roots normally lead to approximate results only, i.e., $f(x)$ is not made exactly zero, but very close to zero. More precisely, an approximate root x fulfills $|f(x)| \leq \epsilon$, where ϵ is a small number. Methods for finding roots are of an iterative nature: we start with a rough approximation to a root and perform a repetitive set of steps that aim to improve the approximation. Our particular method for computing roots, the Bisection method, guarantees to find an approximate root, while other methods, such as the widely used Newton's method (see Sect. A.1.10), can fail to find roots.

The idea of the Bisection method is to start with an interval $[a, b]$ that contains a root of $f(x)$. The interval is halved at $m = (a + b)/2$, and if $f(x)$ changes sign in the left half interval $[a, m]$, one continues with that interval, otherwise one continues with the right half interval $[m, b]$. This procedure is repeated, say n times, and the root is then guaranteed to be inside an interval of length $2^{-n}(b - a)$. The task is to write a program that implements the Bisection method and verify the implementation.

Solution   To implement the Bisection method, we need to translate the description in the previous paragraph to a precise algorithm that can be almost directly translated to computer code. Since the halving of the interval is repeated many times, it is natural to do this inside a loop. We start with the interval $[a, b]$, and adjust a to m if the root must be in the right half of the interval, or we adjust b to m if the root must be in the left half. In a language close to computer code we can express the algorithm precisely as follows:

```
for i in range(0, n+1):
    m = (a + b)/2
    if f(a)*f(m) <= 0:
        b = m  # root is in left half
    else:
        a = m  # root is in right half

# f(x) has a root in [a,b]
```

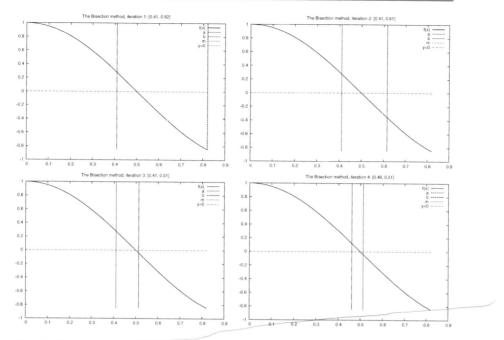

Fig. 4.2 Illustration of the first four iterations of the Bisection algorithm for solving $\cos(\pi x) = 0$. The vertical lines correspond to the current value of a and b

Figure 4.2 displays graphically the first four steps of this algorithm for solving the equation $\cos(\pi x) = 0$, starting with the interval $[0, 0.82]$. The graphs are automatically produced by the program `bisection_movie.py`, which was run as follows for this particular example:

```Terminal
bisection_movie.py 'cos(pi*x)' 0 0.82
```

The first command-line argument is the formula for $f(x)$, the next is a, and the final is b.

In the algorithm listed above, we recompute $f(a)$ in each `if`-test, but this is not necessary if a has not changed since the last $f(a)$ computations. It is a good habit in numerical programming to avoid redundant work. On modern computers the Bisection algorithm normally runs so fast that we can afford to do more work than necessary. However, if $f(x)$ is not a simple formula, but computed by comprehensive calculations in a program, the evaluation of f might take minutes or even hours, and reducing the number of evaluations in the Bisection algorithm is then very important. We will therefore introduce extra variables in the algorithm above to save an $f(m)$ evaluation in each iteration in the `for` loop:

```
f_a = f(a)
for i in range(0, n+1):
    m = (a + b)/2
```

```
    f_m = f(m)
    if f_a*f_m <= 0:
        b = m   # root is in left half
    else:
        a = m   # root is in right half
        f_a = f_m

# f(x) has a root in [a,b]
```

To execute the algorithm above, we need to specify n. Say we want to be sure that the root lies in an interval of maximum extent ϵ. After n iterations the length of our current interval is $2^{-n}(b-a)$, if $[a,b]$ is the initial interval. The current interval is sufficiently small if

$$2^{-n}(b-a) = \epsilon,$$

which implies

$$n = -\frac{\ln \epsilon - \ln(b-a)}{\ln 2}. \tag{4.6}$$

Instead of calculating this n, we may simply stop the iterations when the length of the current interval is less than ϵ. The loop is then naturally implemented as a while loop testing on whether $b-a \leq \epsilon$. To make the algorithm more foolproof, we also insert a test to ensure that $f(x)$ really changes sign in the initial interval. This guarantees a root in $[a,b]$. (However, $f(a)f(b) < 0$ is not a necessary condition if there is an even number of roots in the initial interval.)

Our final version of the Bisection algorithm now becomes

```
f_a=f(a)
if f_a*f(b) > 0:
    # error: f does not change sign in [a,b]

i = 0
while b-a > epsilon:
    i = i + 1
    m = (a + b)/2
    f_m = f(m)
    if f_a*f_m <= 0:
        b = m  # root is in left half
    else:
        a = m  # root is in right half
        f_a = f_m

# if x is the real root, |x-m| < epsilon
```

This is the algorithm we aim to implement in a Python program.

A direct translation of the previous algorithm to a valid Python program is a matter of some minor edits:

```
eps = 1E-5
a, b = 0, 10

fa = f(a)
if fa*f(b) > 0:
```

```
      print 'f(x) does not change sign in [%g,%g].' % (a, b)
      sys.exit(1)

i = 0    # iteration counter
while b-a > eps:
    i += 1
    m = (a + b)/2.0
    fm = f(m)
    if fa*fm <= 0:
        b = m   # root is in left half of [a,b]
    else:
        a = m   # root is in right half of [a,b]
        fa = fm
    print 'Iteration %d: interval=[%g, %g]' % (i, a, b)

x = m            # this is the approximate root
print 'The root is', x, 'found in', i, 'iterations'
print 'f(%g)=%g' % (x, f(x))
```

This program is found in the file bisection_v1.py.

Verification To verify the implementation in bisection_v1.py we choose a very simple $f(x)$ where we know the exact root. One suitable example is a linear function, $f(x) = 2x - 3$ such that $x = 3/2$ is the root of f. As can be seen from the source code above, we have inserted a print statement inside the while loop to control that the program really does the right things. Running the program yields the output

```
Iteration 1: interval=[0, 5]
Iteration 2: interval=[0, 2.5]
Iteration 3: interval=[1.25, 2.5]
Iteration 4: interval=[1.25, 1.875]
...
Iteration 19: interval=[1.5, 1.50002]
Iteration 20: interval=[1.5, 1.50001]
The root is 1.50000572205 found in 20 iterations
f(1.50001)=1.14441e-05
```

It seems that the implementation works. Further checks should include hand calculations for the first (say) three iterations and comparison of the results with the program.

Making a function The previous implementation of the bisection algorithm is fine for many purposes. To solve a new problem $f(x) = 0$ it is just necessary to change the f(x) function in the program. However, if we encounter solving $f(x) = 0$ in another program in another context, we must put the bisection algorithm into that program in the right place. This is simple in practice, but it requires some careful work, and it is easy to make errors. The task of solving $f(x) = 0$ by the bisection algorithm is much simpler and safer if we have that algorithm available as a function in a module. Then we can just import the function and call it. This requires a minimum of writing in later programs.

When you have a "flat" program as shown above, without basic steps in the program collected in functions, you should always consider dividing the code into functions. The reason is that parts of the program will be much easier to reuse in other programs. You save coding, and that is a good rule! A program with functions is also easier to understand, because statements are collected into logical, separate units, which is another good rule! In a mathematical context, functions are particularly important since they naturally split the code into general algorithms (like the bisection algorithm) and a problem-specific part (like a special choice of $f(x)$).

Shuffling statements in a program around to form a new and better designed version of the program is called *refactoring*. We shall now refactor the `bisection_v1.py` program by putting the statements in the bisection algorithm in a function `bisection`. This function naturally takes $f(x)$, a, b, and ϵ as parameters and returns the found root, perhaps together with the number of iterations required:

```python
def bisection(f, a, b, eps):
    fa = f(a)
    if fa*f(b) > 0:
        return None, 0

    i = 0    # iteration counter
    while b-a > eps:
        i += 1
        m = (a + b)/2.0
        fm = f(m)
        if fa*fm <= 0:
            b = m  # root is in left half of [a,b]
        else:
            a = m  # root is in right half of [a,b]
            fa = fm
    return m, i
```

After this function we can have a test program:

```python
def f(x):
    return 2*x - 3    # one root x=1.5

x, iter = bisection(f, a=0, b=10, eps=1E-5)
if x is None:
    print 'f(x) does not change sign in [%g,%g].' % (a, b)
else:
    print 'The root is', x, 'found in', iter, 'iterations'
    print 'f(%g)=%g' % (x, f(x))
```

The complete code is found in file `bisection_v2.py`.

Making a test function   Rather than having a main program as above for verifying the implementation, we should make a test function `test_bisection` as described in Sect. 4.9.4. To this end, we move the statements above inside a function, drop the output, but instead make a boolean variable `success` that is `True` if the test is passed and `False` otherwise. Then we do `assert success, msg`, which will abort the program if the test fails. The `msg` variable is a string with more explanation of what went wrong the test fails. A test function with this structure is easy to integrate into the widely used testing frameworks nose and pytest, and there are no good reasons for not adopting this structure. The code checking that the root is within a distance ϵ to the exact root becomes

```
def test_bisection():
    def f(x):
        return 2*x - 3     # one root x=1.5

    eps = 1E-5
    x_expected = 1.5
    x, iter = bisection(f, a=0, b=10, eps=eps)
    success = abs(x - x_expected) < eps   # test within eps tolerance
    assert success, 'found x=%g != 1.5' % x
```

Making a module A motivating factor for implementing the bisection algorithm as a function `bisection` was that we could import this function in other programs to solve $f(x) = 0$ equations. We therefore need to make a module file `bisection.py` such that we can do, e.g.,

```
from bisection import bisection
x, iter = bisection(lambda x: x**3 + 2*x -1, -10, 10, 1E-5)
```

A module file should not execute a main program, but just define functions, import modules, and define global variables. Any execution of a main program must take place in the test block, otherwise the `import` statement will start executing the main program, resulting in very disturbing statements for another program that wants to solve a different $f(x) = 0$ equation.

The `bisection_v2.py` file had a main program that was just a simple test for checking that the `bisection` algorithm works for a linear function. We took this main program and wrapped in a test function `test_bisection` above. To run the test, we make the call to this function from the test block:

```
if __name__ == '__main__':
    test_bisection()
```

This is all that is demanded to turn the file `bisection_v2.py` into a proper module file `bisection.py`.

Defining a user interface It is nice to have our `bisection` module do more than just test itself: there should be a user interface such that we can solve real problems $f(x) = 0$, where $f(x)$, a, b, and ϵ are defined on the command line by the user. A dedicated function can read from the command line and return the data as Python object. For reading the function $f(x)$ we can either apply `eval` on the command-line argument, or use the more sophisticated `StringFunction` tool from Sect. 4.3.3. With `eval` we need to import functions from the `math` module in case the user have such functions in the expression for $f(x)$. With `StringFunction` this is not necessary.

A `get_input()` for getting input from the command line can be implemented as

```
def get_input():
    """Get f, a, b, eps from the command line."""
    from scitools.std import StringFunction
    try:
        f = StringFunction(sys.argv[1])
        a = float(sys.argv[2])
        b = float(sys.argv[3])
        eps = float(sys.argv[4])
    except IndexError:
        print 'Usage %s: f a b eps' % sys.argv[0]
        sys.exit(1)
    return f, a, b, eps
```

To solve the corresponding $f(x) = 0$ problem, we simply add a branch in the `if` test in the test block:

```
if __name__ == '__main__':
    import sys
    if len(sys.argv) >= 2 and sys.argv[1] == 'test':
        test_bisection()
    else:
        f, a, b, eps = get_input()
        x, iter = bisection(f, a, b, eps)
        print 'Found root x=%g in %d iterations' % (x, iter)
```

Desired properties of a module

Our `bisection.py` code is a complete module file with the following generally desired features of Python modules:

- other programs can import the `bisection` function,
- the module can test itself (with a pytest/nose-compatible test function),
- the module file can be run as a program with a user interface where a general rooting finding problem can be specified in terms of a formula for $f(x)$ along with the parameters a, b, and ϵ.

Using the module Suppose you want to solve $x/(x - 1) = \sin x$ using the `bisection` module. What do you have to do? First, you must reformulate the equation as $f(x) = 0$, i.e., $x/(x - 1) - \sin x = 0$, or maybe multiply by $x - 1$ to get $f(x) = x - (x - 1) \sin x$.

It is required to identify an interval for the root. By evaluating $f(x)$ for some points x one can be trial and error locate an interval. A more convenient approach is to plot the function $f(x)$ and visually inspect where a root is. Chapter 5 describes the techniques, but here we simply state the recipe. We start `ipython -pylab` and write

```
In [1]: x = linspace(-3, 3, 50)   # generate 50 coordinates in [-3,3]

In [2]: y = x - (x-1)*sin(x)

In [3]: plot(x, y)
```

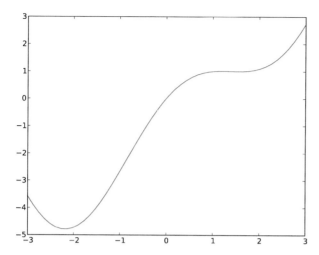

Fig. 4.3 Plot of $f(x) = x - \sin(x)$

Figure 4.3 shows $f(x)$ and we clearly see that, e.g., $[-2, 1]$ is an appropriate interval.

The next step is to run the Bisection algorithm. There are two possibilities:

- make a program where you code $f(x)$ and run the `bisection` function, or
- run the `bisection.py` program directly.

The latter approach is the simplest:

```Terminal
bisection.py "x - (x-1)*sin(x)" -2 1 1E-5
Found root x=-1.90735e-06 in 19 iterations
```

The alternative approach is to make a program:

```
from bisection import bisection
from math import sin

def f(x):
    return x - (x-1)*sin(x)

x, iter = bisection(f, a=-2, b=1, eps=1E-5)
print x, iter
```

Potential problems with the software   Let us solve

- $x = \tanh x$ with start interval $[-10, 10]$ and $\epsilon = 10^{-6}$,
- $x^5 = \tanh(x^5)$ with start interval $[-10, 10]$ and $\epsilon = 10^{-6}$.

Both equations have one root $x = 0$.

Terminal

```
bisection.py "x-tanh(x)" -10 10
Found root x=-5.96046e-07 in 25 iterations

bisection.py "x**5-tanh(x**5)" -10 10
Found root x=-0.0266892 in 25 iterations
```

These results look strange. In both cases we halve the start interval $[-10, 10]$ 25 times, but in the second case we end up with a much less accurate root although the value of ϵ is the same. A closer inspection of what goes on in the bisection algorithm reveals that the inaccuracy is caused by rounding errors. As $a, b, m \rightarrow 0$, raising a small number to the fifth power in the expression for $f(x)$ yields a much smaller result. Subtracting a very small number $\tanh x^5$ from another very small number x^5 may result in a small number with wrong sign, and the sign of f is essential in the bisection algorithm. We encourage the reader to graphically inspect this behavior by running these two examples with the `bisection_plot.py` program using a smaller interval $[-1, 1]$ to better see what is going on. The command-line arguments for the `bisection_plot.py` program are `'x-tanh(x)' -1 1` and `'x**5-tanh(x**5)' -1 1`. The very flat area, in the latter case, where $f(x) \approx 0$ for $x \in [-1/2, 1/2]$ illustrates well that it is difficult to locate an exact root.

Distributing the bisection module to others The Python standard for installing software is to run a `setup.py` program,

---
Terminal

```
Terminal> sudo python setup.py install
```
---

to install the system. The relevant `setup.py` for the `bisection` module arises from substituting the name `interest` by `bisection` in the `setup.py` file listed in Sect. 4.9.8. You can then distribute `bisection.py` and `setup.py` together.

4.12 Exercises

Exercise 4.1: Make an interactive program
Make a program that asks the user for a temperature in Fahrenheit degrees and reads the number; computes the corresponding temperature in Celsius degrees; and prints out the temperature in the Celsius scale.
Filename: `f2c_qa`.

Exercise 4.2: Read a number from the command line
Modify the program from Exercise 4.1 such that the Fahrenheit temperature is read from the command line.
Filename: `f2c_cml`.

Exercise 4.3: Read a number from a file
Modify the program from Exercise 4.1 such that the Fahrenheit temperature is read from a file with the following content:

```
Temperature data
----------------

Fahrenheit degrees: 67.2
```

Hint Create a sample file manually. In the program, skip the first three lines, split the fourth line into words and grab the third word.
Filename: `f2c_file_read`.

Exercise 4.4: Read and write several numbers from and to file
This is a variant of Exercise 4.3 where we have several Fahrenheit degrees in a file and want to read all of them into a list and convert the numbers to Celsius degrees. Thereafter, we want to write out a file with two columns, the left with the Fahrenheit degrees and the right with the Celsius degrees.

An example on the input file format looks like

```
Temperature data
----------------

Fahrenheit degrees: 67.2
Fahrenheit degrees: 66.0
Fahrenheit degrees: 78.9
Fahrenheit degrees: 102.1
Fahrenheit degrees: 32.0
Fahrenheit degrees: 87.8
```

A sample file is `Fdeg.dat`[6].
Filename: `f2c_file_read_write`.

Exercise 4.5: Use exceptions to handle wrong input
Extend the program from Exercise 4.2 with a `try-except` block to handle the potential error that the Fahrenheit temperature is missing on the command line.
Filename: `f2c_cml_exc`.

Exercise 4.6: Read input from the keyboard
Make a program that asks for input from the user, applies `eval` to this input, and prints out the type of the resulting object and its value. Test the program by providing five types of input: an integer, a real number, a complex number, a list, and a tuple.
Filename: `objects_qa`.

Exercise 4.7: Read input from the command line

a) Let a program store the result of applying the `eval` function to the first command-line argument. Print out the resulting object and its type.
b) Run the program with different input: an integer, a real number, a list, and a tuple.

---

[6] http://tinyurl.com/pwyasaa/input/Fdeg.dat

Hint On Unix systems you need to surround the tuple expressions in quotes on the command line to avoid error message from the Unix shell.

c) Try the string `"this is a string"` as a command-line argument. Why does this string cause problems and what is the remedy?

Filename: `objects_cml`.

Exercise 4.8: Try MSWord or LibreOffice to write a program
The purpose of this exercise is to tell you how hard it may be to write Python programs in the standard programs that most people use for writing text.

a) Type the following one-line program in either MSWord or LibreOffice:

```
print "Hello, World!"
```

Both Word and LibreOffice are so "smart" that they automatically edit "print" to "Print" since a sentence should always start with a capital. This is just an example that word processors are made for writing documents, not computer programs.

b) Save the program as a `.docx` (Word) or `.odt` (LibreOffice) file. Now try to run this file as a Python program. What kind of error message do you get? Can you explain why?

c) Save the program as a `.txt` file in Word or LibreOffice and run the file as a Python program. What happened now? Try to find out what the problem is.

Exercise 4.9: Prompt the user for input to a formula
Consider the simplest program for evaluating the formula $y(t) = v_0 t - \frac{1}{2} g t^2$:

```
v0 = 3; g = 9.81; t = 0.6
y = v0*t - 0.5*g*t**2
print y
```

Modify this code so that the program asks the user questions `t=?` and `v0=?`, and then gets `t` and `v0` from the user's input through the keyboard.
Filename: `ball_qa`.

Exercise 4.10: Read parameters in a formula from the command line
Modify the program listed in Exercise 4.9 such that `v0` and `t` are read from the command line.
Filename: `ball_cml`.

Exercise 4.11: Use exceptions to handle wrong input
The program from Exercise 4.10 reads input from the command line. Extend that program with exception handling such that missing command-line arguments are detected. In the `except IndexError` block, use the `raw_input` function to ask the user for missing input data.
Filename: `ball_cml_qa`.

Exercise 4.12: Test validity of input data
Test if the t value read in the program from Exercise 4.10 lies between 0 and $2v_0/g$.
If not, print a message and abort the execution.
Filename: `ball_cml_tcheck`.

Exercise 4.13: Raise an exception in case of wrong input
Instead of printing an error message and aborting the program explicitly, raise
a `ValueError` exception in the `if` test on legal t values in the program from Exercise 4.12. Notify the user about the legal interval for t in the exception message.
Filename: `ball_cml_ValueError`.

Exercise 4.14: Evaluate a formula for data in a file
We consider the formula $y(t) = v_0 t - 0.5gt^2$ and want to evaluate y for a range of
t values found in a file with format

```
v0: 3.00
t:
0.15592   0.28075    0.36807889 0.35 0.57681501876
0.21342619   0.0519085   0.042   0.27   0.50620017 0.528
0.2094294   0.1117   0.53012   0.3729850   0.39325246
0.21385894   0.3464815 0.57982969 0.10262264
0.29584013   0.17383923
```

More precisely, the first two lines are always present, while the next lines contain
an arbitrary number of t values on each line, separated by one or more spaces.

a) Write a function that reads the input file and returns v_0 and a list with the t
   values. A sample file is `ball.dat`[7]
b) Make a test function that generates an input file, calls the function in a) for
   reading the file, and checks that the returned data objects are correct.
c) Write a function that creates a file with two nicely formatted columns containing
   the t values to the left and the corresponding y values to the right. Let the t
   values appear in increasing order (note that the input file does not necessarily
   have the t values sorted).

Filename: `ball_file_read_write`.

Exercise 4.15: Write a function given its test function
A common software development technique in the IT industry is to write the test
function *before* writing the function itself.

a) We want to write a function `halve(x)` that returns the half of its argument x.
   The test function is

```
def test_halve():
    assert halve(5.0) == 2.5   # Real number division
    assert halve(5) == 2       # Integer division
```

---

[7] http://tinyurl.com/pwyasaa/input/ball.dat

Write the associated function `halve`. Call `test_halve` (or run pytest or nose) to verify that `halve` works.

b) We want to write a function `add(a, b)` that returns the sum of its arguments a and b. The test function reads

```
def test_add():
    # Test integers
    assert add(1, 2) == 3

    # Test floating-point numbers with rounding error
    tol = 1E-14
    a = 0.1;  b = 0.2
    computed = add(a, b)
    expected = 0.3
    assert abs(expected - computed) < tol

    # Test lists
    assert add([1,4], [4,7]) == [1,4,4,7]

    # Test strings
    assert add('Hello, ', 'World!') == 'Hello, World!'
```

Write the associated function `add`. Call `test_add` (or run pytest or nose) to verify that `add` works.

c) We want to write a function `equal(a, b)` for determining if two strings a and b are equal. If equal, the function returns `True` and the string a. If not equal, the function returns `False` and a string displaying the differences. This latter string contains the characters common in a and b, but for every difference, the character from a and b are written with a pipe symbol '|' in between. In case a and b are of unequal length, pad the string displaying differences with a * where one of the strings lacks content. For example, `equal('abc', 'aBc')` would return `False`, `'ab|Bc'`, while `equal('abc', 'aBcd')` would return `False`, `'ab|Bc*|d'`. Here is the test function:

```
def test_equal():
    assert equal('abc', 'abc') == (True, 'abc')
    assert equal('abc', 'aBc') == (False, 'ab|Bc')
    assert equal('abc', 'aBcd') == (False, 'ab|Bc*|d')
    assert equal('Hello, World!', 'hello world') == \
           (False, 'H|hello,|  |wW|oo|rr|ll|dd|*!|*')
```

Write the `equal` function (which is handy to detect very small differences between texts).

Filename: `testfunc2func`.

Exercise 4.16: Compute the distance it takes to stop a car

A car driver, driving at velocity v_0, suddenly puts on the brake. What braking distance d is needed to stop the car? One can derive, using Newton's second law of motion or a corresponding energy equation, that

$$d = \frac{1}{2} \frac{v_0^2}{\mu g}.$$ (4.7)

Make a program for computing d in (4.7) when the initial car velocity v_0 and the friction coefficient μ are given on the command line. Run the program for two cases: $v_0 = 120$ and $v_0 = 50$ km/h, both with $\mu = 0.3$ (μ is dimensionless).

Hint Remember to convert the velocity from km/h to m/s before inserting the value in the formula.
Filename: `stopping_length`.

Exercise 4.17: Look up calendar functionality
The purpose of this exercise is to make a program that takes a date, consisting of year (4 digits), month (2 digits), and day (1–31) on the command line and prints the corresponding name of the weekday (Monday, Tuesday, etc.). Python has a module `calendar`, which makes it easy to solve the exercise, but the task is to find out how to use this module.
Filename: `weekday`.

Exercise 4.18: Use the StringFunction tool
Make the program `integrate.py` from Sect. 4.3.2 shorter by using the convenient `StringFunction` tool from Sect. 4.3.3. Write a test function for verifying this new implementation.
Filename: `integrate2`.

Exercise 4.19: Why we test for specific exception types
The simplest way of writing a `try-except` block is to test for *any* exception, for example,

```
try:
    C = float(sys.arg[1])
except:
    print 'C must be provided as command-line argument'
    sys.exit(1)
```

Write the above statements in a program and test the program. What is the problem?
   The fact that a user can forget to supply a command-line argument when running the program was the original reason for using a `try` block. Find out what kind of exception that is relevant for this error and test for this specific exception and re-run the program. What is the problem now? Correct the program.
Filename: `unnamed_exception`.

Exercise 4.20: Make a complete module

a) Make six conversion functions between temperatures in Celsius, Kelvin, and Fahrenheit: C2F, F2C, C2K, K2C, F2K, and K2F.

b) Collect these functions in a module `convert_temp`.

c) Import the module in an interactive Python shell and demonstrate some sample calls on temperature conversions.

d) Insert the session from c) in a triple quoted string at the top of the module file as a doc string for demonstrating the usage.

e) Write a function `test_conversion()` that verifies the implementation. Call this function from the test block if the first command-line argument is `verify`.

Hint Check that `C2F(F2C(f))` is `f`, `K2C(C2K(c))` is `c`, and `K2F(F2K(f))` is `f` — with tolerance. Follow the conventions for test functions outlined in Sects. 4.9.4 and 4.11.2 with a boolean variable that is `False` if a test failed, and `True` if all test are passed, and then an `assert` statement to abort the program when any test fails.

f) Add a user interface to the module such that the user can write a temperature as the first command-line argument and the corresponding temperature scale as the second command-line argument, and then get the temperature in the two other scales as output. For example, `21.3 C` on the command line results in the output `70.3 F 294.4 K`. Encapsulate the user interface in a function, which is called from the test block.

Filename: `convert_temp`.

Exercise 4.21: Organize a previous program as a module

Collect the `f` and `S` functions in the program from Exercise 3.21 in a separate file such that this file becomes a module. Put the statements making the table (i.e., the main program from Exercise 3.21) in a separate function `table(n_values, alpha_values, T)`. Make a test block in the module to read T and a series of n and α values as positional command-line arguments and make a corresponding call to `table`.
Filename: `sinesum2`.

Exercise 4.22: Read options and values from the command line

Let the input to the program in Exercise 4.21 be option-value pairs with the options `-n`, `-alpha`, and `-T`. Provide sensible default values in the module file.

Hint Apply the `argparse` module to read the command-line arguments. Do not copy code from the `sinesum2` module, but make a new file for reading option-value pairs from the command and import the `table` function from the `sinesum2` module.
Filename: `sinesum3`.

Exercise 4.23: Check if mathematical identities hold

Because of rounding errors, it could happen that a mathematical rule like $(ab)^3 = a^3 b^3$ does not hold exactly on a computer. The idea of testing this potential problem is to check such identities for a large number of random numbers. We can make random numbers using the `random` module in Python:

```
import random
a = random.uniform(A, B)
b = random.uniform(A, B)
```

Here, a and b will be random numbers, which are always larger than or equal to A and smaller than B.

a) Make a function `power3_identity(A=-100, B=100, n=1000)` that tests the identity `(a*b)**3 == a**3*b**3` a large number of times, n. Return the fraction of failures.

Hint Inside the loop over n, draw random numbers a and b as described above and count the number of times the test is `True`.

b) We shall now parameterize the expressions to be tested. Make a function

```
equal(expr1, expr2, A=-100, B=100, n=500)
```

where `expr1` and `expr2` are strings containing the two mathematical expressions to be tested. More precisely, the function draws random numbers a and b between A and B and tests if `eval(expr1) == eval(expr2)`. Return the fraction of failures.

Test the function on the identities $(ab)^3 = a^3 b^3$, $e^{a+b} = e^a e^b$, and $\ln a^b = b \ln a$.

Hint Make the `equal` function robust enough to handle illegal a and b values in the mathematical expressions (e.g., $a \leq 0$ in $\ln a$).

c) We want to test the validity of the following set of identities on a computer:
   - $a - b$ and $-(b - a)$
   - a/b and $1/(b/a)$
   - $(ab)^4$ and $a^4 b^4$
   - $(a + b)^2$ and $a^2 + 2ab + b^2$
   - $(a + b)(a - b)$ and $a^2 - b^2$
   - e^{a+b} and $e^a e^b$
   - $\ln a^b$ and $b \ln a$
   - $\ln ab$ and $\ln a + \ln b$
   - ab and $e^{\ln a + \ln b}$
   - $1/(1/a + 1/b)$ and $ab/(a + b)$
   - $a(\sin^2 b + \cos^2 b)$ and a
   - $\sinh(a + b)$ and $(e^a e^b - e^{-a} e^{-b})/2$
   - $\tan(a + b)$ and $\sin(a + b)/\cos(a + b)$
   - $\sin(a + b)$ and $\sin a \cos b + \sin b \cos a$

   Store all the expressions in a list of 2-tuples, where each 2-tuple contains two mathematically equivalent expressions as strings, which can be sent to the `equal` function. Make a nicely formatted table with a pair of equivalent expressions at each line followed by the failure rate. Write this table to a file. Try out A=1 and B=2 as well as A=1 and B=100. Does the failure rate seem to depend on the magnitude of the numbers a and b?

Filename: `math_identities_failures`.

Exercise 4.24: Compute probabilities with the binomial distribution
Consider an uncertain event where there are two outcomes only, typically success or failure. Flipping a coin is an example: the outcome is uncertain and of two types, either head (can be considered as success) or tail (failure). Throwing a die can be another example, if (e.g.) getting a six is considered success and all other outcomes represent failure. Such experiments are called *Bernoulli trials*.

Let the probability of success be p and that of failure $1 - p$. If we perform n experiments, where the outcome of each experiment does not depend on the outcome of previous experiments, the probability of getting success x times, and consequently failure $n - x$ times, is given by

$$B(x, n, p) = \frac{n!}{x!(n-x)!} p^x (1-p)^{n-x} \,. \tag{4.8}$$

This formula (4.8) is called the binomial distribution. The expression $x!$ is the factorial of x: $x! = x(x-1)(x-2) \cdots 1$ and `math.factorial` can do this computation.

a) Implement (4.8) in a function `binomial(x, n, p)`.
b) What is the probability of getting two heads when flipping a coin five times? This probability corresponds to $n = 5$ events, where the success of an event means getting head, which has probability $p = 1/2$, and we look for $x = 2$ successes.
c) What is the probability of getting four ones in a row when throwing a die? This probability corresponds to $n = 4$ events, success is getting one and has probability $p = 1/6$, and we look for $x = 4$ successful events.
d) Suppose cross country skiers typically experience one ski break in one out of 120 competitions. Hence, the probability of breaking a ski can be set to $p = 1/120$. What is the probability b that a skier will experience a ski break during five competitions in a world championship?

Hint This question is a bit more demanding than the other two. We are looking for the probability of 1, 2, 3, 4 or 5 ski breaks, so it is simpler to ask for the probability c of *not* breaking a ski, and then compute $b = 1 - c$. Define *success* as breaking a ski. We then look for $x = 0$ successes out of $n = 5$ trials, with $p = 1/120$ for each trial. Compute b.
Filename: `Bernoulli_trials`.

Exercise 4.25: Compute probabilities with the Poisson distribution
Suppose that over a period of t_m time units, a particular uncertain event happens (on average) νt_m times. The probability that there will be x such events in a time period t is approximately given by the formula

$$P(x, t, \nu) = \frac{(\nu t)^x}{x!} e^{-\nu t} \,. \tag{4.9}$$

This formula is known as the Poisson distribution. (It can be shown that (4.9) arises from (4.8) when the probability p of experiencing the event in a small time interval t/n is $p = \nu t/n$ and we let $n \to \infty$.) An important assumption is that all events are independent of each other and that the probability of experiencing an event does not change significantly over time. This is known as a *Poisson process* in probability theory.

a) Implement (4.9) in a function Poisson(x, t, nu), and make a program that reads x, t, and v from the command line and writes out the probability $P(x,t,v)$. Use this program to solve the problems below.

b) Suppose you are waiting for a taxi in a certain street at night. On average, 5 taxis pass this street every hour at this time of the night. What is the probability of not getting a taxi after having waited 30 minutes? Since we have 5 events in a time period of $t_m = 1$ hour, $v t_m = v = 5$. The sought probability is then $P(0, 1/2, 5)$. Compute this number. What is the probability of having to wait two hours for a taxi? If 8 people need two taxis, that is the probability that two taxis arrive in a period of 20 minutes?

c) In a certain location, 10 earthquakes have been recorded during the last 50 years. What is the probability of experiencing exactly three earthquakes over a period of 10 years in this area? What is the probability that a visitor for one week does not experience any earthquake? With 10 events over 50 years we have $v t_m = v \cdot 50 \text{ years} = 10$ events, which implies $v = 1/5$ event per year. The answer to the first question of having $x = 3$ events in a period of $t = 10$ years is given directly by (4.9). The second question asks for $x = 0$ events in a time period of 1 week, i.e., $t = 1/52$ years, so the answer is $P(0, 1/52, 1/5)$.

d) Suppose that you count the number of misprints in the first versions of the reports you write and that this number shows an average of six misprints per page. What is the probability that a reader of a first draft of one of your reports reads six pages without hitting a misprint? Assuming that the Poisson distribution can be applied to this problem, we have "time" t_m as 1 page and $v \cdot 1 = 6$, i.e., $v = 6$ events (misprints) per page. The probability of no events in a "period" of six pages is $P(0, 6, 6)$.

Filename: Poisson_processes.

Array Computing and Curve Plotting

5

A list object is handy for storing tabular data, such as a sequence of objects or a table of objects. An array is very similar to a list, but less flexible and computationally much more efficient. When using the computer to perform mathematical calculations, we often end up with a huge amount of numbers and associated arithmetic operations. Storing numbers in lists may in such contexts lead to slow programs, while arrays can make the programs run much faster. This is crucial for many advanced applications of mathematics in industry and science, where computer programs may run for hours and days, or even weeks. Any clever idea that reduces the execution time by some factor is therefore paramount.

However, one can argue that programmers of mathematical software have traditionally paid too much attention to efficiency and "clever" program constructs. The resulting software often becomes very hard to maintain and extend. In this book we advocate a focus on clear, well-designed, and easy-to-understand programs that work correctly. Thereafter, one can start thinking about optimization for speed. Fortunately, arrays contribute to clear code, correctness and speed – all at once.

This chapter gives an introduction to arrays: how they are created and what they can be used for. Array computing usually ends up with a lot of numbers. It may be very hard to understand what these numbers mean by just looking at them. Since the human is a visual animal, a good way to understand numbers is to visualize them. In this chapter we concentrate on visualizing curves that reflect functions of one variable; i.e., curves of the form $y = f(x)$. A synonym for curve is graph, and the image of curves on the screen is often called a plot. We will use arrays to store the information about points along the curve. In a nutshell, array computing demands visualization and visualization demands arrays.

All program examples in this chapter can be found as files in the folder `src/plot`[1].

---

[1] http://tinyurl.com/pwyasaa/plot

© Springer-Verlag Berlin Heidelberg 2016
H.P. Langtangen, *A Primer on Scientific Programming with Python*,
Texts in Computational Science and Engineering 6, DOI 10.1007/978-3-662-49887-3_5

5.1 Vectors

This section gives a brief introduction to the vector concept, assuming that you
have heard about vectors in the plane and maybe vectors in space before. This
background will be valuable when we start to work with arrays and curve plotting.

5.1.1 The Vector Concept

Some mathematical quantities are associated with a set of numbers. One example is
a point in the plane, where we need two coordinates (real numbers) to describe the
point mathematically. Naming the two coordinates of a particular point as x and y,
it is common to use the notation (x, y) for the point. That is, we group the numbers
inside parentheses. Instead of symbols we might use the numbers directly: $(0, 0)$
and $(1.5, -2.35)$ are also examples of coordinates in the plane.

A point in three-dimensional space has three coordinates, which we may name
x_1, x_2, and x_3. The common notation groups the numbers inside parentheses:
(x_1, x_2, x_3). Alternatively, we may use the symbols x, y, and z, and write the
point as (x, y, z), or numbers can be used instead of symbols.

From high school you may have a memory of solving two equations with
two unknowns. At the university you will soon meet problems that are formulated as n equations with n unknowns. The solution of such problems contains
n numbers that we can collect inside parentheses and number from 1 to n:
$(x_1, x_2, x_3, \ldots, x_{n-1}, x_n)$.

Quantities such as (x, y), (x, y, z), or $(x_1, \ldots, x_n)$ are known as *vectors* in mathematics. A visual representation of a vector is an arrow that goes from the origin to
a point. For example, the vector (x, y) is an arrow that goes from $(0, 0)$ to the point
with coordinates (x, y) in the plane. Similarly, (x, y, z) is an arrow from $(0, 0, 0)$
to the point (x, y, z) in three-dimensional space.

Mathematicians found it convenient to introduce spaces with higher dimension
than three, because when we have a solution of n equations collected in a vector
$(x_1, \ldots, x_n)$, we may think of this vector as a point in a space with dimension n, or
equivalently, an arrow that goes from the origin $(0, \ldots, 0)$ in n-dimensional space
to the point $(x_1, \ldots, x_n)$. Figure 5.1 illustrates a vector as an arrow, either starting
at the origin, or at any other point. Two arrows/vectors that have the same direction
and the same length are mathematically equivalent.

We say that $(x_1, \ldots, x_n)$ is an n-vector or a vector with n components. Each
of the numbers x_1, x_2, ... is a component or an element. We refer to the first
component (or element), the second component (or element), and so forth.

A Python program may use a list or tuple to represent a vector:

```
v1 = [x, y]          # list of variables
v2 = (-1, 2)         # tuple of numbers
v3 = (x1, x2, x3)    # tuple of variables
from math import exp
v4 = [exp(-i*0.1) for i in range(150)]
```

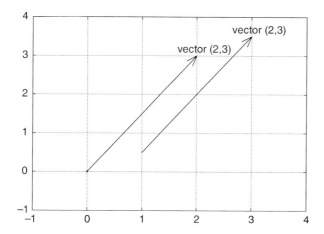

Fig. 5.1 A vector $(2, 3)$ visualized in the standard way as an arrow from the origin to the point $(2, 3)$, and mathematically equivalently, as an arrow from $(1, \frac{1}{2})$ (or any point (a, b)) to $(3, 3\frac{1}{2})$ (or $(a + 2, b + 3)$)

While v1 and v2 are vectors in the plane and v3 is a vector in three-dimensional space, v4 is a vector in a 150-dimensional space, consisting of 150 values of the exponential function. Since Python lists and tuples have 0 as the first index, we may also in mathematics write the vector (x_1, x_2) as (x_0, x_1). This is not at all common in mathematics, but makes the distance from a mathematical description of a problem to its solution in Python shorter.

It is impossible to visually demonstrate how a space with 150 dimensions looks like. Going from the plane to three-dimensional space gives a rough feeling of what it means to add a dimension, but if we forget about the idea of a visual perception of space, the mathematics is very simple: going from a 4-dimensional vector to a 5-dimensional vector is just as easy as adding an element to a list of symbols or numbers.

5.1.2 Mathematical Operations on Vectors

Since vectors can be viewed as arrows having a length and a direction, vectors are extremely useful in geometry and physics. The velocity of a car has a magnitude and a direction, so has the acceleration, and the position of a point in the car is also a vector. An edge of a triangle can be viewed as a line (arrow) with a direction and length.

In geometric and physical applications of vectors, mathematical operations on vectors are important. We shall exemplify some of the most important operations on vectors below. The goal is not to teach computations with vectors, but more to illustrate that such computations are defined by mathematical rules. Given two vectors, (u_1, u_2) and (v_1, v_2), we can add these vectors according to the rule:

$$(u_1, u_2) + (v_1, v_2) = (u_1 + v_1, u_2 + v_2) . \tag{5.1}$$

We can also subtract two vectors using a similar rule:

$$(u_1, u_2) - (v_1, v_2) = (u_1 - v_1, u_2 - v_2). \tag{5.2}$$

A vector can be multiplied by a number. This number, called a below, is usually denoted as a *scalar*:

$$a \cdot (v_1, v_2) = (av_1, av_2). \tag{5.3}$$

The inner product, also called dot product, or scalar product, of two vectors is a number:

$$(u_1, u_2) \cdot (v_1, v_2) = u_1 v_1 + u_2 v_2. \tag{5.4}$$

(From high school mathematics and physics you might recall that the inner or dot product also can be expressed as the product of the lengths of the two vectors multiplied by the cosine of the angle between them, but we will not make use of that formula. There is also a *cross product* defined for 2-vectors or 3-vectors, but we do not list the cross product formula here.)

The length of a vector is defined by

$$||(v_1, v_2)|| = \sqrt{(v_1, v_2) \cdot (v_1, v_2)} = \sqrt{v_1^2 + v_2^2}. \tag{5.5}$$

The same mathematical operations apply to n-dimensional vectors as well. Instead of counting indices from 1, as we usually do in mathematics, we now count from 0, as in Python. The addition and subtraction of two vectors with n components (or elements) read

$$(u_0, \ldots, u_{n-1}) + (v_0, \ldots, v_{n-1}) = (u_0 + v_0, \ldots, u_{n-1} + v_{n-1}), \tag{5.6}$$

$$(u_0, \ldots, u_{n-1}) - (v_0, \ldots, v_{n-1}) = (u_0 - v_0, \ldots, u_{n-1} - v_{n-1}). \tag{5.7}$$

Multiplication of a scalar a and a vector $(v_0, \ldots, v_{n-1})$ equals

$$(av_0, \ldots, av_{n-1}). \tag{5.8}$$

The inner or dot product of two n-vectors is defined as

$$(u_0, \ldots, u_{n-1}) \cdot (v_0, \ldots, v_{n-1}) = u_0 v_0 + \cdots + u_{n-1} v_{n-1} = \sum_{j=0}^{n-1} u_j v_j. \tag{5.9}$$

Finally, the length $||v||$ of an n-vector $v = (v_0, \ldots, v_{n-1})$ is

$$\sqrt{(v_0, \ldots, v_{n-1}) \cdot (v_0, \ldots, v_{n-1})} = \left(v_0^2 + v_1^2 + \cdots + v_{n-1}^2 \right)^{\frac{1}{2}}$$

$$= \left(\sum_{j=0}^{n-1} v_j^2 \right)^{\frac{1}{2}}. \tag{5.10}$$

5.1.3   Vector Arithmetics and Vector Functions

In addition to the operations on vectors in Sect. 5.1.2, which you might recall from high school mathematics, we can define other operations on vectors. This is very useful for speeding up programs. Unfortunately, the forthcoming vector operations are hardly treated in textbooks on mathematics, yet these operations play a significant role in mathematical software, especially in computing environment such as MATLAB, Octave, Python, and R.

Applying a mathematical function of one variable, $f(x)$, to a vector is defined as a vector where f is applied to each element. Let $v = (v_0, \ldots, v_{n-1})$ be a vector. Then

$$f(v) = (f(v_0), \ldots, f(v_{n-1})).$$

For example, the sine of v is

$$\sin(v) = (\sin(v_0), \ldots, \sin(v_{n-1})).$$

It follows that squaring a vector, or the more general operation of raising the vector to a power, can be defined as applying the operation to each element:

$$v^b = (v_0^b, \ldots, v_{n-1}^b).$$

Another operation between two vectors that arises in computer programming of mathematics is the "asterisk" multiplication, defined as

$$u * v = (u_0 v_0, u_1 v_1, \ldots, u_{n-1} v_{n-1}). \tag{5.11}$$

Adding a scalar to a vector or array can be defined as adding the scalar to each component. If a is a scalar and v a vector, we have

$$a + v = (a + v_0, \ldots, a + v_{n-1}).$$

A compound vector expression may look like

$$v^2 * \cos(v) * e^v + 2. \tag{5.12}$$

How do we calculate this expression? We use the normal rules of mathematics, working our way, term by term, from left to right, paying attention to the fact that powers are evaluated before multiplications and divisions, which are evaluated prior to addition and subtraction. First we calculate v^2, which results in a vector we may call u. Then we calculate $\cos(v)$ and call the result p. Then we multiply $u * p$ to get a vector which we may call w. The next step is to evaluate e^v, call the result q, followed by the multiplication $w * q$, whose result is stored as r. Then we add $r + 2$ to get the final result. It might be more convenient to list these operations after each other:

- $u = v^2$
- $p = \cos(v)$
- $w = u * p$

- $q = e^v$
- $r = w * q$
- $s = r + 2$

Writing out the vectors u, w, p, q, and r in terms of a general vector $v = (v_0, \ldots, v_{n-1})$ (do it!) shows that the result of the expression (5.12) is the vector

$$(v_0^2 \cos(v_0)e^{v_0} + 2, \ldots, v_{n-1}^2 \cos(v_{n-1})e^{v_{n-1}} + 2).$$

That is, component no. i in the result vector equals the number arising from applying the formula (5.12) to v_i, where the $*$ multiplication is ordinary multiplication between two numbers.

We can, alternatively, introduce the function

$$f(x) = x^2 \cos(x)e^x + 2$$

and use the result that $f(v)$ means applying f to each element in v. The result is the same as in the vector expression (5.12).

In Python programming it is important for speed (and convenience too) that we can apply functions of one variable, like $f(x)$, to vectors. What this means mathematically is something we have tried to explain in this subsection. Doing Exercises 5.5 and 5.6 may help to grasp the ideas of vector computing, and with more programming experience you will hopefully discover that vector computing is very useful. It is not necessary to have a thorough understanding of vector computing in order to proceed with the next sections.

Arrays are used to represent vectors in a program, but one can do more with arrays than with vectors. Until Sect. 5.8 it suffices to think of arrays as the same as vectors in a program.

5.2   Arrays in Python Programs

This section introduces array programming in Python, but first we create some lists and show how arrays differ from lists.

5.2.1   Using Lists for Collecting Function Data

Suppose we have a function $f(x)$ and want to evaluate this function at a number of x points $x_0, x_1, \ldots, x_{n-1}$. We could collect the n pairs $(x_i, f(x_i))$ in a list, or we could collect all the x_i values, for $i = 0, \ldots, n - 1$, in a list and all the associated $f(x_i)$ values in another list. The following interactive session demonstrates how to create these three types of lists:

```
>>> def f(x):
...     return x**3      # sample function
...
```

```
>>> n = 5                    # no of points along the x axis
>>> dx = 1.0/(n-1)           # spacing between x points in [0,1]
>>> xlist = [i*dx for i in range(n)]
>>> ylist = [f(x) for x in xlist]
>>> pairs = [[x, y] for x, y in zip(xlist, ylist)]
```

Here we have used list comprehensions for achieving compact code. Make sure that you understand what is going on in these list comprehensions (if not, try to write the same code using standard `for` loops and appending new list elements in each pass of the loops).

The list elements consist of objects of the same type: any element in `pairs` is a list of two `float` objects, while any element in `xlist` or `ylist` is a `float`. Lists are more flexible than that, because an element can be an object of any type, e.g.,

```
mylist = [2, 6.0, 'tmp.pdf', [0,1]]
```

Here `mylist` holds an `int`, a `float`, a string, and a list. This combination of diverse object types makes up what is known as *heterogeneous* lists. We can also easily remove elements from a list or add new elements anywhere in the list. This flexibility of lists is in general convenient to have as a programmer, but in cases where the elements are of the same type and the number of elements is fixed, arrays can be used instead. The benefits of arrays are faster computations, less memory demands, and extensive support for mathematical operations on the data. Because of greater efficiency and mathematical convenience, arrays will be used to a large extent in this book. The great use of arrays is also prominent in other programming environments such as MATLAB, Octave, and R, for instance. Lists will be our choice instead of arrays when we need the flexibility of adding or removing elements or when the elements may be of different object types.

5.2.2   Basics of Numerical Python Arrays

An *array* object can be viewed as a variant of a list, but with the following assumptions and features:

- All elements must be of the same type, preferably integer, real, or complex numbers, for efficient numerical computing and storage.
- The number of elements must be known when the array is created.
- Arrays are not part of standard Python – one needs an additional package called *Numerical Python*, often abbreviated as NumPy. The Python name of the package, to be used in `import` statements, is `numpy`.
- With `numpy`, a wide range of mathematical operations can be done directly on complete arrays, thereby removing the need for loops over array elements. This is commonly called *vectorization*
- Arrays with one index are often called vectors. Arrays with two indices are used as an efficient data structure for tables, instead of lists of lists. Arrays can also have three or more indices.

We have two remarks to the above list. First, there is actually an object type called `array` in standard Python, but this data type is not so efficient for mathematical computations, and we will not use it in this book. Second, the number of elements in an array *can* be changed, but at a substantial computational cost.

The following text lists some important functionality of NumPy arrays. A more comprehensive treatment is found in the excellent *NumPy Tutorial, NumPy User Guide, NumPy Reference, Guide to NumPy,* and *NumPy for MATLAB Users,* all accessible at scipy.org[2].

The standard import statement for Numerical Python reads

```
import numpy as np
```

To convert a list `r` to an array, we use the `array` function from `numpy`:

```
a = np.array(r)
```

To create a new array of length n, filled with zeros, we write

```
a = np.zeros(n)
```

The array elements are of a type that corresponds to Python's `float` type. A second argument to `np.zeros` can be used to specify other element types, e.g., `int`. A similar function,

```
a = np.zeros_like(c)
```

generates an array of zeros where the length is that of the array c and the element type is the same as those in c. Arrays with more than one index are treated in Sect. 5.8.

Often one wants an array to have *n* elements with uniformly distributed values in an interval $[p, q]$. The numpy function `linspace` creates such arrays:

```
a = np.linspace(p, q, n)
```

Array elements are accessed by square brackets as for lists: `a[i]`. Slices also work as for lists, for example, `a[1:-1]` picks out all elements except the first and the last, but contrary to lists, `a[1:-1]` is not a copy of the data in a. Hence,

```
b = a[1:-1]
b[2] = 0.1
```

will also change a[3] to 0.1. A slice `a[i:j:s]` picks out the elements starting with index i and stepping s indices at the time up to, but not including, j. Omitting i implies i=0, and omitting j implies j=n if n is the number of elements in the array. For example, `a[0:-1:2]` picks out every two elements up to, but not including, the last element, while `a[::4]` picks out every four elements in the whole array.

---

[2] http://scipy.org

Remarks on importing NumPy
The statement

```
import numpy as np
```

with subsequent prefixing of all NumPy functions and variables by np, has evolved as a standard syntax in the Python scientific computing community. However, to make Python programs look closer to MATLAB and ease the transition to and from that language, one can do

```
from numpy import *
```

to get rid of the prefix (this is evolved as the standard in *interactive* Python shells). This author prefers mathematical functions from numpy to be written without the prefix to make the formulas as close as possible to the mathematics. So, $f(x) = \sinh(x - 1)\sin(wt)$ would be coded as

```
from numpy import sinh, sin

def f(x):
    return sinh(x-1)*sin(w*t)
```

or one may take the less recommended lazy approach `from numpy import *` and fill up the program with *a lot* of functions and variables from numpy.

5.2.3 Computing Coordinates and Function Values

With these basic operations at hand, we can continue the session from the previous section and make arrays out of the lists xlist and ylist:

```
>>> import numpy as np
>>> x2 = np.array(xlist)       # turn list xlist into array x2
>>> y2 = np.array(ylist)
>>> x2
array([ 0.  ,  0.25,  0.5 ,  0.75,  1.  ])
>>> y2
array([ 0.     ,  0.015625,  0.125  ,  0.421875,  1.     ])
```

Instead of first making a list and then converting the list to an array, we can compute the arrays directly. The equally spaced coordinates in x2 are naturally computed by the np.linspace function. The y2 array can be created by np.zeros, to ensure that y2 has the right length len(x2), and then we can run a for loop to fill in all elements in y2 with f values:

```
>>> n = len(xlist)
>>> x2 = np.linspace(0, 1, n)
>>> y2 = np.zeros(n)
```

```
>>> for i in xrange(n):
...     y2[i] = f(x2[i])
...
>>> y2
array([ 0.     ,  0.015625,  0.125   ,  0.421875,  1.      ])
```

Note that we here in the `for` loop have used `xrange` instead of `range`. The former is faster for long loops because it avoids generating *and storing* a list of integers, it just generates the values one by one. Hence, we prefer `xrange` over `range` for loops over long arrays. In Python version 3.x, `range` is the same as `xrange`.

Creating an array of a given length is frequently referred to as *allocating* the array. It simply means that a part of the computer's memory is marked for being occupied by this array. Mathematical computations will often fill up most of the computer's memory by allocating long arrays.

We can shorten the previous code by creating the y2 data in a list comprehension, but list comprehensions produce lists, not arrays, so we need to transform the list object to an array object:

```
>>> x2 = np.linspace(0, 1, n)
>>> y2 = np.array([f(xi) for xi in x2])
```

Nevertheless, there is a much faster way of computing y2 as the next paragraph explains.

5.2.4 Vectorization

Loops over very long arrays may run slowly. A great advantage with arrays is that we can get rid of the loops and apply `f` directly to the whole array:

```
>>> y2 = f(x2)
>>> y2
array([ 0.     ,  0.015625,  0.125   ,  0.421875,  1.      ])
```

The magic that makes `f(x2)` work builds on the vector computing concepts from Sect. 5.1.3. Instead of calling `f(x2)` we can equivalently write the function formula `x2**3` directly.

The point is that numpy implements vector arithmetics *for arrays* of any dimension. Moreover, numpy provides its own versions of mathematical functions like cos, sin, exp, log, etc., which work for array arguments and apply the mathematical function to each element. The following code, computes each array element separately:

```
from math import sin, cos, exp
import numpy as np
x = np.linspace(0, 2, 201)
r = np.zeros(len(x))
for i in xrange(len(x)):
    r[i] = sin(np.pi*x[i])*cos(x[i])*exp(-x[i]**2) + 2 + x[i]**2
```

while here is a corresponding code that operates on arrays directly:

```
r = np.sin(np.pi*x)*np.cos(x)*np.exp(-x**2) + 2 + x**2
```

Many will prefer to see such formulas without the np prefix:

```
from numpy import sin, cos, exp, pi
r = sin(pi*x)*cos(x)*exp(-x**2) + 2 + x**2
```

An important thing to understand is that sin from the math module is different from the sin function provided by numpy. The former does not allow array arguments, while the latter accepts both real numbers and arrays.

Replacing a loop like the one above, for computing r[i], by a vector/array expression like sin(x)*cos(x)*exp(-x**2) + 2 + x**2, is called *vectorization*. The loop version is often referred to as *scalar code*. For example,

```
import numpy as np
import math
x = np.zeros(N);  y = np.zeros(N)
dx = 2.0/(N-1) # spacing of x coordinates
for i in range(N):
    x[i] = -1 + dx*i
    y[i] = math.exp(-x[i])*x[i]
```

is scalar code, while the corresponding vectorized version reads

```
x = np.linspace(-1, 1, N)
y = np.exp(-x)*x
```

We remark that list comprehensions,

```
x = array([-1 + dx*i for i in range(N)])
y = array([np.exp(-xi)*xi for xi in x])
```

result in scalar code because we still have explicit, slow Python for loops operating on scalar quantities. The requirement of vectorized code is that there are no explicit Python for loops. The loops required to compute each array element are performed in fast C or Fortran code in the numpy package.

Most Python functions intended for a scalar argument x, like

```
def f(x):
    return x**4*exp(-x)
```

automatically work for an array argument x:

```
x = np.linspace(-3, 3, 101)
y = f(x)
```

provided that the exp function in the definition of f accepts an array argument. This means that exp must have been imported as from numpy import * or explicitly as from numpy import exp. One can, of course, prefix exp as in np.exp, at the loss of a less attractive mathematical syntax in the formula.

When a Python function f(x) works for an array argument x, we say that the function f is vectorized. Provided that the mathematical expressions in f involve arithmetic operations and basic mathematical functions from the math module, f will be automatically vectorized by just importing the functions from numpy instead of math. However, if the expressions inside f involve if tests, the code needs a rewrite to work with arrays. Section 5.4.1 presents examples where we have to do special actions in order to vectorize functions.

Vectorization is very important for speeding up Python programs that perform heavy computations with arrays. Moreover, vectorization gives more compact code that is easier to read. Vectorization is particularly important for statistical simulations, as demonstrated in Chap. 8.

5.3   Curve Plotting

Visualizing a function $f(x)$ is done by drawing the curve $y = f(x)$ in an xy coordinate system. When we use a computer to do this task, we say that we *plot* the curve. Technically, we plot a curve by drawing straight lines between n points on the curve. The more points we use, the smoother the curve appears.

Suppose we want to plot the function $f(x)$ for $a \leq x \leq b$. First we pick out n x coordinates in the interval $[a, b]$, say we name these $x_0, x_1, \ldots, x_{n-1}$. Then we evaluate $y_i = f(x_i)$ for $i = 0, 1, \ldots, n - 1$. The points (x_i, y_i), $i = 0, 1, \ldots, n - 1$, now lie on the curve $y = f(x)$. Normally, we choose the x_i coordinates to be equally spaced, i.e.,

$$x_i = a + ih, \quad h = \frac{b - a}{n - 1}.$$

If we store the x_i and y_i values in two arrays x and y, we can plot the curve by a command like plot(x,y).

Sometimes the names of the independent variable and the function differ from x and f, but the plotting procedure is the same. Our first example of curve plotting demonstrates this fact by involving a function of t.

5.3.1   MATLAB-Style Plotting with Matplotlib

The standard package for curve plotting in Python is Matplotlib. We first exemplify a usage of this package that is very similar with how you plot in MATLAB as many readers will have MATLAB knowledge of will need to operate MATLAB at some point.

A basic plot Let us plot the curve $y = t^2 \exp(-t^2)$ for t values between 0 and 3. First we generate equally spaced coordinates for t, say 51 values (50 intervals). Then we compute the corresponding y values at these points, before we call the plot(t,y) command to make the curve plot. Here is the complete program:

```
from numpy import *
from matplotlib.pyplot import *

def f(t):
    return t**2*exp(-t**2)

t = linspace(0, 3, 51)      # 51 points between 0 and 3
y = zeros(len(t))           # allocate y with float elements
for i in xrange(len(t)):
    y[i] = f(t[i])

plot(t, y)
show()
```

In this program we pre-allocate the y array and fill it with values, element by element, in a Python loop. Alternatively, we may operate on the whole t array at once, which yields faster and shorter code:

```
y = f(t)
```

To include the plot in electronic documents, we need a hardcopy of the figure in PDF, PNG, or another image format. The savefig function saves the plot to files in various image formats:

```
savefig('tmp1.pdf') # produce PDF
savefig('tmp1.png') # produce PNG
```

The filename extension determines the format: .pdf for PDF and .png for PNG. Figure 5.2 displays the resulting plot.

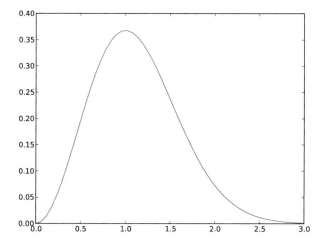

Fig. 5.2 A simple plot in PDF format (Matplotlib)

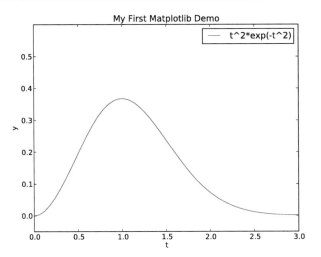

Fig. 5.3 A single curve with label, title, and axis adjusted (Matplotlib)

Decorating the plot The x and y axes in curve plots should have labels, here t and y, respectively. Also, the curve should be identified with a label, or legend as it is often called. A title above the plot is also common. In addition, we may want to control the extent of the axes (although most plotting programs will automatically adjust the axes to the range of the data). All such things are easily added after the `plot` command:

```
plot(t, y)
xlabel('t')
ylabel('y')
legend(['t^2*exp(-t^2)'])
axis([0, 3, -0.05, 0.6])   # [tmin, tmax, ymin, ymax]
title('My First Matplotlib Demo')
savefig('tmp2.pdf')
show()
```

Removing the `show()` call prevents the plot from being shown on the screen, which is advantageous if the program's purpose is to make a large number of plots in PDF or PNG format (you do not want all the plot windows to appear on the screen and then kill all of them manually). This decorated plot is displayed in Fig. 5.3.

Plotting multiple curves A common plotting task is to compare two or more curves, which requires multiple curves to be drawn in the same plot. Suppose we want to plot the two functions $f_1(t) = t^2 \exp(-t^2)$ and $f_2(t) = t^4 \exp(-t^2)$. We can then just issue two `plot` commands, one for each function. To make the syntax resemble MATLAB, we call `hold('on')` after the first `plot` command to indicate that subsequent `plot` commands are to draw the curves in the first plot.

```
def f1(t):
    return t**2*exp(-t**2)
```

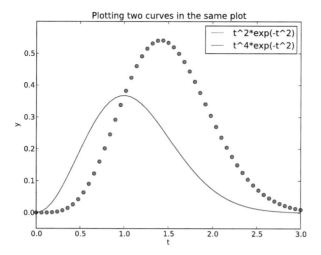

Fig. 5.4 Two curves in the same plot (Matplotlib)

```
def f2(t):
    return t**2*f1(t)

t = linspace(0, 3, 51)
y1 = f1(t)
y2 = f2(t)

plot(t, y1, 'r-')
hold('on')
plot(t, y2, 'bo')
xlabel('t')
ylabel('y')
legend(['t^2*exp(-t^2)', 't^4*exp(-t^2)'])
title('Plotting two curves in the same plot')
show()
```

In these `plot` commands, we have also specified the line type: `r-` means red (`r`) line (`-`), while `bo` means a blue (`b`) circle (`o`) at each data point. Figure 5.4 shows the result. The legends for each curve is specified in a list where the sequence of strings correspond to the sequence of `plot` commands. Doing a `hold('off')` makes the next `plot` command create a new plot.

Placing several plots in one figure We may also put plots together in a figure with `r` rows and `c` columns of plots. The `subplot(r,c,a)` does this, where `a` is a row-wise counter for the individual plots. Here is an example with two rows of plots, and one plot in each row, (see Fig. 5.5):

```
figure()  # make separate figure
subplot(2, 1, 1)
t = linspace(0, 3, 51)
y1 = f1(t)
y2 = f2(t)
```

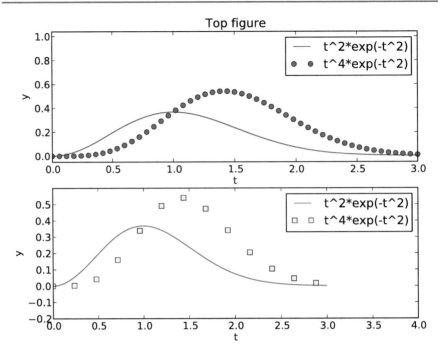

Fig. 5.5 Example on two plots in one figure (Matplotlib)

```
plot(t, y1, 'r-', t, y2, 'bo')
xlabel('t')
ylabel('y')
axis([t[0], t[-1], min(y2)-0.05, max(y2)+0.5])
legend(['t^2*exp(-t^2)', 't^4*exp(-t^2)'])
title('Top figure')

subplot(2, 1, 2)
t3 = t[::4]
y3 = f2(t3)

plot(t, y1, 'b-', t3, y3, 'ys')
xlabel('t')
ylabel('y')
axis([0, 4, -0.2, 0.6])
legend(['t^2*exp(-t^2)', 't^4*exp(-t^2)'])
savefig('tmp4.pdf')
show()
```

The figure() call creates a new plot window on the screen.

All of the examples above on plotting with Matplotlib are collected in the file mpl_pylab_examples.py.

5.3.2   Matplotlib; Pyplot Prefix

The Matplotlib developers do not promote the plotting style we exemplified above. Instead, they recommend to prefix plotting commands by the `matplotlib.pyplot` module and also prefix array computing commands to demonstrate that they come from Numerical Python:

```
import numpy as np
import matplotlib.pyplot as plt
```

The plot in Fig. 5.3 can typically be obtained by prefixing the previously shown plotting commands with `plt`:

```
plt.plot(t, y)
plt.legend(['t^2*exp(-t^2)'])
plt.xlabel('t')
plt.ylabel('y')
plt.axis([0, 3, -0.05, 0.6])   # [tmin, tmax, ymin, ymax]
plt.title('My First Matplotlib Demo')
plt.show()
plt.savefig('tmp2.pdf') # produce PDF
```

Instead of giving plot data and legends separately, it is more common to write

```
plt.plot(t, y, label='t^2*exp(-t^2)')
```

However, in this book we shall stick to the `legend` command since this makes the transition to/from MATLAB easier.

Figure 5.4 can be produced by

```
def f1(t):
    return t**2*np.exp(-t**2)

def f2(t):
    return t**2*f1(t)

t = np.linspace(0, 3, 51)
y1 = f1(t)
y2 = f2(t)

plt.plot(t, y1, 'r-')
plt.plot(t, y2, 'bo')
plt.xlabel('t')
plt.ylabel('y')
plt.legend(['t^2*exp(-t^2)', 't^4*exp(-t^2)'])
plt.title('Plotting two curves in the same plot')
plt.savefig('tmp3.pdf')
plt.show()
```

Putting multiple plots in a figure follows the same set-up with `subplot` as previously shown, except that commands are prefixed by `plt`. The complete example, along with the codes listed above, are found in the file `mpl_pyplot_examples.py`.

Once you have created a basic plot, there are numerous possibilities for fine-tuning the figure, i.e., adjusting tick marks on the axis, inserting text, etc. The Matplotlib website is full of instructive examples on what you can do with this excellent package.

5.3.3   SciTools and Easyviz

Matplotlib has become the *de facto* standard for curve plotting in Python, but there are several other alternative packages, especially if we also consider plotting of 2D/3D scalar and vector fields. Python has interfaces to many leading visualization packages: MATLAB, Gnuplot, Grace, OpenDX, and VTK. Even basic plotting with these packages has very different syntax, and deciding what package and syntax to go with was and still is a challenge. As a response to this challenge, Easyviz was created to provide a common uniform interface to all the mentioned visualization packages (including Matplotlib). The syntax of this interface was made very close to that of MATLAB, since most scientists and engineers have experience with MATLAB or most probably will be using it in some context. (In general, the Python syntax used in the examples in this book is constructed to ease the transition to and from MATLAB.)

Easyviz is part of the SciTools package, which consists of a set of Python tools building on Numerical Python, ScientificPython, the comprehensive SciPy environment, and other packages for scientific computing with Python. SciTools contains in particular software related to the book [13] and the present text. Installation is straightforward as described on the web page https://github.com/hplgit/scitools.

Importing SciTools and Easyviz   A standard import of SciTools is

```
from scitools.std import *
```

The advantage of this statement is that it, with a minimum of typing, imports a lot of useful modules for numerical Python programming: Easyviz for MATLAB-style plotting, all of numpy (from numpy import *), all of scipy (from scipy import *) if available, the StringFunction tool (see Sect. 4.3.3), many mathematical functions and tools in SciTools, plus commonly applied modules such as sys, os, and math. The imported standard mathematical functions (sqrt, sin, asin, exp, etc.) are from numpy.lib.scimath and deal transparently with real and complex input/output (as the corresponding MATLAB functions):

```
>>> from scitools.std import *
>>> a = array([-4., 4])
>>> sqrt(a)                  # need complex output
array([ 0.+2.j,  2.+0.j])
>>> a = array([16., 4])
>>> sqrt(a)                  # can reduce to real output
array([ 4.,  2.])
```

The inverse trigonometric functions have different names in math and numpy, a fact that prevents an expression written for scalars, using math names, to be immediately valid for arrays. Therefore, the `from scitools.std import *` action also imports the names `asin`, `acos`, and `atan` for the `numpy` or `scipy` names `arcsin`, `arccos`, and `arctan` functions, to ease vectorization of mathematical expressions involving inverse trigonometric functions.

The downside of the "star import" from `scitools.std` is twofold. First, it fills up your program or interactive session with the names of several hundred functions. Second, when using a particular function, you do not know the package it comes from. Both problems are solved by doing an import of the type used in Sect. 5.3.2:

```
import scitools.std as st
import numpy as np
```

All of the SciTools and Easyviz functions must then be prefixed by `st`. Although the `numpy` functions are available through the `st` prefix, we recommend using the `np` prefix to clearly see where functionality comes from.

Since the Easyviz syntax for plotting is very close to that of MATLAB, it is also very close to the syntax of Matplotlib shown earlier. This will be demonstrated in the forthcoming examples. The advantage of using Easyviz is that the underlying plotting package, used to create the graphics and known as a *backend*, can trivially be replaced by another package. If users of your Python software have not installed a particular visualization package, the software can still be used with another alternative (which might be considerably easier to install). By default, Easyviz now employs Matplotlib for plotting. Other popular alternatives are Gnuplot and MATLAB. For 2D/3D scalar and vector fields, VTK is a popular backend for Easyviz.

We shall next redo the curve plotting examples from Sect. 5.3.1 using Easyviz syntax.

A basic plot Plotting the curve $y = t^2 \exp(-t^2)$ for $t \in [0, 3]$, using 31 equally spaced points (30 intervals) is performed by like this:

```
from scitools.std import *

def f(t):
    return t**2*exp(-t**2)

t = linspace(0, 3, 31)
y = f(t)
plot(t, y, '-')
```

To save the plot in a file, we use the `savefig` function, which takes the filename as argument:

```
savefig('tmp1.pdf') # produce PDF
savefig('tmp1.eps') # produce PostScript
savefig('tmp1.png') # produce PNG
```

The filename extension determines the format: `.pdf` for PDF, `.ps` or `.eps` for PostScript, and `.png` for PNG. A synonym for the `savefig` function is `hardcopy`.

What if the plot window quickly disappears?

On some platforms, some backends may result in a plot that is shown in just a fraction of a second on the screen before the plot window disappears (the Gnuplot backend on Windows machines and the Matplotlib backend constitute two examples). To make the window stay on the screen, add

```
raw_input('Press the Return key to quit: ')
```

at the end of the program. The plot window is killed when the program terminates, and this statement postpones the termination until the user hits the Return key.

Decorating the plot   Let us plot the same curve, but now with a legend, a plot title, labels on the axes, and specified ranges of the axes:

```
from scitools.std import *

def f(t):
    return t**2*exp(-t**2)

t = linspace(0, 3, 31)
y = f(t)
plot(t, y, '-')
xlabel('t')
ylabel('y')
legend('t^2*exp(-t^2)')
axis([0, 3, -0.05, 0.6])   # [tmin, tmax, ymin, ymax]
title('My First Easyviz Demo')
```

Easyviz has also introduced a more Pythonic `plot` command where all the plot properties can be set at once through keyword arguments:

```
plot(t, y, '-',
     xlabel='t',
     ylabel='y',
     legend='t^2*exp(-t^2)',
     axis=[0, 3, -0.05, 0.6],
     title='My First Easyviz Demo',
     savefig='tmp1.pdf',
     show=True)
```

With `show=False` one can avoid the plot window on the screen and just make the plot file.

Note that we in the curve legend write t square as t^2 (LATEX style) rather than t**2 (program style). Whichever form you choose is up to you, but the LATEX form sometimes looks better in some plotting programs (Matplotlib and Gnuplot are two examples).

Plotting multiple curves   Next we want to compare the two functions $f_1(t) = t^2 \exp(-t^2)$ and $f_2(t) = t^4 \exp(-t^2)$. Writing two plot commands after each other makes two separate plots. To make the second curve appear together with the first one, we need to issue a `hold('on')` call after the first `plot` command. All subsequent `plot` commands will then draw curves in the same plot, until `hold('off')` is called.

```
from scitools.std import *

def f1(t):
    return t**2*exp(-t**2)

def f2(t):
    return t**2*f1(t)

t = linspace(0, 3, 51)
y1 = f1(t)
y2 = f2(t)

plot(t, y1, 'r-')
hold('on')
plot(t, y2, 'b-')

xlabel('t')
ylabel('y')
legend('t^2*exp(-t^2)', 't^4*exp(-t^2)')
title('Plotting two curves in the same plot')
savefig('tmp3.pdf')
```

The sequence of the multiple legends is such that the first legend corresponds to the first curve, the second legend to the second curve, and so forth.

Instead of separate calls to `plot` and the use of `hold('on')`, we can do everything at once and just send several curves to `plot`:

```
plot(t, y1, 'r-', t, y2, 'b-', xlabel='t', ylabel='y',
     legend=('t^2*exp(-t^2)', 't^4*exp(-t^2)'),
     title='Plotting two curves in the same plot',
     savefig='tmp3.pdf')
```

Throughout this book, we very often make use of this type of compact `plot` command, which also only requires an import of the form `from scitools.std import plot`.

Changing backend   Easyviz applies Matplotlib for plotting by default, so the resulting figures so far will be similar to those of Fig. 5.2–5.4.

However, we can use other backends (plotting packages) for creating the graphics. The specification of what package to use is defined in a configuration file (see the heading *Setting Parameters in the Configuration File* in the Easyviz documentation), or on the command line:

Terminal

Terminal> python myprog.py --SCITOOLS_easyviz_backend gnuplot

---

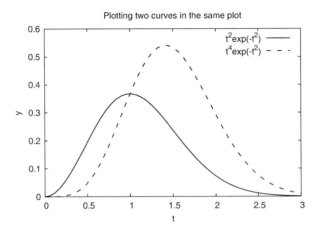

Fig. 5.6 Two curves in the same plot (Gnuplot)

Now, the plotting commands in `myprog.py` will make use of Gnuplot to create the graphics, with a slightly different result than that created by Matplotlib (compare Figs. 5.4 and 5.6). A nice feature of Gnuplot is that the line types are automatically changed if we save a figure to file, such that the lines are easily distinguishable in a black-and-white plot. With Matplotlib one has to carefully set the line types to make them effective on a grayscale.

Placing several plots in one figure Finally, we redo the example from Sect. 5.3.1 where two plots are combined into one figure, using the `subplot` command:

```
figure()
subplot(2, 1, 1)
t = linspace(0, 3, 51)
y1 = f1(t)
y2 = f2(t)
plot(t, y1, 'r-', t, y2, 'bo', xlabel='t', ylabel='y',
     legend=('t^2*exp(-t^2)', 't^4*exp(-t^2)'),
     axis=[t[0], t[-1], min(y2)-0.05, max(y2)+0.5],
     title='Top figure')

subplot(2, 1, 2)
t3 = t[::4]
y3 = f2(t3)

plot(t, y1, 'b-', t3, y3, 'ys',
     xlabel='t', ylabel='y',
     axis=[0, 4, -0.2, 0.6],
     legend=('t^2*exp(-t^2)', 't^4*exp(-t^2)'))
savefig('tmp4.pdf')
```

Note that `figure()` must be used if you want a program to make different plot windows on the screen: each `figure()` call creates a new, separate plot.

All of the Easyviz examples above are found in the file `easyviz_examples.py`. We remark that Easyviz is just a thin layer of code providing access to the most common plotting functionality for curves as well as 2D/3D scalar and vector fields. Fine-tuning of plots, e.g., specifying tick marks on the axes, is not supported, simply because most of the curve plots in the daily work can be made without such functionality. For fine-tuning the plot with special commands, you need to grab an object in Easyviz that communicates directly with the underlying plotting package used to create the graphics. With this object you can issue package-specific commands and do whatever the underlying package allows you do. This is explained in the Easyviz manual[3], which also comes up by running `pydoc scitools.easyviz`. As soon as you have digested the very basics of plotting, you are strongly recommend to read through the curve plotting part of the Easyviz manual.

5.3.4 Making Animations

A sequence of plots can be combined into an animation on the screen and stored in a video file. The standard procedure is to generate a series of individual plots and to show them in sequence to obtain an animation effect. Plots store in files can be combined to a video file.

Example The function

$$f(x; m, s) = (2\pi)^{-1/2} s^{-1} \exp\left[-\frac{1}{2}\left(\frac{x-m}{s}\right)^2\right]$$

is known as the Gaussian function or the probability density function of the normal (or Gaussian) distribution. This bell-shaped function is wide for large s and peak-formed for small s, see Fig. 5.7. The function is symmetric around $x = m$ ($m = 0$ in the figure). Our goal is to make an animation where we see how this function evolves as s is decreased. In Python we implement the formula above as a function `f(x, m, s)`.

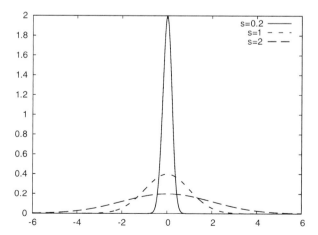

Fig. 5.7 Different shapes of a Gaussian function

---
[3] https://scitools.googlecode.com/hg/doc/easyviz/easyviz.html

The animation is created by varying s in a loop and for each s issue a `plot` command. A moving curve is then visible on the screen. One can also make a video that can be played as any other computer movie using a standard movie player. To this end, each plot is saved to a file, and all the files are combined together using some suitable tool to be explained later. Before going into programming detail there is one key point to emphasize.

Keep the extent of axes fixed during animations!

The underlying plotting program will normally adjust the axis to the maximum and minimum values of the curve if we do not specify the axis ranges explicitly. For an animation such automatic axis adjustment is misleading – any axis range must be fixed to avoid a jumping axis.

The relevant values for the y axis range in the present example is the minimum and maximum value of f. The minimum value is zero, while the maximum value appears for $x = m$ and increases with decreasing s. The range of the y axis must therefore be $[0, f(m; m, \min s)]$.

The function f is defined for all $-\infty < x < \infty$, but the function value is very small already $3s$ away from $x = m$. We may therefore limit the x coordinates to $[m - 3s, m + 3s]$.

Animation in Easyviz We start with using Easyviz for animation since this is almost identical to making standard static plots, and you can choose the plotting engine you want to use, say Gunplot or Matplotlib. The Easyviz recipe for animating the Gaussian function as s goes from 2 to 0.2 looks as follows.

```
from scitools.std import sqrt, pi, exp, linspace, plot, movie
import time

def f(x, m, s):
    return (1.0/(sqrt(2*pi)*s))*exp(-0.5*((x-m)/s)**2)

m = 0
s_min = 0.2
s_max = 2
x = linspace(m -3*s_max, m + 3*s_max, 1000)
s_values = linspace(s_max, s_min, 30)
# f is max for x=m; smaller s gives larger max value
max_f = f(m, m, s_min)

# Show the movie on the screen
# and make hardcopies of frames simultaneously.

counter = 0
for s in s_values:
    y = f(x, m, s)
    plot(x, y, '-', axis=[x[0], x[-1], -0.1, max_f],
        xlabel='x', ylabel='f', legend='s=%4.2f' % s,
        savefig='tmp%04d.png' % counter)
    counter += 1
    #time.sleep(0.2)  # can insert a pause to control movie speed
```

Note that the s values are decreasing (`linspace` handles this automatically if the start value is greater than the stop value). Also note that we, simply because we think it is visually more attractive, let the y axis go from -0.1 although the f function is always greater than zero. The complete code is found in the file `movie1.py`.

Notice

It is crucial to use the single, compound `plot` command shown above, where axis, labels, legends, etc., are set in the same call. Splitting up in individual calls to `plot`, `axis`, and so forth, results in jumping curves and axis. Also, when visualizing more than one animated curve at a time, make sure you send all data to a single `plot` command.

Remark on naming plot files

For each frame (plot) in the movie we store the plot in a file, with the purpose of combining all the files to an ordinary video file. The different files need different names such that various methods for listing the files will list them in the correct order. To this end, we recommend using filenames of the form `tmp0001.png`, `tmp0002.png`, `tmp0003.png`, etc. The printf format 04d pads the integers with zeros such that 1 becomes 0001, 13 becomes 0013 and so on. The expression `tmp*.png` will now expand (by an alphabetic sort) to a list of all files in proper order.

Without the padding with zeros, i.e., names of the form `tmp1.png`, `tmp2.png`, ..., `tmp12.png`, etc., the alphabetic order will give a wrong sequence of frames in the movie. For instance, `tmp12.png` will appear before `tmp2.png`.

Basic animation in Matplotlib Animation is Matplotib requires more than a loop over a parameter and making a plot inside the loop. The set-up that is closest to standard static plots is shown first, while the newer and more widely used tool `FuncAnimation` is explained afterwards.

The first part of the program, where we define `f`, `x`, `s_values`, and so forth, is the same regardless of the animation technique. Therefore, we concentrate on the graphics part here:

```
import matplotlib.pyplot as plt
...
# Make a first plot
plt.ion()
y = f(x, m, s_max)
lines = plt.plot(x, y)
plt.axis([x[0], x[-1], -0.1, max_f])
plt.xlabel('x')
plt.ylabel('f')

# Show the movie, and make hardcopies of frames simulatenously
counter = 0
for s in s_values:
    y = f(x, m, s)
    lines[0].set_ydata(y)
```

```
plt.legend(['s=%4.2f' % s])
plt.draw()
plt.savefig('tmp_%04d.png' % counter)
counter += 1
```

The `plt.ion()` call is important, so is the first plot, where we grab the result of the plot command, which is a list of Matplotlib's Line2D objects. The idea is then to update the data via `lines[0].set_ydata` and show the plot via `plt.draw()` for each frame. For multiple curves we must update the y data for each curve, e.g.,

```
lines = plot(x, y1, x, y2, x, y3)

for parameter in parameters:
    y1 = ...
    y2 = ...
    y3 = ...
    for line, y in zip(lines, [y1, y2, y3]):
        line.set_ydata(y)
    plt.draw()
```

The file `movie1_mpl1.py` contains the complete program for doing animation with native Matplotlib syntax.

Using FuncAnimation in Matplotlib The recommended approach to animation in Matplotlib is to use the `FuncAnimation` tool:

```
import matplotlib.pyplot as plt
from matplotlib.animation import animation

anim = animation.FuncAnimation(
    fig, frame, all_args, interval=150, init_func=init, blit=True)
```

Here, `fig` is the `plt.figure()` object for the current figure, `frame` is a user-defined function for plotting each frame, `all_args` is a list of arguments for `frame`, `interval` is the delay in ms between each frame, `init_func` is a function called for defining the background plot in the animation, and `blit=True` speeds up the animation. For frame number i, `FuncAnimation` will call `frame(all_args[i])`. Hence, the user's task is mostly to write the `frame` function and construct the `all_args` arguments.

After having defined m, s_max, s_min, s_values, and max_f as shown earlier, we have to make a first plot:

```
fig = plt.figure()
plt.axis([x[0], x[-1], -0.1, max_f])
lines = plt.plot([], [])
plt.xlabel('x')
plt.ylabel('f')
```

Notice that we save the return value of `plt.plot` in `lines` such that we can conveniently update the data for the curve(s) in each frame.

The function for defining a background plot draws an empty plot in this example:

```
def init():
    lines[0].set_data([], [])   # empty plot
    return lines
```

The function that defines the individual plots in the animation basically computes y from f and updates the data of the curve:

```
def frame(args):
    frame_no, s, x, lines = args
    y = f(x, m, s)
    lines[0].set_data(x, y)
    return lines
```

Multiple curves can be updated as shown earlier.

We are now ready to call `animation.FuncAnimation`:

```
anim = animation.FuncAnimation(
    fig, frame, all_args, interval=150, init_func=init, blit=True)
```

A common next action is to make a video file, here in the MP4 format with 5 frames per second:

```
anim.save('movie1.mp4', fps=5)    # movie in MP4 format
```

Finally, we must `plt.show()` as always to watch any plots on the screen.

The video making requires additional software on the computer, such as `ffmpeg`, and can fail. One gets more control over the potentially fragile movie making process by explicitly saving plots to file and explicitly running movie making programs like `ffmeg` later. Such programs are explained in Sect. 5.3.5.

The complete code showing the basic use of `FuncAnimation` is available in `movie1_FuncAnimation.py`. There is also a MATLAB Animation Tutorial[4] with more basic information, plus a set of animation examples on http://matplotlib.org/examples.

Remove old plot files!

We strongly recommend removing previously generated plot files before a new set of files is made. Otherwise, the movie may get old and new files mixed up. The following Python code removes all files of the form `tmp*.png`:

```
import glob, os
for filename in glob.glob('tmp*.png'):
    os.remove(filename)
```

These code lines should be inserted at the beginning of programs or functions performing animations.

---

[4] http://jakevdp.github.io/blog/2012/08/18/matplotlib-animation-tutorial/

Instead of deleting the individual plotfiles, one may store all plot files in a subfolder and later delete the subfolder. Here is a suitable code segment:

```
import shutil, os
subdir = 'temp'              # subfolder name for plot files
if os.path.isdir(subdir):    # does the subfolder already exist?
    shutil.rmtree(subdir)    # delete the whole folder
os.mkdir(subdir)             # make new subfolder
os.chdir(subdir)             # move to subfolder
# ... perform all the plotting, make movie ...
os.chdir(os.pardir)          # optional: move up to parent folder
```

Note that Python and many other languages use the word directory instead of folder. Consequently, the name of functions dealing with folders have a name containing dir for directory.

5.3.5   Making Videos

Suppose we have a set of frames in an animation, saved as plot files tmp_*.png. The filenames are generated by the printf syntax 'tmp_%04d.png' % i, using a frame counter i that goes from 0 to some value. The corresponding files are then tmp_0000.png, tmp_0001.png, tmp_0002.png, and so on. Several tools can be used to create videos in common formats from the individual frames in the plot files.

Animated GIF file   The ImageMagick[5] software suite contains a program convert for making animated GIF files:

```
 ─────────────────────────────── Terminal ───────────────────────────────
Terminal> convert -delay 50 tmp_*.png movie.gif
```

The delay between frames, here 50, is measured in units of $1/100$ s. The resulting animated GIF file movie.gif can be viewed by another program in the ImageMagick suite: animate movie.gif, but the most common way of displaying animated GIF files is to include them in web pages. Writing the HTML code

```
<img src="movie.gif">
```

in some file with extension .html and loading this file into a web browser will play the movie repeatedly. You may try this out online[6].

MP4, Ogg, WebM, and Flash videos   The modern video formats that are best suited for being displayed in web browsers are MP4, Ogg, WebM, and Flash. The program ffmpeg, or the almost equivalent avconv, is a common tool to create such movies. Creating a flash video is done by

---

[5] http://www.imagemagick.org/
[6] http://hplgit.github.io/scipro-primer/video/gaussian.html

```
                                    ┌──────────┐
──────────────────────────────────┤ Terminal ├──────────────────────────────
                                    └──────────┘
Terminal> ffmpeg -i tmp_%04d.png -r 5 -vcodec flv movie.flv
──────────────────────────────────────────────────────────────────────────────
```

The -i option specifies the printf string that was used to make the names of the individual plot files, -r specifies the number of frames per second, here 5, -vcodec is the video codec for Flash, which is called flv, and the final argument is the name of the video file. On Debian Linux systems, such as Ubuntu, you use the avconv program instead of ffmpeg.

Other formats are created in the same way, but we need to specify the codec and use the right extension in the video file:

Format	Codec and filename
Flash	-vcodec flv movie.flv
MP4	-vcodec libx264 movie.mp4
Webm	-vcodec libvpx movie.webm
Ogg	-vcodec libtheora movie.ogg

Video files are normally trivial to play in graphical file browser: double lick the filename or right-click and choose a player. On Linux systems there are several players that can be run from the command line, e.g., vlc, mplayer, gxine, and totem.

It is easy to create the video file from a Python program since we can run any operating system command in (e.g.) os.system:

```
cmd = 'convert -delay 50 tmp_*.png movie.gif'
os.system(cmd)
```

It might happen that your downloaded and installed version of ffmpeg fails to generate videos in some of the mentioned formats. The reason is that ffmpeg depends on many other packages that may be missing on your system. Getting ffmpeg to work with the libx264 codec for making MP4 files is often challenging. On Debian-based Linux systems, such as Ubuntu, the installation procedure at the time of this writing goes like

```
                                    ┌──────────┐
──────────────────────────────────┤ Terminal ├──────────────────────────────
                                    └──────────┘
Terminal> sudo apt-get install lib-avtools libavcodec-extra-53 \
          libx264-dev
──────────────────────────────────────────────────────────────────────────────
```

5.3.6 Curve Plots in Pure Text

Sometimes it can be desirable to show a graph in pure ASCII text, e.g., as part of a trial run of a program included in the program itself, or a graph that can be illustrative in a doc string. For such purposes we have slightly extended a module by Imri Goldberg (aplotter.py) and included it as a module in SciTools. Running pydoc on scitools.aplotter describes the capabilities of this type of primitive plotting. Here we just give an example of what it can do:

```
>>> import numpy as np
>>> x = np.linspace(-2, 2, 81)
>>> y = np.exp(-0.5*x**2)*np.cos(np.pi*x)
>>> from scitools.aplotter import plot
>>> plot(x, y)
```

```
                                |
                              -+1
                             // |\
                            /  |  \
                           /   |   \
                          /    |    \
                         /     |     \
                        /      |      \
                       /       |       \
        -------\      /        |        \
  ---+-------\-----------------/---------+--------\----------------/
    -2         \              /          |          \              /
                \            /           |            \           //
                 \          /            |             \          /
                  \        /             |              \        /
                   \      /              |               \      /
                    \    /               |                \    /
                     \  //               |                 \-  //
                      ----               -0.63               ---/
```

5.4 Plotting Difficulties

The previous examples on plotting functions demonstrate how easy it is to make graphs. Nevertheless, the shown techniques might easily fail to plot some functions correctly unless we are careful. Next we address two types of difficult functions: piecewisely defined functions and rapidly varying functions.

5.4.1 Piecewisely Defined Functions

A piecewisely defined function has different function definitions in different intervals along the x axis. The resulting function, made up of pieces, may have discontinuities in the function value or in derivatives. We have to be very careful when plotting such functions, as the next two examples will show. The problem is that the plotting mechanism draws straight lines between coordinates on the function's curve, and these straight lines may not yield a satisfactory visualization of the function. The first example has a discontinuity in the function itself at one point, while the other example has a discontinuity in the derivative at three points.

Example: The Heaviside function Let us plot the Heaviside function

$$H(x) = \begin{cases} 0, & x < 0 \\ 1, & x \geq 0 \end{cases}$$

The most natural way to proceed is first to define the function as

```
def H(x):
    return (0 if x < 0 else 1)
```

The standard plotting procedure where we define a coordinate array x and call y = H(x) will not work for array arguments x, of reasons to be explained in Sect. 5.5.2. However, we may use techniques from that chapter to create a function Hv(x) that works for array arguments. Even with such a function we face difficulties with plotting it.

Since the Heaviside function consists of two flat lines, one may think that we do not need many points along the x axis to describe the curve. Let us try with nine points:

```
x = np.linspace(-10, 10, 9)
from scitools.std import plot
plot(x, Hv(x), axis=[x[0], x[-1], -0.1, 1.1])
```

However, so few x points are not able to describe the jump from 0 to 1 at $x = 0$, as shown by the solid line in Fig. 5.8 (left). Using more points, say 50 between -10 and 10,

```
x2 = np.linspace(-10, 10, 50)
plot(x, Hv(x), 'r', x2, Hv(x2), 'b',
     legend=('5 points', '50 points'),
     axis=[x[0], x[-1], -0.1, 1.1])
```

makes the curve look better. However, the step is still not strictly vertical. More points will improve the situation. Nevertheless, the best is to draw two flat lines directly: from $(-10, 0)$ to $(0, 0)$, then to $(0, 1)$ and then to $(10, 1)$:

```
plot([-10, 0, 0, 10], [0, 0, 1, 1],
     axis=[x[0], x[-1], -0.1, 1.1])
```

The result is shown in Fig. 5.8 (right).

Some will argue that the plot of $H(x)$ should not contain the vertical line from $(0, 0)$ to $(0, 1)$, but only two horizontal lines. To make such a plot, we must draw two distinct curves, one for each horizontal line:

```
plot([-10,0], [0,0], 'r-', [0,10], [1,1], 'r-',
     axis=[x[0], x[-1], -0.1, 1.1])
```

Observe that we must specify the same line style for both lines (curves), otherwise they would by default get different colors on the screen or different line types in a hardcopy of the plot. We remark, however, that discontinuous functions like $H(x)$ are often visualized with vertical lines at the jumps, as we do in Fig. 5.8b.

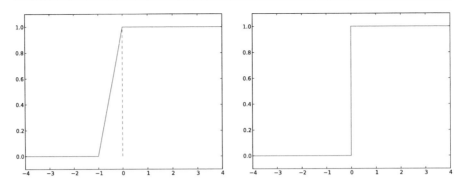

Fig. 5.8 Plot of the Heaviside function using 9 equally spaced x points (*left*) and with a double point at $x = 0$ (*right*)

Example: A hat function Let us plot the hat function $N(x)$, shown as the solid line in Fig. 5.9. This function is a piecewise linear function. The implementation of $N(x)$ must use if tests to locate where we are along the x axis and then evaluate the right linear piece of $N(x)$. A straightforward implementation with plain if tests does not work with array arguments x, but Sect. 5.5.3 explains how to make a vectorized version Nv(x) that works for array arguments as well. Anyway, both the scalar and the vectorized versions face challenges when it comes to plotting.

A first approach to plotting could be

```
x = np.linspace(-2, 4, 6)
plot(x, Nv(x), 'r', axis=[x[0], x[-1], -0.1, 1.1])
```

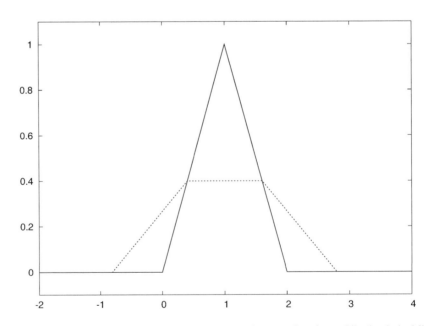

Fig. 5.9 Plot of a hat function. The solid line shows the exact function, while the dashed line arises from using inappropriate points along the x axis

This results in the dashed line in Fig. 5.9. What is the problem? The problem lies in the computation of the x vector, which does not contain the points $x = 1$ and $x = 2$ where the function makes significant changes. The result is that the hat is flattened. Making an x vector with all critical points in the function definitions, $x = 0, 1, 2$, provides the necessary remedy, either

```
x = np.linspace(-2, 4, 7)
```

or the simple

```
x = [-2, 0, 1, 2, 4]
```

Any of these x alternatives and a `plot(x, Nv(x))` will result in the solid line in Fig. 5.9, which is the correct visualization of the $N(x)$ function.

As in the case of the Heaviside function, it is perhaps best to drop using vectorized evaluations and just draw straight lines between the critical points of the function (since the function is linear):

```
x = [-2, 0, 1, 2, 4]
y = [N(xi) for xi in x]
plot(x, y, 'r', axis=[x[0], x[-1], -0.1, 1.1])
```

5.4.2 Rapidly Varying Functions

Let us now visualize the function $f(x) = \sin(1/x)$, using 10 and 1000 points:

```
def f(x):
    return sin(1.0/x)

from scitools.std import linspace, plot
x1 = linspace(-1, 1, 10)
x2 = linspace(-1, 1, 1000)
plot(x1, f(x1), label='%d points' % len(x))
plot(x2, f(x2), label='%d points' % len(x))
```

The two plots are shown in Fig. 5.10. Using only 10 points gives a completely wrong picture of this function, because the function oscillates faster and faster as we approach the origin. With 1000 points we get an impression of these oscillations, but the accuracy of the plot in the vicinity of the origin is still poor. A plot with 100000 points has better accuracy, in principle, but the extremely fast oscillations near the origin just drowns in black ink (you can try it out yourself).

Another problem with the $f(x) = \sin(1/x)$ function is that it is easy to define an x vector containing $x = 0$, such that we get division by zero. Mathematically, the $f(x)$ function has a singularity at $x = 0$: it is difficult to define $f(0)$, so one should exclude this point from the function definition and work with a domain $x \in [-1, -\epsilon] \cup [\epsilon, 1]$, with ϵ chosen small.

The lesson learned from these examples is clear. You must investigate the function to be visualized and make sure that you use an appropriate set of x coordinates along the curve. A relevant first step is to double the number of x coordinates

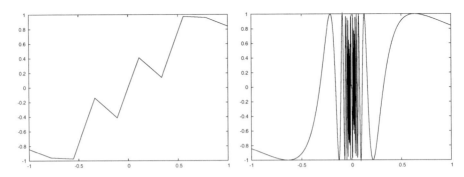

Fig. 5.10 Plot of the function $\sin(1/x)$ with 10 points (*left*) and 1000 points (*right*)

and check if this changes the plot. If not, you probably have an adequate set of x coordinates.

5.5 More Advanced Vectorization of Functions

So far we have seen that vectorization of a Python function f(x) implementing some mathematical function $f(x)$ seems trivial: f(x) works right away with an array argument x and, in that case, returns an array where f is applied to each element in x. When the expression for $f(x)$ is given in terms of a string and the StringFunction tool is used to generate the corresponding Python function f(x), one extra step must be performed to vectorize the Python function. This step is explained in Sect. 5.5.1.

The described vectorization works well as long as the expression $f(x)$ is a mathematical formula without any if test. As soon as we have if tests (conditional mathematical expressions) the vectorization becomes more challenging. Some useful techniques are explained through two examples in Sects. 5.5.2 and 5.5.3. The described techniques are considered advanced material and only necessary when the time spent on evaluating a function at a very large set of points needs to be significantly decreased.

5.5.1 Vectorization of StringFunction Objects

The StringFunction object from scitools.std can convert a formula, available as a string, to a callable Python function (see Sect. 4.3.3). However, the function cannot work with array arguments unless we explicitly tell the StringFunction object to do so. The recipe is very simple. Say f is some StringFunction object. To allow array arguments we are required to call f.vectorize(globals()) once:

```
from numpy import *
x = linspace(0, 1, 30)
# f(x) will in general not work

f.vectorize(globals())
values = f(x)            # f works with array arguments
```

It is important that you import everything from numpy (or scitools.std) *before* calling f.vectorize, exactly as shown.

You may take the f.vectorize call as a magic recipe. Still, some readers want to know what problem f.vectorize solves. Inside the StringFunction module we need to have access to mathematical functions for expressions like sin(x)*exp(x) to be evaluated. These mathematical functions are by default taken from the math module and hence they do not work with array arguments. If the user, in the main program, has imported mathematical functions that work with array arguments, these functions are registered in a dictionary returned from globals(). By the f.vectorize call we supply the StringFunction module with the user's global namespace so that the evaluation of the string expression can make use of the mathematical functions for arrays from the user's program. Unless you use np.sin(x)*np.cos(x) etc. in the string formulas, make sure you do a from numpy import * so that the function names are defined without any prefix.

Even after calling f.vectorize(globals()), a StringFunction object may face problems with vectorization. One example is a piecewise constant function as specified by a string expression '1 if x > 2 else 0'. Section 5.5.2 explains why if tests fail for arrays and what the remedies are.

5.5.2   Vectorization of the Heaviside Function

We consider the widely used Heaviside function defined by

$$H(x) = \begin{cases} 0, & x < 0 \\ 1, & x \geq 0 \end{cases}$$

The most compact way if implementing this function is

```
def H(x):
    return (0 if x < 0 else 1)
```

Trying to call H(x) with an array argument x fails:

```
>>> def H(x): return (0 if x < 0 else 1)
...
>>> import numpy as np
>>> x = np.linspace(-10, 10, 5)
>>> x
array([-10.,  -5.,   0.,   5.,  10.])
>>> H(x)
...
ValueError: The truth value of an array with more than
one element is ambiguous. Use a.any() or a.all()
```

The problem is related to the test x < 0, which results in an array of boolean values, while the if test needs a single boolean value (essentially taking bool(x < 0)):

```
>>> b = x < 0
>>> b
array([ True,   True, False, False, False], dtype=bool)
>>> bool(b)   # evaluate b in a boolean context
...
ValueError: The truth value of an array with more than
one element is ambiguous. Use a.any() or a.all()
>>> b.any()   # True if any element in b is True
True
>>> b.all()   # True if all elements in b are True
False
```

The any and all calls do not help us since we want to take actions element by element depending on whether x[i] < 0 or not.

There are four ways to find a remedy to our problems with the if x < 0 test: (i) we can write an explicit loop for computing the elements, (ii) we can use a tool for automatically vectorize H(x), (iii) we can mix boolean and floating-point calculations, or (iv) we can manually vectorize the H(x) function. All four methods will be illustrated next.

Loop The following function works well for arrays if we insert a simple loop over the array elements (such that H(x) operates on scalars only):

```
def H_loop(x):
    r = np.zeros(len(x))
    for i in xrange(len(x)):
        r[i] = H(x[i])
    return r

# Example:
x = np.linspace(-5, 5, 6)
y = H_loop(x)
```

Automatic vectorization Numerical Python contains a method for automatically vectorizing a Python function H(x) that works with scalars (pure numbers) as x argument:

```
import numpy as np
H_vec = np.vectorize(H)
```

The H_vec(x) function will now work with vector/array arguments x. Unfortunately, such automatically vectorized functions runs at a fairly slow speed compared to the implementations below (see the end of Sect. 5.5.3 for specific timings).

Mixing boolean and floating-point calculations It appears that a very simple solution to vectorizing the H(x) function is to implement it as

```
def H(x):
    return x >= 0
```

The return value is now a `bool` object, not an `int` or `float` as we would math-ematically expect to be the proper type of the result. However, the `bool` object works well in both scalar and vectorized operations as long as we involve the re-turned `H(x)` in some arithmetic expression. The `True` and `False` values are then interpreted as 1 and 0. Here is a demonstration:

```
>>> x = np.linspace(-1, 1, 5)
>>> H(x)
array([False, False,  True,  True,  True], dtype=bool)
>>> 1*H(x)
array([0, 0, 1, 1, 1])
>>> H(x) - 2
array([-2, -2, -1, -1, -1])
>>>
>>> x = 0.2    # test scalar argument
>>> H(x)
True
>>> 1*H(x)
1
>>> H(x) - 2
-1
```

If returning a boolean value is considered undesirable, we can turn the `bool` object into the proper type by

```
def H(x):
    r = x >= 0
    if isinstance(x, (int,float)):
        return int(r)
    elif isinstance(x, np.ndarray):
        return np.asarray(r, dtype=np.int)
```

Manual vectorization By manual vectorization we normally mean translating the algorithm into a set of calls to functions in the `numpy` package such that no loops are visible in the Python code. The last version of the `H(x)` is a manual vectorization, but now we shall look at a more general technique when the result is not necessarily 0 or 1. In general, manual vectorization is non-trivial and requires knowledge of and experience with the underlying library for array computations. Fortunately, there is a simple `numpy` recipe for turning functions of the form

```
def f(x):
    if condition:
        r = <expression1>
    else:
        r = <expression2>
    return r
```

into vectorized form:

```
def f_vectorized(x):
    x1 = <expression1>
    x2 = <expression2>
```

```
    r = np.where(condition, x1, x2)
    return r
```

The `np.where` function returns an array of the same length as `condition`, whose element no. i equals `x1[i]` if `condition[i]` is True, and `x2[i]` otherwise. With Python loops we can express this principle as

```
def my_where(condition, x1, x2):
    r = np.zeros(len(condition))   # result
    for i in xrange(condition):
        r[i] = x1[i] if condition[i] else x2[i]
    return r
```

The `x1` and `x2` variables can be pure numbers too in the call to `np.where`.

In our case we can use the `np.where` function as follows:

```
def Hv(x):
    return np.where(x < 0, 0.0, 1.0)
```

Instead of using `np.where` we can apply *boolean indexing*. The idea is that an array a allows to be indexed by an array b of boolean values: `a[b]`. The result `a[b]` is a new array with all the elements `a[i]` where `b[i]` is True:

```
>>> a
array([ 0. ,   2.5,   5. ,   7.5,  10. ])
>>> b = a > 5
>>> b
array([False, False, False,  True,  True], dtype=bool)
>>> a[b]
array([ 7.5,  10. ])
```

We can assign new values to the elements in a where b is True:

```
>>> a[b]
array([ 7.5,  10. ])
>>> a[b] = np.array([-10, -20], dtype=np.float)
>>> a
array([ 0. ,   2.5,   5. , -10. , -20. ])
>>> a[b] = -4
>>> a
array([ 0. ,   2.5,   5. ,  -4. ,  -4. ])
```

To implement the Heaviside function, we start with an array of zeros and then assign 1 to the elements where x >= 0:

```
def Hv(x):
    r = np.zeros(len(x), dtype=np.int)
    r[x >= 0] = 1
    return r
```

5.5.3   Vectorization of a Hat Function

We now turn the attention to the hat function $N(x)$ defined by

$$N(x) = \begin{cases} 0, & x < 0 \\ x, & 0 \le x < 1 \\ 2 - x, & 1 \le x < 2 \\ 0, & x \ge 2 \end{cases}$$

The corresponding Python implementation N(x) is

```
def N(x):
    if x < 0:
        return 0.0
    elif 0 <= x < 1:
        return x
    elif 1 <= x < 2:
        return 2 - x
    elif x >= 2:
        return 0.0
```

Unfortunately, this N(x) function does not work with array arguments x, because the boolean expressions, like x < 0, are arrays and they cannot yield a single True or False value for the if tests, as explained in Sect. 5.5.2.

The simplest remedy is to use np.vectorize from Sect. 5.5.2:

```
N_vec = np.vectorize(N)
```

It is then important that N(x) returns float and not int values, otherwise the vectorized version will produce int values and hence be incorrect.

A manual rewrite, yielding a faster vectorized function, is more demanding than for the Heaviside function because we now have multiple branches in the if test. One sketch is to replace

```
if condition1:
    r = <expression1>
elif condition2:
    r = <expression2>
elif condition3:
    r = <expression3>
else:
    r = <expression4>
```

by

```
x1 = <expression1>
x2 = <expression2>
x3 = <expression3>
x4 = <expression4>
r = np.where(condition1, x1, x4)   # initialize with "else" expr.
r = np.where(condition2, x2, r)
r = np.where(condition3, x3, r)
```

Alternatively, we can use boolean indexing. Assuming that <expressionX> is some expression involving an array x and coded as a Python function fX(x) (X is 1, 2, 3, or 4), we can write

```
r = f4(x)
r[condition1] = f1(x[condition1])
r[condition2] = f2(x[condition2])
r[condition3] = f2(x[condition3])
```

Specifically, when the function for scalar arguments x reads

```
def N(x):
    if x < 0:
        return 0.0
    elif 0 <= x < 1:
        return x
    elif 1 <= x < 2:
        return 2 - x
    elif x >= 2:
        return 0.0
```

a vectorized attempt would be

```
def Nv(x):
    r = np.where(x < 0,         0.0,  0.0)
    r = np.where(0 <= x < 1, x,      r)
    r = np.where(1 <= x < 2, 2-x,    r)
    r = np.where(x >= 2,        0.0,  r)
    return r
```

The first and last line are not strictly necessary as we could just start with a zero vector (making the insertion of zeros for the first and last condition a redundant operation).

However, any condition like 0 <= x < 1, which is equivalent to 0 <= x and x < 1, does not work because the and operator does not work with array arguments. Fortunately, there is a simple solution to this problem: the function logical_and in numpy. A working Nv function must apply logical_and instead in each condition:

```
def Nv1(x):
    condition1 = x < 0
    condition2 = np.logical_and(0 <= x, x < 1)
    condition3 = np.logical_and(1 <= x, x < 2)
    condition4 = x >= 2

    r = np.where(condition1, 0.0, 0.0)
    r = np.where(condition2, x,   r)
    r = np.where(condition3, 2-x, r)
    r = np.where(condition4, 0.0, r)
    return r
```

With boolean indexing we get the alternative form

```
def Nv2(x):
    condition1 = x < 0
    condition2 = np.logical_and(0 <= x, x < 1)
    condition3 = np.logical_and(1 <= x, x < 2)
    condition4 = x >= 2

    r = np.zeros(len(x))
    r[condition1] = 0.0
    r[condition2] = x[condition2]
    r[condition3] = 2-x[condition3]
    r[condition4] = 0.0
    return r
```

Again, the first and last assignment to r can be omitted in this special case where we start with a zero vector.

The file hat.py implements four vectorized versions of the N(x) function: N_loop, which is a plain loop calling up N(x) for each x[i] element in the array x; N_vec, which is the result of automatic vectorization via np.vectorize; the Nv1 function shown above, which uses the np.where constructions; and the Nv2 function, which uses boolean indexing. With a length of x of 1,000,000, the results on my computer (MacBook Air 11", 2 1.6GHz Intel CPU, running Ubuntu 12.04 in a VMWare virtual machine) became 4.8 s for N_loop, 1 s N_vec, 0.3 s for Nv1, and 0.08 s for Nv2. Boolean indexing is clearly the fastest method.

5.6   More on Numerical Python Arrays

This section lists some more advanced but useful operations with Numerical Python arrays.

5.6.1   Copying Arrays

Let x be an array. The statement a = x makes a refer to the same array as x. Changing a will then also affect x:

```
>>> import numpy as np
>>> x = np.array([1, 2, 3.5])
>>> a = x
>>> a[-1] = 3  # this changes x[-1] too!
>>> x
array([ 1.,  2.,  3.])
```

Changing a without changing x requires a to be a copy of x:

```
>>> a = x.copy()
>>> a[-1] = 9
>>> a
array([ 1.,  2.,  9.])
>>> x
array([ 1.,  2.,  3.])
```

5.6.2   In-Place Arithmetics

Let a and b be two arrays of the same shape. The expression a += b means a = a
+ b, but this is not the complete story. In the statement a = a + b, the sum a +
b is first computed, yielding a new array, and then the name a is bound to this new
array. The old array a is lost unless there are other names assigned to this array. In
the statement a += b, elements of b are added directly into the elements of a (in
memory). There is no hidden intermediate array as in a = a + b. This implies
that a += b is more efficient than a = a + b since Python avoids making an extra
array. We say that the operators +=, *=, and so on, perform *in-place* arithmetics in
arrays.

Consider the compound array expression

```
a = (3*x**4 + 2*x + 4)/(x + 1)
```

The computation actually goes as follows with seven hidden arrays for storing intermediate results:

- r1 = x**4
- r2 = 3*r1
- r3 = 2*x
- r4 = r2 + r3
- r5 = r4 + 4
- r6 = x + 1
- r7 = r5/r6
- a = r7

With in-place arithmetics we can get away with creating three new arrays, at a cost
of a significantly less readable code:

```
a = x.copy()
a **= 4
a *= 3
a += 2*x
a += 4
a /= x + 1
```

The three extra arrays in this series of statement arise from copying x, and computing the right-hand sides 2*x and x+1.

Quite often in computational science and engineering, a huge number of arithmetics is performed on very large arrays, and then saving memory and array allocation time by doing in-place arithmetics is important.

The mix of assignment and in-place arithmetics makes it easy to make unintended changes of more than one array. For example, this code changes x:

```
a = x
a += y
```

since a refers to the same array as x and the change of a is done in-place.

5.6.3 Allocating Arrays

We have already seen that the np.zeros function is handy for making a new array a of a given size. Very often the size and the type of array elements have to match another existing array x. We can then either copy the original array, e.g.,

```
a = x.copy()
```

and fill elements in a with the right new values, or we can say

```
a = np.zeros(x.shape, x.dtype)
```

The attribute x.dtype holds the array element type (dtype for data type), and x.shape is a tuple with the array dimensions. The variable a.ndim holds the number of dimensions.

Sometimes we may want to ensure that an object is an array, and if not, turn it into an array. The np.asarray function is useful in such cases:

```
a = np.asarray(a)
```

Nothing is copied if a already is an array, but if a is a list or tuple, a new array with a copy of the data is created.

5.6.4 Generalized Indexing

Section 5.2.2 shows how slices can be used to extract and manipulate subarrays. The slice f:t:i corresponds to the index set f, f+i, f+2*i, ... up to, but not including, t. Such an index set can be given explicitly too: a[range(f,t,i)]. That is, the integer list from range can be used as a set of indices. In fact, any integer list or integer array can be used as index:

```
>>> a = np.linspace(1, 8, 8)
>>> a
array([ 1.,  2.,  3.,  4.,  5.,  6.,  7.,  8.])
>>> a[[1,6,7]] = 10
>>> a
array([ 1.,  10.,   3.,   4.,   5.,   6.,  10.,  10.])
>>> a[range(2,8,3)] = -2   # same as a[2:8:3] = -2
>>> a
array([ 1.,  10.,  -2.,   4.,   5.,  -2.,  10.,  10.])
```

We can also use boolean arrays to generate an index set. The indices in the set will correspond to the indices for which the boolean array has True values. This functionality allows expressions like a[x<m]. Here are two examples, continuing the previous interactive session:

```
>>> a[a < 0]              # pick out the negative elements of a
array([-2., -2.])
>>> a[a < 0] = a.max()
>>> a
array([ 1.,   10.,   10.,    4.,    5.,   10.,   10.,   10.])
>>> # Replace elements where a is 10 by the first
>>> # elements from another array/list:
>>> a[a == 10] = [10, 20, 30, 40, 50, 60, 70]
>>> a
array([ 1.,   10.,   20.,    4.,    5.,   30.,   40.,   50.])
```

Generalized indexing using integer arrays or lists is important for vectorized initialization of array elements. The syntax for generalized indexing of higher-dimensional arrays is slightly different, see Sect. 5.8.2.

5.6.5   Testing for the Array Type

Inside an interactive Python shell you can easily check an object's type using the type function (see Sect. 1.5.2). In case of a Numerical Python array, the type name is ndarray:

```
>>> a = np.linspace(-1, 1, 3)
>>> a
array([-1.,   0.,   1.])
>>> type(a)
<type 'numpy.ndarray'>
```

Sometimes you need to test if a variable is an ndarray or a float or int. The isinstance function can be used this purpose:

```
>>> isinstance(a, np.ndarray)
True
>>> isinstance(a, (float,int))   # float or int?
False
```

A typical use of isinstance and type to check on object's type is shown next.

Example: Vectorizing a constant function Suppose we have a constant function,

```
def f(x):
    return 2
```

This function accepts an array argument x, but will return a float while a vectorized version of the function should return an array of the same shape as x where each element has the value 2. The vectorized version can be realized as

```
def fv(x):
    return np.zeros(x.shape, x.dtype) + 2
```

The optimal vectorized function would be one that works for both a scalar and an array argument. We must then test on the argument type:

```
def f(x):
    if isinstance(x, (float, int)):
        return 2
    elif isinstance(x, np.ndarray):
        return np.zeros(x.shape, x.dtype) + 2
    else:
        raise TypeError\
        ('x must be int, float or ndarray, not %s' % type(x))
```

5.6.6   Compact Syntax for Array Generation

There is a special compact syntax `r_[f:t:s]` for the `linspace` function:

```
>>> a = r_[-5:5:11j]   # same as linspace(-5, 5, 11)
>>> print a
[-5. -4. -3. -2. -1.  0.  1.  2.  3.  4.  5.]
```

Here, `11j` means 11 coordinates (between -5 and 5, including the upper limit 5). That is, the number of elements in the array is given with the imaginary number syntax.

5.6.7   Shape Manipulation

The `shape` attribute in array objects holds the shape, i.e., the size of each dimension. A function `size` returns the total number of elements in an array. Here are a few equivalent ways of changing the shape of an array:

```
>>> a = np.linspace(-1, 1, 6)
>>> print a
[-1.  -0.6 -0.2  0.2  0.6  1. ]
>>> a.shape
(6,)
>>> a.size
6
>>> a.shape = (2, 3)
>>> a = a.reshape(2, 3)      # alternative
>>> a.shape
(2, 3)
>>> print a
[[-1.  -0.6 -0.2]
 [ 0.2  0.6  1. ]]
>>> a.size                   # total no of elements
6
>>> len(a)                   # no of rows
2
>>> a.shape = (a.size,)      # reset shape
```

Note that `len(a)` always returns the length of the first dimension of an array.

5.7   High-Performance Computing with Arrays

Programs with lots of array calculations may soon consume much time and memory, so it may quickly become crucial to speed up calculations and use as little memory as possible. The main technique for speeding up array calculations is vectorization, i.e., avoiding explicit loops in Python over array elements. To save memory usage, one needs to understand when arrays get allocated and avoid this by in-place array arithmetics. You should review Sect. 5.6.2 about array allocation and in-place arithmetics before reading on.

Example: `axpy` Our computational case study concerns the famous "axpy" operation: $r = ax + y$, where a is a number and x and y are arrays. All implementations and the associated experimentation are found in the file `hpc_axpy.py`.

5.7.1   Scalar Implementation

A naive loop implementation of the "axpy" operation $ax + y$ reads

```
def axpy_loop_newr(a, x, y):
    r = np.zeros_like(x)
    for i in range(r.size):
        r[i] = a*x[i] + y[i]
    return r
```

Classical implementations overwrite y by $ax + y$: $y \leftarrow ax + y$, but we shall make implementations where we either can overwrite y or place $ax + y$ in another array. The function above creates a new array for the result.

Rather than allocating the array inside the function, we can put that burden on the user and provide a result array `r` as input:

```
def axpy_loop(a, x, y, r):
    for i in range(r.size):
        r[i] = a*x[i] + y[i]
    return r
```

The advantage of this version is that we can either overwrite y by $ax + y$ or store $ax + y$ in a separate array:

```
# Classical axpy
y = axpy_loop(a, x, y, y)

# Store axpy result in separate array
r = np.zeros_like(x)
r = axpy_loop(a, x, y, r)
```

Python functions return output data

The call

```
r = axpy_loop(a, x, y, r)
```

can equally well be written

```
axpy_loop(a, x, y, r)
```

This is the typical coding style in Fortran, C, or C++ (where r is then transferred as a reference or pointer to the array data). In Python, there is no need for the function axpy_loop to return r, because the assignment to r[i] inside the loop changes all elements of the r. The array r will therefore be modified after calling axpy_loop(a, x, y, r) anyway. However, it is a good convention in Python that *all input data to a function are arguments and all output data are returned*. With

```
r = axpy_loop(a, x, y, r)
```

we clearly see that r is both input and output.

5.7.2 Vectorized Implementation

The vectorized implementation of the "axpy" operation reads

```
def axpy1(a, x, y):
    r = a*x + y
    return r
```

Note that the result is placed in a new array arrising from the operation a*x+y. The speed up of vectorization is significant, see Fig. 5.11 (made by the function effect_of_vec in the file hpc_axpy.py).

Temporary arrays are needed in the vectorized implementation. One would expect that a*x must be calculated and stored in a temporary array, call it r1, and then r1 + y must be evaluated and stored in another allocated array r, which is returned. It appears that a*x + y only needs the allocation of one array, the one to be returned (we have investigated the memory consumption in detail using the memory_profiler module). Anyway, repeated calls to axpy1 with large arrays lead to an allocation of a new large array in each call.

5.7.3 Memory-Saving Implementation

Applications with large arrays should avoid unnecessary allocation of temporary arrays and instead reuse pre-allocated arrays. Suppose we have allocated an array for the result r = ax + y once and for all. We can pass the r array to the computing function as in the axpy_loop function above and use the memory in r for intermediate calculation. In vectorized code, this requires use of *in-place* array arithmetics (see Sect. 5.6.2)

In-place arithmetics for doing r = a*x + y in a pre-allocated array r will first copy all elements of x into r, then perform elementwise multiplication by a, and finally elementwise addition of y:

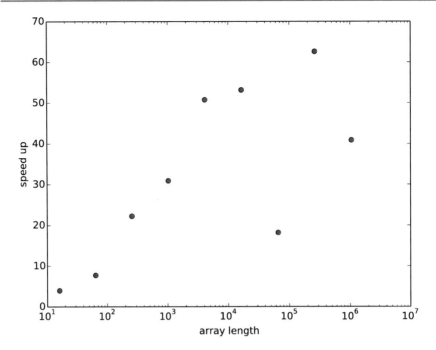

Fig. 5.11 Improved efficiency of vectorizing the "axpy" operation as function of array length

```
def axpy2(a, x, y, r):
    r[:] = x
    r *= a
    r += y
    return r
```

Note that $r[:] = x$ inserts the elements of x into r. A mathematically equivalent construction, $r = x.copy()$, allocates a new object and fills it with the values of x before the name r refers to this object. The array r supplied as argument to the function is then lost, and the returned array is another object.

We can perform repeated calls to the axpy2 function without any extra memory allocation. This can be proved by a small code snippet where we use id(r) to see the unique identity of the r array. If this identity remains constant through calls, we always reuse the pre-allocated r array:

```
r = np.zeros_like(x)
print id(r)
for i in range(10):
    r = axpy2(a, x, y, r)
    print id(r)
```

The output prints the same number, proving that it is physically the same object we feed as r to axpy2 that is also returned.

We can equally well drop returning r and utilize that the changes made by in-place arithmetics is always reflected in the r array we allocate outside the function:

```
def axpy3(a, x, y, r):
    r[:] = x
    r *= a
    r += y
```

The call can now just be

```
axpy3(a, x, y, r)
```

However, as emphasized in Sect. 5.7.1, in Python we usually return the output data (because this is no extra cost, only references to objects are physically transferred back to the calling code).

5.7.4 Analysis of Memory Usage

The module `memory_profiler` is very useful for analyzing the memory usage of every statement in a program. The module is installed by

---
Terminal

```
Terminal> sudo pip install memory_profiler
```

Each function we want to analyze must have (the decorator) `@profile` at the line above, e.g.,

```
@profile
def axpy1(a, x, y:
    r = a*x + y
    return r
```

We can then run a program with name axpy.py by

Terminal

```
Terminal> python -m memory_profiler axpy.py
```

The arrays must be quite large to see a significant increase in memory usage. With 10,000,000 elements in each array we get output like

```
Line #    Mem usage    Increment   Line Contents
================================================
     1   251.977 MiB    0.000 MiB   @profile
     2                              def axpy1(a, x, y:
     3   328.273 MiB   76.297 MiB       r = a*x + y
     4   328.273 MiB    0.000 MiB       return r
```

```
Line #    Mem usage     Increment   Line Contents
================================================
     6  251.977 MiB    0.000 MiB   @profile
     7                             def axpy2(a, x, y, r):
     8  251.977 MiB    0.000 MiB       r[:] = x
     9  251.977 MiB    0.000 MiB       r *= a
    10  251.977 MiB    0.000 MiB       r += y
    11  251.977 MiB    0.000 MiB       return r
```

This demonstrates the larger memory consumption of axpy1 compared with axpy2.

5.7.5 Analysis of the CPU Time

The module line_profiler can time each line of a program. Installation is easily done by sudo pip install line_profiler. As for module_profiler described in the previous section, also line_profiler requires each function to be analyzed to have (the decorator) @profile at the line above the function. The module is installed together with an analysis script kernprof that we use to run the program:

Terminal

```
Terminal> kernprof -l -v axpy.py
```

With an array length of 500,000 we get output like

```
Total time: 0.014291 s
Function: axpy1 at line 1

Line #  Hits      Time  Per Hit   % Time  Line Contents
=======================================================
   1                                      @profile
   2                                      def axpy1(a, x, y):
   3       3     14283   4761.0     99.9      r = a*x + y
   4       3         8      2.7      0.1      return r

Total time: 0.004382 s
Function: axpy2 at line 6

Line #  Hits  Time  Per Hit   % Time  Line Contents
===================================================
   6                                   @profile
   7                                   def axpy2(a, x, y, r):
   8     3     1981    660.3     45.2      r[:] = x
   9     3     1258    419.3     28.7      r *= a
  10     3     1138    379.3     26.0      r += y
  11     3        5      1.7      0.1      return r

Total time: 1.38674 s
Function: axpy_loop at line 26
```

```
Line #  Hits     Time Per Hit  % Time  Line Contents
==============================================================
26                                     @profile
27                                     def axpy_loop(a, x, y, r):
28  1500006  449747   0.3   32.4         for i in range(r.size):
29  1500003  936985   0.6   67.6           r[i] = a*x[i] + y[i]
30        3       10   3.3    0.0         return r
```

We see that the administration of the for loop takes 1/3 of the total cost of the loop.

Both `line_profiler` and `memory_profiler` are very useful tools for spotting inefficient constructions in a code.

5.8 Higher-Dimensional Arrays

5.8.1 Matrices and Arrays

Vectors appeared when mathematicians needed to calculate with a list of numbers. When they needed a table (or a list of lists in Python terminology), they invented the concept of *matrix* (singular) and *matrices* (plural). A table of numbers has the numbers ordered into rows and columns. One example is

$$\begin{bmatrix} 0 & 12 & -1 & 5 \\ -1 & -1 & -1 & 0 \\ 11 & 5 & 5 & -2 \end{bmatrix}$$

This table with three rows and four columns is called a 3×4 matrix (mathematicians may not like this sentence, but it suffices for our purposes). If the symbol A is associated with this matrix, $A_{i,j}$ denotes the number in row number i and column number j. Counting rows and columns from 0, we have, for instance, $A_{0,0} = 0$ and $A_{2,3} = -2$. We can write a general $m \times n$ matrix A as

$$\begin{bmatrix} A_{0,0} & \cdots & A_{0,n-1} \\ \vdots & \ddots & \vdots \\ A_{m-1,0} & \cdots & A_{m-1,n-1} \end{bmatrix}$$

Matrices can be added and subtracted. They can also be multiplied by a scalar (a number), and there is a concept of length or size. The formulas are quite similar to those presented for vectors, but the exact form is not important here.

We can generalize the concept of table and matrix to *array*, which holds quantities with in general d indices. Equivalently we say that the array has rank d. For $d = 3$, an array A has elements with three indices: $A_{p,q,r}$. If p goes from 0 to n_p-1, q from 0 to $n_q - 1$, and r from 0 to $n_r - 1$, the A array has $n_p \times n_q \times n_r$ elements in total. We may speak about the *shape* of the array, which is a d-vector holding the number of elements in each "array direction", i.e., the number of elements for each index. For the mentioned A array, the shape is (n_p, n_q, n_r).

The special case of $d = 1$ is a vector, and $d = 2$ corresponds to a matrix. When we program we may skip thinking about vectors and matrices (if you are not so familiar with these concepts from a mathematical point of view) and instead just work with arrays. The number of indices corresponds to what is convenient in the programming problem we try to solve.

5.8.2   Two-Dimensional Numerical Python Arrays

Consider a nested list `table` of two-pairs `[C, F]` (see Sect. 2.4) constructed by

```
>>> Cdegrees = [-30 + i*10 for i in range(3)]
>>> Fdegrees = [9./5*C + 32 for C in Cdegrees]
>>> table = [[C, F] for C, F in zip(Cdegrees, Fdegrees)]
>>> print table
[[-30, -22.0], [-20, -4.0], [-10, 14.0]]
```

This nested list can be turned into an array,

```
>>> table2 = np.array(table)
>>> print table2
[[-30. -22.]
 [-20.  -4.]
 [-10.  14.]]
>>> type(table2)
<type 'numpy.ndarray'>
```

We say that `table2` is a *two-dimensional* array, or an array of rank 2.

The `table` list and the `table2` array are stored very differently in memory. The `table` variable refers to a list object containing three elements. Each of these elements is a reference to a separate list object with two elements, where each element refers to a separate `float` object. The `table2` variable is a reference to a single array object that again refers to a consecutive sequence of bytes in memory where the six floating-point numbers are stored. The data associated with `table2` are found in one chunk in the computer's memory, while the data associated with `table` are scattered around in memory. On today's machines, it is much more expensive to find data in memory than to compute with the data. Arrays make the data fetching more efficient, and this is major reason for using arrays. However, this efficiency gain is only present for very large arrays, not for a 3×2 array.

Indexing a nested list is done in two steps, first the outer list is indexed, giving access to an element that is another list, and then this latter list is indexed:

```
>>> table[1][0]      # table[1] is [-20,4], whose index 0 holds -20
-20
```

This syntax works for two-dimensional arrays too:

```
>>> table2[1][0]
-20.0
```

but there is another syntax that is more common for arrays:

```
>>> table2[1,0]
-20.0
```

A two-dimensional array reflects a table and has a certain number of rows and columns. We refer to rows as the *first dimension* of the array and columns as the *second dimension*. These two dimensions are available as `table2.shape`:

```
>>> table2.shape
(3, 2)
```

Here, 3 is the number of rows and 2 is the number of columns.

A loop over all the elements in a two-dimensional array is usually expressed as two *nested* `for` loops, one for each index:

```
>>> for i in range(table2.shape[0]):
...     for j in range(table2.shape[1]):
...         print 'table2[%d,%d] = %g' % (i, j, table2[i,j])
...
table2[0,0] = -30
table2[0,1] = -22
table2[1,0] = -20
table2[1,1] = -4
table2[2,0] = -10
table2[2,1] = 14
```

An alternative (but less efficient) way of visiting each element in an array with any number of dimensions makes use of a single `for` loop:

```
>>> for index_tuple, value in np.ndenumerate(table2):
...     print 'index %s has value %g' % \
...           (index_tuple, table2[index_tuple])
...
index (0,0) has value -30
index (0,1) has value -22
index (1,0) has value -20
index (1,1) has value -4
index (2,0) has value -10
index (2,1) has value 14
```

In the same way as we can extract sublists of lists, we can extract subarrays of arrays using slices.

```
table2[0:table2.shape[0], 1]   # 2nd column (index 1)
array([-22.,  -4.,  14.])

>>> table2[0:, 1]              # same
array([-22.,  -4.,  14.])

>>> table2[:, 1]              # same
array([-22.,  -4.,  14.])
```

To illustrate array slicing further, we create a bigger array:

```
>>> t = np.linspace(1, 30, 30).reshape(5, 6)
>>> t
array([[  1.,   2.,   3.,   4.,   5.,   6.],
       [  7.,   8.,   9.,  10.,  11.,  12.],
       [ 13.,  14.,  15.,  16.,  17.,  18.],
       [ 19.,  20.,  21.,  22.,  23.,  24.],
       [ 25.,  26.,  27.,  28.,  29.,  30.]])

>>> t[1:-1:2, 2:]
array([[  9.,  10.,  11.,  12.],
       [ 21.,  22.,  23.,  24.]])
```

To understand the slice, look at the original t array and pick out the two rows corresponding to the first slice 1:-1:2,

```
      [  7.,   8.,   9.,  10.,  11.,  12.]
      [ 19.,  20.,  21.,  22.,  23.,  24.]
```

Among the rows, pick the columns corresponding to the second slice 2:,

```
      [  9.,  10.,  11.,  12.]
      [ 21.,  22.,  23.,  24.]
```

Another example is

```
>>> t[:-2, :-1:2]
array([[  1.,   3.,   5.],
       [  7.,   9.,  11.],
       [ 13.,  15.,  17.]])
```

Generalized indexing as described for one-dimensional arrays in Sect. 5.6.4 requires a more comprehensive syntax for higher-dimensional arrays. Say we want to extract a subarray of t that consists of the rows with indices 0 and 3 and the columns with indices 1 and 2:

```
>>> t[np.ix_([0,3], [1,2])]
array([[  2.,   3.],
       [ 20.,  21.]])
>>> t[np.ix_([0,3], [1,2])] = 0
>>> t
array([[  1.,   0.,   0.,   4.,   5.,   6.],
       [  7.,   8.,   9.,  10.,  11.,  12.],
       [ 13.,  14.,  15.,  16.,  17.,  18.],
       [ 19.,   0.,   0.,  22.,  23.,  24.],
       [ 25.,  26.,  27.,  28.,  29.,  30.]])
```

Recall that slices only gives a view to the array, not a copy of the values:

```
>>> a = t[1:-1:2, 1:-1]
>>> a
array([[  8.,   9.,  10.,  11.],
       [  0.,   0.,  22.,  23.]])
```

```
>>> a[:,:] = -99
>>> a
array([[-99., -99., -99., -99.],
       [-99., -99., -99., -99.]])
>>> t  # is t changed to? yes!
array([[  1.,   0.,   0.,   4.,   5.,   6.],
       [  7., -99., -99., -99., -99.,  12.],
       [ 13.,  14.,  15.,  16.,  17.,  18.],
       [ 19., -99., -99., -99., -99.,  24.],
       [ 25.,  26.,  27.,  28.,  29.,  30.]])
```

5.8.3 Array Computing

The operations on vectors in Sect. 5.1.3 can quite straightforwardly be extended to arrays of any dimension. Consider the definition of applying a function $f(v)$ to a vector v: we apply the function to each element v_i in v. For a two-dimensional array A with elements $A_{i,j}$, $i = 0, \ldots, m$, $j = 0, \ldots, n$, the same definition yields

$$f(A) = (f(A_{0,0}), \ldots, f(A_{m-1,0}), f(A_{1,0}), \ldots, f(A_{m-1,n-1})).$$

For an array B with any rank, $f(B)$ means applying f to each array entry.

The asterisk operation from Sect. 5.1.3 is also naturally extended to arrays: $A * B$ means multiplying an element in A by the corresponding element in B, i.e., element (i, j) in $A * B$ is $A_{i,j} B_{i,j}$. This definition naturally extends to arrays of any rank, provided the two arrays have the same shape.

Adding a scalar to an array implies adding the scalar to each element in the array. Compound expressions involving arrays, e.g., $\exp(-A^2) * A + 1$, work as for vectors. One can in fact just imagine that all the array elements are stored after each other in a long vector (this is actually the way the array elements are stored in the computer's memory), and the array operations can then easily be defined in terms of the vector operations from Sect. 5.1.3.

Remark Readers with knowledge of matrix computations may get confused by the meaning of A^2 in matrix computing and A^2 in array computing. The former is a matrix-matrix product, while the latter means squaring all elements of A. Which rule to apply, depends on the context, i.e., whether we are doing linear algebra or vectorized arithmetics. In mathematical typesetting, A^2 can be written as AA, while the array computing expression A^2 can be alternatively written as $A * A$. In a program, A*A and A**2 are identical computations, meaning squaring all elements (array arithmetics). With NumPy arrays the matrix-matrix product is obtained by dot(A, A). The matrix-vector product Ax, where x is a vector, is computed by dot(A, x). However, with matrix objects (see Sect. 5.8.4) A*A implies the mathematical matrix multiplication AA.

We shall leave this subject of notational confusion between array computing and linear algebra here since this book will not further understanding and the confusion is seldom serious in program code if one has a good overview of the mathematics that is to be carried out.

5.8.4   Matrix Objects

This section only makes sense if you are familiar with basic linear algebra and the matrix concept. The arrays created so far have been of type `ndarray`. NumPy also has a matrix type called `matrix` or `mat` for one- and two-dimensional arrays. One-dimensional arrays are then extended with one extra dimension such that they become matrices, i.e., either a row vector or a column vector:

```
>>> import numpy as np
>>> x1 = np.array([1, 2, 3], float)
>>> x2 = np.matrix(x1)            # or mat(x1)
>>> x2                            # row vector
matrix([[ 1.,   2.,   3.]])
>>> x3 = mat(x1).T                # transpose = column vector
>>> x3
matrix([[ 1.],
        [ 2.],
        [ 3.]])

>>> type(x3)
<class 'numpy.matrixlib.defmatrix.matrix'>
>>> isinstance(x3, np.matrix)
True
```

A special feature of `matrix` objects is that the multiplication operator represents the matrix-matrix, vector-matrix, or matrix-vector product as we know from linear algebra:

```
>>> A = np.eye(3)                 # identity matrix
>>> A
array([[ 1.,   0.,   0.],
       [ 0.,   1.,   0.],
       [ 0.,   0.,   1.]])
>>> A = mat(A)
>>> A
matrix([[ 1.,   0.,   0.],
        [ 0.,   1.,   0.],
        [ 0.,   0.,   1.]])
>>> y2 = x2*A                     # vector-matrix product
>>> y2
matrix([[ 1.,   2.,   3.]])
>>> y3 = A*x3                     # matrix-vector product
>>> y3
matrix([[ 1.],
        [ 2.],
        [ 3.]])
```

One should note here that the multiplication operator between standard `ndarray` objects is quite different!

Readers who are familiar with MATLAB, or intend to use Python and MATLAB together, should seriously think about programming with `matrix` objects instead of `ndarray` objects, because the `matrix` type behaves quite similar to matrices and vectors in MATLAB. Nevertheless, `matrix` cannot be used for arrays of larger dimension than two.

5.9   Some Common Linear Algebra Operations

Python has strong support for numerical linear algebra, much like the functionality found in MATLAB. Some of the most widely used operations are exemplified below.

5.9.1   Inverse, Determinant, and Eigenvalues

We start with showing how to find the inverse and the determinant of a matrix, and how to compute the eigenvalues and eigenvectors:

```
>>> import numpy as np
>>> A = np.array([[2, 0], [0, 5]], dtype=float)

>>> np.linalg.inv(A)    # inverse matrix
array([[ 0.5,  0. ],
       [ 0. ,  0.2]])

>>> np.linalg.det(A)    # determinant
9.9999999999999982

>>> eig_values, eig_vectors = np.linalg.eig(A)
>>> eig_values
array([ 2.,  5.])
>>> eig_vectors
array([[ 1.,  0.],
       [ 0.,  1.]])
```

The eigenvectors are normalized to have unit lengths.

5.9.2   Products

The `np.dot` function is used for scalar or dot product as well as matrix-vector and matrix-matrix products *between array objects*:

```
>>> a = np.array([4, 0])
>>> b = np.array([0, 1])
>>> np.dot(A, a)        # matrix vector product
array([ 8.,  0.])
>>> np.dot(a, b)        # dot product between vectors
0
>>>
>>> B = np.ones((2, 2))  # 2x2 matrix with 1's
>>> np.dot(A, B)        # matrix-matrix product
array([[ 2.,  2.],
       [ 5.,  5.]])
```

Note that using the `matrix` class instead of plain arrays (see Sect. 5.8.4) allows * to be used as operator for matrix-vector and matrix-matrix products.

The cross product $a \times b$, between vectors a and b of length 3, is computed by

```
>>> np.cross([1, 1, 1], [0, 0, 1])
array([ 1, -1,  0])
```

Finding the angle between vectors a and b,

$$\theta = \cos^{-1}\left(\frac{a \cdot b}{||a||\,||b||}\right),$$

goes like

```
>>> np.arccos(np.dot(a, b)/(np.linalg.norm(a)*np.linalg.norm(b)))
1.5707963267948966
```

5.9.3   Norms

Various norms of matrices and vectors are well supported by NumPy. Some common examples are

```
>>> np.linalg.norm(A)        # Frobenius norm for matrices
5.3851648071345037
>>> np.sqrt(np.sum(A**2))    # Frobenius norm: direct formula
5.3851648071345037
>>> np.linalg.norm(a)        # l2 norm for vectors
4.0
```

See `pydoc numpy.linalg.norm` for information on other norms.

5.9.4   Sum and Extreme Values

The sum of all elements or of the elements in a particular row or column is computed by `np.sum`:

```
>>> np.sum(B)        # sum of all elements
2.0
>>> B.sum()          # sum of all elements; alternative syntax
2.0
>>> np.sum(B, axis=0) # sum over index 0 (rows)
array([ 4., -2.])
>>> np.sum(B, axis=1) # sum over index 1 (columns)
array([ 3., -1.])
```

The maximum or minimum value of an array is also often needed:

```
>>> np.max(B)          # max over all elements
3.0
>>> B.max()            # max over all elements, alt. syntax
3.0
>>> np.min(B)          # min over all elements
-4.0
>>> np.abs(B).min()    # min absolute value
1.0
```

A very frequent application of computing the minimum absolute value occurs in
test functions where we want to verify a result, e.g., that $AA^{-1} = I$, where I is the
identity matrix. We then want to check the smallest absolute value in $AA^{-1} - I$:

```
>>> I = np.eye(2)    # identity matrix of size 2
>>> I
array([[ 1.,   0.],
       [ 0.,   1.]])
>>> np.abs(np.dot(A, np.linalg.inv(A)) - I).max()
0.0
```

Never use == when testing real numbers!
It could be tempting to test $AA^{-1} = I$ using the syntax

```
>>> np.dot(A, np.linalg.inv(A)) == np.eye(2)
array([[ True,   True],
       [ True,   True]], dtype=bool)
```

but there are two major problems with this test:

1. the result is a boolean matrix, not suitable for an `if` test
2. using == for matrices with float elements may fail because of rounding errors

The second problem must be solved by computing differences and comparing
them against small tolerances, as we did above. Here is an example where ==
fails:

```
>>> A = np.array([[4, 0], [0, 49]], dtype=float)
>>> np.dot(A, np.linalg.inv(A)) == np.eye(2)
array([[ True,   True],
       [ True, False]], dtype=bool)
```

(`1.0/49*49` is not exactly 1 because of rounding errors.)
     The first problem is solved by using the `C.all()`, which returns one boolean
variable `True` if all elements in the boolean array `C` are `True`, otherwise it returns
`False`, as in the case above:

```
>>> (np.dot(A, np.linalg.inv(A)) == np.eye(2)).all()
False
```

5.9.5 Indexing

Indexing an element is done by A[i,j]. A row or column is extracted as

```
>>> A[0,:]  # first row
array([ 2.,   0.])
>>> A[:,1]  # second column
array([ 0.,   5.])
```

NumPy also supports multiple values for the indices via the np.ix_ function. Here is an example where we grab row 0 and 2, then column 1:

```
>>> C = np.array([[1,2,3],[4,5,6],[7,8,9]])
>>> C[np.ix_([0,2], [1])]  # row 0 and 2, then column 1
array([[2],
       [8]])
```

You can also use the colon notation to pick out other parts of a matrix. If C is a 3×5-matrix,

```
C[1:3, 0:4]
```

gives a sub-matrix consisting of the two rows of C after the first, and the first four columns of C (recall that the upper limits, here 3 and 4, are not included).

Readers familiar with MATLAB should note that the indexing may be a bit unexpected when referring to parts of a matrix: writing C[[0, 2], [0, 2]] one would expect entries residing in rows/columns 0 and 2, but that behavior requires in Python the np.ix_ command:

```
>>> C = np.array([[1, 2, 3], [4, 5, 6], [7, 8, 9]])
>>> C[np.ix_([0, 2], [0, 2])]
[[1 3]
 [7 9]]
>>> # Grab row 0, 2, then column 0 from row 0 and column 2 from row 2
>>> C[[0, 2], [0, 2]]
[1 9]
```

5.9.6 Transpose and Upper/Lower Triangular Parts

The transpose of a matrix B is obtained by B.T:

```
>>> B = np.array([[1, 2], [3, -4]], dtype=float)
>>> B.T                 # the transpose
array([[ 1.,   3.],
       [ 2.,  -4.]])
```

NumPy has rich functionality for doing operations on array objects. For example, one can strip down a matrix to its upper or lower triangular parts:

```
>>> np.triu(B)  # upper triangular part of B
array([[ 1.,   2.],
       [ 0.,  -4.]])
>>> np.tril(B)  # lower triangular part of B
array([[ 1.,   0.],
       [ 3.,  -4.]])
```

5.9.7 Solving Linear Systems

The perhaps most frequent operation in linear algebra is the solution of systems of linear algebraic equations: $Ax = b$, where A is a coefficient matrix, b is a given right-hand side vector, and x is the solution vector. The function `np.linalg.solve(A, b)` does the job:

```
>>> A = np.array([[1, 2], [-2, 2.5]])
>>> x = np.array([-1, 1], dtype=float)    # pick a solution
>>> b = np.dot(A, x)                       # find right-hand side

>>> np.linalg.solve(A, b)                  # will this compute x?
array([-1.,   1.])
```

5.9.8 Matrix Row and Column Operations

Implementing Gaussian elimination constitutes a good pedagogical example on how to perform row and column operations on a matrix. Some needed functionality is

```
A[[i, j]] = A[[j, i]]    # swap rows i and j
A[i] *= k                # multiply row i by a constant k
A[j] += k*A[i]           # add row i, multiplied by k, to row j
```

With these operations, Gaussian elimination is programmed as follows.

```
m, n = shape(A)
for j in range(n - 1):
    for i in range(j + 1, m):
        A[i,j:] -= (A[i,j]/A[j,j])*A[j,j:]
```

Note the special syntax `j:`, which refers to indices from j and up to the end of the array. More generally, when referring to an array a with length n, the following are equivalent:

```
a[0:n]
a[:n]
a[0:]
a[:]
```

In the code for Gaussian elimination, we first eliminate the entries below the diagonal in the first column, by adding a scaled version of the first row to the other rows. Then the same procedure is applied for the second row, and so on. The result is an upper triangular matrix. The code can fail if some of the entries A[j,j] become zero along the way. To avoid this, we can swap rows when the problem arises. The following code implements the idea and will not fail, even if some of the columns are zero.

```
def Gaussian_elimination(A):
    rank = 0
    m, n = np.shape(A)
    i = 0
    for j in range(n):
        p = np.argmax(abs(A[i:m,j]))
        if p > 0: # swap rows
            A[[i,p+i]] = A[[p+i, i]]
        if A[i,j] != 0:
            # j is a pivot column
            rank += 1
            for r in range(i+1, m):
                A[r,j:] -= (A[r,j]/A[i,j])*A[i,j:]
            i += 1
        if i > m:
            break
    return A, rank
```

Note that we stick to the habit of returning all results from a function, here the modified matrix A and its rank.

5.9.9 Computing the Rank of a Matrix

The rank of a matrix equals the number of pivot columns after Gaussian elimination. The variable rank counts these in the code above.

Due to rounding errors, the computed rank may be higher than the actual rank: the rounding errors may imply that A[i,j] != 0 is true, even if Gaussian elimination performed in exact arithmetics gives exactly zero. Such situations can be avoided by replacing if A[i,j] !=0: with if abs(A[i,j]) > tol:, where tol is some small tolerance.

A more reliable way to compute the rank is to compute the singular value decomposition of A, and check how many of the singular values that are larger than a threshold epsilon:

```
>>> A = np.array([[1, 2.01], [2.01, 4.0401]])
>>> U, s, V = np.linalg.svd(A) # s are the singular values of A
# abs(s) > tol gives an array with True and False values
# s.nonzero() lists indices k so that s[k] != 0
>>> shape((abs(s) > tol).nonzero())[1]  # rank
1
>>> A, rank = Gaussian_elimination(A)
>>> rank
2
```

If you use a tolerance check on the form if abs(A[i,j]) > 1E-10: in the function Gaussian_elimination, the code will say that the rank is 1, which is the correct value also found by using the singular value decomposition.

It is known that the determinant is nonzero if and only if the rank equals the number of rows/columns. For the matrix A we used above, the determinant should thus be 0, but also here roundoff errors come into play:

```
>>> A = np.array([[1, 2.01], [2.01, 4.0401]])
>>> A[0, 0]*A[1, 1] - A[0, 1]*A[1, 0]
8.881784197e-16
>>> np.linalg.det(A)
8.92619311799e-16
```

Using our own Gaussian elimination function for computing the rank is less efficient than calling NumPy's singular value decomposition. Here are timings for a random 100×100-matrix:

```
>>> A = np.random.uniform(0, 1, (100, 100))
>>> %timeit U, s, V = np.linalg.svd(A)
100 loops, best of 3: 3.7 ms per loop
>>> %timeit A, rank = Gaussian_elimination(A)
100 loops, best of 3: 22.3 ms per loop
```

5.9.10 Symbolic Linear Algebra

SymPy supports symbolic computations also for linear algebra operations. We may create a matrix and find its inverse and determinant:

```
>>> import sympy as sym
>>> A = sym.Matrix([[2, 0], [0, 5]])

>>> A**-1    # the inverse
Matrix([
[1/2,   0],
[  0, 1/5]])

>>> A.inv()  # the inverse
Matrix([
[1/2,   0],
[  0, 1/5]])

>>> A.det()  # the determinant
10
```

Note that the entries in the inverse matrix are rational numbers (sym.Rational objects to be precise).

Eigenvalues can also be computed exactly:

```
>>> A.eigenvals()
{2: 1, 5: 1}
```

The output is a dictionary meaning here that 2 is an eigenvalue with multiplicity 1 and 5 is an eigenvalue with multiplicity 1. It is more convenient to have the eigenvalues in a list:

```
>>> e = list(A.eigenvals().keys())
>>> e
[2, 5]
```

Eigenvector computations have a somewhat complicated output:

```
>>> A.eigenvects()
[(2, 1, [Matrix([
[1],
[0]])]), (5, 1, [Matrix([
[0],
[1]])])]
```

The output is a list of three-tuples, one for each eigenvalue and eigenvector. The three-tuple contains the eigenvalue, its multiplicity, and the eigenvector as a sym.Matrix object. To isolate the first eigenvector, we can index the list and tuple:

```
>>> v1 = A.eigenvects()[0][2]
>>> v1
Matrix([
[1],
[0]])
```

The vector is a sym.Matrix object with two indices. To extract the vector elements in a plain list, we can do this:

```
>>> v1 = [v1[i,0] for i in range(v1.shape[0])]
>>> v1
[1, 0]
```

The following code extracts all eigenvectors as a list of 2-lists, which may be a convenient data structure for the eigenvectors:

```
>>> v = [[t[2][0][i,0] for i in range(t[2][0].shape[0])]
         for t in A.eigenvects()]
>>> v
[[1, 0], [0, 1]]
```

The norm of a matrix or vector is an exact expression:

```
>>> A.norm()
sqrt(29)
>>> a = sym.Matrix([1, 2])    # vector [1, 2]
>>> a
```

```
Matrix([
[1],
[2]])
>>> a.norm()
sqrt(5)
```

The matrix-vector product and the dot product between vectors are done like this:

```
>>> A*a                        # matrix*vector
Matrix([
[ 2],
[10]])
>>> b = sym.Matrix([2, -1])  # vector [2, -1]
>>> a.dot(b)
0
```

Solving linear systems exactly is also possible:

```
>>> x = sym.Matrix([-1, 1])/2
>>> x
Matrix([
[-1/2],
[ 1/2]])
>>> b = A*x
>>> x = A.LUsolve(b)   # does it compute x?
>>> x                  # x is a matrix object
Matrix([
[-1/2],
[ 1/2]])
```

Sometimes one wants to convert x to a plain numpy array with float values:

```
>>> x = np.array([float(x[i,0].evalf())
                  for i in range(x.shape[0])])
>>> x
array([-0.5,  0.5])
```

Exact row operations can be done as exemplified here:

```
>>> A[1,:] + 2*A[0,:]  # [0,5] + 2*[2,0]
Matrix([[4, 5]])
```

We refer to the online SymPy linear algebra tutorial[7] for more information.

---

[7] http://docs.sympy.org/dev/tutorial/matrices.html

5.10   Plotting of Scalar and Vector Fields

Visualization of scalar and vector fields in Python is commonly done using Matplotlib or Mayavi. Both packages support basic visualization of 2D scalar and vector fields, but Mayavi offers more advanced three-dimensional visualization techniques, especially for 3D scalar and vector fields.

One can also use SciTools for visualizing 2D scalar and vector fields, using either Matplotlib, Gnuplot, or VTK as plotting engines, but this topic is omitted from the present book. However, for fast visualization of large 2D scalar fields, Gnuplot is a viable tool, and the SciTools interface offers a convenient MATLAB-style set of commands to operate Gnuplot.

To exemplify visualization of scalar and vector fields with Matplotlib and Mayavi, we use a common set of examples. A scalar function of x and y is visualized either as a flat two-dimensional plot with contour lines of the field, or as a three-dimensional surface where the height of the surface corresponds to the function value of the field. In the latter case we also add a three-dimensional parameterized curve to the plot.

To illustrate plotting of vector fields, we simply plot the gradient of the scalar field, together with the scalar field. Our convention for variable names goes as follows:

- x, y for one-dimensional coordinates along each axis direction.
- xv, yv for the corresponding vectorized coordinates in a 2D.
- u, v for the components of a vector field at points corresponding to xv, yv.

The following sections contain more mathematical details on the various scalar and vector fields we aim to plot.

5.10.1   Installation

Previously in the book we have explained how to obtain Matplotlib for various platforms. To obtain Mayavi on Ubuntu platforms you can write

```
Terminal
pip install mayavi --upgrade
```

For Mac OS X and Windows, we recommend using Anaconda. To obtain Mayavi for Anaconda you can write

```
Terminal
conda install mayavi
```

5.10.2   Surface Plots

We consider the 2D scalar field defined by

$$h(x, y) = \frac{h_0}{1 + \frac{x^2+y^2}{R^2}}. \tag{5.13}$$

$h(x, y)$ may model the height of an isolated circular mountain, h being the height above sea level, while x and y are Cartesian coordinates on the earth's surface, h_0 the height of the mountain, and R the radius of the mountain. Since mountains are actually quite flat (or more precisely, their heights are small compared to the horizontal extent), we use meter as length unit for vertical distances (z direction) and km as length unit for horizontal distances (x and y coordinates). Prior to all code below we have initialized h_0 and R with the following values: $h_0 = 2277$ m and $R = 4$ km.

Grid for 2D scalar fields   Before we can plot $h(x, y)$, we need to create a rectangular grid in the xy plane with all the points used for plotting. Regardless of which plotting package we will use later on, the grid can be made as follows:

```
x = y = np.linspace(-10., 10., 41)
xv, yv = np.meshgrid(x, y, indexing='ij', sparse=False)

hv = h0/(1 + (xv**2+yv**2)/(R**2))
```

The grid is based on equally spaced coordinates x and y in the interval $[-10, 10]$ km. Note the mysterious extra parameters to meshgrid here, which are needed in order for the coordinates to have the right order such that the arithmetics in the expression for hv becomes correct. The expression computes the surface value at the 41×41 grid points in one vectorized operation.

   A surface plot of a 2D scalar field $h(x, y)$ is a visualization of the surface $z = h(x, y)$ in three-dimensional space. Most plotting packages have functions which can be used to create surface plots of 2D scalar fields. These can be either *wireframe plots*, where only lines connecting the grid points are drawn, or plots where the faces of the surface are colored. In Fig. 5.12 we have shown two such plots of the surface $h(x, y)$. Section 5.11.1 presents the code which generates these plots.

5.10.3   Parameterized Curve

To illustrate the plotting of three-dimensional parameterized curves, we consider a trajectory that represents a circular climb to the top of the mountain:

$$\boldsymbol{r}(t) = \left(10\left(1 - \frac{t}{2\pi}\right)\cos(t)\right)\boldsymbol{i} + \left(10\left(1 - \frac{t}{2\pi}\right)\sin(t)\right)\boldsymbol{j}$$

$$+ \frac{h_0}{1 + \frac{100(1-t/(2\pi))^2}{R^2}}\boldsymbol{k}. \tag{5.14}$$

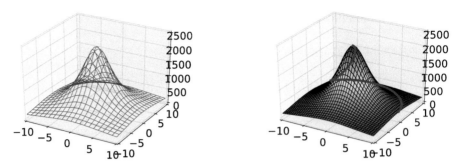

Fig. 5.12 Two different plots of a mountain. The right plot also shows a trajectory to the top of the mountain

Here i, j, and k denote the unit vectors in the x-, y-, and z-directions, respectively. The coordinates of $r(t)$ can be produced by

```
s = np.linspace(0, 2*np.pi, 100)
curve_x = 10*(1 - s/(2*np.pi))*np.cos(s)
curve_y = 10*(1 - s/(2*np.pi))*np.sin(s)
curve_z = h0/(1 + 100*(1 - s/(2*np.pi))**2/(R**2))
```

The parameterized curve is shown together with the surface $h(x, y)$ in the right plot in Fig. 5.12.

5.10.4 Contour Lines

Contour lines are lines defined by the implicit equation $h(x, y) = C$, where C is some constant representing the contour level. Normally, we let C run over some equally spaced values, and very often, the plotting program computes the C values. To distinguish contours, one often associates each contour level C with its own color.

Figure 5.13 shows different ways contour lines can be used to visualize the surface $h(x, y)$. The first and last plot are visualizations utilizing two spatial dimensions. The first draws a small set of contour lines only, while the last one displays the surface as an image, whose colors reflect the values of the field, or equivalently, the height of the surface. The third plot actually combines three different types of contours, each type corresponding to keeping a coordinate constant and projecting the contours on a "wall". The code used to generate these plots is presented in Sect. 5.11.2.

5.10.5 The Gradient Vector Field

The *gradient vector field* ∇h of a 2D scalar field $h(x, y)$ is defined by

$$\nabla h = \frac{\partial h}{\partial x} i + \frac{\partial h}{\partial y} j. \tag{5.15}$$

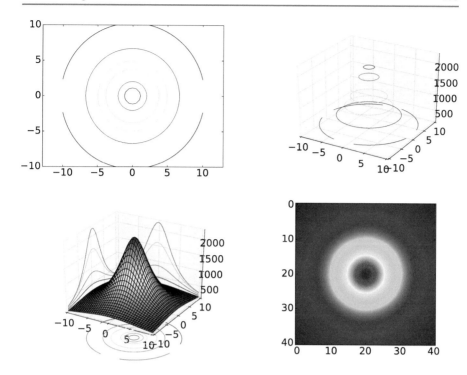

Fig. 5.13 Different types of contour plots of a 2D scalar field in two and three dimensions

One learns in vector calculus that the gradient points in the direction where h increases most, and that the gradients are orthogonal to the contour lines. This is something we can easily illustrate by creating 2D plots of the contours and the gradient field. A challenge in making such plots is to get the right arrow lengths so that the arrows are well visible, but they do not collide and make a cluttered visual impression. Since the arrows are drawn at each point in a 2D grid, one way of controlling the number of arrows is to control the resolution of the grid.

So, let us create a grid with 20 instead of 40 intervals in the horizontal directions:

```
x2 = y2 = np.linspace(-10.,10.,11)
x2v, y2v = np.meshgrid(x2, y2, indexing='ij', sparse=False)
h2v = h0/(1 + (x2v**2 + y2v**2)/(R**2))  # h on coarse grid
```

The gradient vector field of $h(x, y)$ can now be computed using the function np.gradient:

```
dhdx, dhdy = np.gradient(h2v)  # dh/dx, dh/dy
```

The gradient field (5.15) together with the contours appear in Fig. 5.14, from which the orthogonality can be easily seen. Section 5.11.3 explains the code needed to make this plot.

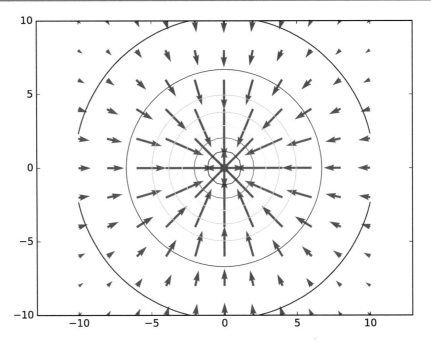

Fig. 5.14 Gradient field with contour plot

5.11 Matplotlib

We import any visualization package under the name `plt`, so for Matplotlib the import is done by

```
import matplotlib.pyplot as plt
```

When creating two-dimensional plots of scalar and vector fields, we shall make use of a Matplotlib `Axes` object, named `ax` and made by

```
fig = plt.figure(1)    # Get current figure
ax = fig.gca()         # Get current axes
```

For three-dimensional visualization, we need the following alternative lines:

```
from mpl_toolkits.mplot3d import Axes3D

fig = plt.figure(1)
ax = fig.gca(projection='3d')
```

5.11.1 Surface Plots

The Matplotlib functions for producing surface plots of 2D scalar fields are `ax.plot_wireframe` and `ax.plot_surface`. The first one produces a wire-

frame plot, and the second one colors the surface. The following code uses the
functions to produce the plots shown in Fig. 5.12, once the grid has been defined as
in Sect. 5.10.2, and the coordinates of the parameterized curve have been computed
as in Sect. 5.10.3.

```
fig = plt.figure(1)
ax = fig.gca(projection='3d')
ax.plot_wireframe(xv, yv, hv, rstride=2, cstride=2)

# Simple plot of mountain and parametric curve
fig = plt.figure(2)
ax = fig.gca(projection='3d')
from matplotlib import cm
ax.plot_surface(xv, yv, hv, cmap=cm.coolwarm,
                rstride=1, cstride=1)

# add the parametric curve. linewidth controls the width of the curve
ax.plot(curve_x, curve_y, curve_z, linewidth=5)
```

Recall that a final `plt.show()` command is necessary to force Matplotlib to show
a plot on the screen.

Note that the second plot in this figure is drawn using a finer grid. This is controlled with the `rstride` and `cstride` parameters, which sets the number of grid
lines in each direction. Setting one of these to 1 means that a grid line is drawn for
every value in the grid in the corresponding direction, and setting to 2 means that
a grid line will be drawn for every two values in the grid. You will normally need
to experiment with such parameters to get a visually attractive plot.

A surface with colors reflecting the height of the surface needs specification of
a *color map*, which is a mapping between function values and colors. Above we
applied the common `coolwarm` scheme which goes from blue ("cool" color for
minimum values) to red ("warm" color for maximum values). There are lots of
colormaps to choose from, and you have to experiment to find appropriate choices
according to your taste and to the problem at hand.

To the latter plot we also added the parameterized curve $r(t)$, defined by (5.14),
using the command `plot`. The attribute `linewidth` is increased here in order to
make the curve thicker and more visible. By default, Matplotlib adds plots to each
other without any need for `plt.hold('on')`, although such a command can indeed
be used.

5.11.2 Contour Plots

The following code exemplifies different types of contour plots. The first two
plots (default two-dimensional and three-dimensional contour plots) are shown in
Fig. 5.13. The next four plots appear in Fig. 5.15. Note that, when we asked Matplotlib to plot 10 contours, the response was, surprisingly, 9 contour lines, where
one of the contours was incomplete. This kind of behavior may also be found in
other plotting packages (such as MATLAB): the package will do its best to plot
the requested number of complete contour lines, but there is no guarantee that this
number is achieved exactly.

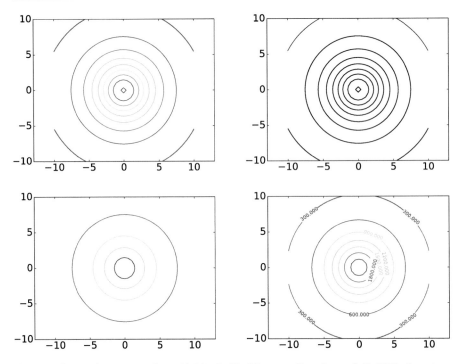

Fig. 5.15 Some other contour plots with Matplotlib: 10 contour lines (*upper left*), 10 black contour lines (*upper right*), specified contour levels (*lower left*), and labeled levels (*lower right*)

```
# Default two-dimensional contour plot with 7 colored lines
fig = plt.figure(3)
ax = fig.gca()
ax.contour(xv, yv, hv)
plt.axis('equal')

# Default three-dimensional contour plot
fig = plt.figure(4)
ax = fig.gca(projection='3d')
ax.contour(xv, yv, hv)

# Plot of mountain and contour lines projected on the
# coordinate planes
fig = plt.figure(5)
ax = fig.gca(projection='3d')
ax.plot_surface(xv, yv, hv, cmap=cm.coolwarm,
                rstride=1, cstride=1)
# zdir is the projection axis
# offset is the offset of the projection plane
ax.contour(xv, yv, hv, zdir='z', offset=-1000, cmap=cm.coolwarm)
ax.contour(xv, yv, hv, zdir='x', offset=-10,   cmap=cm.coolwarm)
ax.contour(xv, yv, hv, zdir='y', offset=10,    cmap=cm.coolwarm)

# View the contours by displaying as an image
fig = plt.figure(6)
ax = fig.gca()
ax.imshow(hv)
```

```
# 10 contour lines (equally spaced contour levels)
fig = plt.figure(7)
ax = fig.gca()
ax.contour(xv, yv, hv, 10)
plt.axis('equal')

# 10 black ('k') contour lines
fig = plt.figure(8)
ax = fig.gca()
ax.contour(xv, yv, hv, 10, colors='k')
plt.axis('equal')

# Specify the contour levels explicitly as a list
fig = plt.figure(9)
ax = fig.gca()
levels = [500., 1000., 1500., 2000.]
ax.contour(xv, yv, hv, levels=levels)
plt.axis('equal')

# Add labels with the contour level for each contour line
fig = plt.figure(10)
ax = fig.gca()
cs = ax.contour(xv, yv, hv)
plt.clabel(cs)
plt.axis('equal')
```

5.11.3 Vector Field Plots

The code for plotting the gradient field (5.15) together with contours goes as explained below, once the grid has been defined as in Sect. 5.10.5. The corresponding plot is shown in Fig. 5.14.

```
fig = plt.figure(11)
ax = fig.gca()
ax.quiver(x2v, y2v, dhdx, dhdy, color='r',
          angles='xy', scale_units='xy')
ax.contour(xv, yv, hv)
plt.axis('equal')
```

5.12 Mayavi

Mayavi is an advanced, free, easy to use, scientific data visualizer, with an emphasis on three-dimensional visualization techniques. The package is written in Python, and uses the Visualization Toolkit (VTK) in C++ for rendering graphics. Since VTK can be configured with different backends, so can Mayavi. Mayavi is cross platform and runs on most platforms, including Mac OS X, Windows, and Linux.

The web page http://docs.enthought.com/mayavi/mayavi/ collects pointers to all relevant documentation of Mayavi. We shall primarily deal with the `mayavi.mlab` module, which provides a simple interface to plotting of 2D scalar and vector fields with commands that mimic those of MATLAB. Let us import this module under our usual name `plt` for a plotting package:

```
import mayavi.mlab as plt
```

The official documentation of the `mlab` module is provided in two places, one for the basic functionality[8] and one for further functionality[9]. Basic figure handling[10] is very similar to the one we know from Matplotlib. Just as for Matplotlib, all plotting commands you do in `mlab` will go into the same figure, until you manually change to a new figure.

5.12.1   Surface Plots

Mayavi has the functions `mesh` and `surf` for producing surface plots. These are similar, but `surf` assumes an orthogonal grid, and uses this assumption to make efficient data structures, while `mesh` makes no such assumptions on the grid. Here we only use orthogonal grids and hence apply `surf`. The following code plots the surface $h(x, y)$ in (5.13), as well as the parameterized curve $r(t)$ in (5.14). The resulting graphics appears in Fig. 5.16.

```
# Create a figure with white background and black foreground
plt.figure(1, fgcolor=(.0, .0, .0), bgcolor=(1.0, 1.0, 1.0))
# 'representation' sets type of plot, here a wireframe plot
plt.surf(xv, yv, hv, extent=(0,1,0,1,0,1),
         representation='wireframe')
# Decorate axes (nb_labels is the number of labels used
# in each direction)
plt.axes(xlabel='x', ylabel='y', zlabel='z', nb_labels=5,
         color=(0., 0., 0.))
# Decorate the plot with a title
plt.title('h(x,y)', size=0.4)

# Simple plot of mountain and parametric curve.
plt.figure(2, fgcolor=(.0, .0, .0), bgcolor=(1.0, 1.0, 1.0))
# Here, representation has default: colored surface elements
plt.surf(xv, yv, hv, extent=(0,1,0,1,0,1))
# Add the parametric curve. tube_radius is the width of the
# curve (use 'extent' for auto-scaling)
plt.plot3d(curve_x, curve_y, curve_z, tube_radius=0.2,
           extent=(0,1,0,1,0,1))

plt.figure(3, fgcolor=(.0, .0, .0), bgcolor=(1.0, 1.0, 1.0))
# Use 'warp_scale' for vertical scaling
plt.surf(xv, yv, hv, warp_scale=0.01, color=(.5, .5, .5))
plt.plot3d(curve_x, curve_y, 0.01*curve_z, tube_radius=0.2)
```

`surf` can produce wireframe plots, as well as plots where the faces of the surface are colored. The parameter `representation` controls this, as exemplified in the first two plots. The first plot was also decorated with axes and a title.

The calls to `plt.figure()` take three parameters: First the usual index for the plot, then two tuples of numbers , representing the RGB-values to be used for the foreground (`fgcolor`) and the background (`bgcolor`). White and black are $(1,1,1)$

---

[8] http://docs.enthought.com/mayavi/mayavi/auto/mlab_helper_functions.html
[9] http://docs.enthought.com/mayavi/mayavi/auto/mlab_other_functions.html
[10] http://docs.enthought.com/mayavi/mayavi/auto/mlab_figure.html

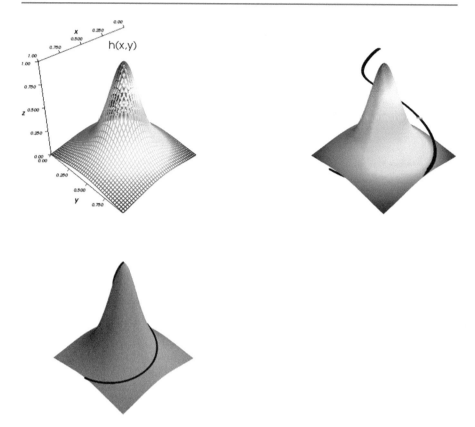

Fig. 5.16 Surface plots produced with the `surf` function of Mayavi: The curve $r(t)$ is also shown in the two last plots

and (0,0,0), respectively. The foreground color is used for text and labels included in the plot. The `color` attribute in `plt.surf` adjusts the surface so that it is colored with small variations from the provided base color, here (`.5, .5, .5`).

The command `plot3d` is used to plot the curve $r(t)$. We have here increased the attribute `tube_radius` to make the curve thicker and more visible.

Mayavi does no auto-scaling of the axes by default (contrary to Matplotlib), so if the magnitudes in the vertical and horizontal directions are very different, as they are for $h(x, y)$, the plots may be very concentrated in one direction. We therefore need to apply some auto-scaling procedure. In Fig. 5.16 two such procedures are exemplified. In the first two plots the parameter `extent` is used. It tells Mayavi to auto-scale the surface and curve to fit the contents described by the six listed values (we will return to what these values mean when we have a more illustrating example). Since the curve and the surface span different areas in space, we see that they are auto-scaled differently in the second plot, with the undesired effect that $r(t)$ is not drawn on the surface. The last plot has avoided this problem by using the `warp_scale` parameter for scaling the vertical direction. Not all Mayavi functions accept this parameter. A remedy for this is to scale the z-coordinates manually, as here exemplified in the last `plot3d`-call. As is seen, the curve is

drawn correctly with respect to the surface in the last plot. In the following we will
use the `warp_scale` parameter to avoid such auto-scaling problems.

Subplots The two plots in Fig. 5.16 were created as separate figures. One can also
create them as subplots within one figure:

```
plt.figure(4, fgcolor=(.0, .0, .0), bgcolor=(1.0, 1.0, 1.0))
plt.mesh(xv, yv, hv, extent=(0, 0.25, 0, 0.25, 0, 0.25),
         colormap='cool')
plt.outline(plt.mesh(
    xv, yv, hv,
    extent=(0.375, 0.625, 0, 0.25, 0, 0.25),
    colormap='Accent'))
plt.outline(plt.mesh(
    xv, yv, hv, extent=(0.75, 1, 0, 0.25, 0, 0.25),
    colormap='prism'), color=(.5, .5, .5))
```

The result is shown in Fig. 5.17. Three separate `mesh` commands are run, each
producing a new plot in the current figure. The commands use different values for
the `colormap` attribute to color the surface in different ways. When this attribute
is not provided, as in the code producing the two first plots in Fig. 5.16, a default
colormap is used.

The `plt.outline` command is used to create a frame around the subplots, and
as seen, we exemplify this possibility for the last two subplots, but not the first one.
We see that one of the two frames has a different color, obtained by setting the
`color` attribute of the `plt.outline` command.

From the computer code it is hopefully clear that the six values listed in `extent`
represent fractions of the cube (0,1,0,1,0,1), where the corresponding plots are
placed. The extents for the three plots are here defined such that they do not overlap.

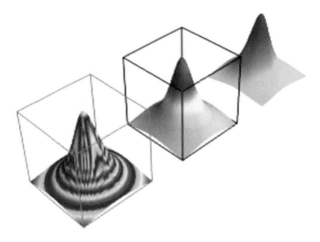

Fig. 5.17 A plot with three subplots created with Mayavi

5.12.2 Contour Plots

The following code exemplifies how one can produce contour plots with Mayavi. The code is very similar to that of Matplotlib, but one difference is that the attribute contours now can represent the number of levels, as well as the levels themselves. The plots are shown in Fig. 5.18.

```
# Default contour plot plotted together with surf.
plt.figure(5, fgcolor=(.0, .0, .0), bgcolor=(1.0, 1.0, 1.0))
plt.surf(xv, yv, hv, warp_scale=0.01)
plt.contour_surf(xv, yv, hv, warp_scale=0.01)

# 10 contour lines (equally spaced contour levels).
plt.figure(6, fgcolor=(.0, .0, .0), bgcolor=(1.0, 1.0, 1.0))
plt.contour_surf(xv, yv, hv, contours=10, warp_scale=0.01)

# 10 contour lines (equally spaced contour levels) together
# with surf. Black color for contour lines.
plt.figure(7, fgcolor=(.0, .0, .0), bgcolor=(1.0, 1.0, 1.0))
plt.surf(xv, yv, hv, warp_scale=0.01)
plt.contour_surf(xv, yv, hv, contours=10, color=(0., 0., 0.),
                 warp_scale=0.01)

# Specify the contour levels explicitly as a list.
plt.figure(8, fgcolor=(.0, .0, .0), bgcolor=(1.0, 1.0, 1.0))
levels = [500., 1000., 1500., 2000.]
plt.contour_surf(xv, yv, hv, contours=levels, warp_scale=0.01)

# View the contours by displaying as an image.
plt.figure(9, fgcolor=(.0, .0, .0), bgcolor=(1.0, 1.0, 1.0))
plt.imshow(hv)
```

Note that there is no function in Mayavi which labels the contours.

Contour plots in Mayavi are shown in three-dimensional space, but you can rotate and look at them from above if you want a two-dimensional plot. Their visual appearance may be enhanced by also including the surface plot itself. We have done this for the top and middle left plots in Fig. 5.18. There is a clear difference in visual impression between these two plots: in the first one, default surface- and contour coloring is used, resulting in less visible contours, but in the middle left plot (plt.figure 6), we set black contours to make them better stand out.

5.12.3 Vector Field Plots

Mayavi supports only vector fields in three-dimensional space. We will therefore visualize the two-dimensional gradient field (5.15) by adding a third component of zero. The following code plots this gradient field together with the contours of h.

```
plt.figure(11, fgcolor=(.0, .0, .0), bgcolor=(1.0, 1.0, 1.0))
plt.contour_surf(xv, yv, hv, contours=20, warp_scale=0.01)
```

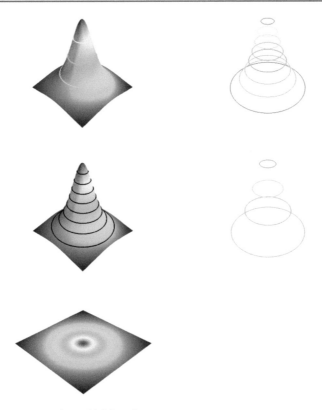

Fig. 5.18 Some contour plots with Mayavi

```
# mode controls the style how vectors are drawn
# color controls the colors of the vectors
# scale_mode='none' ensures that vectors are drawn with the same length
plt.quiver3d(x2v, y2v, 0.01*h2v, dhdx, dhdy, np.zeros_like(dhdx),
             mode='arrow', color=(1,0,0), scale_mode='none')
```

This will produce a 3D view, which we again can rotate to obtain a 2D view. The result is shown in Fig. 5.19, which is similar to Fig. 5.14.

5.12.4 A 3D Scalar Field and Its Gradient Field

Mayavi has functionality for drawing contour surfaces of 3D scalar fields. Let us consider the 3D scalar field

$$g(x, y, z) = z - h(x, y). \tag{5.16}$$

A three-dimensional grid for g can be computed as follows.

```
x = y = np.linspace(-10.,10.,41)
z = np.linspace(0, 50, 41)
```

Fig. 5.19 Gradient field with contour plot

```
xv, yv, zv = np.meshgrid(x, y, z,
                         sparse=False, indexing='ij')
hv = 0.01*h0/(1 + (xv**2+yv**2)/(R**2))
gv = zv - hv
```

The contours are now surfaces defined by the implicit equation $g(x, y, z) = C$, corresponding to vertical shifts of the surface $h(x, y)$.

A corresponding vector field can be calculated:

$$\nabla g = \frac{\partial g}{\partial x}\boldsymbol{i} + \frac{\partial g}{\partial y}\boldsymbol{j} + \frac{\partial g}{\partial z}\boldsymbol{k}. \qquad (5.17)$$

numpy's gradient function can be used to compute a gradient vector field in 3D as well, but you need a three-dimensional grid for the field as input. For the field (5.16), the gradient field is computed as follows.

```
x2 = y2 = np.linspace(-10.,10.,5)
z2 = np.linspace(0, 50, 5)
x2v, y2v, z2v = np.meshgrid(x2, y2, z2,
                           indexing='ij', sparse=False)
h2v = 0.01*h0/(1 + (x2v**2 + y2v**2)/(R**2))
g2v = z2v - h2v
dhdx, dhdy, dhdz = np.gradient(g2v)
```

Again we have used a coarser grid for the vector field.

To visualize the field (5.16) and its gradient field together, we draw enough contours, as we did in the 2D case in Fig. 5.14. The following code can be used.

```
plt.figure(12, fgcolor=(.0, .0, .0), bgcolor=(1.0, 1.0, 1.0))
# opacity controls how contours are visible through each other
plt.contour3d(xv, yv, zv, gv, contours=7, opacity=0.5)
# scale_mode='none': vectors should not be scaled
plt.quiver3d(x2v, y2v, z2v, dhdx, dhdy, dhdz, mode='arrow',
             scale_mode='none', opacity=0.5)
```

The result is shown in Fig. 5.20.

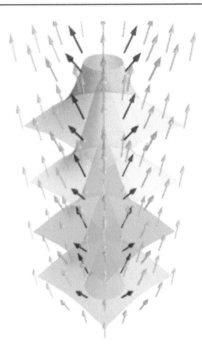

Fig. 5.20   The 3D scalar field (5.16) and its gradient field

This example demonstrates some of the challenges in plotting three-dimensional vector fields. The vectors must not be too dense, and not too long. It is inevitable that contours shadow one another. Fortunately, Mayavi supports an opacity setting, which controls how contours are visible through each other. Visualizing a 3D scalar field is clearly challenging, and we have only touched the subject.

5.12.5   Animations

It is straightforward to create animations with Mayavi. In the following code the function $h(x, y)$ is scaled vertically, for different scaling constants between 0 and 1, and each plot is saved in its own file. The files can then be combined to a standard video file.

```
plt.figure(13, fgcolor=(.0, .0, .0), bgcolor=(1.0, 1.0, 1.0))
s = plt.surf(xv, yv, hv, warp_scale=0.01)

for i in range(10):
    # s.mlab_source.scalars is a handle for the values of the surface,
    # and is updated here
    s.mlab_source.scalars = hv*0.1*(i+1)
    plt.savefig('tmp_%04d.png' % i)
```

5.13 Summary

5.13.1 Chapter Topics

This chapter has introduced computing with arrays and plotting curve data stored in arrays. The Numerical Python package contains lots of functions for array computing, including the ones listed in the table below. Plotting has been done with tools that closely resemble the syntax of MATLAB.

Construction	Meaning
`array(ld)`	copy list data `ld` to a `numpy` array
`asarray(d)`	make array of data `d` (no data copy if already array)
`zeros(n)`	make a `float` vector/array of length `n`, with zeros
`zeros(n, int)`	make an `int` vector/array of length `n` with zeros
`zeros((m,n))`	make a two-dimensional `float` array with shape $(m,'n')$
`zeros_like(x)`	make array of same shape and element type as `x`
`linspace(a,b,m)`	uniform sequence of `m` numbers in $[a,b]$
`a.shape`	tuple containing `a`'s shape
`a.size`	total no of elements in `a`
`len(a)`	length of a one-dim. array `a` (same as `a.shape[0]`)
`a.dtype`	the type of elements in `a`
`a.reshape(3,2)`	return `a` reshaped as 3×2 array
`a[i]`	vector indexing
`a[i,j]`	two-dim. array indexing
`a[1:k]`	slice: reference data with indices `1,2,...,k-1`
`a[1:10:3]`	slice: reference data with indices `1,4,7`
`b = a.copy()`	copy an array
`sin(a), exp(a), ...`	`numpy` functions applicable to arrays
`c = concatenate((a, b))`	`c` contains `a` with `b` appended
`c = where(cond, a1, a2)`	`c[i] = a1[i]` if `cond[i]`, else `c[i] = a2[i]`
`isinstance(a, ndarray)`	is `True` if `a` is an array

Array computing When we apply a Python function `f(x)` to a Numerical Python array `x`, the result is the same as if we apply `f` to each element in `x` separately. However, when `f` contains `if` statements, these are in general invalid if an array `x` enters the boolean expression. We then have to rewrite the function, often by applying the `where` function from Numerical Python.

Plotting curves Sections 5.3.1 and 5.3.2 provide a quick overview of how to plot curves with the aid of Matplotlib. The same examples coded with the Easyviz plotting interface appear in Sect. 5.3.3.

Making movies Each frame in a movie must be a hardcopy of a plot in PNG format. These plot files should have names containing a counter padded with leading zeros. One example may be `tmp_0000.png`, `tmp_0001.png`, `tmp_0002.png`.

Having the plot files with names on this form, we can make an animated GIF movie in the file `movie.gif`, with two frames per second, by

```
os.system('convert -delay 50 tmp_*.png movie.gif')
```

Alternatively, we may combine the plot files to a Flash video:

```
os.system('ffmpeg -r 5 -i tmp_%04d.png -vcodec flv movie.flv')
```

Other formats can be made using other codecs, see Sect. 5.3.5.

Terminology   The important topics in this chapter are

- array computing
- vectorization
- plotting
- animations

5.13.2   Example: Animating a Function

Problem   In this chapter's summarizing example we shall visualize how the temperature varies downward in the earth as the surface temperature oscillates between high day and low night values. One question may be: What is the temperature change 10 m down in the ground if the surface temperature varies between 2 C in the night and 15 C in the day?

Let the z axis point downwards, towards the center of the earth, and let $z = 0$ correspond to the earth's surface. The temperature at some depth z in the ground at time t is denoted by $T(z, t)$. If the surface temperature has a periodic variation around some mean value T_0, according to

$$T(0, t) = T_0 + A \cos(\omega t),$$

one can find, from a mathematical model for heat conduction, that the temperature at an arbitrary depth is

$$T(z, t) = T_0 + A e^{-az} \cos(\omega t - az), \quad a = \sqrt{\frac{\omega}{2k}}. \tag{5.18}$$

The parameter k reflects the ground's ability to conduct heat (k is called the *thermal diffusivity* or the *heat conduction coefficient*).

The task is to make an animation of how the temperature profile in the ground, i.e., T as a function of z, varies in time. Let ω correspond to a time period of 24 hours. The mean temperature T_0 is taken as 10 C, and the maximum variation A is assumed to be 10 C. The heat conduction coefficient k may be set as 1 mm^2/s (which is 10^{-6} m^2/s in proper SI units).

Solution To animate $T(z,t)$ in time, we need to make a loop over points in time, and in each pass in the loop we must save a plot of T, as a function of z, to file. The plot files can then be combined to a movie. The algorithm becomes

- for $t_i = i\,\Delta t$, $i = 0, 1, 2\ldots,n$:
  - plot the curve $y(z) = T(z,t_i)$
  - store the plot in a file
- combine all the plot files into a movie

It can be wise to make a general `animate` function where we just feed in some $f(x,t)$ function and make all the plot files. If `animate` has arguments for setting the labels on the axis and the extent of the y axis, we can easily use `animate` also for a function $T(z,t)$ (we just use z as the name for the x axis and T as the name for the y axis in the plot). Recall that it is important to fix the extent of the y axis in a plot when we make animations, otherwise most plotting programs will automatically fit the extent of the axis to the current data, and the tick marks on the y axis will jump up and down during the movie. The result is a wrong visual impression of the function.

The names of the plot files must have a common stem appended with some frame number, and the frame number should have a fixed number of digits, such as 0001, 0002, etc. (if not, the sequence of the plot files will not be correct when we specify the collection of files with an asterisk for the frame numbers, e.g., as in `tmp*.png`). We therefore include an argument to `animate` for setting the name stem of the plot files. By default, the stem is `tmp_`, resulting in the filenames `tmp_0000.png`, `tmp_0001.png`, `tmp_0002.png`, and so forth. Other convenient arguments for the `animate` function are the initial time in the plot, the time lag Δt between the plot frames, and the coordinates along the x axis. The `animate` function then takes the form

```
def animate(tmax, dt, x, function, ymin, ymax, t0=0,
            xlabel='x', ylabel='y', filename='tmp_'):
    t = t0
    counter = 0
    while t <= tmax:
        y = function(x, t)
        plot(x, y, '-',
             axis=[x[0], x[-1], ymin, ymax],
             title='time=%2d h' % (t/3600.0),
             xlabel=xlabel, ylabel=ylabel,
             savefig=filename + '%04d.png' % counter)
        savefig('tmp_%04d.pdf' % counter)
        t += dt
        counter += 1
```

The $T(z,t)$ function is easy to implement, but we need to decide whether the parameters A, ω, T_0, and k shall be arguments to the Python implementation of $T(z,t)$ or if they shall be global variables. Since the `animate` function expects that the function to be plotted has only two arguments, we must implement $T(z,t)$ as `T(z,t)` in Python and let the other parameters be global variables (Sects. 7.1.1 and 7.1.2 explain this problem in more detail and present a better implementation). The `T(z,t)` implementation then reads

```
def T(z, t):
    # T0, A, k, and omega are global variables
    a = sqrt(omega/(2*k))
    return T0 + A*exp(-a*z)*cos(omega*t - a*z)
```

Suppose we plot $T(z,t)$ at n points for $z \in [0, D]$. We make such plots for $t \in [0, t_{max}]$ with a time lag Δt between the them. The frames in the movie are now made by

```
# set T0, A, k, omega, D, n, tmax, dt
z = linspace(0, D, n)
animate(tmax, dt, z, T, T0-A, T0+A, 0, 'z', 'T')
```

We have here set the extent of the y axis in the plot as $[T_0 - A, T_0 + A]$, which is in accordance with the $T(z,t)$ function.

The call to `animate` above creates a set of files with names of the form `tmp_*.png`. Out of these files we can create an animated GIF movie or a video in, e.g., Flash format by running operating systems commands with `convert` and `avconv` (or `ffmpeg`):

```
os.system('convert -delay 50 tmp_*.png movie.gif')
os.system('avconv -i tmp_%04d.png -r 5 -vcodec flv movie.flv')
```

See Sect. 5.3.5 for how to create videos in other formats.

It now remains to assign proper values to all the global variables in the program: n, D, T0, A, omega, dt, tmax, and k. The oscillation period is 24 hours, and ω is related to the period P of the cosine function by $\omega = 2\pi/P$ (realize that $\cos(t 2\pi/P)$ has period P). We then express $P = 24$ h as $24 \cdot 60 \cdot 60$ s and compute ω as $2\pi/P \approx 7 \cdot 10^{-5}$ s^{-1}. The total simulation time can be 3 periods, i.e., $t_{max} = 3P$. The $T(z,t)$ function decreases exponentially with the depth z so there is no point in having the maximum depth D larger than the depth where T is visually zero, say 0.001. We have that $e^{-aD} = 0.001$ when $D = -a^{-1} \ln 0.001$, so we can use this estimate in the program. The proper initialization of all parameters can then be expressed as follows:

```
k = 1E-6          # thermal diffusivity (in m*m/s)
P = 24*60*60.     # oscillation period of 24 h (in seconds)
omega = 2*pi/P
dt = P/24         # time lag: 1 h
tmax = 3*P        # 3 day/night simulation
T0 = 10           # mean surface temperature in Celsius
A = 10            # amplitude of the temperature variations in Celsius
a = sqrt(omega/(2*k))
D = -(1/a)*log(0.001) # max depth
n = 501           # no of points in the z direction
```

Note that it is very important to use consistent units. Here we express all units in terms of meter, second, and Kelvin or Celsius.

We encourage you to run the program `heatwave.py` to see the movie. The hardcopy of the movie is in the file `movie.gif`. Figure 5.21 displays two snapshots in time of the $T(z,t)$ function.

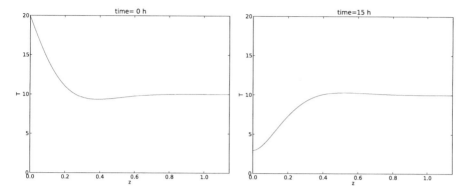

Fig. 5.21 Plot of the temperature $T(z,t)$ in the ground for two different t values

Scaling In this example, as in many other scientific problems, it was easier to write the code than to assign proper physical values to the input parameters in the program. To learn about the physical process, here how heat propagates from the surface and down in the ground, it is often advantageous to scale the variables in the problem so that we work with dimensionless variables. Through the scaling procedure we normally end up with much fewer physical parameters that must be assigned values. Let us show how we can take advantage of scaling the present problem.

Consider a variable x in a problem with some dimension. The idea of scaling is to introduce a new variable $\bar{x} = x/x_c$, where x_c is a *characteristic size* of x. Since x and x_c have the same dimension, the dimension cancels in $\bar{x}$ such that $\bar{x}$ is dimensionless. Choosing x_c to be the expected maximum value of x, ensures that $\bar{x} \leq 1$, which is usually considered a good idea. That is, we try to have all dimensionless variables varying between zero and one. For example, we can introduce a dimensionless z coordinate: $\bar{z} = z/D$, and now $\bar{z} \in [0, 1]$. Doing a proper scaling of a problem is challenging so for now it is sufficient to just follow the steps below – and not worry why we choose a certain scaling.

In the present problem we introduce these dimensionless variables:

$$\bar{z} = z/D$$
$$\bar{T} = \frac{T - T_0}{A}$$
$$\bar{t} = \omega t$$

We now insert $z = \bar{z}D$ and $t = \bar{t}/\omega$ in the expression for $T(z,t)$ and get

$$T = T_0 + Ae^{-b\bar{z}} \cos(\bar{t} - b\bar{z}), \quad b = aD$$

or

$$\bar{T}(\bar{z}, \bar{t}) = \frac{T - T_0}{A} = e^{-b\bar{z}} \cos(\bar{t} - b\bar{z}) .$$

We see that $\bar{T}$ depends on only *one* dimensionless parameter b in addition to the independent dimensionless variables $\bar{z}$ and $\bar{t}$. It is common practice at this stage of the scaling to just drop the bars and write

$$T(z,t) = e^{-bz} \cos(t - bz). \tag{5.19}$$

This function is much simpler to plot than the one with lots of physical parameters, because now we know that T varies between -1 and 1, t varies between 0 and 2π for one period, and z varies between 0 and 1. The scaled temperature has only one parameter b in addition to the independent variable. That is, the shape of the graph is completely determined by b.

   In our previous movie example, we used specific values for D, ω, and k, which then implies a certain $b = D\sqrt{\omega/(2k)}$ (≈ 6.9). However, we can now run different b values and see the effect on the heat propagation. Different b values will in our problems imply different periods of the surface temperature variation and/or different heat conduction values in the ground's composition of rocks. Note that doubling ω and k leaves the same b – it is only the fraction ω/k that influences the value of b.

   We can reuse the `animate` function also in the scaled case, but we need to make a new $T(z,t)$ function and, e.g., a main program where b can be read from the command line:

```
def T(z, t):
    return exp(-b*z)*cos(t - b*z)  # b is global

b = float(sys.argv[1])
n = 401
z = linspace(0, 1, n)
animate(3*2*pi, 0.05*2*pi, z, T, -1.2, 1.2, 0, 'z', 'T')
movie('tmp_*.png', encoder='convert', fps=2,
      output_file='tmp_heatwave.gif')
os.system('convert -delay 50 tmp_*.png movie.gif')
```

   Running the program, found as the file `heatwave_scaled.py`, for different b values shows that b governs how deep the temperature variations on the surface $z = 0$ penetrate. A large b makes the temperature changes confined to a thin layer close to the surface, while a small b leads to temperature variations also deep down in the ground. You are encouraged to run the program with $b = 2$ and $b = 20$ to experience the major difference, or just view the ready-made animations[11].

   We can understand the results from a physical perspective. Think of increasing ω, which means reducing the oscillation period so we get a more rapid temperature variation. To preserve the value of b we must increase k by the same factor. Since a large k means that heat quickly spreads down in the ground, and a small k implies the opposite, we see that more rapid variations at the surface requires a larger k to more quickly conduct the variations down in the ground. Similarly, slow temperature variations on the surface can penetrate deep in the ground even if the ground's ability to conduct (k) is low.

---

[11] http://hplgit.github.io/scipro-primer/video/heatwave.html

5.14 Exercises

Exercise 5.1: Fill lists with function values
Define

$$h(x) = \frac{1}{\sqrt{2\pi}} e^{-\frac{1}{2}x^2} . \qquad (5.20)$$

Fill lists `xlist` and `hlist` with x and $h(x)$ values for 41 uniformly spaced x coordinates in $[-4, 4]$.

Hint You may adapt the example in Sect. 5.2.1.
Filename: `fill_lists`.

Exercise 5.2: Fill arrays; loop version
The aim is to fill two arrays x and y with x and $h(x)$ values, respectively, where $h(x)$ is defined in (5.20). Let the x values be as in Exercise 5.1. Create empty x and y arrays and compute each element in x and y with a `for` loop.
Filename: `fill_arrays_loop`.

Exercise 5.3: Fill arrays; vectorized version
Vectorize the code in Exercise 5.2 by creating the x values using the `linspace` function from the numpy package and by evaluating $h(x)$ for an array argument.
Filename: `fill_arrays_vectorized`.

Exercise 5.4: Plot a function
Make a plot of the function in Exercise 5.1 for $x \in [-4, 4]$.
Filename: `plot_Gaussian`.

Exercise 5.5: Apply a function to a vector
Given a vector $v = (2, 3, -1)$ and a function $f(x) = x^3 + xe^x + 1$, apply f to each element in v. Then calculate by hand $f(v)$ as the NumPy expression `v**3 + v*exp(v) + 1` using vector computing rules. Demonstrate that the two results are equal.
Filename: `apply_vecfunc`.

Exercise 5.6: Simulate by hand a vectorized expression
Suppose x and t are two arrays of the same length, entering a vectorized expression

```
y = cos(sin(x)) + exp(1/t)
```

If x holds two elements, 0 and 2, and t holds the elements 1 and 1.5, calculate by hand (using a calculator) the y array. Thereafter, write a program that mimics the series of computations you did by hand (typically a sequence of operations of the kind we listed in Sect. 5.1.3 – use explicit loops, but at the end you can use Numerical Python functionality to check the results).
Filename: `simulate_vector_computing`.

Exercise 5.7: Demonstrate array slicing
Create an array `w` with values $0, 0.1, 0.2, \ldots, 3$. Write out `w[:]`, `w[:-2]`, `w[::5]`, `w[2:-2:6]`. Convince yourself in each case that you understand which elements of the array that are printed.
Filename: `slicing`.

Exercise 5.8: Replace list operations by array computing
The data analysis problem in Sect. 2.6.2 is solved by list operations. Convert the list to a two-dimensional array and perform the computations using array operations (i.e., no explicit loops, but you need a loop to make the printout).
Filename: `sun_data_vec`.

Exercise 5.9: Plot a formula
Make a plot of the function $y(t) = v_0 t - \frac{1}{2} g t^2$ for $v_0 = 10$, $g = 9.81$, and $t \in [0, 2v_0/g]$. Set the axes labels as `time (s)` and `height (m)`.
Filename: `plot_ball1`.

Exercise 5.10: Plot a formula for several parameters
Make a program that reads a set of v_0 values from the command line and plots the corresponding curves $y(t) = v_0 t - \frac{1}{2} g t^2$ in the same figure, with $t \in [0, 2v_0/g]$ for each curve. Set $g = 9.81$.

Hint You need a different vector of t coordinates for each curve.
Filename: `plot_ball2`.

Exercise 5.11: Specify the extent of the axes in a plot
Extend the program from Exercises 5.10 such that the minimum and maximum t and y values are computed, and use the extreme values to specify the extent of the axes. Add some space above the highest curve to make the plot look better.
Filename: `plot_ball3`.

Exercise 5.12: Plot exact and inexact Fahrenheit-Celsius conversion formulas
A simple rule to quickly compute the Celsius temperature from the Fahrenheit degrees is to subtract 30 and then divide by 2: $C = (F - 30)/2$. Compare this curve against the exact curve $C = (F - 32)5/9$ in a plot. Let F vary between -20 and 120.
Filename: `f2c_shortcut_plot`.

Exercise 5.13: Plot the trajectory of a ball
The formula for the trajectory of a ball is given by

$$f(x) = x \tan \theta - \frac{1}{2 v_0^2} \frac{g x^2}{\cos^2 \theta} + y_0, \tag{5.21}$$

where x is a coordinate along the ground, g is the acceleration of gravity, v_0 is the size of the initial velocity, which makes an angle θ with the x axis, and $(0, y_0)$ is the initial position of the ball.

In a program, first read the input data y_0, θ, and v_0 from the command line. Then plot the trajectory $y = f(x)$ for $y \geq 0$.
Filename: `plot_trajectory`.

Exercise 5.14: Plot data in a two-column file
The file `src/plot/xy.dat`[12] contains two columns of numbers, corresponding to x and y coordinates on a curve. The start of the file looks as this:

```
-1.0000      -0.0000
-0.9933      -0.0087
-0.9867      -0.0179
-0.9800      -0.0274
-0.9733      -0.0374
```

Make a program that reads the first column into a list x and the second column into a list y. Plot the curve. Print out the mean y value as well as the maximum and minimum y values.

Hint Read the file line by line, split each line into words, convert to `float`, and append to x and y. The computations with y are simpler if the list is converted to an array.
Filename: `read_2columns`.

Remarks The function `loadtxt` in `numpy` can read files with tabular data (any number of columns) and return the data in a two-dimensional array:

```
import numpy as np
# Read table of floats
data = np.loadtxt('xy.dat', dtype=np.float)
# Extract one-dim arrays from two-dim data
x = data[:,0]  # column with index 0
y = data[:,1]  # column with index 1
```

The present exercise asks you to implement a simplified version of `loadtxt`, but for later loading of a file with tabular data into an array you will certainly use `loadtxt`.

Exercise 5.15: Write function data to file
We want to dump x and $f(x)$ values to a file, where the x values appear in the first column and the $f(x)$ values appear in the second. Choose n equally spaced x values in the interval $[a, b]$. Provide f, a, b, n, and the filename as input data on the command line.

Hint You may use the `StringFunction` tool (see Sects. 4.3.3 and 5.5.1) to turn the textual expression for f into a Python function. (Note that the program from Exercise 5.14 can be used to read the file generated in the present exercise into arrays again for visualization of the curve $y = f(x)$.)
Filename: `write_cml_function`.

---

[12] http://tinyurl.com/pwyasaa/plot/xy.dat

Exercise 5.16: Plot data from a file

The files `density_water.dat` and `density_air.dat` files in the folder `src/plot`[13] contain data about the density of water and air (respectively) for different temperatures. The data files have some comment lines starting with # and some lines are blank. The rest of the lines contain density data: the temperature in the first column and the corresponding density in the second column. The goal of this exercise is to read the data in such a file and plot the density versus the temperature as distinct (small) circles for each data point. Let the program take the name of the data file as command-line argument. Apply the program to both files.
Filename: `read_density_data`.

Exercise 5.17: Write table to file

Given a function of two parameters x and y, we want to create a file with a table of function values. The left column of the table contains y values in decreasing order as we go down the rows, and the last row contains the x values in increasing order. That is, the first column and the last row act like numbers on an x and y axis in a coordinate system. The rest of the table cells contains function values corresponding to the x and y values for the respective rows and columns. For example, if the function formula is $x + 2y$, x runs from 0 to 2 in steps of 0.5, and y run from -1 to 2 in steps of 1, the table looks as follows:

```
  2     4  4.5    5  5.5    6
  1     2  2.5    3  3.5    4
  0     0  0.5    1  1.5    2
 -1    -2 -1.5   -1 -0.5    0

        0  0.5    1  1.5    2
```

The task is to write a function

```python
def write_table_to_file(f, xmin, xmax, nx, ymin, ymax, ny,
                         width=10, decimals=None,
                         filename='table.dat'):
```

where f is the formula, given as a Python function; xmin, xmax, ymin, and ymax are the minimum and maximum x and y values; nx is the number of intervals in the x coordinates (the number of steps in x direction is then (xmax-xmin)/nx); ny is the number of intervals in the y coordinates; width is the width of each column in the table (a positive integer); decimals is the number of decimals used when writing out the numbers (None means no decimal specification), and filename is the name of the output file. For example, width=10 and decimals=1 gives the output format %10.1g, while width=5 and decimals=None implies %5g.

Here is a test function which you should use to verify the implementation:

```
def test_write_table_to_file():
    filename = 'tmp.dat'
    write_table_to_file(f=lambda x, y: x + 2*y,
                        xmin=0, xmax=2, nx=4,
                        ymin=-1, ymax=2, ny=3,
                        width=5, decimals=None,
                        filename=filename)
    # Load text in file and compare with expected results
    with open(filename, 'r') as infile:
        computed = infile.read()
    expected = """\
 2    4  4.5    5  5.5    6
 1    2  2.5    3  3.5    4
 0    0  0.5    1  1.5    2
-1   -2 -1.5   -1 -0.5    0

     0  0.5    1  1.5    2"""
    assert computed == expected
```

Filename: `write_table_to_file`.

Exercise 5.18: Fit a polynomial to data points

The purpose of this exercise is to find a simple mathematical formula for how the density of water or air depends on the temperature. The idea is to load density and temperature data from file as explained in Exercise 5.16 and then apply some NumPy utilities that can find a polynomial that approximates the density as a function of the temperature.

NumPy has a function `polyfit(x, y, deg)` for finding a best fit of a polynomial of degree `deg` to a set of data points given by the array arguments `x` and `y`. The `polyfit` function returns a list of the coefficients in the fitted polynomial, where the first element is the coefficient for the term with the highest degree, and the last element corresponds to the constant term. For example, given points in `x` and `y`, `polyfit(x, y, 1)` returns the coefficients `a`, `b` in a polynomial `a*x + b` that fits the data in the best way. (More precisely, a line $y = ax + b$ is a best fit to the data points (x_i, y_i), $i = 0, \ldots, n-1$ if a and b are chosen to make the sum of squared errors $R = \sum_{j=0}^{n-1}(y_j - (ax_j + b))^2$ as small as possible. This approach is known as *least squares approximation* to data and proves to be extremely useful throughout science and technology.)

NumPy also has a utility `poly1d`, which can take the tuple or list of coefficients calculated by, e.g., `polyfit` and return the polynomial as a Python function that can be evaluated. The following code snippet demonstrates the use of `polyfit` and `poly1d`:

```
coeff = polyfit(x, y, deg)
p = poly1d(coeff)
print p                 # prints the polynomial expression
y_fitted = p(x)         # computes the polynomial at the x points
# Use red circles for data points and a blue line for the polyn.
plot(x, y, 'ro', x, y_fitted, 'b-',
     legend=('data', 'fitted polynomial of degree %d' % deg))
```

a) Write a function fit(x, y, deg) that creates a plot of data in x and y arrays along with polynomial approximations of degrees collected in the list deg as explained above.
b) We want to call fit to make a plot of the density of water versus temperature and another plot of the density of air versus temperature. In both calls, use deg=[1,2] such that we can compare linear and quadratic approximations to the data.
c) From a visual inspection of the plots, can you suggest simple mathematical formulas that relate the density of air to temperature and the density of water to temperature?

Filename: fit_density_data.

Exercise 5.19: Fit a polynomial to experimental data
Suppose we have measured the oscillation period T of a simple pendulum with a mass m at the end of a massless rod of length L. We have varied L and recorded the corresponding T value. The measurements are found in a file src/plot/pendulum.dat[14]. The first column in the file contains L values and the second column has the corresponding T values.

a) Plot L versus T using circles for the data points.
b) We shall assume that L as a function of T is a polynomial. Use the NumPy utilities polyfit and poly1d, as explained in Exercise 5.18, to fit polynomials of degree 1, 2, and 3 to the L and T data. Visualize the polynomial curves together with the experimental data. Which polynomial fits the measured data best?

Filename: fit_pendulum_data.

Exercise 5.20: Read acceleration data and find velocities
A file src/plot/acc.dat[15] contains measurements $a_0, a_1, \ldots, a_{n-1}$ of the acceleration of an object moving along a straight line. The measurement a_k is taken at time point $t_k = k\Delta t$, where Δt is the time spacing between the measurements. The purpose of the exercise is to load the acceleration data into a program and compute the velocity $v(t)$ of the object at some time t.

In general, the acceleration $a(t)$ is related to the velocity $v(t)$ through $v'(t) = a(t)$. This means that

$$v(t) = v(0) + \int_0^t a(\tau)d\tau. \tag{5.22}$$

If $a(t)$ is only known at some discrete, equally spaced points in time, $a_0, \ldots, a_{n-1}$ (which is the case in this exercise), we must compute the integral in (5.22) numerically, for example by the Trapezoidal rule:

$$v(t_k) \approx \Delta t \left(\frac{1}{2}a_0 + \frac{1}{2}a_k + \sum_{i=1}^{k-1} a_i \right), \quad 1 \le k \le n-1. \tag{5.23}$$

We assume $v(0) = 0$ so that also $v_0 = 0$.

---

[14] http://tinyurl.com/pwyasaa/plot/pendulum.dat
[15] http://tinyurl.com/pwyasaa/plot/acc.dat

Read the values $a_0, \ldots, a_{n-1}$ from file into an array, plot the acceleration versus time, and use (5.23) to compute one $v(t_k)$ value, where Δt and $k \geq 1$ are specified on the command line.
Filename: `acc2vel_v1`.

Exercise 5.21: Read acceleration data and plot velocities
The task in this exercise is the same as in Exercise 5.20, except that we now want to compute $v(t_k)$ for all time points $t_k = k\Delta t$ and plot the velocity versus time. Now only Δt is given on the command line, and the $a_0, \ldots, a_{n-1}$ values must be read from file as in Exercise 5.20.

Hint Repeated use of (5.23) for all k values is very inefficient. A more efficient formula arises if we add the area of a new trapezoid to the previous integral (see also Sect. A.1.7):

$$v(t_k) = v(t_{k-1}) + \int_{t_{k-1}}^{t_k} a(\tau)d\tau \approx v(t_{k-1}) + \Delta t \frac{1}{2}(a_{k-1} + a_k), \tag{5.24}$$

for $k = 1, 2, \ldots, n - 1$, while $v_0 = 0$. Use this formula to fill an array v with velocity values.
Filename: `acc2vel`.

Exercise 5.22: Plot a trip's path and velocity from GPS coordinates
A GPS device measures your position at every s seconds. Imagine that the positions corresponding to a specific trip are stored as (x, y) coordinates in a file `src/plot/pos.dat`[16] with an x and y number on each line, except for the first line, which contains the value of s.

a) Plot the two-dimensional curve of corresponding to the data in the file.

Hint Load s into a `float` variable and then the x and y numbers into two arrays. Draw a straight line between the points, i.e., plot the y coordinates versus the x coordinates.

b) Plot the velocity in x direction versus time in one plot and the velocity in y direction versus time in another plot.

Hint If $x(t)$ and $y(t)$ are the coordinates of the positions as a function of time, we have that the velocity in x direction is $v_x(t) = dx/dt$, and the velocity in y direction is $v_y = dy/dt$. Since x and y are only known for some discrete times, $t_k = ks, k = 0, \ldots, n - 1$, we must use numerical differentiation. A simple (forward) formula is

$$v_x(t_k) \approx \frac{x(t_{k+1}) - x(t_k)}{s}, \quad v_y(t_k) \approx \frac{y(t_{k+1}) - y(t_k)}{s}, \quad k = 0, \ldots, n - 2.$$

---

[16] http://tinyurl.com/pwyasaa/plot/pos.dat

Compute arrays vx and vy with velocities based on the formulas above for $v_x(t_k)$ and $v_y(t_k)$, $k = 0, \ldots, n - 2$.
Filename: `position2velocity`.

Exercise 5.23: Vectorize the Midpoint rule for integration
The Midpoint rule for approximating an integral can be expressed as

$$\int_a^b f(x)dx \approx h \sum_{i=1}^n f(a - \frac{1}{2}h + ih), \tag{5.25}$$

where $h = (b - a)/n$.

a) Write a function `midpointint(f, a, b, n)` to compute Midpoint rule. Use a plain Python `for` loop to implement the sum.
b) Make a vectorized implementation of the Midpoint rule where you compute the sum by Python's built-in function `sum`.
c) Make another vectorized implementation of the Midpoint rule where you compute the sum by the `sum` function in the `numpy` package.
d) Organize the three implementations above in a module file `midpoint_vec.py`. Equip the module with one test function for verifying the three implementations. Use the integral $\int_2^4 2x\,dx = 12$ as test case since the Midpoint rule will integrate such a linear integrand exactly.
e) Start IPython, import the functions from `midpoint_vec.py`, define some Python implementation of a mathematical function $f(x)$ to integrate, and use the `%timeit` feature of IPython to measure the efficiency of the three alternative implementations.

Hint The `%timeit` feature is described in Sect. H.8.1.
Filename: `midpoint_vec`.

Remarks The lesson learned from the experiments in e) is that `numpy.sum` is much more efficient than Python's built-in function `sum`. Vectorized implementations must always make use of `numpy.sum` to compute sums.

Exercise 5.24: Vectorize a function for computing the area of a polygon
The area of a polygon is given by (3.17) in Exercise 3.19. Vectorize this formula such that there are no Python loops in the implementation. Make a test function that compares the scalar implementation in the referred exercise with the new vectorized implementation for some chosen polygons (the scalar version must then be available in a module so that the function can be imported).

Hint Observe that the formula $x_1 y_2 + x_2 y_3 + \cdots + x_{n-1} y_n = \sum_{i=0}^{n-1} x_i y_{i+1}$ is the dot product of two vectors, `x[:-1]` and `y[1:]`, which can be computed as `numpy.dot(x[:-1], y[1:])`.
Filename: `polygon_area_vec`.

Exercise 5.25: Implement Lagrange's interpolation formula

Imagine we have $n + 1$ measurements of some quantity y that depends on x: $(x_0, y_0), (x_1, y_1), \ldots, (x_n, y_n)$. We may think of y as a function of x and ask what y is at some arbitrary point x not coinciding with any of the points $x_0, \ldots, x_n$. It is not clear how y varies between the measurement points, but we can make assumptions or models for this behavior. Such a problem is known as *interpolation*.

One way to solve the interpolation problem is to fit a continuous function that goes through all the $n + 1$ points and then evaluate this function for any desired x. A candidate for such a function is the polynomial of degree n that goes through all the points. It turns out that this polynomial can be written

$$p_L(x) = \sum_{k=0}^{n} y_k L_k(x), \tag{5.26}$$

where

$$L_k(x) = \prod_{i=0, i \neq k}^{n} \frac{x - x_i}{x_k - x_i}. \tag{5.27}$$

The $\prod$ notation corresponds to $\sum$, but the terms are multiplied. For example,

$$\prod_{i=0, i \neq k}^{n} x_i = x_0 x_1 \cdots x_{k-1} x_{k+1} \cdots x_n.$$

The polynomial $p_L(x)$ is known as Lagrange's interpolation formula, and the points $(x_0, y_0), \ldots, (x_n, y_n)$ are called interpolation points.

a) Make functions p_L(x, xp, yp) and L_k(x, k, xp, yp) that evaluate $p_L(x)$ and $L_k(x)$ by (5.26) and (5.27), respectively, at the point x. The arrays xp and yp contain the x and y coordinates of the $n + 1$ interpolation points, respectively. That is, xp holds $x_0, \ldots, x_n$, and yp holds $y_0, \ldots, y_n$.

b) To verify the program, we observe that $L_k(x_k) = 1$ and that $L_k(x_i) = 0$ for $i \neq k$, implying that $p_L(x_k) = y_k$. That is, the polynomial p_L goes through all the points $(x_0, y_0), \ldots, (x_n, y_n)$. Write a function test_p_L(xp, yp) that computes $|p_L(x_k) - y_k|$ at all the interpolation points (x_k, y_k) and checks that the value is approximately zero. Call test_p_L with xp and yp corresponding to 5 equally spaced points along the curve $y = \sin(x)$ for $x \in [0, \pi]$. Thereafter, evaluate $p_L(x)$ for an x in the middle of two interpolation points and compare the value of $p_L(x)$ with the exact one.

Filename: Lagrange_poly1.

Exercise 5.26: Plot Lagrange's interpolating polynomial

a) Write a function

```
def graph(f, n, xmin, xmax, resolution=1001):
```

for plotting $p_L(x)$ in Exercise 5.25, based on interpolation points taken from some mathematical function $f(x)$ represented by the argument f. The argument n denotes the number of interpolation points sampled from the $f(x)$ function, and `resolution` is the number of points between xmin and xmax used to plot $p_L(x)$. The x coordinates of the n interpolation points can be uniformly distributed between xmin and xmax. In the graph, the interpolation points $(x_0, y_0), \ldots, (x_n, y_n)$ should be marked by small circles. Test the graph function by choosing 5 points in $[0, \pi]$ and f as $\sin x$.

b) Make a module `Lagrange_poly2` containing the p_L, L_k, test_p_L, and graph functions. The call to `test_p_L` described in Exercise 5.25 and the call to graph described above should appear in the module's test block.

Hint Section 4.9 describes how to make a module. In particular, a test block is explained in Sect. 4.9.3, test functions like `test_p_L` are demonstrated in Sect. 4.9.4 and also in Sect. 3.4.2, and how to combine `test_p_L` and graph calls in the test block is exemplified in Sect. 4.9.5.
Filename: `Lagrange_poly2`.

Exercise 5.27: Investigate the behavior of Lagrange's interpolating polynomials

Unfortunately, the polynomial $p_L(x)$ defined and implemented in Exercise 5.25 can exhibit some undesired oscillatory behavior that we shall explore graphically in this exercise. Call the graph function from Exercise 5.26 with $f(x) = |x|$, $x \in [-2, 2]$, for $n = 2, 4, 6, 10$. All the graphs of $p_L(x)$ should appear in the same plot for comparison. In addition, make a new figure with results from calls to graph for $n = 13$ and $n = 20$. All the code necessary for solving this exercise should appear in some separate program file, which imports the `Lagrange_poly2` module made in Exercise 5.26.
Filename: `Lagrange_poly2b`.

Remarks The purpose of the $p_L(x)$ function is to compute (x, y) between some given (often measured) data points $(x_0, y_0), \ldots, (x_n, y_n)$. We see from the graphs that for a small number of interpolation points, $p_L(x)$ is quite close to the curve $y = |x|$ we used to generate the data points, but as n increases, $p_L(x)$ starts to oscillate, especially toward the end points (x_0, y_0) and (x_n, y_n). Much research has historically been focused on methods that do not result in such strange oscillations when fitting a polynomial to a set of points.

Exercise 5.28: Plot a wave packet

The function

$$f(x, t) = e^{-(x-3t)^2} \sin\left(3\pi(x - t)\right) \tag{5.28}$$

describes for a fixed value of t a wave localized in space. Make a program that visualizes this function as a function of x on the interval $[-4, 4]$ when $t = 0$.
Filename: `plot_wavepacket`.

Exercise 5.29: Judge a plot

Assume you have the following program for plotting a parabola:

```
import numpy as np
x = np.linspace(0, 2, 20)
y = x*(2 - x)
import matplotlib.pyplot as plt
plt.plot(x, y)
plt.show()
```

Then you switch to the function $\cos(18\pi x)$ by altering the computation of y to y = cos(18*pi*x). Judge the resulting plot. Is it correct? Display the $\cos(18\pi x)$ function with 1000 points in the same plot.
Filename: judge_plot.

Exercise 5.30: Plot the viscosity of water
The viscosity of water, μ, varies with the temperature T (in Kelvin) according to

$$\mu(T) = A \cdot 10^{B/(T-C)}, \tag{5.29}$$

where $A = 2.414 \cdot 10^{-5}$ Pa s, $B = 247.8$ K, and $C = 140$ K. Plot $\mu(T)$ for T between 0 and 100 degrees Celsius. Label the x axis with 'temperature (C)' and the y axis with 'viscosity (Pa s)'. Note that T in the formula for μ must be in Kelvin.
Filename: water_viscosity.

Exercise 5.31: Explore a complicated function graphically
The wave speed c of water surface waves depends on the length λ of the waves. The following formula relates c to λ:

$$c(\lambda) = \sqrt{\frac{g\lambda}{2\pi}\left(1 + s\frac{4\pi^2}{\rho g\lambda^2}\right)\tanh\left(\frac{2\pi h}{\lambda}\right)}. \tag{5.30}$$

Here, g is the acceleration of gravity (9.81 m/s²), s is the air-water surface tension ($7.9 \cdot 10^{-2}$ N/m), ρ is the density of water (can be taken as 1000 kg/m³), and h is the water depth. Let us fix h at 50 m. First make a plot of $c(\lambda)$ (in m/s) for small λ (0.001 m to 0.1 m). Then make a plot $c(\lambda)$ for larger λ (1 m to 2 km.
Filename: water_wave_velocity.

Exercise 5.32: Plot Taylor polynomial approximations to sin x
The sine function can be approximated by a polynomial according to the following formula:

$$\sin x \approx S(x; n) = \sum_{j=0}^{n}(-1)^j\frac{x^{2j+1}}{(2j+1)!}. \tag{5.31}$$

The expression $(2j + 1)!$ is the factorial (math.factorial can compute this quantity). The error in the approximation $S(x; n)$ decreases as n increases and in the limit we have that $\lim_{n\to\infty} S(x; n) = \sin x$. The purpose of this exercise is to visualize the quality of various approximations $S(x; n)$ as n increases.

a) Write a Python function S(x, n) that computes $S(x; n)$. Use a straightforward approach where you compute each term as it stands in the formula, i.e.,

$(-1)^j x^{2j+1}$ divided by the factorial $(2j + 1)!$. (We remark that Exercise A.14 outlines a much more efficient computation of the terms in the series.)

b) Plot $\sin x$ on $[0, 4\pi]$ together with the approximations $S(x; 1)$, $S(x; 2)$, $S(x; 3)$, $S(x; 6)$, and $S(x; 12)$.

Filename: `plot_Taylor_sin`.

Exercise 5.33: Animate a wave packet
Display an animation of the function $f(x, t)$ in Exercise 5.28 by plotting f as a function of x on $[-6, 6]$ for a set of t values in $[-1, 1]$. Also make an animated GIF file.

Hint A suitable resolution can be 1000 intervals (1001 points) along the x axis, 60 intervals (61 points) in time, and 6 frames per second in the animated GIF file. Use the recipe in Sect. 5.3.4 and remember to remove the family of old plot files in the beginning of the program.
Filename: `plot_wavepacket_movie`.

Exercise 5.34: Animate a smoothed Heaviside function
Visualize the smoothed Heaviside function $H_\epsilon(x)$, defined in 3.26), as an animation where ϵ starts at 2 and then goes to zero.
Filename: `smoothed_Heaviside_movie`.

Exercise 5.35: Animate two-scale temperature variations
We consider temperature oscillations in the ground as addressed in Sect. 5.13.2. Now we want to visualize daily and annual variations. Let A_1 be the amplitude of annual variations and A_2 the amplitude of the day/night variations. Let also $P_1 = 365$ days and $P_2 = 24$ h be the periods of the annual and the daily oscillations. The temperature at time t and depth z is then given by

$$T(z, t) = T_0 + A_1 e^{-a_1 z} \sin(\omega_1 t - a_1 z) + A_2 e^{-a_2 z} \sin(\omega_2 t - a_2 z), \qquad (5.32)$$

where

$$\omega_1 = 2\pi P_1,$$
$$\omega_2 = 2\pi P_2,$$
$$a_1 = \sqrt{\frac{\omega_1}{2k}},$$
$$a_2 = \sqrt{\frac{\omega_2}{2k}}.$$

Choose $k = 10^{-6}$ m^2/s, $A_1 = 15$ C, $A_2 = 7$ C, and the resolution Δt as $P_2/10$. Modify the `heatwave.py` program in order to animate this new temperature function.
Filename: `heatwave2`.

Remarks We assume in this problem that the temperature T equals the reference temperature T_0 at $t = 0$, resulting in a sine variation rather than the cosine variation in (5.18).

Exercise 5.36: Use non-uniformly distributed coordinates for visualization

Watching the animation in Exercise 5.35 reveals that there are rapid oscillations in a small layer close to $z = 0$. The variations away from $z - 0$ are much smaller in time and space. It would therefore be wise to use more z coordinates close to $z = 0$ than for larger z values. Given a set $x_0 < x_1 < \cdots < x_n$ of uniformly spaced coordinates in $[a, b]$, we can compute new coordinates $\bar{x}_i$, stretched toward $x = a$, by the formula

$$\bar{x}_i = a + (b - a) \left(\frac{x_i - a}{b - a} \right)^s,$$

for some $s > 1$. In the present example, we can use this formula to stretch the z coordinates to the left.

a) Experiment with $s \in [1.2, 3]$ and few points (say 15) and visualize the curve as a line with circles at the points so that you can easily see the distribution of points toward the left end. Identify a suitable value of s.
b) Run the animation with no circles and (say) 501 points with the found s value.

Filename: `heatwave2a`.

Exercise 5.37: Animate a sequence of approximations to π

Exercise 3.18 outlines an idea for approximating π as the length of a polygon inside the circle. Wrap the code from that exercise in a function `pi_approx(N)`, which returns the approximation to π using a polygon with $N + 1$ equally distributed points. The task of the present exercise is to visually display the polygons as a movie, where each frame shows the polygon with $N + 1$ points together with the circle and a title reflecting the corresponding error in the approximate value of π. The whole movie arises from letting N run through $4, 5, 6, \ldots, K$, where K is some (large) prescribed value. Let there be a pause of $0.3\,\mathrm{s}$ between each frame in the movie. By playing the movie you will see how the polygons move closer and closer to the circle and how the approximation to π improves.
Filename: `pi_polygon_movie`.

Exercise 5.38: Animate a planet's orbit

A planet's orbit around a star has the shape of an ellipse. The purpose of this exercise is to make an animation of the movement along the orbit. One should see a small disk, representing the planet, moving along an elliptic curve. An evolving solid line shows the development of the planet's orbit as the planet moves and the title displays the planet's instantaneous velocity magnitude. As a test, run the special case of a circle and verify that the magnitude of the velocity remains constant as the planet moves.

Hint 1 The points (x, y) along the ellipse are given by the expressions

$$x = a \cos(\omega t), \quad y = b \sin(\omega t),$$

where a is the semi-major axis of the ellipse, b is the semi-minor axis, ω is an angular velocity of the planet around the star, and t denotes time. One complete orbit corresponds to $t \in [0, 2\pi/\omega]$. Let us discretize time into time points $t_k = k\Delta t$, where $\Delta t = 2\pi/(\omega n)$. Each frame in the movie corresponds to (x, y) points along the curve with t values $t_0, t_1, \ldots, t_i$, i representing the frame number $(i = 1, \ldots, n)$.

Hint 2 The velocity vector is

$$(\frac{dx}{dt}, \frac{dy}{dt}) = (-\omega a \sin(\omega t), \omega b \cos(\omega t)),$$

and the magnitude of this vector becomes $\omega \sqrt{a^2 \sin^2(\omega t) + b^2 \cos^2(\omega t)}$. Filename: `planet_orbit`.

Exercise 5.39: Animate the evolution of Taylor polynomials
A general series approximation (to a function) can be written as

$$S(x; M, N) = \sum_{k=M}^{N} f_k(x).$$

For example, the Taylor polynomial of degree N for e^x equals $S(x; 0, N)$ with $f_k(x) = x^k/k!$. The purpose of the exercise is to make a movie of how $S(x; M, N)$ develops and improves as an approximation as we add terms in the sum. That is, the frames in the movie correspond to plots of $S(x; M, M)$, $S(x; M, M + 1)$, $S(x; M, M + 2)$, ..., $S(x; M, N)$.

a) Make a function

```
animate_series(fk, M, N, xmin, xmax, ymin, ymax, n, exact)
```

for creating such animations. The argument `fk` holds a Python function implementing the term $f_k(x)$ in the sum, `M` and `N` are the summation limits, the next arguments are the minimum and maximum x and y values in the plot, `n` is the number of x points in the curves to be plotted, and `exact` holds the function that $S(x)$ aims at approximating.

Hint Here is some more information on how to write the `animate_series` function. The function must accumulate the $f_k(x)$ terms in a variable s, and for each k value, s is plotted against x together with a curve reflecting the exact function. Each plot must be saved in a file, say with names `tmp_0000.png`, `tmp_0001.png`, and so on (these filenames can be generated by `tmp_%04d.png`, using an appropriate counter). Use the `movie` function to combine all the plot files into a movie in a desired movie format.

In the beginning of the `animate_series` function, it is necessary to remove all old plot files of the form `tmp_*.png`. This can be done by the `glob` module and the `os.remove` function as exemplified in Sect. 5.3.4.

b) Call the `animate_series` function for the Taylor series for $\sin x$, where $f_k(x) = (-1)^k x^{2k+1}/(2k+1)!$, and $x \in [0, 13\pi]$, $M = 0$, $N = 40$, $y \in [-2, 2]$.

c) Call the `animate_series` function for the Taylor series for e^{-x}, where $f_k(x) = (-x)^k/k!$, and $x \in [0, 15]$, $M = 0$, $N = 30$, $y \in [-0.5, 1.4]$.

Filename: `animate_Taylor_series`.

Exercise 5.40: Plot the velocity profile for pipeflow
A fluid that flows through a (very long) pipe has zero velocity on the pipe wall and a maximum velocity along the centerline of the pipe. The velocity v varies through the pipe cross section according to the following formula:

$$v(r) = \left(\frac{\beta}{2\mu_0}\right)^{1/n} \frac{n}{n+1} \left(R^{1+1/n} - r^{1+1/n}\right), \qquad (5.33)$$

where R is the radius of the pipe, β is the pressure gradient (the force that drives the flow through the pipe), μ_0 is a viscosity coefficient (small for air, larger for water and even larger for toothpaste), n is a real number reflecting the viscous properties of the fluid ($n = 1$ for water and air, $n < 1$ for many modern plastic materials), and r is a radial coordinate that measures the distance from the centerline ($r = 0$ is the centerline, $r = R$ is the pipe wall).

a) Make a Python function that evaluates $v(r)$.

b) Plot $v(r)$ as a function of $r \in [0, R]$, with $R = 1$, $\beta = 0.02$, $\mu_0 = 0.02$, and $n = 0.1$.

c) Make an animation of how the $v(r)$ curves varies as n goes from 1 and down to 0.01. Because the maximum value of $v(r)$ decreases rapidly as n decreases, each curve can be normalized by its $v(0)$ value such that the maximum value is always unity.

Filename: `plot_velocity_pipeflow`.

Exercise 5.41: Plot sum-of-sines approximations to a function
Exercise 3.21 defines the approximation $S(t; n)$ to a function $f(t)$. Plot $S(t; 1)$, $S(t; 3)$, $S(t; 20)$, $S(t; 200)$, and the exact $f(t)$ function in the same plot. Use $T = 2\pi$.
Filename: `sinesum1_plot`.

Exercise 5.42: Animate the evolution of a sum-of-sine approximation to a function
First perform Exercise 5.41. A natural next step is to animate the evolution of $S(t; n)$ as n increases. Create such an animation and observe how the discontinuity in $f(t)$ is poorly approximated by $S(t; n)$, even when n grows large (plot $f(t)$ in each frame). This is a well-known deficiency, called Gibb's phenomenon, when approximating discontinuous functions by sine or cosine (Fourier) series.
Filename: `sinesum1_movie`.

Exercise 5.43: Plot functions from the command line

For quickly getting a plot of a function $f(x)$ for $x \in [x_{\min}, x_{\max}]$ it could be nice to a have a program that takes the minimum amount of information from the command line and produces a plot on the screen and saves the plot to a file `tmp.png`. The usage of the program goes as follows:

```
                                      Terminal
plotf.py "f(x)" xmin xmax
```

Plotting $e^{-0.2x} \sin(2\pi x)$ for $x \in [0, 4\pi]$ is then specified as

```
                                      Terminal
plotf.py "exp(-0.2*x)*sin(2*pi*x)" 0 4*pi
```

Write the `plotf.py` program with as short code as possible (we leave it to Exercise 5.44 to test for valid input).

Hint Make x coordinates from the second and third command-line arguments and then use `eval` (or `StringFunction` from `scitools.std`, see Sects. 4.3.3 and 5.5.1) on the first argument.
Filename: `plotf`.

Exercise 5.44: Improve command-line input

Equip the program from Exercise 5.43 with tests on valid input on the command line. Also allow an optional fourth command-line argument for the number of points along the function curve. Set this number to 501 if it is not given.
Filename: `plotf2`.

Exercise 5.45: Demonstrate energy concepts from physics

The vertical position $y(t)$ of a ball thrown upward is given by $y(t) = v_0 t - \frac{1}{2}gt^2$, where g is the acceleration of gravity and v_0 is the velocity at $t = 0$. Two important physical quantities in this context are the potential energy, obtained by doing work against gravity, and the kinetic energy, arising from motion. The potential energy is defined as $P = mgy$, where m is the mass of the ball. The kinetic energy is defined as $K = \frac{1}{2}mv^2$, where v is the velocity of the ball, related to y by $v(t) = y'(t)$.

Make a program that can plot $P(t)$ and $K(t)$ in the same plot, along with their sum $P + K$. Let $t \in [0, 2v_0/g]$. Read m and v_0 from the command line. Run the program with various choices of m and v_0 and observe that $P + K$ is always constant in this motion. (In fact, it turns out that $P + K$ is constant for a large class of motions, and this is a very important result in physics.)
Filename: `energy_physics`.

Exercise 5.46: Plot a w-like function

Define mathematically a function that looks like the "w" character. Plot the function. Also write a formal test function that verifies the implementation.
Filename: `plot_w`.

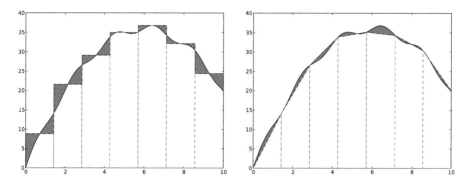

Fig. 5.22 Visualization of numerical integration rules, with the Midpoint rule to the left and the Trapezoidal rule to the right. The filled areas illustrate the deviations in the approximation of the area under the curve

Exercise 5.47: Plot a piecewise constant function
Consider the piecewise constant function defined in Exercise 3.32. Make a Python function `plot_piecewise(data, xmax)` that draws a graph of the function, where `data` is the nested list explained in mentioned exercise and `xmax` is the maximum x coordinate. Use ideas from Sect. 5.4.1.
Filename: `plot_piecewise_constant`.

Exercise 5.48: Vectorize a piecewise constant function
Consider the piecewise constant function defined in Exercise 3.32. Make a vectorized implementation `piecewise_constant_vec(x, data, xmax)` of such a function, where `x` is an array.

Hint You can use ideas from the `Nv1` function in Sect. 5.5.3. However, since the number of intervals is not known, it is necessary to store the various intervals and conditions in lists.
Filename: `piecewise_constant_vec`.

Remarks Plotting the array returned from `piecewise_constant_vec` faces the same problems as encountered in Sect. 5.4.1. It is better to make a custom plotting function that simply draws straight horizontal lines in each interval (Exercise 5.47).

Exercise 5.49: Visualize approximations in the Midpoint integration rule
Consider the midpoint rule for integration from Exercise 3.12. Use Matplotlib to make an illustration of the midpoint rule as shown to the left in Fig. 5.22.
The $f(x)$ function used in Fig. 5.22 is

$$f(x) = x(12 - x) + \sin(\pi x), \quad x \in [0, 10].$$

Hint Look up the documentation of the Matplotlib function `fill_between` and use this function to create the filled areas between $f(x)$ and the approximating rectangles.
Note that the `fill_between` requires the two curves to have the same number of points. For accurate visualization of $f(x)$ you need quite many x coordinates,

and the rectangular approximation to $f(x)$ must be drawn using the same set of x coordinates.

Filename: `viz_midpoint`.

Exercise 5.50: Visualize approximations in the Trapezoidal integration rule
Redo Exercise 5.49 for the Trapezoidal rule from Exercise 3.11 to produce the graph shown to the right in Fig. 5.22.
Filename: `viz_trapezoidal`.

Exercise 5.51: Experience overflow in a function
We are give the mathematical function

$$v(x) = \frac{1 - e^{x/\mu}}{1 - e^{1/\mu}},$$

where μ is a parameter.

a) Make a Python function `v(x, mu=1E-6, exp=math.exp)` for calculating the formula for $v(x)$ using `exp` as a possibly user-given exponential function. Let the v function return the nominator and denominator in the formula as well as the fraction.

b) Call the v function for various x values between 0 and 1 in a `for` loop, let `mu` be `1E-3`, and have an inner `for` loop over two different `exp` functions: `math.exp` and `numpy.exp`. The output will demonstrate how the denominator is subject to overflow and how difficult it is to calculate this function on a computer.

c) Plot $v(x)$ for $\mu = 1, 0.01, 0.001$ on $[0, 1]$ using 10,000 points to see what the function looks like.

d) Convert x and eps to a higher precision representation of real numbers, with the aid of the NumPy type `float96`, before calling `v`:

```
import numpy
x = numpy.float96(x); mu = numpy.float96(e)
```

Repeat point b) with these type of variables and observe how much better results we get with `float96` compared with the standard `float` value, which is `float64` (the number reflects the number of bits in the machine's representation of a real number).

e) Call the v function with x and mu as `float32` variables and report how the function now behaves.

Filename: `boundary_layer_func1`.

Remarks When an object (ball, car, airplane) moves through the air, there is a very, very thin layer of air close to the object's surface where the air velocity varies dramatically, from the same value as the velocity of the object at the object's surface to zero a few centimeters away. This layer is called a *boundary layer*. The physics in the boundary layer is important for air resistance and cooling/heating of objects. The change in velocity in the boundary layer is quite abrupt and can be modeled by the functiion $v(x)$, where $x = 1$ is the object's surface, and $x = 0$ is some distance away where one cannot notice any wind velocity v because of the passing object ($v = 0$). The wind velocity coincides with the velocity of the object at $x = 1$, here set to $v = 1$. The parameter μ is very small and related to the viscosity of air. With a small value of μ, it becomes difficult to calculate $v(x)$ on a computer. The exercise demonstrates the difficulties and provides a remedy.

Exercise 5.52: Apply a function to a rank 2 array
Let A be the two-dimensional array

$$\begin{bmatrix} 0 & 2 & -1 \\ -1 & -1 & 0 \\ 0 & 5 & 0 \end{bmatrix}$$

Apply the function f from Exercise 5.5 to each element in A. Then calculate the result of the array expression `A**3 + A*exp(A) + 1`, and demonstrate that the end result of the two methods are the same.
Filename: `apply_arrayfunc`.

Exercise 5.53: Explain why array computations fail
The following loop computes the array y from x:

```
>>> import numpy as np
>>> x = np.linspace(0, 1, 3)
>>> y = np.zeros(len(x))
>>> for i in range(len(x)):
...     y[i] = x[i] + 4
```

However, the alternative loop

```
>>> for xi, yi in zip(x, y):
...     yi = xi + 5
```

leaves y unchanged. Why? Explain in detail what happens in each pass of this loop and write down the contents of `xi`, `yi`, `x`, and `y` as the loop progresses.
Filename: `find_errors_arraycomp`.

Exercise 5.54: Verify linear algebra results
When we want to verify that a mathematical result is true, we often generate matrices or vectors with random elements and show that the result holds for these "arbitrary" mathematical objects. As an example, consider testing that $A + B = B + A$ for matrices A and B:

```
def test_addition():
    n = 4   # matrix size
    A = matrix(random.rand(n, n))
    B = matrix(random.rand(n, n))
```

```
tol = 1E-14
result1 = A + B
result2 = B + A
assert abs(result1 - result2).max() < tol
```

Use this technique to write test functions for the following mathematical results:

1. $(A + B)C = AC + BC$
2. $(AB)C = A(BC)$
3. $\operatorname{rank} A = \operatorname{rank} A^T$
4. $\det(AB) = \det A \det B$
5. The eigenvalues if A equals the eigenvalues of A^T when A is square.

Filename: `verify_linalg`.

Dictionaries and Strings

6

The present chapter addresses many techniques for interpreting information in files and storing the data in convenient Python objects for further data analysis. A particularly handy object for many purposes is the dictionary, which maps objects to objects, very often strings to various kinds of data that later can be looked up through the strings. Section 6.1 is devoted to dictionaries.

Information in files often appear as pure text, so to interpret and extract data from files it is sometimes necessary to carry out sophisticated operations on the text. Python strings have many methods for performing such operations, and the most important functionality is described in Sect. 6.2.

The World Wide Web is full of information and scientific data that may be useful to access from a program. Section 6.3 tells you how to read web pages from a program and interpret the contents using string operations.

Working with data often involves spreadsheets. Python programs not only need to extract data from spreadsheet files, but it can be advantageous and convenient to actually to the data processing in a Python program rather than in a spreadsheet program like Microsoft Excel or LibreOffice. Section 6.4 goes through relevant techniques for reading and writing files in the common CSV format for spreadsheets.

The present chapter builds on fundamental programming concepts such as loops, lists, arrays, `if` tests, command-line arguments, and curve plotting. The folder `src/files`[1] contains all the relevant program example files and associated data files.

6.1 Dictionaries

So far in the book we have stored information in various types of objects, such as numbers, strings, list, and arrays. A *dictionary* is a very flexible object for storing various kind of information, and in particular when reading files. It is therefore time to introduce the popular dictionary type.

A list is a collection of objects indexed by an integer going from 0 to the number of elements minus one. Instead of looking up an element through an integer index,

---

[1] http://tinyurl.com/pwyasaa/files

© Springer-Verlag Berlin Heidelberg 2016
H.P. Langtangen, *A Primer on Scientific Programming with Python*,
Texts in Computational Science and Engineering 6, DOI 10.1007/978-3-662-49887-3_6

it can be more handy to use a text. Roughly speaking, a list where the index can be a text is called a dictionary in Python. Other computer languages use other names for the same thing: HashMap, hash, associative array, or map.

6.1.1 Making Dictionaries

Suppose we need to store the temperatures from three cities: Oslo, London, and Paris. For this purpose we can use a list,

```
temps = [13, 15.4, 17.5]
```

but then we need to remember the sequence of cities, e.g., that index 0 corresponds to Oslo, index 1 to London, and index 2 to Paris. That is, the London temperature is obtained as `temps[1]`. A dictionary with the city name as index is more convenient, because this allows us to write `temps['London']` to look up the temperature in London. Such a dictionary is created by one of the following two statements

```
temps = {'Oslo': 13, 'London': 15.4, 'Paris': 17.5}
# or
temps = dict(Oslo=13, London=15.4, Paris=17.5)
```

Additional text-value pairs can be added when desired. We can, for instance, write

```
temps['Madrid'] = 26.0
```

The `temps` dictionary has now four text-value pairs, and a `print temps` yields

```
{'Oslo': 13, 'London': 15.4, 'Paris': 17.5, 'Madrid': 26.0}
```

6.1.2 Dictionary Operations

The string "indices" in a dictionary are called *keys*. To loop over the keys in a dictionary d, one writes `for key in d:` and works with `key` and the corresponding value `d[key]` inside the loop. We may apply this technique to write out the temperatures in the `temps` dictionary from the previous paragraph:

```
>>> for city in temps:
...     print 'The temperature in %s is %g' % (city, temps[city])
...
The temperature in Paris is 17.5
The temperature in Oslo is 13
The temperature in London is 15.4
The temperature in Madrid is 26
```

We can check if a key is present in a dictionary by the syntax if key in d:

```
>>> if 'Berlin' in temps:
...     print 'Berlin:', temps['Berlin']
... else:
...     print 'No temperature data for Berlin'
...
No temperature data for Berlin
```

Writing key in d yields a standard boolean expression, e.g.,

```
>>> 'Oslo' in temps
True
```

The keys and values can be extracted as lists from a dictionary:

```
>>> temps.keys()
['Paris', 'Oslo', 'London', 'Madrid']
>>> temps.values()
[17.5, 13, 15.4, 26.0]
```

An important feature of the keys method in dictionaries is that the order of the returned list of keys is unpredictable. If you need to traverse the keys in a certain order, you can sort the keys. A loop over the keys in the temps dictionary in alphabetic order is written as

```
>>> for city in sorted(temps):
...     print city
...
London
Madrid
Oslo
Paris
```

Python also has a special dictionary type OrderedDict where the key-value pairs has a specific order, see Sect. 6.1.4.

A key-value pair can be removed by del d[key]:

```
>>> del temps['Oslo']
>>> temps
{'Paris': 17.5, 'London': 15.4, 'Madrid': 26.0}
>>> len(temps)  # no of key-value pairs in dictionary
3
```

Sometimes we need to take a copy of a dictionary:

```
>>> temps_copy = temps.copy()
>>> del temps_copy['Paris']   # this does not affect temps
>>> temps_copy
{'London': 15.4, 'Madrid': 26.0}
>>> temps
{'Paris': 17.5, 'London': 15.4, 'Madrid': 26.0}
```

Note that if two variables refer to the same dictionary and we change the contents of the dictionary through either of the variables, the change will be seen in both variables:

```
>>> t1 = temps
>>> t1['Stockholm'] = 10.0      # change t1
>>> temps                       # temps is also changed
{'Stockholm': 10.0, 'Paris': 17.5, 'London': 15.4, 'Madrid': 26.0}
```

To avoid that `temps` is affected by adding a new key-value pair to `t1`, `t1` must be a copy of `temps`.

Remark In Python version 2.x, `temps.keys()` returns a list object while in Python version 3.x, `temps.keys()` only enables iterating over the keys. To write code that works with both versions one can use `list(temps.keys())` in the cases where a list is really needed and just `temps.keys()` in a `for` loop over the keys.

6.1.3   Example: Polynomials as Dictionaries

Python objects that cannot change their contents are known as *immutable* data types and consist of `int`, `float`, `complex`, `str`, and `tuple`. Lists and dictionaries can change their contents and are called *mutable* objects.

The keys in a dictionary are not restricted to be strings. In fact, any immutable Python object can be used as key. For example, if you want a list as key, it cannot be used since lists can change their contents are hence mutable objects, but a tuple will do, since it is immutable.

A common type of key in dictionaries is integers. Next we shall explain how dictionaries with integers as key provide a handy way of representing polynomials. Consider the polynomial

$$p(x) = -1 + x^2 + 3x^7 .$$

The data associated with this polynomial can be viewed as a set of power-coefficient pairs, in this case the coefficient -1 belongs to power 0, the coefficient 1 belongs to power 2, and the coefficient 3 belongs to power 7. A dictionary can be used to map a power to a coefficient:

```
p = {0: -1, 2: 1, 7: 3}
```

A list can, of course, also be used, but in this case we must fill in all the zero coefficients too, since the index must match the power:

```
p = [-1, 0, 1, 0, 0, 0, 0, 3]
```

The advantage with a dictionary is that we need to store only the non-zero coefficients. For the polynomial $1 + x^{100}$ the dictionary holds two elements while the list holds 101 elements (see Exercise 6.10).

The following function can be used to evaluate a polynomial represented as a dictionary:

```
def eval_poly_dict(poly, x):
    sum = 0.0
    for power in poly:
        sum += poly[power]*x**power
    return sum
```

The `poly` argument must be a dictionary where `poly[power]` holds the coefficient associated with the term `x**power`.

A more compact implementation can make use of Python's `sum` function to sum the elements of a list:

```
def eval_poly_dict2(poly, x):
    return sum([poly[power]*x**power for power in poly])
```

That is, we first make a list of the terms in the polynomial using a list comprehension, and then we feed this list to the `sum` function. We can in fact drop the brackets and storing all the `poly[power]*x**power` numbers in a list, because `sum` can directly add elements of an iterator (like `for power in poly`):

```
def eval_poly_dict2(poly, x):
    return sum(poly[power]*x**power for power in poly)
```

Be careful with redefining variables!
The name `sum` appears in both `eval_poly_dict` and `eval_poly_dict2`. In the former, `sum` is a `float` object, and in the latter, `sum` is a built-in Python function. When we set `sum=0.0` in the first implementation, we bind the name `sum` to a new `float` object, and the built-in Python function associated with the name `sum` is then no longer accessible inside the `eval_poly_dict` function. (Actually, this is not strictly correct, because `sum` is a local variable while the built-in Python `sum` function is associated with a global name `sum`, which can always be reached through `globals()['sum']`.) Outside the `eval_poly_dict` function, nevertheless, `sum` will be Python's summation function and the local `sum` variable inside the `eval_poly_dict` function is destroyed.

As a rule of thumb, avoid using `sum` or other names associated with frequently used functions as new variables unless you are in a very small function (like `eval_poly_dict`) where there is no danger that you need the original meaning of the name.

With a list instead of dictionary for representing the polynomial, a slightly different evaluation function is needed:

```
def eval_poly_list(poly, x):
    sum = 0
    for power in range(len(poly)):
        sum += poly[power]*x**power
    return sum
```

If there are many zeros in the `poly` list, `eval_poly_list` must perform all the multiplications with the zeros, while `eval_poly_dict` computes with the non-zero coefficients only and is hence more efficient.

Another major advantage of using a dictionary to represent a polynomial rather than a list is that negative powers are easily allowed, e.g.,

```
p = {-3: 0.5, 4: 2}
```

can represent $\frac{1}{2}x^{-3} + 2x^4$. With a list representation, negative powers require much more book-keeping. We may, for example, set

```
p = [0.5, 0, 0, 0, 0, 0, 0, 2]
```

and remember that `p[i]` is the coefficient associated with the power `i-3`. In particular, the `eval_poly_list` function will no longer work for such lists, while the `eval_poly_dict` function works also for dictionaries with negative keys (powers).

There is a dictionary counterpart to list comprehensions, called *dictionary comprehensions*, for quickly generating parameterized key-value pairs with a `for` loop. Such a construction is convenient to generate the coefficients in a polynomial:

```
from math import factorial
d = {k: (-1)**k/float(factorial(k)) for k in range(n+1)}
```

The `d` dictionary now contains the power-coefficient pairs of the Taylor polynomial of degree `n` for e^{-x}. (Note the use of `float` to avoid integer division.)

You are now encouraged to solve Exercise 6.11 to become more familiar with the concept of dictionaries.

6.1.4 Dictionaries with Default Values and Ordering

Dictionaries with default values Looking up keys that are not present in the dictionary requires special treatment. Consider a polynomial dictionary of the type introduced in Sect. 6.1.3. Say we have $2x^{-3} - 1.5x^{-1} - 2x^2$ represented by

```
p1 = {-3: 2, -1: -1.5, 2: -2}
```

If the code tries to look up `p1[1]`, this operation results in a `KeyError` since 1 is not a registered key in `p1`. We therefore need to do either

```
if key in p1:
    value = p1[key]
```

or use

```
value = p1.get(key, 0.0)
```

where p1.get returns p1[key] if key in p1 and the default value 0.0 if not. A third possibility is to work with a dictionary with a default value:

```
from collections import defaultdict

def polynomial_coeff_default():
    # default value for polynomial dictionary
    return 0.0

p2 = defaultdict(polynomial_coeff_default)
p2.update(p1)
```

The p2 can be indexed by any key, and for unregistered keys the polynomial_coeff_default function is called to provide a value. This must be a function without arguments. Usually, a separate function is never made, but either a type is inserted or a lambda function. The example above is equivalent to

```
p2 = defaultdict(lambda: 0.0)
p2 = defaultdict(float)
```

In the latter case float() is called for each unknown key, and float() returns a float object with zero value. Now we can look up p2[1] and get the default value 0. It must be remarked that this key is then a part of the dictionary:

```
>>> p2 = defaultdict(lambda: 0.0)
>>> p2.update({2: 8})  # only one key
>>> p2[1]
0.0
>>> p2[0]
0.0
>>> p2[-2]
0.0
>>> print p2
{0: 0.0, 1: 0.0, 2: 8, -2: 0.0}
```

Ordered dictionaries The elements of a dictionary have an undefined order. For example,

```
>>> p1 = {-3: 2, -1: -1.5, 2: -2}
>>> print p1
{2: -2, -3: 2, -1: -1.5}
```

One can control the order by sorting the keys, either by the default sorting (alphabetically for string keys, ascending order for number keys):

```
>>> for key in sorted(p1):
...     print key, p1[key]
...
-3 2
-1 -1.5
2 -2
```

The `sorted` function also accept an optional argument where the user can supply a function that sorts two keys (see Exercise 3.39).

However, Python features a dictionary type that preserves the order of the keys as they were registered:

```
>>> from collections import OrderedDict
>>> p2 = OrderedDict({-3: 2, -1: -1.5, 2: -2})
>>> print p2
OrderedDict([(2, -2), (-3, 2), (-1, -1.5)])
>>> p2[-5] = 6
>>> for key in p2:
...     print key, p2[key]
...
2 -2
-3 2
-1 -1.5
-5 6
```

Here is an example with dates as keys where the order is important.

```
>>> data = {'Jan 2': 33, 'Jan 16': 0.1, 'Feb 2': 2}
>>> for date in data:
...     print date, data[date]
...
Feb 2 2
Jan 2 33
Jan 16 0.1
```

The order of the keys in the loop is not the right registered order, but this is easily achieved by `OrderedDict`

```
>>> data = OrderedDict()
>>> data['Jan 2'] = 33
>>> data['Jan 16'] = 0.1
>>> data['Feb 2'] = 2
>>> for date in data:
...     print date, data[date]
...
Jan2 33
Jan 16 0.1
Feb 2 2
```

A comment on alternative solutions should be made here. Trying to sort the `data` dictionary when it is an ordinary `dict` object does not help, as by default the sorting will be alphabetically, resulting in the sequence 'Feb 2', 'Jan 16', and 'Jan 2'. What does help, however, is to use Python's `datetime` objects as keys reflecting dates, since these objects will be correctly sorted. A `datetime` object can be created from a string like 'Jan 2, 2017' using a special syntax (see the module documentation). The relevant code is

```
>>> import datetime
>>> data = {}
>>> d = datetime.datetime.strptime  # short form
>>> data[d('Jan 2, 2017', '%b %d, %Y')] = 33
>>> data[d('Jan 16, 2017', '%b %d, %Y')] = 0.1
>>> data[d('Feb 2, 2017', '%b %d, %Y')] = 2
```

Printing out in sorted order gives the right sequence of dates:

```
>>> for date in sorted(data):
...     print date, data[date]
...
2017-01-02 00:00:00 33
2017-01-16 00:00:00 0.1
2017-02-02 00:00:00 2
```

The time is automatically part of a datetime object and set to 00:00:00 when not specified.

While OrderedDict provides a simpler and shorter solution to keeping keys (here dates) in the right order in a dictionary, using datetime objects for dates has many advantages: dates can be formatted and written out in various ways, counting days between two dates is easy (see Sect. A.1.1), calculating the corresponding week number and name of the weekday is supported, to mention some functionality.

6.1.5 Example: Storing File Data in Dictionaries

Problem The file files/densities.dat contains a table of densities of various substances measured in g/cm^3:

```
air           0.0012
gasoline      0.67
ice           0.9
pure water    1.0
seawater      1.025
human body    1.03
limestone     2.6
granite       2.7
iron          7.8
silver        10.5
mercury       13.6
gold          18.9
platinium     21.4
Earth mean    5.52
Earth core    13
Moon          3.3
Sun mean      1.4
Sun core      160
proton        2.3E+14
```

In a program we want to access these density data. A dictionary with the name of the substance as key and the corresponding density as value seems well suited for storing the data.

Solution We can read the `densities.dat` file line by line, split each line into words, use a float conversion of the last word as density value, and the remaining one or two words as key in the dictionary.

```
def read_densities(filename):
    infile = open(filename, 'r')
    densities = {}
    for line in infile:
        words = line.split()
        density = float(words[-1])

        if len(words[:-1]) == 2:
            substance = words[0] + ' ' + words[1]
        else:
            substance = words[0]

        densities[substance] = density
    infile.close()
    return densities

densities = read_densities('densities.dat')
```

This code is found in the file `density.py`. With string operations from Sect. 6.2.1 we can avoid the special treatment of one or two words in the name of the substance and achieve simpler and more general code, see Exercise 6.3.

6.1.6   Example: Storing File Data in Nested Dictionaries

Problem   We are given a data file with measurements of some properties with given names (here A, B, C ...). Each property is measured a given number of times. The data are organized as a table where the rows contain the measurements and the columns represent the measured properties:

	A	B	C	D
1	11.7	0.035	2017	99.1
2	9.2	0.037	2019	101.2
3	12.2	no	no	105.2
4	10.1	0.031	no	102.1
5	9.1	0.033	2009	103.3
6	8.7	0.036	2015	101.9

The word no stands for no data, i.e., we lack a measurement. We want to read this table into a dictionary data so that we can look up measurement no. i of (say) property C as data['C'][i]. For each property p, we want to compute the mean of all measurements and store this as data[p]['mean'].

Algorithm The algorithm for creating the data dictionary goes as follows:

```
examine the first line: split it into words and
initialize a dictionary with the property names
as keys and empty dictionaries {} as values

for each of the remaining lines in the file:
    split the line into words
    for each word after the first:
        if the word is not 'no':
            transform the word to a real number and store
            the number in the relevant dictionary
```

- examine the first line: split it into words and initialize a dictionary with the property names as keys and empty dictionaries as values
- for each of the remaining lines in the file
  - split the line into words
  - for each word after the first
    * if the word is not no:
      · transform the word to a real number and store the number in the relevant dictionary

Implementation A new aspect needed in the solution is *nested dictionaries*, that is, dictionaries of dictionaries. The latter topic is first explained, via an example:

```
>>> d = {'key1': {'key1': 2, 'key2': 3}, 'key2': 7}
```

Observe here that the value of d['key1'] is a dictionary, which we can index with its keys key1 and key2:

```
>>> d['key1']              # this is a dictionary
{'key2': 3, 'key1': 2}
>>> type(d['key1'])        # proof
<type 'dict'>
>>> d['key1']['key1']      # index a nested dictionary
2
>>> d['key1']['key2']
3
```

In other words, repeated indexing works for nested dictionaries as for nested lists. The repeated indexing does not apply to d['key2'] since that value is just an integer:

```
>>> d['key2']['key1']
    ...
TypeError: unsubscriptable object
>>> type(d['key2'])
<type 'int'>
```

When we have understood the concept of nested dictionaries, we are in a position to present a complete code that solves our problem of loading the tabular data in the file `table.dat` into a nested dictionary `data` and computing mean values. First, we list the program, stored in the file `table2dict.py`, and display the program's output. Thereafter, we dissect the code in detail.

```
infile = open('table.dat', 'r')
lines = infile.readlines()
infile.close()
data = {}   # data[property][measurement_no] = propertyvalue
first_line = lines[0]
properties = first_line.split()
for p in properties:
    data[p] = {}

for line in lines[1:]:
    words = line.split()
    i = int(words[0])        # measurement number
    values = words[1:]       # values of properties
    for p, v in zip(properties, values):
        if v != 'no':
            data[p][i] = float(v)

# Compute mean values
for p in data:
    values = data[p].values()
    data[p]['mean'] = sum(values)/len(values)

for p in sorted(data):
    print 'Mean value of property %s = %g' % (p, data[p]['mean'])
```

The corresponding output from this program becomes

```
Mean value of property A = 10.1667
Mean value of property B = 0.0344
Mean value of property C = 2015
Mean value of property D = 102.133
```

To view the nested `data` dictionary, we may insert

```
import scitools.pprint2; scitools.pprint2.pprint(data)
```

which produces something like

```
{'A': {1: 11.7, 2: 9.2, 3: 12.2, 4: 10.1, 5: 9.1, 6: 8.7,
       'mean': 10.1667},
 'B': {1: 0.035, 2: 0.037, 4: 0.031, 5: 0.033, 6: 0.036,
       'mean': 0.0344},
 'C': {1: 2017, 2: 2019, 5: 2009, 6: 2015, 'mean': 2015},
 'D': {1: 99.1,
       2: 101.2,
       3: 105.2,
       4: 102.1,
       5: 103.3,
       6: 101.9,
       'mean': 102.133}}
```

Dissection To understand a computer program, you need to understand what the result of every statement is. Let us work through the code, almost line by line, and see what it does.

First, we load all the lines of the file into a list of strings called `lines`. The `first_line` variable refers to the string

```
'       A       B       C       D'
```

We split this line into a list of words, called `properties`, which then contains

```
['A', 'B', 'C', 'D']
```

With each of these property names we associate a dictionary with the measurement number as key and the property value as value, but first we must create these "inner" dictionaries as empty before we can add the measurements:

```
for p in properties:
    data[p] = {}
```

The first pass in the `for` loop picks out the string

```
'1      11.7    0.035    2017    99.1'
```

as the `line` variable. We split this line into words, the first word (`words[0]`) is the measurement number, while the rest `words[1:]` is a list of property values, here named `values`. To pair up the right properties and values, we loop over the `properties` and `values` lists simultaneously:

```
for p, v in zip(properties, values):
    if v != 'no':
        data[p][i] = float(v)
```

Recall that some values may be missing and we drop to record that value (we could, alternatively, set the value to `None`). Because the `values` list contains strings (words) read from the file, we need to explicitly transform each string to a `float` number before we can compute with the values.

After the `for line in lines[1:]` loop, we have a dictionary `data` of dictionaries where all the property values are stored for each measurement number and property name. Figure 6.1 shows a graphical representation of the `data` dictionary.

It remains to compute the average values. For each property name p, i.e., key in the `data` dictionary, we can extract the recorded values as the list `data[p]. values()` and simply send this list to Python's `sum` function and divide by the number of measured values for this property, i.e., the length of the list:

```
for p in data:
    values = data[p].values()
    data[p]['mean'] = sum(values)/len(values)
```

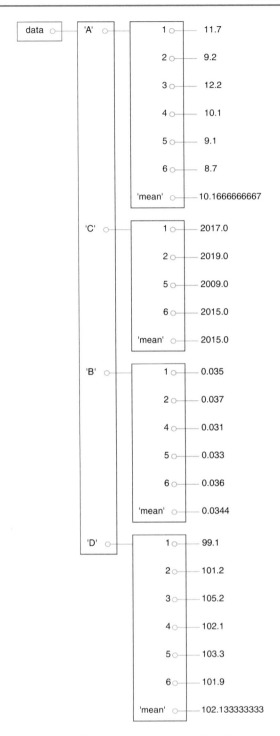

Fig. 6.1 Illustration of the nested dictionary created in the `table2dict.py` program

Alternatively, we can write an explicit loop to compute the average:

```
for p in data:
    sum_values = 0
    for value in data[p]:
        sum_values += value
    data[p]['mean'] = sum_values/len(data[p])
```

When we want to look up a measurement no. n of property B, we must recall that this particular measurement may be missing so we must do a test if n is key in the dictionary `data[p]`:

```
if n in data['B']:
    value = data['B'][n]

# alternative:
value = data['B'][n] if n in data['B'] else None
```

6.1.7 Example: Reading and Plotting Data Recorded at Specific Dates

Problem We want to compare the evolution of the stock prices of some giant companies in the computer industry: Microsoft, Apple, and Google. Relevant data files for stock prices can be downloaded from http://finance.yahoo.com. Fill in the company's name and click on *Search Finance* in the top bar of this page and choose *Historical Prices* in the left pane. On the resulting web page one can specify start and end dates for the historical prices of the stock. The default values were used in this example. Ticking off *Monthly* values and clicking *Get Prices* result in a table of stock prices for each month since the stock was introduced. The table can be downloaded as a spreadsheet file in CSV format, typically looking like

```
Date,Open,High,Low,Close,Volume,Adj Close
2014-02-03,502.61,551.19,499.30,545.99,12244400,545.99
2014-01-02,555.68,560.20,493.55,500.60,15698500,497.62
2013-12-02,558.00,575.14,538.80,561.02,12382100,557.68
2013-11-01,524.02,558.33,512.38,556.07,9898700,552.76
2013-10-01,478.45,539.25,478.28,522.70,12598400,516.57
...
1984-11-01,25.00,26.50,21.87,24.75,5935500,2.71
1984-10-01,25.00,27.37,22.50,24.87,5654600,2.73
1984-09-07,26.50,29.00,24.62,25.12,5328800,2.76
```

The file format is simple: columns are separated by comma, the first line contains column headings, and the data lines have the date in the first column and various measures of stock prices in the next columns. Reading about the meaning of the various data on the Yahoo! web pages reveals that our interest concerns the final column (as these prices are adjusted for splits and dividends). Three relevant data files can be found in the folder `src/files`[2] with the names `stockprices_X.csv`, where X is `Microsoft`, `Apple`, or `Google`.

The task is visually illustrate the historical, relative stock market value of these companies. For this purpose it is natural to scale the prices of a company's stock to start at a unit value when the most recent company entered the market. Since the date of entry varies, the oldest data point can be skipped such that all data points correspond to the first trade day every month.

Solution   There are two major parts of this problem: reading the file and plotting the data. The reading part is quite straightforward, while the plotting part needs some special considerations since the x values in the plot are dates and not real numbers. In the forthcoming text we solve the individual subproblems one by one, showing the relevant Python snippets. The complete program is found in the file `stockprices.py`.

We start with the reading part. Since the reading will be repeated for several companies, we create a function for extracting the relevant data for a specific company. These data cover the dates in column 1 and the stock prices in the last column. Since we want to plot prices versus dates, it will be convenient to turn the dates into `date` objects. In more detail the algorithms has the following points:

1. open the file
2. create two empty lists, dates and prices, for collecting the data
3. read the first line (of no interest)
4. for each line in the rest of the file:
   (a) split the line wrt. comma into words
   (b) append the first word to the dates list
   (c) append the last word to the prices list
5. reverse the lists (oldest date first)
6. convert date strings to datetime objects
7. convert prices list to float array for computations
8. return dates and prices, except for the first (oldest) data point

There are a couple of additional points to consider. First, the words on a line are strings, and at least the prices (last word) should be converted to a float. Second, the recipe for converting dates like '2008-02-04' to `date` (or `datetime`) objects goes as

```
from datetime import datetime
datefmt = '%Y-%m-%d'   # date format YYYY-MM-DD used in datetime
strdate = '2008-02-04'
datetime_object = datetime.strptime(strdate, datefmt)
date_object = datetime_object.date()
```

The nice thing with `date` and `datetime` object is that we can compute with them and in particular used them in plotting with Matplotlib.

---

[2] http://tinyurl.com/pwyasaa/files

We can now translate the algorithm to Python code:

```
from datetime import datetime

def read_file(filename):
    infile = open(filename, 'r')
    infile.readline()  # read column headings
    dates = []; prices = []
    for line in infile:
        words = line.split(',')
        dates.append(words[0])
        prices.append(float(words[-1]))
    infile.close()
    dates.reverse()
    prices.reverse()
    # Convert dates on the form 'YYYY-MM-DD' to date objects
    datefmt = '%Y-%m-%d'
    dates = [datetime.strptime(_date, datefmt).date()
             for _date in dates]
    prices = np.array(prices)
    return dates[1:], prices[1:]
```

Although we work with three companies in this example, it is easy and almost always a good idea to generalize the program to an arbitrary number of companies. All we assume is that their stock prices are in files with names of the form stockprices_X.csv, where X is the company name. With aid of the function call glob.glob('stockprices_*.csv') we get a list of all such files. By looping over this list, extracting the company name, and calling read_file, we can store the dates and corresponding prices in dictionaries dates and prices, indexed by the company name:

```
dates = {}; prices = {}
import glob, numpy as np
filenames = glob.glob('stockprices_*.csv')
companies = []
for filename in filenames:
    company = filename[12:-4]
    d, p = read_file(filename)
    dates[company] = d
    prices[company] = p
```

The next step is to normalize the prices such that they coincide on a certain date. We pick this date as the first month we have data for the youngest company. In lists of date or datetime objects, we can use Python's max and min functions to extract the newest and oldest date.

```
first_months = [dates[company][0] for company in dates]
normalize_date = max(first_months)
for company in dates:
    index = dates[company].index(normalize_date)
    prices[company] /= prices[company][index]
```

```
# Plot log of price versus years

import matplotlib.pyplot as plt
from matplotlib.dates import YearLocator, MonthLocator, DateFormatter

fig, ax = plt.subplots()
legends = []
for company in prices:
    ax.plot_date(dates[company], np.log(prices[company]),
                 '-', label=company)
    legends.append(company)
ax.legend(legends, loc='upper left')
ax.set_ylabel('logarithm of normalized value')

# Format the ticks
years   = YearLocator(5)    # major ticks every 5 years
months  = MonthLocator(6)   # minor ticks every 6 months
yearsfmt = DateFormatter('%Y')
ax.xaxis.set_major_locator(years)
ax.xaxis.set_major_formatter(yearsfmt)
ax.xaxis.set_minor_locator(months)
ax.autoscale_view()
fig.autofmt_xdate()

plt.savefig('tmp.pdf'); plt.savefig('tmp.png')
plt.show()
```

The normalized prices vary a lot, so to see the development over 30 years better, we decide to take the logarithm of the prices. The plotting procedure is somewhat involved so the reader should take the coming code more as a recipe than as a sequence of statement to really understand:

```
import matplotlib.pyplot as plt
from matplotlib.dates import YearLocator, MonthLocator, DateFormatter

fig, ax = plt.subplots()
legends = []
for company in prices:
    ax.plot_date(dates[company], np.log(prices[company]),
                 '-', label=company)
    legends.append(company)
ax.legend(legends, loc='upper left')
ax.set_ylabel('logarithm of normalized value')

# Format the ticks
years   = YearLocator(5)    # major ticks every 5 years
months  = MonthLocator(6)   # minor ticks every 6 months
yearsfmt = DateFormatter('%Y')
ax.xaxis.set_major_locator(years)
ax.xaxis.set_major_formatter(yearsfmt)
ax.xaxis.set_minor_locator(months)
ax.autoscale_view()
fig.autofmt_xdate()

plt.savefig('tmp.pdf'); plt.savefig('tmp.png')
```

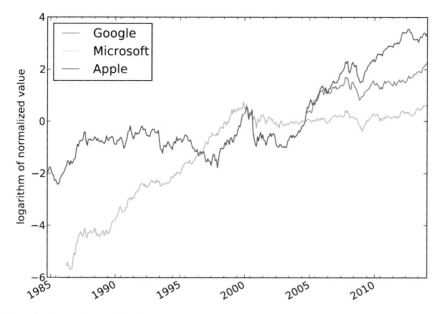

Fig. 6.2 The evolution of stock prices for three companies

Figure 6.2 shows the resulting plot. We observe that the normalized prices coincide when Google entered the market, here at Sep 1, 2004. Note that there is a log scale on the vertical axis. You may want to plot the real normalized prices to get a stronger impression of the significant recent rise in value, especially for Apple.

6.2 Strings

Many programs need to manipulate text. For example, when we read the contents of a file into a string or list of strings (lines), we may want to change parts of the text in the string(s) – and maybe write out the modified text to a new file. So far in this chapter we have converted parts of the text to numbers and computed with the numbers. Now it is time to learn how to manipulate the text strings themselves.

6.2.1 Common Operations on Strings

Python has a rich set of operations on string objects. Some of the most common operations are listed below.

Substring specification The expression `s[i:j]` extracts the substring starting with character number `i` and ending with character number `j-1` (similarly to lists, 0 is the index of the first character):

```
>>> s = 'Berlin: 18.4 C at 4 pm'
>>> s[8:]      # from index 8 to the end of the string
'18.4 C at 4 pm'
>>> s[8:12]    # index 8, 9, 10 and 11 (not 12!)
'18.4'
```

A negative upper index counts, as usual, from the right such that s[-1] is the last element, s[-2] is the next last element, and so on.

```
>>> s[8:-1]
'18.4 C at 4 p'
>>> s[8:-8]
'18.4 C'
```

Searching for substrings The call s.find(s1) returns the index where the substring s1 first appears in s. If the substring is not found, −1 is returned.

```
>>> s.find('Berlin')  # where does 'Berlin' start?
0
>>> s.find('pm')
20
>>> s.find('Oslo')    # not found
-1
```

Sometimes the aim is to just check if a string is contained in another string, and then we can use the syntax:

```
>>> 'Berlin' in s:
True
>>> 'Oslo' in s:
False
```

Here is a typical use of the latter construction in an if test:

```
>>> if 'C' in s:
...       print 'C found'
... else:
...       print 'no C'
...
C found
```

Two other convenient methods for checking if a string starts with or ends with a specified string are startswith and endswith:

```
>>> s.startswith('Berlin')
True
>>> s.endswith('am')
False
```

Substitution The call `s.replace(s1, s2)` replaces substring `s1` by `s2` everywhere in `s`:

```
>>> s.replace(' ', '_')
'Berlin:_18.4_C__at_4_pm'
>>> s.replace('Berlin', 'Bonn')
'Bonn: 18.4 C at 4 pm'
```

A variant of the last example, where several string operations are put together, consists of replacing the text before the first colon:

```
>>> s.replace(s[:s.find(':')], 'Bonn')
'Bonn: 18.4 C at 4 pm'
```

Take a break at this point and convince yourself that you understand how we specify the substring to be replaced!

String splitting The call `s.split()` splits the string `s` into words separated by whitespace (space, tabulator, or newline):

```
>>> s.split()
['Berlin:', '18.4', 'C', 'at', '4', 'pm']
```

Splitting a string `s` into words separated by a text `t` can be done by `s.split(t)`. For example, we may split with respect to colon:

```
>>> s.split(':')
['Berlin', ' 18.4 C at 4 pm']
```

We know that `s` contains a city name, a colon, a temperature, and then `C`:

```
>>> s = 'Berlin: 18.4 C at 4 pm'
```

With `s.splitlines()`, a multi-line string is split into lines (very useful when a file has been read into a string and we want a list of lines):

```
>>> t = '1st line\n2nd line\n3rd line'
>>> print t
1st line
2nd line
3rd line
>>> t.splitlines()
['1st line', '2nd line', '3rd line']
```

Upper and lower case `s.lower()` transforms all characters to their lower case equivalents, and `s.upper()` performs a similar transformation to upper case letters:

```
>>> s.lower()
'berlin: 18.4 c at 4 pm'
>>> s.upper()
'BERLIN: 18.4 C AT 4 PM'
```

Strings are constant   A string cannot be changed, i.e., any change always results in a new string. Replacement of a character is not possible:

```
>>> s[18] = 5
...
TypeError: 'str' object does not support item assignment
```

If we want to replace s[18], a new string must be constructed, for example by keeping the substrings on either side of s[18] and inserting a '5' in between:

```
>>> s[:18] + '5' + s[19:]
'Berlin: 18.4 C at 5 pm'
```

Strings with digits only   One can easily test whether a string contains digits only or not:

```
>>> '214'.isdigit()
True
>>> '  214 '.isdigit()
False
>>> '2.14'.isdigit()
False
```

Whitespace   We can also check if a string contains spaces only by calling the isspace method. More precisely, isspace tests for *whitespace*, which means the space character, newline, or the TAB character:

```
>>> '     '.isspace()   # blanks
True
>>> ' \n'.isspace()     # newline
True
>>> ' \t '.isspace()    # TAB
True
>>> ''.isspace()        # empty string
False
```

The isspace is handy for testing for blank lines in files. An alternative is to strip first and then test for an empty string:

```
>>> line = '   \n'
>>> line.strip() == ''
True
```

Stripping off leading and/or trailing spaces in a string is sometimes useful:

```
>>> s = '  text with leading/trailing space   \n'
>>> s.strip()
'text with leading/trailing space'
>>> s.lstrip()   # left strip
'text with leading/trailing space   \n'
>>> s.rstrip()   # right strip
'  text with leading/trailing space'
```

Joining strings The opposite of the `split` method is `join`, which joins elements in a list of strings with a specified delimiter in between. That is, the following two types of statements are inverse operations:

```
t = delimiter.join(words)
words = t.split(delimiter)
```

An example on using `join` may be

```
>>> strings = ['Newton', 'Secant', 'Bisection']
>>> t = ', '.join(strings)
>>> t
'Newton, Secant, Bisection'
```

As an illustration of the usefulness of `split` and `join`, we want to remove the first two words on a line. This task can be done by first splitting the line into words and then joining the words of interest:

```
>>> line = 'This is a line of words separated by space'
>>> words = line.split()
>>> line2 = ' '.join(words[2:])
>>> line2
'a line of words separated by space'
```

There are many more methods in string objects. All methods are described in the String Methods[3] section of the Python Standard Library online document.

6.2.2 Example: Reading Pairs of Numbers

Problem Suppose we have a file consisting of pairs of real numbers, i.e., text of the form (a, b), where a and b are real numbers. This notation for a pair of numbers is often used for points in the plane, vectors in the plane, and complex numbers. A sample file may look as follows:

```
(1.3,0)    (-1,2)    (3,-1.5)
(0,1)      (1,0)     (1,1)
(0,-0.01)  (10.5,-1) (2.5,-2.5)
```

The file can be found as `read_pairs1.dat`. Our task is to read this text into a nested list `pairs` such that `pairs[i]` holds the pair with index `i`, and this pair is a tuple of two `float` objects. We assume that there are no blanks inside the parentheses of a pair of numbers (we rely on a split operation, which would otherwise not work).

Solution To solve this programming problem, we can read in the file line by line; for each line: split the line into words (i.e., split with respect to whitespace); for

---

[3] http://docs.python.org/2/library/stdtypes.html#string-methods

each word: strip off the parentheses, split with respect to comma, and convert the resulting two words to floats. Our brief algorithm can be almost directly translated to Python code:

```
# Load the file into list of lines
with open('read_pairs1.dat', 'r') as infile:
    lines = infile.readlines()

# Analyze the contents of each line
pairs = []   # list of (n1, n2) pairs of numbers
for line in lines:
    words = line.split()
    for word in words:
        word = word[1:-1]  # strip off parenthesis
        n1, n2 = word.split(',')
        n1 = float(n1);  n2 = float(n2)
        pair = (n1, n2)
        pairs.append(pair)  # add 2-tuple to last row
```

This code is available in the file `read_pairs1.py`. The `with` statement is the modern Python way of reading files, see Sect. 4.5.2, with the advantage that we do not need to think about closing the file. Figure 6.3 shows a snapshot of the state of the variables in the program after having treated the first line. You should explain each line in the program to yourself, and compare your understanding with the figure.

The output from the program becomes

```
[(1.3, 0.0),
 (-1.0, 2.0),
 (3.0, -1.5),
 (0.0, 1.0),
 (1.0, 0.0),
 (1.0, 1.0),
 (0.0, -0.01),
 (10.5, -1.0),
 (2.5, -2.5)]
```

We remark that our solution to this programming problem relies heavily on the fact that spaces inside the parentheses are not allowed. If spaces were allowed, the simple split to obtain the pairs on a line as words would not work. What can we then do?

We can first strip off all blanks on a line, and then observe that the pairs are separated by the text `')('`. The first and last pair on a line will have an extra parenthesis that we need to remove. The rest of code is similar to the previous code and can be found in `read_pairs2.py`:

```
with open('read_pairs2.dat', 'r') as infile:
    lines = infile.readlines()

# Analyze the contents of each line
pairs = []   # list of (n1, n2) pairs of numbers
```

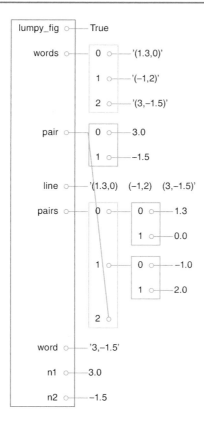

Fig. 6.3 Illustration of the variables in the `read_pairs.py` program after the first pass in the loop over words in the first line of the data file

```
for line in lines:
    line = line.strip()  # remove whitespace such as newline
    line = line.replace(' ', '')  # remove all blanks
    words = line.split(')(')
    # strip off leading/trailing parenthesis in first/last word:
    words[0] = words[0][1:]     # (-1,3  ->  -1,3
    words[-1] = words[-1][:-1]  # 8.5,9) ->  8.5,9
    for word in words:
        n1, n2 = word.split(',')
        n1 = float(n1);  n2 = float(n2)
        pair = (n1, n2)
        pairs.append(pair)
```

The program can be tested on the file `read_pairs2.dat`:

```
   (1.3 , 0)    (-1 , 2 )    (3, -1.5)
(0 , 1)      ( 1, 0)      ( 1 , 1 )
(0,-0.01)  (10.5,-1) (2.5, -2.5)
```

A third approach is to notice that if the pairs were separated by commas,

```
(1, 3.0),    (-1, 2),    (3, -1.5),
(0, 1),      (1, 0),     (1, 1),
```

the file text is very close to the Python syntax of a list of 2-tuples. We only miss the enclosing brackets:

```
[(1, 3.0),    (-1, 2),    (3, -1.5),
(0, 1),      (1, 0),     (1, 1),]
```

Running eval on this text will automatically produce the list object we want to construct! All we need to do is to read the file into a string, add a comma after every right parenthesis, add opening and closing bracket, and call eval (program read_pairs3.py):

```
with open('read_pairs2.dat', 'r') as infile:
    text = infile.read()
text = text.replace(')', '),')
text = '[' + text + ']'
pairs = eval(text)
```

In general, it is a good idea to construct file formats that are as close as possible to valid Python syntax such that one can take advantage of the eval or exec functions to turn text into "live objects".

6.2.3 Example: Reading Coordinates

Problem Suppose we have a file with coordinates (x, y, z) in three-dimensional space. The file format looks as follows:

```
x=-1.345    y= 0.1112    z= 9.1928
x=-1.231    y=-0.1251    z= 1001.2
x= 0.100    y= 1.4344E+6 z=-1.0100
x= 0.200    y= 0.0012    z=-1.3423E+4
x= 1.5E+5   y=-0.7666    z= 1027
```

The goal is to read this file and create a list with (x,y,z) 3-tuples, and thereafter convert the nested list to a two-dimensional array with which we can compute.

Note that there is sometimes a space between the = signs and the following number and sometimes not. Splitting with respect to space and extracting every second word is therefore not an option. We shall present three solutions.

Solution 1: substring extraction The file format looks very regular with the x=, y=, and z= texts starting in the same columns at every line. By counting characters, we realize that the x= text starts in column 2, the y= text starts in column 16, while the z= text starts in column 31. Introducing

```
x_start = 2
y_start = 16
z_start = 31
```

the three numbers in a `line` string are obtained as the substrings

```
x = line[x_start+2:y_start]
y = line[y_start+2:z_start]
z = line[z_start+2:]
```

The following code, found in file `file2coor_v1.py`, creates the `coor` array with shape $(n, 3)$, where n is the number of (x, y, z) coordinates.

```
infile = open('xyz.dat', 'r')
coor = []   # list of (x,y,z) tuples
for line in infile:
    x_start = 2
    y_start = 16
    z_start = 31
    x = line[x_start+2:y_start]
    y = line[y_start+2:z_start]
    z = line[z_start+2:]
    print 'debug: x="%s", y="%s", z="%s"' % (x,y,z)
    coor.append((float(x), float(y), float(z)))
infile.close()

import numpy as np
coor = np.array(coor)
print coor.shape, coor
```

The `print` statement inside the loop is always wise to include when doing string manipulations, simply because counting indices for substring limits quickly leads to errors. Running the program, the output from the loop looks like this

```
debug: x="-1.345   ", y=" 0.1112   ", z=" 9.1928
"
```

for the first line in the file. The double quotes show the exact extent of the extracted coordinates. Note that the last quote appears on the next line. This is because `line` has a newline at the end (this newline must be there to define the end of the line), and the substring `line[z_start:]` contains the newline at the of `line`. Writing `line[z_start:-1]` would leave the newline out of the z coordinate. However, this has no effect in practice since we transform the substrings to `float`, and an extra newline or other blanks make no harm.

The `coor` object at the end of the program has the value

```
[[ -1.34500000e+00   1.11200000e-01   9.19280000e+00]
 [ -1.23100000e+00  -1.25100000e-01   1.00120000e+03]
 [  1.00000000e-01   1.43440000e+06  -1.01000000e+00]
 [  2.00000000e-01   1.20000000e-03  -1.34230000e+04]
 [  1.50000000e+05  -7.66600000e-01   1.02700000e+03]]
```

Solution 2: string search One problem with the solution approach above is that the program will not work if the file format is subject to a change in the column positions of x=, y=, or z=. Instead of hardcoding numbers for the column positions, we can use the `find` method in string objects to locate these column positions:

```
x_start = line.find('x=')
y_start = line.find('y=')
z_start = line.find('z=')
```

The rest of the code is similar to the complete program listed above, and the complete code is stored in the file `file2coor_v2.py`.

Solution 3: string split String splitting is a powerful tool, also in the present case. Let us split with respect to the equal sign. The first line in the file then gives us the words

```
['x', '-1.345    y', ' 0.1112    z', ' 9.1928']
```

We throw away the first word, and strip off the last character in the next word. The final word can be used as is. The complete program is found in the file `file2coor_v3.py` and looks like

```
infile = open('xyz.dat', 'r')
coor = []  # list of (x,y,z) tuples
for line in infile:
    words = line.split('=')
    x = float(words[1][:-1])
    y = float(words[2][:-1])
    z = float(words[3])
    coor.append((x, y, z))
infile.close()

import numpy as np
coor = np.array(coor)
print coor.shape, coor
```

More sophisticated examples of string operations appear in Sect. 6.3.4.

6.3 Reading Data from Web Pages

Python has a module `urllib` which makes it possible to read data from a web page as easily as we can read data from an ordinary file. (In principle this is true, but in practice the text in web pages tend to be much more complicated than the text in the files we have treated so far.) Before we do this, a few concepts from the Internet world must be touched.

6.3.1 About Web Pages

Web pages are viewed with a web browser. There are many browsers: Firefox, Internet Explorer, Safari, Opera, and Google Chrome to mention the most famous. Any web page you visit is associated with an address, usually something like

```
http://www.some.where.net/some/file.html
```

This type of web address is called a URL (Uniform Resource Locator) or URI (Uniform Resource Identifier). (We stick to the term URL in this book because Python's tools for accessing resources on the Internet have `url` as part of module and function names.) The graphics you see in a web browser, i.e., the web page you see with your eyes, is produced by a series of commands that specifies the text on the page, the images, buttons to be pressed, etc. Roughly speaking, these commands are like statements in computer programs. The commands are stored in a text file and follow rules in a language, exactly as you are used to when writing statements in a programming language.

The common language for defining web pages is HTML. A web page is then simply a text file with text containing HTML commands. Instead of a physical file, the web page can also be the output text from a program. In that case the URL is the name of the program file.

The web browser interprets the text and the HTML commands, and then decides how to display the information visually. Let us demonstrate this for a very simple web page shown in Fig. 6.4. This page was produced by the following text with embedded HTML commands:

```
<html>
<body bgcolor="orange">
<h1>A Very Simple HTML Page</h1> <!-- headline -->
Web pages are written in a language called
<a href="http://www.w3.org/MarkUp/Guide/">HTML</a>.
Ordinary text is written as ordinary text, but when we
need links, headlines, lists,
<ul>
<li><em>emphasized words</em>, or
```

Fig. 6.4 Example of what a very simple HTML file looks like in a web browser

```
<li> <b>boldface text</b>,
</ul>
we need to embed the text inside HTML tags. We can also
insert GIF or PNG images, taken from other Internet sites,
if desired.
<hr> <!-- horizontal line -->
<img src="http://www.simula.no/simula_logo.gif">
</body>
</html>
```

A typical HTML command consists of an opening and a closing *tag*. For example, emphasized text is specified by enclosing the text inside em (emphasize) tags:

```
<em>emphasized words</em>
```

The opening tag is enclosed in less than and greater than signs, while the closing tag has an additional forward slash before the tag name.

In the HTML file we see an opening and closing `html` tag around the whole text in the file. Similarly, there is a pair of `body` tags, where the first one also has a parameter `bgcolor` which can be used to specify a background color in the web page. Section headlines are specified by enclosing the headline text inside `h1` tags. Subsection headlines apply `h2` tags, which results in a smaller font compared with `h1` tags. Comments appear inside `<!--` and `->`. Links to other web pages are written inside a tags, with an argument `href` for the link's web address. Lists apply the `ul` (unordered list) tag, while each item is written with just an opening tag `li` (list item), but no closing tag is necessary. Images are also specified with just an opening tag having name `img`, and the image file is given as a file name or URL of a file, enclosed in double quotes, as the `src` parameter.

The ultra-quick HTML course in the previous paragraphs gives a glimpse of how web pages can be constructed. One can either write the HTML file by hand in a pure text editor, or one can use programs such as Dream Weaver to help design the page graphically in a user-friendly environment, and then the program can automatically generate the right HTML syntax in files.

6.3.2 How to Access Web Pages in Programs

Why is it useful to know some HTML and how web pages are constructed? The reason is that the web is full of information that we can get access to through programs and use in new contexts. What we can get access to is not the visual web page you see, but the underlying HTML file. The information you see on the screen appear in text form in the HTML file, and by extracting text, we can get hold of the text's information in a program.

Given the URL as a string stored in a variable, there are two ways of accessing the HTML text in a Python program:

Alternative 1 Download the HTML file and store it as a local file with a given name, say `webpage.html`:

```
import urllib
url = 'http://www.simula.no/research/scientific/cbc'
urllib.urlretrieve(url, filename='webpage.html')
```

Alternative 2 Open the HTML file as a file-like object:

```
infile = urllib.urlopen(url)
```

This `infile` object has methods such as `read`, `readline`, and `readlines`.

6.3.3   Example: Reading Pure Text Files

Some web pages are just pure text files. Extracting the data from such pages are as easy as reading ordinary text files. Here is an example of historic weather data from the UK:

```
http://www.metoffice.gov.uk/climate/uk/stationdata/
```

We may choose a station, say Oxford, which directs us to the page

```
http://www.metoffice.gov.uk/climate/uk/stationdata/oxforddata.txt
```

We can download this data file by

```
import urllib
url = \
'http://www.metoffice.gov.uk/climate/uk/stationdata/oxforddata.txt'
urllib.urlretrieve(url, filename='Oxford.txt')
```

The files looks as follows:

```
Oxford
Location: 4509E 2072N, 63 metres amsl
Estimated data is marked with a * after the value.
Missing data (more than 2 days missing in month) is marked by   ---.
Sunshine data taken from an automatic ...
   yyyy  mm   tmax    tmin      af    rain     sun
              degC    degC    days      mm   hours
   1853   1    8.4     2.7       4    62.8     ---
   1853   2    3.2    -1.8      19    29.3     ---
   1853   3    7.7    -0.6      20    25.9     ---
   1853   4   12.6     4.5       0    60.1     ---
   1853   5   16.8     6.1       0    59.5     ---

...

   2010   1    4.7    -1.0      17    56.4    68.2
   2010   2    7.1     1.3       7    79.8    59.3
   2010   3   11.3     3.2       8    47.6   130.2
   2010   4   15.8     4.9       0    25.5   209.5
   2010   5   17.6     7.3       0    28.6   207.4
```

```
2010    6    23.0    11.1       0    34.5    230.5
2010    7    23.3*   14.1*      0*   24.4*   184.4*   Provisional
2010    8    21.4    12.0       0   146.2    123.8    Provisional
2010    9    19.0    10.0       0    48.1    118.6    Provisional
2010   10    14.6     7.4       2    43.5    128.8    Provisional
```

After the 7 header lines the data consists of 7 or 8 columns of numbers, the 8th being of no interest. Some numbers may have * or # appended to them, but this character must be stripped off before using the number. The columns contain the year, the month number (1–12), average maximum temperature, average minimum temperature, total number of days of air frost (af) during the month, total rainfall during the month, and the total number of hours with sun during the month. The temperature averages are taken over the maximum and minimum temperatures for all days in the month. Unavailable data are marked by three dashes.

The data can be conveniently stored in a dictionary with, e.g., three main keys: place (name), location (the info on the 2nd), and data. The latter is a dictionary with two keys: year and month.

The following program creates the data dictionary:

```python
infile = open(local_file, 'r')
data = {}
data['place'] = infile.readline().strip()
data['location'] = infile.readline().strip()
# Skip the next 5 lines
for i in range(5):
    infile.readline()

data['data'] ={}
for line in infile:
    columns = line.split()

    year = int(columns[0])
    month = int(columns[1])

    if columns[-1] == 'Provisional':
        del columns[-1]
    for i in range(2, len(columns)):
        if columns[i] == '---':
            columns[i] = None
        elif columns[i][-1] == '*' or columns[i][-1] == '#':
            # Strip off trailing character
            columns[i] = float(columns[i][:-1])
        else:
            columns[i] = float(columns[i])

    tmax, tmin, air_frost, rain, sun = columns[2:]

    if not year in data['data']:
        data['data'][year] = {}
    data['data'][year][month] = {'tmax': tmax,
                                 'tmin': tmin,
                                 'air frost': air_frost,
                                 'sun': sun}
```

The code is available in the file historic_weather.py.

With a few lines of code, we can extract the data we want, say a two-dimensional array of the number of sun hours in a month (these data are available from year 1929):

```
sun = [[data['data'][y][m]['sun'] for m in range(1,13)] \
       for y in range(1929, 2010)]
import numpy as np
sun = np.array(sun)
```

One can now do analysis of the data as exemplified in Sect. 2.6.2 and Exercise 5.8.

6.3.4   Example: Extracting Data from HTML

Very often, interesting data in a web page appear inside HTML code. We then need to interpret the text using string operations and store the data in variables. An example will clarify the principle.

The web site `www.worldclimate.com` contains data on temperature and rainfall in a large number of cities around the world. For example,

```
http://www.worldclimate.com/cgi-bin/data.pl?ref=N38W009+2100+08535W
```

contains a table of the average rainfall for each month of the year in the town Lisbon, Portugal. Our task is to download this web page and extract the tabular data (rainfall per month) in a list.

Downloading the file is done with `urllib` as explained in Sects. 6.3.2 and 6.3.3. Before attempting to read and interpret the text in the file, we need to look at the HTML code to find the interesting parts and determine how we can extract the data. The table with the rainfall data appears in the middle of the file. A sketch of the relevant HTML code goes as follows:

```
<p>Weather station <strong>LISBOA</strong> ...
<tr><th align=right><th>  Jan<th>  Feb<th> ...  <br>
<tr><td> mm <td align=right> 95.2 <td align=right> 86.7 ...<br>
<tr><td>inches <td align=right>3.7<td align=right>3.4 ...<br>
```

Our task is to walk through the file line by line and stop for processing the first and third line above:

```
infile = open('Lisbon_rainfall.html', 'r')
rainfall = []
for line in infile:
    if 'Weather station' in line:
        station = line.split('</strong>')[0].split('<strong>')[1]
    if '<td> mm <td' in line:
        data = line.split('<td align=right>')
```

The resulting `data` list looks like

```
['<tr><td> mm ', ' 95.2 ', ..., '702.4<br> \n']
```

To process this list further, we strip off the `<br>`... part of the last element:

```
data[-1] = data[-1].split('<br>')[0]
```

Then we drop the first element and convert the others to `float` objects:

```
data = [float(x) for x in data[1:]]
```

Now we have the rainfall data for each month as a list of real numbers. The complete program appears in the file `Lisbon_rainfall.py`. The recipe provided in this example can be used to interpret many other types of web pages where HTML code and data are wired together.

6.3.5   Handling Non-English Text

By default, Python only accepts English characters in a program file. Comments and strings in other languages, containing non-English characters, requires a special comment line before any non-English characters appears:

```
# -*- coding: utf-8 -*-
```

This line specifies that the file applies the UTF-8 encoding. Alternative encodings are UTF-16 and latin-1, depending on what your computer system supports. UTF-8 is most common nowadays.

There are two types of strings in Python: plain strings (known as byte strings) with type `str` and unicode strings with type `unicode`. Plain strings suffice as long as you are writing English text only. A string is then just a series of bytes representing integers between 0 and 255. The first characters corresponding to the numbers 0 to 127 constitute the ASCII set. These can be printed out:

```
for i in range(0, 128):
    print i, chr(i)
```

The keys on an English keyboard can be recognized from i=32 to i=126. The next numbers are used to represent non-English characters.

Texts with non-English characters are recommended to be represented by unicode strings. This is the default string type in Python 3, while in Python 2 we need to explicitly annotate a string as unicode by a u prefix as in s = u'my text'.

We shall now explore plain strings and unicode strings and will for that purpose need a help function for displaying a string in the terminal window, printing the type of string, dumping the exact content of the string, and telling us the length of the string in bytes:

```
def check(s):
    print '%s, %s: %s (%d)' % \
          (s, s.__class__.__name__, repr(s), len(s))
```

Let us start with a German character typed with a German keyboard:

```
>>> Gauss = 'C. F. ßGau'
>>> check(Gauss)
C. F. ßGau, str: 'C. F. Gau\xc3\x9f' (11)
```

Observe that there are 10 characters in the string, but `len(Gauss)` is 11. We can write each character:

```
>>> for char in Gauss:
...     print ord(char),
...
67 46 32 70 46 32 71 97 117 195 159
```

The last character in the `Gauss` object, the special German character, is represented by two bytes: 195 and 159. The other characters are in the range 0–127.

The `Gauss` object above is a plain Python 2 (byte) string. We can define the string as unicode in Python 2:

```
>>> Gauss = u'C. F. ßGau'
>>> check(Gauss)
C. F. ßGau, unicode: u'C. F. Gau\xdf' (10)
```

This time the unicode representation is as long as the expected number of characters, and the special German ß looks like `\xdf`. In fact, this character has unicode representation DF and we can use this code directly when we define the string, instead of a German keyboard:

```
>>> Gauss = u'C. F. Gau\xdf'
>>> check(Gauss)
C. F. ßGau, unicode: u'C. F. Gau\xdf' (10)
```

The string can be defined through the UTF-8 bytecode counterpart to ß, which is C3 9F:

```
>>> Gauss = 'C. F. Gau\xc3\x9f'  # plain string
>>> check(Gauss)
C. F. ßGau, str: 'C. F. Gau\xc3\x9f' (11)
```

Mixing UTF-8 bytecode in unicode strings, as in u'C. F. Gau\xc3\x9f', gives and unreadable output.

We can convert from a unicode representation to UTF-8 bytecode and back again:

```
>>> Gauss = u'C. F. Gau\xdf'
>>> repr(Gauss.encode('utf-8'))  # convert to UTF-8 bytecode
'C. F. Gau\xc3\x9f'
>>> unicode(Gauss.encode('utf-8'), 'utf-8')  # convert back again
u'C. F. Gau\xdf'
```

Other encodings are UTF-16 and latin-1:

```
>>> repr(Gauss.encode('utf-16'))
'\xff\xfeC\x00.\x00 \x00F\x00.\x00 \x00G\x00a\x00u\x00\xdf\x00'
>>> repr(Gauss.encode('latin-1'))
'C. F. Gau\xdf'
```

Writing the unicode variable Gauss to file, a la f.write(Gauss), leads to a UnicodeEncodeError in Python 2, saying that 'ascii' codec can't encode character u'\xdf' in position 9. The UTF-8 bytecode representation of strings does not pose any problems with file writing. The solution for unicode strings is to use the codecs module and explicitly work with a file object that converts unicode to UTF-8:

```
import codecs
with codecs.open('tmp.txt', 'w', 'utf-8') as f:
    f.write(Gauss)
```

This is not necessary with Python 3, so if you use non-English characters, Python 3 has a clear advantage over Python 2.

To summarize, non-English character can be input with a non-English keyboard and stored either as a plain (byte) string or as a unicode string:

```
>>> name = 'Åsmund Øådegrd'  # plain string
>>> check(name)Å
smund Øådegrd, str: '\xc3\x85smund \xc3\x98deg\xc3\xa5rd' (17)
>>> name = u'Åsmund Øådegrd' # unicode
>>> check(name)Å
smund Øådegrd, unicode: u'\xc5smund \xd8deg\xe5rd' (14)
```

Alternatively, the non-English characters can be specified with special codes, depending on whether the representation is a plain UTF-8 string or a unicode string. Using a table[4] with conversion between unicode and UTF-8 representation we find that in UTF-8, Å has the code C3 85, Ø is C3 98, and å is C3 A5:

```
>>> name = '\xc3\x85smund \xc3\x98deg\xc3\xa5rd'
>>> check(name)Å
smund Øådegrd, str: '\xc3\x85smund \xc3\x98deg\xc3\xa5rd' (17)
```

In unicode, Å is C5, Ø is D8, å is E5:

```
>>> name = u'\xc5smund \xd8deg\xe5rd'
>>> check(name)Å
smund Øådegrd, unicode: u'\xc5smund \xd8deg\xe5rd' (14)
```

The examples above have been collected in the file unicode_utf8.py.

---

[4] http://www.utf8-chartable.de/

6.4   Reading and Writing Spreadsheet Files

From school you are probably used to spreadsheet programs such as Microsoft Excel or LibreOffice. This type of program is used to represent a table of numbers and text. Each table entry is known as a *cell*, and one can easily perform calculations with cells that contain numbers. The application of spreadsheet programs for mathematical computations and graphics is steadily growing.

Also Python may be used to do spreadsheet-type calculations on tabular data. The advantage of using Python is that you can easily extend the calculations far beyond what a spreadsheet program can do. However, even if you can view Python as a substitute for a spreadsheet program, it may be beneficial to combine the two. Suppose you have some data in a spreadsheet. How can you read these data into a Python program, perform calculations on the data, and thereafter read the data back to the spreadsheet program? This is exactly what we will explain below through an example. With this example, you should understand how easy it is to combine Excel or LibreOffice with your own Python programs.

6.4.1   CSV Files

The table of data in a spreadsheet can be saved in so-called CSV files, where CSV stands for *comma separated values*. The CSV file format is very simple: each row in the spreadsheet table is a line in the file, and each cell in the row is separated by a comma or some other specified separation character. CSV files can easily be read into Python programs, and the table of cell data can be stored in a nested list (table, see Sect. 2.4), which can be processed as we desire. The modified table of cell data can be written back to a CSV file and read into the spreadsheet program for further processing.

Figure 6.5 shows a simple spreadsheet in the LibreOffice program. The table contains 4×4 cells, where the first row contains column headings and the first column contains row headings. The remaining 3×3 subtable contains numbers that

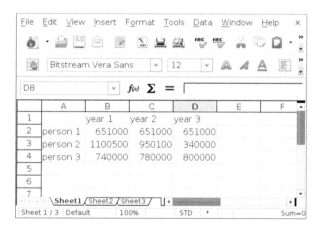

Fig. 6.5 A simple spreadsheet in LibreOffice

we may compute with. Let us save this spreadsheet to a file in the CSV format. The complete file will typically look as follows:

```
,"year 1","year 2","year 3"
"person 1",651000,651000,651000
"person 2",1100500,950100,340000
"person 3",740000,780000,800000
```

Our primary task is now to load these data into a Python program, compute the sum of each column, and write the data out again in the CSV format.

6.4.2 Reading CSV Files

We start with loading the data into a table, represented as a nested list, with aid of the csv module from Python's standard library. This approach gives us complete control of all details. Later, we will use more high-level numpy functionality for accomplishing the same thing with less lines.

The csv module offers functionality for reading one line at a time from a CSV file:

```
infile = open('budget.csv', 'r')  # CSV file
import csv
table = []
for row in csv.reader(infile):
    table.append(row)
infile.close()
```

The row variable is a list of column values that are read from the file by the csv module. The three lines computing table can be condensed to one using a list comprehension:

```
table = [row for row in csv.reader(infile)]
```

We can easily print table,

```
import pprint
pprint.pprint(table)
```

to see what the spreadsheet looks like when it is represented as a nested list in a Python program:

```
[['', 'year 1', 'year 2', 'year 3'],
 ['person 1', '651000', '651000', '651000'],
 ['person 2', '1100500', '950100', '340000'],
 ['person 3', '740000', '780000', '800000']]
```

Observe now that all entries are surrounded by quotes, which means that all entries are string (str) objects. This is a general rule: the csv module reads all cells into string objects. To compute with the numbers, we need to transform the string objects to `float` objects. The transformation should not be applied to the first row and first column, since the cells here hold text. The transformation from strings to numbers therefore applies to the indices r and c in `table` (`table[r][c]`), such that the row counter r goes from 1 to `len(table)-1`, and the column counter c goes from 1 to `len(table[0])-1` (`len(table[0])` is the length of the first row, assuming the lengths of all rows are equal to the length of the first row). The relevant Python code for this transformation task becomes

```
for r in range(1,len(table)):
    for c in range(1, len(table[0])):
        table[r][c] = float(table[r][c])
```

A `pprint.pprint(table)` statement after this transformation yields

```
[['', 'year 1', 'year 2', 'year 3'],
 ['person 1', 651000.0, 651000.0, 651000.0],
 ['person 2', 1100500.0, 950100.0, 340000.0],
 ['person 3', 740000.0, 780000.0, 800000.0]]
```

The numbers now have a decimal and no quotes, indicating that the numbers are `float` objects and hence ready for mathematical calculations.

6.4.3 Processing Spreadsheet Data

Let us perform a very simple calculation with `table`, namely adding a final row with the sum of the numbers in the columns:

```
row = [0.0]*len(table[0])
row[0] = 'sum'
for c in range(1, len(row)):
    s = 0
    for r in range(1, len(table)):
        s += table[r][c]
    row[c] = s
```

As seen, we first create a list `row` consisting of zeros. Then we insert a text in the first column, before we invoke a loop over the numbers in the table and compute the sum of each column. The `table` list now represents a spreadsheet with four columns and five rows:

```
[['', 'year 1', 'year 2', 'year 3'],
 ['person 1', 651000.0, 651000.0, 651000.0],
 ['person 2', 1100500.0, 950100.0, 340000.0],
 ['person 3', 740000.0, 780000.0, 800000.0],
 ['sum', 2491500.0, 2381100.0, 1791000.0]]
```

6.4.4   Writing CSV Files

Our final task is to write the modified `table` list back to a CSV file so that the data can be loaded in a spreadsheet program. The write task is done by the code segment

```
outfile = open('budget2.csv', 'w')
writer = csv.writer(outfile)
for row in table:
    writer.writerow(row)
outfile.close()
```

The `budget2.csv` looks like this:

```
year 1,year 2,year 3
person 1,651000.0,651000.0,651000.0
person 2,1100500.0,950100.0,340000.0
person 3,740000.0,780000.0,800000.0
sum,2491500.0,2381100.0,1791000.0
```

The final step is to read `budget2.csv` into a spreadsheet. The result is displayed in Fig. 6.6 (in LibreOffice one must specify in the *Open* dialog that the spreadsheet data are separated by commas, i.e., that the file is in CSV format).

The complete program reading the `budget.csv` file, processing its data, and writing the `budget2.csv` file can be found in `rw_csv.py`. With this example at hand, you should be in a good position to combine spreadsheet programs with your own Python programs.

Remark   You may wonder why we used the `csv` module to read and write CSV files when such files have comma separated values, which we can extract by splitting lines with respect to the comma (this technique is used in Sect. 6.1.7 to read a CSV file):

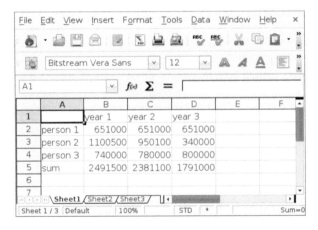

Fig. 6.6   A spreadsheet processed in a Python program and loaded back into LibreOffice

```
infile = open('budget.csv', 'r')
for line in infile:
    row = line.split(',')
```

This works well for the present budget.csv file, but the technique breaks down when a text in a cell contains a comma, for instance "Aug 8, 2007". The line.split(',') will split this cell text, while the csv.reader functionality is smart enough to avoid splitting text cells with a comma.

6.4.5   Representing Number Cells with Numerical Python Arrays

Instead of putting the whole spreadsheet into a single nested list, we can make a Python data structure more tailored to the data at hand. What we have are two headers (for rows and columns, respectively) and a subtable of numbers. The headers can be represented as lists of strings, while the subtable could be a two-dimensional Numerical Python array. The latter makes it easier to implement various mathematical operations on the numbers. A dictionary can hold all the three items: two header lists and one array. The relevant code for reading, processing, and writing the data is shown below and can be found in the file rw_csv_numpy.py:

```
infile = open('budget.csv', 'r')
import csv
table = [row for row in csv.reader(infile)]
infile.close()

# Convert subtable of numbers (string to float)
import numpy
subtable = [[float(c) for c in row[1:]] for row in table[1:]]

data = {'column headings': table[0][1:],
        'row headings': [row[0] for row in table[1:]],
        'array': numpy.array(subtable)}

# Add a new row with sums
data['row headings'].append('sum')
a = data['array']    # short form
data['column sum'] = [sum(a[:,c]) for c in range(a.shape[1])]

outfile = open('budget2.csv', 'w')
writer = csv.writer(outfile)
# Turn data dictionary into a nested list first (for easy writing)
table = a.tolist()    # transform array to nested list
table.append(data['column sum'])
table.insert(0, data['column headings'])
# Extend table with row headings (a new column)
[table[r+1].insert(0, data['row headings'][r])
 for r in range(len(table)-1)]
for row in table:
    writer.writerow(row)
outfile.close()
```

The code makes heavy use of list comprehensions, and the transformation between a nested list, for file reading and writing, and the `data` dictionary, for representing the data in the Python program, is non-trivial. If you manage to understand every line in this program, you have digested a lot of topics in Python programming!

6.4.6 Using More High-Level Numerical Python Functionality

The previous program can be shortened significantly by applying the `genfromtxt` function from Numerical Python:

```
import numpy as np
arr = np.genfromtxt('budget.csv', delimiter=',', dtype=str)

data = {'column headings': arr[0,1:].tolist(),
        'row headings': arr[1:,0].tolist(),
        'array': np.asarray(arr[1:,1:], dtype=np.float)}

data['row headings'].append('sum')
data['column sum'] = np.sum(data['array'], axis=1).tolist()
```

Doing a `repr(arr)` on the array returned from `genfromtxt` results in

```
array([['', '"year 1"', '"year 2"', '"year 3"'],
       ['"person 1"', '651000', '651000', '651000'],
       ['"person 2"', '1100500', '950100', '340000'],
       ['"person 3"', '740000', '780000', '800000']],
      dtype='|S10')
```

That is, the data in the CSV file are available as an array of strings. The code shows how we can easily use slices to extract the row and column headings, convert the numbers to a floating-point array for computations, compute the sums, and store various object in the `data` dictionary. Then we may write a CSV file as described in the previous example (see `rw_csv_numpy2.py`) or we may take a different approach and extend the `arr` array with an extra row and fill in the row heading and the sums (see `rw_csv_numpy3.py` for the complete code):

```
arr = np.genfromtxt('budget.csv', delimiter=',', dtype=str)

# Add row for sum of columns
arr.resize((arr.shape[0]+1, arr.shape[1]))
arr[-1,0] = '"sum"'
subtable = np.asarray(arr[1:-1,1:], dtype=np.float)
sum_row = np.sum(subtable, axis=1)
arr[-1,1:] = np.asarray(sum_row, dtype=str)

# numpy.savetxt writes table with a delimiter between entires
np.savetxt('budget2c.csv', arr, delimiter=',', fmt='%s')
```

Observe how we extract the numbers in `subtable`, compute with them, and put the results back into the `arr` array as strings. The `savetxt` function saves a two-dimensional array as a plain table in a text file, here with comma as delimiter. The function suffices in this example, none of the approaches with `genfromtxt` and `savetxt` work with column or row headings containing a comma. Then we need to use the `csv` module.

6.5 Examples from Analyzing DNA

We shall here continue the bioinformatics applications started in Sect. 3.3. Analysis of DNA sequences is conveniently done in Python, much with the aid of lists, dictionaries, `numpy` arrays, strings, and files. This will be illustrated through a series of examples.

6.5.1 Computing Frequencies

Your genetic code is essentially the same from you are born until you die, and the same in your blood and your brain. Which genes that are turned on and off make the difference between the cells. This regulation of genes is orchestrated by an immensely complex mechanism, which we have only started to understand. A central part of this mechanism consists of molecules called transcription factors that float around in the cell and attach to DNA, and in doing so turn nearby genes on or off. These molecules bind preferentially to specific DNA sequences, and this binding preference pattern can be represented by a table of frequencies of given symbols at each position of the pattern. More precisely, each row in the table corresponds to the bases A, C, G, and T, while column j reflects how many times the base appears in position j in the DNA sequence.

For example, if our set of DNA sequences are TAG, GGT, and GGG, the table becomes

base	0	1	2
A	0	1	0
C	0	0	0
G	2	2	2
T	1	0	1

From this table we can read that base A appears once in index 1 in the DNA strings, base C does not appear at all, base G appears twice in all positions, and base T appears once in the beginning and end of the strings.

In the following we shall present different data structures to hold such a table and different ways of computing them. The table is known as a *frequency matrix* in bioinformatics and this is the term used here too.

Separate frequency lists Since we know that there are only four rows in the frequency matrix, an obvious data structure would be four lists, each holding a row. A function computing these lists may look like

```
def freq_lists(dna_list):
    n = len(dna_list[0])
    A = [0]*n
    T = [0]*n
    G = [0]*n
    C = [0]*n
    for dna in dna_list:
        for index, base in enumerate(dna):
            if base == 'A':
                A[index] += 1
            elif base == 'C':
                C[index] += 1
            elif base == 'G':
                G[index] += 1
            elif base == 'T':
                T[index] += 1
    return A, C, G, T
```

We need to initialize the lists with the right length and a zero for each element, since each list element is to be used as a counter. Creating a list of length n with object x in all positions is done by [x]*n. Finding the proper length is here carried out by inspecting the length of the first element in dna_list, the list of all DNA strings to be counted, assuming that all elements in this list have the same length.

In the for loop we apply the enumerate function, which is used to extract both the element value *and* the element index when iterating over a sequence. For example,

```
>>> for index, base in enumerate(['t', 'e', 's', 't']):
...     print index, base
...
0 t
1 e
2 s
3 t
```

Here is a test,

```
dna_list = ['GGTAG', 'GGTAC', 'GGTGC']
A, C, G, T = freq_lists(dna_list)
print A
print C
print G
print T
```

with output

```
[0, 0, 0, 2, 0]
[0, 0, 0, 0, 2]
[3, 3, 0, 1, 1]
[0, 0, 3, 0, 0]
```

Nested list The frequency matrix can also be represented as a nested list M such that M[i][j] is the frequency of base i in position j in the set of DNA strings.

Here i is an integer, where 0 corresponds to A, 1 to T, 2 to G, and 3 to C. The
frequency is the number of times base i appears in position j in a set of DNA
strings. Sometimes this number is divided by the number of DNA strings in the set
so that the frequency is between 0 and 1. Note that all the DNA strings must have
the same length.

The simplest way to make a nested list is to insert the A, C, G, and T lists into
another list:

```
>>> frequency_matrix = [A, C, G, T]
>>> frequency_matrix[2][3]
2
>>> G[3]   # same element
2
```

Alternatively, we can illustrate how to compute this type of nested list directly:

```
def freq_list_of_lists_v1(dna_list):
    # Create empty frequency_matrix[i][j] = 0
    # i=0,1,2,3 corresponds to A,T,G,C
    # j=0,...,length of dna_list[0]
    frequency_matrix = [[0 for v in dna_list[0]] for x in 'ACGT']

    for dna in dna_list:
      for index, base in enumerate(dna):
          if base == 'A':
              frequency_matrix[0][index] +=1
          elif base == 'C':
              frequency_matrix[1][index] +=1
          elif base == 'G':
              frequency_matrix[2][index] +=1
          elif base == 'T':
              frequency_matrix[3][index] +=1

    return frequency_matrix
```

As in the case with individual lists we need to initialize all elements in the nested
list to zero.

A call and printout,

```
dna_list = ['GGTAG', 'GGTAC', 'GGTGC']
frequency_matrix = freq_list_of_lists_v1(dna_list)
print frequency_matrix
```

results in

```
[[0, 0, 0, 2, 0], [0, 0, 0, 0, 2], [3, 3, 0, 1, 1], [0, 0, 3, 0, 0]]
```

Dictionary for more convenient indexing   The series of if tests in the Python
function freq_list_of_lists_v1 are somewhat cumbersome, especially if we
want to extend the code to other bioinformatics problems where the alphabet is
larger. What we want is a mapping from base, which is a character, to the corresponding index 0, 1, 2, or 3. A Python dictionary may represent such mappings:

```
>>> base2index = {'A': 0, 'C': 1, 'G': 2, 'T': 3}
>>> base2index['G']
2
```

With the `base2index` dictionary we do not need the series of `if` tests and the
alphabet `'ATGC'` could be much larger without affecting the length of the code:

```
def freq_list_of_lists_v2(dna_list):
    frequency_matrix = [[0 for v in dna_list[0]] for x in 'ACGT']
    base2index = {'A': 0, 'C': 1, 'G': 2, 'T': 3}
    for dna in dna_list:
        for index, base in enumerate(dna):
            frequency_matrix[base2index[base]][index] += 1

    return frequency_matrix
```

Numerical Python array As long as each sublist in a list of lists has the same
length, a list of lists can be replaced by a Numerical Python (numpy) array. Processing of such arrays is often much more efficient than processing of the nested
list data structure. To initialize a two-dimensional numpy array we need to know
its size, here 4 times `len(dna_list[0])`. Only the first line in the function
`freq_list_of_lists_v2` needs to be changed in order to utilize a numpy array:

```
import numpy as np

def freq_numpy(dna_list):
    frequency_matrix = np.zeros((4, len(dna_list[0])), dtype=int)
    base2index = {'A': 0, 'C': 1, 'G': 2, 'T': 3}
    for dna in dna_list:
        for index, base in enumerate(dna):
            frequency_matrix[base2index[base]][index] += 1

    return frequency_matrix
```

The resulting `frequency_matrix` object can be indexed as `[b][i]` or `[b,i]`,
with integers b and i. Typically, b will be something line `base2index['C']`.

Dictionary of lists Instead of going from a character to an integer index via
`base2index`, we may prefer to index `frequency_matrix` by the base name and
the position index directly, like in `['C'][14]`. This is the most natural syntax
for a user of the frequency matrix. The relevant Python data structure is then
a dictionary of lists. That is, `frequency_matrix` is a dictionary with keys `'A'`,
`'C'`, `'G'`, and `'T'`. The value for each key is a list. Let us now also extend the
flexibility such that `dna_list` can have DNA strings of different lengths. The lists
in `frequency_list` will have lengths equal to the longest DNA string. A relevant
function is

```
def freq_dict_of_lists_v1(dna_list):
    n = max([len(dna) for dna in dna_list])
    frequency_matrix = {
        'A': [0]*n,
        'C': [0]*n,
        'G': [0]*n,
        'T': [0]*n,
        }
    for dna in dna_list:
        for index, base in enumerate(dna):
            frequency_matrix[base][index] += 1

    return frequency_matrix
```

Running the test code

```
frequency_matrix = freq_dict_of_lists_v1(dna_list)
import pprint    # for nice printout of nested data structures
pprint.pprint(frequency_matrix)
```

results in the output

```
{'A': [0, 0, 0, 2, 0],
 'C': [0, 0, 0, 0, 2],
 'G': [3, 3, 0, 1, 1],
 'T': [0, 0, 3, 0, 0]}
```

The initialization of `frequency_matrix` in the above code can be made more compact by using a dictionary comprehension:

```
dict = {key: value for key in some_sequence}
```

In our example we set

```
frequency_matrix = {base: [0]*n for base in 'ACGT'}
```

Adopting this construction in the `freq_dict_of_lists_v1` function leads to a slightly more compact version:

```
def freq_dict_of_lists_v2(dna_list):
    n = max([len(dna) for dna in dna_list])
    frequency_matrix = {base: [0]*n for base in 'ACGT'}
    for dna in dna_list:
        for index, base in enumerate(dna):
            frequency_matrix[base][index] += 1

    return frequency_matrix
```

As an additional comment on computing the maximum length of the DNA strings can be made as there are several alternative ways of doing this. The classical use of max is to apply it to a list as done above:

```
n = max([len(dna) for dna in dna_list])
```

However, for very long lists it is possible to avoid the memory demands of storing the result of the list comprehension, i.e., the list of lengths. Instead max can work with the lengths as they are computed:

```
n = max(len(dna) for dna in dna_list)
```

It is also possible to write

```
n = max(dna_list, key=len)
```

Here, len is applied to each element in dna_list, and the maximum of the resulting values is returned.

Dictionary of dictionaries.

The dictionary of lists data structure can alternatively be replaced by a dictionary of dictionaries object, often just called a dict of dicts object. That is, frequency_matrix[base] is a dictionary with key i and value equal to the added number of occurrences of base in dna[i] for all dna strings in the list dna_list. The indexing frequency_matrix['C'][i] and the values are exactly as in the last example; the only difference is whether frequency_matrix['C'] is a list or dictionary.

Our function working with frequency_matrix as a dict of dicts is written as

```
def freq_dict_of_dicts_v1(dna_list):
    n = max([len(dna) for dna in dna_list])
    frequency_matrix = {base: {index: 0 for index in range(n)}
                        for base in 'ACGT'}
    for dna in dna_list:
        for index, base in enumerate(dna):
            frequency_matrix[base][index] += 1

    return frequency_matrix
```

Using dictionaries with default values The manual initialization of each subdictionary to zero,

```
    frequency_matrix = {base: {index: 0 for index in range(n)}
                        for base in 'ACGT'}
```

can be simplified by using a dictionary with default values for any key. The construction defaultdict(lambda: obj) makes a dictionary with obj as default value. This construction simplifies the previous function a bit:

```
from collections import defaultdict

def freq_dict_of_dicts_v2(dna_list):
    n = max([len(dna) for dna in dna_list])
    frequency_matrix = {base: defaultdict(lambda: 0)
                        for base in 'ACGT'}
    for dna in dna_list:
        for index, base in enumerate(dna):
            frequency_matrix[base][index] += 1

    return frequency_matrix
```

Remark Dictionary comprehensions were new in Python 2.7 and 3.1, but can be simulated in earlier versions by making (key, value) tuples via list comprehensions. A dictionary comprehension

```
d = {key: value for key in sequence}
```

is then constructed as

```
d = dict([(key, value) for key in sequence])
```

Using arrays and vectorization The `frequency_matrix` dict of lists for can easily be changed to a dict of numpy arrays: just replace the initialization `[0]*n` by `np.zeros(n, dtype=np.int)`. The indexing remains the same:

```
def freq_dict_of_arrays_v1(dna_list):
    n = max([len(dna) for dna in dna_list])
    frequency_matrix = {base: np.zeros(n, dtype=np.int)
                        for base in 'ACGT'}
    for dna in dna_list:
        for index, base in enumerate(dna):
            frequency_matrix[base][index] += 1

    return frequency_matrix
```

Having `frequency_matrix[base]` as a numpy array instead of a list does not give any immediate advantage, as the storage and CPU time is about the same. The loop over the `dna` string and the associated indexing is what consumes all the CPU time. However, the numpy arrays provide a potential for increasing efficiency through vectorization, i.e., replacing the element-wise operations on `dna` and `frequency_matrix[base]` by operations on the entire arrays at once.

Let us use the interactive Python shell to explore the possibilities of vectorization. We first convert the string to a numpy array of characters:

```
>>> dna = 'ACAT'
>>> dna = np.array(dna, dtype='c')
>>> dna
array(['A', 'C', 'A', 'T'],
      dtype='|S1')
```

For a given base, say A, we can in one vectorized operation find which locations in dna that contain A:

```
>>> b = dna == 'A'
>>> b
array([ True, False,  True, False], dtype=bool)
```

By converting b to an integer array i we can update the frequency counts for all indices by adding i to frequency_matrix['A']:

```
>>> i = np.asarray(b, dtype=np.int)
>>> i
array([1, 0, 1, 0])
>>> frequency_matrix['A'] = frequency_matrix['A'] + i
```

This recipe can be repeated for all bases:

```
for dna in dna_list:
    dna = np.array(dna, dtype='c')
    for base in 'ACGT':
        b = dna == base
        i = np.asarray(b, dtype=np.int)
        frequency_matrix[base] = frequency_matrix[base] + i
```

It turns out that we do not need to convert the boolean array b to an integer array i, because doing arithmetics with b directly is possible: False is interpreted as 0 and True as 1 in arithmetic operations. We can also use the += operator to update all elements of frequency_matrix[base] directly, without first computing the sum of two arrays frequency_matrix[base] + i and then assigning this result to frequency_matrix[base]. Collecting all these ideas in one function yields the code

```
def freq_dict_of_arrays_v2(dna_list):
    n = max([len(dna) for dna in dna_list])
    frequency_matrix = {base: np.zeros(n, dtype=np.int)
                        for base in 'ACGT'}
    for dna in dna_list:
        dna = np.array(dna, dtype='c')
        for base in 'ACCT':
            frequency_matrix[base] += dna == base

    return frequency_matrix
```

This vectorized function runs almost 10 times as fast as the (scalar) counterpart freq_list_of_arrays_v1!

6.5.2   Analyzing the Frequency Matrix

Having built a frequency matrix out of a collection of DNA strings, it is time to use it for analysis. The short DNA strings that a frequency matrix is built out of, is typically a set of substrings of a larger DNA sequence, which shares some common purpose. An example of this is to have a set of substrings that serves as a kind of anchors/magnets at which given molecules attach to DNA and perform biological functions (like turning genes on or off). With the frequency matrix constructed from a limited set of known anchor locations (substrings), we can now scan for other similar substrings that have the potential to perform the same function. The simplest way to do this is to first determine the most typical substring according to the frequency matrix, i.e., the substring having the most frequent nucleotide at each position. This is referred to as the consensus string of the frequency matrix. We can then look for occurrences of the consensus substring in a larger DNA sequence, and consider these occurrences as likely candidates for serving the same function (e.g., as anchor locations for molecules).

For instance, given three substrings ACT, CCA and AGA, the frequency matrix would be (list of lists, with rows corresponding to A, C, G, and T):

```
[[2, 0, 2]
 [1, 2, 0]
 [0, 1, 0]
 [0, 0, 1]]
```

We see that for position 0, which corresponds to the left-most column in the table, the symbol A has the highest frequency (2). The maximum frequencies for the other positions are seen to be C for position 1, and A for position 2. The consensus string is therefore ACA. Note that the consensus string does not need to be equal to any of the substrings that formed the basis of the frequency matrix (this is indeed the case for the above example).

List of lists frequency matrix   Let `frequency_matrix` be a list of lists. For each position i we run through the rows in the frequency matrix and keep track of the maximum frequency value and the corresponding letter. If two or more letters have the same frequency value we use a dash to indicate that this position in the consensus string is undetermined.

The following function computes the consensus string:

```python
def find_consensus_v1(frequency_matrix):
    base2index = {'A': 0, 'C': 1, 'G': 2, 'T': 3}
    consensus = ''
    dna_length = len(frequency_matrix[0])

    for i in range(dna_length):  # loop over positions in string
        max_freq = -1            # holds the max freq. for this i
        max_freq_base = None     # holds the corresponding base

        for base in 'ATGC':
            if frequency_matrix[base2index[base]][i] > max_freq:
                max_freq = frequency_matrix[base2index[base]][i]
                max_freq_base = base
            elif frequency_matrix[base2index[base]][i] \
                    == max_freq:
                max_freq_base = '-'  # more than one base as max
```

```
        consensus += max_freq_base  # add new base with max freq
    return consensus
```

Since this code requires `frequency_matrix` to be a list of lists we should insert a test and raise an exception if the type is wrong:

```
def find_consensus_v1(frequency_matrix):
    if isinstance(frequency_matrix, list) and \
       isinstance(frequency_matrix[0], list):
        pass # right type
    else:
        raise TypeError('frequency_matrix must be list of lists')
    ...
```

Dict of dicts frequency matrix How must the `find_consensus_v1` function be altered if `frequency_matrix` is a dict of dicts?

1. The `base2index` dict is no longer needed.
2. Access of sublist, `frequency_matrix[0]`, to test for type and length of the strings, must be replaced by `frequency_matrix['A']`.

The updated function becomes

```
def find_consensus_v3(frequency_matrix):
    if isinstance(frequency_matrix, dict) and \
       isinstance(frequency_matrix['A'], dict):
        pass # right type
    else:
        raise TypeError('frequency_matrix must be dict of dicts')

    consensus = ''
    dna_length = len(frequency_matrix['A'])

    for i in range(dna_length):  # loop over positions in string
        max_freq = -1             # holds the max freq. for this i
        max_freq_base = None      # holds the corresponding base

        for base in 'ACGT':
            if frequency_matrix[base][i] > max_freq:
                max_freq = frequency_matrix[base][i]
                max_freq_base = base
            elif frequency_matrix[base][i] == max_freq:
                max_freq_base = '-' # more than one base as max

        consensus += max_freq_base  # add new base with max freq
    return consensus
```

Here is a test:

```
frequency_matrix = freq_dict_of_dicts_v1(dna_list)
pprint.pprint(frequency_matrix)
print find_consensus_v3(frequency_matrix)
```

with output

```
{'A': {0: 0, 1: 0, 2: 0, 3: 2, 4: 0},
 'C': {0: 0, 1: 0, 2: 0, 3: 0, 4: 2},
 'G': {0: 3, 1: 3, 2: 0, 3: 1, 4: 1},
 'T': {0: 0, 1: 0, 2: 3, 3: 0, 4: 0}}
Consensus string: GGTAC
```

Let us try `find_consensus_v3` with the dict of defaultdicts as input (`freq_dicts_of_dicts_v2`). The code runs fine, but the output string is just G! The reason is that `dna_length` is 1, and therefore that the length of the A dict in `frequency_matrix` is 1. Printing out `frequency_matrix` yields

```
{'A': defaultdict(X, {3: 2}),
 'C': defaultdict(X, {4: 2}),
 'G': defaultdict(X, {0: 3, 1: 3, 3: 1, 4: 1}),
 'T': defaultdict(X, {2: 3})}
```

where our X is a short form for text like

```
'<function <lambda> at 0xfaede8>'
```

We see that the length of a defaultdict will only count the nonzero entries. Hence, to use a defaultdict our function must get the length of the DNA string to build as an extra argument:

```
def find_consensus_v4(frequency_matrix, dna_length):
    ...
```

Exercise 6.16 suggests to make a unified `find_consensus` function which works with all of the different representations of `frequency_matrix` that we have used.

The functions making and using the frequency matrix are found in the file `freq.py`.

6.5.3 Finding Base Frequencies

DNA consists of four molecules called nucleotides, or bases, and can be represented as a string of the letters A, C, G, and T. But this does not mean that all four nucleotides need to be similarly frequent. Are some nucleotides more frequent than others, say in yeast, as represented by the first chromosome of yeast? Also, DNA is really not a single thread, but two threads wound together. This wounding is based on an A from one thread binding to a T of the other thread, and C binding to G (that is, A will only bind with T, not with C or G). Could this fact force groups of the four symbol frequencies to be equal? The answer is that the A-T and G-C binding does not in principle force certain frequencies to be equal, but in practice they usually become so because of evolutionary factors related to this pairing.

Our first programming task now is to compute the frequencies of the bases A, C, G, and T. That is, the number of times each base occurs in the DNA string, divided by the length of the string. For example, if the DNA string is ACGGAAA, the length is 7, A appears 4 times with frequency 4/7, C appears once with frequency 1/7, G appears twice with frequency 2/7, and T does not appear so the frequency is 0.

From a coding perspective we may create a function for counting how many times A, C, G, and T appears in the string and then another function for computing the frequencies. In both cases we want dictionaries such that we can index with the character and get the count or the frequency out. Counting is done by

```
def get_base_counts(dna):
    counts = {'A': 0, 'T': 0, 'G': 0, 'C': 0}
    for base in dna:
        counts[base] += 1
    return counts
```

This function can then be used to compute the base frequencies:

```
def get_base_frequencies_v1(dna):
    counts = get_base_counts(dna)
    return {base: count*1.0/len(dna)
            for base, count in counts.items()}
```

Since we learned at the end of Sect. 3.3.2 that dna.count(base) was much faster than the various manual implementations of counting, we can write a faster and simpler function for computing all the base frequencies:

```
def get_base_frequencies_v2(dna):
    return {base: dna.count(base)/float(len(dna))
            for base in 'ATGC'}
```

A little test,

```
dna = 'ACCAGAGT'
frequencies = get_base_frequencies_v2(dna)

def format_frequencies(frequencies):
    return ', '.join(['%s: %.2f' % (base, frequencies[base])
                      for base in frequencies])

print "Base frequencies of sequence '%s':\n%s" % \
      (dna, format_frequencies(frequencies))
```

gives the result

```
Base frequencies of sequence 'ACCAGAGT':
A: 0.38, C: 0.25, T: 0.12, G: 0.25
```

The format_frequencies function was made for nice printout of the frequencies with 2 decimals. The one-line code is an effective combination of a dictionary, list comprehension, and the `join` functionality. The latter is used to get a comma correctly inserted between the items in the result. Lazy programmers would probably just do a `print frequencies` and live with the curly braces in the output and (in general) 16 disturbing decimals.

We can try the frequency computation on real data. The file

```
http://hplgit.github.com/bioinf-py/data/yeast_chr1.txt
```

contains the DNA for yeast. We can download this file from the Internet by

```
urllib.urlretrieve(url, filename=name_of_local_file)
```

where `url` is the Internet address of the file and `name_of_local_file` is a string containing the name of the file on the computer where the file is downloaded. To avoid repeated downloads when the program is run multiple times, we insert a test on whether the local file exists or not. The call `os.path.isfile(f)` returns `True` if a file with name `f` exists in the current working folder.

The appropriate download code then becomes

```
import urllib, os
urlbase = 'http://hplgit.github.com/bioinf-py/data/'
yeast_file = 'yeast_chr1.txt'
if not os.path.isfile(yeast_file):
    url = urlbase + yeast_file
    urllib.urlretrieve(url, filename=yeast_file)
```

A copy of the file on the Internet is now in the current working folder under the name `yeast_chr1.txt`. (See Sect. 6.3.2 for more information about `urllib` and downloading files from the Internet.)

The `yeast_chr1.txt` files contains the DNA string split over many lines. We therefore need to read the lines in this file, strip each line to remove the trailing newline, and join all the stripped lines to recover the DNA string:

```
def read_dnafile_v1(filename):
    lines = open(filename, 'r').readlines()
    # Remove newlines in each line (line.strip()) and join
    dna = ''.join([line.strip() for line in lines])
    return dna
```

As usual, an alternative programming solution can be devised:

```
def read_dnafile_v2(filename):
    dna = ''
    for line in open(filename, 'r'):
        dna += line.strip()
    return dna

dna = read_dnafile_v2(yeast_file)
yeast_freq = get_base_frequencies_v2(dna)
print "Base frequencies of yeast DNA (length %d):\n%s" % \
      (len(dna), format_frequencies(yeast_freq))
```

The output becomes

```
Base frequencies of yeast DNA (length 230208):
A: 0.30, C: 0.19, T: 0.30, G: 0.20
```

The varying frequency of different nucleotides in DNA is referred to as nucleotide bias. The nucleotide bias varies between organisms, and have a range of biological implications. For many organisms the nucleotide bias has been highly optimized through evolution and reflects characteristics of the organisms and their environments, for instance the typical temperature the organism is adapted to.

The functions computing base frequencies are available in the file `basefreq.py`.

6.5.4   Translating Genes into Proteins

An important usage of DNA is for cells to store information on their arsenal of proteins. Briefly, a gene is, in essence, a region of the DNA, consisting of several coding parts (called exons), interspersed by non-coding parts (called introns). The coding parts are concatenated to form a string called mRNA, where also occurrences of the letter T in the coding parts are substituted by a U. A triplet of mRNA letters code for a specific amino acid, which are the building blocks of proteins. Consecutive triplets of letters in mRNA define a specific sequence of amino acids, which amounts to a certain protein.

Here is an example of using the mapping from DNA to proteins to create the Lactase protein (LPH), using the DNA sequence of the Lactase gene (LCT) as underlying code. An important functional property of LPH is in digesting Lactose, which is found most notably in milk. Lack of the functionality of LPH leads to digestive problems referred to as lactose intolerance. Most mammals and humans lose their expression of LCT and therefore their ability to digest milk when they stop receiving breast milk.

The file

```
http://hplgit.github.com/bioinf-py/doc/src/data/genetic_code.tsv
```

contains a mapping of genetic codes to amino acids. The file format looks like

```
UUU    F      Phe    Phenylalanine
UUC    F      Phe    Phenylalanine
UUA    L      Leu    Leucine
UUG    L      Leu    Leucine
CUU    L      Leu    Leucine
CUC    L      Leu    Leucine
CUA    L      Leu    Leucine
CUG    L      Leu    Leucine
AUU    I      Ile    Isoleucine
AUC    I      Ile    Isoleucine
AUA    I      Ile    Isoleucine
AUG    M      Met    Methionine (Start)
```

The first column is the genetic code (triplet in mRNA), while the other columns represent various ways of expressing the corresponding amino acid: a 1-letter symbol, a 3-letter name, and the full name.

Downloading the `genetic_code.tsv` file can be done by this robust function:

```
def download(urlbase, filename):
    if not os.path.isfile(filename):
        url = urlbase + filename
        try:
            urllib.urlretrieve(url, filename=filename)
        except IOError as e:
            raise IOError('No Internet connection')
        # Check if downloaded file is an HTML file, which
        # is what github.com returns if the URL is not existing
        f = open(filename, 'r')
        if 'DOCTYPE html' in f.readline():
            raise IOError('URL %s does not exist' % url)
```

We want to make a dictionary of this file that maps the code (first column) on to the 1-letter name (second column):

```
def read_genetic_code_v1(filename):
    infile = open(filename, 'r')
    genetic_code = {}
    for line in infile:
        columns = line.split()
        genetic_code[columns[0]] = columns[1]
    return genetic_code
```

Downloading the file, reading it, and making the dictionary are done by

```
urlbase = 'http://hplgit.github.com/bioinf-py/data/'
genetic_code_file = 'genetic_code.tsv'
download(urlbase, genetic_code_file)
code = read_genetic_code_v1(genetic_code_file)
```

Not surprisingly, the `read_genetic_code_v1` can be made much shorter by collecting the first two columns as list of 2-lists and then converting the 2-lists to key-value pairs in a dictionary:

```
def read_genetic_code_v2(filename):
    return dict([line.split()[0:2]
                 for line in open(filename, 'r')])
```

Creating a mapping of the code onto all the three variants of the amino acid name is also of interest. For example, we would like to make look ups like ['CUU']['3-letter'] or ['CUU']['amino acid']. This requires a dictionary of dictionaries:

```
def read_genetic_code_v3(filename):
    genetic_code = {}
    for line in open(filename, 'r'):
        columns = line.split()
        genetic_code[columns[0]] = {}
        genetic_code[columns[0]]['1-letter']   = columns[1]
        genetic_code[columns[0]]['3-letter']   = columns[2]
        genetic_code[columns[0]]['amino acid'] = columns[3]
    return genetic_code
```

An alternative way of writing the last function is

```
def read_genetic_code_v4(filename):
    genetic_code = {}
    for line in open(filename, 'r'):
        c = line.split()
        genetic_code[c[0]] = {
        '1-letter': c[1], '3-letter': c[2], 'amino acid': c[3]}
    return genetic_code
```

To form mRNA, we need to grab the exon regions (the coding parts) of the lactase gene. These regions are substrings of the lactase gene DNA string, corresponding to the start and end positions of the exon regions. Then we must replace T by U, and combine all the substrings to build the mRNA string.

Two straightforward subtasks are to load the lactase gene and its exon positions into variables. The file lactase_gene.txt, at the same Internet location as the other files, stores the lactase gene. The file has the same format as yeast_chr1.txt. Using the download function and the previously shown read_dnafile_v1, we can easily load the data in the file into the string lactase_gene.

The exon regions are described in a file lactase_exon.tsv, also found at the same Internet site as the other files. The file is easily transferred to your computer by calling download. The file format is very simple in that each line holds the start and end positions of an exon region:

```
0        651
3990     4070
7504     7588
13177    13280
15082    15161
```

We want to have this information available in a list of (start, end) tuples. The following function does the job:

```
def read_exon_regions_v1(filename):
    positions = []
    infile = open(filename, 'r')
    for line in infile:
        start, end = line.split()
        start, end = int(start), int(end)
        positions.append((start, end))
```

```
    infile.close()
    return positions
```

Readers favoring compact code will appreciate this alternative version of the function:

```
def read_exon_regions_v2(filename):
    return [tuple(int(x) for x in line.split())
            for line in open(filename, 'r')]

lactase_exon_regions = read_exon_regions_v2(lactase_exon_file)
```

For simplicity's sake, we shall consider mRNA as the concatenation of exons, although in reality, additional base pairs are added to each end. Having the lactase gene as a string and the exon regions as a list of (start, end) tuples, it is straightforward to extract the regions as substrings, replace T by U, and add all the substrings together:

```
def create_mRNA(gene, exon_regions):
    mrna = ''
    for start, end in exon_regions:
        mrna += gene[start:end].replace('T','U')
    return mrna

mrna = create_mRNA(lactase_gene, lactase_exon_regions)
```

We would like to store the mRNA string in a file, using the same format as lactase_gene.txt and yeast_chr1.txt, i.e., the string is split on multiple lines with, e.g., 70 characters per line. An appropriate function doing this is

```
def tofile_with_line_sep_v1(text, filename, chars_per_line=70):
    outfile = open(filename, 'w')
    for i in xrange(0, len(text), chars_per_line):
        start = i
        end = start + chars_per_line
        outfile.write(text[start:end] + '\n')
    outfile.close()
```

It might be convenient to have a separate folder for files that we create. Python has good support for testing if a folder exists, and if not, make a folder:

```
output_folder = 'output'
if not os.path.isdir(output_folder):
    os.mkdir(output_folder)
filename = os.path.join(output_folder, 'lactase_mrna.txt')
tofile_with_line_sep_v1(mrna, filename)
```

Python's term for folder is directory, which explains why isdir is the function name for testing on a folder existence. Observe especially that the combination of a folder and a filename is done via os.path.join rather than just inserting a forward slash, or backward slash on Windows: `os.path.join` will insert the right slash, forward or backward, depending on the current operating system.

Occasionally, the output folder is nested, say

```
output_folder = os.path.join('output', 'lactase')
```

In that case, `os.mkdir(output_folder)` may fail because the intermediate folder output is missing. Making a folder and also all missing intermediate folders is done by `os.makedirs`. We can write a more general file writing function that takes a folder name and file name as input and writes the file. Let us also add some flexibility in the file format: one can either write a fixed number of characters per line, or have the string on just one long line. The latter version is specified through `chars_per_line='inf'` (for infinite number of characters per line). The flexible file writing function then becomes

```
def tofile_with_line_sep_v2(text, foldername, filename,
                            chars_per_line=70):
    if not os.path.isdir(foldername):
        os.makedirs(foldername)
    filename = os.path.join(foldername, filename)
    outfile = open(filename, 'w')

    if chars_per_line == 'inf':
        outfile.write(text)
    else:
        for i in xrange(0, len(text), chars_per_line):
            start = i
            end = start + chars_per_line
            outfile.write(text[start:end] + '\n')
    outfile.close()
```

To create the protein, we replace the triplets of the mRNA strings by the corresponding 1-letter name as specified in the `genetic_code.tsv` file.

```
def create_protein(mrna, genetic_code):
    protein = ''
    for i in xrange(len(mrna)/3):
        start = i * 3
        end = start + 3
        protein += genetic_code[mrna[start:end]]
    return protein

genetic_code = read_genetic_code_v1('genetic_code.tsv')
protein = create_protein(mrna, genetic_code)
tofile_with_line_sep_v2(protein, 'output',
```

Unfortunately, this first try to simulate the translation process is incorrect. The problem is that the translation always begins with the amino acid Methionine, code AUG, and ends when one of the stop codons is met. We must thus check for the correct start and stop criteria. A fix is

```
def create_protein_fixed(mrna, genetic_code):
    protein_fixed = ''
    trans_start_pos = mrna.find('AUG')
    for i in range(len(mrna[trans_start_pos:])/3):
        start = trans_start_pos + i*3
        end = start + 3
        amino = genetic_code[mrna[start:end]]
        if amino == 'X':
            break
        protein_fixed += amino
    return protein_fixed

protein = create_protein_fixed(mrna, genetic_code)
tofile_with_line_sep_v2(protein, 'output',
                        'lactase_protein_fixed.txt', 70)

print '10 last amino acids of the correct lactase protein: ', \
      protein[-10:]
print 'Lenght of the correct protein: ', len(protein)
```

The output, needed below for comparison, becomes

```
10 last amino acids of the correct lactase protein:  QQELSPVSSF
Lenght of the correct protein:  1927
```

6.5.5   Some Humans Can Drink Milk, While Others Cannot

One type of lactose intolerance is called *Congenital lactase deficiency*. This is a rare genetic disorder that causes lactose intolerance from birth, and is particularly common in Finland. The disease is caused by a mutation of the base in position 30049 (0-based) of the lactase gene, a mutation from T to A. Our goal is to check what happens to the protein if this base is mutated. This is a simple task using the previously developed tools:

```
def congential_lactase_deficiency(
    lactase_gene,
    genetic_code,
    lactase_exon_regions,
    output_folder=os.curdir,
    mrna_file=None,
    protein_file=None):

    pos = 30049
    mutated_gene = lactase_gene[:pos] + 'A' + lactase_gene[pos+1:]
    mutated_mrna = create_mRNA(mutated_gene, lactase_exon_regions)

    if mrna_file is not None:
        tofile_with_line_sep_v2(
            mutated_mrna, output_folder, mrna_file)

    mutated_protein = create_protein_fixed(
        mutated_mrna, genetic_code)
```

```
    if protein_file:
        tofile_with_line_sep_v2(
            mutated_protein, output_folder, protein_file)
    return mutated_protein

mutated_protein = congential_lactase_deficiency(
    lactase_gene, genetic_code, lactase_exon_regions,
    output_folder='output',
    mrna_file='mutated_lactase_mrna.txt',
    protein_file='mutated_lactase_protein.txt')

print '10 last amino acids of the mutated lactase protein:', \
    mutated_protein[-10:]
print 'Lenght of the mutated lactase protein:', \
    len(mutated_protein)
```

The output, to be compared with the non-mutated gene above, is now

```
10 last amino acids of the mutated lactase protein: GFIWSAASAA
Lenght of the mutated lactase protein: 1389
```

As we can see, the translation stops prematurely, creating a much smaller protein, which will not have the required characteristics of the lactase protein.

A couple of mutations in a region for LCT located in front of LCT (actually in the introns of another gene) is the reason for the common lactose intolerance. That is, the one that sets in for adults only. These mutations control the expression of the LCT gene, i.e., whether that the gene is turned on or off. Interestingly, different mutations have evolved in different regions of the world, e.g., Africa and Northern Europe. This is an example of convergent evolution: the acquisition of the same biological trait in unrelated lineages. The prevalence of lactose intolerance varies widely, from around 5 % in northern Europe, to close to 100 % in south-east Asia.

The functions analyzing the lactase gene are found in the file genes2proteins. py.

6.6 Making Code that is Compatible with Python 2 and 3

Some basic differences between Python 2 and 3 are covered Sect. 4.10. With the additional constructions met in this chapter, there are some important additional differences between the two versions of Python.

6.6.1 More Basic Differences Between Python 2 and 3

xrange **in Python 2 is** range **in Python 3** The range function in Python 2 generates a list of integers, and for very long loops this list may consume significant computer memory. The xrange function in Python was therefore made to just generate a series of integers without storing them. In Python 3, range is the xrange function from Python 2. If one wants a list of integers in Python 3, one has to do list(range(5)) to store the output from range in a list.

Python 3 often avoids returning lists and dictionaries The Python 3 idea of letting range just generate one object at a time instead of storing all of them applies to many other constructions too. Let d be a dictionary. In Python 2, d.keys() returns a list of the keys in the dictionary, while in Python 3, d.keys() just enables iteration over the keys in a for loop. Similarly, d.values() and d.items() returns lists of values or key-value pairs in Python 2, while in Python 3 we can only iterate over the values in a for loop. A simple loop like

```
for key in d.keys():
    ...
```

works well for both Python versions, but

```
keys = d.keys()
```

in Python 2, where we want the keys as a list, needs a modification in Python 3:

```
keys = list(d.keys())
```

We should add that for key in d.keys() is not the preferred syntax anyway – use for key in d. Also, if we just want a for loop over all key-value pairs, we can use d.iteritems() which does not return any list, neither in Python 2 nor in Python 3.

Library modules have different names We have used the urllib module in Sects. 6.3.2 and 6.3.3. Python 3 has some different names for this module:

```
# Python 2
import urllib
with urllib.urlopen('http://google.com') as webfile:
    text = webfile.read()
urllib.urlretrieve('http://google.com', filename='tmp.html')

# Python 3
import urllib.request as urllibr
with urllibr.urlopen('http://google.com') as webfile:
    text = webfile.read()
urllibr.urlretrieve('http://google.com', filename='tmp.html')
```

A lot of other modules have also changed names, but the futurize program (see below) help you to find the right new names.

Python 3 has unicode and byte strings A standard Python 2 string, s = 'abc', is a sequence of bytes, called byte string in Python 3, declared as s = b'abc' in Python 3. The assignment s = 'abc' in Python 3 leads to a unicode string and is equivalent to s = u'abc' in Python 2. To convert a byte string in Python 3 to an ordinary (unicode) string, do s.decode('utf-8'). String handling is often the most tricky task when converting Python 2 code to Python 3.

Python 3 has different relative import syntax inside packages If you work with
Python packages[5], relative imports *inside* a package has slightly different syntax
in Python 2 and 3. Say you want to import `somefunc` from a module `somemod` in
some other module at the same level (same subfolder) in the package. Python 2 syntax would be `from somemod import somefunc`, while Python 3 demands `from
.somemod import somefunc`. The leading dot in the module name indicates that
`somemod` is a module located in the same subfolder as the file containing this import
statement. The alternative import, `import somemod`, in Python 2 must read `from
. import somemod` in Python 3.

6.6.2 Turning Python 2 Code into Python 3 Code

As demonstrated in Sect. 4.10, one can use the `futurize` program to turn a Python
2 program into a version that works with both Python 2 and 3. For the programs at
this stage in the book, and also for more advanced programs, we recommend to run
the command

```Terminal
Terminal> futurize --all-imports -w -n -o py23 prog.py
```

which generates the new version of the program `prog.py` in the subfolder `py23`.
Sometimes manual changes are needed in addition, but this depends on the complexity of `prog.py`.

By frequently running just `futurize prog.py` to see what needs to be changed,
you can learn a lot of the differences between Python 2 and 3 and also change
your programming style in Python 2 so that it comes even closer to Python 3. The
`python-future` documentation has a very useful list of difference between Python
2 and 3[6] and recipes on how to make common code for both versions.

Porting of larger programs from Python 2 to 3 is recommended to use `futurize`
in a two-stage fashion[7].

6.7 Summary

6.7.1 Chapter Topics

Dictionaries Array or list-like objects with text or other (fixed-valued) Python objects as indices are called dictionaries. They are very useful for storing general
collections of objects in a single data structure. The table below displays some of
the most important dictionary operations.

[5] https://docs.python.org/3/tutorial/modules.html#packages
[6] http://python-future.org/compatible_idioms.html
[7] http://python-future.org/futurize.html#forwards-conversion-stage1

Construction	Meaning
`a = {}`	initialize an empty dictionary
`a = {'point': [0,0.1], 'value': 7}`	initialize a dictionary
`a = dict(point=[2,7], value=3)`	initialize a dictionary w/string keys
`a.update(b)`	add/update key-value pairs from b in a
`a.update(key1=value1, key2=value2)`	add/update key-value pairs in a
`a['hide'] = True`	add new key-value pair to a
`a['point']`	get value corresponding to key `point`
`for key in a:`	loop over keys in unknown order
`for key in sorted(a):`	loop over keys in alphabetic order
`'value' in a`	`True` if string `value` is a key in a
`del a['point']`	delete a key-value pair from a
`list(a.keys())`	list of keys
`list(a.values())`	list of values
`len(a)`	number of key-value pairs in a
`isinstance(a, dict)`	is `True` if a is a dictionary

Strings Some of the most useful functionalities in a string object s are listed below.

Split the string into substrings separated by `delimiter`:

```
words = s.split(delimiter)
```

Join elements in a list of strings:

```
newstring = delimiter.join(words[i:j])
```

Extract a **substring**:

```
substring = s[2:n-4]
```

Replace a substring `substr` by new a string `replacement`:

```
s_new = s.replace(substr, replacement)
```

Check if a substring **is contained** within another string:

```
if 'some text' in s:
    ...
```

Find the index where some text starts:

```
index = s.find(text)
if index == -1:
    print 'Could not find "%s" in "%s" (text, s)
else:
    substring = s[index:]  # strip off chars before text
```

Extend a string:

```
s += another_string    # append at the end
s = another_string + s # append at the beginning
```

Check if a string contains **whitespace only**:

```
if s.isspace():
    ...
```

Note: you cannot change the characters in a string like you can change elements in a list (a string is in this sense like a tuple). You have to make a new string:

```
>>> filename = 'myfile1.txt'
>>> filename[6] = '2'
Traceback (most recent call last):
  ...
TypeError: 'str' object does not support item assignment
>>> filename.replace('1', '2')
'myfile2.txt'
>>> filename[:6] + '2' + filename[7:]   # 'myfile' + '2' + '.txt'
'myfile2.txt'
```

Downloading Internet files Internet files can be downloaded if we know their URL:

```
import urllib
url = 'http://www.some.where.net/path/thing.html'
urllib.urlretrieve(url, filename='thing.html')
```

The downloaded information is put in the local file `thing.html` in the current working folder. Alternatively, we can open the URL as a file object:

```
webpage = urllib.urlopen(url)
```

HTML files are often messy to interpret by string operations.

Terminology The important computer science topics in this chapter are

- dictionaries
- strings and string operations
- CSV files
- HTML files

6.7.2   Example: A File Database

Problem We have a file containing information about the courses that students have taken. The file format consists of blocks with student data, where each block starts with the student's name (`Name:`), followed by the courses that the student has taken. Each course line starts with the name of the course, then comes the semester when the exam was taken, then the size of the course in terms of credit points, and finally the grade is listed (letters `A` to `F`). Here is an example of a file with three student entries:

```
Name: John Doe
Astronomy                       2003 fall 10 A
Introductory Physics            2003 fall 10 C
Calculus I                      2003 fall 10 A
Calculus II                     2004 spring 10 B
Linear Algebra                  2004 spring 10 C
Quantum Mechanics I             2004 fall 10 A
Quantum Mechanics II            2005 spring 10 A
Numerical Linear Algebra        2004 fall 5 E
Numerical Methods               2004 spring 20 C

Name: Jan Modaal
Calculus I                      2005 fall 10 A
Calculus II                     2006 spring 10 A
Introductory C++ Programming    2005 fall 15 D
Introductory Python Programming 2006 spring 5 A
Astronomy                       2005 fall 10 A
Basic Philosophy                2005 fall 10 F

Name: Kari Nordmann
Introductory Python Programming 2006 spring 5 A
Astronomy                       2005 fall 10 D
```

Our problem consists of reading this file into a dictionary `data` with the student name as key and a list of courses as value. Each element in the list of courses is a dictionary holding the course name, the semester, the credit points, and the grade. A value in the `data` dictionary may look as

```
'Kari Nordmann': [{'credit': 5,
                   'grade': 'A',
                   'semester': '2006 spring',
                   'title': 'Introductory Python Programming'},
                  {'credit': 10,
                   'grade': 'D',
                   'semester': '2005 fall',
                   'title': 'Astronomy'}],
```

Having the `data` dictionary, the next task is to print out the average grade of each student.

Solution We divide the problem into two major tasks: loading the file data into the `data` dictionary, and computing the average grades. These two tasks are naturally placed in two functions.

We need to have a strategy for reading the file and interpreting the contents. It will be natural to read the file line by line, and for each line check if this is a line containing a new student's name, a course information line, or a blank line. In the latter case we jump to the next pass in the loop. When a new student name is encountered, we initialize a new entry in the `data` dictionary to an empty list. In the case of a line about a course, we must interpret the contents on that line, which we postpone a bit.

We can now sketch the algorithm described above in terms of some unfinished Python code, just to get the overview:

```
def load(studentfile):
    infile = open(studentfile, 'r')
    data = {}
    for line in infile:
        i = line.find('Name:')
        if i != -1:
            # line contains 'Name:', extract the name.
            ...
        elif line.isspace():      # Blank line?
            continue              # Yes, go to next loop iteration.
        else:
            # This must be a course line, interpret the line.
            ...
    infile.close()
    return data
```

If we find `'Name:'` as a substring in `line`, we must extract the name. This can be done by the substring `line[i+5:]`. Alternatively, we can split the line with respect to colon and strip off the first word:

```
words = line.split(':')
name = ' '.join(words[1:])
```

We have chosen the former strategy of extracting the name as a substring in the final program.

Each course line is naturally split into words for extracting information:

```
words = line.split()
```

The name of the course consists of a number of words, but we do not know how many. Nevertheless, we know that the final words contain the semester, the credit points, and the grade. We can hence count from the right and extract information, and when we are finished with the semester information, the rest of the `words` list holds the words in the name of the course. The code goes as follows:

```
grade = words[-1]
credit = int(words[-2])
semester = ' '.join(words[-4:-2])
course_name = ' '.join(words[:-4])
data[name].append({'title':    course_name,
                   'semester': semester,
                   'credit':   credit,
                   'grade':    grade})
```

This code is a good example of the usefulness of split and join operations when extracting information from a text.

Now to the second task of computing the average grade. Since the grades are letters we cannot compute with them. A natural way to proceed is to convert the letters to numbers, compute the average number, and then convert that number back to a letter. Conversion between letters and numbers is easily represented by a dictionary:

```
grade2number = {'A': 5, 'B': 4, 'C': 3, 'D': 2, 'E': 1, 'F': 0}
```

To convert from numbers to grades, we construct the inverse dictionary:

```
number2grade = {}
for grade in grade2number:
    number2grade[grade2number[grade]] = grade
```

In the computation of the average grade we should use a weighted sum such that larger courses count more than smaller courses. The weighted mean value of a set of numbers r_i with weights w_i, $i = 0, \ldots, n - 1$, is given by

$$\frac{\sum_{i=0}^{n-1} w_i r_i}{\sum_{i=0}^{n-1} w_i}.$$

This weighted mean value must then be rounded to the nearest integer, which can be used as key in number2grade to find the corresponding grade expressed as a letter. The weight w_i is naturally taken as the number of credit points in the course with grade r_i. The whole process is performed by the following function:

```
def average_grade(data, name):
    sum = 0; weights = 0
    for course in data[name]:
        weight = course['credit']
        grade  = course['grade']
        sum += grade2number[grade]*weight
        weights += weight
    avg = sum/float(weights)
    return number2grade[round(avg)]
```

The complete code is found in the file students.py. Running this program gives the following output of the average grades:

```
John Doe: B
Kari Nordmann: C
Jan Modaal: C
```

One feature of the students.py code is that the output of the names are sorted after the last name. How can we accomplish that? A straight for name in data loop will visit the keys in an unknown (random) order. To visit the keys in alphabetic order, we must use

```
for name in sorted(data):
```

This default sort will sort with respect to the first character in the name strings. We want a sort according to the last part of the name. A tailored sort function can then be written (see Exercise 3.39 for an introduction to tailored sort functions). In this function we extract the last word in the names and compare them:

```
def sort_names(name1, name2):
    last_name1 = name1.split()[-1]
    last_name2 = name2.split()[-1]
    if last_name1 < last_name2:
        return -1
    elif last_name1 > last_name2:
        return 1
    else:
        return 0
```

We can now pass on sort_names to the sorted function to get a sequence that is sorted with respect to the last word in the students' names:

```
for name in sorted(data, sort_names):
    print '%s: %s' % (name, average_grade(data, name))
```

6.8 Exercises

Exercise 6.1: Make a dictionary from a table
The file src/dictstring/constants.txt[8] contains a table of the values and the dimensions of some fundamental constants from physics. We want to load this table into a dictionary constants, where the keys are the names of the constants. For example, constants['gravitational constant'] holds the value of the gravitational constant $(6.67259 \cdot 10^{-11})$ in Newton's law of gravitation. Make a function that reads and interprets the text in the file, and finally returns the dictionary.
Filename: fundamental_constants.

Exercise 6.2: Explore syntax differences: lists vs. dicts
Consider this code:

```
t1 = {}
t1[0] = -5
t1[1] = 10.5
```

Explain why the lines above work fine while the ones below do not:

```
t2 = []
t2[0] = -5
t2[1] = 10.5
```

What must be done in the last code snippet to make it work properly?
Filename: list_vs_dict.

---

[8] http://tinyurl.com/pwyasaa/dictstring/constants.txt

Exercise 6.3: Use string operations to improve a program
Consider the program density.py from Sect. 6.1.5. One problem with this program is that the name of the substance can contain only one or two words, while more comprehensive tables may have substances with names consisting of several words. The purpose of this exercise is to use string operations to shorten the code and make it more general and elegant.

a) Make a Python function that lets the name substance consist of all the words that line is split into, but not the last (which is the value of the corresponding density). Use the join method in string objects to combine the words that make up the name of the substance.
b) Observe that all the density values in the file densities.dat start in the same column. Write an alternative function that makes use of substring indexing to divide line into two parts (substance and density).

Hint Remember to strip the first part such that, e.g., the density of ice is obtained as densities['ice'] and not densities['ice '].

c) Make a test function that calls the two other functions and tests that they produce the same result.

Filename: density_improved.

Exercise 6.4: Interpret output from a program
The program src/funcif/lnsum.py produces, among other things, this output:

```
epsilon: 1e-04, exact error: 8.18e-04, n=55
epsilon: 1e-06, exact error: 9.02e-06, n=97
epsilon: 1e-08, exact error: 8.70e-08, n=142
epsilon: 1e-10, exact error: 9.20e-10, n=187
epsilon: 1e-12, exact error: 9.31e-12, n=233
```

Redirect the output to a file (by python lnsum.py > file). Write a Python program that reads the file and extracts the numbers corresponding to epsilon, exact error, and n. Store the numbers in three arrays and plot epsilon and the exact error versus n. Use a logarithmic scale on the y axis.

Hint The function semilogy is an alternative to plot and gives logarithmic scale on y axis.
Filename: read_error.

Exercise 6.5: Make a dictionary
Based on the stars data in Exercise 3.39, make a dictionary where the keys contain the names of the stars and the values correspond to the luminosity.
Filename: stars_data_dict1.

Exercise 6.6: Make a nested dictionary
Store the data about stars from Exercise 3.39 in a nested dictionary such that we can look up the distance, the apparent brightness, and the luminosity of a star with name N by

```
stars[N]['distance']
stars[N]['apparent brightness']
stars[N]['luminosity']
```

Hint Initialize the data by just copying the stars.txt[9] text into the program.
Filename: stars_data_dict2.

Exercise 6.7: Make a nested dictionary from a file
The file src/dictstring/human_evolution.txt[10] holds information about various human species and their height, weight, and brain volume. Make a program that reads this file and stores the tabular data in a nested dictionary humans. The keys in humans correspond to the specie name (e.g., homo erectus), and the values are dictionaries with keys for height, weight, brain volume, and when (the latter for when the specie lived). For example, humans['homo neanderthalensis']['mass'] should equal '55-70'. Let the program write out the humans dictionary in a nice tabular form similar to that in the file.
Filename: humans.

Exercise 6.8: Make a nested dictionary from a file
The viscosity μ of gases depends on the temperature. For some gases the following formula is relevant:

$$\mu(T) = \mu_0 \frac{T_0 - C}{T + C} \left(\frac{T}{T_0}\right)^{1.5},$$

where the values of the constants C, T_0, and μ_0 are found in the file src/dictstring/viscosity_of_gases.dat[11]. The temperature is measured in Kelvin.

a) Load the file into a nested dictionary mu_data such that we can look up C, T_0, and μ_0 for a gas with name name by mu_data[name][X], where X is 'C' for C, 'T_0' for T_0, and 'mu_0' for μ_0.
b) Make a function mu(T, gas, mu_data) for computing $\mu(T)$ for a gas with name gas (according to the file) and information about constants C, T_0, and μ_0 in mu_data.
c) Plot $\mu(T)$ for air, carbon dioxide, and hydrogen with $T \in [223, 373]$.

Filename: viscosity_of_gases.

[9] http://tinyurl.com/pwyasaa/funcif/stars.txt
[10] http://tinyurl.com/pwyasaa/dictstring/human_evolution.txt
[11] http://tinyurl.com/pwyasaa/dictstring/viscosity_of_gases.txt

Exercise 6.9: Compute the area of a triangle
The purpose of this exercise is to write an `area` function as in Exercise 3.16, but now we assume that the vertices of the triangle is stored in a dictionary and not a list. The keys in the dictionary correspond to the vertex number (1, 2, or 3) while the values are 2-tuples with the x and y coordinates of the vertex. For example, in a triangle with vertices $(0, 0)$, $(1, 0)$, and $(0, 2)$ the `vertices` argument becomes

```
{1: (0,0), 2: (1,0), 3: (0,2)}
```

Filename: `area_triangle_dict`.

Exercise 6.10: Compare data structures for polynomials
Write a code snippet that uses both a list and a dictionary to represent the polynomial $-\frac{1}{2} + 2x^{100}$. Print the list and the dictionary, and use them to evaluate the polynomial for $x = 1.05$.

Hint You can apply the `eval_poly_dict` and `eval_poly_list` functions from Sect. 6.1.3).
Filename: `poly_repr`.

Exercise 6.11: Compute the derivative of a polynomial
A polynomial can be represented by a dictionary as explained in Sect. 6.1.3. Write a function `diff` for differentiating such a polynomial. The `diff` function takes the polynomial as a dictionary argument and returns the dictionary representation of the derivative. Here is an example of the use of the function `diff`:

```
>>> p = {0: -3, 3: 2, 5: -1}    # -3 + 2*x**3 - x**5
>>> diff(p)                      # should be 6*x**2 - 5*x**4
{2: 6, 4: -5}
```

Hint Recall the formula for differentiation of polynomials:

$$\frac{d}{dx}\sum_{j=0}^{n} c_j x^j = \sum_{j=1}^{n} jc_j x^{j-1}. \qquad (6.1)$$

This means that the coefficient of the x^{j-1} term in the derivative equals j times the coefficient of x^j term of the original polynomial. With p as the polynomial dictionary and dp as the dictionary representing the derivative, we then have dp[j-1] = j*p[j] for j running over all keys in p, except when j equals 0.
Filename: `poly_diff`.

Exercise 6.12: Specify functions on the command line
Explain what the following two code snippets do and give an example of how they can be used.

Hint Read about the `StringFunction` tool in Sect. 4.3.3 and about a variable number of keyword arguments in Sect. H.7.

a)

```
import sys
from scitools.StringFunction import StringFunction
parameters = {}
for prm in sys.argv[4:]:
    key, value = prm.split('=')
    parameters[key] = eval(value)
f = StringFunction(sys.argv[1], independent_variables=sys.argv[2],
                   **parameters)
var = float(sys.argv[3])
print f(var)
```

b)

```
import sys
from scitools.StringFunction import StringFunction
f = eval('StringFunction(sys.argv[1], ' + \
         'independent_variables=sys.argv[2], %s)' % \
         (', '.join(sys.argv[4:])))
var = float(sys.argv[3])
print f(var)
```

Filename: `cml_functions`.

Exercise 6.13: Interpret function specifications
To specify arbitrary functions $f(x_1, x_2, \ldots; p_1, p_2, \ldots)$ with independent variables $x_1, x_2, \ldots$ and a set of parameters $p_1, p_2, \ldots$, we allow the following syntax on the command line or in a file:

```
<expression> is function of <list1> with parameter <list2>
```

where `<expression>` denotes the function formula, `<list1>` is a comma-separated list of the independent variables, and `<list2>` is a comma-separated list of name=value parameters. The part `with parameters <list2>` is omitted if there are no parameters. The names of the independent variables and the parameters can be chosen freely as long as the names can be used as Python variables. Here are four different examples of what we can specify on the command line using this syntax:

```
sin(x) is a function of x
sin(a*y) is a function of y with parameter a=2
sin(a*x-phi) is a function of x with parameter a=3, phi=-pi
exp(-a*x)*cos(w*t) is a function of t with parameter a=1,w=pi,x=2
```

Create a Python function that takes such function specifications as input and returns an appropriate `StringFunction` object. This object must be created from the function expression and the list of independent variables and parameters. For example, the last function specification above leads to the following `StringFunction` creation:

```
f = StringFunction('exp(-a*x)*cos(w*t)',
                   independent_variables=['t'],
                   a=1, w=pi, x=2)
```

Write a test function for verifying the implementation (fill `sys.argv` with appropriate content prior to each individual test).

Hint Use string operations to extract the various parts of the string. For example, the expression can be split out by calling `split('is a function of')`. Typically, you need to extract `<expression>`, `<list1>`, and `<list2>`, and create a string like

```
StringFunction(<expression>, independent_variables=[<list1>],
               <list2>)
```

and sending it to `eval` to create the object.
Filename: `text2func`.

Exercise 6.14: Compare average temperatures in cities
The tarfile `src/misc/city_temp.tar.gz`[12] contains a set of files with temperature data for a large number of cities around the world. The files are in text format with four columns, containing the month number, the date, the year, and the temperature, respectively. Missing temperature observations are represented by the value −99. The mapping between the names of the text files and the names of the cities are defined in an HTML file `citylistWorld.htm`.

a) Write a function that can read the `citylistWorld.htm` file and create a dictionary with mapping between city and filenames.
b) Write a function that takes this dictionary and a city name as input, opens the corresponding text file, and loads the data into an appropriate data structure (dictionary of arrays and city name is a suggestion).
c) Write a function that can take a number of data structures and the corresponding city names to create a plot of the temperatures over a certain time period.

Filename: `temperature_data`.

Exercise 6.15: Generate an HTML report with figures
The goal of this exercise is to let a program write a report in HTML format containing the solution to Exercise 5.33. First, include the program from that exercise, with additional explaining text if necessary. Program code can be placed inside `<pre>` and `</pre>` tags. Second, insert three plots of the $f(x, t)$ function for three different t values (find suitable t values that illustrate the displacement of the wave packet). Third, add an animated GIF file with the movie of $f(x, t)$. Insert headlines (`<h1>` tags) wherever appropriate.
Filename: `wavepacket_report`.

---

[12] http://tinyurl.com/pwyasaa/misc/city_temp.tar.gz

Exercise 6.16: Allow different types for a function argument
Consider the family of find_consensus_v* functions from Sect. 6.5.2. The different versions work on different representations of the frequency matrix. Make a unified find_consensus function that accepts different data structures for the frequency_matrix. Test on the type of data structure and perform the necessary actions.
Filename: find_consensus.

Exercise 6.17: Make a function more robust
Consider the function get_base_counts(dna) from Sect. 6.5.3, which counts how many times A, C, G, and T appears in the string dna:

```
def get_base_counts(dna):
    counts = {'A': 0, 'T': 0, 'G': 0, 'C': 0}
    for base in dna:
        counts[base] += 1
    return counts
```

Unfortunately, this function crashes if other letters appear in dna. Write an enhanced function get_base_counts2 which solves this problem. Test it on a string like 'ADLSTTLLD'.
Filename: get_base_counts2.

Exercise 6.18: Find proportion of bases inside/outside exons
Consider the lactase gene as described in Sects. 6.5.4 and 6.5.5. What is the proportion of base A inside and outside exons of the lactase gene?

Hint Write a function get_exons, which returns all the substrings of the exon regions concatenated. Also write a function get_introns, which returns all the substrings between the exon regions concatenated. The function get_base_frequencies from Sect. 6.5.3 can then be used to analyze the frequencies of bases A, C, G, and T in the two strings.
Filename: prop_A_exons.

Introduction to Classes

7

A class packs a set of data (variables) together with a set of functions operating on the data. The goal is to achieve more modular code by grouping data and functions into manageable (often small) units. Most of the mathematical computations in this book can easily be coded without using classes, but in many problems, classes enable either more elegant solutions or code that is easier to extend at a later stage. In the non-mathematical world, where there are no mathematical concepts and associated algorithms to help structure the problem solving, software development can be very challenging. Classes may then improve the understanding of the problem and contribute to simplify the modeling of data and actions in programs. As a consequence, almost all large software systems being developed in the world today are heavily based on classes.

Programming with classes is offered by most modern programming languages, also Python. In fact, Python employs classes to a very large extent, but one can use the language for lots of purposes without knowing what a class is. However, one will frequently encounter the class concept when searching books or the World Wide Web for Python programming information. And more important, classes often provide better solutions to programming problems. This chapter therefore gives an introduction to the class concept with emphasis on applications to numerical computing. More advanced use of classes, including inheritance and object orientation, is treated in Chap. 9.

The folder `src/class`[1] contains all the program examples from the present chapter.

7.1 Simple Function Classes

Classes can be used for many things in scientific computations, but one of the most frequent programming tasks is to represent mathematical functions that have a set of parameters in addition to one or more independent variables. Section 7.1.1 explains why such mathematical functions pose difficulties for programmers, and Sect. 7.1.2 shows how the class idea meets these difficulties. Sections 7.1.4 presents another example where a class represents a mathematical function. More advanced material

---

[1] http://tinyurl.com/pwyasaa/class

© Springer-Verlag Berlin Heidelberg 2016
H.P. Langtangen, *A Primer on Scientific Programming with Python*,
Texts in Computational Science and Engineering 6, DOI 10.1007/978-3-662-49887-3_7

about classes, which for some readers may clarify the ideas, but which can also be skipped in a first reading, appears in Sects. 7.1.5 and Sect. 7.1.6.

7.1.1 Challenge: Functions with Parameters

To motivate for the class concept, we will look at functions with parameters. One example is $y(t) = v_0 t - \frac{1}{2} g t^2$. Conceptually, in physics, the y quantity is viewed as a function of t, but y also depends on two other parameters, v_0 and g, although it is not natural to view y as a *function* of these parameters. We may write $y(t; v_0, g)$ to indicate that t is the independent variable, while v_0 and g are parameters. Strictly speaking, g is a fixed parameter (as long as we are on the surface of the earth and can view g as constant), so only v_0 and t can be arbitrarily chosen in the formula. It would then be better to write $y(t; v_0)$.

In the general case, we may have a function of x that has n parameters $p_1, \ldots, p_n$: $f(x; p_1, \ldots, p_n)$. One example could be

$$g(x; A, a) = A e^{-ax}.$$

How should we implement such functions? One obvious way is to have the independent variable and the parameters as arguments:

```
def y(t, v0):
    g = 9.81
    return v0*t - 0.5*g*t**2

def g(x, a, A):
    return A*exp(-a*x)
```

Problem There is one major problem with this solution. Many software tools we can use for mathematical operations on functions assume that a function of one variable has only one argument in the computer representation of the function. For example, we may have a tool for differentiating a function $f(x)$ at a point x, using the approximation

$$f'(x) \approx \frac{f(x+h) - f(x)}{h} \tag{7.1}$$

coded as

```
def diff(f, x, h=1E-5):
    return (f(x+h) - f(x))/h
```

The `diff` function works with any function f that takes one argument:

```
def h(t):
    return t**4 + 4*t

dh = diff(h, 0.1)

from math import sin, pi
x = 2*pi
dsin = diff(sin, x, h=1E-6)
```

Unfortunately, `diff` will not work with our `y(t, v0)` function. Calling `diff(y, t)` leads to an error inside the `diff` function, because it tries to call our `y` function with only one argument while the `y` function requires two.

Writing an alternative `diff` function for `f` functions having two arguments is a bad remedy as it restricts the set of admissible `f` functions to the very special case of a function with one independent variable and one parameter. A fundamental principle in computer programming is to strive for software that is as general and widely applicable as possible. In the present case, it means that the `diff` function should be applicable to all functions `f` of one variable, and letting `f` take one argument is then the natural decision to make.

The mismatch of function arguments, as outlined above, is a major problem because a lot of software libraries are available for operations on mathematical functions of one variable: integration, differentiation, solving $f(x) = 0$, finding extrema, etc. All these libraries will try to call the mathematical function we provide with only one argument. When our function has more arguments, the code inside the library aborts in the call to our function, and such errors may not always be easy to track down.

A bad solution: global variables The requirement is thus to define Python implementations of mathematical functions of one variable with one argument, the independent variable. The two examples above must then be implemented as

```
def y(t):
    g = 9.81
    return v0*t - 0.5*g*t**2

def g(t):
    return A*exp(-a*x)
```

These functions work only if `v0`, `A`, and `a` are global variables, initialized before one attempts to call the functions. Here are two sample calls where `diff` differentiates `y` and `g`:

```
v0 = 3
dy = diff(y, 1)

A = 1; a = 0.1
dg = diff(g, 1.5)
```

The use of global variables is in general considered bad programming. Why global variables are problematic in the present case can be illustrated when there is need to work with several versions of a function. Suppose we want to work with two versions of $y(t; v_0)$, one with $v_0 = 1$ and one with $v_0 = 5$. Every time we call `y` we must remember which version of the function we work with, and set `v0` accordingly prior to the call:

```
v0 = 1; r1 = y(t)
v0 = 5; r2 = y(t)
```

Another problem is that variables with simple names like v0, a, and A may easily be used as global variables in other parts of the program. These parts may change our v0 in a context different from the y function, but the change affects the correctness of the y function. In such a case, we say that changing v0 has *side effects*, i.e., the change affects other parts of the program in an unintentional way. This is one reason why a golden rule of programming tells us to limit the use of global variables as much as possible.

Another solution to the problem of needing two v_0 parameters could be to introduce two y functions, each with a distinct v_0 parameter:

```
def y1(t):
    g = 9.81
    return v0_1*t - 0.5*g*t**2

def y2(t):
    g = 9.81
    return v0_2*t - 0.5*g*t**2
```

Now we need to initialize v0_1 and v0_2 once, and then we can work with y1 and y2. However, if we need 100 v_0 parameters, we need 100 functions. This is tedious to code, error prone, difficult to administer, and simply a really bad solution to a programming problem.

So, is there a good remedy? The answer is yes: the class concept solves all the problems described above!

7.1.2 Representing a Function as a Class

A class contains a set of variables (data) and a set of functions, held together as one unit. The variables are visible in all the functions in the class. That is, we can view the variables as "global" in these functions. These characteristics also apply to modules, and modules can be used to obtain many of the same advantages as classes offer (see comments in Sect. 7.1.6). However, classes are technically very different from modules. You can also make many copies of a class, while there can be only one copy of a module. When you master both modules and classes, you will clearly see the similarities and differences. Now we continue with a specific example of a class.

Consider the function $y(t; v_0) = v_0 t - \frac{1}{2}gt^2$. We may say that v_0 and g, represented by the variables v0 and g, constitute the data. A Python function, say value(t), is needed to compute the value of $y(t; v_0)$ and this function must have access to the data v0 and g, while t is an argument.

A programmer experienced with classes will then suggest to collect the data v0 and g, and the function value(t), together as a class. In addition, a class usually has another function, called *constructor* for initializing the data. The constructor is always named __init__. Every class must have a name, often starting with a capital, so we choose Y as the name since the class represents a mathematical function with name y. Figure 7.1 sketches the contents of class Y as a so-called UML diagram, here created with aid of the program class_Y_v1_UML.py. The

Fig. 7.1 UML diagram with function and data in the simple class Y for representing a mathematical function $y(t; v_0)$

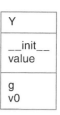

UML diagram has two "boxes", one where the functions are listed, and one where the variables are listed. Our next step is to implement this class in Python.

Implementation The complete code for our class Y looks as follows in Python:

```
class Y:
    def __init__(self, v0):
        self.v0 = v0
        self.g = 9.81

    def value(self, t):
        return self.v0*t - 0.5*self.g*t**2
```

A puzzlement for newcomers to Python classes is the `self` parameter, which may take some efforts and time to fully understand.

Usage and dissection Before we dig into what each line in the class implementation means, we start by showing how the class can be used to compute values of the mathematical function $y(t; v_0)$.

A class creates a new data type, here of name Y, so when we use the class to make objects, those objects are of type Y. (Actually, all the standard Python objects, such as lists, tuples, strings, floating-point numbers, integers, etc., are built-in Python classes, with names `list`, `tuple`, `str`, `float`, `int`, etc.) An object of a user-defined class (like Y) is usually called an *instance*. We need such an instance in order to use the data in the class and call the `value` function. The following statement constructs an instance bound to the variable name y:

```
y = Y(3)
```

Seemingly, we call the class Y as if it were a function. Actually, `Y(3)` is automatically translated by Python to a call to the constructor `__init__` in class Y. The arguments in the call, here only the number 3, are always passed on as arguments to `__init__` *after* the `self` argument. That is, v0 gets the value 3 and `self` is just dropped in the call. This may be confusing, but it is a rule that the `self` argument is never used in calls to functions in classes.

With the instance y, we can compute the value $y(t = 0.1; v_0 = 3)$ by the statement

```
v = y.value(0.1)
```

Here also, the `self` argument is dropped in the call to `value`. To access functions and variables in a class, we must prefix the function and variable names by the name of the instance and a dot: the `value` function is reached as `y.value`, and the variables are reached as `y.v0` and `y.g`. We can, for example, print the value of v0 in the instance y by writing

```
print y.v0
```

The output will in this case be 3.

We have already introduced the term "instance" for the object of a class. Functions in classes are commonly called *methods*, and variables (data) in classes are called *data attributes*. Methods are also known as *method attributes*. From now on we will use this terminology. In our sample class Y we have two methods or method attributes, `__init__` and `value`, two data attributes, v0 and g, and four attributes in total (`__init__`, `value`, v0, and g). The names of attributes can be chosen freely, just as names of ordinary Python functions and variables. However, the constructor must have the name `__init__`, otherwise it is not automatically called when we create new instances.

You can do whatever you want in whatever method, but it is a common convention to use the constructor for initializing the variables in the class.

Extension of the class We can have as many attributes as we like in a class, so let us add a new method to class Y. This method is called `formula` and prints a string containing the formula of the mathematical function y. After this formula, we provide the value of v_0. The string can then be constructed as

```
'v0*t - 0.5*g*t**2; v0=%g' % self.v0
```

where `self` is an instance of class Y. A call of `formula` does not need any arguments:

```
print y.formula()
```

should be enough to create, return, and print the string. However, even if the `formula` method does not need any arguments, it must have a `self` argument, which is left out in the call but needed inside the method to access the attributes. The implementation of the method is therefore

```
def formula(self):
    return 'v0*t - 0.5*g*t**2; v0=%g' % self.v0
```

For completeness, the whole class now reads

```
class Y:
    def __init__(self, v0):
        self.v0 = v0
        self.g = 9.81
```

```
    def value(self, t):
        return self.v0*t - 0.5*self.g*t**2

    def formula(self):
        return 'v0*t - 0.5*g*t**2; v0=%g' % self.v0
```

Example on use may be

```
y = Y(5)
t = 0.2
v = y.value(t)
print 'y(t=%g; v0=%g) = %g' % (t, y.v0, v)
print y.formula()
```

with the output

```
y(t=0.2; v0=5) = 0.8038
v0*t - 0.5*g*t**2; v0=5
```

Be careful with indentation in class programming

A common mistake done by newcomers to the class construction is to place the code that applies the class at the same indentation as the class methods. This is illegal. Only method definitions and assignments to so-called static data attributes (Sect. 7.6) can appear in the indented block under the `class` headline. Ordinary data attribute assignment must be done inside methods. The main program using the class must appear with the same indent as the `class` headline.

Using methods as ordinary functions We may create several y functions with different values of v_0:

```
y1 = Y(1)
y2 = Y(1.5)
y3 = Y(-3)
```

We can treat y1.value, y2.value, and y3.value as ordinary Python functions of t, and then pass them on to any Python function that expects a function of one variable. In particular, we can send the functions to the diff(f, x) function from Sect. 7.1.1:

```
dy1dt = diff(y1.value, 0.1)
dy2dt = diff(y2.value, 0.1)
dy3dt = diff(y3.value, 0.2)
```

Inside the diff(f, x) function, the argument f now behaves as a function of one variable that automatically carries with it two variables v0 and g. When f refers to (e.g.) y3.value, Python actually knows that f(x) means y3.value(x), and inside the y3.value method self is y3, and we have access to y3.v0 and y3.g.

New-style classes versus classic classes When use Python version 2 and write a class like

```
class V:
    ...
```

we get what is known as an old-style or classic class. A revised implementation of classes in Python came in version 2.2 with *new-style* classes. The specification of a new-style class requires (object) after the class name:

```
class V(object):
    ...
```

New-style classes have more functionality, and it is in general recommended to work with new-style classes. We shall therefore from now write V(object) rather than just V. In Python 3, all classes are new-style whether we write V or V(object).

Doc strings A function may have a doc string right after the function definition, see Sect. 3.1.11. The aim of the doc string is to explain the purpose of the function and, for instance, what the arguments and return values are. A class can also have a doc string, it is just the first string that appears right after the class headline. The convention is to enclose the doc string in triple double quotes """:

```
class Y(object):
    """The vertical motion of a ball."""

    def __init__(self, v0):
        ...
```

More comprehensive information can include the methods and how the class is used in an interactive session:

```
class Y(object):
    """
    Mathematical function for the vertical motion of a ball.

    Methods:
        constructor(v0): set initial velocity v0.
        value(t): compute the height as function of t.
        formula(): print out the formula for the height.

    Data attributes:
        v0: the initial velocity of the ball (time 0).
        g: acceleration of gravity (fixed).

    Usage:
    >>> y = Y(3)
    >>> position1 = y.value(0.1)
    >>> position2 = y.value(0.3)
    >>> print y.formula()
    v0*t - 0.5*g*t**2; v0=3
    """
```

7.1.3 The Self Variable

Now we will provide some more explanation of the `self` parameter and how the class methods work. Inside the constructor `__init__`, the argument `self` is a variable holding the new instance to be constructed. When we write

```
self.v0 = v0
self.g = 9.81
```

we define two new data attributes in this instance. The `self` parameter is invisibly returned to the calling code. We can imagine that Python translates the syntax `y = Y(3)` to a call written as

```
Y.__init__(y, 3)
```

Now, `self` becomes the new instance `y` we want to create, so when we do `self.v0 = v0` in the constructor, we actually assign v0 to `y.v0`. The prefix with `Y.` illustrates how to reach a class method with a syntax similar to reaching a function in a module (just like `math.exp`). If we prefix with `Y.`, we need to explicitly feed in an instance for the `self` argument, like `y` in the code line above, but if we prefix with `y.` (the instance name) the `self` argument is dropped in the syntax, and Python will automatically assign the `y` instance to the `self` argument. It is the latter "instance name prefix" which we shall use when computing with classes. (`Y.__init__(y, 3)` will not work since `y` is undefined and supposed to be an `Y` object. However, if we first create `y = Y(2)` and then call `Y.__init__(y, 3)`, the syntax works, and `y.v0` is 3 after the call.)

Let us look at a call to the `value` method to see a similar use of the `self` argument. When we write

```
value = y.value(0.1)
```

Python translates this to a call

```
value = Y.value(y, 0.1)
```

such that the `self` argument in the `value` method becomes the `y` instance. In the expression inside the `value` method,

```
self.v0*t - 0.5*self.g*t**2
```

`self` is `y` so this is the same as

```
y.v0*t - 0.5*y.g*t**2
```

The use of `self` may become more apparent when we have multiple class instances. We can make a class that just has one parameter so we can easily identify a class instance by printing the value of this parameter. In addition, every Python object `obj` has a unique identifier obtained by `id(obj)` that we can also print to track what `self` is.

```
class SelfExplorer(object):
    def __init__(self, a):
        self.a = a
        print 'init: a=%g, id(self)=%d' % (self.a, id(self))

    def value(self, x):
        print 'value: a=%g, id(self)=%d' % (self.a, id(self))
        return self.a*x
```

Here is an interactive session with this class:

```
>>> s1 = SelfExplorer(1)
init: a=1, id(self)=38085696
>>> id(s1)
38085696
```

We clearly see that `self` inside the constructor is the same object as `s1`, which we want to create by calling the constructor.

A second object is made by

```
>>> s2 = SelfExplorer(2)
init: a=2, id(self)=38085192
>>> id(s2)
38085192
```

Now we can call the `value` method using the standard syntax `s1.value(x)` and the "more pedagogical" syntax `SelfExplorer.value(s1, x)`. Using both `s1` and `s2` illustrates how `self` take on different values, while we may look at the method `SelfExplorer.value` as a single function that just operates on different `self` and `x` objects:

```
>>> s1.value(4)
value: a=1, id(self)=38085696
4
>>> SelfExplorer.value(s1, 4)
value: a=1, id(self)=38085696
4

>>> s2.value(5)
value: a=2, id(self)=38085192
10
>>> SelfExplorer.value(s2, 5)
value: a=2, id(self)=38085192
10
```

Hopefully, these illustrations help to explain that `self` is just the instance used in the method call prefix, here `s1` or `s2`. If not, patient work with class programming in Python will over time reveal an understanding of what `self` really is.

Rules regarding `self`

- Any class method must have `self` as first argument. (The name can be any valid variable name, but the name `self` is a widely established convention in Python.)
- `self` represents an (arbitrary) instance of the class.
- To access any class attribute inside class methods, we must prefix with `self`, as in `self.name`, where `name` is the name of the attribute.
- `self` is dropped as argument in calls to class methods.

7.1.4 Another Function Class Example

Let us apply the ideas from the Y class to the function

$$v(r) = \left(\frac{\beta}{2\mu_0}\right)^{1/n} \frac{n}{n+1} \left(R^{1+1/n} - r^{1+1/n}\right),$$

where r is the independent variable. We may write this function as $v(r; \beta, \mu_0, n, R)$ to explicitly indicate that there is one primary independent variable (r) and four physical parameters β, μ_0, n, and R. Exercise 5.40 describes a physical interpretation of v as the velocity of a fluid. The class typically holds the physical parameters as variables and provides an `value(r)` method for computing the v function:

```
class V(object):
    def __init__(self, beta, mu0, n, R):
        self.beta, self.mu0, self.n, self.R = beta, mu0, n, R

    def value(self, r):
        beta, mu0, n, R = self.beta, self.mu0, self.n, self.R
        n = float(n)  # ensure float divisions
        v = (beta/(2.0*mu0))**(1/n)*(n/(n+1))*\
            (R**(1+1/n) - r**(1+1/n))
        return v
```

There is seemingly one new thing here in that we initialize several variables on the same line:

```
        self.beta, self.mu0, self.n, self.R = beta, mu0, n, R
```

The comma-separated list of variables on the right-hand side forms a tuple so this assignment is just the a valid construction where a set of variables on the left-hand side is set equal to a list or tuple on the right-hand side, element by element. An equivalent multi-line code is

```
        self.beta = beta
        self.mu0 = mu0
        self.n = n
        self.R = R
```

In the `value` method it is convenient to avoid the `self.` prefix in the mathematical formulas and instead introduce the local short names `beta`, `mu0`, `n`, and `R`. This is in general a good idea, because it makes it easier to read the implementation of the formula and check its correctness.

Remark   Another solution to the problem of sending functions with parameters to a general library function such as `diff` is provided in Sect. H.7. The remedy there is to transfer the parameters as arguments "through" the `diff` function. This can be done in a general way as explained in that appendix.

7.1.5   Alternative Function Class Implementations

To illustrate class programming further, we will now realize class Y from Sect. 7.1.2 in a different way. You may consider this section as advanced and skip it, but for some readers the material might improve the understanding of class Y and give some insight into class programming in general.

It is a good habit always to have a constructor in a class and to initialize the data attributes in the class here, but this is not a requirement. Let us drop the constructor and make v0 an optional argument to the `value` method. If the user does not provide v0 in the call to `value`, we use a v0 value that must have been provided in an earlier call and stored as a data attribute `self.v0`. We can recognize if the user provides v0 as argument or not by using `None` as default value for the keyword argument and then test if `v0 is None`.

Our alternative implementation of class Y, named Y2, now reads

```
class Y2(object):
    def value(self, t, v0=None):
        if v0 is not None:
            self.v0 = v0
        g = 9.81
        return self.v0*t - 0.5*g*t**2
```

This time the class has only one method and one data attribute as we skipped the constructor and let g be a local variable in the `value` method.

But if there is no constructor, how is an instance created? Python fortunately creates an empty constructor. This allows us to write

```
y = Y2()
```

to make an instance y. Since nothing happens in the automatically generated empty constructor, y has no data attributes at this stage. Writing

```
print y.v0
```

therefore leads to the exception

```
AttributeError: Y2 instance has no attribute 'v0'
```

By calling

```
v = y.value(0.1, 5)
```

we create an attribute `self.v0` inside the `value` method. In general, we can create any attribute `name` in any method by just assigning a value to `self.name`. Now trying a

```
print y.v0
```

will print 5. In a new call,

```
v = y.value(0.2)
```

the previous v0 value (5) is used inside `value` as `self.v0` unless a v0 argument is specified in the call.

The previous implementation is not foolproof if we fail to initialize v0. For example, the code

```
y = Y2()
v = y.value(0.1)
```

will terminate in the `value` method with the exception

```
AttributeError: Y2 instance has no attribute 'v0'
```

As usual, it is better to notify the user with a more informative message. To check if we have an attribute v0, we can use the Python function `hasattr`. Calling `hasattr(self, 'v0')` returns True only if the instance `self` has an attribute with name 'v0'. An improved `value` method now reads

```
def value(self, t, v0=None):
    if v0 is not None:
        self.v0 = v0
    if not hasattr(self, 'v0'):
        print 'You cannot call value(t) without first '\
              'calling value(t,v0) to set v0'
        return None
    g = 9.81
    return self.v0*t - 0.5*g*t**2
```

Alternatively, we can try to access `self.v0` in a `try-except` block, and perhaps raise an exception `TypeError` (which is what Python raises if there are not enough arguments to a function or method):

```
def value(self, t, v0=None):
    if v0 is not None:
        self.v0 = v0
    g = 9.81
    try:
        value = self.v0*t - 0.5*g*t**2
    except AttributeError:
        msg = 'You cannot call value(t) without first '
              'calling value(t,v0) to set v0'
        raise TypeError(msg)
    return value
```

Note that Python detects an `AttributeError`, but from a user's point of view, not enough parameters were supplied in the call so a `TypeError` is more appropriate to communicate back to the calling code.

We think class Y is a better implementation than class Y2, because the former is simpler. As already mentioned, it is a good habit to include a constructor and set data here rather than "recording data on the fly" as we try to in class Y2. The whole purpose of class Y2 is just to show that Python provides great flexibility with respect to defining attributes, and that there are no requirements to what a class *must* contain.

7.1.6 Making Classes Without the Class Construct

Newcomers to the class concept often have a hard time understanding what this concept is about. The present section tries to explain in more detail how we can introduce classes without having the class construct in the computer language. This information may or may not increase your understanding of classes. If not, programming with classes will definitely increase your understanding with time, so there is no reason to worry. In fact, you may safely jump to Sect. 7.3 as there are no important concepts in this section that later sections build upon.

A class contains a collection of variables (data) and a collection of methods (functions). The collection of variables is unique to each instance of the class. That is, if we make ten instances, each of them has its own set of variables. These variables can be thought of as a dictionary with keys equal to the variable names. Each instance then has its own dictionary, and we may roughly view the instance as this dictionary. (The instance can also contain static data attributes (Sect. 7.6), but these are to be viewed as global variables in the present context.)

On the other hand, the methods are shared among the instances. We may think of a method in a class as a standard global function that takes an instance in the form of a dictionary as first argument. The method has then access to the variables in the instance (dictionary) provided in the call. For the Y class from Sect. 7.1.2 and an instance y, the methods are ordinary functions with the following names and arguments:

```
Y.value(y, t)
Y.formula(y)
```

The class acts as a *namespace*, meaning that all functions must be prefixed by the namespace name, here Y. Two different classes, say C1 and C2, may have functions with the same name, say value, but when the value functions belong to different namespaces, their names C1.value and C2.value become distinct. Modules are also namespaces for the functions and variables in them (think of math.sin, cmath.sin, numpy.sin).

The only peculiar thing with the class construct in Python is that it allows us to use an alternative syntax for method calls:

```
y.value(t)
y.formula()
```

This syntax coincides with the traditional syntax of calling class methods and providing arguments, as found in other computer languages, such as Java, C#, C++, Simula, and Smalltalk. The dot notation is also used to access variables in an instance such that we inside a method can write self.v0 instead of self['v0'] (self refers to y through the function call).

We could easily implement a simple version of the class concept without having a class construction in the language. All we need is a dictionary type and ordinary functions. The dictionary acts as the instance, and methods are functions that take this dictionary as the first argument such that the function has access to all the variables in the instance. Our Y class could now be implemented as

```
def value(self, t):
    return self['v0']*t - 0.5*self['g']*t**2

def formula(self):
    print 'v0*t - 0.5*g*t**2; v0=%g' % self['v0']
```

The two functions are placed in a module called Y. The usage goes as follows:

```
import Y
y = {'v0': 4, 'g': 9.81}    # make an "instance"
y1 = Y.value(y, t)
```

We have no constructor since the initialization of the variables is done when declaring the dictionary y, but we could well include some initialization function in the Y module

```
def init(v0):
    return {'v0': v0, 'g': 9.81}
```

The usage is now slightly different:

```
import Y
y = Y.init(4)        # make an "instance"
y1 = Y.value(y, t)
```

This way of implementing classes with the aid of a dictionary and a set of ordinary functions actually forms the basis for class implementations in many languages. Python and Perl even have a syntax that demonstrates this type of implementation. In fact, every class instance in Python has a dictionary `__dict__` as attribute, which holds all the variables in the instance. Here is a demo that proves the existence of this dictionary in class Y:

```
>>> y = Y(1.2)
>>> print y.__dict__
{'v0': 1.2, 'g': 9.8100000000000005}
```

To summarize: A Python class can be thought of as some variables collected in a dictionary, and a set of functions where this dictionary is automatically provided as first argument such that functions always have full access to the class variables.

First remark We have in this section provided a view of classes *from a technical point of view*. Others may view a class as a way of modeling the world in terms of data and operations on data. However, in sciences that employ the language of mathematics, the modeling of the world is usually done by mathematics, and the mathematical structures provide understanding of the problem and structure of programs. When appropriate, mathematical structures can conveniently be mapped on to classes in programs to make the software simpler and more flexible.

Second remark The view of classes in this section neglects very important topics such as inheritance and dynamic binding (explained in Chap. 9). For more completeness of the present section, we therefore briefly describe how our combination of dictionaries and global functions can deal with inheritance and dynamic binding (but this will not make sense unless you know what inheritance is).

Data inheritance can be obtained by letting a subclass dictionary do an `update` call with the superclass dictionary as argument. In this way all data in the superclass are also available in the subclass dictionary. Dynamic binding of methods is more complicated, but one can think of checking if the method is in the subclass module (using `hasattr`), and if not, one proceeds with checking super class modules until a version of the method is found.

7.1.7 Closures

This section follows up the discussion in Sect. 7.1.6 and presents a more advanced construction that may serve as alternative to class constructions in some cases.

Our motivating example is that we want a Python implementation of a mathematical function $y(t; v_0) = v_0 t - \frac{1}{2} g t^2$ to have t as the only argument, but also have access to the parameter v_0. Consider the following function, which returns a function:

```
>>> def generate_y():
...     v0 = 5
...     g = 9.81
...     def y(t):
...         return v0*t - 0.5*g*t**2
...     return y
...
>>> y = generate_y()
>>> y(1)
0.09499999999999975
```

The remarkable property of the y function is that it remembers the value of v0 and g, although these variables are not local to the parent function `generate_y` and not local in y. In particular, we can specify v0 as argument to `generate_y`:

```
>>> def generate_y(v0):
...     g = 9.81
...     def y(t):
...         return v0*t - 0.5*g*t**2
...     return y
...
>>> y1 = generate_y(v0=1)
>>> y2 = generate_y(v0=5)
>>> y1(1)
-3.9050000000000002
>>> y2(1)
0.09499999999999975
```

Here, `y1(t)` has access to v0=1 while `y2(t)` has access to v0=5.

The function `y(t)` we construct and return from `generate_y` is called a *closure* and it remembers the value of the surrounding local variables in the parent function (at the time we create the y function). Closures are very convenient for many purposes in mathematical computing. Examples appear in Sect. 7.3.2. Closures are also central in a programming style called *functional programming*.

Generating multiple closures in a function

As soon as you get the idea of a closure, you will probably use it a lot because it is a convenient way to pack a function with extra data. However, there are some pitfalls. The biggest is illustrated below, but this is considered advanced material!

Let us generate a series of functions `v(t)` for various values of a parameter v0. Each function just returns a tuple `(v0, t)` such that we can easily see what the argument and the parameter are. We use `lambda` to quickly define each function, and we place the functions in a list:

```
>>> def generate():
...     return [lambda t: (v0, t) for v0 in [0, 1, 5, 10]]
...
>>> funcs = generate()
```

Now, `funcs` is a list of functions with one argument. Calling each function and printing the return values v0 and t gives

```
>>> for func in funcs:
...         print func(1)
...
(10, 1)
(10, 1)
(10, 1)
(10, 1)
```

As we see, all functions have v0=10, i.e., they stored the most recent value of v0 before return. This is not what we wanted.

The trick is to let v0 be a keyword argument in each function, because the value of a keyword argument is frozen at the time the function is defined:

```
>>> def generate():
...         return [lambda t, v0=v0: (v0, t)
...                 for v0 in [0, 1, 5, 10]]
...
>>> funcs = generate()
>>> for func in funcs:
...         print func(1)
...
(0, 1)
(1, 1)
(5, 1)
(10, 1)
```

7.2   More Examples on Classes

The use of classes to solve problems from mathematical and physical sciences may not be so obvious. On the other hand, in many administrative programs for managing interactions between objects in the real world the objects themselves are natural candidates for being modeled by classes. Below we give some examples on what classes can be used to model.

7.2.1   Bank Accounts

The concept of a bank account in a program is a good candidate for a class. The account has some data, typically the name of the account holder, the account number, and the current balance. Three things we can do with an account is withdraw money, put money into the account, and print out the data of the account. These actions are modeled by methods. With a class we can pack the data and actions together into a new data type so that one account corresponds to one variable in a program.

Class `Account` can be implemented as follows:

```
class Account(object):
    def __init__(self, name, account_number, initial_amount):
        self.name = name
        self.no = account_number
        self.balance = initial_amount

    def deposit(self, amount):
        self.balance += amount

    def withdraw(self, amount):
        self.balance -= amount

    def dump(self):
        s = '%s, %s, balance: %s' % \
            (self.name, self.no, self.balance)
        print s
```

Here is a simple test of how class `Account` can be used:

```
>>> from classes import Account
>>> a1 = Account('John Olsson', '19371554951', 20000)
>>> a2 = Account('Liz Olsson',  '19371564761', 20000)
>>> a1.deposit(1000)
>>> a1.withdraw(4000)
>>> a2.withdraw(10500)
>>> a1.withdraw(3500)
>>> print "a1's balance:", a1.balance
a1's balance: 13500
>>> a1.dump()
John Olsson, 19371554951, balance: 13500
>>> a2.dump()
Liz Olsson, 19371564761, balance: 9500
```

The author of this class does not want users of the class to operate on the attributes directly and thereby change the name, the account number, or the balance. The intention is that users of the class should only call the constructor, the `deposit`, `withdraw`, and `dump` methods, and (if desired) inspect the `balance` attribute, but never change it. Other languages with class support usually have special keywords that can restrict access to attributes, but Python does not. Either the author of a Python class has to rely on correct usage, or a special convention can be used: any name starting with an underscore represents an attribute that should never be touched. One refers to names starting with an underscore as *protected* names. These can be freely used inside methods in the class, but not outside.

In class `Account`, it is natural to protect access to the `name`, `no`, and `balance` attributes by prefixing these names by an underscore. For *reading* only of the `balance` attribute, we provide a new method `get_balance`. The user of the class should now only call the methods in the class and not access any data attributes directly.

The new "protected" version of class `Account`, called `AccountP`, reads

```
class AccountP(object):
    def __init__(self, name, account_number, initial_amount):
        self._name = name
        self._no = account_number
        self._balance = initial_amount

    def deposit(self, amount):
        self._balance += amount

    def withdraw(self, amount):
        self._balance -= amount

    def get_balance(self):
        return self._balance

    def dump(self):
        s = '%s, %s, balance: %s' % \
            (self._name, self._no, self._balance)
        print s
```

We can technically access the data attributes, but we then break the convention that names starting with an underscore should never be touched outside the class. Here is class `AccountP` in action:

```
>>> a1 = AccountP('John Olsson', '19371554951', 20000)
>>> a1.deposit(1000)
>>> a1.withdraw(4000)
>>> a1.withdraw(3500)
>>> a1.dump()
John Olsson, 19371554951, balance: 13500
>>> print a1._balance       # it works, but a convention is broken
13500
print a1.get_balance()      # correct way of viewing the balance
13500
>>> a1._no = '19371554955'  # this is a "serious crime"
```

Python has a special construct, called *properties*, that can be used to protect data attributes from being changed. This is very useful, but the author considers properties a bit too complicated for this introductory book.

7.2.2 Phone Book

You are probably familiar with the phone book on your mobile phone. The phone book contains a list of persons. For each person you can record the name, telephone numbers, email address, and perhaps other relevant data. A natural way of representing such personal data in a program is to create a class, say class `Person`. The data attributes of the class hold information like the name, mobile phone number, office phone number, private phone number, and email address. The constructor may initialize some of the data about a person. Additional data can be specified later by calling methods in the class. One method can print the data. Other methods can register additional telephone numbers and an email address. In addition we initialize some of the data attributes in a constructor method. The attributes that are not initialized when constructing a Person instance can be added later by calling appropriate methods. For example, adding an office number is done by calling add_office_number.

Class Person may look as

```python
class Person(object):
    def __init__(self, name,
                 mobile_phone=None, office_phone=None,
                 private_phone=None, email=None):
        self.name = name
        self.mobile = mobile_phone
        self.office = office_phone
        self.private = private_phone
        self.email = email

    def add_mobile_phone(self, number):
        self.mobile = number

    def add_office_phone(self, number):
        self.office = number

    def add_private_phone(self, number):
        self.private = number

    def add_email(self, address):
        self.email = address
```

Note the use of None as default value for various data attributes: the object None is commonly used to indicate that a variable or attribute is defined, but yet not with a sensible value.

A quick demo session of class Person may go as follows:

```python
>>> p1 = Person('Hans Hanson',
...             office_phone='767828283', email='h@hanshanson.com')
>>> p2 = Person('Ole Olsen', office_phone='767828292')
>>> p2.add_email('olsen@somemail.net')
>>> phone_book = [p1, p2]
```

It can be handy to add a method for printing the contents of a Person instance in a nice fashion:

```python
class Person(object):
    ...
    def dump(self):
        s = self.name + '\n'
        if self.mobile is not None:
            s += 'mobile phone:   %s\n' % self.mobile
        if self.office is not None:
            s += 'office phone:   %s\n' % self.office
```

```
        if self.private is not None:
            s += 'private phone:  %s\n' % self.private
        if self.email is not None:
            s += 'email address:  %s\n' % self.email
        print s
```

With this method we can easily print the phone book:

```
>>> for person in phone_book:
...     person.dump()
...
Hans Hanson
office phone:    767828283
email address:   h@hanshanson.com

Ole Olsen
office phone:    767828292
email address:   olsen@somemail.net
```

A phone book can be a list of `Person` instances, as indicated in the examples above. However, if we quickly want to look up the phone numbers or email address for a given name, it would be more convenient to store the `Person` instances in a dictionary with the name as key:

```
>>> phone_book = {'Hanson': p1, 'Olsen': p2}
>>> for person in sorted(phone_book):  # alphabetic order
...     phone_book[person].dump()
```

The current example of `Person` objects is extended in Sect. 7.3.5.

7.2.3   A Circle

Geometric figures, such as a circle, are other candidates for classes in a program. A circle is uniquely defined by its center point (x_0, y_0) and its radius R. We can collect these three numbers as data attributes in a class. The values of x_0, y_0, and R are naturally initialized in the constructor. Other methods can be `area` and `circumference` for calculating the area πR^2 and the circumference $2\pi R$:

```
class Circle(object):
    def __init__(self, x0, y0, R):
        self.x0, self.y0, self.R = x0, y0, R

    def area(self):
        return pi*self.R**2

    def circumference(self):
        return 2*pi*self.R
```

An example of using class `Circle` goes as follows:

```
>>> c = Circle(2, -1, 5)
>>> print 'A circle with radius %g at (%g, %g) has area %g' % \
...          (c.R, c.x0, c.y0, c.area())
A circle with radius 5 at (2, -1) has area 78.5398
```

The ideas of class `Circle` can be applied to other geometric objects as well: rectangles, triangles, ellipses, boxes, spheres, etc. Exercise 7.4 tests if you are able to adapt class `Circle` to a rectangle and a triangle.

Verification We should include a test function for checking that the implementation of class `Circle` is correct:

```
def test_Circle():
    R = 2.5
    c = Circle(7.4, -8.1, R)

    from math import pi
    expected_area = pi*R**2
    computed_area = c.area()
    diff = abs(expected_area - computed_area)
    tol = 1E-14
    assert diff < tol, 'bug in Circle.area, diff=%s' % diff

    expected_circumference = 2*pi*R
    computed_circumference = c.circumference()
    diff = abs(expected_circumference - computed_circumference)
    assert diff < tol, 'bug in Circle.circumference, diff=%s' % diff
```

The `test_Circle` function is written in a way that it can be used in a pytest or nose testing framework (see Sect. H.9, or the brief examples in Sects. 3.3.3, 3.4.2, and 4.9.4). The necessary conventions are that the function name starts with `test_`, the function takes no arguments, and all tests are of the form `assert success` or `assert success, msg` where `success` is a boolean condition for the test and `msg` is an optional message to be written if the test fails (`success` is `False`). It is a good habit to write such test functions to verify the implementation of classes.

Remark There are usually many solutions to a programming problem. Representing a circle is no exception. Instead of using a class, we could collect x_0, y_0, and R in a list and create global functions `area` and `circumference` that take such a list as argument:

```
x0, y0, R = 2, -1, 5
circle = [x0, y0, R]

def area(c):
    R = c[2]
    return pi*R**2

def circumference(c):
    R = c[2]
    return 2*pi*R
```

Alternatively, the circle could be represented by a dictionary with keys 'center' and 'radius':

```
circle = {'center': (2, -1), 'radius': 5}

def area(c):
    R = c['radius']
    return pi*R**2

def circumference(c):
    R = c['radius']
    return 2*pi*R
```

7.3   Special Methods

Some class methods have names starting and ending with a double underscore. These methods allow a special syntax in the program and are called *special methods*. The constructor `__init__` is one example. This method is automatically called when an instance is created (by calling the class as a function), but we do not need to explicitly write `__init__`. Other special methods make it possible to perform arithmetic operations with instances, to compare instances with >, >=, !=, etc., to call instances as we call ordinary functions, and to test if an instance evaluates to True or False, to mention some possibilities.

7.3.1   The Call Special Method

Computing the value of the mathematical function represented by class Y from Sect. 7.1.2, with y as the name of the instance, is performed by writing y.value(t). If we could write just y(t), the y instance would look as an ordinary function. Such a syntax is indeed possible and offered by the special method named `__call__`. Writing y(t) implies a call

```
y.__call__(t)
```

if class Y has the method `__call__` defined. We may easily add this special method:

```
class Y(object):
    ...
    def __call__(self, t):
        return self.v0*t - 0.5*self.g*t**2
```

The previous value method is now redundant. A good programming convention is to include a `__call__` method in all classes that represent a mathematical function. Instances with `__call__` methods are said to be *callable* objects, just as plain functions are callable objects as well. The call syntax for callable objects is the same, regardless of whether the object is a function or a class instance. Given an object a,

```
if callable(a):
```

tests whether a behaves as a callable, i.e., if a is a Python function or an instance with a `__call__` method.

In particular, an instance of class Y can be passed as the f argument to the `diff` function from Sect. 7.1.1:

```
y = Y(v0=5)
dydt = diff(y, 0.1)
```

Inside `diff`, we can test that f is not a function but an instance of class Y. However, we only use f in calls, like `f(x)`, and for this purpose an instance with a `__call__` method works as a plain function. This feature is very convenient.

The next section demonstrates a neat application of the call operator `__call__` in a numerical algorithm.

7.3.2 Example: Automagic Differentiation

Problem Given a Python implementation `f(x)` of a mathematical function $f(x)$, we want to create an object that behaves as a Python function for computing the derivative $f'(x)$. For example, if this object is of type `Derivative`, we should be able to write something like

```
>>> def f(x):
        return x**3
...
>>> dfdx = Derivative(f)
>>> x = 2
>>> dfdx(x)
12.000000992884452
```

That is, `dfdx` behaves as a straight Python function for implementing the derivative $3x^2$ of x^3 (well, the answer is only approximate, with an error in the 7th decimal, but the approximation can easily be improved).

Maple, Mathematica, and many other software packages can do exact symbolic mathematics, including differentiation and integration. The Python package `sympy` for symbolic mathematics (see Sect. 1.7) makes it trivial to calculate the exact derivative of a large class of functions $f(x)$ and turn the result into an ordinary Python function. However, mathematical functions that are defined in an algorithmic way (e.g., solution of another mathematical problem), or functions with branches, random numbers, etc., pose fundamental problems to symbolic differentiation, and then numerical differentiation is required. Therefore we base the computation of derivatives in `Derivative` instances on finite difference formulas. Use of exact symbolic differentiation via SymPy is also possible.

Solution The most basic (but not the best) formula for a numerical derivative is

$$f'(x) \approx \frac{f(x+h) - f(x)}{h} . \qquad (7.2)$$

The idea is that we make a class to hold the function to be differentiated, call it f, and a step size h to be used in (7.2). These variables can be set in the constructor. The __call__ operator computes the derivative with aid of (7.1). All this can be coded in a few lines:

```
class Derivative(object):
    def __init__(self, f, h=1E-5):
        self.f = f
        self.h = float(h)

    def __call__(self, x):
        f, h = self.f, self.h      # make short forms
        return (f(x+h) - f(x))/h
```

Note that we turn h into a `float` to avoid potential integer division.

Below follows an application of the class to differentiate two functions $f(x) = \sin x$ and $g(t) = t^3$:

```
>>> from math import sin, cos, pi
>>> df = Derivative(sin)
>>> x = pi
>>> df(x)
-1.000000082740371
>>> cos(x)   # exact
-1.0
>>> def g(t):
...        return t**3
...
>>> dg = Derivative(g)
>>> t = 1
>>> dg(t)   # compare with 3 (exact)
3.000000248221113
```

The expressions df(x) and dg(t) look as ordinary Python functions that evaluate the derivative of the functions sin(x) and g(t). Class Derivative works for (almost) any function $f(x)$.

Verification It is a good programming habit to include a test function for verifying the implementation of a class. We can construct a test based on the fact that the approximate differentiation formula (7.2) is exact for linear functions:

```
def test_Derivative():
    # The formula is exact for linear functions, regardless of h
    f = lambda x: a*x + b
    a = 3.5; b = 8
    dfdx = Derivative(f, h=0.5)
    diff = abs(dfdx(4.5) - a)
    assert diff < 1E-14, 'bug in class Derivative, diff=%s' % diff
```

We have here used a lambda function for compactly defining a function f, see Sect. 3.1.14. A special feature of f is that it remembers the variables a and b when f is sent to class Derivative (it is a closure, see Sect. 7.1.7). Note that the test function above follows the conventions for test functions outlined in Sect. 7.2.3.

Application: Newton's method In what situations will it be convenient to automatically produce a Python function df(x) which is the derivative of another Python function f(x)? One example arises when solving nonlinear algebraic equations $f(x) = 0$ with Newton's method and we, because of laziness, lack of time, or lack of training do not manage to derive $f'(x)$ by hand. Consider a function Newton for solving $f(x) = 0$: Newton(f, x, dfdx, epsilon=1.0E-7, N=100). Section A.1.10 presents a specific implementation in a module file Newton.py. The arguments are a Python function f for $f(x)$, a float x for the initial guess (start value) of x, a Python function dfdx for $f'(x)$, a float epsilon for the accuracy ϵ of the root: the algorithms iterates until $|f(x)| < \epsilon$, and an int N for the maximum number of iterations that we allow. All arguments are easy to provide, except dfdx, which requires computing $f'(x)$ by hand then implementation of the formula in a Python function. Suppose our target equation reads

$$f(x) = 10^5 (x - 0.9)^2 (x - 1.1)^3 = 0.$$

The function $f(x)$ is plotted in Fig. 7.2. The following session employs the Derivative class to quickly make a derivative so we can call Newton's method:

```
>>> from classes import Derivative
>>> from Newton import Newton
>>> def f(x):
...     return 100000*(x - 0.9)**2 * (x - 1.1)**3
...
>>> df = Derivative(f)
>>> Newton(f, 1.01, df, epsilon=1E-5)
(1.0987610068093443, 8, -7.5139644257961411e-06)
```

The output 3-tuple holds the approximation to a root, the number of iterations, and the value of f at the approximate root (a measure of the error in the equation).

The exact root is 1.1, and the convergence toward this value is very slow. (Newton's method converges very slowly when the derivative of f is zero at the roots of f. Even slower convergence appears when higher-order derivatives also are zero, like in this example. Notice that the error in x is much larger than the error in the equation (epsilon). For example, an epsilon tolerance of 10^{-10} requires 18 iterations with an error of 10^{-3}.) Using an exact derivative gives almost the same result:

```
>>> def df_exact(x):
...     return 100000*(2*(x-0.9)*(x-1.1)**3 + \
...                    (x-0.9)**2*3*(x-1.1)**2)
...
>>> Newton(f, 1.01, df_exact, epsilon=1E-5)
(1.0987610065618421, 8, -7.5139689100699629e-06)
```

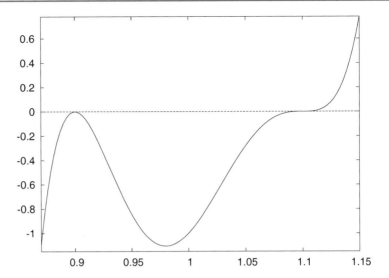

Fig. 7.2 Plot of $y = 10^5(x - 0.9)^2(x - 1.1)^3$

This example indicates that there are hardly any drawbacks in using a "smart" inexact general differentiation approach as in the `Derivative` class. The advantages are many – most notably, `Derivative` avoids potential errors from possibly incorrect manual coding of possibly lengthy expressions of possibly wrong hand-calculations. The errors in the involved approximations can be made smaller, usually much smaller than other errors, like the tolerance in Newton's method in this example or the uncertainty in physical parameters in real-life problems.

Solution utilizing SymPy Class `Derivative` is based on numerical differentiation, but it is possible to make an equally short class that can do exact differentiation. In SymPy, one can perform symbolic differentiation of an expression e with respect to a symbolic independent variable x by `diff(e, x)` (see Sect. 1.7.1). Assuming that the user's f function can be evaluated for a symbolic independent variable x, we can call `f(x)` to get the SymPy expression for the formula in f and then use `diff` to calculate the exact derivative. Thereafter, we turn the symbolic expression of the derivative into an ordinary Python function (via `lambdify`) and define this function as the `__call__` method. The proper Python code is very short:

```
class Derivative_sympy(object):
    def __init__(self, f):
        from sympy import Symbol, diff, lambdify
        x = Symbol('x')
        sympy_f = f(x)   # make sympy expression
        sympy_dfdx = diff(sympy_f, x)
        self.__call__ = lambdify([x], sympy_dfdx)
```

Note how the `__call__` method is defined by assigning a function to it (even though the function returned by `lambdify` is a function of x only, it works to call `obj(x)` for an instance `obj` of type `Derivative_sympy`).

Both demonstration of the class and verification of the implementation can be placed in a test function:

```
def test_Derivative_sympy():
    def g(t):
        return t**3

    dg = Derivative_sympy(g)
    t = 2
    exact = 3*t**2
    computed = dg(t)
    tol = 1E-14
    assert abs(exact - computed) < tol

    def h(y):
        return exp(-y)*sin(2*y)

    from sympy import exp, sin
    dh = Derivative_sympy(h)
    from math import pi, exp, sin, cos
    y = pi
    exact = -exp(-y)*sin(2*y) + exp(-y)*2*cos(2*y)
    computed = dh(y)
    assert abs(exact - computed) < tol
```

The example with the g(t) should be straightforward to understand. In the constructor of class Derivative_sympy, we call g(x), with the symbol x, and g returns the SymPy expression x**3. The __call__ method then becomes a function lambda x: 3*x**2.

The h(y) function, however, deserves more explanation. When then constructor of class Derivative_sympy makes the call h(x), with the symbol x, the h function will return the SymPy expression exp(-x)*sin(2*x), provided exp and sin are SymPy functions. Since we do from sympy import exp, sin prior to calling the constructor in class Derivative_sympy, the names exp and sin are defined in the test function, and our local h function will have access to all local variables, as it is a closure as mentioned above and in Sect. 7.1.7 (see also Sect. 9.2.6). This means that h has access to sympy.sin and sympy.cos when the constructor in class Derivative_sympy calls h. Thereafter, we want to do some numerical computing and need exp, sin, and cos from the math module. If we had tried to do Derivative_sympy(h) after the import from math, h would then call math.exp and math.sin with a SymPy symbol as argument, and would cause a TypeError since math.exp expects a float, not a Symbol object from SymPy.

Although the Derivative_sympy class is small and compact, its construction and use as explained here bring up more advanced topics than class Derivative and its plain numerical computations. However, it may be interesting to see that a class for exact differentiation of a Python function can be realized in very few lines.

7.3.3   Example: Automagic Integration

We can apply the ideas from Sect. 7.3.2 to make a class for computing the integral of a function numerically. Given a function $f(x)$, we want to compute

$$F(x; a) = \int_a^x f(t)dt \,.$$

The computational technique consists of using the Trapezoidal rule with n intervals ($n + 1$ points):

$$\int_a^x f(t)dt = h\left(\frac{1}{2}f(a) + \sum_{i=1}^{n-1} f(a + ih) + \frac{1}{2}f(x) \right), \qquad (7.3)$$

where $h = (x - a)/n$. In an application program, we want to compute $F(x; a)$ by a simple syntax like

```
def f(x):
    return exp(-x**2)*sin(10*x)

a = 0; n = 200
F = Integral(f, a, n)
print F(x)
```

Here, `f(x)` is the Python function to be integrated, and `F(x)` behaves as a Python function that calculates values of $F(x; a)$.

A simple implementation Consider a straightforward implementation of the Trapezoidal rule in a Python function:

```
def trapezoidal(f, a, x, n):
    h = (x-a)/float(n)
    I = 0.5*f(a)
    for i in range(1, n):
        I += f(a + i*h)
    I += 0.5*f(x)
    I *= h
    return I
```

Class `Integral` must have some data attributes and a `__call__` method. Since the latter method is supposed to take `x` as argument, the other parameters `a`, `f`, and `n` must be data attributes. The implementation then becomes

```
class Integral(object):
    def __init__(self, f, a, n=100):
        self.f, self.a, self.n = f, a, n

    def __call__(self, x):
        return trapezoidal(self.f, self.a, x, self.n)
```

Observe that we just reuse the `trapezoidal` function to perform the calculation. We could alternatively have copied the body of the `trapezoidal` function into the `__call__` method. However, if we already have this algorithm implemented and tested as a function, it is better to call the function. The class is then known as a *wrapper* of the underlying function. A wrapper allows something to be called with alternative syntax.

An application program computing $\int_0^{2\pi} \sin x \, dx$ might look as follows:

```
from math import sin, pi

G = Integral(sin, 0, 200)
value = G(2*pi)
```

An equivalent calculation is

```
value = trapezoidal(sin, 0, 2*pi, 200)
```

Verification via symbolic computing We should always provide a test function for verification of the implementation. To avoid dealing with unknown approximation errors of the Trapezoidal rule, we use the obvious fact that linear functions are integrated exactly by the rule. Although it is really easy to pick a linear function, integrate it, and figure out what an integral is, we can also demonstrate how to automate such a process by SymPy. Essentially, we define an expression in SymPy, ask SymPy to integrate it, and then turn the resulting symbolic integral to a plain Python function for computing:

```
>>> import sympy as sp
>>> x = sp.Symbol('x')
>>> f_expr = sp.cos(x) + 5*x
>>> f_expr
5*x + cos(x)
>>> F_expr = sp.integrate(f_expr, x)
>>> F_expr
5*x**2/2 + sin(x)
>>> F = sp.lambdify([x], F_expr)   # turn f_expr to F(x) func.
>>> F(0)
0.0
>>> F(1)
3.3414709848078967
```

Using such functionality to do exact integration, we can write our test function as

```
def test_Integral():
    # The Trapezoidal rule is exact for linear functions
    import sympy as sp
    x = sp.Symbol('x')
    f_expr = 2*x + 5
    # Turn sympy expression into plain Python function f(x)
    f = sp.lambdify([x], f_expr)
```

```
# Find integral of f_expr and turn into plain Python function F
F_expr = sp.integrate(f_expr, x)
F = sp.lambdify([x], F_expr)

a = 2
x = 6
exact = F(x) - F(a)
computed = Integral(f, a, n=4)
diff = abs(exact - computed)
tol = 1E-15
assert diff < tol, 'bug in class Integral, diff=%s' % diff
```

If you think it is overkill to use SymPy for integrating linear functions, you can equally well do it yourself and define f = lambda x: 2*x + 5 and F = lambda x: x**2 + 5*x.

Remark Class Integral is inefficient (but probably more than fast enough) for plotting $F(x; a)$ as a function x. Exercise 7.22 suggests to optimize the class for this purpose.

7.3.4 Turning an Instance into a String

Another useful special method is __str__. It is called when a class instance needs to be converted to a string. This happens when we say print a, and a is an instance. Python will then look into the a instance for a __str__ method, which is supposed to return a string. If such a special method is found, the returned string is printed, otherwise just the name of the class is printed. An example will illustrate the point. First we try to print an y instance of class Y from Sect. 7.1.2 (where there is no __str__ method):

```
>>> print y
<__main__.Y instance at 0xb751238c>
```

This means that y is an Y instance in the __main__ module (the main program or the interactive session). The output also contains an address telling where the y instance is stored in the computer's memory.

If we want print y to print out the y instance, we need to define the __str__ method in class Y:

```
class Y(object):
    ...
    def __str__(self):
        return 'v0*t - 0.5*g*t**2; v0=%g' % self.v0
```

Typically, __str__ replaces our previous formula method and __call__ replaces our previous value method. Python programmers with the experience that we now have gained will therefore write class Y with special methods only:

```
class Y(object):
    def __init__(self, v0):
        self.v0 = v0
        self.g = 9.81

    def __call__(self, t):
        return self.v0*t - 0.5*self.g*t**2

    def __str__(self):
        return 'v0*t - 0.5*g*t**2; v0=%g' % self.v0
```

Let us see the class in action:

```
>>> y = Y(1.5)
>>> y(0.2)
0.1038
>>> print y
v0*t - 0.5*g*t**2; v0=1.5
```

What have we gained by using special methods? Well, we can still only evaluate the formula and write it out, but many users of the class will claim that the syntax is more attractive since y(t) in code means $y(t)$ in mathematics, and we can do a print y to view the formula. The bottom line of using special methods is to achieve a more user-friendly syntax. The next sections illustrate this point further.

Note that the __str__ method is called whenever we do str(a), and print a is effectively print str(a), i.e., print a.__str__().

7.3.5 Example: Phone Book with Special Methods

Let us reconsider class Person from Sect. 7.2.2. The dump method in that class is better implemented as a __str__ special method. This is easy: we just change the method name and replace print s by return s.

Storing Person instances in a dictionary to form a phone book is straightforward. However, we make the dictionary a bit easier to use if we wrap a class around it. That is, we make a class PhoneBook which holds the dictionary as an attribute. An add method can be used to add a new person:

```
class PhoneBook(object):
    def __init__(self):
        self.contacts = {}    # dict of Person instances

    def add(self, name, mobile=None, office=None,
            private=None, email=None):
        p = Person(name, mobile, office, private, email)
        self.contacts[name] = p
```

A `__str__` can print the phone book in alphabetic order:

```
def __str__(self):
    s = ''
    for p in sorted(self.contacts):
        s += str(self.contacts[p]) + '\n'
    return s
```

To retrieve a `Person` instance, we use the `__call__` with the person's name as argument:

```
def __call__(self, name):
    return self.contacts[name]
```

The only advantage of this method is simpler syntax: for a `PhoneBook` b we can get data about `NN` by calling `b('NN')` rather than accessing the internal dictionary `b.contacts['NN']`.

We can make a simple demo code for a phone book with three names:

```
b = PhoneBook()
b.add('Ole Olsen', office='767828292',
      email='olsen@somemail.net')
b.add('Hans Hanson',
      office='767828283', mobile='995320221')
b.add('Per Person', mobile='906849781')
print b('Per Person')
print b
```

The output becomes

```
Per Person
mobile phone:    906849781

Hans Hanson
mobile phone:    995320221
office phone:    767828283

Ole Olsen
office phone:    767828292
email address:   olsen@somemail.net

Per Person
mobile phone:    906849781
```

You are strongly encouraged to work through this last demo program by hand and simulate what the program does. That is, jump around in the code and write down on a piece of paper what various variables contain after each statement. This is an important and good exercise! You enjoy the happiness of mastering classes if you get the same output as above. The complete program with classes `Person` and `PhoneBook` and the test above is found in the file `PhoneBook.py`. You can run this program, statement by statement, either in the Online Python Tutor[2] or in a debugger (see Sect. F.1) to control that your understanding of the program flow is correct.

Remark Note that the names are sorted with respect to the first names. The reason is that strings are sorted after the first character, then the second character, and so on. We can supply our own tailored sort function, as explained in Exercise 3.39. One possibility is to split the name into words and use the last word for sorting:

```
def last_name_sort(name1, name2):
    lastname1 = name1.split()[-1]
    lastname2 = name2.split()[-1]
    if lastname1 < lastname2:
        return -1
    elif lastname1 > lastname2:
        return 1
    else: # equality
        return 0

for p in sorted(self.contacts, last_name_sort):
    ...
```

7.3.6 Adding Objects

Let a and b be instances of some class C. Does it make sense to write a + b? Yes, this makes sense if class C has defined a special method __add__:

```
class C(object):
    ...
    __add__(self, other):
        ...
```

The __add__ method should add the instances self and other and return the result as an instance. So when Python encounters a + b, it will check if class C has an __add__ method and interpret a + b as the call a.__add__(b). The next example will hopefully clarify what this idea can be used for.

7.3.7 Example: Class for Polynomials

Let us create a class Polynomial for polynomials. The coefficients in the polynomial can be given to the constructor as a list. Index number i in this list represents the coefficients of the x^i term in the polynomial. That is, writing Polynomial([1,0,-1,2]) defines a polynomial

$$1 + 0 \cdot x - 1 \cdot x^2 + 2 \cdot x^3 = 1 - x^2 + 2x^3 \,.$$

---

[2] http://www.pythontutor.com/

Polynomials can be added (by just adding the coefficients corresponding to the same powers) so our class may have an `__add__` method. A `__call__` method is natural for evaluating the polynomial, given a value of x. The class is listed below and explained afterwards.

```python
class Polynomial(object):
    def __init__(self, coefficients):
        self.coeff = coefficients

    def __call__(self, x):
        """Evaluate the polynomial."""
        s = 0
        for i in range(len(self.coeff)):
            s += self.coeff[i]*x**i
        return s

    def __add__(self, other):
        """Return self + other as Polynomial object."""
        # Two cases:
        #
        # self:   X X X X X X
        # other:  X X X
        #
        # or:
        #
        # self:   X X X X X
        # other:  X X X X X X X X

        # Start with the longest list and add in the other
        if len(self.coeff) > len(other.coeff):
            result_coeff = self.coeff[:]  # copy!
            for i in range(len(other.coeff)):
                result_coeff[i] += other.coeff[i]
        else:
            result_coeff = other.coeff[:] # copy!
            for i in range(len(self.coeff)):
                result_coeff[i] += self.coeff[i]
        return Polynomial(result_coeff)
```

Implementation Class `Polynomial` has one data attribute: the list of coefficients. To evaluate the polynomial, we just sum up coefficient no. i times x^i for $i = 0$ to the number of coefficients in the list.

The `__add__` method looks more advanced. The goal is to add the two lists of coefficients. However, it may happen that the lists are of unequal length. We therefore start with the longest list and add in the other list, element by element. Observe that `result_coeff` starts out as a *copy* of `self.coeff`: if not, changes in `result_coeff` as we compute the sum will be reflected in `self.coeff`. This means that `self` would be the sum of itself and the `other` instance, or in other words, adding two instances, p1+p2, changes p1 – this is not what we want! An alternative implementation of class `Polynomial` is found in Exercise 7.24.

A subtraction method `__sub__` can be implemented along the lines of `__add__`, but is slightly more complicated and left as Exercise 7.25. You are strongly encouraged to do this exercise as it will help increase the understanding of the interplay between mathematics and programming in class `Polynomial`.

A more complicated operation on polynomials, from a mathematical point of view, is the multiplication of two polynomials. Let $p(x) = \sum_{i=0}^{M} c_i x^i$ and $q(x) = \sum_{j=0}^{N} d_j x^j$ be the two polynomials. The product becomes

$$\left(\sum_{i=0}^{M} c_i x^i \right) \left(\sum_{j=0}^{N} d_j x^j \right) = \sum_{i=0}^{M} \sum_{j=0}^{N} c_i d_j x^{i+j} .$$

The double sum must be implemented as a double loop, but first the list for the resulting polynomial must be created with length $M+N+1$ (the highest exponent is $M+N$ and then we need a constant term). The implementation of the multiplication operator becomes

```
def __mul__(self, other):
    c = self.coeff
    d = other.coeff
    M = len(c) - 1
    N = len(d) - 1
    result_coeff = numpy.zeros(M+N+1)
    for i in range(0, M+1):
        for j in range(0, N+1):
            result_coeff[i+j] += c[i]*d[j]
    return Polynomial(result_coeff)
```

We could also include a method for differentiating the polynomial according to the formula

$$\frac{d}{dx} \sum_{i=0}^{n} c_i x^i = \sum_{i=1}^{n} i c_i x^{i-1} .$$

If c_i is stored as a list c, the list representation of the derivative, say its name is dc, fulfills `dc[i-1] = i*c[i]` for i running from 1 to the largest index in c. Note that dc has one element less than c.

There are two different ways of implementing the differentiation functionality, either by changing the polynomial coefficients, or by returning a new `Polynomial` instance from the method such that the original polynomial instance is intact. We let `p.differentiate()` be an implementation of the first approach, i.e., this method does not return anything, but the coefficients in the `Polynomial` instance p are altered. The other approach is implemented by `p.derivative()`, which returns a new `Polynomial` object with coefficients corresponding to the derivative of p.

The complete implementation of the two methods is given below:

```
class Polynomial(object):
    ...
    def differentiate(self):
        """Differentiate this polynomial in-place."""
        for i in range(1, len(self.coeff)):
            self.coeff[i-1] = i*self.coeff[i]
        del self.coeff[-1]
```

```
def derivative(self):
    """Copy this polynomial and return its derivative."""
    dpdx = Polynomial(self.coeff[:])  # make a copy
    dpdx.differentiate()
    return dpdx
```

The Polynomial class with a differentiate method and not a derivative method would be mutable (i.e., the object's content can change) and allow in-place changes of the data, while the Polynomial class with derivative and not differentiate would yield an immutable object where the polynomial initialized in the constructor is never altered. (Technically, it is possible to grab the coeff variable in a class instance and alter this list. By starting coeff with an under-score, a Python programming convention tells programmers that this variable is for internal use in the class only, and not to be altered by users of the instance, see Sects. 7.2.1 and 7.5.2.) A good rule is to offer only one of these two functions such that a Polynomial object is either mutable or immutable (if we leave out differentiate, its function body must of course be copied into derivative since derivative now relies on that code). However, since the main purpose of this class is to illustrate various types of programming techniques, we keep both versions.

Usage   As a demonstration of the functionality of class Polynomial, we introduce the two polynomials

$$p_1(x) = 1 - x, \quad p_2(x) = x - 6x^4 - x^5.$$

```
>>> p1 = Polynomial([1, -1])
>>> p2 = Polynomial([0, 1, 0, 0, -6, -1])
>>> p3 = p1 + p2
>>> print p3.coeff
[1, 0, 0, 0, -6, -1]
>>> p4 = p1*p2
>>> print p4.coeff
[0, 1, -1, 0, -6, 5, 1]
>>> p5 = p2.derivative()
>>> print p5.coeff
[1, 0, 0, -24, -5]
```

One verification of the implementation may be to compare p3 at (e.g.) $x = 1/2$ with $p_1(x) + p_2(x)$:

```
>>> x = 0.5
>>> p1_plus_p2_value = p1(x) + p2(x)
>>> p3_value = p3(x)
>>> print p1_plus_p2_value - p3_value
0.0
```

Note that p1 + p2 is very different from p1(x) + p2(x). In the former case, we add two instances of class Polynomial, while in the latter case we add two instances of class float (since p1(x) and p2(x) imply calling __call__ and that method returns a float object).

Pretty print of polynomials The `Polynomial` class can also be equipped with a `__str__` method for printing the polynomial to the screen. A first, rough implementation could simply add up strings of the form `+ self.coeff[i]*x^i`:

```
class Polynomial(object):
    ...
    def __str__(self):
        s = ''
        for i in range(len(self.coeff)):
            s += ' + %g*x^%d' % (self.coeff[i], i)
        return s
```

However, this implementation leads to ugly output from a mathematical viewpoint. For instance, a polynomial with coefficients `[1,0,0,-1,-6]` gets printed as

```
 + 1*x^0 + 0*x^1 + 0*x^2 + -1*x^3 + -6*x^4
```

A more desired output would be

```
1 - x^3 - 6*x^4
```

That is, terms with a zero coefficient should be dropped; a part `'+ -'` of the output string should be replaced by `'- '`; unit coefficients should be dropped, i.e., `'1*'` should be replaced by space `' '`; unit power should be dropped by replacing `'x^1 '` by `'x '`; zero power should be dropped and replaced by 1, initial spaces should be fixed, etc. These adjustments can be implemented using the `replace` method in string objects and by composing slices of the strings. The new version of the `__str__` method below contains the necessary adjustments. If you find this type of string manipulation tricky and difficult to understand, you may safely skip further inspection of the improved `__str__` code since the details are not essential for your present learning about the class concept and special methods.

```
class Polynomial(object):
    ...
    def __str__(self):
        s = ''
        for i in range(0, len(self.coeff)):
            if self.coeff[i] != 0:
                s += ' + %g*x^%d' % (self.coeff[i], i)
        # Fix layout
        s = s.replace('+ -', '- ')
        s = s.replace('x^0', '1')
        s = s.replace(' 1*', ' ')
        s = s.replace('x^1 ', 'x ')
        if s[0:3] == ' + ':   # remove initial +
            s = s[3:]
        if s[0:3] == ' - ':   # fix spaces for initial -
            s = '-' + s[3:]
        return s
```

Programming sometimes turns into coding (what one think is) a general solution followed by a series of special cases to fix caveats in the "general" solution, just as we experienced with the `__str__` method above. This situation often calls for additional future fixes and is often a sign of a suboptimal solution to the programming problem.

Pretty print of `Polynomial` instances can be demonstrated in an interactive session:

```
>>> p1 = Polynomial([1, -1])
>>> print p1
1 - x^1
>>> p2 = Polynomial([0, 1, 0, 0, -6, -1])
>>> p2.differentiate()
>>> print p2
1 - 24*x^3 - 5*x^4
```

Verifying the implementation   It is always a good habit to include a test function `test_Polynomial()` for verifying the functionality in class `Polynomial`. To this end, we construct some examples of addition, multiplication, and differentiation of polynomials by hand and make tests that class `Polynomial` reproduces the correct results. Testing the `__str__` method is left as Exercise 7.26.

Rounding errors may be an issue in class `Polynomial`: `__add__`, `derivative`, and `differentiate` will lead to integer coefficients if the polynomials to be added have integer coefficients, while `__mul__` always results in a polynomial with the coefficients stored in a `numpy` array with `float` elements. Integer coefficients in lists can be compared using `==` for lists, while coefficients in `numpy` arrays must be compared with a tolerance. One can either subtract the `numpy` arrays and use the `max` method to find the largest deviation and compare this with a tolerance, or one can use `numpy.allclose(a, b, rtol=tol)` for comparing the arrays a and b with a (relative) tolerance `tol`.

Let us pick polynomials with integer coefficients as test cases such that `__add__`, `derivative`, and `differentiate` can be verified by testing equality (`==`) of the `coeff` lists. Multiplication in `__mul__` must employ `numpy.allclose`.

We follow the convention that all tests are on the form `assert success`, where `success` is a boolean expression for the test. (The actual version of the test function in the file `Polynomial.py` adds an error message `msg` to the test: `assert success, msg`.) Another part of the convention is that the function starts with `test_` and the function takes no arguments.

Our test function now becomes

```
def test_Polynomial():
    p1 = Polynomial([1, -1])
    p2 = Polynomial([0, 1, 0, 0, -6, -1])

    p3 = p1 + p2
    p3_exact = Polynomial([1, 0, 0, 0, -6, -1])
    assert p3.coeff == p3_exact.coeff

    p4 = p1*p2
    p4_exact = Polynomial(numpy.array([0, 1, -1, 0, -6, 5, 1]))
    assert numpy.allclose(p4.coeff, p4_exact.coeff, rtol=1E-14)
```

```
p5 = p2.derivative()
p5_exact = Polynomial([1, 0, 0, -24, -5])
assert p5.coeff == p5_exact.coeff

p6 = Polynomial([0, 1, 0, 0, -6, -1])  # p2
p6.differentiate()
p6_exact = p5_exact
assert p6.coeff == p6_exact.coeff
```

7.3.8 Arithmetic Operations and Other Special Methods

Given two instances a and b, the standard binary arithmetic operations with a and b are defined by the following special methods:

- a + b : a.__add__(b)
- a - b : a.__sub__(b)
- a*b : a.__mul__(b)
- a/b : a.__div__(b)
- a**b : a.__pow__(b)

Some other special methods are also often useful:

- the length of a, len(a): a.__len__()
- the absolute value of a, abs(a): a.__abs__()
- a == b : a.__eq__(b)
- a > b : a.__gt__(b)
- a >= b : a.__ge__(b)
- a < b : a.__lt__(b)
- a <= b : a.__le__(b)
- a != b : a.__ne__(b)
- -a : a.__neg__()
- evaluating a as a boolean expression (as in the test if a:) implies calling the special method a.__bool__(), which must return True or False – if __bool__ is not defined, __len__ is called to see if the length is zero (False) or not (True)

We can implement such methods in class Polynomial, see Exercise 7.25. Section 7.4 contains examples on implementing the special methods listed above.

7.3.9 Special Methods for String Conversion

Look at this class with a __str__ method:

```
>>> class MyClass(object):
...     def __init__(self):
...         self.data = 2
...     def __str__(self):
```

```
...                   return 'In __str__: %s' % str(self.data)
...
>>> a = MyClass()
>>> print a
In __str__: 2
```

Hopefully, you understand well why we get this output (if not, go back to Sect. 7.3.4).

But what will happen if we write just a at the command prompt in an interactive shell?

```
>>> a
<__main__.MyClass instance at 0xb75125ac>
```

When writing just a in an interactive session, Python looks for a special method __repr__ in a. This method is similar to __str__ in that it turns the instance into a string, but there is a convention that __str__ is a pretty print of the instance contents while __repr__ is a complete representation of the contents of the instance. For a lot of Python classes, including int, float, complex, list, tuple, and dict, __repr__ and __str__ give identical output. In our class MyClass the __repr__ is missing, and we need to add it if we want

```
>>> a
```

to write the contents like print a does.

Given an instance a, str(a) implies calling a.__str__() and repr(a) implies calling a.__repr__(). This means that

```
>>> a
```

is actually a repr(a) call and

```
>>> print a
```

is actually a print str(a) statement.

A simple remedy in class MyClass is to define

```
def __repr__(self):
    return self.__str__()  # or return str(self)
```

However, as we explain below, the __repr__ is best defined differently.

Recreating objects from strings The Python function eval(e) evaluates a valid Python expression contained in the string e, see Sect. 4.3.1. It is a convention that __repr__ returns a string such that eval applied to the string recreates the instance. For example, in case of the Y class from Sect. 7.1.2, __repr__ should return 'Y(10)' if the v0 variable has the value 10. Then eval('Y(10)') will be the same as if we had coded Y(10) directly in the program or an interactive session.

Below we show examples of `__repr__` methods in classes Y (Sect. 7.1.2), Polynomial (Sect. 7.3.7), and MyClass (above):

```
class Y(object):
    ...
    def __repr__(self):
        return 'Y(v0=%s)' % self.v0

class Polynomial(object):
    ...
    def __repr__(self):
        return 'Polynomial(coefficients=%s)' % self.coeff

class MyClass(object):
    ...
    def __repr__(self):
        return 'MyClass()'
```

With these definitions, `eval(repr(x))` recreates the object x if it is of one of the three types above. In particular, we can write x to file and later recreate the x from the file information:

```
# somefile is some file object
somefile.write(repr(x))
somefile.close()
...
data = somefile.readline()
x2 = eval(data)  # recreate object
```

Now, x2 will be equal to x (`x2 == x` evaluates to `True`).

7.4 Example: Class for Vectors in the Plane

This section explains how to implement two-dimensional vectors in Python such that these vectors act as objects we can add, subtract, form inner products with, and do other mathematical operations on. To understand the forthcoming material, it is necessary to have digested Sect. 7.3, in particular Sects. 7.3.6 and 7.3.8.

7.4.1 Some Mathematical Operations on Vectors

Vectors in the plane are described by a pair of real numbers, (a, b). In Sect. 5.1.2 we present mathematical rules for adding and subtracting vectors, multiplying two vectors (the inner or dot or scalar product), the length of a vector, and multiplication by a scalar:

$$(a, b) + (c, d) = (a + c, b + d), \tag{7.4}$$

$$(a, b) - (c, d) = (a - c, b - d), \tag{7.5}$$

$$(a, b) \cdot (c, d) = ac + bd, \tag{7.6}$$

$$||(a, b)|| = \sqrt{(a, b) \cdot (a, b)}. \tag{7.7}$$

Moreover, two vectors (a, b) and (c, d) are equal if $a = c$ and $b = d$.

7.4.2   Implementation

We may create a class for plane vectors where the above mathematical operations are implemented by special methods. The class must contain two data attributes, one for each component of the vector, called x and y below. We include special methods for addition, subtraction, the scalar product (multiplication), the absolute value (length), comparison of two vectors (== and !=), as well as a method for printing out a vector.

```
class Vec2D(object):
    def __init__(self, x, y):
        self.x = x
        self.y = y

    def __add__(self, other):
        return Vec2D(self.x + other.x, self.y + other.y)

    def __sub__(self, other):
        return Vec2D(self.x - other.x, self.y - other.y)

    def __mul__(self, other):
        return self.x*other.x + self.y*other.y

    def __abs__(self):
        return math.sqrt(self.x**2 + self.y**2)

    def __eq__(self, other):
        return self.x == other.x and self.y == other.y

    def __str__(self):
        return '(%g, %g)' % (self.x, self.y)

    def __ne__(self, other):
        return not self.__eq__(other)   # reuse __eq__
```

The `__add__`, `__sub__`, `__mul__`, `__abs__`, and `__eq__` methods should be quite straightforward to understand from the previous mathematical definitions of these operations. The last method deserves a comment: here we simply reuse the equality operator `__eq__`, but precede it with a not. We could also have implemented this method as

```
    def __ne__(self, other):
        return self.x != other.x or self.y != other.y
```

Nevertheless, this implementation requires us to write more, and it has the danger of introducing an error in the logics of the boolean expressions. A more reliable approach, when we know that the `__eq__` method works, is to reuse this method and observe that `a != b` means `not (a == b)`.

A word of warning is in place regarding our implementation of the equality operator (`==` via `__eq__`). We test for equality of each component, which is correct from a mathematical point of view. However, each vector component is a floating-point number that may be subject to rounding errors both in the representation on the computer and from previous (inexact) floating-point calculations. Two mathematically equal components may be different in their inexact representations on the computer. The remedy for this problem is to avoid testing for equality, but instead check that the difference between the components is sufficiently small. The function `numpy.allclose` can be used for this purpose:

```
if a == b:
```

by

```
if numpy.allclose(a, b):
```

A more reliable equality operator can now be implemented:

```
class Vec2D(object):
    ...
    def __eq__(self, other):
        return numpy.allclose(self.x, other.x) and \
               numpy.allclose(self.y, other.y)
```

As a rule of thumb, you should never apply the `==` test to two `float` objects.

The special method `__len__` could be introduced as a synonym for `__abs__`, i.e., for a `Vec2D` instance named `v`, `len(v)` is the same as `abs(v)`, because the absolute value of a vector is mathematically the same as the length of the vector. However, if we implement

```
def __len__(self):
    # Reuse implementation of __abs__
    return abs(self)  # equiv. to self.__abs__()
```

we will run into trouble when we compute `len(v)` and the answer is (as usual) a `float`. Python will then complain and tell us that `len(v)` must return an `int`. Therefore, `__len__` cannot be used as a synonym for the length of the vector in our application. On the other hand, we could let `len(v)` mean the number of components in the vector:

```
def __len__(self):
    return 2
```

This is not a very useful function, though, as we already know that all our `Vec2D` vectors have just two components. For generalizations of the class to vectors with n components, the `__len__` method is of course useful.

7.4.3   Usage

Let us play with some Vec2D objects:

```
>>> u = Vec2D(0,1)
>>> v = Vec2D(1,0)
>>> w = Vec2D(1,1)
>>> a = u + v
>>> print a
(1, 1)
>>> a == w
True
>>> a = u - v
>>> print a
(-1, 1)
>>> a = u*v
>>> print a
0
>>> print abs(u)
1.0
>>> u == v
False
>>> u != v
True
```

When you read through this interactive session, you should check that the calculation is mathematically correct, that the resulting object type of a calculation is correct, and how each calculation is performed in the program. The latter topic is investigated by following the program flow through the class methods. As an example, let us consider the expression u != v. This is a boolean expression that is True since u and v are different vectors. The resulting object type should be bool, with values True or False. This is confirmed by the output in the interactive session above. The Python calculation of u != v leads to a call to

```
u.__ne__(v)
```

which leads to a call to

```
u.__eq__(v)
```

The result of this last call is False, because the special method will evaluate the boolean expression

```
0 == 1 and 1 == 0
```

which is obviously False. When going back to the __ne__ method, we end up with a return of not False, which evaluates to True.

Comment For real computations with vectors in the plane, you would probably just use a Numerical Python array of length 2. However, one thing such objects cannot do is evaluating u*v as a scalar product. The multiplication operator for Numerical Python arrays is not defined as a scalar product (it is rather defined as $(a, b) \cdot (c, d) = (ac, bd)$). Another difference between our Vec2D class and Numerical Python arrays is the abs function, which computes the length of the vector in class Vec2D, while it does something completely different with Numerical Python arrays.

7.5   Example: Class for Complex Numbers

Imagine that Python did not already have complex numbers. We could then make a class for such numbers and support the standard mathematical operations. This exercise turns out to be a very good pedagogical example of programming with classes and special methods, so we shall make our own class for complex numbers and go through all the details of the implementation.

The class must contain two data attributes: the real and imaginary part of the complex number. In addition, we would like to add, subtract, multiply, and divide complex numbers. We would also like to write out a complex number in some suitable format. A session involving our own complex numbers may take the form

```
>>> u = Complex(2,-1)
>>> v = Complex(1)       # zero imaginary part
>>> w = u + v
>>> print w
(3, -1)
>>> w != u
True
>>> u*v
Complex(2, -1)
>>> u < v
illegal operation "<" for complex numbers
>>> print w + 4
(7, -1)
>>> print 4 - w
(1, 1)
```

We do not manage to use exactly the same syntax with j as imaginary unit as in Python's built-in complex numbers so to specify a complex number we must create a Complex instance.

7.5.1   Implementation

Here is the complete implementation of our class for complex numbers:

```
class Complex(object):
    def __init__(self, real, imag=0.0):
        self.real = real
        self.imag = imag
```

```
def __add__(self, other):
    return Complex(self.real + other.real,
                   self.imag + other.imag)

def __sub__(self, other):
    return Complex(self.real - other.real,
                   self.imag - other.imag)

def __mul__(self, other):
    return Complex(self.real*other.real - self.imag*other.imag,
                   self.imag*other.real + self.real*other.imag)

def __div__(self, other):
    sr, si, or, oi = self.real, self.imag, \
                     other.real, other.imag # short forms
    r = float(or**2 + oi**2)
    return Complex((sr*or+si*oi)/r, (si*or-sr*oi)/r)

def __abs__(self):
    return sqrt(self.real**2 + self.imag**2)

def __neg__(self):   # defines -c (c is Complex)
    return Complex(-self.real, -self.imag)

def __eq__(self, other):
    return self.real == other.real and self.imag == other.imag

def __ne__(self, other):
    return not self.__eq__(other)

def __str__(self):
    return '(%g, %g)' % (self.real, self.imag)

def __repr__(self):
    return 'Complex' + str(self)

def __pow__(self, power):
    raise NotImplementedError\
        ('self**power is not yet impl. for Complex')
```

The special methods for addition, subtraction, multiplication, division, and the absolute value follow easily from the mathematical definitions of these operations for complex numbers (see Sect. 1.6). What -c means when c is of type `Complex`, is also easy to define and implement. The `__eq__` method needs a word of caution: the method is mathematically correct, but comparison of real numbers on a computer should always employ a tolerance. The version of `__eq__` shown above is about compact code and equivalence to the mathematics. Any real-world numerical computations should employ a test that abs(`self.real` - `other.real`) < eps *and* abs(`self.imag` - `other.imag`) < eps, where eps is some small tolerance, say eps = 1E-14.

The final `__pow__` method exemplifies a way to introduce a method in a class, while we postpone its implementation. The simplest way to do this is by inserting an empty function body using the pass ("do nothing") statement:

```
class Polynomial(object):
    ...
    def __pow__(self, power):
        # Postpone implementation of self**power
        pass
```

However, the preferred method is to raise a `NotImplementedError` exception so that users writing power expressions are notified that this operation is not available. The simple `pass` will just silently bypass this serious fact!

7.5.2   Illegal Operations

Some mathematical operations, like the comparison operators >, >=, etc., do not have a meaning for complex numbers. By default, Python allows us to use these comparison operators for our `Complex` instances, but the boolean result will be mathematical nonsense. Therefore, we should implement the corresponding special methods and give a sensible error message that the operations are not available for complex numbers. Since the messages are quite similar, we make a separate method to gather common operations:

```
def _illegal(self, op):
    print 'illegal operation "%s" for complex numbers' % op
```

Note the underscore prefix: this is a Python convention telling that the `_illegal` method is local to the class in the sense that it is not supposed to be used outside the class, just by other class methods. In computer science terms, we say that names starting with an underscore are not part of the *application programming interface*, known as the API. Other programming languages, such as Java, C++, and C#, have special keywords, like `private` and `protected` that can be used to technically hide both data and methods from users of the class. Python will never restrict anybody who tries to access data or methods that are considered private to the class, but the leading underscore in the name reminds any user of the class that she now touches parts of the class that are not meant to be used "from the outside".

Various special methods for comparison operators can now call up `_illegal` to issue the error message:

```
def __gt__(self, other):  self._illegal('>')
def __ge__(self, other):  self._illegal('>=')
def __lt__(self, other):  self._illegal('<')
def __le__(self, other):  self._illegal('<=')
```

7.5.3   Mixing Complex and Real Numbers

The implementation of class `Complex` is far from perfect. Suppose we add a complex number and a real number, which is a mathematically perfectly valid operation:

```
w = u + 4.5
```

This statement leads to an exception,

```
AttributeError: 'float' object has no attribute 'real'
```

In this case, Python sees u + 4.5 and tries to use u.__add__(4.5), which causes trouble because the other argument in the __add__ method is 4.5, i.e., a float object, and float objects do not contain an attribute with the name real (other.real is used in our __add__ method, and accessing other.real is what causes the error).

One idea for a remedy could be to set

```
other = Complex(other)
```

since this construction turns a real number other into a Complex object. However, when we add two Complex instances, other is of type Complex, and the constructor simply stores this Complex instance as self.real (look at the method __init__). This is not what we want!

A better idea is to test for the type of other and perform the right conversion to Complex:

```
def __add__(self, other):
    if isinstance(other, (float,int)):
        other = Complex(other)
    return Complex(self.real + other.real,
                   self.imag + other.imag)
```

We could alternatively drop the conversion of other and instead implement two addition rules, depending on the type of other:

```
def __add__(self, other):
    if isinstance(other, (float,int)):
        return Complex(self.real + other, self.imag)
    else:
        return Complex(self.real + other.real,
                       self.imag + other.imag)
```

A third way is to look for what we require from the other object, and check that this demand is fulfilled. Mathematically, we require other to be a complex or real number, but from a programming point of view, all we demand (in the original __add__ implementation) is that other has real and imag attributes. To check if an object a has an attribute with name stored in the string attr, one can use the function

```
hasattr(a, attr)
```

In our context, we need to perform the test

```
if hasattr(other, 'real') and hasattr(other, 'imag'):
```

Our third implementation of the `__add__` method therefore becomes

```
def __add__(self, other):
    if isinstance(other, (float,int)):
        other = Complex(other)
    elif not (hasattr(other, 'real') and \
              hasattr(other, 'imag')):
        raise TypeError('other must have real and imag attr.')
    return Complex(self.real + other.real,
                   self.imag + other.imag)
```

The advantage with this third alternative is that we may add instances of class `Complex` and Python's own complex class (`complex`), since all we need is an object with `real` and `imag` attributes.

7.5.4 Dynamic, Static, Strong, Weak, and Duck Typing

The presentations of alternative implementations of the `__add__` actually touch some very important computer science topics. In Python, function arguments can refer to objects of any type, and the type of an argument can change during program execution. This feature is known as *dynamic typing* and supported by languages such as Python, Perl, Ruby, and Tcl. Many other languages, C, C++, Java, and C# for instance, restrict a function argument to be of one type, which must be known when we write the program. Any attempt to call the function with an argument of another type is flagged as an error. One says that the language employs *static typing*, since the type cannot change as in languages having dynamic typing. The code snippet

```
a = 6     # a is integer
a = 'b'   # a is string
```

is valid in a language with dynamic typing, but not in a language with static typing.

Our next point is easiest illustrated through an example. Consider the code

```
a = 6
b = '9'
c = a + b
```

The expression a + b adds an integer and a string, which is illegal in Python. However, since b is the string '9', it is natural to interpret a + b as 6 + 9. That is, if the string b is converted to an integer, we may calculate a + b. Languages performing this conversion automatically are said to employ *weak typing*, while languages that require the programmer to explicit perform the conversion, as in

```
c = a + float(b)
```

are known to have *strong typing*. Python, Java, C, and C# are examples of languages with strong typing, while Perl and C++ allow weak typing. However, in our third implementation of the `__add__` method, certain types – `int` and `float` – are automatically converted to the right type `Complex`. The programmer has therefore imposed a kind of weak typing in the behavior of the addition operation for complex numbers.

There is also something called *duck typing* where the code only imposes a requirement of some data or methods in the object, rather than demanding the object to be of a particular type. The explanation of the term duck typing is the principle: *if it walks like a duck, and quacks like a duck, it's a duck.* An operation a + b may be valid if a and b have certain properties that make it possible to add the objects, regardless of the type of a or b. To enable a + b in our third implementation of the `__add__` method, it is sufficient that b has `real` and `imag` attributes. That is, objects with `real` and `imag` look like `Complex` objects. Whether they really are of type `Complex` is not considered important in this context.

There is a continuously ongoing debate in computer science which kind of typing that is preferable: dynamic versus static, and weak versus strong. Static and strong typing, as found in Java and C#, support coding safety and reliability at the expense of long and sometimes repetitive code, while dynamic and weak typing support programming flexibility and short code. Many will argue that short code is more readable and reliable than long code, so there is no simple conclusion.

7.5.5 Special Methods for "Right" Operands

What happens if we add a `float` and a `Complex` in that order?

```
w = 4.5 + u
```

This statement causes the exception

```
TypeError: unsupported operand type(s) for +: 'float' and 'instance'
```

This time Python cannot find any definition of what the plus operation means with a `float` on the left-hand side and a `Complex` object on the right-hand side of the plus sign. The `float` class was created many years ago without any knowledge of our `Complex` objects, and we are not allowed to extend the `__add__` method in the `float` class to handle `Complex` instances. Nevertheless, Python has a special method `__radd__` for the case where the class instance (`self`) is on the right-hand side of the operator and the `other` object is on the left-hand side. That is, we may implement a possible `float` or `int` plus a `Complex` by

```
def __radd__(self, other):      # defines other + self
    return self.__add__(other)  # other + self = self + other
```

Similar special methods exist for subtraction, multiplication, and division. For the subtraction operator, observe that `other` - `self`, which is the operation assumed to implemented in `__rsub__`, can be realized by `other.__sub__(self)`. A possible implementation is

```
    def __sub__(self, other):
        print 'in sub, self=%s, other=%s' % (self, other)
        if isinstance(other, (float,int)):
            other = Complex(other)
        return Complex(self.real - other.real,
                       self.imag - other.imag)

    def __rsub__(self, other):
        print 'in rsub, self=%s, other=%s' % (self, other)
        if isinstance(other, (float,int)):
            other = Complex(other)
        return other.__sub__(self)
```

The `print` statements are inserted to better understand how these methods are visited. A quick test demonstrates what happens:

```
>>> w = u - 4.5
in sub, self=(2, -1), other=4.5
>>> print w
(-2.5, -1)
>>> w = 4.5 - u
in rsub, self=(2, -1), other=4.5
in sub, self=(4.5, 0), other=(2, -1)
>>> print w
(2.5, 1)
```

Remark As you probably realize, there is quite some code to be implemented and lots of considerations to be resolved before we have a class `Complex` for professional use in the real world. Fortunately, Python provides its `complex` class, which offers everything we need for computing with complex numbers. This fact reminds us that it is important to know what others already have implemented, so that we avoid "reinventing the wheel". In a learning process, however, it is a probably a very good idea to look into the details of a class `Complex` as we did above.

7.5.6 Inspecting Instances

The purpose of this section is to explain how we can easily look at the contents of a class instance, i.e., the data attributes and the methods. As usual, we look at an example – this time involving a very simple class:

```
class A(object):
    """A class for demo purposes."""
    def __init__(self, value):
        self.v = value

    def dump(self):
        print self.__dict__
```

The `self.__dict__` attribute is briefly mentioned in Sect. 7.1.6. Every instance is automatically equipped with this attribute, which is a dictionary that stores all the ordinary attributes of the instance (the variable names are keys, and the object references are values). In class A there is only one data attribute, so the `self.__dict__` dictionary contains one key, `'v'`:

```
>>> a = A([1,2])
>>> a.dump()
{'v': [1, 2]}
```

Another way of inspecting what an instance a contains is to call `dir(a)`. This Python function writes out the names of all methods and variables (and more) of an object:

```
>>> dir(a)
'__doc__', '__init__', '__module__', 'dump', 'v']
```

The `__doc__` variable is a docstring, similar to docstrings in functions (see Sect. 3.1.11), i.e., a description of the class appearing as a first string right after the `class` headline:

```
>>> a.__doc__
'A class for demo purposes.'
```

The `__module__` variable holds the name of the module in which the class is defined. If the class is defined in the program itself and not in an imported module, `__module__` equals `'__main__'`.

The rest of the entries in the list returned from `dir(a)` correspond to attribute names defined by the programmer of the class, in this example the method attributes `__init__` and dump, and the data attribute v.

Now, let us try to add new variables to an existing instance:

```
>>> a.myvar = 10
>>> a.dump()
{'myvar': 10, 'v': [1, 2]}
>>> dir(a)
['__doc__', '__init__', '__module__', 'dump', 'myvar', 'v']
```

The output of `a.dump()` and `dir(a)` show that we were successful in adding a new variable to this instance on the fly. If we make a new instance, it contains only the variables and methods that we find in the definition of class A:

```
>>> b = A(-1)
>>> b.dump()
{'v': -1}
>>> dir(b)
['__doc__', '__init__', '__module__', 'dump', 'v']
```

We may also add new methods to an instance, but this will not be shown here.

Adding or removing attributes may sound scary and highly illegal to C, C++, and Java programmers, but more dynamic classes is natural and legal in many other languages – and often useful.

Python classes are dynamic and their contents can be inspected
As seen by the examples above,

1. a class instance is dynamic and allows attributes to be added or removed while the program is running,
2. the contents of an instance can be inspected by the `dir` function, and the data attributes are available through the `__dict__` dictionary.

There is a special module, `inspect`, doing more detailed inspection of Python objects. One can, for example, get the arguments of functions or methods and even inspect the code of the object.

7.6   Static Methods and Attributes

Up to now, each instance has its own copy of data attributes. Sometimes it can be natural to have data attributes that are shared among all instances. For example, we may have an attribute that counts how many instances that have been made so far. We can exemplify how to do this in a little class for points (x, y, z) in space:

```
>>> class SpacePoint(object):
...     counter = 0
...     def __init__(self, x, y, z):
...         self.p = (x, y, z)
...         SpacePoint.counter += 1
```

The `counter` data attribute is initialized at the same indentation level as the methods in the class, and the attribute is not prefixed by `self`. Such attributes declared outside methods are shared among all instances and called *static attributes*. To access the `counter` attribute, we must prefix by the classname `SpacePoint` instead of `self`: `SpacePoint.counter`. In the constructor we increase this common counter by 1, i.e., every time a new instance is made the counter is updated to keep track of how many objects we have created so far:

```
>>> p1 = SpacePoint(0,0,0)
>>> SpacePoint.counter
1
>>> for i in range(400):
...     p = SpacePoint(i*0.5, i, i+1)
...
>>> SpacePoint.counter
401
```

The methods we have seen so far must be called through an instance, which is fed in as the `self` variable in the method. We can also make class methods that can be called without having an instance. The method is then similar to a plain Python function, except that it is contained inside a class and the method name must be prefixed by the classname. Such methods are known as *static methods*. Let us illustrate the syntax by making a very simple class with just one static method `write`:

```
>>> class A(object):
...     @staticmethod
...     def write(message):
...         print message
...
>>> A.write('Hello!')
Hello!
```

As demonstrated, we can call `write` without having any instance of class `A`, we just prefix with the class name. Also note that `write` does not take a `self` argument. Since this argument is missing inside the method, we can never access non-static attributes since these always must be prefixed by an instance (i.e., `self`). However, we can access static attributes, prefixed by the classname.

If desired, we can make an instance and call `write` through that instance too:

```
>>> a = A()
>>> a.write('Hello again')
Hello again
```

Static methods are used when you want a global function, but find it natural to let the function belong to a class and be prefixed with the classname.

7.7 Summary

7.7.1 Chapter Topics

Classes A class contains attributes, which are variables (data attributes) and functions (method attributes, also called just methods). A first rough overview of a class can be to just list the attributes, e.g., in a UML diagram.

Below is a sample class with three data attributes (m, M, and G) and three methods (a constructor, `force`, and `visualize`). The class represents the gravity force between two masses. This force is computed by the `force` method, while the `visualize` method plots the force as a function of the distance between the masses.

```
class Gravity(object):
    """Gravity force between two physical objects."""

    def __init__(self, m, M):
        self.m = m            # mass of object 1
        self.M = M            # mass of object 2
        self.G = 6.67428E-11 # gravity constant, m**3/kg/s**2

    def force(self, r):
        G, m, M = self.G, self.m, self.M
        return G*m*M/r**2

    def visualize(self, r_start, r_stop, n=100):
        from scitools.std import plot, linspace
        r = linspace(r_start, r_stop, n)
        g = self.force(r)
        title='Gravity force: m=%g, M=%g' % (self.m, self.M)
        plot(r, g, title=title)
```

Note that to access attributes inside the `force` method, and to call the `force` method inside the `visualize` method, we must prefix with `self`. Also recall that all methods must take `self`, "this" instance, as first argument, but the argument is left out in calls. The assignment of a data attributes to a local variable (e.g., `G = self.G`) inside methods is not necessary, but here it makes the mathematical formula easier to read and compare with standard mathematical notation.

This class (found in file `Gravity.py`) can be used to find the gravity force between the Moon and the Earth:

```
mass_moon = 7.35E+22;  mass_earth = 5.97E+24
gravity = Gravity(mass_moon, mass_earth)
r = 3.85E+8  # Earth-Moon distance in meters
Fg = gravity.force(r)
print 'force:', Fg
```

Special methods A collection of special methods, with two leading and trailing underscores in the method names, offers special syntax in Python programs.

The table below provides an overview of the most important special methods.

Construction	Meaning2
`a.__init__(self, args)`	constructor: `a = A(args)`
`a.__del__(self)`	destructor: `del a`
`a.__call__(self, args)`	call as function: `a(args)`
`a.__str__(self)`	pretty print: `print a`, `str(a)`
`a.__repr__(self)`	representation: `a = eval(repr(a))`
`a.__add__(self, b)`	`a + b`
`a.__sub__(self, b)`	`a - b`
`a.__mul__(self, b)`	`a*b`
`a.__div__(self, b)`	`a/b`
`a.__radd__(self, b)`	`b + a`
`a.__rsub__(self, b)`	`b - a`
`a.__rmul__(self, b)`	`b*a`
`a.__rdiv__(self, b)`	`b/a`
`a.__pow__(self, p)`	`a**p`
`a.__lt__(self, b)`	`a < b`
`a.__gt__(self, b)`	`a > b`
`a.__le__(self, b)`	`a <= b`
`a.__ge__(self, b)`	`a => b`
`a.__eq__(self, b)`	`a == b`
`a.__ne__(self, b)`	`a != b`
`a.__bool__(self)`	boolean expression, as in `if a:`
`a.__len__(self)`	length of a (`int`): `len(a)`
`a.__abs__(self)`	`abs(a)`

Terminology The important computer science topics in this chapter are

- classes
- attributes
- methods

- constructor (`__init__`)
- special methods (`__add__`, `__str__`, `__ne__`, etc.)

7.7.2 Example: Interval Arithmetic

Input data to mathematical formulas are often subject to uncertainty, usually because physical measurements of many quantities involve measurement errors, or because it is difficult to measure a parameter and one is forced to make a qualified guess of the value instead. In such cases it could be more natural to specify an input parameter by an interval $[a, b]$, which is guaranteed to contain the true value of the parameter. The size of the interval expresses the uncertainty in this parameter. Suppose all input parameters are specified as intervals, what will be the interval, i.e., the uncertainty, of the output data from the formula? This section develops a tool for computing this output uncertainty in the cases where the overall computation consists of the standard arithmetic operations.

To be specific, consider measuring the acceleration of gravity by dropping a ball and recording the time it takes to reach the ground. Let the ground correspond to $y = 0$ and let the ball be dropped from $y = y_0$. The position of the ball, $y(t)$, is then

$$y(t) = y_0 - \frac{1}{2}gt^2 .$$

If T is the time it takes to reach the ground, we have that $y(T) = 0$, which gives the equation $\frac{1}{2}gT^2 = y_0$, with solution

$$g = 2y_0 T^{-2} .$$

In such experiments we always introduce some measurement error in the start position y_0 and in the time taking (T). Suppose y_0 is known to lie in $[0.99, 1.01]$ m and T in $[0.43, 0.47]$ s, reflecting a 2 % measurement error in position and a 10 % error from using a stop watch. What is the error in g? With the tool to be developed below, we can find that there is a 22 % error in g.

Problem Assume that two numbers p and q are guaranteed to lie inside intervals,

$$p = [a, b], \quad q = [c, d] .$$

The sum $p + q$ is then guaranteed to lie inside an interval $[s, t]$ where $s = a + c$ and $t = b + d$. Below we list the rules of *interval arithmetic*, i.e., the rules for addition, subtraction, multiplication, and division of two intervals:

- $p + q = [a + c, b + d]$
- $p - q = [a - d, b - c]$
- $pq = [\min(ac, ad, bc, bd), \max(ac, ad, bc, bd)]$
- $p/q = [\min(a/c, a/d, b/c, b/d), \max(a/c, a/d, b/c, b/d)]$ provided that $[c, d]$ does not contain zero

For doing these calculations in a program, it would be natural to have a new type for quantities specified by intervals. This new type should support the operators +, -, *, and / according to the rules above. The task is hence to implement a class for interval arithmetics with special methods for the listed operators. Using the class, we should be able to estimate the uncertainty of two formulas:

- The acceleration of gravity, $g = 2y_0 T^{-2}$, given a 2 % uncertainty in y_0: $y_0 = [0.99, 1.01]$, and a 10 % uncertainty in T: $T = [T_m \cdot 0.95, T_m \cdot 1.05]$, with $T_m = 0.45$.
- The volume of a sphere, $V = \frac{4}{3}\pi R^3$, given a 20 % uncertainty in R: $R = [R_m \cdot 0.9, R_m \cdot 1.1]$, with $R_m = 6$.

Solution The new type is naturally realized as a class `IntervalMath` whose data consist of the lower and upper bound of the interval. Special methods are used to implement arithmetic operations and printing of the object. Having understood class Vec2D from Sect. 7.4, it should be straightforward to understand the class below:

```
class IntervalMath(object):
    def __init__(self, lower, upper):
        self.lo = float(lower)
        self.up = float(upper)

    def __add__(self, other):
        a, b, c, d = self.lo, self.up, other.lo, other.up
        return IntervalMath(a + c, b + d)

    def __sub__(self, other):
        a, b, c, d = self.lo, self.up, other.lo, other.up
        return IntervalMath(a - d, b - c)

    def __mul__(self, other):
        a, b, c, d = self.lo, self.up, other.lo, other.up
        return IntervalMath(min(a*c, a*d, b*c, b*d),
                            max(a*c, a*d, b*c, b*d))

    def __div__(self, other):
        a, b, c, d = self.lo, self.up, other.lo, other.up
        # [c,d] cannot contain zero:
        if c*d <= 0:
            raise ValueError\
                    ('Interval %s cannot be denominator because '\
                    'it contains zero' % other)
        return IntervalMath(min(a/c, a/d, b/c, b/d),
                            max(a/c, a/d, b/c, b/d))

    def __str__(self):
        return '[%g, %g]' % (self.lo, self.up)
```

The code of this class is found in the file `IntervalMath.py`. A quick demo of the class can go as

```
I = IntervalMath
a = I(-3,-2)
b = I(4,5)
expr = 'a+b', 'a-b', 'a*b', 'a/b'
for e in expr:
    print '%s =' % e, eval(e)
```

The output becomes

```
a+b = [1, 3]
a-b = [-8, -6]
a*b = [-15, -8]
a/b = [-0.75, -0.4]
```

This gives the impression that with very short code we can provide a new type that enables computations with interval arithmetic and thereby with uncertain quantities. However, the class above has severe limitations as shown next.

Consider computing the uncertainty of aq if a is expressed as an interval $[4, 5]$ and q is a number (float):

```
a = I(4,5)
q = 2
b = a*q
```

This does not work so well:

```
  File "IntervalMath.py", line 15, in __mul__
    a, b, c, d = self.lo, self.up, other.lo, other.up
AttributeError: 'float' object has no attribute 'lo'
```

The problem is that a*q is a multiplication between an IntervalMath object a and a float object q. The __mul__ method in class IntervalMath is invoked, but the code there tries to extract the lo attribute of q, which does not exist since q is a float.

We can extend the __mul__ method and the other methods for arithmetic operations to allow for a number as operand – we just convert the number to an interval with the same lower and upper bounds:

```
def __mul__(self, other):
    if isinstance(other, (int, float)):
        other = IntervalMath(other, other)
    a, b, c, d = self.lo, self.up, other.lo, other.up
    return IntervalMath(min(a*c, a*d, b*c, b*d),
                        max(a*c, a*d, b*c, b*d))
```

Looking at the formula $g = 2y_0 T^{-2}$, we run into a related problem: now we want to multiply 2 (int) with y_0, and if y_0 is an interval, this multiplication is not defined among int objects. To handle this case, we need to implement an __rmul__(self, other) method for doing other*self, as explained in Sect. 7.5.5:

```
def __rmul__(self, other):
    if isinstance(other, (int, float)):
        other = IntervalMath(other, other)
    return other*self
```

Similar methods for addition, subtraction, and division must also be included in the class.

Returning to $g = 2y_0 T^{-2}$, we also have a problem with T^{-2} when T is an interval. The expression T**(-2) invokes the power operator (at least if we do not rewrite the expression as 1/(T*T)), which requires a __pow__ method in class IntervalMath. We limit the possibility to have integer powers, since this is easy to compute by repeated multiplications:

```
def __pow__(self, exponent):
    if isinstance(exponent, int):
        p = 1
        if exponent > 0:
            for i in range(exponent):
                p = p*self
        elif exponent < 0:
            for i in range(-exponent):
                p = p*self
            p = 1/p
        else:    # exponent == 0
            p = IntervalMath(1, 1)
        return p
    else:
        raise TypeError('exponent must int')
```

Another natural extension of the class is the possibility to convert an interval to a number by choosing the midpoint of the interval:

```
>>> a = IntervalMath(5,7)
>>> float(a)
6
```

float(a) calls a.__float__(), which we implement as

```
def __float__(self):
    return 0.5*(self.lo + self.up)
```

A __repr__ method returning the right syntax for recreating the present instance is also natural to include in any class:

```
def __repr__(self):
    return '%s(%g, %g)' % \
            (self.__class__.__name__, self.lo, self.up)
```

We are now in a position to test out the extended class IntervalMath.

```
>>> g = 9.81
>>> y_0 = I(0.99, 1.01)        # 2% uncertainty
>>> Tm = 0.45                  # mean T
>>> T = I(Tm*0.95, Tm*1.05)   # 10% uncertainty
>>> print T
[0.4275, 0.4725]
>>> g = 2*y_0*T**(-2)
>>> g
IntervalMath(8.86873, 11.053)
>>> # Compute with mean values
>>> T = float(T)
>>> y = 1
>>> g = 2*y_0*T**(-2)
>>> print '%.2f' % g
9.88
```

Another formula, the volume $V = \frac{4}{3}\pi R^3$ of a sphere, shows great sensitivity to uncertainties in R:

```
>>> Rm = 6
>>> R = I(Rm*0.9, Rm*1.1)    # 20 % error
>>> V = (4./3)*pi*R**3
>>> V
IntervalMath(659.584, 1204.26)
>>> print V
[659.584, 1204.26]
>>> print float(V)
931.922044761
>>> # Compute with mean values
>>> R = float(R)
>>> V = (4./3)*pi*R**3
>>> print V
904.778684234
```

Here, a 20 % uncertainty in R gives almost 60 % uncertainty in V, and the mean of the V interval is significantly different from computing the volume with the mean of R.

The complete code of class `IntervalMath` is found in `IntervalMath.py`. Compared to the implementations shown above, the real implementation in the file employs some ingenious constructions and help methods to save typing and repeating code in the special methods for arithmetic operations. You can read more about interval arithmetics on Wikipedia[3].

7.8   Exercises

Exercise 7.1: Make a function class
Make a class F that implements the function

$$f(x; a, w) = e^{-ax} \sin(wx) .$$

---

[3] http://en.wikipedia.org/wiki/Interval_arithmetic

A value(x) method computes values of f, while a and w are data attributes. Test the class in an interactive session:

```
>>> from F import F
>>> f = F(a=1.0, w=0.1)
>>> from math import pi
>>> print f.value(x=pi)
0.013353835137
>>> f.a = 2
>>> print f.value(pi)
0.00057707154012
```

Filename: F.

Exercise 7.2: Add a data attribute to a class
Add a data attribute transactions to the Account class from Sect. 7.2.1. The new attribute counts the number of transactions done in the deposit and withdraw methods. Print the total number of transactions in the dump method. Write a test function test_Account() for testing that the implementation of the extended class Account is correct.
Filename: Account2.

Exercise 7.3: Add functionality to a class
In class AccountP from Sect. 7.2.1, introduce a list self._transactions, where each element holds a dictionary with the amount of a transaction and the point of time the transaction took place. Remove the _balance attribute and use instead the _transactions list to compute the balance in the method get_balance. Print out a nicely formatted table of all transactions, their amounts, and their time in a method print_transactions.

Hint Use the time or datetime module to get the date and local time.
Filename: Account3.

Remarks Observe that the computation of the balance is implemented in a different way in the present version of class AccountP compared to the version in Sect. 7.2.1, but the usage of the class, especially the get_balance method, remains the same. This is one of the great advantages of class programming: users are supposed to use the methods only, and the implementation of data structures and computational techniques inside methods can be changed without affecting existing programs that just call the methods.

Exercise 7.4: Make classes for a rectangle and a triangle
The purpose of this exercise is to create classes like class Circle from Sect. 7.2.3 for representing other geometric figures: a rectangle with width W, height H, and lower left corner (x_0, y_0); and a general triangle specified by its three vertices (x_0, y_0), (x_1, y_1), and (x_2, y_2) as explained in Exercise 3.16. Provide three methods: __init__ (to initialize the geometric data), area, and perimeter. Write test functions test_Rectangle() and test_Triangle() for checking that the results

produced by area and perimeter coincide with exact values within a small tolerance.
Filename: geometric_shapes.

Exercise 7.5: Make a class for quadratic functions

Consider a quadratic function $f(x; a, b, c) = ax^2 + bx + c$. Make a class Quadratic for representing f, where a, b, and c are data attributes, and the methods are

- __init__ for storing the attributes a, b, and c,
- value for computing a value of f at a point x,
- table for writing out a table of x and f values for n x values in the interval $[L, R]$,
- roots for computing the two roots.

The file with class Quadratic and corresponding demonstrations and/or tests should be organized as a module such that other programs can do a from Quadratic import Quadratic to use the class. Also equip the file with a test function for verifying the implementation of value and roots.
Filename: Quadratic.

Exercise 7.6: Make a class for straight lines

Make a class Line whose constructor takes two points p1 and p2 (2-tuples or 2-lists) as input. The line goes through these two points (see function line in Sect. 3.1.11 for the relevant formula of the line). A value(x) method computes a value on the line at the point x. Also make a function test_Line() for verifying the implementation. Here is a demo in an interactive session:

```
>>> from Line import Line, test_Line
>>> line = Line((0,-1), (2,4))
>>> print line.value(0.5), line.value(0), line.value(1)
0.25 -1.0 1.5
>>> test_Line()
```

Filename: Line.

Exercise 7.7: Flexible handling of function arguments

The constructor in class Line in Exercise 7.6 takes two points as arguments. Now we want to have more flexibility in the way we specify a straight line: we can give two points, a point and a slope, or a slope and the line's interception with the y axis. Write this extended class and a test function for checking that the increased flexibility does work.

Hint Let the constructor take two arguments p1 and p2 as before, and test with isinstance whether the arguments are float versus tuple or list to determine what kind of data the user supplies:

```
if isinstance(p1, (tuple,list)) and isinstance(p2, (float,int)):
    # p1 is a point and p2 is slope
    self.a = p2
    self.b = p1[1] - p2*p1[0]
elif ...
```

Filename: Line2.

Exercise 7.8: Wrap functions in a class

The purpose of this exercise is to make a class interface to an already existing set of functions implementing Lagrange's interpolation method from Exercise 5.25. We want to construct a class LagrangeInterpolation with a typical usage like:

```
import numpy as np
# Compute some interpolation points along y=sin(x)
xp = np.linspace(0, np.pi, 5)
yp = np.sin(xp)

# Lagrange's interpolation polynomial
p_L = LagrangeInterpolation(xp, yp)
x = 1.2
print 'p_L(%g)=%g' % (x, p_L(x)),
print 'sin(%g)=%g' % (x, np.sin(x))
p_L.plot()    # show graph of p_L
```

The plot method visualizes $p_L(x)$ for x between the first and last interpolation point (xp[0] and xp[-1]). In addition to writing the class itself, you should write code to verify the implementation.

Hint The class does not need much code as it can call the functions p_L from Exercise 5.25 and graph from Exercise 5.26, available in the Lagrange_poly2 module made in the latter exercise.
Filename: Lagrange_poly3.

Exercise 7.9: Flexible handling of function arguments

Instead of manually computing the interpolation points, as demonstrated in Exercise 7.8, we now want the constructor in class LagrangeInterpolation to also accept some Python function f(x) for computing the interpolation points. Typically, we would like to write this code:

```
from numpy import exp, sin, pi

def myfunction(x):
    return exp(-x/2.0)*sin(x)

p_L = LagrangeInterpolation(myfunction, x=[0, pi], n=11)
```

With such a code, $n = 11$ uniformly distributed x points between 0 and π are computed, and the corresponding y values are obtained by calling myfunction. The Lagrange interpolation polynomial is then constructed from these points. Note that the previous types of calls, `LangrangeInterpolation(xp, yp)`, must still be valid.

Hint The constructor in class `LagrangeInterpolation` must now accept two different sets of arguments: `xp, yp` vs. `f, x, n`. You can use the `isinstance(a, t)` function to test if object `a` is of type `t`. Declare the constructor with three arguments `arg1`, `arg2`, and `arg3=None`. Test if `arg1` and `arg2` are arrays (`isinstance(arg1, numpy.ndarray)`), and in that case, set `xp=arg1` and `yp=arg2`. On the other hand, if `arg1` is a function (`callable(arg1)` is `True`), `arg2` is a list or tuple (`isinstance(arg2, (list,tuple))`), and `arg3` is an integer, set `f=arg1`, `x=arg2`, and `n=arg3`.
Filename: `Lagrange_poly4`.

Exercise 7.10: Deduce a class implementation
Write a class `Hello` that behaves as illustrated in the following session:

```
>>> a = Hello()
>>> print a('students')
Hello, students!
>>> print a
Hello, World!
```

Filename: `Hello`.

Exercise 7.11: Implement special methods in a class
Modify the class from Exercise 7.1 such that the following interactive session can be run:

```
>>> from F import F
>>> f = F(a=1.0, w=0.1)
>>> from math import pi
>>> print f(x=pi)
0.013353835137
>>> f.a = 2
>>> print f(pi)
0.00057707154012
>>> print f
exp(-a*x)*sin(w*x)
```

Filename: `F2`.

Exercise 7.12: Make a class for summation of series
The task in this exercise is to calculate a sum $S(x) = \sum_{k=M}^{N} f_k(x)$, where $f_k(x)$ is some user-given formula for the terms in the sum. The following snippet demonstrates the typical use and functionality of a class `Sum` for computing $S(x) = \sum_{k=0}^{N}(-x)^k$:

```
def term(k, x):
    return (-x)**k

S = Sum(term, M=0, N=3)
x = 0.5
print S(x)
print S.term(k=4, x=x)   # (-0.5)**4
```

a) Implement class Sum such that the code snippet above works.
b) Implement a test function `test_Sum()` for verifying the results of the various
    methods in class Sum for a specific choice of $f_k(x)$.
c) Apply class Sum to compute the Taylor polynomial approximation to $\sin x$ for
    $x = \pi$ and some chosen x and N.

Filename: Sum.

Exercise 7.13: Apply a numerical differentiation class

Isolate class Derivative from Sect. 7.3.2 in a module file. Also isolate class Y
from Sect. 7.1.2 in a module file. Make a program that imports class Derivative
and class Y and applies the former to differentiate the function $y(t) = v_0 t - \frac{1}{2} g t^2$
represented by class Y. Compare the computed derivative with the exact value for
$t = 0, \frac{1}{2} v_0/g, v_0/g$.
Filenames: dYdt.py, Derivative.py, Y.py.

Exercise 7.14: Implement an addition operator

An anthropologist was asking a primitive tribesman about arithmetic. When the
anthropologist asked, *What does two and two make?* the tribesman replied, *Five.*
Asked to explain, the tribesman said, *If I have a rope with two knots, and another
rope with two knots, and I join the ropes together, then I have five knots.*

a) Make a class Rope for representing a rope with a given number of knots. Imple-
    ment the addition operator in this class such that we can join two ropes together
    in the way the tribesman described:

```
>>> from Rope import Rope
>>> rope1 = Rope(2)
>>> rope2 = Rope(2)
>>> rope3 = rope1 + rope2
>>> print rope3
5
```

   As seen, the class also features a `__str__` method for returning the number of
   knots on the rope.
b) Equip the module file with a test function for verifying the implementation of
    the addition operator.

Filename: Rope.py.

Exercise 7.15: Implement in-place += and -= operators
As alternatives to the `deposit` and `withdraw` methods in class `Account` class from
Sect. 7.2.1, we could use the operation += for `deposit` and -= for `withdraw`. Implement the += and -= operators, a `__str__` method, and preferably a `__repr__`
method in class `Account`. Write a `test_Account()` function to verify the implementation of all functionality in class `Account`.

Hint The special methods `__iadd__` and `__isub__` implement the += and -= operators, respectively. For instance, a -= p implies a call to a.`__isub__`(p). One
important feature of `__iadd__` and `__isub__` is that they must return `self` to work
properly, see the documentation of these methods in Chapter 3 of the Python Language Reference[4].
Filename: `Account4`.

Exercise 7.16: Implement a class for numerical differentiation
A widely used formula for numerical differentiation of a function $f(x)$ takes the
form

$$f'(x) \approx \frac{f(x+h) - f(x-h)}{2h} . \qquad (7.8)$$

This formula usually gives more accurate derivatives than (7.1) because it applies
a centered, rather than a one-sided, difference.

The goal of this exercise is to use the formula (7.8) to automatically differentiate a mathematical function $f(x)$ implemented as a Python function `f(x)`. More
precisely, the following code should work:

```
def f(x):
    return 0.25*x**4

df = Central(f)  # make function-like object df
# df(x) computes the derivative of f(x) approximately
x = 2
print 'df(%g)=%g' % (x, df(x))
print 'exact:', x**3
```

a) Implement class `Central` and test that the code above works. Include an optional argument h to the constructor in class `Central` so that h in the approximation (7.8) can be specified.
b) Write a test function `test_Central()` to verify the implementation. Utilize the
   fact that the formula (7.8) is exact for quadratic polynomials (provided h is not
   too small, then rounding errors in (7.8) require use of a (much) larger tolerance
   than the expected machine precision).
c) Write a function `table(f, x, h=1E-5)` that prints a table of errors in the numerical derivative (7.8) applied to a function f at some points x. The argument
   f is a `sympy` expression for a function. This f object can be transformed to
   a Python function and fed to the constructor of class `Central`, and f can be
   used to compute the exact derivative symbolically. The argument x is a list or
   array of points x, and h is the h in (7.8).

[4] http://docs.python.org/2/reference/

Hint The following session demonstrates how `sympy` can differentiate a mathematical expression and turn the result into a Python function:

```
>>> import sympy
>>> x = sympy.Symbol('x')
>>> f_expr = 'x*sin(2*x)'
>>> df_expr = sympy.diff(f_expr)
>>> df_expr
2*x*cos(2*x) + sin(2*x)
>>> df = sympy.lambdify([x], df_expr)  # make Python function
>>> df(0)
0.0
```

d) Organize the file with the class and functions such that it can be used a module.

Filename: `Central`.

Exercise 7.17: Examine a program

Consider this program file for computing a backward difference approximation to the derivative of a function `f(x)`:

```
from math import *

class Backward(object):
    def __init__(self, f, h=e-9):
        self.f, self.h = f, h
    def __call__(self, x):
        h, f = self.h, self.f
        return (f(x) - f(x-h))/h  # finite difference

dsin = Backward(sin)
e = dsin(0) - cos(0); print 'error:', e
dexp = Backward(exp, h=e-7)
e = dexp(0) - exp(0); print 'error:', e
```

The output becomes

```
error: -1.00023355634
error: 371.570909212
```

Is the approximation that bad, or are there bugs in the program?
Filename: `find_errors_class`.

Exercise 7.18: Modify a class for numerical differentiation

Make the two data attributes `h` and `f` of class `Derivative` from Sect. 7.3.2 protected as explained in Sect. 7.2.1. That is, prefix `h` and `f` with an underscore to tell users that these attributes should not be accessed directly. Add two methods `get_precision()` and `set_precision(h)` for reading and changing `h`. Make a separate test function for checking that the new class works as intended.
Filename: `Derivative_protected`.

Exercise 7.19: Make a class for the Heaviside function

a) Use a class to implement the discontinuous Heaviside function (3.25) from Exercise 3.29 and the smoothed continuous version (3.26) from Exercise 3.30 such that the following code works:

```
H = Heaviside()            # original discontinous Heaviside function
print H(0.1)
H = Heaviside(eps=0.8)  # smoothed continuous Heaviside function
print H(0.1)
```

b) Extend class Heaviside such that array arguments are allowed:

```
H = Heaviside()            # original discontinous Heaviside function
x = numpy.linspace(-1, 1, 11)
print H(x)
H = Heaviside(eps=0.8)  # smoothed Heaviside function
print H(x)
```

Hint Use ideas from Sect. 5.5.2.

c) Extend class Heaviside such that it supports plotting:

```
H = Heaviside()
x, y = H.plot(xmin=-4, xmax=4)  # x in [-4, 4]
from matplotlib.pyplot import plot
plot(x, y)

H = Heaviside(eps=1)
x, y = H.plot(xmin=-4, xmax=4)
plot(x, y)
```

Hint Techniques from Sect. 5.4.1 must in the first case be used to return arrays x and y such that the discontinuity is exactly reproduced. In the continuous (smoothed) case, one needs to compute a sufficiently fine resolution (x) based on the eps parameter, e.g., $201/\epsilon$ points in the interval $[-\epsilon, \epsilon]$, with a coarser set of coordinates outside this interval where the smoothed Heaviside function is almost constant, 0 or 1.

d) Write a test function test_Heaviside() for verifying the result of the various methods in class Heaviside.

Filename: Heaviside_class.

Exercise 7.20: Make a class for the indicator function
The purpose of this exercise is the make a class implementation of the indicator function from Exercise 3.31. Let the implementation be based on expressing the indicator function in terms of Heaviside functions. Allow for an ϵ parameter in the calls to the Heaviside function, such that we can easily choose between a discontinuous and a smoothed, continuous version of the indicator function:

```
I = Indicator(a, b)          # indicator function on [a,b]
print I(b+0.1), I((a+b)/2.0)
I = Indicator(0, 2, eps=1)   # smoothed indicator function on [0,2]
print I(0), I(1), I(1.9)
```

Note that if you build on the version of class Heaviside in Exercise 7.19b, any Indicator instance will accept array arguments too.
Filename: Indicator.

Exercise 7.21: Make a class for piecewise constant functions
The purpose of this exercise is to make a class implementation of a piecewise constant function, as defined in Exercise 3.32.

a) Implement the minimum functionality such that the following code works:

```
f = PiecewiseConstant([(0.4, 1), (0.2, 1.5), (0.1, 3)], xmax=4)
print f(1.5), f(1.75), f(4)

x = np.linspace(0, 4, 21)
print f(x)
```

b) Add a plot method to class PiecewiseConstant such that we can easily plot the graph of the function:

```
x, y = f.plot()
from matplotlib.pyplot import plot
plot(x, y)
```

Filename: PiecewiseConstant.

Exercise 7.22: Speed up repeated integral calculations
The observant reader may have noticed that our Integral class from Sect. 7.3.3 is very inefficient if we want to tabulate or plot a function $F(x) = \int_a^x f(x)$ for several consecutive values of x: $x_0 < x_1 < \cdots < x_m$. Requesting $F(x_k)$ will recompute the integral computed for $F(x_{k-1})$, and this is of course waste of computer work. Use the ideas from Sect. A.1.7 to modify the __call__ method such that if x is an array, assumed to contain coordinates of increasing value: $x_0 < x_1 < \cdots < x_m$, the method returns an array with $F(x_0), F(x_1), \ldots, F(x_m)$ with the minimum computational work. Also write a test function to verify that the implementation is correct.

Hint The n (n) parameter in the constructor of the Integral class can be taken as the total number of trapezoids (intervals) that are to be used to compute the final $F(x_m)$ value. The integral over an interval $[x_k, x_{k+1}]$ can then be computed by the trapezoidal function (or an Integral object) using an appropriate fraction of the n total trapezoids. This fraction can be $(x_{k+1} - x_k)/(x_m - a)$ (i.e., $n_k = n(x_{k+1} - x_k)/(x_m - a)$) or one may simply use a constant $n_k = n/m$ number of trapezoids for all the integrals over $[x_k, x_{k+1}]$, $k = 0, \ldots, m - 1$.
Filename: Integral_eff.

Exercise 7.23: Apply a class for polynomials
The Taylor polynomial of degree N for the exponential function e^x is given by

$$p(x) = \sum_{k=0}^{N} \frac{x^k}{k!}.$$

Make a program that (i) imports class Polynomial from Sect. 7.3.7, (ii) reads x and a series of N values from the command line, (iii) creates a Polynomial object for each N value for computing with the given Taylor polynomial, and (iv) prints the values of $p(x)$ for all the given N values as well as the exact value e^x. Try the program out with $x = 0.5, 3, 10$ and $N = 2, 5, 10, 15, 25$.
Filename: Polynomial_exp.

Exercise 7.24: Find a bug in a class for polynomials
Go through this alternative implementation of class Polynomial from Sect. 7.3.7 and explain each line in detail:

```
class Polynomial(object):
    def __init__(self, coefficients):
        self.coeff = coefficients

    def __call__(self, x):
        return sum([c*x**i for i, c in enumerate(self.coeff)])

    def __add__(self, other):
        maxlength = max(len(self), len(other))
        # Extend both lists with zeros to this maxlength
        self.coeff += [0]*(maxlength - len(self.coeff))
        other.coeff += [0]*(maxlength - len(other.coeff))
        result_coeff = self.coeff
        for i in range(maxlength):
            result_coeff[i] += other.coeff[i]
        return Polynomial(result_coeff)
```

The enumerate function, used in the __call__ method, enables us to iterate over a list somelist with both list indices and list elements: for index, element in enumerate(somelist). Write the code above in a file, and demonstrate that adding two polynomials does not work. Find the bug and correct it.
Filename: Polynomial_error.

Exercise 7.25: Implement subtraction of polynomials
Implement the special method __sub__ in class Polynomial from Sect. 7.3.7. Add a test for this functionality in function test_Polynomial.

Hint Study the __add__ method in class Polynomial and treat the two cases, where the lengths of the lists in the polynomials differs, separately.
Filename: Polynomial_sub.

Exercise 7.26: Test the functionality of pretty print of polynomials
Verify the functionality of the __str__ method in class Polynomial from Sect. 7.3.7 by writing a new test function test_Polynomial_str().
Filename: Polynomial_test_str.

Exercise 7.27: Vectorize a class for polynomials
Introducing an array instead of a list in class Polynomial does not enhance the efficiency of the implementation unless the mathematical computations are also vectorized. That is, all explicit Python loops must be substituted by vectorized expressions.

a) Go through class Polynomial.py and make sure the coeff attribute is always a numpy array with float elements.

b) Update the test function test_Polynomial to make use of the fact that the coeff attribute is always a numpy array with float elements. Run test_Polynomial to check that the new implementation is correct.

c) Vectorize the __add__ method by adding the common parts of the coefficients arrays and then appending the rest of the longest array to the result.

Hint Appending an array a to an array b can be done by concatenate(a, b).

d) Vectorize the __call__ method by observing that evaluation of a polynomial, $\sum_{i=0}^{n-1} c_i x^i$, can be computed as the inner product of two arrays: $(c_0, \ldots, c_{n-1})$ and $(x^0, x^1, \ldots, x^{n-1})$. The latter array can be computed by x**p, where p is an array with powers $0, 1, \ldots, n-1$, and x is a scalar.

e) The differentiate method can be vectorized by the statements

```
n = len(self.coeff)
self.coeff[:-1] = linspace(1, n-1, n-1)*self.coeff[1:]
self.coeff = self.coeff[:-1]
```

Show by hand calculations in a case where n is 3 that the vectorized statements produce the same result as the original differentiate method.

Filename: Polynomial_vec.

Remarks The __mul__ method is more challenging to vectorize so you may leave this unaltered. Check that the vectorized versions of __add__, __call__, and differentiate work as intended by calling the test_Polynomial function.

Exercise 7.28: Use a dict to hold polynomial coefficients
Use a dictionary (instead of a list) for the coeff attribute in class Polynomial from Sect. 7.3.7 such that self.coeff[k] holds the coefficient of the x^k term. The advantage with a dictionary is that only the nonzero coefficients in a polynomial need to be stored.

a) Implement a constructor and the __call__ method for evaluating the polynomial. The following demonstration code should work:

```
from Polynomial_dict import Polynomial
p1_dict = {4: 1, 2: -2, 0: 3}  # polynomial x^4 - 2*x^2 + 3
p1 = Polynomial(p1_dict)
print p1(2)  # prints 11 (16-8+3)
```

b) Implement the `__add__` method. The following demonstration code should work:

```
p1 = Polynomial({4: 1, 2: -2, 0: 3})  # x^4 - 2*x^2 + 3
p2 = Polynomial({0: 4, 1: 3}          # 4 + 3*x
p3 = p1 + p2                          # x^4 - 2*x^2 + 3*x + 7
print p3.coeff  # prints {0: 7, 1: 3, 2: -2, 4: 1}
```

Hint The structure of `__add__` may be

```
class Polynomial(object):
    ...
    def __add__(self, other):
        """Return self + other as a Polynomial object."""
        result = self.coeff.copy()
        for exponent in result:
            if exponent in other.coeff:
                # add other's term to result's term
            else:
                result[exponent] = other[exponent]
        # return Polynomial object based on result dict
```

c) Implement the `__sub__` method. The following demonstration code should work:

```
p1 = Polynomial({4: 1, 2: -2, 0: 3})  # x^4 - 2*x^2 + 3
p2 = Polynomial({0: 4, 1: 3}          # 4 + 3*x
p3 = p1 - p2                          # x^4 - 2*x^2 - 3*x - 1
print p3.coeff  # prints {0: -1, 1: -3, 2: -2, 4: 1}
```

d) Implement the `__mul__` method. The following demonstration code should work:

```
p1 = Polynomial({0: 1, 3: 1})  # 1 + x^3
p2 = Polynomial({1: -2, 2: 3}) # -2*x + 3*x^2
p3 = p1*p3
print p3.coeff  # prints {1: -2, 2: 3, 4: -2, 5: 3}
```

Hint Study the `__mul__` method in class `Polynomial` based on a list representation of the data in the polynomial and adapt to a dictionary representation.

e) Write a test function for each of the methods `__call__`, `__add__`, and `__mul__`.

Filename: `Polynomial_dict`.

Exercise 7.29: Extend class Vec2D to work with lists/tuples
The Vec2D class from Sect. 7.4 supports addition and subtraction, but only addition and subtraction of two Vec2D objects. Sometimes we would like to add or subtract a point that is represented by a list or a tuple:

```
u = Vec2D(-2, 4)
v = u + (1,1.5)
w = [-3, 2] - v
```

That is, a list or a tuple must be allowed in the right or left operand. Implement such an extension of class Vec2D.

Hint Ideas are found in Sects. 7.5.3 and 7.5.5.
Filename: Vec2D_lists.

Exercise 7.30: Extend class Vec2D to 3D vectors
Extend the implementation of class Vec2D from Sect. 7.4 to a class Vec3D for vectors in three-dimensional space. Add a method cross for computing the cross product of two 3D vectors.
Filename: Vec3D.

Exercise 7.31: Use NumPy arrays in class Vec2D
The internal code in class Vec2D from Sect. 7.4 can be valid for vectors in any space dimension if we represent the vector as a NumPy array in the class instead of separate variables x and y for the vector components. Make a new class Vec where you apply NumPy functionality in the methods. The constructor should be able to treat all the following ways of initializing a vector:

```
a = array([1, -1, 4], float)   # numpy array
v = Vec(a)
v = Vec([1, -1, 4])            # list
v = Vec((1, -1, 4))            # tuple
v = Vec(1, -1)                 # coordinates
```

Hint In the constructor, use variable number of arguments as described in Sect. H.7. All arguments are then available as a tuple, and if there is only one element in the tuple, it should be an array, list, or tuple you can send through asarray to get a NumPy array. If there are many arguments, these are coordinates, and the tuple of arguments can be transformed by array to a NumPy array. Assume in all operations that the involved vectors have equal dimension (typically that other has the same dimension as self). Recall to return Vec objects from all arithmetic operations, not NumPy arrays, because the next operation with the vector will then not take place in Vec but in NumPy. If self.v is the attribute holding the vector as a NumPy array, the addition operator will typically be implemented as

```
class Vec(object):
    ...
    def __add__(self, other):
        return Vec(selv.v + other.v)
```

Filename: Vec.

Exercise 7.32: Impreciseness of interval arithmetics
Consider the function $f(x) = x/(1 + x)$ on $[1, 2]$. Find the variation of f over
$[1, 2]$. Use interval arithmetics from Sect. 7.7.2 to compute the variation of f when
$x \in [1, 2]$.
Filename: `interval_arithmetics`.

Remarks In this case, interval arithmetics overestimates the variation in f. The
reason is that x occurs more than once in the formula for f (the so-called dependency problem[5]).

Exercise 7.33: Make classes for students and courses
Use classes to reimplement the summarizing problem in Sect. 6.7.2. More precisely,
introduce a class `Student` and a class `Course`. Find appropriate attributes. The
classes should have a `__str__` method for pretty-printing of the contents.
Filename: `Student_Course`.

Exercise 7.34: Find local and global extrema of a function
Extreme points of a function $f(x)$ are normally found by solving $f'(x) = 0$.
A much simpler method is to evaluate $f(x)$ for a set of discrete points in the interval $[a, b]$ and look for local minima and maxima among these points. We work
with $n + 1$ equally spaced points $a = x_0 < x_1 < \cdots < x_n = b$, $x_i = a + ih$,
$h = (b - a)/n$.

First we find all local extreme points in the interior of the domain. Local minima
are recognized by

$$f(x_{i-1}) > f(x_i) < f(x_{i+1}), \quad i = 1, \ldots, n - 1.$$

Similarly, at a local maximum point x_i we have

$$f(x_{i-1}) < f(x_i) > f(x_{i+1}), \quad i = 1, \ldots, n - 1.$$

Let P_{min} be the set of x values for local minima and F_{min} the set of the corresponding
$f(x)$ values at these minima. Two sets P_{max} and F_{max} are defined correspondingly
for the maxima.

The boundary points $x = a$ and $x = b$ are for algorithmic simplicity also defined
as local extreme points: $x = a$ is a local minimum if $f(a) < f(x_1)$, and a local
maximum otherwise. Similarly, $x = b$ is a local minimum if $f(b) < f(x_{n-1})$, and
a local maximum otherwise. The end points a and b and the corresponding function
values must be added to the sets $P_{min}, P_{max}, F_{min}, F_{max}$.

The global maximum point is defined as the x value corresponding to the maximum value in F_{max}. The global minimum point is the x value corresponding to the
minimum value in F_{min}.

a) Make a class `MinMax` with the following functionality:
   • `__init__` takes $f(x)$, a, b, and n as arguments, and calls a method
     `_find_extrema` to compute the local and global extreme points.

[5] http://en.wikipedia.org/wiki/Interval_arithmetic#Dependency_problem

- `_find_extrema` implements the algorithm above for finding local and global extreme points, and stores the sets P_{min}, P_{max}, F_{min}, F_{max} as list attributes in the (`self`) instance.
- `get_global_minimum` returns the global minimum point as a pair $(x, f(x))$.
- `get_global_maximum` returns the global maximum point as a pair $(x, f(x))$.
- `get_all_minima` returns a list or array of all $(x, f(x))$ minima.
- `get_all_maxima` returns a list or array of all $(x, f(x))$ maxima.
- `__str__` returns a string where a nicely formatted table of all the min/max points are listed, plus the global extreme points.

Here is a sample code using class `MinMax`:

```
def f(x):
    return x**2*exp(-0.2*x)*sin(2*pi*x)

m = MinMax(f, 0, 4, 5001)
print m
```

The output becomes

```
All minima: 0.8056, 1.7736, 2.7632, 3.7584, 0
All maxima: 0.3616, 1.284, 2.2672, 3.2608, 4
Global minimum: 3.7584
Global maximum: 3.2608
```

Make sure that the program also works for functions without local extrema, e.g., linear functions $f(x) = ax + b$.

b) The algorithm sketched above finds local extreme points x_i, but all we know is that the true extreme point is in the interval (x_{i-1}, x_{i+1}). A more accurate algorithm may take this interval as a starting point and run a Bisection method (see Sect. 4.11.2) to find the extreme point $\bar{x}$ such that $f'(\bar{x}) = 0$. Add a method `_refine_extrema` in class `MinMax`, which goes through all the interior local minima and maxima and solves $f'(\bar{x}) = 0$. Compute $f'(x)$ using the `Derivative` class (Sect. 7.3.2 with $h \ll x_{i+1} - x_{i-1}$.

Filename: `minmaxf`.

Exercise 7.35: Find the optimal production for a company
The company PROD produces two different products, P_1 and P_2, based on three different raw materials, M_1, M_2 and M_3. The following table shows how much of each raw material M_i that is required to produce *a single unit* of each product P_j:

	P_1	P_2
M_1	2	1
M_2	5	3
M_3	0	4

For instance, to produce one unit of P_2 one needs 1 unit of M_1, 3 units of M_2 and 4 units of M_3. Furthermore, PROD has available 100, 80 and 150 units of material

M_1, M_2 and M_3 respectively (for the time period considered). The revenue per produced unit of product P_1 is 150 NOK, and for one unit of P_2 it is 175 NOK. On the other hand the raw materials M_1, M_2 and M_3 cost 10, 17 and 25 NOK per unit, respectively. The question is: how much should PROD produce of each product? We here assume that PROD wants to maximize its net revenue (which is revenue minus costs).

a) Let x and y be the number of units produced of product P_1 and P_2, respectively. Explain why the total revenue $f(x, y)$ is given by

$$f(x, y) = 150x - (10 \cdot 2 + 17 \cdot 5)x + 175y - (10 \cdot 1 + 17 \cdot 3 + 25 \cdot 4)y$$

and simplify this expression. The function $f(x, y)$ is *linear* in x and y (make sure you know what linearity means).

b) Explain why PROD's problem may be stated mathematically as follows:

$$\begin{array}{lrcrcl}
\text{maximize} & & & f(x, y) & & \\
\text{subject to} & & & & & \\
& 2x & + & y & \leq & 100 \\
& 5x & + & 3y & \leq & 80 \\
& & & 4y & \leq & 150 \\
& x \geq 0, & y \geq 0. & & &
\end{array}$$

(7.9)

This is an example of a *linear optimization problem.*

c) The production (x, y) may be considered as a point in the plane. Illustrate geometrically the set T of all such points that satisfy the constraints in model (7.9). Every point in this set is called a *feasible point.*

Hint For every inequality determine first the straight line obtained by replacing the inequality by equality. Then, find the points satisfying the inequality (a half-plane), and finally, intersect these half-planes.

d) Make a program for drawing the straight lines defined by the inequalities. Each line can be written as $ax + by = c$. Let the program read each line from the command line as a list of the a, b, and c values. In the present case the command-line arguments will be

```
'[2,1,100]' '[5,3,80]' '[0,4,150]' '[1,0,0]' '[0,1,0]'
```

Hint Perform an `eval` on the elements of `sys.argv[1:]` to get a, b, and c for each line as a list in the program.

e) Let α be a positive number and consider the *level set* of the function f, defined as the set

$$L_\alpha = \{(x, y) \in T : f(x, y) = \alpha\}.$$

This set consists of all feasible points having the same net revenue α. Extend the program with two new command-line arguments holding p and q for a function

$f(x, y) = px + qy$. Use this information to compute the level set lines $y = \alpha/q - px/q$, and plot the level set lines for some different values of α (use the α value in the legend for each line).

f) Use what you saw in e) to solve the problem (7.9) geometrically. This solution is called an *optimal solution*.

Hint How large can you choose α such that L_α is nonempty?

g) Assume that we have other values on the revenues and costs than the actual numbers in a). Explain why (7.9), with these new parameter values, still has an optimal solution lying in a corner point of T. Extend the program to calculate all the corner points of a region T in the plane determined by the linear inequalities like those listed above. Moreover, the program shall compute the maximum of a given linear function $f(x, y) = ax + by$ over T by calculating the function values in the corner points and finding the smallest function value.

Filename: `optimization`.

Random Numbers and Simple Games

8

Random numbers have many applications in science and computer programming, especially when there are significant uncertainties in a phenomenon of interest. The purpose of this chapter is to look at some practical problems involving random numbers and learn how to program with such numbers. We shall make several games and also look into how random numbers can be used in physics. You need to be familiar with basic programming concepts such as loops, lists, arrays, vectorization, curve plotting, and command-line arguments in order to study the present chapter. This means that Chaps. 1–5 of the present book should be digested. A few examples and exercises will require familiarity with the class concept from Chap. 7.

The key idea in computer simulations with random numbers is first to formulate an algorithmic description of the phenomenon we want to study. This description frequently maps directly onto a quite simple and short Python program, where we use random numbers to mimic the uncertain features of the phenomenon. The program needs to perform a large number of repeated calculations, and the final answers are "only" approximate, but the accuracy can usually be made good enough for practical purposes. Most programs related to the present chapter produce their results within a few seconds. In cases where the execution times become large, we can vectorize the code. Vectorized computations with random numbers is definitely the most demanding topic in this chapter, but is not mandatory for seeing the power of mathematical modeling via random numbers.

All files associated with the examples in this chapter are found in the folder `src/random`[1].

8.1 Drawing Random Numbers

Python has a module `random` for generating random numbers. The function call `random.random()` generates a random number in the half open interval $[0, 1)$ (recall that in the half open interval $[0, 1)$ the lower limit is included, but the upper limit is not). We can try it out:

---

[1] http://tinyurl.com/pwyasaa/random

© Springer-Verlag Berlin Heidelberg 2016

H.P. Langtangen, *A Primer on Scientific Programming with Python*,

Texts in Computational Science and Engineering 6, DOI 10.1007/978-3-662-49887-3_8

```
>>> import random
>>> random.random()
0.81550546885338104
>>> random.random()
0.44913326809029852
>>> random.random()
0.88320653116367454
```

All computations of random numbers are based on deterministic algorithms (see
Exercise 8.20 for an example), so the sequence of numbers cannot be truly random.
However, the sequence of numbers appears to lack any pattern, and we can therefore
view the numbers as random.

8.1.1   The Seed

Every time we import random, the subsequent sequence of random.random() calls
will yield different numbers. For debugging purposes it is useful to get the same
sequence of random numbers every time we run the program. This functionality is
obtained by setting a *seed* before we start generating numbers. With a given value
of the seed, one and only one sequence of numbers is generated. The seed is an
integer and set by the random.seed function:

```
>>> random.seed(121)
```

Let us generate two series of random numbers at once, using a list comprehension
and a format with two decimals only:

```
>>> random.seed(2)
>>> ['%.2f' % random.random() for i in range(7)]
['0.96', '0.95', '0.06', '0.08', '0.84', '0.74', '0.67']
>>> ['%.2f' % random.random() for i in range(7)]
['0.31', '0.61', '0.61', '0.58', '0.16', '0.43', '0.39']
```

If we set the seed to 2 again, the sequence of numbers is regenerated:

```
>>> random.seed(2)
>>> ['%.2f' % random.random() for i in range(7)]
['0.96', '0.95', '0.06', '0.08', '0.84', '0.74', '0.67']
```

If we do not give a seed, the random module sets a seed based on the current time.
That is, the seed will be different each time we run the program and consequently
the sequence of random numbers will also be different from run to run. This is
what we want in most applications. However, we always recommend setting a seed
during program development to simplify debugging and verification.

8.1.2 Uniformly Distributed Random Numbers

The numbers generated by random.random() tend to be equally distributed between 0 and 1, which means that there is no part of the interval $[0, 1)$ with more random numbers than other parts. We say that the distribution of random numbers in this case is *uniform*. The function random.uniform(a,b) generates uniform random numbers in the half open interval $[a, b)$, where the user can specify a and b. With the following program (in file uniform_numbers0.py) we may generate lots of random numbers in the interval $[-1, 1)$ and visualize how they are distributed:

```
import random
random.seed(42)
N = 500  # no of samples
x = range(N)
y = [random.uniform(-1,1) for i in x]

import scitools.std as st
st.plot(x, y, '+', axis=[0,N-1,-1.2,1.2])
```

Figure 8.1 shows the values of these 500 numbers, and as seen, the numbers appear to be random and uniformly distributed between -1 and 1.

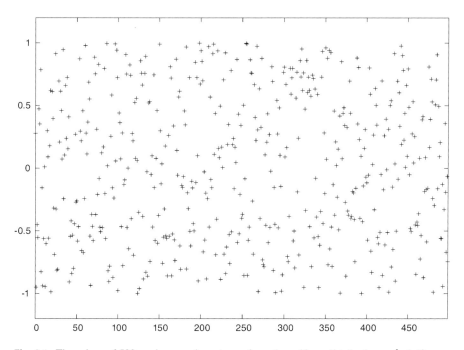

Fig. 8.1 The values of 500 random numbers drawn from the uniform distribution on $[-1, 1)$

8.1.3   Visualizing the Distribution

It is of interest to see how N random numbers in an interval $[a, b]$ are distributed throughout the interval, especially as $N \rightarrow \infty$. For example, when drawing numbers from the uniform distribution, we expect that no parts of the interval get more numbers than others. To visualize the distribution, we can divide the interval into subintervals and display how many numbers there are in each subinterval.

Let us formulate this method more precisely. We divide the interval $[a, b)$ into n equally sized subintervals, each of length $h = (b - a)/n$. These subintervals are called *bins*. We can then draw N random numbers by calling `random.random()` N times. Let $\hat{H}(i)$ be the number of random numbers that fall in bin no. i, $[a + ih, a + (i + 1)h]$, $i = 0, \ldots, n - 1$. If N is small, the value of $\hat{H}(i)$ can be quite different for the different bins, but as N grows, we expect that $\hat{H}(i)$ varies little with i.

Ideally, we would be interested in how the random numbers are distributed as $N \rightarrow \infty$ and $n \rightarrow \infty$. One major disadvantage is that $\hat{H}(i)$ increases as N increases, and it decreases with n. The quantity $\hat{H}(i)/N$, called the frequency count, will reach a finite limit as $N \rightarrow \infty$. However, $\hat{H}(i)/N$ will be smaller and smaller as we increase the number of bins. The quantity $H(i) = \hat{H}(i)/(Nh)$ reaches a finite limit as $N, n \rightarrow \infty$. The probability that a random number lies inside subinterval no. i is then $\hat{H}(i)/N = H(i)h$.

We can visualize $H(i)$ as a bar diagram (see Fig. 8.2), called a *normalized histogram*. We can also define a piecewise constant function $p(x)$ from $H(i)$: $p(x) = H(i)$ for $x \in [a + ih, a + (i + 1)h)$, $i = 0, \ldots, n - 1$. As $n, N \rightarrow \infty$, $p(x)$ approaches the probability density function of the distribution in question. For example, `random.uniform(a,b)` draws numbers from the uniform distribution on $[a, b)$, and the probability density function is constant, equal to $1/(b - a)$. As we increase n and N, we therefore expect $p(x)$ to approach the constant $1/(b - a)$.

The function `compute_histogram` from `scitools.std` returns two arrays `x` and `y` such that `plot(x,y)` plots the piecewise constant function $p(x)$. The plot is hence the histogram of the set of random samples. The program below exemplifies the usage:

```
from scitools.std import plot, compute_histogram
import random
samples = [random.random() for i in range(100000)]
x, y = compute_histogram(samples, nbins=20)
plot(x, y)
```

Figure 8.2 shows two plots corresponding to N taken as 10^3 and 10^6. For small N, we see that some intervals get more random numbers than others, but as N grows, the distribution of the random numbers becomes more and more equal among the intervals. In the limit $N \rightarrow \infty$, $p(x) \rightarrow 1$, which is illustrated by the plot.

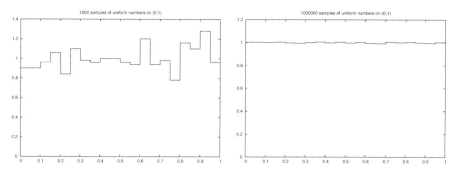

Fig. 8.2 The histogram of uniformly distributed random numbers in 20 bins

8.1.4 Vectorized Drawing of Random Numbers

There is a `random` module in the Numerical Python package, which can be used to efficiently draw a possibly large array of random numbers:

```
import numpy as np
r = np.random.random()              # one number between 0 and 1
r = np.random.random(size=10000)    # array with 10000 numbers
r = np.random.uniform(-1, 10)       # one number between -1 and 10
r = np.random.uniform(-1, 10, size=10000)  # array
```

There are thus two `random` modules to be aware of: one in the standard Python library and one in `numpy`. For drawing uniformly distributed numbers, the two `random` modules have the same interface, except that the functions from `numpy`'s `random` module has an extra `size` parameter. Both modules also have a `seed` function for fixing the seed.

Vectorized drawing of random numbers using `numpy`'s `random` module is efficient because all the numbers are drawn "at once" in fast C code. You can measure the efficiency gain with the `time.clock()` function as explained in Sect. 8.5.3 and H.8.1.

Warning It is easy to do an `import random` followed by a `from numpy import *` or maybe `from scitools.std import *` without realizing that the latter two import statements import a name `random` from `numpy` that overwrites the same name that was imported in `import random`. The result is that the effective random module becomes the one from `numpy`. A possible solution to this problem is to introduce a different name for Python's `random` module, say

```
import random as random_number
```

Another solution is to do `import numpy as np` and work explicitly with `np.random`.

8.1.5   Computing the Mean and Standard Deviation

You probably know the formula for the mean or average of a set of n numbers $x_0, x_1, \ldots, x_{n-1}$:

$$x_m = \frac{1}{n} \sum_{j=0}^{n-1} x_j .$$ (8.1)

The amount of spreading of the x_i values around the mean x_m can be measured by the *variance*,

$$x_v = \frac{1}{n} \sum_{j=0}^{n-1} (x_j - x_m)^2 .$$ (8.2)

Textbooks in statistics teach you that it is more appropriate to divide by $n-1$ instead of n, but we are not going to worry about that fact in this book. A variant of (8.2) reads

$$x_v = \frac{1}{n} \left(\sum_{j=0}^{n-1} x_j^2 \right) - x_m^2 .$$ (8.3)

The good thing with this latter formula is that one can, as a statistical experiment progresses and n increases, record the sums

$$s_m = \sum_{j=0}^{q-1} x_j, \quad s_v = \sum_{j=0}^{q-1} x_j^2$$ (8.4)

and then, when desired, efficiently compute the most recent estimate on the mean value and the variance after q samples by

$$x_m = s_m/q, \quad x_v = s_v/q - s_m^2/q^2 .$$ (8.5)

The *standard deviation*

$$x_s = \sqrt{x_v}$$ (8.6)

is often used as an alternative to the variance, because the standard deviation has the same unit as the measurement itself. A common way to express an uncertain quantity x, based on a data set $x_0, \ldots, x_{n-1}$, from simulations or physical measurements, is $x_m \pm x_s$. This means that x has an uncertainty of one standard deviation x_s to either side of the mean value x_m. With probability theory and statistics one can provide many other, more precise measures of the uncertainty, but that is the topic of a different course.

Below is an example where we draw numbers from the uniform distribution on $[-1, 1)$ and compute the evolution of the mean and standard deviation 10 times during the experiment, using the formulas (8.1) and (8.3)–(8.6):

```
import sys
N = int(sys.argv[1])
import random
from math import sqrt
sm = 0; sv = 0
for q in range(1, N+1):
    x = random.uniform(-1, 1)
    sm += x
    sv += x**2

    # Write out mean and st.dev. 10 times in this loop
    if q % (N/10) == 0:
        xm = sm/q
        xs = sqrt(sv/q - xm**2)
        print '%10d mean: %12.5e  stdev: %12.5e' % (q, xm, xs)
```

The `if` test applies the *mod* function, see Sect. 3.4.2, for checking if a number can be divided by another without any remainder. The particular `if` test here is `True` when i equals 0, N/10, 2*N/10, ..., N, i.e., 10 times during the execution of the loop. The program is available in the file `mean_stdev_uniform1.py`. A run with $N = 10^6$ gives the output

```
 100000 mean:  1.86276e-03  stdev:  5.77101e-01
 200000 mean:  8.60276e-04  stdev:  5.77779e-01
 300000 mean:  7.71621e-04  stdev:  5.77753e-01
 400000 mean:  6.38626e-04  stdev:  5.77944e-01
 500000 mean: -1.19830e-04  stdev:  5.77752e-01
 600000 mean:  4.36091e-05  stdev:  5.77809e-01
 700000 mean: -1.45486e-04  stdev:  5.77623e-01
 800000 mean:  5.18499e-05  stdev:  5.77633e-01
 900000 mean:  3.85897e-05  stdev:  5.77574e-01
1000000 mean: -1.44821e-05  stdev:  5.77616e-01
```

We see that the mean is getting smaller and approaching zero as expected since we generate numbers between -1 and 1. The theoretical value of the standard deviation, as $N \to \infty$, equals $\sqrt{1/3} \approx 0.57735$.

We have also made a corresponding vectorized version of the code above using numpy's `random` module and the ready-made functions `mean`, `var`, and `std` for computing the mean, variance, and standard deviation (respectively) of an array of numbers:

```
import sys
N = int(sys.argv[1])
import numpy as np
x = np.random.uniform(-1, 1, size=N)
xm = np.mean(x)
xv = np.var(x)
xs = np.std(x)
print '%10d mean: %12.5e  stdev: %12.5e' % (N, xm, xs)
```

This program can be found in the file `mean_stdev_uniform2.py`.

8.1.6 The Gaussian or Normal Distribution

In some applications we want random numbers to cluster around a specific value m. This means that it is more probable to generate a number close to m than far away from m. A widely used distribution with this qualitative property is the Gaussian or normal distribution. For example, the statistical distribution of the height or the blood pressure among adults of one gender are well described by a normal distribution. The normal distribution has two parameters: the mean value m and the standard deviation s. The latter measures the width of the distribution, in the sense that a small s makes it less likely to draw a number far from the mean value, and a large s makes more likely to draw a number far from the mean value.

Single random numbers from the normal distribution can be generated by

```
import random
r = random.normalvariate(m, s)
```

while efficient generation of an array of length N is enabled by

```
import numpy as np
r = np.random.normal(m, s, size=N)
r = np.random.randn(N)   # mean=0, std.dev.=1
```

The following program draws N random numbers from the normal distribution, computes the mean and standard deviation, and plots the histogram:

```
import sys
N = int(sys.argv[1])
m = float(sys.argv[2])
s = float(sys.argv[3])

import numpy as np
np.random.seed(12)
samples = np.random.normal(m, s, N)
print np.mean(samples), np.std(samples)

import scitools.std as st
x, y = st.compute_histogram(samples, 20, piecewise_constant=True)
st.plot(x, y, savefig='tmp.pdf',
        title ='%d samples of Gaussian/normal numbers on (0,1)' % N)
```

The corresponding program file is `normal_numbers1.py`, which gives a mean of -0.00253 and a standard deviation of 0.99970 when run with N as 1 million, m as 0, and s equal to 1. Figure 8.3 shows that the random numbers cluster around the mean $m = 0$ in a histogram. This normalized histogram will, as N goes to infinity, approach the famous, bell-shaped, normal distribution probability density function[2].

---

[2] http://en.wikipedia.org/wiki/Normal_distribution

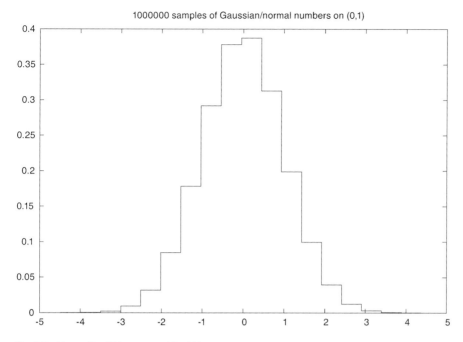

Fig. 8.3 Normalized histogram of 1 million random numbers drawn from the normal distribution

8.2   Drawing Integers

Suppose we want to draw a random integer among the values 1, 2, 3, and 4, and that each of the four values is equally probable. One possibility is to draw real numbers from the uniform distribution on, e.g., $[0, 1)$ and divide this interval into four equal subintervals:

```
import random
r = random.random()
if 0 <=  r < 0.25:
    r = 1
elif 0.25 <= r < 0.5:
    r = 2
elif 0.5 <= r < 0.75:
    r = 3
else:
    r = 4
```

Nevertheless, the need for drawing uniformly distributed integers occurs quite frequently, so there are special functions for returning random integers in a specified interval $[a, b]$.

8.2.1   Random Integer Functions

Python's `random` module has a built-in function `randint(a,b)` for drawing an integer in $[a, b]$, i.e., the return value is among the numbers a, a+1, ..., b-1, b.

```
import random
r = random.randint(a, b)
```

The `numpy.random.randint(a, b, N)` function has a similar functionality for vectorized drawing of an array of length N of random integers in $[a, b)$. The upper limit b is not among the drawn numbers, so if we want to draw from a, a+1, ..., b-1, b, we must write

```
import numpy as np
r = np.random.randint(a, b+1, N)
```

Another function, `random_integers(a, b, N)`, also in `numpy.random`, includes the upper limit b in the possible set of random integers:

```
r = np.random.random_integers(a, b, N)
```

8.2.2   Example: Throwing a Die

Scalar version We can make a function that lets the computer throw a die N times and returns the fraction of the throws the die shows six eyes:

```
def six_eyes(N):
    M = 0                          # no of times we get 6 eyes
    for i in xrange(N):
        outcome = random.randint(1, 6)
        if outcome == 6:
            M += 1
    return float(M)/N
```

We use `xrange` instead of `range` because the former avoids storing N numbers in memory, which can be an important feature when N is large.

Vectorized version Too speed up the experiments, we can vectorize the drawing of the random numbers and the counting of the number successful experiments:

```
import numpy as np

def six_eyes_vec(N):
    eyes = np.random.randint(1, 7, N)
    success = eyes == 6        # True/False array
    M = np.sum(success)        # treats True as 1, False as 0
    return float(M)/N
```

The eyes == 6 construction results in an array with True or False values, and np.sum applied to this array treats True as 1 and False as 0 (the integer equivalents to the boolean values), so the sum is the number of elements in eyes that equals 6. A very important point here for computational efficiency is to use np.sum and not the standard sum function that is available in standard Python. With np.sum function, the vectorized version runs about 60 times faster than the scalar version. With Python's standard sum function, the vectorized versions is in fact twice as slow as the scalar version (!). We can illustrate the gain and loss in efficiency by the follow IPython session:

```
In [1]: from roll_die import six_eyes, six_eyes_vec

In [2]: %timeit six_eyes(100000)
1 loops, best of 3: 250 ms per loop

In [3]: %timeit six_eyes_vec(100000)
100 loops, best of 3: 4.11 ms per loop

In [4]: 250/4.11               # performance fraction
Out[4]: 60.8272506082725

In [5]: from roll_die import np

In [6]: np.sum = sum   # fool numpy to use built-in Python sum

In [7]: %timeit six_eyes_vec(100000)
1 loops, best of 3: 543 ms per loop
```

(Note how we can bind Python's built-in sum function to the np.sum name such that np.sum in six_eyes_vec applies Python's sum function instead of the original np.sum.)

Vectorized version with batches The disadvantage with the vectorized version is that all the random numbers must be stored in the computer's memory. A large N may cause the program to run out of memory and raise a MemoryError. Instead of drawing all random numbers at once, we can draw them in batches of size arraysize. There will be N//arraysize such batches, plus a rest. Note the double slash in N//arraysize: here we indeed want integer division, which is explicitly instructed in Python by the double forward slash. The rest is obtained by the mod operator: rest = N % arraysize. The size of the batches can be stored in a list:

```
rest = N % arraysize
batch_sizes = [arraysize]*(N//arraysize) + [rest]
```

We can now make one batch of random numbers at a time and count how many times we get six:

```
def six_eyes_vec2(N, arraysize=1000000):
    # Split all experiments into batches of size arraysize,
    # plus a final batch of size rest
    # (note: N//arraysize is integer division)
    rest = N % arraysize
    batch_sizes = [arraysize]*(N//arraysize) + [rest]

    M = 0
    for batch_size in batch_sizes:
        eyes = np.random.randint(1, 7, batch_size)
        success = eyes == 6        # True/False array
        M += np.sum(success)       # treats True as 1, False as 0
    return float(M)/N
```

Because we fix the seed, the computed f will always be the same in this function.

Verification of the scalar version Verifying computations with random numbers requires the seed to be fixed. When we believe the scalar version in function six_eyes is correct, mainly by observing that the return value approaches 1/6 as the number of experiments, N, grows, we can call the function with a small N and record the return value for use in a test function:

```
def test_scalar():
    random.seed(3)
    f = six_eyes(100)
    f_exact = 0.26
    assert abs(f_exact - f) < 1E-15
```

Verification of all versions Since we have three alternative functions for computing the same quantity, a verification can be based on comparing the output of all three functions. This is somewhat problematic since the scalar and vectorized versions apply different random number generators. Fixing the seed of Python's random module and numpy.random does not help as these two tools will generate different sequences of random integers. Nevertheless, we can fool the scalar version in the six_eyes function to use np.random.randint instead of random.randit: this is just a matter of setting random = np.random (with a declaration global random) before calling six_eyes. The problem is that the call np.random.randint(1, 6) in six_eyes will then generate the numbers up to *but not including* 6, so M will always be zero. A little trick, can solve the problem: we redefine random.randit to be a function that calls np.random.randint:

```
random.randint = lambda a, b: np.random.randint(a, b+1, 1)[0]
```

The call random.randint(1, 6) in six_eyes now becomes np.random.randint(1, 7, 1)[0], i.e., we generate an array of 1 random integer and extract the first element so the result is a scalar number as before.

A test function can call all three functions, with the same fixed seed, and compare the returned values:

```
def test_all():
    # Use np.random as random number generator for all three
    # functions and make sure all of them applies the same seed
    N = 100
    arraysize = 40
    random.randint = lambda a, b: np.random.randint(a, b+1, 1)[0]
    tol = 1E-15

    np.random.seed(3)
    f_scalar = six_eyes(N)
    np.random.seed(3)
    f_vec = six_eyes_vec(N)
    assert abs(f_scalar - f_vec) < tol

    np.random.seed(3)
    f_vec2 = six_eyes_vec2(N, arraysize=80)
    assert abs(f_vec - f_vec2) < tol
```

All the functions above are found in the file `roll_die.py`[3].

8.2.3 Drawing a Random Element from a List

Given a list a, the statement

```
re = random.choice(a)
```

picks out an element of a at random, and re refers to this element. The shown call
to `random.choice` is the same as

```
re = a[random.randint(0, len(a)-1)]
```

There is also a function `shuffle` that performs a random permutation of the list
elements:

```
random.shuffle(a)
```

Picking now a[0], for instance, has the same effect as `random.choice` on the
original, unshuffled list. Note that `shuffle` changes the list given as argument.

The `numpy.random` module has also a `shuffle` function with the same functionality.

A small session illustrates the various methods for picking a random element
from a list:

```
>>> awards = ['car', 'computer', 'ball', 'pen']
>>> import random
>>> random.choice(awards)
'car'
```

---

[3] http://tinyurl.com/pwyasaa/random/roll_die

```
>>> awards[random.randint(0, len(awards)-1)]
'pen'
>>> random.shuffle(awards)
>>> awards[0]
'computer'
```

8.2.4 Example: Drawing Cards from a Deck

The following function creates a deck of cards, where each card is represented as a string, and the deck is a list of such strings:

```
def make_deck():
    ranks = ['A', '2', '3', '4', '5', '6', '7',
             '8', '9', '10', 'J', 'Q', 'K']
    suits = ['C', 'D', 'H', 'S']
    deck = []
    for s in suits:
        for r in ranks:
            deck.append(s + r)
    random.shuffle(deck)
    return deck
```

Here, 'A' means an ace, 'J' represents a jack, 'Q' represents a queen, 'K' represents a king, 'C' stands for clubs, 'D' stands for diamonds, 'H' means hearts, and 'S' means spades. The computation of the list deck can alternatively (and more compactly) be done by a one-line list comprehension:

```
deck = [s+r for s in suits for r in ranks]
```

We can draw a card at random by

```
deck = make_deck()
card = deck[0]
del deck[0]
# or better:
card = deck.pop(0)  # return and remove element with index 0
```

Drawing a hand of n cards from a shuffled deck is accomplished by

```
def deal_hand(n, deck):
    hand = [deck[i] for i in range(n)]
    del deck[:n]
    return hand, deck
```

Note that we must return deck to the calling code since this list is changed. Also note that the n first cards of the deck are random cards if the deck is shuffled (and any deck made by make_deck is shuffled).

The following function deals cards to a set of players:

```
def deal(cards_per_hand, no_of_players):
    deck = make_deck()
    hands = []
    for i in range(no_of_players):
        hand, deck = deal_hand(cards_per_hand, deck)
        hands.append(hand)
    return hands
```

```
players = deal(5, 4)
import pprint; pprint.pprint(players)
```

The `players` list may look like

```
[['D4', 'CQ', 'H10', 'DK', 'CK'],
 ['D7', 'D6', 'SJ', 'S4', 'C5'],
 ['C3', 'DQ', 'S3', 'C9', 'DJ'],
 ['H6', 'H9', 'C6', 'D5', 'S6']]
```

The next step is to analyze a hand. Of particular interest is the number of pairs, three of a kind, four of a kind, etc. That is, how many combinations there are of `n_of_a_kind` cards of the same rank (e.g., `n_of_a_kind=2` finds the number of pairs):

```
def same_rank(hand, n_of_a_kind):
    ranks = [card[1:] for card in hand]
    counter = 0
    already_counted = []
    for rank in ranks:
        if rank not in already_counted and \
               ranks.count(rank) == n_of_a_kind:
            counter += 1
            already_counted.append(rank)
    return counter
```

Note how convenient the `count` method in list objects is for counting how many copies there are of one element in the list.

Another analysis of the hand is to count how many cards there are of each suit. A dictionary with the suit as key and the number of cards with that suit as value, seems appropriate to return. We pay attention only to suits that occur more than once:

```
def same_suit(hand):
    suits = [card[0] for card in hand]
    counter = {}   # counter[suit] = how many cards of suit
    for suit in suits:
        count = suits.count(suit)
        if count > 1:
            counter[suit] = count
    return counter
```

For a set of players we can now analyze their hands:

```
for hand in players:
    print """\
The hand %s
    has %d pairs, %s 3-of-a-kind and %s cards of the same suit."""%\
    (', '.join(hand), same_rank(hand, 2),
     same_rank(hand, 3),
     '+'.join([str(s) for s in same_suit(hand).values()]))
```

The values we feed into the printf string undergo some massage: we join the card values with comma and put a plus in between the counts of cards with the same suit. (The `join` function requires a string argument. That is why the integer counters of cards with the same suit, returned from `same_suit`, must be converted to strings.) The output of the `for` loop becomes

```
The hand D4, CQ, H10, DK, CK
    has 1 pairs, 0 3-of-a-kind and 2+2 cards of the same suit.
The hand D7, D6, SJ, S4, C5
    has 0 pairs, 0 3-of-a-kind and 2+2 cards of the same suit.
The hand C3, DQ, S3, C9, DJ
    has 1 pairs, 0 3-of-a-kind and 2+2 cards of the same suit.
The hand H6, H9, C6, D5, S6
    has 0 pairs, 1 3-of-a-kind and 2 cards of the same suit.
```

The file `cards.py` contains the functions `make_deck`, `hand`, `same_rank`, `same_suit`, and the test snippets above. With the `cards.py` file one can start to implement real card games.

8.2.5 Example: Class Implementation of a Deck

To work with a deck of cards with the code from the previous section one needs to shuffle a global variable `deck` in and out of functions. A set of functions that update global variables (like `deck`) is a primary candidate for a class: the global variables are stored as data attributes and the functions become class methods. This means that the code from the previous section is better implemented as a class. We introduce class `Deck` with a list of cards, `deck`, as data attribute, and methods for dealing one or several hands and for putting back a card:

```
class Deck(object):
    def __init__(self):
        ranks = ['A', '2', '3', '4', '5', '6', '7',
                 '8', '9', '10', 'J', 'Q', 'K']
        suits = ['C', 'D', 'H', 'S']
        self.deck = [s+r for s in suits for r in ranks]
        random.shuffle(self.deck)
```

```
    def hand(self, n=1):
        """Deal n cards. Return hand as list."""
        hand = [self.deck[i] for i in range(n)]   # pick cards
        del self.deck[:n]                          # remove cards
        return hand

    def deal(self, cards_per_hand, no_of_players):
        """Deal no_of_players hands. Return list of lists."""
        return [self.hand(cards_per_hand) \
                for i in range(no_of_players)]

    def putback(self, card):
        """Put back a card under the rest."""
        self.deck.append(card)

    def __str__(self):
        return str(self.deck)
```

This class is found in the module file Deck.py. Dealing a hand of five cards to p players is coded as

```
from Deck import Deck
deck = Deck()
print deck
players = deck.deal(5, 4)
```

Here, players become a nested list as shown in Sect. 8.2.4.

One can go a step further and make more classes for assisting card games. For example, a card has so far been represented by a plain string, but we may well put that string in a class Card:

```
class Card(object):
    """Representation of a card as a string (suit+rank)."""
    def __init__(self, suit, rank):
        self.card = suit + str(rank)

    def __str__(self):   return self.card
    def __repr__(self):  return str(self)
```

Note that str(self) is equivalent to self.__str__().

A Hand contains a set of Card instances and is another natural abstraction, and hence a candidate for a class:

```
class Hand(object):
    """Representation of a hand as a list of Card objects."""
    def __init__(self, list_of_cards):
        self.hand = list_of_cards

    def __str__(self):   return str(self.hand)
    def __repr__(self):  return str(self)
```

With the aid of classes Card and Hand, class Deck can be reimplemented as

```
class Deck(object):
    """Representation of a deck as a list of Card objects."""

    def __init__(self):
        ranks = ['A', '2', '3', '4', '5', '6', '7',
                 '8', '9', '10', 'J', 'Q', 'K']
        suits = ['C', 'D', 'H', 'S']
        self.deck = [Card(s,r) for s in suits for r in ranks]
        random.shuffle(self.deck)

    def hand(self, n=1):
        """Deal n cards. Return hand as a Hand object."""
        hand = Hand([self.deck[i] for i in range(n)])
        del self.deck[:n]          # remove cards
        return hand

    def deal(self, cards_per_hand, no_of_players):
        """Deal no_of_players hands. Return list of Hand obj."""
        return [self.hand(cards_per_hand) \
                   for i in range(no_of_players)]

    def putback(self, card):
        """Put back a card under the rest."""
        self.deck.append(card)

    def __str__(self):
        return str(self.deck)

    def __repr__(self):
        return str(self)

    def __len__(self):
        return len(self.deck)
```

The module file Deck2.py contains this implementation. The usage of the two Deck classes is the same,

```
from Deck2 import Deck
deck = Deck()
players = deck.deal(5, 4)
```

with the exception that players in the last case holds a list of Hand instances, and each Hand instance holds a list of Card instances.

We stated in Sect. 7.3.9 that the __repr__ method should return a string such that one can recreate the object from this string by the aid of eval. However, we did not follow this rule in the implementation of classes Card, Hand, and Deck. Why? The reason is that we want to print a Deck instance. Python's print or pprint on a list applies repr(e) to print an element e in the list. Therefore, if we had implemented

```
class Card(object):
    ...
    def __repr__(self):
        return "Card('%s', %s)" % (self.card[0], self.card[1:])

class Hand(object):
    ...
    def __repr__(self): return 'Hand(%s)' % repr(self.hand)
```

a plain printing of the deck list of Hand instances would lead to output like

```
[Hand([Card('C', '10'), Card('C', '4'), Card('H', 'K'), ...]),
 ...,
 Hand([Card('D', '7'), Card('C', '5'), ..., Card('D', '9')])]
```

This output is harder to read than

```
[[C10, C4, HK, DQ, HQ],
 [SA, S8, H3, H10, C2],
 [HJ, C7, S2, CQ, DK],
 [D7, C5, DJ, S3, D9]]
```

That is why we let `__repr__` in classes Card and Hand return the same pretty print string as `__str__`, obtained by returning `str(self)`.

8.3 Computing Probabilities

With the mathematical rules from *probability theory* one may compute the probability that a certain event happens, say the probability that you get one black ball when drawing three balls from a hat with four black balls, six white balls, and three green balls. Unfortunately, theoretical calculations of probabilities may soon become hard or impossible if the problem is slightly changed. There is a simple numerical way of computing probabilities that is generally applicable to problems with uncertainty. The principal ideas of this approximate technique is explained below, followed by three examples of increasing complexity.

8.3.1 Principles of Monte Carlo Simulation

Assume that we perform N experiments where the outcome of each experiment is random. Suppose that some event takes place M times in these N experiments. An estimate of the probability of the event is then M/N. The estimate becomes more accurate as N is increased, and the exact probability is assumed to be reached in the limit as $N \to \infty$. (Note that in this limit, $M \to \infty$ too, so for rare events, where M may be small in a program, one must increase N such that M is sufficiently large for M/N to become a good approximation to the probability.)

Programs that run a large number of experiments and record the outcome of events are often called *simulation programs*. (Note that this term is also applied for programs that solve equations arising in mathematical models in general, but it is even more common to use the term when random numbers are used to estimate probabilities.) The mathematical technique of letting the computer perform lots of experiments based on drawing random numbers is commonly called *Monte Carlo simulation*. This technique has proven to be extremely useful throughout science and industry in problems where there is uncertain or random behavior is involved.

> *As far as the laws of mathematics refer to reality, they are not certain, as far as they are certain, they do not refer to reality.* Albert Einstein, physicist, 1879–1955.

For example, in finance the stock market has a random variation that must be taken into account when trying to optimize investments. In offshore engineering, environmental loads from wind, currents, and waves show random behavior. In nuclear and particle physics, random behavior is fundamental according to quantum mechanics and statistical physics. Many probabilistic problems can be calculated exactly by mathematics from probability theory, but very often Monte Carlo simulation is the only way to solve statistical problems. Sections 8.3.2–8.3.5 applies examples to explain the essence of Monte Carlo simulation in problems with inherent uncertainty. However, also deterministic problems, such as integration of functions, can be computed by Monte Carlo simulation (see Sect. 8.5).

It appears that Monte Carlo simulation programmed in pure Python is a computationally feasible approach, even on small laptops, in all the forthcoming examples. Significant speed-up can be achieved by vectorizing the code, which is explained in detail for many of the examples. However, large-scale Monte Carlo simulations and other heavy computations run slowly in pure Python, and the core of the computations should be moved to a compiled language such as C. In Appendix G, you can find a Monte Carlo application that is implemented in pure Python, in vectorized numpy Python, in the extended (and very closely related) Cython language, as well as in pure C code. Various ways of combining Python with C are also illustrated.

8.3.2 Example: Throwing Dice

You throw two dice, one black and one green. What is the probability that the number of eyes on the black die is larger than the number of eyes on the green die?

Straightforward solution We can simulate N throws of two dice in a program. For each throw we see if the event is successful, and if so, increase M by one:

```
import sys
N = int(sys.argv[1])                # no of experiments

import random
M = 0                               # no of successful events
```

```
for i in range(N):
    black = random.randint(1, 6)    # throw black
    green = random.randint(1, 6)    # throw brown
    if black > green:               # success?
        M += 1
p = float(M)/N
print 'probability:', p
```

This program is named `black_gt_green.py`.

Vectorization Although the `black_gt_green.py` program runs $N = 10^6$ in a few seconds, Monte Carlo simulation programs can quickly require quite some simulation time so speeding up the algorithm by vectorization is often desired. Let us vectorize the code shown above. The idea is to draw all the random numbers ($2N$) at once. We make an array of random numbers between 1 and 6 with 2 rows and N columns. The first row can be taken as the number of eyes on the black die in all the experiments, while the second row are the corresponding eyes on the green die:

```
r = np.random.random_integers(1, 6, size=(2, N))
black = r[0,:]          # eyes for all throws with black
green = r[1,:]          # eyes for all throws with green
```

The condition `black > green` results in an array of length N of boolean values: True when the element in `black` is greater than the corresponding element in `green`, and False if not. The number of True elements in the boolean array `black > green` is then M. This number can be computed by summing up all the boolean values. In arithmetic operations, True is 1 and False i 0, so the sum equals M. Fast summation of arrays requires `np.sum` and not Python's standard sum function. The code goes like

```
success = black > green    # success[i] is true if black[i]>green[i]
M = np.sum(success)        # sum up all successes
p = float(M)/N
print 'probability:', p
```

The code, found in the file `black_gt_green_vec.py`, runs over 10 times faster than the corresponding scalar code in `black_gt_green.py`.

Exact solution In this simple example we can quite easily compute the exact solution. To this end, we set up all the outcomes of the experiment, i.e., all the possible combinations of eyes on two dice:

```
combinations = [(black, green)
                for black in range(1, 7)
                for green in range(1, 7)]
```

Then we count how many of the (black, green) pairs that have the property black > green:

```
success = [black > green for black, green in combinations]
M = sum(success)
```

It turns out that M is 15, giving a probability $15/36 \approx 0.41667$ since there are 36 combinations in total. Running the Monte Carlo simulations with $N = 10^6$ typically gives probabilities in $[0.416, 0.417]$.

A game Suppose a games is constructed such that you have to pay 1 euro to throw the two dice. You win 2 euros if there are more eyes on the black than on the green die. Should you play this game? We can easily simulate the game directly (file `black_gt_green_game.py`):

```
import sys
N = int(sys.argv[1])                  # no of experiments

import random
start_capital = 10
money = start_capital
for i in range(N):
    money -= 1                        # pay for the game
    black = random.randint(1, 6)      # throw black
    green = random.randint(1, 6)      # throw brown
    if black > green:                 # success?
        money += 2                    # get award

net_profit_total = money - start_capital
net_profit_per_game = net_profit_total/float(N)
print 'Net profit per game in the long run:', net_profit_per_game
```

Experimenting with a few N shows that the net profit per game is always negative. That is, you should *not* play this game.

A vectorized version is beneficial of efficiency reasons (the corresponding file is `black_gt_green_game_vec.py`):

```
import sys
N = int(sys.argv[1])     # no of experiments

import numpy as np
r = np.random.random_integers(1, 6, size=(2, N))

money = 10 - N           # capital after N throws
black = r[0,:]           # eyes for all throws with black
green = r[1,:]           # eyes for all throws with green
success = black > green  # success[i] is true if black[i]>green[i]
M = np.sum(success)      # sum up all successes
money += 2*M             # add all awards for winning
print 'Net profit per game in the long run:', (money-10)/float(N)
```

Decide if a game is fair Suppose the cost of playing a game once is q and that the award for winning is r. The net income in a winning game is $r - q$. Winning

M out of N games means that the cost is Nq and the income is Mr, making a net profit $s = Mr - Nq$. Now $p = M/N$ is the probability of winning the game so $s = (pr - q)N$. A fair game means that we neither win nor lose in the long run: $s = 0$, which implies that $r = q/p$. That is, given the cost q and the probability p of winning, the award paid for winning the game must be $r = q/p$ in a fair game.

When somebody comes up with a game you can use Monte Carlo simulation to estimate p and then conclude that you should not play the game of $r < q/p$. The example above has $p = 15/36$ (exact) and $q = 1$, so $r = 2.4$ makes a fair game.

The reasoning above is based on common sense and an intuitive interpretation of probability. More precise reasoning from probability theory will introduce the game as an experiment with two outcomes, either you win with probability p and or lose with probability $1 - p$. The expected payment is then the sum of the probabilities times the corresponding net income for each event: $-q(1 - p) + (r - q)p$ (recall that the net income in a winning game is $r - q$). A fair game has zero expected payment, which leads to $r = q/p$.

8.3.3 Example: Drawing Balls from a Hat

Suppose there are 12 balls in a hat: four black, four red, and four blue. We want to make a program that draws three balls at random from the hat. It is natural to represent the collection of balls as a list. Each list element can be an integer 1, 2, or 3, since we have three different types of balls, but it would be easier to work with the program if the balls could have a color instead of an integer number. This is easily accomplished by defining color names:

```
colors = 'black', 'red', 'blue'   # (tuple of strings)
hat = []
for color in colors:
    for i in range(4):
        hat.append(color)
```

Drawing a ball at random is performed by

```
import random
color = random.choice(hat)
print color
```

Drawing n balls without replacing the drawn balls requires us to remove an element from the hat when it is drawn. There are three ways to implement the procedure: (i) we perform a hat.remove(color), (ii) we draw a random index with randint from the set of legal indices in the hat list, and then we do a del hat[index] to remove the element, or (iii) we can compress the code in (ii) to hat.pop(index).

```
def draw_ball(hat):
    color = random.choice(hat)
    hat.remove(color)
    return color, hat
```

```
def draw_ball(hat):
    index = random.randint(0, len(hat)-1)
    color = hat[index]
    del hat[index]
    return color, hat

def draw_ball(hat):
    index = random.randint(0, len(hat)-1)
    color = hat.pop(index)
    return color, hat

# Draw n balls from the hat
balls = []
for i in range(n):
    color, hat = draw_ball(hat)
    balls.append(color)
print 'Got the balls', balls
```

We can extend the experiment above and ask the question: what is the probability of drawing two or more black balls from a hat with 12 balls, four black, four red, and four blue? To this end, we perform N experiments, count how many times M we get two or more black balls, and estimate the probability as M/N. Each experiment consists of making the hat list, drawing a number of balls, and counting how many black balls we got. The latter task is easy with the count method in list objects: hat.count('black') counts how many elements with value 'black' we have in the list hat. A complete program for this task is listed below. The program appears in the file balls_in_hat.py.

```
import random

def draw_ball(hat):
    """Draw a ball using list index."""
    index = random.randint(0, len(hat)-1)
    color = hat.pop(index)
    return color, hat

def draw_ball(hat):
    """Draw a ball using list index."""
    index = random.randint(0, len(hat)-1)
    color = hat[index]
    del hat[index]
    return color, hat

def draw_ball(hat):
    """Draw a ball using list element."""
    color = random.choice(hat)
    hat.remove(color)
    return color, hat
```

```
def new_hat():
    colors = 'black', 'red', 'blue'   # (tuple of strings)
    hat = []
    for color in colors:
        for i in range(4):
            hat.append(color)
    return hat

n = int(raw_input('How many balls are to be drawn? '))
N = int(raw_input('How many experiments? '))

# Run experiments
M = 0  # no of successes
for e in range(N):
    hat = new_hat()
    balls = []              # the n balls we draw
    for i in range(n):
        color, hat = draw_ball(hat)
        balls.append(color)
    if balls.count('black') >= 2:  # at least two black balls?
        M += 1
print 'Probability:', float(M)/N
```

Running the program with $n = 5$ (drawing 5 balls each time) and $N = 4000$ gives a probability of 0.57. Drawing only 2 balls at a time reduces the probability to about 0.09.

One can with the aid of probability theory derive theoretical expressions for such probabilities, but it is much simpler to let the computer perform a large number of experiments to estimate an approximate probability.

A class version of the code in this section is better than the code presented, because we avoid shuffling the `hat` variable in and out of functions. Exercise 8.21 asks you to design and implement a class `Hat`.

8.3.4 Random Mutations of Genes

A simple mutation model A fundamental principle of biological evolution is that DNA undergoes mutation. Since DNA can be represented as a string consisting of the letters A, C, G, and T, as explained in Sect. 3.3, mutation of DNA is easily modeled by replacing the letter in a randomly chosen position of the DNA by a randomly chosen letter from the alphabet A, C, G, and T. A function for replacing the letter in a randomly selected position (index) by a random letter among A, C, G, and T is most straightforwardly implemented by converting the DNA string to a list of letters, since changing a character in a Python string is impossible without constructing a new string. However, an element in a list can be changed in-place:

```
import random

def mutate_v1(dna):
    dna_list = list(dna)
    mutation_site = random.randint(0, len(dna_list) - 1)
    dna_list[mutation_site] = random.choice(list('ATCG'))
    return ''.join(dna_list)
```

Using `get_base_frequencies_v2` and `format_frequencies` from Sect. 6.5.3, we can easily mutate a gene a number of times and see how the frequencies of the bases A, C, G, and T change:

```
dna = 'ACGGAGATTTCGGTATGCAT'
print 'Starting DNA:', dna
print format_frequencies(get_base_frequencies_v2(dna))

nmutations = 10000
for i in range(nmutations):
    dna = mutate_v1(dna)

print 'DNA after %d mutations:' % nmutations, dna
print format_frequencies(get_base_frequencies_v2(dna))
```

Here is the output from a run:

```
Starting DNA: ACGGAGATTTCGGTATGCAT
A: 0.25, C: 0.15, T: 0.30, G: 0.30
DNA after 10000 mutations: AACCAATCCGACGAGGAGTG
A: 0.35, C: 0.25, T: 0.10, G: 0.30
```

Vectorized version The efficiency of the `mutate_v1` function with its surrounding loop can be significantly increased up by performing all the mutations at once using numpy arrays. This speed-up is of interest for long dna strings and many mutations. The idea is to draw all the mutation sites at once, and also all the new bases at these sites at once. The `np.random` module provides functions for drawing several random numbers at a time, but only integers and real numbers can be drawn, not characters from the alphabet A, C, G, and T. We therefore have to simulate these four characters by the numbers (say) 0, 1, 2, and 3. Afterwards we can translate the integers to letters by some clever vectorized indexing.

Drawing N mutation sites is a matter of drawing N random integers among the legal indices:

```
import numpy as np
mutation_sites = np.random.random_integers(0, len(dna)-1, size=N)
```

Drawing N bases, represented as the integers 0–3, is similarly done by

```
new_bases_i = np.random.random_integers(0, 3, N)
```

Converting say the integers 1 to the base symbol C is done by picking out the indices (in a boolean array) where `new_bases_i` equals 1, and inserting the character `'C'` in a companion array of characters:

```
new_bases_c = np.zeros(N, dtype='c')
indices = new_bases_i == 1
new_bases_c[indices] = 'C'
```

We must do this integer-to-letter conversion for all four integers/letters. Thereafter, `new_bases_c` must be inserted in `dna` for all the indices corresponding to the randomly drawn mutation sites,

```
dna[mutation_sites] = new_bases_c
```

The final step is to convert the `numpy` array of characters `dna` back to a standard string by first converting `dna` to a list and then joining the list elements: `''.join(dna.tolist())`.

The complete vectorized function can now be expressed as follows:

```
import numpy as np
# Use integers in random numpy arrays and map these
# to characters according to
i2c = {0: 'A', 1: 'C', 2: 'G', 3: 'T'}

def mutate_v2(dna, N):
    dna = np.array(dna, dtype='c')  # array of characters
    mutation_sites = np.random.random_integers(
        0, len(dna) - 1, size=N)
    # Must draw bases as integers
    new_bases_i = np.random.random_integers(0, 3, size=N)
    # Translate integers to characters
    new_bases_c = np.zeros(N, dtype='c')
    for i in i2c:
        new_bases_c[new_bases_i == i] = i2c[i]
    dna[mutation_sites] = new_bases_c
    return ''.join(dna.tolist())
```

It is of interest to time `mutate_v2` versus `mutate_v1`. For this purpose we need a long test string. A straightforward generation of random letters is

```
def generate_string_v1(N, alphabet='ACGT'):
    return ''.join([random.choice(alphabet) for i in xrange(N)])
```

A vectorized version of this function can also be made, using the ideas explored above for the `mutate_v2` function:

```
def generate_string_v2(N, alphabet='ACGT'):
    # Draw random integers 0,1,2,3 to represent bases
    dna_i = np.random.random_integers(0, 3, N)
```

```
# Translate integers to characters
dna = np.zeros(N, dtype='c')
for i in i2c:
    dna[dna_i == i] = i2c[i]
return ''.join(dna.tolist())
```

The `time_mutate` function in the file `mutate.py` performs timing of the generation of test strings and the mutations. To generate a DNA string of length 100,000 the vectorized function is about 8 times faster. When performing 10,000 mutations on this string, the vectorized version is almost 3000 times faster! These numbers stay approximately the same also for larger strings and more mutations. Hence, this case study on vectorization is a striking example on the fact that a straightforward and convenient function like `mutate_v1` might occasionally be very slow for large-scale computations.

A Markov chain mutation model The observed rate at which mutations occur at a given position in the genome is not independent of the type of nucleotide (base) at that position, as was assumed in the previous simple mutation model. We should therefore take into account that the rate of transition depends on the base.

There are a number of reasons why the observed mutation rates vary between different nucleotides. One reason is that there are different mechanisms generating transitions from one base to another. Another reason is that there are extensive repair process in living cells, and the efficiency of this repair mechanism varies for different nucleotides.

Mutation of nucleotides may be modeled using distinct probabilities for the transitions from each nucleotide to every other nucleotide. For example, the probability of replacing A by C may be prescribed as (say) 0.2. In total we need 4×4 probabilities since each nucleotide can transform into itself (no change) or three others. The sum of all four transition probabilities for a given nucleotide must sum up to one. Such statistical evolution, based on probabilities for transitioning from one state to another, is known as a Markov process or Markov chain.

First we need to set up the probability matrix, i.e., the 4×4 table of probabilities where each row corresponds to the transition of A, C, G, or T into A, C, G, or T. Say the probability transition from A to A is 0.2, from A to C is 0.1, from A to G is 0.3, and from A to T is 0.4.

Rather than just prescribing some arbitrary transition probabilities for test purposes, we can use random numbers for these probabilities. To this end, we generate three random numbers to divide the interval $[0, 1]$ into four intervals corresponding to the four possible transitions. The lengths of the intervals give the transition probabilities, and their sum is ensured to be 1. The interval limits, 0, 1, and three random numbers must be sorted in ascending order to form the intervals. We use the function `random.random()` to generate random numbers in $[0, 1)$:

```
slice_points = sorted(
    [0] + [random.random() for i in range(3)] + [1])
transition_probabilities = [slice_points[i+1] - slice_points[i]
                            for i in range(4)]
```

The transition probabilities are handy to have available as a dictionary:

```
markov_chain['A'] = {'A': ..., 'C': ..., 'G': ..., 'T': ...}
```

which can be computed by

```
markov_chain['A'] = {base: p for base, p in
                        zip('ACGT', transition_probabilities)}
```

To select a transition, we need to draw a random letter (A, C, G, or T) according to the probabilities `markov_chain[b]` where b is the base at the current position. Actually, this is a very common operation, namely drawing a random value from a *discrete probability distribution* (`markov_chain[b]`). The natural approach is therefore write a general function for drawing from any discrete probability distribution given as a dictionary:

```
def draw(discrete_probdist):
    """
    Draw random value from discrete probability distribution
    represented as a dict: P(x=value) = discrete_probdist[value].
    """
    # Method:
    # http://en.wikipedia.org/wiki/Pseudo-random_number_sampling
    limit = 0
    r = random.random()
    for value in discrete_probdist:
        limit += discrete_probdist[value]
        if r < limit:
            return value
```

Basically, the algorithm divides [0, 1] into intervals of lengths equal to the probabilities of the various outcomes and checks which interval is hit by a random variable in [0, 1]. The corresponding value is the random choice.

A complete function creating all the transition probabilities and storing them in a dictionary of dictionaries takes the form

```
def create_markov_chain():
    markov_chain = {}
    for from_base in 'ATGC':
        # Generate random transition probabilities by dividing
        # [0,1] into four intervals of random length
        slice_points = sorted(
            [0] + [random.random()for i in range(3)] + [1])
        transition_probabilities = \
            [slice_points[i+1] - slice_points[i] for i in range(4)]
        markov_chain[from_base] = {base: p for base, p
                        in zip('ATGC', transition_probabilities)}
    return markov_chain

mc = create_markov_chain()
print mc
print mc['A']['T'] # probability of transition from A to T
```

It is natural to develop a function for checking that the generated probabilities are consistent. The transition from a particular base into one of the four bases happens with probability 1, which means that the probabilities in a row must sum up to 1:

```
def check_transition_probabilities(markov_chain):
    for from_base in 'ATGC':
        s = sum(markov_chain[from_base][to_base]
                for to_base in 'ATGC')
        if abs(s - 1) > 1E-15:
            raise ValueError('Wrong sum: %s for "%s"' % \
                             (s, from_base))
```

Another test is to check that `draw` actually draws random values in accordance with the underlying probabilities. To this end, we draw a large number of values, N, count the frequencies of the various values, divide by N and compare the empirical normalized frequencies with the probabilities:

```
def check_draw_approx(discrete_probdist, N=1000000):
    """
    See if draw results in frequencies approx equal to
    the probability distribution.
    """
    frequencies = {value: 0 for value in discrete_probdist}
    for i in range(N):
        value = draw(discrete_probdist)
        frequencies[value] += 1
    for value in frequencies:
        frequencies[value] /= float(N)
    print ', '.join(['%s: %.4f (exact %.4f)' % \
                     (v, frequencies[v], discrete_probdist[v])
                     for v in frequencies])
```

This test is only approximate, but does bring evidence to the correctness of the implementation of the `draw` function.

A vectorized version of `draw` can also be made. We refer to the source code file `mutate.py` for details (the function is relatively complicated).

Now we have all the tools needed to run the Markov chain of transitions for a randomly selected position in a DNA sequence:

```
def mutate_via_markov_chain(dna, markov_chain):
    dna_list = list(dna)
    mutation_site = random.randint(0, len(dna_list) - 1)
    from_base = dna[mutation_site]
    to_base = draw(markov_chain[from_base])
    dna_list[mutation_site] = to_base
    return ''.join(dna_list)
```

Exercise 8.47 suggests some efficiency enhancements of simulating mutations via these functions.

Here is a simulation of mutations using the method based on Markov chains:

```
dna = 'TTACGGAGATTTCGGTATGCAT'
print 'Starting DNA:', dna
print format_frequencies(get_base_frequencies_v2(dna))

mc = create_markov_chain()
import pprint
print 'Transition probabilities:\n', pprint.pformat(mc)
nmutations = 10000
for i in range(nmutations):
    dna = mutate_via_markov_chain(dna, mc)

print 'DNA after %d mutations (Markov chain):' % nmutations, dna
print format_frequencies(get_base_frequencies_v2(dna))
```

The output will differ each time the program is run unless `random.seed(i)` is called in the beginning of the program for some integer i. This call makes the sequence of random numbers the same every time the program is run and is very useful for debugging. An example on the output may look like

```
Starting DNA: TTACGGAGATTTCGGTATGCAT
A: 0.23, C: 0.14, T: 0.36, G: 0.27
Transition probabilities:
{'A': {'A': 0.4288890546751146,
       'C': 0.4219086988655296,
       'G': 0.00668870644455688,
       'T': 0.14251354001479888},
 'C': {'A': 0.24999667668640035,
       'C': 0.04718309085408834,
       'G': 0.6250440975238185,
       'T': 0.0777761349356928},
 'G': {'A': 0.16022955651881965,
       'C': 0.34652746609882423,
       'G': 0.1328031742612512,
       'T': 0.3604398031211049},
 'T': {'A': 0.20609823213950174,
       'C': 0.17641112746655452,
       'G': 0.010267621176125452,
       'T': 0.6072230192178183}}
DNA after 10000 mutations (Markov chain): GGTTTAAGTCAGCTATGATTCT
A: 0.23, C: 0.14, T: 0.41, G: 0.23
```

The various functions performing mutations are located in the file `mutate.py`.

8.3.5   Example: Policies for Limiting Population Growth

China has for many years officially allowed only one child per couple. However, the success of the policy has been somewhat limited. One challenge is the current over-representation of males in the population (families have favored sons to live up). An alternative policy is to allow each couple to continue getting children until they get a son. We can simulate both policies and see how a population will develop under the *one child* and the *one son* policies. Since we expect to work with a large population over several generations, we aim at vectorized code at once.

Suppose we have a collection of n individuals, called `parents`, consisting of males and females randomly drawn such that a certain portion (`male_portion`) constitutes males. The `parents` array holds integer values, 1 for male and 2 for females. We can introduce constants, MALE=1 and FEMALE=2, to make the code easier to read. Our task is to see how the `parents` array develop from one generation to the next under the two policies. Let us first show how to draw the random integer array `parents` where there is a probability `male_portion` of getting the value MALE:

```
import numpy as np
r = np.random.random(n)
parents = np.zeros(n, int)
MALE = 1; FEMALE=2
parents[r <  male_portion] = MALE
parents[r >= male_portion] = FEMALE
```

The number of potential couples is the minimum of males and females. However, only a fraction (`fertility`) of the couples will actually get a child. Under the perfect one child policy, these couples can have one child each:

```
males = len(parents[parents==MALE])
females = len(parents) - males
couples = min(males, females)
n = int(fertility*couples)  # couples that get a child

# The next generation, one child per couple
r = random.random(n)
children = np.zeros(n, int)
children[r <  male_portion] = MALE
children[r >= male_portion] = FEMALE
```

The code for generating a new population will be needed in every generation. Therefore, it is natural to collect the last statements in a separate function such that we can repeat the statements when needed.

```
def get_children(n, male_portion, fertility):
    n = int(fertility*n)
    r = random.random(n)
    children = zeros(n, int)
    children[r <  male_portion] = MALE
    children[r >= male_portion] = FEMALE
    return children
```

Under the one son policy, the families can continue getting a new child until they get the first son:

```
# First try
children = get_children(couples, male_portion, fertility)
```

```
# Continue with getting a new child for each daughter
daughters = children[children == FEMALE]
while len(daughters) > 0:
    new_children = get_children(len(daughters),
                                male_portion, fertility)
    children = np.concatenate((children, new_children))
    daughters = new_children[new_children == FEMALE]
```

The program `birth_policy.py` organizes the code segments above for the two
policies into a function `advance_generation`, which we can call repeatedly to
see the evolution of the population.

```
def advance_generation(parents, policy='one child',
                       male_portion=0.5, fertility=1.0):
    males = len(parents[parents==MALE])
    females = len(parents) - males
    couples = min(males, females)

    if policy == 'one child':
        children = get_children(couples, male_portion, fertility)
    elif policy == 'one son':
        # First try at getting a child
        children = get_children(couples, male_portion, fertility)
        # Continue with getting a new child for each daughter
        daughters = children[children == FEMALE]
        while len(daughters) > 0:
            new_children = get_children(len(daughters),
                                        male_portion, fertility)
            children = np.concatenate((children, new_children))
            daughters = new_children[new_children == FEMALE]
    return children
```

The simulation is then a matter of repeated calls to `advance_generation`:

```
N = 1000000                     # population size
male_portion = 0.51
fertility = 0.92
# Start with a "perfect" generation of parents
parents = get_children(N, male_portion=0.5, fertility=1.0)
print 'one son policy, start: %d' % len(parents)
for i in range(10):
    parents = advance_generation(parents, 'one son',
                                 male_portion, fertility)
    print '%3d: %d' % (i+1, len(parents))
```

Under ideal conditions with unit `fertility` and a `male_portion` of 0.5, the
program predicts that the one child policy halves the population from one generation to the next, while the one son policy, where we expect each couple to get
one daughter and one son on average, keeps the population constant. Increasing
`male_portion` slightly and decreasing `fertility`, which corresponds more to
reality, will in both cases lead to a reduction of the population. You can try the
program out with various values of these input parameters.

An obvious extension is to incorporate the effect that a portion of the population does not follow the policy and get c children on average. The program

`birth_policy.py` can account for the effect, which is quite dramatic: if a fraction 0.01 of the population does not follow the one son policy and get 4 children on average, the population grows with a factor 1.5 over 10 generations (`male_portion` and `fertility` kept at the ideal values 0.5 and 1, respectively).

Normally, simple models like the difference equations (A.9) and (A.12), from Sects. A.1.4 and A.1.5, or the differential equations (C.11) or (C.23), are used to model population growth. However, these models track the number of individuals through time with a very simple growth factor from one generation to the next. The model above tracks each individual in the population and applies rules involving random actions to each individual. Such a detailed and much more computer-time consuming model can be used to see the effect of different policies. Using the results of this detailed model, we can (sometimes) estimate growth factors for simpler models so that these mimic the overall effect on the population size. Exercise 8.26 asks you to investigate if a certain realization of the one son policy leads to simple exponential growth.

8.4 Simple Games

This section presents the implementation of some simple games based on drawing random numbers. The games can be played by two humans, but here we consider a human versus the computer.

8.4.1 Guessing a Number

The game The computer determines a secret number, and the player shall guess the number. For each guess, the computer tells if the number is too high or too low.

The implementation We let the computer draw a random integer in an interval known to the player, let us say $[1, 100]$. In a `while` loop the program prompts the player for a guess, reads the guess, and checks if the guess is higher or lower than the drawn number. An appropriate message is written to the screen. We think the algorithm can be expressed directly as executable Python code:

```
import random
number = random.randint(1, 100)
attempts = 0  # count no of attempts to guess the number
guess = 0
while guess != number:
    guess = eval(raw_input('Guess a number: '))
    attempts += 1
    if guess == number:
        print 'Correct! You used', attempts, 'attempts!'
        break
    elif guess < number:
        print 'Go higher!'
    else:
        print 'Go lower!'
```

The program is available as the file `guess_number.py`. Try it out! Can you come up with a strategy for reducing the number of attempts? See Exercise 8.27 for an automatic investigation of two possible strategies.

8.4.2 Rolling Two Dice

The game The player is supposed to roll two dice, and beforehand guess the sum of the eyes. If the guess on the sum is n and it turns out to be right, the player earns n euros. Otherwise, the player must pay 1 euro. The machine plays in the same way, but the machine's guess of the number of eyes is a uniformly distributed number between 2 and 12. The player determines the number of rounds, r, to play, and receives r euros as initial capital. The winner is the one that has the largest amount of euros after r rounds, or the one that avoids to lose all the money.

The implementation There are three actions that we can naturally implement as functions: (i) roll two dice and compute the sum; (ii) ask the player to guess the number of eyes; (iii) draw the computer's guess of the number of eyes. One soon realizes that it is as easy to implement this game for an arbitrary number of dice as it is for two dice. Consequently we can introduce `ndice` as the number of dice. The three functions take the following forms:

```python
import random

def roll_dice_and_compute_sum(ndice):
    return sum([random.randint(1, 6) \
                for i in range(ndice)])

def computer_guess(ndice):
    return random.randint(ndice, 6*ndice)

def player_guess(ndice):
    return input('Guess the sum of the no of eyes '\
                 'in the next throw: ')
```

We can now implement one round in the game for the player or the computer. The round starts with a capital, a guess is performed by calling the right function for guessing, and the capital is updated:

```python
def play_one_round(ndice, capital, guess_function):
    guess = guess_function(ndice)
    throw = roll_dice_and_compute_sum(ndice)
    if guess == throw:
        capital += guess
    else:
        capital -= 1
    return capital, throw, guess
```

Here, `guess_function` is either `computer_guess` or `player_guess`.

With the `play_one_round` function we can run a number of rounds involving both players:

```
def play(nrounds, ndice=2):
    player_capital = computer_capital = nrounds  # start capital

    for i in range(nrounds):
        player_capital, throw, guess = \
            play_one_round(ndice, player_capital, player_guess)
        print 'YOU guessed %d, got %d' % (guess, throw)

        computer_capital, throw, guess = \
            play_one_round(ndice, computer_capital, computer_guess)

        print 'Machine guessed %d, got %d' % (guess, throw)

        print 'Status: you have %d euros, machine has %d euros' % \
            (player_capital, computer_capital)

        if player_capital == 0 or computer_capital == 0:
            break

    if computer_capital > player_capital:
        winner = 'Machine'
    else:
        winner = 'You'
    print winner, 'won!'
```

The name of the program is `ndice.py`.

Example Here is a session (with a fixed seed of 20):

```
Guess the sum of the no of eyes in the next throw: 7
YOU guessed 7, got 11
Machine guessed 10, got 8
Status: you have 9 euros, machine has 9 euros

Guess the sum of the no of eyes in the next throw: 9
YOU guessed 9, got 10
Machine guessed 11, got 6
Status: you have 8 euros, machine has 8 euros

Guess the sum of the no of eyes in the next throw: 9
YOU guessed 9, got 9
Machine guessed 3, got 8
Status: you have 17 euros, machine has 7 euros
```

Exercise 8.12 asks you to perform simulations to determine whether a certain strategy can make the player win over the computer in the long run.

A class version We can cast the previous code segment in a class. Many will argue that a class-based implementation is closer to the problem being modeled and hence easier to modify or extend.

A natural class is Dice, which can throw n dice:

```
class Dice(object):
    def __init__(self, n=1):
        self.n = n    # no of dice

    def throw(self):
        return [random.randint(1,6) \
                for i in range(self.n)]
```

Another natural class is Player, which can perform the actions of a player. Functions can then make use of Player to set up a game. A Player has a name, an initial capital, a set of dice, and a Dice object to throw the object:

```
class Player(object):
    def __init__(self, name, capital, guess_function, ndice):
        self.name = name
        self.capital = capital
        self.guess_function = guess_function
        self.dice = Dice(ndice)

    def play_one_round(self):
        self.guess = self.guess_function(self.dice.n)
        self.throw = sum(self.dice.throw())
        if self.guess == self.throw:
            self.capital += self.guess
        else:
            self.capital -= 1
        self.message()
        self.broke()

    def message(self):
        print '%s guessed %d, got %d' % \
              (self.name, self.guess, self.throw)

    def broke(self):
        if self.capital == 0:
            print '%s lost!' % self.name
            sys.exit(0)   # end the program
```

The guesses of the computer and the player are specified by functions:

```
def computer_guess(ndice):
    # All guesses have the same probability
    return random.randint(ndice, 6*ndice)

def player_guess(ndice):
    return input('Guess the sum of the no of eyes '\
                 'in the next throw: ')
```

The key function to play the whole game, utilizing the Player class for the computer and the user, can be expressed as

```
def play(nrounds, ndice=2):
    player = Player('YOU', nrounds, player_guess, ndice)
    computer = Player('Computer', nrounds, computer_guess, ndice)

    for i in range(nrounds):
        player.play_one_round()
        computer.play_one_round()
        print 'Status: user has %d euro, machine has %d euro\n' % \
              (player.capital, computer.capital)

    if computer.capital > player.capital:
        winner = 'Machine'
    else:
        winner = 'You'
    print winner, 'won!'
```

The complete code is found in the file `ndice2.py`. There is no new functionality compared to the `ndice.py` implementation, just a new and better structuring of the code.

8.5 Monte Carlo Integration

One of the earliest applications of random numbers was numerical computation of integrals, which is actually non-random (deterministic) problem. Computing integrals with the aid of random numbers is known as Monte Carlo integration and is one of the most powerful and widely used mathematical technique throughout science and engineering.

Our main focus here will be integrals of the type $\int_a^b f(x)dx$ for which the Monte Carlo integration is not a competitive technique compared to simple methods such as the Trapezoidal method, the Midpoint method, or Simpson's method. However, for integration of functions of *many variables*, the Monte Carlo approach is the best method we have. Such integrals arise all the time in quantum physics, financial engineering, and when estimating the uncertainty of mathematical computations. What you learn about Monte Carlo integration of functions of one variable ($\int_a^b f(x)dx$) is directly transferable to the important application cases where there are many variables involved.

8.5.1 Derivation of Monte Carlo Integration

There are two ways to introduce Monte Carlo integration, one based on calculus and one based on probability theory. The goal is to compute a numerical approximation to

$$\int_a^b f(x)dx \ .$$

The calculus approach via the mean-value theorem   The mean-value theorem from calculus states that

$$\int_a^b f(x)dx = (b-a)\bar{f}, \tag{8.7}$$

where $\bar{f}$ is the mean value of f, defined as

$$\bar{f} = \frac{1}{b-a}\int_a^b f(x)dx.$$

One way of using (8.7) to define a numerical method for integration is to approximate $\bar{f}$ by taking the average of f at n points $x_0, \ldots, x_{n-1}$:

$$\bar{f} \approx \frac{1}{n}\sum_{i=0}^{n-1} f(x_i). \tag{8.8}$$

(We let the numbering of the points go from 0 to $n-1$ because these numbers will later be indices in Python arrays, which have to start at 0.)

There is freedom in how to choose the points $x_0, \ldots, x_{n-1}$. We could, for example, make them uniformly spaced in the interval $[a, b]$. The particular choice

$$x_0 = a + ih + \frac{1}{2}h, \quad i = 0, \ldots, n-1, \quad h = \frac{b-a}{n-1},$$

correspond to the famous Midpoint rule for numerical integration. Intuitively, we anticipate that the more points we use, the better is the approximation $\frac{1}{n}\sum_i f(x_i)$ to the exact mean value $\bar{f}$. For the Midpoint rule one can show mathematically, or numerically estimate through examples, that the error in the approximation depends on n as n^{-2}. That is, doubling the number of points reduces the error by a factor 1/4.

A slightly different set of uniformly distributed points is

$$x_i = a + ih, \quad i = 0, \ldots, n-1, \ldots, \quad h = \frac{b-1}{n-2}.$$

These points might look more intuitive to many since they start at a and end at b ($x_0 = a$, $x_{n-1} = b$), but the error in the integration rule now goes as n^{-1}: doubling the number of points just halves the error. That is, computing more function values to get a better integration estimate is less effective with this set points than with the slight displaced points used in the Midpoint rule.

One could also throw in the idea of using a set of *random* points, uniformly distributed in $[a, b]$. This is the Monte Carlo integration technique. The error now goes like $n^{-1/2}$, which means that many more points and corresponding function evaluations are needed to reduce the error compared to using the Midpoint method. The surprising fact, however, is that using random points for many variables (in high-dimensional vector spaces) yields a very effective integration technique, dramatically more effective than extending the ideas of the Midpoint rule to many variables.

The probability theory approach People who are into probability theory usually like to interpret integrals as mathematical expectations of a random variable. (If you are not one of those, you can safely jump to Sect. 8.5.2 where we just program the simple sum in the Monte Carlo integration method.) More precisely, the integral $\int_a^b f(x)dx$ can be expressed as a mathematical expectation of $f(x)$ if x is a uniformly distributed random variable on $[a, b]$. This expectation can be estimated by the average of random samples, which results in the Monte Carlo integration method. To see this, we start with the formula for the probability density function for a uniformly distributed random variable X on $[a, b]$:

$$p(x) = \begin{cases} (b-a)^{-1}, & x \in [a, b] \\ 0, & \text{otherwise} \end{cases}$$

Now we can write the standard formula for the mathematical expectation $\mathrm{E}(f(X))$:

$$\mathrm{E}(f(X)) = \int_{-\infty}^{\infty} f(x)p(x)dx = \int_a^b f(x)\frac{1}{b-a}dx = (b-a)\int_a^b f(x)dx \, .$$

The last integral is exactly what we want to compute. An expectation is usually estimated from a lot of samples, in this case uniformly distributed random numbers $x_0, \ldots, x_{n-1}$ in $[a, b]$, and computing the sample mean:

$$\mathrm{E}(f(X)) \approx \frac{1}{n}\sum_{i=0}^{n-1} f(x_i) \, .$$

The integral can therefore be estimated by

$$\int_a^b f(x)dx \approx (b-a)\frac{1}{n}\sum_{i=0}^{n-1} f(x_i),$$

which is nothing but the Monte Carlo integration method.

8.5.2 Implementation of Standard Monte Carlo Integration

To summarize, Monte Carlo integration consists in generating n uniformly distributed random numbers x_i in $[a, b]$ and then compute

$$(b-a)\frac{1}{n}\sum_{i=0}^{n-1} f(x_i) \, . \tag{8.9}$$

We can implement (8.9) in a small function:

```
import random

def MCint(f, a, b, n):
    s = 0
    for i in xrange(n):
        x = random.uniform(a, b)
        s += f(x)
    I = (float(b-a)/n)*s
    return I
```

One normally needs a large n to obtain good results with this method, so a faster vectorized version of the `MCint` function is very useful:

```
import numpy as np

def MCint_vec(f, a, b, n):
    x = np.random.uniform(a, b, n)
    s = np.sum(f(x))
    I = (float(b-a)/n)*s
    return I
```

The functions above are available in the module file `MCint.py`. We can test the gain in efficiency with `%timeit` in an IPython session:

```
In [1]: from MCint import MCint, MCint_vec

In [2]: from math import sin, pi

In [3]: %timeit MCint(sin, 0, pi, 1000000)
1 loops, best of 3: 1.19 s per loop

In [4]: from numpy import sin

In [5]: %timeit MCint_vec(sin, 0, pi, 1000000)
1 loops, best of 3: 173 ms per loop

In [6]: 1.19/0.173              # relative performance
Out[6]: 6.878612716763006
```

Note that we use `sin` from `math` in the scalar function `MCint` because this function is significantly faster than `sin` from `numpy`:

```
In [7]: from math import sin

In [8]: %timeit sin(1.2)
10000000 loops, best of 3: 179 ns per loop

In [9]: from numpy import sin

In [10]: %timeit sin(1.2)
100000 loops, best of 3: 3.22 microsec per loop

In [11]: 3.22E-6/179E-9         # relative performance
Out[11]: 17.988826815642458
```

(A similar test reveals that `math.sin` is 1.3 times slower calling `sin` without a prefix. The differences are much smaller between `numpy.sin` and the same function without the prefix.)

The increase in efficiency by using `MCint_vec` instead of `MCint` is in the test above a factor of 6–7, which is not dramatic. Moreover, the vectorized version needs to store n random numbers and n function values in memory. A better vectorized implementation for large n is to split the x and f(x) arrays into chunks of a given size `arraysize` such that we can control the memory usage. Mathematically, it means to split the sum $\frac{1}{n}\sum_i f(x_i)$ into a sum of smaller sums. An appropriate implementation reads

```
def MCint_vec2(f, a, b, n, arraysize=1000000):
    s = 0
    # Split sum into batches of size arraysize
    # + a sum of size rest (note: n//arraysize is integer division)
    rest = n % arraysize
    batch_sizes = [arraysize]*(n//arraysize) + [rest]
    for batch_size in batch_sizes:
        x = np.random.uniform(a, b, batch_size)
        s += np.sum(f(x))
    I = (float(b-a)/n)*s
    return I
```

With 100 million points, `MCint_vec2` is about 10 faster than `MCint`. (Note that the latter function must use `xrange` and not `range` for so large n, otherwise the array returned by `range` may become to large to be stored in memory in a small computer. The `xrange` function generates one i at a time without the need to store all the i values.)

Example Let us try the Monte Carlo integration method on a simple linear function $f(x) = 2 + 3x$, integrated from 1 to 2. Most other numerical integration methods will integrate such a linear function exactly, regardless of the number of function evaluations. This is not the case with Monte Carlo integration.

It would be interesting to see how the quality of the Monte Carlo approximation increases with n. To plot the evolution of the integral approximation we must store intermediate I values. This requires a slightly modified `MCint` method:

```
def MCint2(f, a, b, n):
    s = 0
    # Store the intermediate integral approximations in an
    # array I, where I[k-1] corresponds to k function evals.
    I = np.zeros(n)
    for k in range(1, n+1):
        x = random.uniform(a, b)
        s += f(x)
        I[k-1] = (float(b-a)/k)*s
    return I
```

Note that we let k go from 1 to n, such that k reflects the actual number of points used in the method. Since n can be very large, the I array may consume more memory than what we have on the computer. Therefore, we decide to store only every N values of the approximation. Determining if a value is to be stored or not can be computed by the mod function: k % N gives the remainder when k is divided by N. In our case we can store when this reminder is zero,

```
for k in range(1, n+1):
    ...
    if k % N == 0:
        # store
```

This recipe of doing something every N-th pass in long loops has lots of applications in scientific computing! The complete function now takes the following form:

```
def MCint3(f, a, b, n, N=100):
    s = 0
    # Store every N intermediate integral approximations in an
    # array I and record the corresponding k value.
    I_values = []
    k_values = []
    for k in range(1, n+1):
        x = random.uniform(a, b)
        s += f(x)
        if k % N == 0:
            I = (float(b-a)/k)*s
            I_values.append(I)
            k_values.append(k)
    return k_values, I_values

def demo():
```

Our sample application goes like

```
def f1(x):
    return 2 + 3*x

k, I = MCint3(f1, 1, 2, 1000000, N=10000)
error = 6.5 - np.array(I)
```

Figure 8.4 shows a plot of error versus the number of function evaluation k.

Remark We claimed that the Monte Carlo method is slow for integration functions of one variable. There are, however, many techniques that apply smarter ways of drawing random numbers, so called variance reduction techniques, and thereby increase the computational efficiency.

8.5.3 Area Computing by Throwing Random Points

Think of some geometric region G in the plane and a surrounding bounding box B with geometry $[x_L, x_H] \times [y_L, y_H]$. One way of computing the area of G is to draw N random points inside B and count how many of them, M, that lie inside G. The area of G is then the fraction M/N (G's fraction of B's area) times the area of B,

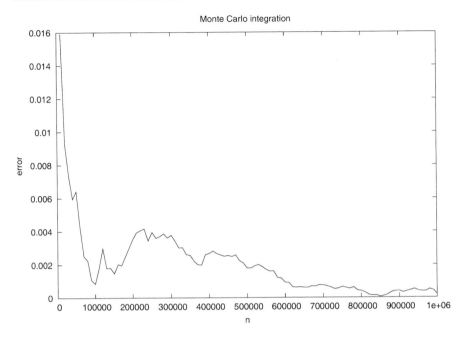

Fig. 8.4 The convergence of Monte Carlo integration applied to $\int_1^2 (2 + 3x)dx$

$(x_H - x_L)(y_H - y_L)$. Phrased differently, this method is a kind of dart game where you record how many hits there are inside G if every throw hits uniformly within B.

Let us formulate this method for computing the integral $\int_a^b f(x)dx$. The important observation is that this integral is the area under the curve $y = f(x)$ and above the x axis, between $x = a$ and $x = b$. We introduce a rectangle B,

$$B = \{(x, y) \mid a \leq x \leq b,\ 0 \leq y \leq m\},$$

where $m \leq \max_{x \in [a,b]} f(x)$. The algorithm for computing the area under the curve is to draw N random points inside B and count how many of them, M, that are above the x axis and below the $y = f(x)$ curve, see Fig. 8.5. The area or integral is then estimated by

$$\frac{M}{N}m(b - a).$$

First we implement the "dart method" by a simple loop over points:

```
def MCint_area(f, a, b, n, m):
    below = 0   # counter for no of points below the curve
    for i in range(n):
        x = random.uniform(a, b)
        y = random.uniform(0, m)
        if y <= f(x):
            below += 1
    area = below/float(n)*m*(b-a)
    return area
```

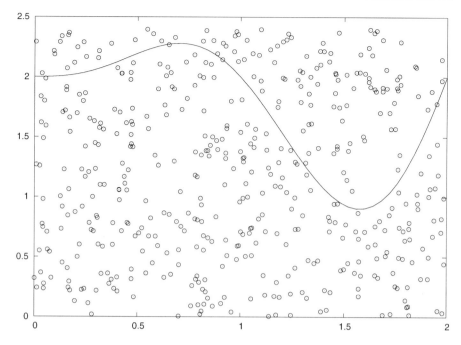

Fig. 8.5 The "dart" method for computing integrals. When M out of N random points in the rectangle $[0, 2] \times [0, 2.4]$ lie under the curve, the area under the curve is estimated as the M/N fraction of the area of the rectangle, i.e., $(M/N)2 \cdot 2.4$

Note that this method draws twice as many random numbers as the previous method.

A vectorized implementation reads

```
import numpy as np

def MCint_area_vec(f, a, b, n, m):
    x = np.random.uniform(a, b, n)
    y = np.random.uniform(0, m, n)
    below = np.sum(y < f(x))
    area = below/float(n)*m*(b-a)
    return area
```

The only non-trivial line here is the expression `y[y < f(x)]`, which applies boolean indexing (Sect. 5.5.2) to extract the y values that are below the `f(x)` curve. The sum of the boolean values (interpreted as 0 and 1) in `y < f(x)` gives the number of points below the curve.

Even for 2 million random numbers the plain loop version is not that slow as it executes within some seconds on a slow laptop. Nevertheless, if you need the integration being repeated many times inside another calculation, the superior efficiency of the vectorized version may be important. We can quantify the efficiency gain by the aid of the timer `time.clock()` in the following way (see Sect. H.8.1):

```
import time
t0 = time.clock()
print MCint_area(f1, a, b, n, fmax)
t1 = time.clock()    # time of MCint_area is t1-t0
print MCint_area_vec(f1, a, b, n, fmax)
t2 = time.clock()    # time of MCint_area_vec is t2-t1
print 'loop/vectorized fraction:', (t1-t0)/(t2-t1)
```

With $n = 10^6$ the author achieved a factor of about 8 in favor of the vectorized version.

8.6   Random Walk in One Space Dimension

In this section we shall simulate a collection of particles that move around in a random fashion. This type of simulations are fundamental in physics, biology, chemistry as well as other sciences and can be used to describe many phenomena. Some application areas include molecular motion, heat conduction, quantum mechanics, polymer chains, population genetics, brain research, hazard games, and pricing of financial instruments.

Imagine that we have some particles that perform random moves, either to the right or to the left. We may flip a coin to decide the movement of each particle, say head implies movement to the right and tail means movement to the left. Each move is one unit length. Physicists use the term *random walk* for this type of movement of a particle. You may try this yourself: flip the coin and make one step to the left or right, and repeat this process.

The movement is also known as *drunkard's walk*. You may have experienced this after a very wet night on a pub: you step forward and backward in a random fashion. Since these movements on average make you stand still, and since you know that you normally reach home within reasonable time, the model is not good for a real walk. We need to add a *drift* to the walk, so the probability is greater for going forward than backward. This is an easy adjustment, see Exercise 8.32. What may come as a surprise is the following fact: even when there is equal probability of going forward and backward, one can prove mathematically that the drunkard will always reach his home. Or more precisely, he will get home in finite time ("almost surely" as the mathematicians must add to this statement). Exercise 8.33 asks you to experiment with this fact. For many practical purposes, "finite time" does not help much as there might be more steps involved than the time it takes to get sufficiently sober to remove the completely random component of the walk.

8.6.1   Basic Implementation

How can we implement n_s random steps of n_p particles in a program? Let us introduce a coordinate system where all movements are along the x axis. An array of x values then holds the positions of all particles. We draw random numbers to simulate flipping a coin, say we draw from the integers 1 and 2, where 1 means head (movement to the right) and 2 means tail (movement to the left). We think the algorithm is conveniently expressed directly as a complete Python program:

```
import random
import numpy
np = 4                          # no of particles
ns = 100                        # no of steps
positions = numpy.zeros(np)     # all particles start at x=0
HEAD = 1;   TAIL = 2            # constants

for step in range(ns):
    for p in range(np):
        coin = random.randint(1,2)  # flip coin
        if coin == HEAD:
            positions[p] += 1   # one unit length to the right
        elif coin == TAIL:
            positions[p] -= 1   # one unit length to the left
```

This program is found in the file walk1D.py.

8.6.2   Visualization

We may add some visualization of the movements by inserting a plot command at the end of the step loop and a little pause to better separate the frames in the animation:

```
plot(positions, y, 'ko3', axis=[xmin, xmax, -0.2, 0.2])
time.sleep(0.2)   # pause
```

These two statements require from scitools.std import plot and import time.

It is very important that the extent of the axis are kept fixed in animations, otherwise one gets a wrong visual impression. We know that in n_s steps, no particle can move longer than n_s unit lengths to the right or to the left so the extent of the x axis becomes $[-n_s, n_s]$. However, the probability of reaching these lower or upper limit is very small. To be specific, the probability is 2^{-n_s}, which becomes about 10^{-9} for 30 steps. Most of the movements will take place in the center of the plot. We may therefore shrink the extent of the axis to better view the movements. It is known that the expected extent of the particles is of the order $\sqrt{n_s}$, so we may take the maximum and minimum values in the plot as $\pm 2\sqrt{n_s}$. However, if a position of a particle exceeds these values, we extend xmax and xmin by $2\sqrt{n_s}$ in positive and negative x direction, respectively.

The y positions of the particles are taken as zero, but it is necessary to have some extent of the y axis, otherwise the coordinate system collapses and most plotting packages will refuse to draw the plot. Here we have just chosen the y axis to go from -0.2 to 0.2. You can find the complete program in walk1Dp.py. The np and ns parameters can be set as the first two command-line arguments:

———————————————— | Terminal | ————————————————

walk1Dp.py 6 200

It is hard to claim that this program has astonishing graphics. In Sect. 8.7, where we let the particles move in two space dimensions, the graphics gets much more exciting.

8.6.3   Random Walk as a Difference Equation

The random walk process can easily be expressed in terms of a difference equation (see Appendix A for an introduction to difference equations). Let x_n be the position of the particle at time n. This position is an evolution from time $n-1$, obtained by adding a random variable s to the previous position x_{n-1}, where $s = 1$ has probability 1/2 and $s = -1$ has probability 1/2. In statistics, the expression *probability of event A* is written $P(A)$. We can therefore write $P(s = 1) = 1/2$ and $P(s = -1) = 1/2$. The difference equation can now be expressed mathematically as

$$x_n = x_{n-1} + s, \quad x_0 = 0, \quad P(s = 1) = P(s = -1) = 1/2. \tag{8.10}$$

This equation governs the motion of one particle. For a collection m of particles we introduce $x_n^{(i)}$ as the position of the i-th particle at the n-th time step. Each $x_n^{(i)}$ is governed by (8.10), and all the s values in each of the m difference equations are independent of each other.

8.6.4   Computing Statistics of the Particle Positions

Scientists interested in random walks are in general not interested in the graphics of our `walk1D.py` program, but more in the statistics of the positions of the particles at each step. We may therefore, at each step, compute a histogram of the distribution of the particles along the x axis, plus estimate the mean position and the standard deviation. These mathematical operations are easily accomplished by letting the SciTools function `compute_histogram` and the `numpy` functions `mean` and `std` operate on the `positions` array (see Sect. 8.1.5):

```
mean_pos  = numpy.mean(positions)
stdev_pos = numpy.std(positions)
pos, freq = compute_histogram(positions, nbins=int(xmax),
                              piecewise_constant=True)
```

The number of bins in the histogram is just based on the extent of the particles. It could also have been a fixed number.

We can plot the particles as circles, as before, and add the histogram and vertical lines for the mean and the positive and negative standard deviation (the latter indicates the "width" of the distribution of particles). The vertical lines can be defined by the six lists

```
xmean, ymean  = [mean_pos, mean_pos],   [yminv, ymaxv]
xstdv1, ystdv1 = [stdev_pos, stdev_pos], [yminv, ymaxv]
xstdv2, ystdv2 = [-stdev_pos, -stdev_pos], [yminv, ymaxv]
```

where `yminv` and `ymaxv` are the minimum and maximum y values of the vertical lines. The following command plots the position of every particle as circles, the histogram as a curve, and the vertical lines with a thicker line:

```
plot(positions, y, 'ko3',       # particles as circles
     pos, freq, 'r',            # histogram
     xmean, ymean, 'r2',        # mean position as thick line
     xstdv1, ystdv1, 'b2',      # +1 standard dev.
     xstdv2, ystdv2, 'b2',      # -1 standard dev.
     axis=[xmin, xmax, ymin, ymax],
     title='random walk of %d particles after %d steps' %
           (np, step+1))
```

This plot is then created at every step in the random walk. By observing the graphics, one will soon realize that the computation of the extent of the y axis in the plot needs some considerations. We have found it convenient to base `ymax` on the maximum value of the histogram (`max(freq)`), plus some space (chosen as 10 percent of `max(freq)`). However, we do not change the `ymax` value unless it is more than 0.1 different from the previous `ymax` value (otherwise the axis "jumps" too often). The minimum value, `ymin`, is set to `ymin=-0.1*ymax` every time we change the `ymax` value. The complete code is found in the file `walk1Ds.py`. If you try out 2000 particles and 30 steps, the final graphics becomes like that in Fig. 8.6. As the number of steps is increased, the particles are dispersed in the positive and negative x direction, and the histogram gets flatter and flatter. Letting $\hat{H}(i)$ be the histogram value in interval number i, and each interval having width Δx, the probability of finding a particle in interval i is $\hat{H}(i)\Delta x$. It can be shown mathematically that the histogram is an approximation to the probability density function of the normal distribution with mean zero and standard deviation $s \sim \sqrt{n}$, where n is the step number.

8.6.5 Vectorized Implementation

There is no problem with the speed of our one-dimensional random walkers in the `walk1Dp.py` or `walk1Ds.py` programs, but in real-life applications of such simulation models, we often have a very large number of particles performing a very large number of steps. It is then important to make the implementation as efficient as possible. Two loops over all particles and all steps, as we have in the programs above, become very slow compared to a vectorized implementation.

A vectorized implementation of a one-dimensional walk should utilize the functions `randint` or `random_integers` from `numpy`'s `random` module. A first idea may be to draw steps for all particles at a step simultaneously. Then we repeat this process in a loop from 0 to $n_s - 1$. However, these repetitions are just new vectors of random numbers, and we may avoid the loop if we draw $n_p \times n_s$ random numbers at once:

```
moves = numpy.random.randint(1, 3, size=np*ns)
# or
moves = numpy.random.random_integers(1, 2, size=np*ns)
```

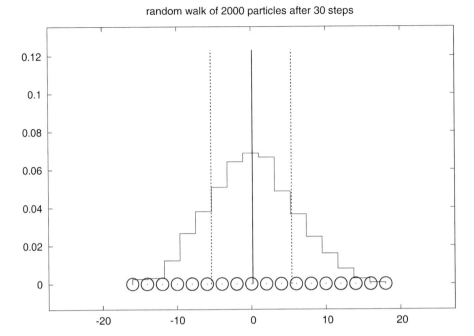

Fig. 8.6 Particle positions (circles), histogram (piecewise constant curve), and vertical lines indicating the mean value and the standard deviation from the mean after a one-dimensional random walk of 2000 particles for 30 steps

The values are now either 1 or 2, but we want -1 or 1. A simple scaling and translation of the numbers transform the 1 and 2 values to -1 and 1 values:

```
moves = 2*moves - 3
```

Then we can create a two-dimensional array out of `moves` such that `moves[i,j]` is the `i`-th step of particle number `j`:

```
moves.shape = (ns, np)
```

It does not make sense to plot the evolution of the particles and the histogram in the vectorized version of the code, because the point with vectorization is to speed up the calculations, and the visualization takes much more time than drawing random numbers, even in the `walk1Dp.py` and `walk1Ds.py` programs from Sect. 8.6.4. We therefore just compute the positions of the particles inside a loop over the steps and some simple statistics. At the end, after n_s steps, we plot the histogram of the particle distribution along with circles for the positions of the particles. The rest of the program, found in the file `walk1Dv.py`, looks as follows:

```
positions = numpy.zeros(np)
for step in range(ns):
    positions += moves[step, :]
```

```
    mean_pos = numpy.mean(positions)
    stdev_pos = numpy.std(positions)
    print mean_pos, stdev_pos

 nbins = int(3*sqrt(ns))      # no of intervals in histogram
 pos, freq = compute_histogram(positions, nbins,
                               piecewise_constant=True)

 plot(positions, zeros(np), 'ko3',
      pos, freq, 'r',
      axis=[min(positions), max(positions), -0.01, 1.1*max(freq)],
      savefig='tmp.pdf')
```

8.7   Random Walk in Two Space Dimensions

A random walk in two dimensions performs a step either to the north, south, west, or east, each one with probability $1/4$. To demonstrate this process, we introduce x and y coordinates of n_p particles and draw random numbers among 1, 2, 3, or 4 to determine the move. The positions of the particles can easily be visualized as small circles in an xy coordinate system.

8.7.1   Basic Implementation

The algorithm described above is conveniently expressed directly as a complete working program:

```
 def random_walk_2D(np, ns, plot_step):
     xpositions = numpy.zeros(np)
     ypositions = numpy.zeros(np)
     # extent of the axis in the plot:
     xymax = 3*numpy.sqrt(ns); xymin = -xymax

     NORTH = 1;  SOUTH = 2;  WEST = 3;  EAST = 4  # constants

     for step in range(ns):
         for i in range(np):
             direction = random.randint(1, 4)
             if direction == NORTH:
                 ypositions[i] += 1
             elif direction == SOUTH:
                 ypositions[i] -= 1
             elif direction == EAST:
                 xpositions[i] += 1
             elif direction == WEST:
                 xpositions[i] -= 1

         # Plot just every plot_step steps
         if (step+1) % plot_step == 0:
             plot(xpositions, ypositions, 'ko',
                  axis=[xymin, xymax, xymin, xymax],
```

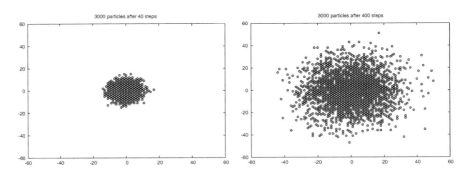

Fig. 8.7 Location of 3000 particles starting at the origin and performing a random walk, with 40 steps (*left*) and 400 steps (*right*)

```
              title='%d particles after %d steps' %
                    (np, step+1),
              savefig='tmp_%03d.pdf' % (step+1))
    return xpositions, ypositions

# main program:
import random
random.seed(10)
import sys
import numpy
from scitools.std import plot

np        = int(sys.argv[1])   # number of particles
ns        = int(sys.argv[2])   # number of steps
plot_step = int(sys.argv[3])   # plot every plot_step steps
x, y = random_walk_2D(np, ns, plot_step)
```

The program is found in the file `walk2D.py`. Figure 8.7 shows two snapshots of the distribution of 3000 particles after 40 and 400 steps. These plots were generated with command-line arguments 3000 400 20, the latter implying that we visualize the particles every 20 time steps only.

To get a feeling for the two-dimensional random walk you can try out only 30 particles for 400 steps and let each step be visualized (i.e., command-line arguments 30 400 1). The update of the movements is now fast.

The `walk2D.py` program dumps the plots to PDF files with names of the form `tmp_xxx.pdf`, where xxx is the step number. We can create a movie out of these individual files using the program `convert` from the ImageMagick suite:

Terminal

```
Terminal> convert -delay 50 -loop 1000 tmp_*.pdf movie.gif
```

All the plots are now put after each other as frames in a movie, with a delay of 50 ms between each frame. The movie will run in a loop 1000 times. The resulting movie file is named `movie.gif`, which can be viewed by the `animate` program (also from the ImageMagick program suite), just write `animate movie.gif`. Making and showing the movie are slow processes if a large number of steps are included in the movie. The alternative is to make a true video file in, e.g., the Flash format:

```
Terminal
Terminal> avconv -r 5 -i tmp_%04d.png -c:v flv movie.flv
```

This requires the plot files to be in PNG format.

8.7.2   Vectorized Implementation

The `walk2D.py` program is quite slow. Now the visualization is much faster than the movement of the particles. Vectorization may speed up the `walk2D.py` program significantly. As in the one-dimensional phase, we draw all the movements at once and then invoke a loop over the steps to update the x and y coordinates. We draw $n_s \times n_p$ numbers among 1, 2, 3, and 4. We then reshape the vector of random numbers to a two-dimensional array `moves[i, j]`, where i counts the steps, j counts the particles. The `if` test on whether the current move is to the north, south, east, or west can be vectorized using the `where` function. For example, if the random numbers for all particles in the current step are accessible in an array `this_move`, we could update the x positions by

```
xpositions += np.where(this_move == EAST, 1, 0)
xpositions -= np.where(this_move == WEST, 1, 0)
```

provided EAST and WEST are constants, equal to 3 and 4, respectively. A similar construction can be used for the y moves.

The complete program is listed below:

```
def random_walk_2D(np, ns, plot_step):
    xpositions = numpy.zeros(np)
    ypositions = numpy.zeros(np)
    moves = numpy.random.random_integers(1, 4, size=ns*np)
    moves.shape = (ns, np)

    # Estimate max and min positions
    xymax = 3*numpy.sqrt(ns); xymin = -xymax

    NORTH = 1;   SOUTH = 2;   WEST = 3;   EAST = 4  # constants

    for step in range(ns):
        this_move = moves[step,:]
        ypositions += numpy.where(this_move == NORTH, 1, 0)
        ypositions -= numpy.where(this_move == SOUTH, 1, 0)
        xpositions += numpy.where(this_move == EAST,  1, 0)
        xpositions -= numpy.where(this_move == WEST,  1, 0)
```

```
      # Just plot every plot_step steps
      if (step+1) % plot_step == 0:
          plot(xpositions, ypositions, 'ko',
               axis=[xymin, xymax, xymin, xymax],
               title='%d particles after %d steps' %
                     (np, step+1),
               savefig='tmp_%03d.pdf' % (step+1))
  return xpositions, ypositions

# Main program
from scitools.std import plot
import numpy, sys
numpy.random.seed(11)

np = int(sys.argv[1])   # number of particles
ns = int(sys.argv[2])   # number of steps
plot_step = int(sys.argv[3])  # plot each plot_step step
x, y = random_walk_2D(np, ns, plot_step)
```

You will easily experience that this program, found in the file walk2Dv.py, runs significantly faster than the walk2D.py program.

8.8 Summary

8.8.1 Chapter Topics

Drawing random numbers Random numbers can be scattered throughout an interval in various ways, specified by the *distribution* of the numbers. We have considered a uniform distribution (Sect. 8.1.2) and a normal (or Gaussian) distribution (Sect. 8.1.6).

The table below shows the syntax for generating random numbers of these two distributions, using either the standard scalar random module in Python or the vectorized numpy.random module. Here, N is the array length in vectorized drawing, while m and s represent the mean and standard deviation values of a normal distribution. Functions from the standard random module appear in the middle column, while the corresponding functions from numpy.random are listed in the right column.

Functionality	Python's random	numpy.random
uniform numbers in $[0, 1)$	random()	random(N)
uniform numbers in $[a, b)$	uniform(a, b)	uniform(a, b, N)
integers in $[a, b]$	randint(a, b)	randint(a, b+1, N)
integers in $[a, b]$		random_integers(a, b, N)
Gaussian numbers	gauss(m, s)	normal(m, s, N)
set seed (i)	seed(i)	seed(i)
shuffle list in-place	shuffle(a)	shuffle(a)
choose a random element in list	choice(a)	

Typical probability computation via Monte Carlo simulation Many programs performing probability computations draw a large number N of random numbers and count how many times M a random number leads to some true condition (success):

```
import random
M = 0
for i in xrange(N):
    r = random.randint(a, b)
    if success:
        M += 1
print 'Probability estimate:', float(M)/N
```

For example, if we seek the probability that we get at least four eyes when throwing a dice, we choose the random number to be the number of eyes, i.e., an integer in the interval $[1, 6]$ (a=1, b=6) and success becomes r >= 4.

For large N we can speed up such programs by vectorization, i.e., drawing all random numbers at once in a big array and use operations on the array to find M. The similar vectorized version of the program above looks like

```
import numpy as np
r = np.random.uniform(a, b, N)
M = np.sum(condition)
# or
M = np.sum(where(condition, 1, 0))
print 'Probability estimate:', float(M)/N
```

(Combinations of boolean expressions in the condition argument to where requires special constructs as outlined in Exercise 8.17.) Make sure you use np.sum when operating on large arrays and not the much slower built-in sum function in Python.

Statistical measures Given an array of random numbers, the following code computes the mean, variance, and standard deviation of the numbers and finally displays a plot of the histogram, which reflects how the numbers are statistically distributed:

```
from scitools.std import compute_histogram, plot
import numpy as np
m = np.mean(numbers)
v = np.var(numbers)
s = np.std(numbers)
x, y = compute_histogram(numbers, 50, piecewise_constant=True)
plot(x, y)
```

Terminology The important topics in this chapter are

- random numbers
- random number distribution
- Monte Carlo simulation
- Monte Carlo integration
- random walk

8.8.2   Example: Random Growth

Section A.1.1 presents simple mathematical models for how an investment grows when there is an interest rate being added to the investment at certain intervals. The models can easily allow for a time-varying interest rate, but for forecasting the growth of an investment, it is difficult to predict the future interest rate. One commonly used method is to build a probabilistic model for the development of the interest rate, where the rate is chosen randomly at random times. This gives a random growth of the investment, but by simulating many random scenarios we can compute the mean growth and use the standard deviation as a measure of the uncertainty of the predictions.

Problem   Let p be the annual interest rate in a bank in percent. Suppose the interest is added to the investment q times per year. The new value of the investment, x_n, is given by the previous value of the investment, x_{n-1}, plus the p/q percent interest:

$$x_n = x_{n-1} + \frac{p}{100q} x_{n-1} \,.$$

Normally, the interest is added daily ($q = 360$ and n counts days), but for efficiency in the computations later we shall assume that the interest is added monthly, so $q = 12$ and n counts months.

   The basic assumption now is that p is random and varies with time. Suppose p increases with a random amount γ from one month to the next:

$$p_n = p_{n-1} + \gamma \,.$$

A typical size of p adjustments is 0.5. However, the central bank does not adjust the interest every month. Instead this happens every M months on average. The probability of a $\gamma \neq 0$ can therefore be taken as $1/M$. In a month where $\gamma \neq 0$, we may say that $\gamma = m$ with probability 1/2 or $\gamma = -m$ with probability 1/2 if it is equally likely that the rate goes up as down (this is not a good assumption, but a more complicated change in γ is postponed now).

Solution   First we must develop the precise formulas to be implemented. The difference equations for x_n and p_n are simple in the present case, but the details of computing γ must be worked out.

   In a program, we can draw two random numbers to estimate γ: one for deciding if $\gamma \neq 0$ and the other for determining the sign of the change. Since the probability for $\gamma \neq 0$ is $1/M$, we can draw a number r_1 among the integers $1, \ldots, M$ and if $r_1 = 1$ we continue with drawing a second number r_2 among the integers 1 and 2. If $r_2 = 1$ we set $\gamma = m$, and if $r_2 = 2$ we set $\gamma = -m$. We must also assure that p_n does not take on unreasonable values, so we choose $p_n < 1$ and $p_n > 15$ as cases where p_n is not changed.

   The mathematical model for the investment must track both x_n and p_n. Below we express with precise mathematics the equations for x_n and p_n and the computation of the random γ quantity:

$$x_n = x_{n-1} + \frac{p_{n-1}}{12 \cdot 100} x_{n-1}, \quad i = 1, \ldots, N \tag{8.11}$$

$$r_1 = \text{random integer in } [1, M] \tag{8.12}$$

$$r_2 = \text{random integer in } [1, 2] \tag{8.13}$$

$$\gamma = \begin{cases} m, & \text{if } r_1 = 1 \text{ and } r_2 = 1, \\ -m, & \text{if } r_1 = 1 \text{ and } r_2 = 2, \\ 0, & \text{if } r_1 \neq 1 \end{cases} \tag{8.14}$$

$$p_n = p_{n-1} + \begin{cases} \gamma, & \text{if } p_{n-1} + \gamma \in [1, 15], \\ 0, & \text{otherwise} \end{cases} \tag{8.15}$$

We remark that the evolution of p_n is much like a random walk process (Sect. 8.6), the only differences is that the plus/minus steps are taken at some random points among the times $0, 1, 2, \ldots, N$ rather than at all times $0, 1, 2, \ldots, N$. The random walk for p_n also has barriers at $p = 1$ and $p = 15$, but that is common in a standard random walk too.

Each time we calculate the x_n sequence in the present application, we get a different development because of the random numbers involved. We say that one development of $x_0, \ldots, x_n$ is a *path* (or realization, but since the realization can be viewed as a curve x_n or p_n versus n in this case, it is common to use the word path). Our Monte Carlo simulation approach consists of computing a large number of paths, as well as the sum of the path and the sum of the paths squared. From the latter two sums we can compute the mean and standard deviation of the paths to see the average development of the investment and the uncertainty of this development. Since we are interested in complete paths, we need to store the complete sequence of x_n for each path. We may also be interested in the statistics of the interest rate so we store the complete sequence p_n too.

Programs should be built in pieces so that we can test each piece before testing the whole program. In the present case, a natural piece is a function that computes one path of x_n and p_n with N steps, given M, m, and the initial conditions x_0 and p_0. We can then test this function before moving on to calling the function a large number of times. An appropriate code may be

```
def simulate_one_path(N, x0, p0, M, m):
    x = np.zeros(N+1)
    p = np.zeros(N+1)
    index_set = range(0, N+1)

    x[0] = x0
    p[0] = p0

    for n in index_set[1:]:
        x[n] = x[n-1] + p[n-1]/(100.0*12)*x[n-1]
```

```
        # Update interest rate p
        r = random.randint(1, M)
        if r == 1:
            # Adjust gamma
            r = random.randint(1, 2)
            gamma = m if r == 1 else -m
        else:
            gamma = 0
        pn = p[n-1] + gamma
        p[n] = pn if 1 <= pn <= 15 else p[n-1]
    return x, p
```

Testing such a function is challenging because the result is different each time because of the random numbers. A first step in verifying the implementation is to turn off the randomness ($m = 0$) and check that the deterministic parts of the difference equations are correctly computed:

```
x, p = simulate_one_path(3, 1, 10, 1, 0)
print x
```

The output becomes

```
[ 1.           1.00833333   1.01673611   1.02520891]
```

These numbers can quickly be checked against the famous formula

$$x_n = x_0 \left(1 + \frac{p}{12 \cdot 100}\right)^n$$

in an interactive session:

```
>>> def g(x0, n, p):
...     return x0*(1 + p/(12.*100))**n
...
>>> g(1, 1, 10)
1.0083333333333333
>>> g(1, 2, 10)
1.0167361111111111
>>> g(1, 3, 10)
1.0252089120370369
```

We can conclude that our function works well when there is no randomness. A next step is to carefully examine the code that computes gamma and compare with the mathematical formulas.

Simulating many paths and computing the average development of x_n and p_n is a matter of calling simulate_one_path repeatedly, use two arrays xm and pm to collect the sum of x and p, respectively, and finally obtain the average path by dividing xm and pm by the number of paths we have computed:

```
def simulate_n_paths(n, N, L, p0, M, m):
    xm = np.zeros(N+1)
    pm = np.zeros(N+1)
    for i in range(n):
        x, p = simulate_one_path(N, L, p0, M, m)
        # Accumulate paths
        xm += x
        pm += p
    # Compute average
    xm /= float(n)
    pm /= float(n)
    return xm, pm
```

We can also compute the standard deviation of the paths using formulas (8.3) and (8.6), with x_j as either an x or a p array. It might happen that small rounding errors generate a small *negative* variance, which mathematically should have been slightly greater than zero. Taking the square root will then generate complex arrays and problems with plotting. To avoid this problem, we therefore replace all negative elements by zeros in the variance arrays before taking the square root. The new lines for computing the standard deviation arrays xs and ps are indicated below:

```
def simulate_n_paths(n, N, x0, p0, M, m):
    ...
    xs = np.zeros(N+1)   # standard deviation of x
    ps = np.zeros(N+1)   # standard deviation of p
    for i in range(n):
        x, p = simulate_one_path(N, x0, p0, M, m)
        # Accumulate paths
        xm += x
        pm += p
        xs += x**2
        ps += p**2

    ...
    # Compute standard deviation
    xs = xs/float(n) - xm*xm   # variance
    ps = ps/float(n) - pm*pm   # variance
    # Remove small negative numbers (round off errors)
    xs[xs < 0] = 0
    ps[ps < 0] = 0
    xs = np.sqrt(xs)
    ps = np.sqrt(ps)
    return xm, xs, pm, ps
```

A remark regarding the efficiency of array operations is appropriate here. The statement xs += x**2 could equally well, from a mathematical point of view, be written as xs = xs + x**2. However, in this latter statement, two extra arrays are created (one for the squaring and one for the sum), while in the former only one array (x**2) is made. Since the paths can be long and we make many simulations, such optimizations can be important.

One may wonder whether x**2 is wise in the sense that squaring is detected and computed as x*x, not as a general (slow) power function. This is indeed the case for arrays, as we have investigated in the little test program `smart_power.py`. This program applies time measurement methods from Sect. H.8.2.

Our `simulate_n_paths` function generates four arrays that are natural to visualize. Having a mean and a standard deviation curve, it is often common to plot the mean curve with one color or linetype and then two curves, corresponding to plus one and minus one standard deviation, with another less visible color. This gives an indication of the mean development and the uncertainty of the underlying process. We therefore make two plots: one with `xm`, `xm+xs`, and `xm-xs`, and one with `pm`, `pm+ps`, and `pm-ps`.

Both for debugging and curiosity it is handy to have some plots of a few actual paths. We may pick out 5 paths from the simulations and visualize these:

```
def simulate_n_paths(n, N, x0, p0, M, m):
    ...
    for i in range(n):
        ...
        # Show 5 random sample paths
        if i % (n/5) == 0:
            figure(1)
            plot(x, title='sample paths of investment')
            hold('on')
            figure(2)
            plot(p, title='sample paths of interest rate')
            hold('on')
    figure(1); savefig('tmp_sample_paths_investment.pdf')
    figure(2); savefig('tmp_sample_paths_interestrate.pdf')
    ...
    return ...
```

Note the use of `figure`: we need to hold on both figures to add new plots and switch between the figures, both for screen plotting and calls to `savefig`.

After the visualization of sample paths we make the mean ± standard deviation plots by this code:

```
xm, xs, pm, ps = simulate_n_paths(n, N, x0, p0, M, m)
figure(3)
months = range(len(xm))   # indices along the x axis
plot(months, xm, 'r',
     months, xm-xs, 'y',
     months, xm+xs, 'y',
     title='Mean +/- 1 st.dev. of investment',
     savefig='tmp_mean_investment.pdf')
figure(4)
plot(months, pm, 'r',
     months, pm-ps, 'y',
     months, pm+ps, 'y',
     title='Mean +/- 1 st.dev. of annual interest rate',
     savefig='tmp_mean_interestrate.pdf')
```

The complete program for simulating the investment development is found in the file `growth_random.py`.

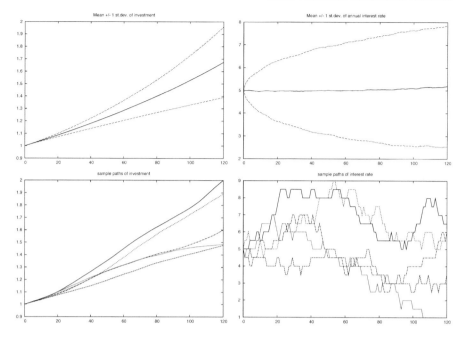

Fig. 8.8 Development of an investment with random jumps of the interest rate at random points of time. *Top left*: mean value of investment ± one standard deviation. *Top right*: mean value of the interest rate ± one standard deviation. *Bottom left*: five paths of the investment development. *Bottom right*: five paths of the interest rate development

Running the program with the input data

```
x0 = 1            # initial investment
p0 = 5            # initial interest rate
N = 10*12         # number of months
M = 3             # p changes (on average) every M months
n = 1000          # number of simulations
m = 0.5           # adjustment of p
```

and initializing the seed of the random generator to 1, we get four plots, which are shown in Fig. 8.8.

8.9 Exercises

Exercise 8.1: Flip a coin times
Make a program that simulates flipping a coin N times. Print out `tail` or `head` for each flip and let the program count and print the number of heads.

Hint Use `r = random.random()` and define head as `r <= 0.5`, or draw an integer among $\{0, 1\}$ with `r = random.randint(0,1)` and define head when r is 0.
Filename: `flip_coin`.

Exercise 8.2: Compute a probability

What is the probability of getting a number between 0.5 and 0.6 when drawing uniformly distributed random numbers from the interval $[0, 1)$? To answer this question empirically, let a program draw N such random numbers using Python's standard `random` module, count how many of them, M, that fall in the interval $[0.5, 0.6]$, and compute the probability as M/N. Run the program with the four values $N = 10^i$ for $i = 1, 2, 3, 6$.
Filename: `compute_prob`.

Exercise 8.3: Choose random colors

Suppose we have eight different colors. Make a program that chooses one of these colors at random and writes out the color.

Hint Use a list of color names and use the `choice` function in the `random` module to pick a list element.
Filename: `choose_color`.

Exercise 8.4: Draw balls from a hat

Suppose there are 40 balls in a hat, of which 10 are red, 10 are blue, 10 are yellow, and 10 are purple. What is the probability of getting two blue and two purple balls when drawing 10 balls at random from the hat?
Filename: `draw_10balls`.

Exercise 8.5: Computing probabilities of rolling dice

This exercise deals with four questions:

1. You throw a die. What is the probability of getting a 6?
2. You throw a die four times in a row. What is the probability of getting 6 all the times?
3. Suppose you have thrown the die three times with 6 coming up all times. What is the probability of getting a 6 in the fourth throw?
4. Suppose you have thrown the die 100 times and experienced a 6 in every throw. What do you think about the probability of getting a 6 in the next throw?

First try to solve the questions from a theoretical or common sense point of view. Thereafter, make functions for simulating cases 1, 2, and 3.
Filename: `rolling_dice`.

Exercise 8.6: Estimate the probability in a dice game

Make a program for estimating the probability of getting at least one die with six eyes when throwing n dice. Read n and the number of experiments from the command line.

As a partial verification, compare the Monte Carlo simulation results to the exact answer 11/36 for $n = 2$ and observe that the approximate probabilities approach the exact probability as the number of simulations grow.
Filename: `one6_ndice`.

Exercise 8.7: Compute the probability of hands of cards
Use the Deck.py module (see Sect. 8.2.5) and the `same_rank` and `same_suit` functions from the `cards` module (see Sect. 8.2.4) to compute the following probabilities by Monte Carlo simulation:

- exactly two pairs among five cards,
- four or five cards of the same suit among five cards,
- four-of-a-kind among five cards.

Filename: `card_hands`.

Exercise 8.8: Decide if a dice game is fair
Somebody suggests the following game. You pay 1 euro and are allowed to throw four dice. If the sum of the eyes on the dice is less than 9, you get paid r euros back, otherwise you lose the 1 euro investment. Assume $r = 10$. Will you then in the long run win or lose money by playing this game? Answer the question by making a program that simulates the game. Read r and the number of experiments N from the command line.
Filename: `sum_4dice`.

Exercise 8.9: Adjust a game to make it fair
It turns out that the game in Exercise 8.8 is not fair, since you lose money in the long run. The purpose of this exercise is to adjust the winning award so that the game becomes fair, i.e., that you neither lose nor win money in the long run.

Make a Python function that computes the probability p of getting a sum less than s when rolling n dice. Use the reasoning in Sect. 8.3.2 to find the award per game, r, that makes the game fair. Run the program from Exercise 8.8 with this r on the command line and verify that the game is now (approximately) fair.
Filename: `sum_ndice_fair`.

Exercise 8.10: Make a test function for Monte Carlo simulation
We consider the Python function in Exercise 8.9 for computing a probability p that the sum of the eyes on n dice is less than s. The aim is to write a test function for verifying the computation of p.

a) Find some combinations of n and s that must result in $p = 0$ and $p = 1$ and make the appropriate code in the test function.
b) Fix the seed of the random number generator and record the first eight random numbers to 16 digits. Set $n = 2$, perform four experiments, and compute by hand what the probability estimate becomes (choose any appropriate s). Write the necessary code in the test function to compare this manually calculated result with the what is produced by the function from Exercise 8.9.

Filename: `test_sum_ndice`.

Exercise 8.11: Generalize a game
Consider the game in Sect. 8.3.2. A generalization is to think as follows: you throw one die until the number of eyes is less than or equal to the previous throw. Let m be the number of throws in a game.

a) Use Monte Carlo simulation to compute the probability of getting $m = 2, 3, 4, \ldots$.

Hint For $m \geq 6$ the throws must be exactly $1, 2, 3, 4, 5, 6, 6, 6, \ldots$, and the probability of each is 1/6, giving the total probability 6^{-m}. Use $N = 10^6$ experiments as this should suffice to estimate the probabilities for $m \leq 5$, and beyond that we have the analytical expression.

b) If you pay 1 euro to play this game, what is the fair amount to get paid when win? Answer this question for each of the cases $m = 2, 3, 4, 5$.

Filename: `incr_eyes`.

Exercise 8.12: Compare two playing strategies
Suggest a player strategy for the game in Sect. 8.4.2. Remove the question in the `player_guess` function in the file `ndice2.py`, and implement the chosen strategy instead. Let the program play a large number of games, and record the number of times the computer wins. Which strategy is best in the long run: the computer's or yours?
Filename: `simulate_strategies1`.

Exercise 8.13: Investigate strategies in a game
Extend the program from Exercise 8.12 such that the computer and the player can use a different number of dice. Let the computer choose a random number of dice between 2 and 20. Experiment to find out if there is a favorable number of dice for the player.
Filename: `simulate_strategies2`.

Exercise 8.14: Investigate the winning chances of some games
An amusement park offers the following game. A hat contains 20 balls: 5 red, 5 yellow, 3 green, and 7 brown. At a cost of $2n$ euros you can draw $4 \leq n \leq 10$ balls at random from the hat (without putting them back). Before you are allowed to look at the drawn balls, you must choose one of the following options:

1. win 60 euros if you have drawn exactly three red balls
2. win $7 + 5\sqrt{n}$ euros if you have drawn at least three brown balls
3. win $n^3 - 26$ euros if you have drawn exactly one yellow ball and one brown ball
4. win 23 euros if you have drawn at least one ball of each color

For each of the $4n$ different types of games you can play, compute the net income (per play) and the probability of winning. Is there any of the games (i.e., any combinations of n and the four options above) where you will win money in the long run?
Filename: `draw_balls`.

Exercise 8.15: Compute probabilities of throwing two dice
Throw two dice a large number of times in a program. Record the sum of the eyes each time and count how many times each of the possibilities for the sum (2, 3, . . ., 12) appear. Compute the corresponding probabilities and compare them with the exact values. (To find the exact probabilities, set up all the 6×6 possible outcomes of throwing two dice, and then count how many of them that has a sum s for $s = 2, 3, \ldots, 12$.)
Filename: `freq_2dice`.

Exercise 8.16: Vectorize flipping a coin
Simulate flipping a coin N times and write out the number of tails. The code should be vectorized, i.e., there must be no loops in Python.

Hint Constructions like `numpy.where(r<=0.5, 1, 0)` combined with `numpy.sum`, or `r[r<=0.5].size`, are useful, where `r` is an array of random numbers between 0 and 1.
Filename: `flip_coin_vec`.

Exercise 8.17: Vectorize a probablility computation
The purpose of this exercise is to speed up the code in Exercise 8.2 by vectorization.

Hint For an array `r` of uniformly distributed random numbers on $[0, 1)$, make use of `r1 = r[r>0.5]` and `r1[r1<0.6]`. An alternative is `numpy.where` combine with a compound boolean expression with `numpy.logical_and(0.5>=r, r<=0.6)`. See the discussion of this topic in Sect. 5.5.3.
Filename: `compute_prob_vec`.

Exercise 8.18: Throw dice and compute a small probability
Use Monte Carlo simulation to compute the probability of getting 6 eyes on all dice when rolling 7 dice.

Hint You need a large number of experiments in this case because of the small probability (see the first paragraph of Sect. 8.3), so a vectorized implementation may be important.
Filename: `roll_7dice`.

Exercise 8.19: Is democracy reliable as a decision maker?
A democracy takes decisions based on majority votes. We shall investigate if this is a good idea or if a single person would produce better decisions.

We shall ask about pure facts, not opinions. This means that the question to be answered by a population has a definite "yes" or "no" answer. For example, "Can Python lists contain tuples as elements?" The correct answer is "yes". Asking a population such a question and relying on the majority of votes, is a reliable procedure if the competence level in the population is sufficiently high.

a) Assume that the competence level in a population can be modeled by a probability p such that if you ask N people a question, $M = pN$ of them will give the correct answer (as $N \to \infty$). Here we make the questionable assumption of a homogeneous population, in the sense that p is the same for every individual. Make a function `homogeneous(p, N)` for simulating whether the majority vote of a population of N individuals arrives at the right answer, if the probability of answering correctly is p for an individual. Make another function `homogeneous_ex()` that runs 10 tests the specific case of $N = 5$ (as when relying on the majority of a student group) and 10 tests when asking a whole city of $N = 1,000,000$ voters. Try $p = 0.49$, $p = 0.51$, and $p = 0.8$. Are the results as you would expect from intuition?

Hint Asking one individual is like flipping a biased coin that has probability p of giving head (right answer) and probability $1 - p$ of giving tail (wrong answer).

b) The problem in a) can be exactly solved, since each question is a Bernoulli trial with success probability p, and the probability of a correct majority vote is the same as the probability of getting $N/2$ *or more* successes in N trials. For large N, the probability of M successes in N trials can be well approximated by a normal (Gaussian) density function:

$$g(M) = (\sqrt{2\pi} Np(1 - p))^{-1} \exp\left(-\frac{1}{2}(M - Np)^2 / (Np(1 - p))\right).$$

The majority vote is correct when $M > N/2$, and the probability of this event is given by $1 - \Phi(N/2)$, where Φ is the cumulative normal distribution with mean Np and variance $Np(1 - p)$.

Plot the probability of being right against p.

Say 5 questions are of importance. What competence level p does a king need to have all 5 right compared to the population having all 5 right.

c) We shall now simulate voting in a *heterogeneous* population. The probability that an individual no. i answers correctly is p_i, where p_i is drawn from a normal (Gaussian) probability density with mean p and standard deviation s. The competence level varies between individuals, with s expressing the spreading of knowledge and p the mean competence level.

Make function `heterogeneous(p, N, s)` for returning whether the majority vote is right or wrong in the heterogeneous case. Rerun the examples from a) with $s = 0.2$.

d) With a somewhat large variation of the population, i.e., s somewhat large, there will be some individuals that always provide wrong or right answers according to this model. To learn about reasonable values s we can investigate unreasonable large amounts of people who are *always* right or wrong.

The probability of always being wrong is the probability of $p_i < 0$. This is given by $\Phi(-p/s)$, where Φ is the cumulative normal distribution with mean zero and unit standard deviation. It can be reached in Python as `scipy.stats.norm.cdf`. The probability of always being right is the probability of $p_i > 1$, which can be computed as $1 - \Phi((1 - p)/s)$. Plot curves of the probability of always being right and always wrong against $s \in [0.1, 0.6]$. Perform this curve plotting in a function `extremes(p)`.

Filename: `democracy`.

Exercise 8.20: Difference equation for random numbers
Simple random number generators are based on simulating difference equations. Here is a typical set of two equations:

$$x_n = (ax_{n-1} + c) \bmod m, \tag{8.16}$$

$$y_n = x_n/m, \tag{8.17}$$

for $n = 1, 2, \ldots$. A seed x_0 must be given to start the sequence. The numbers $y_1, y_2, \ldots$ represent the random numbers and $x_0, x_1, \ldots$ are "help" numbers. Although y_n is completely deterministic from (8.16)–(8.17), the sequence y_n *appears* random. The mathematical expression $p \bmod q$ is coded as p % q in Python.

Use $a = 8121$, $c = 28411$, and $m = 134456$. Solve the system (8.16)–(8.17) in a function to generate N random numbers. Make a histogram to examine the distribution of the numbers (the y_n numbers are uniformly distributed if the histogram is approximately flat).
Filename: `diffeq_random`.

Exercise 8.21: Make a class for drawing balls from a hat
Consider the example about drawing colored balls from a hat in Sect. 8.3.3. It could be handy to have an object that acts as a hat:

```
hat = Hat(red=3, blue=4, green=6)
balls = hat.draw(3)
if balls.count('red') == 1 and balls.count('green') == 2:
    ...
```

a) Write such a class Hat with the shown functionality.

Hint 1 The flexible syntax in the constructor, where the colors of the balls and the number of balls of each color are freely specified, requires use of a dictionary (**kwargs) for handling a variable number of keyword arguments, see Sect. H.7.2.

Hint 2 You can borrow useful code from the `balls_in_hat.py` program and ideas from Sect. 8.2.5.

b) Apply class Hat to compute the probability of getting 2 brown and 2 blue galls when drawing 6 balls from a hat with 6 blue, 8 brown, and 3 green balls.

Filename: Hat.

Exercise 8.22: Independent versus dependent random numbers

a) Generate a sequence of N independent random variables with values 0 or 1 and print out this sequence without space between the numbers (i.e., as 001011010110111010).
b) The purpose now is to generate random zeros and ones that are dependent. If the last generated number was 0, the probability of generating a new 0 is p and a new 1 is $1 - p$. Conversely, if the last generated was 1, the probability of

generating a new 1 is p and a new 0 is $1 - p$. Since the new value depends on the last one, we say the variables are dependent. Implement this algorithm in a function returning an array of N zeros and ones. Print out this array in the condense format as described above.

c) Choose $N = 80$ and try the probabilities $p = 0.5$, $p = 0.8$ and $p = 0.9$. Can you by visual inspection of the output characterize the differences between sequences of independent and dependent random variables?

Filename: `dependent_random_numbers`.

Exercise 8.23: Compute the probability of flipping a coin

a) Simulate flipping a coin N times.

Hint Draw N random integers 0 and 1 using `numpy.random.randint`.

b) Look at a subset $N_1 \leq N$ of the experiments in a) and compute the probability of getting a head (M_1/N_1, where M_1 is the number of heads in N_1 experiments). Choose $N = 1000$ and print out the probability for $N_1 = 10, 100, 500, 1000$. Generate just N numbers once in the program. How do you think the accuracy of the computed probability vary with N_1? Is the output compatible with this expectation?

c) Now we want to study the probability of getting a head, p, as a function of N_1, i.e., for $N_1 = 1, \ldots, N$. A first try to compute the probability array for p is

```
import numpy as np
h = np.where(r <= 0.5, 1, 0)
p = np.zeros(N)
for i in range(N):
    p[i] = np.sum(h[:i+1])/float(i+1)
```

Implement these computations in a function.

d) An array `q[i] = np.sum(h([:i]))` reflects a *cumulative sum* and can be efficiently generated by `np.cumsum`: `q = np.cumsum(h)`. Thereafter we can compute p by `q/I`, where `I[i]=i+1` and I can be computed by `np.arange(1,N+1)` or `r_[1:N+1]` (integers 1, 2, ..., up to but not including N+1). Use `cumsum` to make an alternative vectorized version of the function in c).

e) Write a test function that verifies that the implementations in c) and d) give the same results.

Hint Use `numpy.allclose` to compare two arrays.

f) Make a function that applies the `time` module to measure the relative efficiency of the implementations in c) and d).

g) Plot p against I for the case where $N = 10000$. Annotate the axis and the plot with relevant text.

Filename: `flip_coin_prob`.

Exercise 8.24: Simulate binomial experiments

Exercise 4.24 describes some problems that can be solved exactly using the formula (4.8), but we can also simulate these problems and find approximate numbers for the probabilities. That is the task of this exercise.

Make a general function `simulate_binomial(p, n, x)` for running n experiments, where each experiment have two outcomes, with probabilities p and $1 - p$. The n experiments constitute a *success* if the outcome with probability p occurs exactly x times. The `simulate_binomial` function must repeat the n experiments N times. If M is the number of successes in the N experiments, the probability estimate is M/N. Let the function return this probability estimate together with the error (the exact result is (4.8)). Simulate the three cases in Exercise 4.24 using this function.
Filename: `simulate_binomial`.

Exercise 8.25: Simulate a poker game

Make a program for simulating the development of a poker (or simplified poker) game among n players. Use ideas from Sect. 8.2.4.
Filename: `poker`.

Exercise 8.26: Estimate growth in a simulation model

The simulation model in Sect. 8.3.5 predicts the number of individuals from generation to generation. Make a simulation of the one son policy with 10 generations, a male portion of 0.51 among newborn babies, set the fertility to 0.92, and assume that a fraction 0.06 of the population will break the law and want 6 children in a family. These parameters implies a significant growth of the population. See if you can find a factor r such that the number of individuals in generation n fulfills the difference equation

$$x_n = (1 + r)x_{n-1} .$$

Hint Compute r for two consecutive generations x_{n-1} and x_n ($r = x_n/x_{n-1} - 1$) and see if r is approximately constant as n increases.
Filename: `estimate_growth`.

Exercise 8.27: Investigate guessing strategies

In the game from Sect. 8.4.1 it is smart to use the feedback from the program to track an interval $[p, q]$ that must contain the secret number. Start with $p = 1$ and $q = 100$. If the user guesses at some number n, update p to $n + 1$ if n is less than the secret number (no need to care about numbers smaller than $n + 1$), or update q to $n - 1$ if n is larger than the secret number (no need to care about numbers larger than $n - 1$).

Are there any smart strategies to pick a new guess $s \in [p, q]$? To answer this question, investigate two possible strategies: s as the midpoint in the interval $[p, q]$, or s as a uniformly distributed random integer in $[p, q]$. Make a program that implements both strategies, i.e., the player is not prompted for a guess but the computer computes the guess based on the chosen strategy. Let the program run a large number of games and see if one of the strategies can be considered as superior in the long run.
Filename: `strategies4guess`.

Exercise 8.28: Vectorize a dice game
Vectorize the simulation program from Exercise 8.8 with the aid of the module numpy.random and the numpy.sum function.
Filename: sum9_4dice_vec.

Exercise 8.29: Compute π by a Monte Carlo method
Use the method in Sect. 8.5.3 to compute π by computing the area of a circle. Choose G as the circle with its center at the origin and with unit radius, and choose B as the rectangle $[-1, 1] \times [-1, 1]$. A point (x, y) lies within G if $x^2 + y^2 < 1$. Compare the approximate π with math.pi.
Filename: MC_pi.

Exercise 8.30: Compute π by a Monte Carlo method
This exercise has the same purpose of computing π as in Exercise 8.29, but this time you should choose G as a circle with center at $(2, 1)$ and radius 4. Select an appropriate rectangle B. A point (x, y) lies within a circle with center at (x_c, y_c) and with radius R if $(x - x_c)^2 + (y - y_c)^2 < R^2$.
Filename: MC_pi2.

Exercise 8.31: Compute π by a random sum

a) Let $x_0, \ldots, x_N$ be $N + 1$ uniformly distributed random numbers between 0 and 1. Explain why the random sum $S_N = (N + 1)^{-1} \sum_{i=0}^{N} 2(1 - x_i^2)^{-1/2}$ is an approximation to π.

Hint Interpret the sum as Monte Carlo integration and compute the corresponding integral by hand or sympy.

b) Compute $S_0, S_1, \ldots, S_N$ (using just one set of $N + 1$ random numbers). Plot this sequence versus N. Also plot the horizontal line corresponding to the value of π. Choose N large, e.g., $N = 10^6$.

Filename: MC_pi_plot.

Exercise 8.32: 1D random walk with drift
Modify the walk1D.py program such that the probability of going to the right is r and the probability of going to the left is $1 - r$ (draw numbers in $[0, 1)$ rather than integers in $\{1, 2\}$). Compute the average position of n_p particles after 100 steps, where n_p is read from the command line. Mathematically one can show that the average position approaches $rn_s - (1 - r)n_s$ as $n_p \to \infty$ (n_s is the number of walks). Write out this exact result together with the computed mean position with a finite number of particles.
Filename: walk1D_drift.

Exercise 8.33: 1D random walk until a point is hit
Set np=1 in the walk1Dv.py program and modify the program to measure how many steps it takes for one particle to reach a given point $x = x_p$. Give x_p on the command line. Report results for $x_p = 5, 50, 5000, 50000$.
Filename: walk1Dv_hit_point.

Exercise 8.34: Simulate making a fortune from gaming

A man plays a game where the probability of winning is p and that of losing is consequently $1 - p$. When winning he earns 1 euro and when losing he loses 1 euro. Let x_i be the man's fortune from playing this game i number of times. The starting fortune is x_0. We assume that the man gets a necessary loan if $x_i < 0$ such that the gaming can continue. The target is a fortune F, meaning that the playing stops when $x = F$ is reached.

a) Explain why x_i is a 1D random walk.
b) Modify one of the 1D random walk programs to simulate the average number of games it takes to reach the target fortune $x = F$. This average must be computed by running a large number of random walks that start at x_0 and reach F. Use $x_0 = 10$, $F = 100$, and $p = 0.49$ as example.
c) Suppose the average number of games to reach $x = F$ is proportional to $(F - x_0)^r$, where r is some exponent. Try to find r by experimenting with the program. The r value indicates how difficult it is to make a substantial fortune by playing this game. Note that the *expected* earning is negative when $p < 0.5$, but there is still a small probability for hitting $x = F$.

Filename: `game_as_walk1D`.

Exercise 8.35: Simulate pollen movements as a 2D random walk

The motion of single particles can often be described as random walks. On a water surface, 1000 grains of pollen are placed in a single point. The movement of the pollen grains can be modeled by a random walk model, where for each second each grain will move a random distance, along a two-dimensional vector, whose two components are independently normally distributed with expectation 0 mm and standard deviation 0.05 mm.

a) Make a function that implements this kind of 2D random walk. Return an array with the position of each grain for each step.
b) Make a movie that shows the position of the pollen grains from 0 to 100 seconds.
c) Make a plot of the mean distance from the origin versus time. What do you see?

Filename: `pollen`.

Exercise 8.36: Make classes for 2D random walk

The purpose of this exercise is to reimplement the `walk2D.py` program from Sect. 8.7.1 with the aid of classes.

a) Make a class `Particle` with the coordinates (x, y) and the time step number of a particle as data attributes. A method `move` moves the particle in one of the four directions and updates the (x, y) coordinates. Another class, `Particles`, holds a list of `Particle` objects and a `plotstep` parameter (as in `walk2D.py`). A method `move` moves all the particles one step, a method `plot` can make a plot of all particles, while a method `moves` performs a loop over time steps and calls `move` and `plot` in each step.

b) Equip the `Particle` and `Particles` classes with print functionality such that one can print out all particles in a nice way by saying `print p` (for a `Particles` instance p) or `print self` (inside a method).

Hint In `__str__`, apply the `pformat` function from the `pprint` module to the list of particles, and make sure that `__repr__` just reuse `__str__` in both classes so the output looks nice.

c) Make a test function that compares the first three positions of four particles with the corresponding results computed by the `walk2D.py` program. The seed of the random number generator must of course be fixed identically in the two programs.

d) Organize the complete code as a module such that the classes `Particle` and `Particles` can be reused in other programs. The test block should read the number of particles from the command line and perform a simulation.

e) Compare the efficiency of the class version against the vectorized version in `walk2Dv.py`, using the techniques in Sect. H.8.1.

f) The program developed above cannot be vectorized as long as we base the implementation on class `Particle`. However, if we remove that class and focus on class `Particles`, the latter can employ arrays for holding the positions of all particles and vectorized updates of these positions in the `moves` method. Use ideas from the `walk2Dv.py` program to make a new class `Particles_vec` which vectorizes `Particles`.

g) Verify the code against the `walk2Dv.py` program as explained in c). Automate the verification in a test function.

h) Write a Python function that measures the computational efficiency the vectorized class `Particles_vec` and the scalar class `Particles`.

Filename: `walk2D_class`.

Exercise 8.37: 2D random walk with walls; scalar version
Modify the `walk2D.py` or `walk2Dc.py` programs from Exercise 8.36 so that the walkers cannot walk outside a rectangular area $A = [x_L, x_H] \times [y_L, y_H]$. Do not move the particle if its new position is outside A.
Filename: `walk2D_barrier`.

Exercise 8.38: 2D random walk with walls; vectorized version
Modify the `walk2Dv.py` program so that the walkers cannot walk outside a rectangular area $A = [x_L, x_H] \times [y_L, y_H]$.

Hint First perform the moves of one direction. Then test if new positions are outside A. Such a test returns a boolean array that can be used as index in the position arrays to pick out the indices of the particles that have moved outside A and move them back to the relevant boundary of A.
Filename: `walk2Dv_barrier`.

Exercise 8.39: Simulate mixing of gas molecules
Suppose we have a box with a wall dividing the box into two equally sized parts. In
one part we have a gas where the molecules are uniformly distributed in a random
fashion. At $t = 0$ we remove the wall. The gas molecules will now move around
and eventually fill the whole box.

This physical process can be simulated by a 2D random walk inside a fixed
area A as introduced in Exercises 8.37 and 8.38 (in reality the motion is three-
dimensional, but we only simulate the two-dimensional part of it since we already
have programs for doing this). Use the program from either Exercises 8.37 or 8.38
to simulate the process for $A = [0, 1] \times [0, 1]$. Initially, place 10000 particles at
uniformly distributed random positions in $[0, 1/2] \times [0, 1]$. Then start the random
walk and visualize what happens. Simulate for a long time and make a hardcopy of
the animation (an animated GIF file, for instance). Is the end result what you would
expect?
Filename: `disorder1`.

Remarks Molecules tend to move randomly because of collisions and forces between molecules. We do not model collisions between particles in the random
walk, but the nature of this walk, with random movements, simulates the effect
of collisions. Therefore, the random walk can be used to model molecular motion
in many simple cases. In particular, the random walk can be used to investigate how
a quite ordered system, where one gas fills one half of a box, evolves through time
to a more disordered system.

Exercise 8.40: Simulate slow mixing of gas molecules
Solve Exercise 8.39 when the wall dividing the box is not completely removed, but
instead has a small hole.
Filename: `disorder2`.

Exercise 8.41: Guess beer brands
You are presented n glasses of beer, each containing a different brand. You are
informed that there are $m \geq n$ possible brands in total, and the names of all brands
are given. For each glass, you can pay p euros to taste the beer, and if you guess
the right brand, you get $q \geq p$ euros back. Suppose you have done this before and
experienced that you typically manage to guess the right brand T times out of 100,
so that your probability of guessing the right brand is $b = T/100$.

Make a function `simulate(m, n, p, q, b)` for simulating the beer tasting
process. Let the function return the amount of money earned and how many correct
guesses ($\leq n$) you made. Call `simulate` a large number of times and compute the
average earnings and the probability of getting full score in the case $m = n = 4$,
$p = 3, q = 6$, and $b = 1/m$ (i.e., four glasses with four brands, completely random
guessing, and a payback of twice as much as the cost). How much more can you
earn from this game if your ability to guess the right brand is better, say $b = 1/2$?
Filename: `simulate_beer_tasting`.

Exercise 8.42: Simulate stock prices
A common mathematical model for the evolution of stock prices can be formulated
as a difference equation

$$x_n = x_{n-1} + \Delta t \mu x_{n-1} + \sigma x_{n-1} \sqrt{\Delta t} \, r_{n-1}, \qquad (8.18)$$

where x_n is the stock price at time t_n, Δt is the time interval between two time
levels ($\Delta t = t_n - t_{n-1}$), μ is the growth rate of the stock price, σ is the volatility
of the stock price, and $r_0, \ldots, r_{n-1}$ are normally distributed random numbers with
mean zero and unit standard deviation. An initial stock price x_0 must be prescribed
together with the input data μ, σ, and Δt.

We can make a remark that (8.18) is a Forward Euler discretization of a stochastic
differential equation for a continuous price function $x(t)$:

$$\frac{dx}{dt} = \mu x + \sigma N(t),$$

where $N(t)$ is a so-called white noise random time series signal. Such equations
play a central role in modeling of stock prices.

Make R realizations of (8.18) for $n = 0, \ldots, N$ for $N = 5000$ steps over a time
period of $T = 180$ days with a step size $\Delta t = T/N$.
Filename: `stock_prices`.

Exercise 8.43: Compute with option prices in finance
In this exercise we are going to consider the pricing of so-called Asian options.
An Asian option is a financial contract where the owner earns money when certain
market conditions are satisfied.

The contract is specified by a *strike price* K and a maturity time T. It is written
on the average price of the underlying stock, and if this average is bigger than the
strike K, the owner of the option will earn the difference. If, on the other hand,
the average becomes less, the owner receives nothing, and the option matures in the
value zero. The average is calculated from the last trading price of the stock for
each day.

From the theory of options in finance, the price of the Asian option will be the
expected present value of the payoff. We assume the stock price dynamics given as,

$$S(t + 1) = (1 + r)S(t) + \sigma S(t)\epsilon(t), \qquad (8.19)$$

where r is the interest-rate, and σ is the volatility of the stock price. The time t
is supposed to be measured in days, $t = 0, 1, 2, \ldots$, while $\epsilon(t)$ are independent
identically distributed normal random variables with mean zero and unit standard
deviation. To find the option price, we must calculate the expectation

$$p = (1 + r)^{-T} \mathrm{E}\left[\max\left(\frac{1}{T}\sum_{t=1}^{T} S(t) - K, 0\right)\right]. \qquad (8.20)$$

The price is thus given as the expected discounted payoff. We will use Monte Carlo
simulations to estimate the expectation. Typically, r and σ can be set to $r = 0.0002$
and $\sigma = 0.015$. Assume further $S(0) = 100$.

a) Make a function that simulates a path of $S(t)$, that is, the function computes $S(t)$ for $t = 1, \ldots, T$ for a given T based on the recursive definition in (8.19). The function should return the path as an array.

b) Create a function that finds the average of $S(t)$ from $t = 1$ to $t = T$. Make another function that calculates the price of the Asian option based on N simulated averages. You may choose $T = 100$ days and $K = 102$.

c) Plot the price p as a function of N. You may start with $N = 1000$.

d) Plot the error in the price estimation as a function N (assume that the p value corresponding to the largest N value is the "right" price). Try to fit a curve of the form $c/\sqrt{N}$ for some c to this error plot. The purpose is to show that the error is reduced as $1/\sqrt{N}$.

Filename: `option_price`.

Remarks If you wonder where the values for r and σ come from, you will find the explanation in the following. A reasonable level for the yearly interest-rate is around 5 percent, which corresponds to a daily rate $0.05/250 = 0.0002$. The number 250 is chosen because a stock exchange is on average open this amount of days for trading. The value for σ is calculated as the volatility of the stock price, corresponding to the standard deviation of the daily returns of the stock defined as $(S(t + 1) - S(t))/S(t)$. "Normally", the volatility is around 1.5 percent a day. Finally, there are theoretical reasons why we assume that the stock price dynamics is driven by r, meaning that we consider the *risk-neutral* dynamics of the stock price when pricing options. There is an exciting theory explaining the appearance of r in the dynamics of the stock price. If we want to simulate a stock price dynamics mimicing what we see in the market, r in (8.19) must be substituted with μ, the expected return of the stock. Usually, μ is higher than r.

Exercise 8.44: Differentiate noise measurements

In a laboratory experiment waves are generated through the impact of a model slide into a wave tank. (The intention of the experiment is to model a future tsunami event in a fjord, generated by loose rocks that fall into the fjord.) At a certain location, the elevation of the surface, denoted by η, is measured at discrete points in time using an ultra-sound wave gauge. The result is a time series of vertical positions of the water surface elevations in meter: $\eta(t_0), \eta(t_1), \eta(t_2), \ldots, \eta(t_n)$. There are 300 observations per second, meaning that the time difference between two neighboring measurement values $\eta(t_i)$ and $\eta(t_{i+1})$ is $h = 1/300$ second.

a) Read the η values in the file `gauge.dat`[4] into an array `eta`. Read h from the command line.

b) Plot `eta` versus the time values.

c) Compute the velocity v of the surface by the formula

$$v_i \approx (\eta_{i+1} - \eta_{i-1})/(2h), \quad i = 1, \ldots, n - 1.$$

Plot v versus time values in a separate plot.

---

[4] http://tinyurl.com/pwyasaa/random/gauge.dat

d) Compute the acceleration a of the surface by the formula

$$a_i \approx (\eta_{i+1} - 2\eta_i + \eta_{i-1})/h^2, \quad i = 1, \ldots, n - 1.$$

Plot a versus the time values in a separate plot.

e) If we have a noisy signal η_i, where $i = 0, \ldots, n$ counts time levels, the noise can be reduced by computing a new signal where the value at a point is a weighted average of the values at that point and the neighboring points at each side. More precisely, given the signal η_i, $i = 0, \ldots, n$, we compute a filtered (averaged) signal with values $\eta_i^{(1)}$ by the formula

$$\eta_i^{(1)} = \frac{1}{4}(\eta_{i+1} + 2\eta_i + \eta_{i-1}), \quad i = 1, \ldots, n - 1, \ \eta_0^{(1)} = \eta_0, \ \eta_n^{(1)} = \eta_n.$$
(8.21)

Make a function `filter` that takes the η_i values in an array `eta` as input and returns the filtered $\eta_i^{(1)}$ values in an array.

f) Let $\eta_i^{(k)}$ be the signal arising by applying the `filtered` function k times to the same signal. Make a plot with curves η_i and the filtered $\eta_i^{(k)}$ values for $k = 1, 10, 100$. Make similar plots for the velocity and acceleration where these are made from both the original, measured η data and the filtered data. Discuss the results.

Filename: `labstunami`.

Exercise 8.45: Differentiate noisy signals
The purpose of this exercise is to look into numerical differentiation of time series signals that contain measurement errors. This insight might be helpful when analyzing the noise in real data from a laboratory experiment in Exercise 8.44.

a) Compute a signal

$$\bar{\eta}_i = A \sin(\frac{2\pi}{T} t_i), \quad t_i = i\frac{T}{40}, \ i = 0, \ldots, 200.$$

Display $\bar{\eta}_i$ versus time t_i in a plot. Choose $A = 1$ and $T = 2\pi$. Store the $\bar{\eta}$ values in an array `etabar`.

b) Compute a signal with random noise E_i,

$$\eta_i = \bar{\eta}_i + E_i,$$

E_i is drawn from the normal distribution with mean zero and standard deviation $\sigma = 0.04A$. Plot this η_i signal as circles in the same plot as $\bar{\eta}_i$. Store the E_i in an array `E` for later use.

c) Compute the first derivative of $\bar{\eta}_i$ by the formula

$$\frac{\bar{\eta}_{i+1} - \bar{\eta}_{i-1}}{2h}, \quad i = 1, \ldots, n - 1,$$

and store the values in an array `detabar`. Display the graph.

d) Compute the first derivative of the error term by the formula

$$\frac{E_{i+1} - E_{i-1}}{2h}, \quad i = 1, \ldots, n-1,$$

and store the values in an array dE. Calculate the mean and the standard deviation of dE.

e) Plot detabar and detabar + dE. Use the result of the standard deviation calculations to explain the qualitative features of the graphs.

f) The second derivative of a time signal η_i can be computed by

$$\frac{\eta_{i+1} - 2\eta_i + \eta_{i-1}}{h^2}, \quad i = 1, \ldots, n-1.$$

Use this formula on the etabar data and save the result in d2etabar. Also apply the formula to the E data and save the result in d2E. Plot d2etabar and d2etabar + d2E. Compute the standard deviation of d2E and compare with the standard deviation of dE and E. Discuss the plot in light of these standard deviations.

Filename: sine_noise.

Exercise 8.46: Model noise in a time signal

We assume that the measured data can be modeled as a smooth time signal $\bar{\eta}(t)$ plus a random variation $E(t)$. Computing the velocity of $\eta = \bar{\eta} + E$ results in a smooth velocity from the $\bar{\eta}$ term and a noisy signal from the E term.

a) We can estimate the level of noise in the first derivative of E as follows. The random numbers $E(t_i)$ are assumed to be independent and normally distributed with mean zero and standard deviation σ. It can then be shown that

$$\frac{E_{i+1} - E_{i-1}}{2h}$$

produces numbers that come from a normal distribution with mean zero and standard deviation $2^{-1/2}h^{-1}\sigma$. How much is the original noise, reflected by σ, magnified when we use this numerical approximation of the velocity?

b) The fraction

$$\frac{E_{i+1} - 2E_i + E_{i-1}}{h^2}$$

will also generate numbers from a normal distribution with mean zero, but this time with standard deviation $2h^{-2}\sigma$. Find out how much the noise is magnified in the computed acceleration signal.

c) The numbers in the gauge.dat file in Exercise 8.44 are given with 5 digits. This is no certain indication of the accuracy of the measurements, but as a test we may assume σ is of the order 10^{-4}. Check if the visual results for the velocity and acceleration are consistent with the standard deviation of the noise in these signals as modeled above.

Exercise 8.47: Speed up Markov chain mutation

The functions `transition` and `mutate_via_markov_chain` from Sect. 8.3.4 were made for being easy to read and understand. Upon closer inspection, we realize that the `transition` function constructs the `interval_limits` every time a random transition is to be computed, and we want to run a large number of transitions. By merging the two functions, pre-computing interval limits for each `from_base`, and adding a loop over N mutations, one can reduce the computation of interval limits to a minimum. Perform such an efficiency enhancement. Measure the CPU time of this new function versus the `mutate_via_markov_chain` function for 1 million mutations.

Filename: `markov_chain_mutation2`.

Object-Oriented Programming

9

This chapter introduces the basic ideas of object-oriented programming. Different people put different meanings into the term object-oriented programming: some use the term for programming with objects in general, while others use the term for programming with class hierarchies. The author applies the second meaning, which is the most widely accepted one in computer science. The first meaning is better named *object-based* programming. Since everything in Python is an object, we do object-based programming all the time, yet one usually reserves this term for the case when classes different from Python's basic types (`int`, `float`, `str`, `list`, `tuple`, `dict`) are involved.

Necessary background for the present chapter includes basic knowledge about classes in Python, at least concepts such as attributes (method attributes, data attributes), methods, constructors, the `self` object, and the `__call__` special method. Suitable material for this background is Sects. 7.1, 7.2, and 7.3.1. For Sects. 9.2 and 9.3 one must know the most basic methods for numerical differentiation and integration, for example from Appendix B. During an initial reading of the chapter, it can be beneficial to skip the more advanced material in Sects. 9.2.4–9.2.7.

All the programs associated with this chapter are found in the folder `src/oo`[1].

9.1 Inheritance and Class Hierarchies

Most of this chapter tells you how to put related classes together in families such that the family can be viewed as one unit. This idea helps to hide details in a program, and makes it easier to modify or extend the program.

A family of classes is known as a *class hierarchy*. As in a biological family, there are parent classes and child classes. Child classes can *inherit* data and methods from parent classes, they can modify these data and methods, and they can add their own data and methods. This means that if we have a class with some functionality, we can extend this class by creating a child class and simply add the functionality we need. The original class is still available and the separate child class is small, since it does not need to repeat the code in the parent class.

---

[1] http://tinyurl.com/pwyasaa/oo

© Springer-Verlag Berlin Heidelberg 2016
H.P. Langtangen, *A Primer on Scientific Programming with Python*,
Texts in Computational Science and Engineering 6, DOI 10.1007/978-3-662-49887-3_9

The magic of object-oriented programming is that other parts of the code do not need to distinguish whether an object is the parent or the child – all generations in a family tree can be treated as a unified object. In other words, one piece of code can work with all members in a class family or hierarchy. This principle has revolutionized the development of large computer systems. As an illustration, two of the most widely used computer languages today are Java and C#, and both of them force programs to be written in an object-oriented style.

The concepts of classes and object-oriented programming first appeared in the Simula programming language in the 1960s. Simula was invented by the Norwegian computer scientists Ole-Johan Dahl and Kristen Nygaard, and the impact of the language is particularly evident in C++, Java, and C#, three of the most dominating programming languages in the world today. The invention of object-oriented programming was a remarkable achievement, and the professors Dahl and Nygaard received two very prestigious prizes: the von Neumann medal and the Turing prize (popularly known as the Nobel prize of computer science).

A parent class is usually called *base class* or *superclass*, while the child class is known as a *subclass* or *derived class*. We shall use the terms superclass and subclass from now on.

9.1.1 A Class for Straight Lines

Assume that we have written a class for straight lines, $y = c_0 + c_1 x$:

```
class Line(object):
    def __init__(self, c0, c1):
        self.c0 = c0
        self.c1 = c1

    def __call__(self, x):
        return self.c0 + self.c1*x

    def table(self, L, R, n):
        """Return a table with n points for L <= x <= R."""
        s = ''
        import numpy as np
        for x in np.linspace(L, R, n):
            y = self(x)
            s += '%12g %12g\n' % (x, y)
        return s
```

The constructor `__init__` initializes the coefficients c_0 and c_1 in the expression for the straight line: $y = c_0 + c_1 x$. The call operator `__call__` evaluates the function $c_1 x + c_0$, while the `table` method samples the function at n points and creates a table of x and y values.

9.1.2 A First Try on a Class for Parabolas

A parabola $y = c_0 + c_1 x + c_2 x^2$ contains a straight line as a special case ($c_2 = 0$). A class for parabolas will therefore be similar to a class for straight lines. All we have do to is to add the new term $c_2 x^2$ in the function evaluation and store c_2 in the constructor:

```
class Parabola(object):
    dof __init__(self, c0, c1, c2):
        self.c0 = c0
        self.c1 = c1
        self.c2 = c2

    def __call__(self, x):
        return self.c2*x**2 + self.c1*x + self.c0

    def table(self, L, R, n):
        """Return a table with n points for L <= x <= R."""
        s = ''
        import numpy as np
        for x in np.linspace(L, R, n):
            y = self(x)
            s += '%12g %12g\n' % (x, y)
        return s
```

Observe that we can copy the `table` method from class `Line` without any modifications.

9.1.3 A Class for Parabolas Using Inheritance

Python and other languages that support object-oriented programming have a special construct, so that class `Parabola` does not need to repeat the code that we have already written in class `Line`. We can specify that class `Parabola` *inherits* all code from class `Line` by adding `(Line)` in the class headline:

```
class Parabola(Line):
```

Class `Parabola` now automatically gets all the code from class `Line`. Exercise 9.1 asks you to explicitly demonstrate the validity of this assertion. We say that class `Parabola` is *derived* from class `Line`, or equivalently, that class `Parabola` is a subclass of its superclass `Line`.

Now, class `Parabola` should not be identical to class `Line`: it needs to add data in the constructor (for the new term) and to modify the call operator (because of the new term), but the `table` method can be inherited as it is. If we implement the constructor and the call operator in class `Parabola`, these will *override* the inherited versions from class `Line`. If we do not implement a `table` method, the one inherited from class `Line` is available as if it were coded visibly in class `Parabola`.

Class `Parabola` must first have the statements from the class `Line` methods `__call__` and `__init__`, and then add extra code in these methods. An important

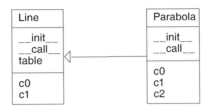

Fig. 9.1 UML diagram for the class hierarchy with superclass Line and subclass Parabola

principle in computer programming is to avoid repeating code. We should therefore
call up functionality in class Line instead of copying statements from class Line
methods to Parabola methods. Any method in the superclass Line can be called
using the syntax

```
Line.methodname(self, arg1, arg2, ...)
# or
super(Parabola, self).methodname(arg1, arg2, ...)
```

The latter construction only works if the super class is derived from Python's general super class object (i.e., class Line must be a new-style class).

Let us now show how to write class Parabola as a subclass of class Line, and
implement just the new additional code that we need and that is not already written
in the superclass:

```
class Parabola(Line):
    def __init__(self, c0, c1, c2):
        Line.__init__(self, c0, c1)   # let Line store c0 and c1
        self.c2 = c2

    def __call__(self, x):
        return Line.__call__(self, x) + self.c2*x**2
```

This short implementation of class Parabola provides exactly the same functionality as the first version of class Parabola that we showed in Sect. 9.1.2 and that did
not inherit from class Line. Figure 9.1 shows the class hierarchy in UML fashion.
The arrow from one class to another indicates inheritance.

A quick demo of the Parabola class in a main program,

```
p = Parabola(1, -2, 2)
p1 = p(x=2.5)
print p1
print p.table(0, 1, 3)
```

gives this output:

```
8.5
            0              1
          0.5            0.5
            1              1
```

Program flow The program flow can be somewhat complicated when we work with class hierarchies. Consider the code segment

```
p = Parabola(1, -1, 2)
p1 = p(x=2.5)
```

Let us explain the program flow in detail for these two statements. As always, you can monitor the program flow in a debugger as explained in Sect. F.1 or you can invoke the very illustrative Online Python Tutor[2].

Calling Parabola(1, -1, 2) leads to a call to the constructor method `__init__`, where the arguments c0, c1, and c2 in this case are int objects with values 1, -1, and 2. The self argument in the constructor is the object that will be returned and referred to by the variable p. Inside the constructor in class Parabola we call the constructor in class Line. In this latter method, we create two data attributes in the self object. Printing out dir(self) will explicitly demonstrate what self contains so far in the construction process. Back in class Parabola's constructor, we add a third attribute c2 to the same self object. Then the self object is invisibly returned and referred to by p.

The other statement, p1 = p(x=2.5), has a similar program flow. First we enter the p.`__call__` method with self as p and x as a float object with value 2.5. The program flow jumps to the `__call__` method in class Line for evaluating the linear part $c_1 x + c_0$ of the expression for the parabola, and then the flow jumps back to the `__call__` method in class Parabola where we add the new quadratic term.

9.1.4 Checking the Class Type

Python has the function isinstance(i,t) for checking if an instance i is of class type t:

```
>>> l = Line(-1, 1)
>>> isinstance(l, Line)
True
>>> isinstance(l, Parabola)
False
```

A Line is not a Parabola, but is a Parabola a Line?

```
>>> p = Parabola(-1, 0, 10)
>>> isinstance(p, Parabola)
True
>>> isinstance(p, Line)
True
```

Yes, from a class hierarchy perspective, a Parabola instance is regarded as a Line instance too, since it contains everything that a Line instance contains.

---

[2] http://www.pythontutor.com/

Every instance has an attribute `__class__` that holds the type of class:

```
>>> p.__class__
<class __main__.Parabola at 0xb68f108c>
>>> p.__class__ == Parabola
True
>>> p.__class__.__name__    # string version of the class name
'Parabola'
```

Note that `p.__class__` is a class object (or class definition one may say), while `p.__class__.__name__` is a string. These two variables can be used as an alternative test for the class type:

```
if p.__class__.__name__ == 'Parabola':
    ...
# or
if p.__class__ == Parabola:
    ...
```

However, `isinstance(p, Parabola)` is the recommended programming style for checking the type of an object.

A function `issubclass(c1, c2)` tests if class c1 is a subclass of class c2, e.g.,

```
>>> issubclass(Parabola, Line)
True
>>> issubclass(Line, Parabola)
False
```

The superclasses of a class are stored as a tuple in the `__bases__` attribute of the class object:

```
>>> p.__class__.__bases__
(<class __main__.Line at 0xb7c5d2fc>,)
>>> p.__class__.__bases__[0].__name__    # extract name as string
'Line'
```

9.1.5   Attribute vs Inheritance: has-a vs is-a Relationship

Instead of letting class Parabola inherit from a class Line, we may let it *contain* a class Line instance as a data attribute:

```
class Parabola(object):
    def __init__(self, c0, c1, c2):
        self.line = Line(c0, c1)  # let Line store c0 and c1
        self.c2 = c2

    def __call__(self, x):
        return self.line(x) + self.c2*x**2
```

Whether to use inheritance or an attribute depends on the problem being solved.

If it is natural to say that class Parabola *is* a Line object, we say that Parabola has an *is-a relationship* with class Line. Alternatively, if it is natural to think that class Parabola *has a* Line object, we speak about a *has-a relationship* with class Line. In the present example, we may argue that technically the expression for the parabola *is a* straight line plus another term and hence claim an is-a relationship, but we can also view a parabola as a quantity that *has a* line plus an extra term, which makes the *has-a* relationship relevant.

From a mathematical point of view, many will say that a parabola *is not* a line, but that a line is a special case of a parabola. Adopting this reasoning reverses the dependency of the classes: now it is more natural to let Line is a subclass of Parabola (Line *is a* Parabola). This easy, and all we have to do is

```
class Parabola(object):
    def __init__(self, c0, c1, c2):
        self.c0, self.c1, self.c2 = c0, c2, c2

    def __call__(self, x):
        return self.c0 + self.c1*x + self.c2*x**2

    def table(self, L, R, n):  # implemented as shown above

class Line(Parabola):
    def __init__(self, c0, c1):
        Parabola.__init__(self, c0, c1, 0)
```

The inherited `__call__` method from class Parabola will work since the c2 coefficient is zero. Exercises 9.4 suggests deriving Parabola from a general class Polynomial and asks you to discuss the alternative class designs.

Extension and restriction of a superclass

In the example where Parabola as a subclass of Line, we used inheritance to *extend* the functionality of the superclass. The case where Line is a subclass of Parabola is an example on *restricting* the superclass functionality in a subclass.

How classes depend on each other is influenced by two factors: sharing of code and logical relations. From a sharing of code perspective, many will say that class Parabola is naturally a subclass of Line, the former adds code to the latter. On the other hand, Line is naturally a subclass of Parabola from the logical relations in mathematics. Computational efficiency is a third perspective when we implement mathematics. When Line is a subclass of Parabola we always evaluate the $c_2 x^2$ term in the parabola although this term is zero. Nevertheless, when Parabola is a subclass of Line, we call Line.`__call__` to evaluate the linear part of the second-degree polynomial, and this call is costly in Python. From a pure efficiency point of view, we would reprogram the linear part in Parabola.`__call__` (which is against the programming habit we have been arguing for!). This little discussion here highlights the many different considerations that come into play when establishing class relations.

9.1.6   Superclass for Defining an Interface

As another example of class hierarchies, we now want to represent functions by classes, as described in Sect. 7.1.2, but in addition to the `__call__` method, we also want to provide methods for the first and second derivative. The class can be sketched as

```
class SomeFunc(object):
    def __init__(self, parameter1, parameter2, ...)
        # Store parameters
    def __call__(self, x):
        # Evaluate function
    def df(self, x):
        # Evaluate the first derivative
    def ddf(self, x):
        # Evaluate the second derivative
```

For a given function, the analytical expressions for first and second derivative must be manually coded. However, we could think of inheriting general functions for computing these derivatives numerically, such that the only thing we must always implement is the function itself. To realize this idea, we create a superclass

```
class FuncWithDerivatives(object):
    def __init__(self, h=1.0E-5):
        self.h = h  # spacing for numerical derivatives

    def __call__(self, x):
        raise NotImplementedError\
        ('__call__ missing in class %s' % self.__class__.__name__)

    def df(self, x):
        """Return the 1st derivative of self.f."""
        # Compute first derivative by a finite difference
        h = self.h
        return (self(x+h) - self(x-h))/(2.0*h)

    def ddf(self, x):
        """Return the 2nd derivative of self.f."""
        # Compute second derivative by a finite difference:
        h = self.h
        return (self(x+h) - 2*self(x) + self(x-h))/(float(h)**2)
```

This class is only meant as a superclass of other classes. For a particular function, say $f(x) = \cos(ax) + x^3$, we represent it by a subclass:

```
class MyFunc(FuncWithDerivatives):
    def __init__(self, a):
        self.a = a

    def __call__(self, x):
        return cos(self.a*x) + x**3
```

```
    def df(self, x):
        a = self.a
        return -a*sin(a*x) + 3*x**2

    def ddf(self, x):
        a = self.a
        return -a*a*cos(a*x) + 6*x
```

The superclass constructor is never called, hence h is never initialized, and there are no possibilities for using numerical approximations via the superclass methods df and ddf. Instead, we override all the inherited methods and implement our own versions.

Tip

Many think it is a good programming style to always call the superclass constructor in a subclass constructor, even in simple classes where we do not need the functionality of the superclass constructor.

For a more complicated function, e.g., $f(x) = \ln |p \tanh(qx \cos rx)|$, we may skip the analytical derivation of the derivatives, and just code $f(x)$ and rely on the difference approximations inherited from the superclass to compute the derivatives:

```
class MyComplicatedFunc(FuncWithDerivatives):
    def __init__(self, p, q, r, h=1.0E-5):
        FuncWithDerivatives.__init__(self, h)
        self.p, self.q, self.r = p, q, r

    def __call__(self, x):
        return log(abs(self.p*tanh(self.q*x*cos(self.r*x))))
```

That's it! We are now ready to use this class:

```
>>> f = MyComplicatedFunc(1, 1, 1)
>>> x = pi/2
>>> f(x)
-36.880306514638988
>>> f.df(x)
-60.593693618216086
>>> f.ddf(x)
3.3217246931444789e+19
```

Class MyComplicatedFunc inherits the df and ddf methods from the superclass FuncWithDerivatives. These methods compute the first and second derivatives approximately, provided that we have defined a __call__ method. If we fail to define this method, we will inherit __call__ from the superclass, which just raises an exception, saying that the method is not properly implemented in class MyComplicatedFunc.

The important message in this subsection is that we introduced a super class to mainly define an *interface*, i.e., the operations (in terms of methods) that one can do with a class in this class hierarchy. The superclass itself is of no direct use, since it does not implement any function evaluation in the __call__ method. However, it stores a variable common to all subclasses (h), and it implements general methods df and ddf that any subclass can make use of. A specific mathematical function must be represented as a subclass, where the programmer can decide whether analytical derivatives are to be used, or if the more lazy approach of inheriting general functionality (df and ddf) for computing numerical derivatives is satisfactory.

In object-oriented programming, the superclass very often defines an interface, and instances of the superclass have no applications on their own – only instances of subclasses can do anything useful.

To digest the present material on inheritance, we recommend doing Exercises 9.1–9.4 before reading the next section.

9.2   Class Hierarchy for Numerical Differentiation

Section 7.3.2 presents a class Derivative that (approximately) differentiate any mathematical function represented by a callable Python object. The class employs the simplest possible numerical derivative. There are a lot of other numerical formulas for computing approximations to $f'(x)$:

$$f'(x) = \frac{f(x + h) - f(x)}{h} + \mathcal{O}(h), \quad \text{(1st-order forward diff.)} \tag{9.1}$$

$$f'(x) = \frac{f(x) - f(x - h)}{h} + \mathcal{O}(h), \quad \text{(1st-order backward diff.)} \tag{9.2}$$

$$f'(x) = \frac{f(x + h) - f(x - h)}{2h} + \mathcal{O}(h^2), \quad \text{(2nd-order central diff.)} \tag{9.3}$$

$$f'(x) = \frac{4}{3}\frac{f(x + h) - f(x - h)}{2h} - \frac{1}{3}\frac{f(x + 2h) - f(x - 2h)}{4h} + \mathcal{O}(h^4),$$
$$\text{(4th-order central diff.)} \tag{9.4}$$

$$f'(x) = \frac{3}{2}\frac{f(x + h) - f(x - h)}{2h} - \frac{3}{5}\frac{f(x + 2h) - f(x - 2h)}{4h} +$$
$$\frac{1}{10}\frac{f(x + 3h) - f(x - 3h)}{6h} + \mathcal{O}(h^6),$$
$$\text{(6th-order central diff.)} \tag{9.5}$$

$$f'(x) = \frac{1}{h}\left(-\frac{1}{6}f(x + 2h) + f(x + h) - \frac{1}{2}f(x) - \frac{1}{3}f(x - h)\right) + \mathcal{O}(h^3),$$
$$\text{(3rd-order forward diff.)} \tag{9.6}$$

The key ideas about the implementation of such a family of formulas are explained in Sect. 9.2.1. For the interested reader, Sects. 9.2.4–9.2.7 contains more advanced additional material that can well be skipped in a first reading. However, the additional material puts the basic solution in Sect. 9.2.1 into a wider perspective, which may increase the understanding of object orientation.

9.2.1 Classes for Differentiation

It is argued in Sect. 7.3.2 that it is wise to implement a numerical differentiation formula as a class where $f(x)$ and h are data attributes and a `__call__` method makes class instances behave as ordinary Python functions. Hence, when we have a collection of different numerical differentiation formulas, like (9.1)–(9.6), it makes sense to implement each one of them as a class.

Doing this implementation (see Exercise 7.16), we realize that the constructors are identical because their task in the present case to store f and h. Object-orientation is now a natural next step: we can avoid duplicating the constructors by letting all the classes inherit the common constructor code. To this end, we introduce a superclass `Diff` and implement the different numerical differentiation rules in subclasses of `Diff`. Since the subclasses inherit their constructor, all they have to do is to provide a `__call__` method that implements the relevant differentiation formula.

Let us show what the superclass `Diff` looks like and how three subclasses implement the formulas (9.1)–(9.3):

```python
class Diff(object):
    def __init__(self, f, h=1E-5):
        self.f = f
        self.h = float(h)

class Forward1(Diff):
    def __call__(self, x):
        f, h = self.f, self.h
        return (f(x+h) - f(x))/h

class Backward1(Diff):
    def __call__(self, x):
        f, h = self.f, self.h
        return (f(x) - f(x-h))/h

class Central2(Diff):
    def __call__(self, x):
        f, h = self.f, self.h
        return (f(x+h) - f(x-h))/(2*h)
```

These small classes demonstrates an important feature of object-orientation: code common to many different classes are placed in a superclass, and the subclasses add just the code that differs among the classes.

We can easily implement the formulas (9.4)–(9.6) by following the same method:

```python
class Central4(Diff):
    def __call__(self, x):
        f, h = self.f, self.h
        return (4./3)*(f(x+h)   - f(x-h))  /(2*h) - \
               (1./3)*(f(x+2*h) - f(x-2*h))/(4*h)
```

```
class Central6(Diff):
    def __call__(self, x):
        f, h = self.f, self.h
        return (3./2) *(f(x+h)   - f(x-h))  /(2*h) - \
                (3./5) *(f(x+2*h) - f(x-2*h))/(4*h) + \
                (1./10)*(f(x+3*h) - f(x-3*h))/(6*h)

class Forward3(Diff):
    def __call__(self, x):
        f, h = self.f, self.h
        return (-(1./6)*f(x+2*h) + f(x+h) - 0.5*f(x) - \
                (1./3)*f(x-h))/h
```

We have placed all the classes in a module file Diff.py. Here is a short interactive example using the module to numerically differentiate the sine function:

```
>>> from Diff import *
>>> from math import sin
>>> mycos = Central4(sin)
>>> mycos(pi)                  # compute sin'(pi)
-1.000000082740371
```

Instead of a plain Python function we may use an object with a __call__ method, here exemplified through the function $f(t; a, b, c) = at^2 + bt + c$:

```
class Poly2(object):
    def __init__(self, a, b, c):
        self.a, self.b, self.c = a, b, c
    def __call__(self, t):
        return self.a*t**2 + self.b*t + self.c

f = Poly2(1, 0, 1)
dfdt = Central4(f)
t = 2
print "f'(%g)=%g" % (t, dfdt(t))
```

Let us examine the program flow. When Python encounters dfdt = Central4(f), it looks for the constructor in class Central4, but there is no constructor in that class. Python then examines the superclasses of Central4, listed in Central4.__bases__. The superclass Diff contains a constructor, and this method is called. When Python meets the dfdt(t) call, it looks for __call__ in class Central4 and finds it, so there is no need to examine the superclass. This process of looking up methods of a class is called *dynamic binding*.

Computer science remark Dynamic binding means that a name is bound to a function while the program is running. Normally, in computer languages, a function name is static in the sense that it is hardcoded as part of the function body and will not change during the execution of the program. This principle is known as static binding of function/method names. Object orientation offers the technical means to associate different functions with the same name, which yields a kind of magic for increased flexibility in programs. The particular function that the name refs to can be set at run-time, i.e., when the program is running, and therefore known as dynamic binding.

In Python, dynamic binding is a natural feature since names (variables) can refer to functions and therefore be dynamically bound during execution, just as any ordinary variable. To illustrate this point, let `func1` and `func2` be two Python functions of one argument, and consider the code

```
if input == 'func1':
    f = func1
elif input == 'func2':
    f = func2
y = f(x)
```

Here, the name `f` is bound to one of the `func1` and `func2` function objects while the program is running. This is a result of two features: (i) dynamic typing (so the contents of `f` can change), and (ii) functions being ordinary objects. The bottom line is that dynamic binding comes natural in Python, while it appears more like convenient magic in languages like C++, Java, and C#.

9.2.2   Verification

We have several alternative numerical methods for differentiation implemented in the `Diff` hierarchy, and the `Diff` module should contain one or more test functions for verifying the implementations. The fundamental problem is that even if we know the exact derivative of a function, we do not know what the numerical error in one of the subclass methods is. This fact prevents us from comparing the numerical and the exact derivative.

Fortunately, numerical differentiation formulas of the type we have encountered above are able to differentiate lower order polynomials exactly. All of them are capable of computing $f'(x) = a$, where $f(x) = ax + b$, without approximation errors for any h. We can use this knowledge to construct a test function:

```
def test_Central2():
    def f(x):
        return a*x + b

    def df_exact(x):
        return a

    a = 0.2; b = -4
    df = Central2(f, h=0.55)
    x = 6.2
    msg = 'method Central2 failed: df/dx=%g != %g' % \
          (df(x), df_exact(x))
    tol = 1E-14
    assert abs(df_exact(x) - df(x)) < tol
```

It will be boring to write such a test function for each class in the hierarchy. Therefore, we parameterize the class name and rewrite `test_Central` such that it can be reused for any class in the `Diff` hierarchy:

```
def _test_one_method(method):
    """Test method in string 'method' on a linear function."""
    f = lambda x: a*x + b
    df_exact = lambda x: a
    a = 0.2; b = -4
    df = eval(method)(f, h=0.55)
    x = 6.2
    msg = 'method %s failed: df/dx=%g != %g' % \
          (method, df(x), df_exact(x))
    tol = 1E-14
    assert abs(df_exact(x) - df(x)) < tol
```

Some comments are needed to explain this function:

- All our test functions are intended for the pytest and nose testing frameworks. (See Sect. H.9 for more information on such test functions.) The function name must then start with `test_` and no arguments are allowed. For the helper function `_test_one_method` with an argument, the function name cannot start with `test`, and that is why an underscore is added.
- Lambda functions (see Sect. 3.1.14) are used to save code in the definitions of `f` and `df_exact`.
- The subclass to be tested is given as a string `method`. Calling the constructor must then be done by `eval(method)(f)`.

It remains to make a loop over all the implemented subclasses and call `_test_one_method` for each of them. As always, we try to find a way to automate boring work, which here consists of listing all the subclasses (and remembering to update the list when new subclasses are added). All global variables in a file is available from the dictionary returned by `globals()`. The key is a variable name and the value is the corresponding object. For example, `print globals()` reveals that all the defined classes are in `globals()`, e.g.,

```
'Central2': <class Diff.Central2 at 0x1a87c80>,
'Central4': <class Diff.Central4 at 0x1a87f58>,
'Diff': <class Diff.Diff at 0x1a870b8>,
```

To find all the relevant classes to test, we grab all names from the `globals()` dictionary, look for names that starts with upper case, and find the names that correspond to a subclass of `Diff` (drop `Diff` itself as this class cannot compute anything and therefore cannot be tested). Translating this algorithm to code gives us a test function that can test all subclasses in the `Diff` hierarchy:

```
def test_all_methods():
    """Call _test_one_method for all subclasses of Diff."""
    print globals()
    names = list(globals().keys())  # all names in this module
    for name in names:
        if name[0].isupper():
            if issubclass(eval(name), Diff):
                if name != 'Diff':
                    _test_one_method(name)
```

9.2.3 A flexible Main Program

As a demonstration of the power of Python programming, we shall now write a main program for our `Diff` module that accepts a function on the command-line, together with information about the difference type (centered, backward, or forward), the order of the approximation, and a value of the independent variable. The corresponding output is the derivative of the given function. An example of the usage of the program goes like this:

`Terminal`

```
Diff.py 'exp(sin(x))' Central 2 3.1
-1.04155573055
```

Here, we asked the program to differentiate $f(x) = e^{\sin x}$ at $x = 3.1$ with a central scheme of order 2 (using the `Central2` class in the `Diff` hierarchy).

We can provide any expression with x as input and request any scheme from the `Diff` hierarchy, and the derivative will be (approximately) computed. One great thing with Python is that the code is very short:

```python
from math import *  # make all math functions available to main

def main():
    from scitools.StringFunction import StringFunction
    import sys

    try:
        formula = sys.argv[1]
        difftype = sys.argv[2]
        difforder = sys.argv[3]
        x = float(sys.argv[4])
    except IndexError:
        print 'Usage:  Diff.py formula difftype difforder x'
        print 'Example: Diff.py "sin(x)*exp(-x)" Central 4 3.14'
        sys.exit(1)

    classname = difftype + difforder
    f = StringFunction(formula)
    df = eval(classname)(f)
    print df(x)

if __name__ == '__main__':
    main()
```

Read the code line by line, and convince yourself that you understand what is going on. You may need to review Sects. 4.3.1 and 4.3.3.

One disadvantage is that the code above is limited to x as the name of the independent variable. If we allow a 5th command-line argument with the name of the independent variable, we can pass this name on to the `StringFunction` constructor, and suddenly our program works with any name for the independent variable!

```python
varname = sys.argv[5]
f = StringFunction(formula, independent_variables=varname)
```

Of course, the program crashes if we do not provide five command-line arguments, and the program does not work properly if we are not careful with ordering of the command-line arguments. There is some way to go before the program is really user friendly, but that is beyond the scope of this chapter.

Many other popular programming languages (C++, Java, C#) cannot perform the eval operation while the program is running. The result is that one needs if tests to turn the information in difftype and difforder into creation of subclass instances. Such type of code would look like this in Python:

```
if classname == 'Forward1':
    df = Forward1(f)
elif classname == 'Backward1':
    df = Backward1(f)
...
```

and so forth. This piece of code is very common in object-oriented systems and often put in a function that is referred to as a *factory function*. Thanks to eval in Python, factory functions are usually only a matter of applying eval to a string.

9.2.4 Extensions

The great advantage of sharing code via inheritance becomes obvious when we want to extend the functionality of a class hierarchy. It is possible to do this by adding more code to the superclass only. Suppose we want to be able to assess the accuracy of the numerical approximation to the derivative by comparing with the exact derivative, if available. All we need to do is to allow an extra argument in the constructor and provide an additional superclass method that computes the error in the numerical derivative. We may add this code to class Diff, or we may add it in a subclass Diff2 and let the other classes for various numerical differentiation formulas inherit from class Diff2. We follow the latter approach:

```
class Diff2(Diff):
    def __init__(self, f, h=1E-5, dfdx_exact=None):
        Diff.__init__(self, f, h)
        self.exact = dfdx_exact

    def error(self, x):
        if self.exact is not None:
            df_numerical = self(x)
            df_exact = self.exact(x)
            return df_exact - df_numerical

class Forward1(Diff2):
    def __call__(self, x):
        f, h = self.f, self.h
        return (f(x+h) - f(x))/h
```

The other subclasses, Backward1, Central2, and so on, must also be derived from Diff2 to equip all subclasses with new functionality for perfectly assessing

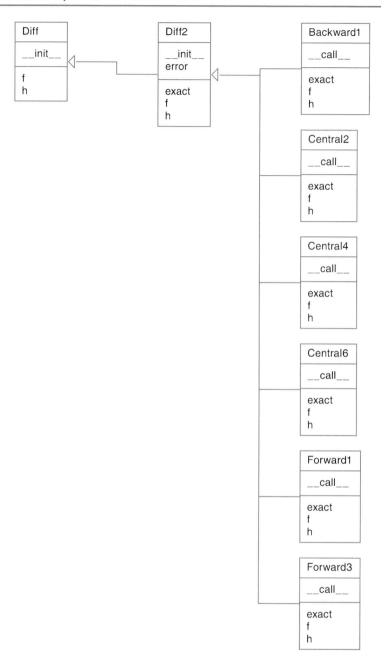

Fig. 9.2 UML diagram of the `Diff` hierarchy for a series of differentiation formulas (`Backward1`, `Central2`, etc.)

the accuracy of the approximation. No other modifications are necessary in this example, since all the subclasses can inherit the superclass constructor and the `error` method. Figure 9.2 shows a UML diagram of the new `Diff` class hierarchy.

Here is an example of usage:

```
mycos = Forward1(sin, dfdx_exact=cos)
print 'Error in derivative is', mycos.error(x=pi)
```

The program flow of the `mycos.error(x=pi)` call can be interesting to follow. We first enter the `error` method in class `Diff2`, which then calls `self(x)`, i.e., the `__call__` method in class `Forward1`, which jumps out to the `self.f` function, i.e., the `sin` function in the `math` module in the present case. After returning to the `error` method, the next call is to `self.exact`, which is the `cos` function (from `math`) in our case.

Application We can apply the methods in the `Diff2` hierarchy to get some insight into the accuracy of various difference formulas. Let us write out a table where the rows correspond to different h values, and the columns correspond to different approximation methods (except the first column, which reflects the h value). The values in the table can be the numerically computed $f'(x)$ or the error in this approximation if the exact derivative is known. The following function writes such a table:

```
def table(f, x, h_values, methods, dfdx=None):
    # Print headline (h and class names for the methods)
    print '      h      ',
    for method in methods:
        print '%-15s' % method.__name__,
    print   # newline
    # Print table
    for h in h_values:
        print '%10.2E' % h,
        for method in methods:
            if dfdx is not None:      # write error
                d = method(f, h, dfdx)
                output = d.error(x)
            else:                     # write value
                d = method(f, h)
                output = d(x)
            print '%15.8E' % output,
        print   # newline
```

The next lines tries three approximation methods on $f(x) = e^{-10x}$ for $x = 0$ and with $h = 1, 1/2, 1/4, 1/16, \ldots, 1/512$:

```
from Diff2 import *
from math import exp

def f1(x):
    return exp(-10*x)

def df1dx(x):
    return -10*exp(-10*x)

table(f1, 0, [2**(-k) for k in range(10)],
      [Forward1, Central2, Central4], df1dx)
```

Note how convenient it is to make a list of class names – class names can be used as ordinary variables, and to print the class name as a string we just use the __name__ attribute. The output of the main program above becomes

```
    h         Forward1         Central2          Central4
1.00E+00 -9.00004540E+00  1.10032329E+04 -4.04157586E+07
5.00E-01 -8.01347589E+00  1.38406421E+02 -3.48320240E+03
2.50E-01 -6.32833999E+00  1.42008179E+01 -2.72010498E+01
1.25E-01 -4.29203837E+00  2.81535264E+00 -9.79802452E-01
6.25E-02 -2.56418286E+00  6.63876231E-01 -5.32825724E-02
3.12E-02 -1.41170013E+00  1.63556996E-01 -3.21608292E-03
1.56E-02 -7.42100948E-01  4.07398036E-02 -1.99260429E-04
7.81E-03 -3.80648092E-01  1.01756309E-02 -1.24266603E-05
3.91E-03 -1.92794011E-01  2.54332554E-03 -7.76243120E-07
1.95E-03 -9.70235594E-02  6.35795004E-04 -4.85085874E-08
```

From one row to the next, h is halved, and from about the 5th row and onwards, the Forward1 errors are also halved, which is consistent with the error $\mathcal{O}(h)$ of this method. Looking at the 2nd column, we see that the errors are reduced to 1/4 when going from one row to the next, at least after the 5th row. This is also according to the theory since the error is proportional to h^2. For the last row with a 4th-order scheme, the error is reduced by 1/16, which again is what we expect when the error term is $\mathcal{O}(h^4)$. What is also interesting to observe, is the benefit of using a higher-order scheme like Central4: with, for example, $h = 1/128$ the Forward1 scheme gives an error of -0.7, Central2 improves this to 0.04, while Central4 has an error of -0.0002. More accurate formulas definitely give better results. (Strictly speaking, it is the fraction of the work and the accuracy that counts: Central4 needs four function evaluations, while Central2 and Forward1 only needs two.) The test example shown here is found in the file Diff2_examples.py.

9.2.5   Alternative Implementation via Functions

Could we implement the functionality offered by the Diff hierarchy of objects by using plain functions and no object orientation? The answer is "yes, almost". What we have to pay for a pure function-based solution is a less friendly user interface to the differentiation functionality: more arguments must be supplied in function calls, because each difference formula, now coded as a straight Python function, must get $f(x)$, x, and h as arguments. In the class version we first store f and h as data attributes in the constructor, and every time we want to compute the derivative, we just supply x as argument.

A Python function for implementing numerical differentiation reads

```python
def central2_func(f, x, h=1.0E-5):
    return (f(x+h) - f(x-h))/(2*h)
```

The usage demonstrates the difference from the class solution:

```python
mycos = central2_func(sin, pi, 1E-6)
# Compute sin'(pi):
print "g'(%g)=%g (exact value is %g)" % (pi, mycos, cos(pi))
```

Now, `mycos` is a number, not a callable object. The nice thing with the class solution is that `mycos` appeared to be a standard Python function whose mathematical values equal the derivative of the Python function `sin(x)`. But does it matter whether `mycos` is a function or a number? Yes, it matters if we want to apply the difference formula twice to compute the second-order derivative. When `mycos` is a callable object of type `Central2`, we just write

```
mysin = Central2(mycos)
# or
mysin = Central2(Central2(sin))

# Compute g''(pi):
print "g''(%g)=%g" % (pi, mysin(pi))
```

With the `central2_func` function, this composition will not work. Moreover, when the derivative is an object, we can send this object to any algorithm that expects a mathematical function, and such algorithms include numerical integration, differentiation, interpolation, ordinary differential equation solvers, and finding zeros of equations, so the applications are many.

9.2.6 Alternative Implementation via Functional Programming

As a conclusion of the previous section, the great benefit of the object-oriented solution in Sect. 9.2.1 is that one can have some subclass instance d from the `Diff` (or `Diff2`) hierarchy and write `d(x)` to evaluate the derivative at a point x. The `d(x)` call behaves as if d were a standard Python function containing a manually coded expression for the derivative.

The `d(x)` interface to the derivative can also be obtained by other and perhaps more direct means than object-oriented programming. In programming languages where functions are ordinary objects that can be referred to by variables, as in Python, one can make a function that returns the right `d(x)` function according to the chosen numerical derivation rule. The code looks as this (see `Diff_functional.py` for the complete code):

```
def differentiate(f, method, h=1.0E-5):
    h = float(h)  # avoid integer division

    if method == 'Forward1':
        def Forward1(x):
            return (f(x+h) - f(x))/h
    return Forward1

    elif method == 'Backward1':
        def Backward1(x):
            return (f(x) - f(x-h))/h
    return Backward1
    ...
```

And the usage goes like

```
mycos = differentiate(sin,   'Forward1')
mysin = differentiate(mycos, 'Forward1')
x = pi
print mycos(x), cos(x), mysin, -sin(x)
```

The surprising thing is that when we call `mycos(x)` we provide only x, while the function itself looks like

```
def Forward1(x):
    return (f(x+h) - f(x))/h
return Forward1
```

How do the parameters `f` and `h` get their values when we call `mycos(x)`? There is some magic attached to the `Forward1` function, or literally, there are some variables attached to `Forward1`: this function remembers the values of `f` and `h` that existed as local variables in the `differentiate` function when the `Forward1` function was defined.

In computer science terms, `Forward1` always has access to variables in the *scope* in which the function was defined. The `Forward1` function is call a *closure* and explained in Sect. 7.1.7. Closures are much used in a programming style called *functional programming*. Two key features of functional programming is operations on lists (like list comprehensions) and returning functions from functions. Python supports functional programming, but we will not consider this programming style further in this book.

9.2.7 Alternative Implementation via a Single Class

Instead of making many classes or functions for the many different differentiation schemes, the basic information about the schemes can be stored in one table. With a single method in one single class can use the table information, and for a given scheme, compute the derivative. To do this, we need to reformulate the mathematical problem (actually by using ideas from Sect. 9.3.1).

A family of numerical differentiation schemes can be written

$$f'(x) \approx h^{-1} \sum_{i=-r}^{r} w_i f(x_i), \tag{9.7}$$

where w_i are weights and x_i are points. The $2r + 1$ points are symmetric around some point x:

$$x_i = x + ih, \quad i = -r, \ldots, r.$$

The weights depend on the differentiation scheme. For example, the Midpoint scheme (9.3) has

$$w_{-1} = -1, \quad w_0 = 0, \quad w_1 = 1.$$

The table below lists the values of w_i for different difference formulas. The type of difference is abbreviated with c for central, f for forward, and b for backward. The number after the nature of a scheme denotes the order of the schemes (for example, "c 2" is a central difference of 2nd order). We have set $r = 4$, which is sufficient for the schemes written up in this book.

	$x - 4h$	$x - 3h$	$x - 2h$	$x - h$	x	$x + h$	$x + 2h$	$x + 3h$	$x + 4h$
c 2	0	0	0	$-\frac{1}{2}$	0	$\frac{1}{2}$	0	0	0
c 4	0	0	$\frac{1}{12}$	$-\frac{2}{3}$	0	$\frac{2}{3}$	$-\frac{1}{12}$	0	0
c 6	0	$-\frac{1}{60}$	$\frac{3}{20}$	$-\frac{3}{4}$	0	$\frac{3}{4}$	$-\frac{3}{20}$	$\frac{1}{60}$	0
c 8	$\frac{1}{280}$	$-\frac{4}{105}$	$\frac{12}{60}$	$-\frac{4}{5}$	0	$\frac{4}{5}$	$-\frac{12}{60}$	$\frac{4}{105}$	$-\frac{1}{280}$
f 1	0	0	0	0	1	1	0	0	0
f 3	0	0	0	$-\frac{2}{6}$	$-\frac{1}{2}$	1	$-\frac{1}{6}$	0	0
b 1	0	0	0	-1	1	0	0	0	0

Given a table of the w_i values, we can use (9.7) to compute the derivative. A faster, vectorized computation can have the x_i, w_i, and $f(x_i)$ values as stored in three vectors. Then $h^{-1} \sum_i w_i f(x_i)$ can be interpreted as a dot product between the two vectors with components w_i and $f(x_i)$, respectively.

A class with the table of weights as a static variable, a constructor, and a `__call__` method for evaluating the derivative via $h^{-1} \sum_i w_i f(x_i)$ looks as follows:

```
class Diff3(object):
    table = {
    ('forward', 1):
    [0, 0, 0, 0, 1, 1, 0, 0, 0],
    ('central', 2):
    [0, 0, 0, -1./2, 0, 1./2, 0, 0, 0],
    ('central', 4):
    [ 0, 0, 1./12, -2./3, 0, 2./3, -1./12, 0, 0],
    ...
    }
    def __init__(self, f, h=1.0E-5, type='central', order=2):
        self.f, self.h, self.type, self.order = f, h, type, order
        self.weights = np.array(Diff2.table[(type, order)])

    def __call__(self, x):
        f_values = np.array([f(self.x+i*self.h) \
                            for i in range(-4,5)])
        return np.dot(self.weights, f_values)/self.h
```

Here we used numpy's `dot(x, y)` function for computing the inner or dot product between two arrays x and y.

Class `Diff3` can be found in the file `Diff3.py`. Using class `Diff3` to differentiate the sine function goes like this:

```
import Diff3
mycos = Diff3.Diff3(sin, type='central', order=4)
print "sin'(pi):", mycos(pi)
```

Remark The downside of class `Diff3`, compared with the other implementation techniques, is that the sum $h^{-1} \sum_i w_i f(x_i)$ contains many multiplications by zero for lower-order schemes. These multiplications are known to yield zero in advance so we waste computer resources on trivial calculations. Once upon a time, programmers would have been extremely careful to avoid wasting multiplications this way, but today arithmetic operations are quite cheap, especially compared to fetching data from the computer's memory. Lots of other factors also influence the computational efficiency of a program, but this is beyond the scope of this book.

9.3   Class Hierarchy for Numerical Integration

There are many different numerical methods for integrating a mathematical function, just as there are many different methods for differentiating a function. It is thus obvious that the idea of object-oriented programming and class hierarchies can be applied to numerical integration formulas in the same manner as we did in Sect. 9.2.

9.3.1   Numerical Integration Methods

First, we list some different methods for integrating $\int_a^b f(x)dx$ using n evaluation points. All the methods can be written as

$$\int_a^b f(x)dx \approx \sum_{i=0}^{n-1} w_i f(x_i), \tag{9.8}$$

where w_i are weights and x_i are evaluation points, $i = 0, \ldots, n-1$. The Midpoint method has

$$x_i = a + \frac{h}{2} + ih, \quad w_i = h, \quad h = \frac{b-a}{n}, \quad i = 0, \ldots, n-1. \tag{9.9}$$

The Trapezoidal method has the points

$$x_i = a + ih, \quad h = \frac{b-a}{n-1}, \quad i = 0, \ldots, n-1, \tag{9.10}$$

and the weights

$$w_0 = w_{n-1} = \frac{h}{2}, \ w_i = h, \quad i = 1, \ldots, n-2. \tag{9.11}$$

Simpson's rule has the same evaluation points as the Trapezoidal rule, but

$$h = 2\frac{b-a}{n-1}, \quad w_0 = w_{n-1} = \frac{h}{6}, \tag{9.12}$$

$$w_i = \frac{h}{3} \quad \text{for } i = 2, 4, \ldots, n-3, \tag{9.13}$$

$$w_i = \frac{2h}{3} \quad \text{for } i = 1, 3, 5, \ldots, n-2. \tag{9.14}$$

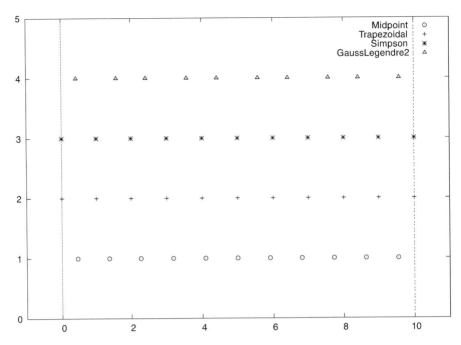

Fig. 9.3 Illustration of the distribution of points for various numerical integration methods. The Gauss-Legendre method has 10 points, while the other methods have 11 points in [0, 10]

Note that n must be odd in Simpson's rule. A Two-Point Gauss-Legendre method takes the form

$$x_i = a + \left(i + \frac{1}{2}\right) h - \frac{1}{\sqrt{3}} \frac{h}{2} \quad \text{for } i = 0, 2, 4, \ldots, n-2, \tag{9.15}$$

$$x_i = a + \left(i + \frac{1}{2}\right) h + \frac{1}{\sqrt{3}} \frac{h}{2} \quad \text{for } i = 1, 3, 5, \ldots, n-1, \tag{9.16}$$

with $h = 2(b-a)/n$. Here n must be even. All the weights have the same value: $w_i = h/2$, $i = 0, \ldots, n-1$. Figure 9.3 illustrates how the points in various integration rules are distributed over a few intervals.

9.3.2 Classes for Integration

We may store x_i and w_i in two NumPy arrays and compute the integral as $\sum_{i=0}^{n-1} w_i f(x_i)$. This operation can also be vectorized as a dot (inner) product between the w_i vector and the $f(x_i)$ vector, provided $f(x)$ is implemented in a vectorizable form.

We argued in Sect. 7.3.3 that it pays off to implement a numerical integration formula as a class. If we do so with the different methods from the previous section, a typical class looks like this:

```
class SomeIntegrationMethod(object):
    def __init__(self, a, b, n):
        # Compute self.points and self.weights

    def integrate(self, f):
        s = 0
        for i in range(len(self.weights)):
            s += self.weights[i]*f(self.points[i])
        return s
```

Making such classes for many different integration methods soon reveals that all the classes contain common code, namely the `integrate` method for computing $\sum_{i=0}^{n-1} w_i f(x_i)$. Therefore, this common code can be placed in a superclass, and subclasses can just add the code that is specific to a certain numerical integration formula, namely the definition of the weights w_i and the points x_i.

Let us start with the superclass:

```
class Integrator(object):
    def __init__(self, a, b, n):
        self.a, self.b, self.n = a, b, n
        self.points, self.weights = self.construct_method()

    def construct_method(self):
        raise NotImplementedError('no rule in class %s' %
                                  self.__class__.__name__)

    def integrate(self, f):
        s = 0
        for i in range(len(self.weights)):
            s += self.weights[i]*f(self.points[i])
        return s
```

As we have seen, we store the a, b, and n data about the integration method in the constructor. Moreover, we compute arrays or lists `self.points` for the x_i points and `self.weights` for the w_i weights. All this code can now be inherited by all subclasses.

The initialization of points and weights is put in a separate method, `construct_method`, which is supposed to be implemented in each subclass, but the superclass provides a default implementation, which tells the user that the method is not implemented. What happens is that when subclasses redefine a method, that method overrides the method inherited from the superclass. Hence, if we forget to redefine `construct_method` in a subclass, we will inherit the one from the superclass, and this method issues an error message. The construction of this error message is quite clever in the sense that it will tell in which class the `construct_method` method is missing (`self` will be the subclass instance and its `__class__.__name__` is a string with the corresponding subclass name).

In computer science one usually speaks about *overloading* a method in a subclass, but the words redefining and overriding are also used. A method that is overloaded is said to be *polymorphic*. A related term, *polymorphism*, refers to coding with polymorphic methods. Very often, a superclass provides some default

implementation of a method, and a subclass overloads the method with the purpose
of tailoring the method to a particular application.

The `integrate` method is common for all integration rules, i.e., for all subclasses, so it can be inherited as it is. A vectorized version can also be added in the
superclass to make it automatically available also in all subclasses:

```
def vectorized_integrate(self, f):
    return np.dot(self.weights, f(self.points))
```

Let us then implement a subclass. Only the `construct_method` method needs
to be written. For the Midpoint rule, this is a matter of translating the formulas in
(9.9) to Python:

```
class Midpoint(Integrator):
    def construct_method(self):
        a, b, n = self.a, self.b, self.n  # quick forms
        h = (b-a)/float(n)
        x = np.linspace(a + 0.5*h, b - 0.5*h, n)
        w = np.zeros(len(x)) + h
        return x, w
```

Observe that we implemented directly a vectorized code. We could also have used
(slow) loops and explicit indexing:

```
x = np.zeros(n)
w = np.zeros(n)
for i in range(n):
    x[i] = a + 0.5*h + i*h
    w[i] = h
```

Before we continue with other subclasses for other numerical integration formulas, we will have a look at the program flow when we use class `Midpoint`. Suppose
we want to integrate $\int_0^2 x^2 dx$ using 101 points:

```
def f(x): return x*x
m = Midpoint(0, 2, 101)
print m.integrate(f)
```

How is the program flow? The assignment to m invokes the constructor in class
`Midpoint`. Since this class has no constructor, we invoke the inherited one
from the superclass `Integrator`. Here data attributes are stored, and then the
`construct_method` method is called. Since `self` is a `Midpoint` instance, it
is the `construct_method` in the `Midpoint` class that is invoked, even if there
is a method with the same name in the superclass. Class `Midpoint` overloads
`construct_method` in the superclass. In a way, we "jump down" from the constructor in class `Integrator` to the `construct_method` in the `Midpoint` class.
The next statement, `m.integrate(f)`, just calls the inherited `integral` method
that is common to all subclasses.

The points and weights for a Trapezoidal rule can be implemented in a vectorized
way in another subclass with name `Trapezoidal`:

```
class Trapezoidal(Integrator):
    def construct_method(self):
        x = np.linspace(self.a, self.b, self.n)
        h = (self.b - self.a)/float(self.n - 1)
        w = np.zeros(len(x)) + h
        w[0]  /= 2
        w[-1] /= 2
        return x, w
```

Observe how we divide the first and last weight by 2, using index 0 (the first) and
-1 (the last) and the /= operator (a /= b is equivalent to a = a/b). We could also
have implemented a scalar version with loops. The relevant code is in function
trapezoidal in Sect. 7.3.3.

Class Simpson has a slightly more demanding rule, at least if we want to vectorize the expression, since the weights are of two types.

```
class Simpson(Integrator):
    def construct_method(self):
        if self.n % 2 != 1:
            print 'n=%d must be odd, 1 is added' % self.n
            self.n += 1
        x = np.linspace(self.a, self.b, self.n)
        h = (self.b - self.a)/float(self.n - 1)*2
        w = np.zeros(len(x))
        w[0:self.n:2] = h*1.0/3
        w[1:self.n-1:2] = h*2.0/3
        w[0]  /= 2
        w[-1] /= 2
        return x, w
```

We first control that we have an odd number of points, by checking that the remainder of self.n divided by two is 1. If not, an exception could be raised, but
for smooth operation of the class, we simply increase n so it becomes odd. Such
automatic adjustments of input is not a rule to be followed in general. Wrong input
is best notified explicitly. However, sometimes it is user friendly to make small adjustments of the input, as we do here, to achieve a smooth and successful operation.
(In cases like this, a user might become uncertain whether the answer can be trusted
if she (later) understands that the input should not yield a correct result. Therefore,
do the adjusted computation, and provide a notification to the user about what has
taken place.)

The computation of the weights w in class Simpson applies slices with stride
(jump/step) 2 such that the operation is vectorized for speed. Recall that the upper
limit of a slice is not included in the set, so self.n-1 is the largest index in the first
case, and self.n-2 is the largest index in the second case. Instead of the vectorized
operation of slices for computing w, we could use (slower) straight loops:

```
for i in range(0, self.n, 2):
    w[i] = h*1.0/3
for i in range(1, self.n-1, 2):
    w[i] = h*2.0/3
```

The points in the Two-Point Gauss-Legendre rule are slightly more complicated to calculate, so here we apply straight loops to make a safe first implementation:

```python
class GaussLegendre2(Integrator):
    def construct_method(self):
        if self.n % 2 != 0:
            print 'n=%d must be even, 1 is subtracted' % self.n
            self.n -= 1
        nintervals = int(self.n/2.0)
        h = (self.b - self.a)/float(nintervals)
        x = np.zeros(self.n)
        sqrt3 = 1.0/math.sqrt(3)
        for i in range(nintervals):
            x[2*i]   = self.a + (i+0.5)*h - 0.5*sqrt3*h
            x[2*i+1] = self.a + (i+0.5)*h + 0.5*sqrt3*h
        w = np.zeros(len(x)) + h/2.0
        return x, w
```

A vectorized calculation of x is possible by observing that the (i+0.5)*h expression can be computed by np.linspace, and then we can add the remaining two terms:

```python
m = np.linspace(0.5*h, (nintervals-1+0.5)*h, nintervals)
x[0:self.n-1:2] = m + self.a - 0.5*sqrt3*h
x[1:self.n:2]   = m + self.a + 0.5*sqrt3*h
```

The array on the right-hand side has half the length of x $(n/2)$, but the length matches exactly the slice with stride 2 on the left-hand side.

The code snippets above are found in the module file integrate.py.

9.3.3   Verification

To verify the implementation we use the fact that all the subclasses implement methods that can integrate a linear function exactly. A suitable test function is therefore

```python
def test_Integrate():
    """Check that linear functions are integrated exactly."""
    def f(x):
        return x + 2

    def F(x):
        """Integral of f."""
        return 0.5*x**2 + 2*x

    a = 2; b = 3; n = 4      # test data
    I_exact = F(b) - F(a)
    tol = 1E-15
```

```
methods = [Midpoint, Trapezoidal, Simpson, GaussLegendre2,
           GaussLegendre2_vec]
for method in methods:
    integrator = method(a, b, n)

    I = integrator.integrate(f)
    assert abs(I_exact - I) < tol

    I_vec = integrator.vectorized_integrate(f)
    assert abs(I_exact - I_vec) < tol
```

A stronger method of verification is to compute how the error varies with n. Exercise 9.15 explains the details.

9.3.4 Using the Class Hierarchy

To verify the implementation, we first try to integrate a linear function. All methods should compute the correct integral value regardless of the number of evaluation points:

```
def f(x):
    return x + 2

a = 2; b = 3; n = 4
for Method in Midpoint, Trapezoidal, Simpson, GaussLegendre2:
    m = Method(a, b, n)
    print m.__class__.__name__, m.integrate(f)
```

Observe how we simply list the class names as a tuple (comma-separated objects), and Method will in the for loop attain the values Midpoint, Trapezoidal, and so forth. For example, in the first pass of the loop, Method(a, b, n) is identical to Midpoint(a, b, n).

The output of the test above becomes

```
Midpoint 4.5
Trapezoidal 4.5
n=4 must be odd, 1 is added
Simpson 4.5
GaussLegendre2 4.5
```

Since $\int_2^3 (x + 2)dx = \frac{9}{2} = 4.5$, all methods passed this simple test.

A more challenging integral, from a numerical point of view, is

$$\int_0^1 \left(1 + \frac{1}{m}\right) t^{\frac{1}{m}} dt = 1.$$

To use any subclass in the Integrator hierarchy, the integrand must be a function of one variable only. For the present integrand, which depends on t and m, we use a class to represent it:

```
class F(object):
    def __init__(self, m):
        self.m = float(m)

    def __call__(self, t):
        m = self.m
        return (1 + 1/m)*t**(1/m)
```

We now ask the question: how much is the error in the integral reduced as we increase the number of integration points (n)? It appears that the error decreases exponentially with n, so if we want to plot the errors versus n, it is best to plot the logarithm of the error versus $\ln n$. We expect this graph to be a straight line, and the steeper the line is, the faster the error goes to zero as n increases. A common conception is to regard one numerical method as better than another if the error goes faster to zero as we increase the computational work (here n).

For a given m and method, the following function computes two lists containing the logarithm of the n values, and the logarithm of the corresponding errors in a series of experiments:

```
def error_vs_n(f, exact, n_values, Method, a, b):
    log_n = []   # log of actual n values (Method may adjust n)
    log_e = []   # log of corresponding errors
    for n_value in n_values:
        method = Method(a, b, n_value)
        error = abs(exact - method.integrate(f))
        log_n.append(log(method.n))
        log_e.append(log(error))
    return log_n, log_e
```

We can plot the error versus n for several methods in the same plot and make one plot for each m value. The loop over m below makes such plots:

```
n_values = [10, 20, 40, 80, 160, 320, 640]
for m in 1./4, 1./8., 2, 4, 16:
    f = F(m)
    figure()
    for Method in Midpoint, Trapezoidal, \
            Simpson, GaussLegendre2:
        n, e = error_vs_n(f, 1, n_values, Method, 0, 1)
        plot(n, e); legend(Method.__name__); hold('on')
    title('m=%g' % m); xlabel('ln(n)'); ylabel('ln(error)')
```

The code snippets above are collected in a function `test` in the `integrate.py` file.

The plots for $m > 1$ look very similar. The plots for $0 < m < 1$ are also similar, but different from the $m > 1$ cases. Let us have a look at the results for $m = 1/4$ and $m = 2$. The first, $m = 1/4$, corresponds to $\int_0^1 5x^4 dx$. Figure 9.4 shows that the error curves for the Trapezoidal and Midpoint methods converge more slowly compared to the error curves for Simpson's rule and the Gauss-Legendre method. This is the usual situation for these methods, and mathematical analysis of the methods can confirm the results in Fig. 9.4.

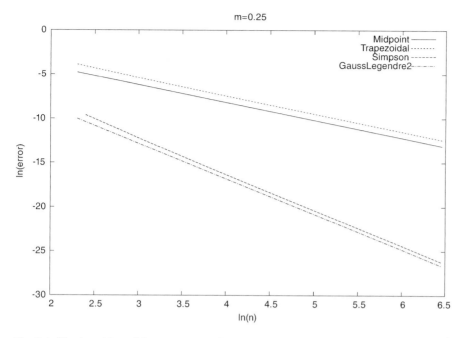

Fig. 9.4 The logarithm of the error versus the logarithm of integration points for integral $5x^4$ computed by the Trapezoidal and Midpoint methods (*upper two lines*), and Simpson's rule and the Gauss-Legendre methods (*lower two lines*)

However, when we consider the integral $\int_0^1 \frac{3}{2}\sqrt{x}\,dx$, ($m = 2$) and $m > 1$ in general, all the methods converge with the same speed, as shown in Fig. 9.5. Our integral is difficult to compute numerically when $m > 1$, and the theoretically better methods (Simpson's rule and the Gauss-Legendre method) do not converge faster than the simpler methods. The difficulty is due to the infinite slope (derivative) of the integrand at $x = 0$.

9.3.5   About Object-Oriented Programming

From an implementational point of view, the advantage of class hierarchies in Python is that we can save coding by inheriting functionality from a superclass. In programming languages where each variable must be specified with a fixed type, class hierarchies are particularly useful because a function argument with a special type also works with all subclasses of that type. Suppose we have a function where we need to integrate:

```
def do_math(arg1, arg2, integrator):
    ...
    I = integrator.integrate(myfunc)
    ...
```

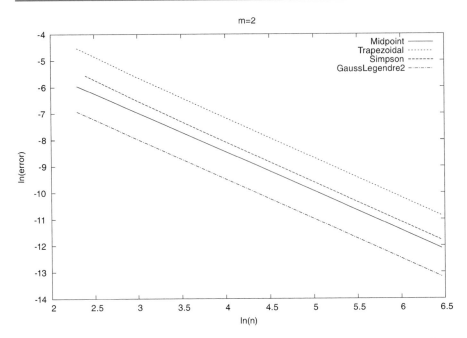

Fig. 9.5 The logarithm of the error versus the logarithm of integration points for integral $\frac{3}{2}\sqrt{x}$ computed by the Trapezoidal method and Simpson's rule (*upper two lines*), and Midpoint and Gauss-Legendre methods (*lower two lines*)

That is, `integrator` must be an instance of some class, or a module, such that the syntax `integrator.integrate(myfunc)` corresponds to a function call, but nothing more (like having a particular type) is demanded.

This Python code will run as long as `integrator` has a method `integrate` taking one argument. In other languages, the function arguments are specified with a type, say in Java we would write

```
void do_math(double arg1, int arg2, Simpson integrator)
```

A compiler will examine all calls to `do_math` and control that the arguments are of the right type. Instead of specifying the integration method to be of type `Simpson`, one can in Java and other object-oriented languages specify `integrator` to be of the superclass type `Integrator`:

```
void do_math(double arg1, int arg2, Integrator integrator)
```

Now it is allowed to pass an object of any subclass type of `Integrator` as the third argument. That is, this method works with `integrator` of type `Midpoint`, `Trapezoidal`, `Simpson`, etc., not just one of them. Class hierarchies and object-oriented programming are therefore important means for parameterizing away types in languages like Java, C++, and C#. We do not need to parameterize types in Python, since arguments are not declared with a fixed type. Object-oriented programming is hence not so technically important in Python as in other languages for providing increased flexibility in programs.

Is there then any use for object-oriented programming beyond inheritance? The answer is yes! For many code developers object-oriented programming is not just a technical way of sharing code, but it is more a way of modeling the world, and understanding the problem that the program is supposed to solve. In mathematical applications we already have objects, defined by the mathematics, and standard programming concepts such as functions, arrays, lists, and loops are often sufficient for solving simpler problems. In the non-mathematical world the concept of objects is very useful because it helps to structure the problem to be solved. As an example, think of the phone book and message list software in a mobile phone. Class `Person` can be introduced to hold the data about one person in the phone book, while class `Message` can hold data related to an SMS message. Clearly, we need to know who sent a message so a `Message` object will have an associated `Person` object, or just a phone number if the number is not registered in the phone book. Classes help to structure both the problem and the program. The impact of classes and object-oriented programming on modern software development can hardly be exaggerated.

A good, real-world, pedagogical example on inheritance is the class hierarchy for numerical methods for ordinary differential equations described in Sect. E.2.

9.4 Class Hierarchy for Making Drawings

Implementing a drawing program provides a very good example on the usefulness of object-oriented programming. In the following we shall develop the simpler parts of a relatively small and compact drawing program for making sketches of the type shown in Fig. 9.6. This is a typical *principal sketch* of a physics problem, here involving a rolling wheel on an inclined plane. The sketch is made up many individual elements: a rectangle filled with a pattern (the inclined plane), a hollow circle with color (the wheel), arrows with labels (the N and Mg forces, and the x axis), an angle with symbol θ, and a dashed line indicating the starting location of the wheel.

Drawing software and plotting programs can produce such figures quite easily in principle, but the amount of details the user needs to control with the mouse can be substantial. Software more tailored to producing sketches of this type would work with more convenient abstractions, such as circle, wall, angle, force arrow, axis, and so forth. And as soon we start *programming* to construct the figure we get a range of other powerful tools at disposal. For example, we can easily translate and rotate parts of the figure and make an animation that illustrates the physics of the problem. Programming as a superior alternative to interactive drawing is the mantra of this section.

Classes are very suitable for implementing the various components that build up a sketch. In particular, we shall demonstrate that as soon as some classes are established, more are easily added. Enhanced functionality for all the classes is also easy to implement in common, generic code that can immediately be shared by all present and future classes.

The fundamental data structure involved in this case study is a hierarchical tree, and much of the material on implementation issues targets how to traverse tree

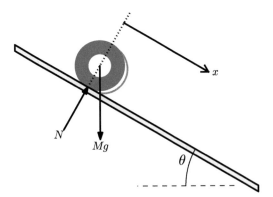

Fig. 9.6 Sketch of a physics problem

structures with recursive function calls. This topic is of key relevance in a wide range of other applications as well.

9.4.1 Using the Object Collection

We start by demonstrating a convenient user interface for making sketches of the type in Fig. 9.6. However, it is more appropriate to start with a significantly simpler example as depicted in Fig. 9.7. This toy sketch consists of several elements: two circles, two rectangles, and a "ground" element.

Basic drawing A typical program creating these five elements is shown next. After importing the `pysketcher` package, the first task is always to define a coordinate system:

```
from pysketcher import *

drawing_tool.set_coordinate_system(
    xmin=0, xmax=10, ymin=-1, ymax=8)
```

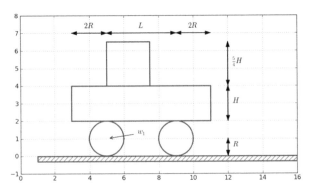

Fig. 9.7 Sketch of a simple figure

Instead of working with lengths expressed by specific numbers it is highly recommended to use variables to parameterize lengths as this makes it easier to change dimensions later. Here we introduce some key lengths for the radius of the wheels, distance between the wheels, etc.:

```
R = 1     # radius of wheel
L = 4     # distance between wheels
H = 2     # height of vehicle body
w_1 = 5   # position of front wheel
drawing_tool.set_coordinate_system(xmin=0, xmax=w_1 + 2*L + 3*R,
                                   ymin=-1, ymax=2*R + 3*H)
```

With the drawing area in place we can make the first `Circle` object in an intuitive fashion:

```
wheel1 = Circle(center=(w_1, R), radius=R)
```

to change dimensions later.

To translate the geometric information about the `wheel1` object to instructions for the plotting engine (in this case Matplotlib), one calls the `wheel1.draw()`. To display all drawn objects, one issues `drawing_tool.display()`. The typical steps are hence:

```
wheel1 = Circle(center=(w_1, R), radius=R)
wheel1.draw()

# Define other objects and call their draw() methods
drawing_tool.display()
drawing_tool.savefig('tmp.png')  # store picture
```

The next wheel can be made by taking a copy of `wheel1` and translating the object to the right according to a displacement vector $(L, 0)$:

```
wheel2 = wheel1.copy()
wheel2.translate((L,0))
```

The two rectangles are also made in an intuitive way:

```
under = Rectangle(lower_left_corner=(w_1-2*R, 2*R),
                  width=2*R + L + 2*R, height=H)
over  = Rectangle(lower_left_corner=(w_1, 2*R + H),
                  width=2.5*R, height=1.25*H)
```

Groups of objects Instead of calling the `draw` method of every object, we can group objects and call `draw`, or perform other operations, for the whole group. For example, we may collect the two wheels in a `wheels` group and the `over` and `under` rectangles in a `body` group. The whole vehicle is a composition of its `wheels` and `body` groups. The code goes like

```
wheels  = Composition({'wheel1': wheel1, 'wheel2': wheel2})
body    = Composition({'under': under, 'over': over})

vehicle = Composition({'wheels': wheels, 'body': body})
```

The ground is illustrated by an object of type `Wall`, mostly used to indicate walls in sketches of mechanical systems. A `Wall` takes the x and y coordinates of some curve, and a `thickness` parameter, and creates a thick curve filled with a simple pattern. In this case the curve is just a flat line so the construction is made of two points on the ground line ($(w_1 - L, 0)$ and $(w_1 + 3L, 0)$):

```
ground = Wall(x=[w_1 - L, w_1 + 3*L], y=[0, 0], thickness=-0.3*R)
```

The negative thickness makes the pattern-filled rectangle appear below the defined line, otherwise it appears above.

We may now collect all the objects in a "top" object that contains the whole figure:

```
fig = Composition({'vehicle': vehicle, 'ground': ground})
fig.draw()  # send all figures to plotting backend
drawing_tool.display()
drawing_tool.savefig('tmp.png')
```

The `fig.draw()` call will visit all subgroups, their subgroups, and so forth in the hierarchical tree structure of figure elements, and call `draw` for every object.

Changing line styles and colors Controlling the line style, line color, and line width is fundamental when designing figures. The `pysketcher` package allows the user to control such properties in single objects, but also set global properties that are used if the object has no particular specification of the properties. Setting the global properties are done like

```
drawing_tool.set_linestyle('dashed')
drawing_tool.set_linecolor('black')
drawing_tool.set_linewidth(4)
```

At the object level the properties are specified in a similar way:

```
wheels.set_linestyle('solid')
wheels.set_linecolor('red')
```

and so on.

Geometric figures can be specified as *filled*, either with a color or with a special visual pattern:

```
# Set filling of all curves
drawing_tool.set_filled_curves(color='blue', pattern='/')

# Turn off filling of all curves
drawing_tool.set_filled_curves(False)
```

```
# Fill the wheel with red color
wheel1.set_filled_curves('red')
```

The figure composition as an object hierarchy   The composition of objects making up the figure is hierarchical, similar to a family, where each object has a parent and a number of children. Do a `print fig` to display the relations:

```
ground
    wall
vehicle
    body
        over
            rectangle
        under
            rectangle
    wheels
        wheel1
            arc
        wheel2
            arc
```

The indentation reflects how deep down in the hierarchy (family) we are. This output is to be interpreted as follows:

- `fig` contains two objects, `ground` and `vehicle`
- `ground` contains an object `wall`
- `vehicle` contains two objects, `body` and `wheels`
- `body` contains two objects, `over` and `under`
- `wheels` contains two objects, `wheel1` and `wheel2`

In this listing there are also objects not defined by the programmer: `rectangle` and `arc`. These are of type `Curve` and automatically generated by the classes `Rectangle` and `Circle`.

More detailed information can be printed by

```
print fig.show_hierarchy('std')
```

yielding the output

```
ground (Wall):
    wall (Curve): 4 coords fillcolor='white' fillpattern='/'
vehicle (Composition):
    body (Composition):
        over (Rectangle):
            rectangle (Curve): 5 coords
        under (Rectangle):
            rectangle (Curve): 5 coords
    wheels (Composition):
        wheel1 (Circle):
            arc (Curve): 181 coords
        wheel2 (Circle):
            arc (Curve): 181 coords
```

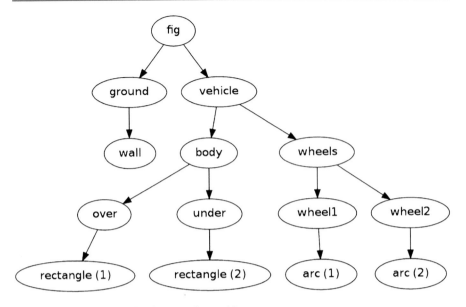

Fig. 9.8 Hierarchical relation between figure objects

Here we can see the class type for each figure object, how many coordinates that are involved in basic figures (`Curve` objects), and special settings of the basic figure (fillcolor, line types, etc.). For example, `wheel2` is a `Circle` object consisting of an `arc`, which is a `Curve` object consisting of 181 coordinates (the points needed to draw a smooth circle). The `Curve` objects are the only objects that really holds specific coordinates to be drawn. The other object types are just compositions used to group parts of the complete figure.

One can also get a graphical overview of the hierarchy of figure objects that build up a particular figure `fig`. Just call `fig.graphviz_dot('fig')` to produce a file `fig.dot` in the *dot format*. This file contains relations between parent and child objects in the figure and can be turned into an image, as in Fig. 9.8, by running the dot program:

Terminal

```
Terminal> dot -Tpng -o fig.png fig.dot
```

The call `fig.graphviz_dot('fig', classname=True)` makes a `fig.dot` file where the class type of each object is also visible, see Fig. 9.9. The ability to write out the object hierarchy or view it graphically can be of great help when working with complex figures that involve layers of subfigures.

Any of the objects can in the program be reached through their names, e.g.,

```
fig['vehicle']
fig['vehicle']['wheels']
fig['vehicle']['wheels']['wheel2']
fig['vehicle']['wheels']['wheel2']['arc']
```

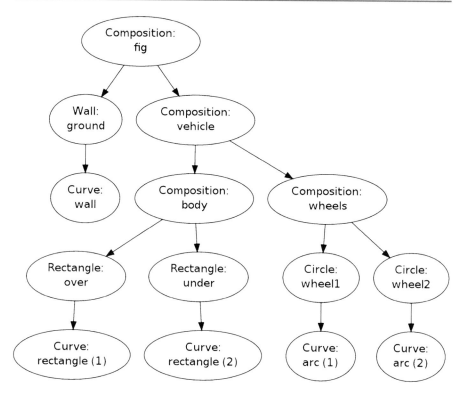

Fig. 9.9 Hierarchical relation between figure objects, including their class names

```
fig['vehicle']['wheels']['wheel2']['arc'].x   # x coords
fig['vehicle']['wheels']['wheel2']['arc'].y   # y coords
fig['vehicle']['wheels']['wheel2']['arc'].linestyle
fig['vehicle']['wheels']['wheel2']['arc'].linetype
```

Grabbing a part of the figure this way is handy for changing properties of that part, for example, colors, line styles (see Fig. 9.10):

```
fig['vehicle']['wheels'].set_filled_curves('blue')
fig['vehicle']['wheels'].set_linewidth(6)
fig['vehicle']['wheels'].set_linecolor('black')

fig['vehicle']['body']['under'].set_filled_curves('red')

fig['vehicle']['body']['over'].set_filled_curves(pattern='/')
fig['vehicle']['body']['over'].set_linewidth(14)
fig['vehicle']['body']['over']['rectangle'].linewidth = 4
```

The last line accesses the Curve object directly, while the line above, accesses the Rectangle object, which will then set the linewidth of its Curve object, and other objects if it had any. The result of the actions above is shown in Fig. 9.10.

We can also change position of parts of the figure and thereby make animations, as shown next.

Fig. 9.10 *Left*: Basic line-based drawing. *Right*: Thicker lines and filled parts

Animation: translating the vehicle Can we make our little vehicle roll? A first attempt will be to fake rolling by just displacing all parts of the vehicle. The relevant parts constitute the `fig['vehicle']` object. This part of the figure can be translated, rotated, and scaled. A translation along the ground means a translation in x direction, say a length L to the right:

```
fig['vehicle'].translate((L,0))
```

You need to erase, draw, and display to see the movement:

```
drawing_tool.erase()
fig.draw()
drawing_tool.display()
```

Without erasing, the old drawing of the vehicle will remain in the figure so you get two vehicles. Without `fig.draw()` the new coordinates of the vehicle will not be communicated to the drawing tool, and without calling display the updated drawing will not be visible.

A figure that moves in time is conveniently realized by the function `animate`:

```
animate(fig, tp, action)
```

Here, `fig` is the entire figure, `tp` is an array of time points, and `action` is a user-specified function that changes `fig` at a specific time point. Typically, `action` will move parts of `fig`.

In the present case we can define the movement through a velocity function `v(t)` and displace the figure `v(t)*dt` for small time intervals `dt`. A possible velocity function is

```
def v(t):
    return -8*R*t*(1 - t/(2*R))
```

Our action function for horizontal displacements `v(t)*dt` becomes

```
def move(t, fig):
    x_displacement = dt*v(t)
    fig['vehicle'].translate((x_displacement, 0))
```

Since our velocity is negative for $t \in [0, 2R]$ the displacement is to the left.

The `animate` function will for each time point t in `tp` erase the drawing, call `action(t, fig)`, and show the new figure by `fig.draw()` and `drawing_tool.display()`. Here we choose a resolution of the animation corresponding to 25 time points in the time interval $[0, 2R]$:

```
import numpy
tp = numpy.linspace(0, 2*R, 25)
dt = tp[1] - tp[0]  # time step

animate(fig, tp, move, pause_per_frame=0.2)
```

The `pause_per_frame` adds a pause, here 0.2 seconds, between each frame in the animation.

We can also ask `animate` to store each frame in a file:

```
files = animate(fig, tp, move_vehicle, moviefiles=True,
                pause_per_frame=0.2)
```

The `files` variable, here `'tmp_frame_%04d.png'`, is the printf-specification used to generate the individual plot files. We can use this specification to make a video file via `ffmpeg` (or `avconv` on Debian-based Linux systems such as Ubuntu). Videos in the Flash and WebM formats can be created by

```
Terminal
Terminal> ffmpeg -r 12 -i tmp_frame_%04d.png -vcodec flv mov.flv
Terminal> ffmpeg -r 12 -i tmp_frame_%04d.png -vcodec libvpx mov.webm
```

An animated GIF movie can also be made using the `convert` program from the ImageMagick software suite:

```
Terminal
Terminal> convert -delay 20 tmp_frame*.png mov.gif
Terminal> animate mov.gif  # play movie
```

The delay between frames, in units of 1/100 s, governs the speed of the movie. To play the animated GIF file in a web page, simply insert `<img src="mov.gif">` in the HTML code.

The individual PNG frames can be directly played in a web browser by running

```
Terminal
Terminal> scitools movie output_file=mov.html fps=5 tmp_frame*
```

or calling

```
from scitools.std import movie
movie(files, encoder='html', output_file='mov.html')
```

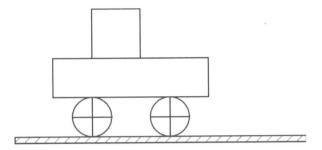

Fig. 9.11 Wheels with spokes to illustrate rolling

in Python. Load the resulting file `mov.html` into a web browser to play the movie.

Try to run `vehicle0.py` and then load `mov.html` into a browser, or play one of the `mov.*` video files. Alternatively, you can view a ready-made movie[3].

Animation: rolling the wheels It is time to show rolling wheels. To this end, we add spokes to the wheels, formed by two crossing lines, see Fig. 9.11. The construction of the wheels will now involve a circle and two lines:

```
wheel1 = Composition({
    'wheel': Circle(center=(w_1, R), radius=R),
    'cross': Composition({'cross1': Line((w_1,0),     (w_1,2*R)),
                          'cross2': Line((w_1-R,R), (w_1+R,R))})})
wheel2 = wheel1.copy()
wheel2.translate((L,0))
```

Observe that `wheel1.copy()` copies all the objects that make up the first wheel, and `wheel2.translate` translates all the copied objects.

The `move` function now needs to displace all the objects in the entire vehicle and also rotate the `cross1` and `cross2` objects in both wheels. The rotation angle follows from the fact that the arc length of a rolling wheel equals the displacement of the center of the wheel, leading to a rotation angle

```
angle = - x_displacement/R
```

With `w_1` tracking the x coordinate of the center of the front wheel, we can rotate that wheel by

```
w1 = fig['vehicle']['wheels']['wheel1']
from math import degrees
w1.rotate(degrees(angle), center=(w_1, R))
```

The `rotate` function takes two parameters: the rotation angle (in degrees) and the center point of the rotation, which is the center of the wheel in this case. The other wheel is rotated by

---

[3] http://tinyurl.com/oou9lp7/mov-tut/vehicle0.html

```
w2 = fig['vehicle']['wheels']['wheel2']
w2.rotate(degrees(angle), center=(w_1 + L, R))
```

That is, the angle is the same, but the rotation point is different. The update of the center point is done by `w_1 += x_displacement`. The complete `move` function with translation of the entire vehicle and rotation of the wheels then becomes

```
w_1 = w_1 + L   # start position

def move(t, fig):
    x_displacement = dt*v(t)
    fig['vehicle'].translate((x_displacement, 0))

    # Rotate wheels
    global w_1
    w_1 += x_displacement
    # R*angle = -x_displacement
    angle = - x_displacement/R
    w1 = fig['vehicle']['wheels']['wheel1']
    w1.rotate(degrees(angle), center=(w_1, R))
    w2 = fig['vehicle']['wheels']['wheel2']
    w2.rotate(degrees(angle), center=(w_1 + L, R))
```

The complete example is found in the file `vehicle1.py`. You may run this file or watch a ready-made movie[4].

The advantages with making figures this way, through programming rather than using interactive drawing programs, are numerous. For example, the objects are parameterized by variables so that various dimensions can easily be changed. Subparts of the figure, possible involving a lot of figure objects, can change color, linetype, filling or other properties through a *single* function call. Subparts of the figure can be rotated, translated, or scaled. Subparts of the figure can also be copied and moved to other parts of the drawing area. However, the single most important feature is probably the ability to make animations governed by mathematical formulas or data coming from physics simulations of the problem, as shown in the example above.

9.4.2 Example of Classes for Geometric Objects

We shall now explain how we can, quite easily, realize software with the capabilities demonstrated in the previous examples. Each object in the figure is represented as a class in a class hierarchy. Using inheritance, classes can inherit properties from parent classes and add new geometric features.

We introduce class `Shape` as superclass for all specialized objects in a figure. This class does not store any data, but provides a series of functions that add functionality to all the subclasses. This will be shown later.

Simple geometric objects One simple subclass is `Rectangle`, specified by the coordinates of the lower left corner and its width and height:

---

[4] http://tinyurl.com/oou9lp7/mov-tut/vehicle1.html

```
class Rectangle(Shape):
    def __init__(self, lower_left_corner, width, height):
        p = lower_left_corner  # short form
        x = [p[0], p[0] + width,
             p[0] + width, p[0], p[0]]
        y = [p[1], p[1], p[1] + height,
             p[1] + height, p[1]]
        self.shapes = {'rectangle': Curve(x,y)}
```

Any subclass of Shape will have a constructor that takes geometric information about the shape of the object and creates a dictionary self.shapes with the shape built of simpler shapes. The most fundamental shape is Curve, which is just a collection of (x, y) coordinates in two arrays x and y. Drawing the Curve object is a matter of plotting y versus x. For class Rectangle the x and y arrays contain the corner points of the rectangle in counterclockwise direction, starting and ending with in the lower left corner.

Class Line is also a simple class:

```
class Line(Shape):
    def __init__(self, start, end):
        x = [start[0], end[0]]
        y = [start[1], end[1]]
        self.shapes = {'line': Curve(x, y)}
```

Here we only need two points, the start and end point on the line. However, we may want to add some useful functionality, e.g., the ability to give an x coordinate and have the class calculate the corresponding y coordinate:

```
    def __call__(self, x):
        """Given x, return y on the line."""
        x, y = self.shapes['line'].x, self.shapes['line'].y
        self.a = (y[1] - y[0])/(x[1] - x[0])
        self.b = y[0] - self.a*x[0]
        return self.a*x + self.b
```

Unfortunately, this is too simplistic because vertical lines cannot be handled (infinite self.a). The true source code of Line therefore provides a more general solution at the cost of significantly longer code with more tests.

A circle implies a somewhat increased complexity. Again we represent the geometric object by a Curve object, but this time the Curve object needs to store a large number of points on the curve such that a plotting program produces a visually smooth curve. The points on the circle must be calculated manually in the constructor of class Circle. The formulas for points (x, y) on a curve with radius R and center at (x_0, y_0) are given by

$$x = x_0 + R \cos(t),$$
$$y = y_0 + R \sin(t),$$

where $t \in [0, 2\pi]$. A discrete set of t values in this interval gives the corresponding set of (x, y) coordinates on the circle. The user must specify the resolution as the number of t values. The circle's radius and center must of course also be specified.

We can write the `Circle` class as

```
class Circle(Shape):
    def __init__(self, center, radius, resolution=180):
        self.center, self.radius = center, radius
        self.resolution = resolution

        t = linspace(0, 2*pi, resolution+1)
        x0 = center[0];   y0 = center[1]
        R = radius
        x = x0 + R*cos(t)
        y = y0 + R*sin(t)
        self.shapes = {'circle': Curve(x, y)}
```

As in class `Line` we can offer the possibility to give an angle θ (equivalent to t in the formulas above) and then get the corresponding x and y coordinates:

```
    def __call__(self, theta):
        """Return (x, y) point corresponding to angle theta."""
        return self.center[0] + self.radius*cos(theta), \
               self.center[1] + self.radius*sin(theta)
```

There is one flaw with this method: it yields illegal values after a translation, scaling, or rotation of the circle.

A part of a circle, an arc, is a frequent geometric object when drawing mechanical systems. The arc is constructed much like a circle, but t runs in $[\theta_s, \theta_s + \theta_a]$. Giving θ_s and θ_a the slightly more descriptive names `start_angle` and `arc_angle`, the code looks like this:

```
class Arc(Shape):
    def __init__(self, center, radius,
                 start_angle, arc_angle,
                 resolution=180):
        self.start_angle = radians(start_angle)
        self.arc_angle = radians(arc_angle)

        t = linspace(self.start_angle,
                     self.start_angle + self.arc_angle,
                     resolution+1)
        x0 = center[0];   y0 = center[1]
        R = radius
        x = x0 + R*cos(t)
        y = y0 + R*sin(t)
        self.shapes = {'arc': Curve(x, y)}
```

Having the `Arc` class, a `Circle` can alternatively be defined as a subclass specializing the arc to a circle:

```
class Circle(Arc):
    def __init__(self, center, radius, resolution=180):
        Arc.__init__(self, center, radius, 0, 360, resolution)
```

Class curve Class `Curve` sits on the coordinates to be drawn, but how is that done? The constructor of class `Curve` just stores the coordinates, while a method `draw` sends the coordinates to the plotting program to make a graph. Or more precisely, to avoid a lot of (e.g.) Matplotlib-specific plotting commands in class `Curve` we have created a small layer with a simple programming interface to plotting programs. This makes it straightforward to change from Matplotlib to another plotting program. The programming interface is represented by the `drawing_tool` object and has a few functions:

- `plot_curve` for sending a curve in terms of x and y coordinates to the plotting program,
- `set_coordinate_system` for specifying the graphics area,
- `erase` for deleting all elements of the graph,
- `set_grid` for turning on a grid (convenient while constructing the figure),
- `set_instruction_file` for creating a separate file with all plotting commands (Matplotlib commands in our case),
- a series of `set_X` functions where X is some property like `linecolor`, `linestyle`, `linewidth`, `filled_curves`.

This is basically all we need to communicate to a plotting program.

Any class in the `Shape` hierarchy inherits `set_X` functions for setting properties of curves. This information is propagated to all other shape objects in the `self.shapes` dictionary. Class `Curve` stores the line properties together with the coordinates of its curve and propagates this information to the plotting program. When saying `vehicle.set_linewidth(10)`, all objects that make up the `vehicle` object will get a `set_linewidth(10)` call, but only the `Curve` object at the end of the chain will actually store the information and send it to the plotting program.

A rough sketch of class `Curve` reads

```
class Curve(Shape):
    """General curve as a sequence of (x,y) coordintes."""
    def __init__(self, x, y):
        self.x = asarray(x, dtype=float)
        self.y = asarray(y, dtype=float)

    def draw(self):
        drawing_tool.plot_curve(
            self.x, self.y,
            self.linestyle, self.linewidth, self.linecolor, ...)

    def set_linewidth(self, width):
        self.linewidth = width

    det set_linestyle(self, style):
        self.linestyle = style
    ...
```

Compound geometric objects The simple classes `Line`, `Arc`, and `Circle` could can the geometric shape through just one `Curve` object. More complicated shapes are built from instances of various subclasses of Shape. Classes used for professional drawings soon get quite complex in composition and have a lot of geometric details, so here we prefer to make a very simple composition: the already drawn vehicle from Fig. 9.7. That is, instead of composing the drawing in a Python program as shown above, we make a subclass Vehicle0 in the Shape hierarchy for doing the same thing.

The Shape hierarchy is found in the pysketcher package, so to use these classes or derive a new one, we need to import pysketcher. The constructor of class Vehicle0 performs approximately the same statements as in the example program we developed for making the drawing in Fig. 9.7.

```
from pysketcher import *

class Vehicle0(Shape):
    def __init__(self, w_1, R, L, H):
        wheel1 = Circle(center=(w_1, R), radius=R)
        wheel2 = wheel1.copy()
        wheel2.translate((L,0))

        under = Rectangle(lower_left_corner=(w_1-2*R, 2*R),
                          width=2*R + L + 2*R, height=H)
        over  = Rectangle(lower_left_corner=(w_1, 2*R + H),
                          width=2.5*R, height=1.25*H)

        wheels = Composition(
            {'wheel1': wheel1, 'wheel2': wheel2})
        body = Composition(
            {'under': under, 'over': over})

        vehicle = Composition({'wheels': wheels, 'body': body})
        xmax = w_1 + 2*L + 3*R
        ground = Wall(x=[R, xmax], y=[0, 0], thickness=-0.3*R)

        self.shapes = {'vehicle': vehicle, 'ground': ground}
```

Any subclass of Shape *must* define the shapes attribute, otherwise the inherited draw method (and a lot of other methods too) will not work.

The painting of the vehicle, as shown in the right part of Fig. 9.10, could in class Vehicle0 be offered by a method:

```
    def colorful(self):
        wheels = self.shapes['vehicle']['wheels']
        wheels.set_filled_curves('blue')
        wheels.set_linewidth(6)
        wheels.set_linecolor('black')
        under = self.shapes['vehicle']['body']['under']
        under.set_filled_curves('red')
        over = self.shapes['vehicle']['body']['over']
        over.set_filled_curves(pattern='/')
        over.set_linewidth(14)
```

The usage of the class is simple: after having set up an appropriate coordinate system as previously shown, we can do

```
vehicle = Vehicle0(w_1, R, L, H)
vehicle.draw()
drawing_tool.display()
```

and go on the make a painted version by

```
drawing_tool.erase()
vehicle.colorful()
vehicle.draw()
drawing_tool.display()
```

A complete code defining and using class `Vehicle0` is found in the file `vehicle2.py`.

The `pysketcher` package contains a wide range of classes for various geometrical objects, particularly those that are frequently used in drawings of mechanical systems.

9.4.3   Adding Functionality via Recursion

The really powerful feature of our class hierarchy is that we can add much functionality to the superclass `Shape` and to the "bottom" class `Curve`, and then all other classes for various types of geometrical shapes immediately get the new functionality. To explain the idea we may look at the `draw` method, which all classes in the `Shape` hierarchy must have. The inner workings of the `draw` method explain the secrets of how a series of other useful operations on figures can be implemented.

Basic principles of recursion Note that we work with two types of hierarchies in the present documentation: one Python *class hierarchy*, with `Shape` as superclass, and one *object hierarchy* of figure elements in a specific figure. A subclass of `Shape` stores its figure in the `self.shapes` dictionary. This dictionary represents the object hierarchy of figure elements for that class. We want to make one `draw` call for an instance, say our class `Vehicle0`, and then we want this call to be propagated to *all* objects that are contained in `self.shapes` and all is nested subdictionaries. How is this done?

The natural starting point is to call `draw` for each `Shape` object in the `self.shapes` dictionary:

```
def draw(self):
    for shape in self.shapes:
        self.shapes[shape].draw()
```

This general method can be provided by class `Shape` and inherited in subclasses like `Vehicle0`. Let v be a `Vehicle0` instance. Seemingly, a call v.draw() just calls

```
v.shapes['vehicle'].draw()
v.shapes['ground'].draw()
```

However, in the former call we call the draw method of a Composition object whose self.shapes attributed has two elements: wheels and body. Since class Composition inherits the same draw method, this method will run through self.shapes and call wheels.draw() and body.draw(). Now, the wheels object is also a Composition with the same draw method, which will run through self.shapes, now containing the wheel1 and wheel2 objects. The wheel1 object is a Circle, so calling wheel1.draw() calls the draw method in class Circle, but this is the same draw method as shown above. This method will therefore traverse the circle's shapes dictionary, which we have seen consists of one Curve element.

The Curve object holds the coordinates to be plotted so here draw really needs to do something "physical", namely send the coordinates to the plotting program. The draw method is outlined in the short listing of class Curve shown previously.

We can go to any of the other shape objects that appear in the figure hierarchy and follow their draw calls in the similar way. Every time, a draw call will invoke a new draw call, until we eventually hit a Curve object at the "bottom" of the figure hierarchy, and then that part of the figure is really plotted (or more precisely, the coordinates are sent to a plotting program).

When a method calls itself, such as draw does, the calls are known as *recursive* and the programming principle is referred to as *recursion*. This technique is very often used to traverse hierarchical structures like the figure structures we work with here. Even though the hierarchy of objects building up a figure are of different types, they all inherit the same draw method and therefore exhibit the same behavior with respect to drawing. Only the Curve object has a different draw method, which does not lead to more recursion.

Explaining recursion Understanding recursion is usually a challenge. To get a better idea of how recursion works, we have equipped class Shape with a method recurse that just visits all the objects in the shapes dictionary and prints out a message for each object. This feature allows us to trace the execution and see exactly where we are in the hierarchy and which objects that are visited.

The recurse method is very similar to draw:

```
def recurse(self, name, indent=0):
    # print message where we are (name is where we come from)
    for shape in self.shapes:
        # print message about which object to visit
        self.shapes[shape].recurse(indent+2, shape)
```

The indent parameter governs how much the message from this recurse method is intended. We increase indent by 2 for every level in the hierarchy, i.e., every row of objects in Fig. 9.12. This indentation makes it easy to see on the printout how far down in the hierarchy we are.

A typical message written by recurse when name is 'body' and the shapes dictionary has the keys 'over' and 'under', will be

```
Composition: body.shapes has entries 'over', 'under'
call body.shapes["over"].recurse("over", 6)
```

The number of leading blanks on each line corresponds to the value of `indent`. The code printing out such messages looks like

```
def recurse(self, name, indent=0):
    space = ' '*indent
    print space, '%s: %s.shapes has entries' % \
        (self.__class__.__name__, name), \
        str(list(self.shapes.keys()))[1:-1]

    for shape in self.shapes:
        print space,
        print 'call %s.shapes["%s"].recurse("%s", %d)' % \
            (name, shape, shape, indent+2)
        self.shapes[shape].recurse(shape, indent+2)
```

Let us follow a `v.recurse('vehicle')` call in detail, `v` being a `Vehicle0` instance. Before looking into the output from `recurse`, let us get an overview of the figure hierarchy in the `v` object (as produced by `print v`)

```
ground
    wall
vehicle
    body
        over
            rectangle
        under
            rectangle
    wheels
        wheel1
            arc
        wheel2
            arc
```

The `recurse` method performs the same kind of traversal of the hierarchy, but writes out and explains a lot more.

The data structure represented by `v.shapes` is known as a *tree*. As in physical trees, there is a *root*, here the `v.shapes` dictionary. A graphical illustration of the tree (upside down) is shown in Fig. 9.12. From the root there are one or more branches, here two: `ground` and `vehicle`. Following the `vehicle` branch, it has two new branches, `body` and `wheels`. Relationships as in family trees are often used to describe the relations in object trees too: we say that `vehicle` is the parent of `body` and that `body` is a child of `vehicle`. The term *node* is also often used to describe an element in a tree. A node may have several other nodes as *descendants*.

Recursion is the principal programming technique to traverse tree structures. Any object in the tree can be viewed as a root of a subtree. For example, `wheels` is the root of a subtree that branches into `wheel1` and `wheel2`. So when processing an object in the tree, we imagine we process the root and then recurse into a subtree, but the first object we recurse into can be viewed as the root of the subtree, so the processing procedure of the parent object can be repeated.

A recommended next step is to simulate the `recurse` method by hand and carefully check that what happens in the visits to `recurse` is consistent with the output

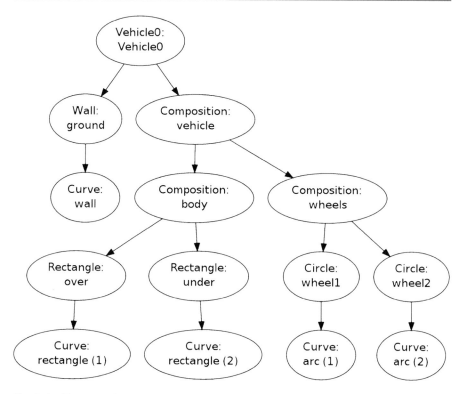

Fig. 9.12 Hierarchy of figure elements in an instance of class `Vehicle0`

listed below. Although tedious, this is a major exercise that guaranteed will help to
demystify recursion.

   A part of the printout of `v.recurse('vehicle')` looks like

```
Vehicle0: vehicle.shapes has entries 'ground', 'vehicle'
call vehicle.shapes["ground"].recurse("ground", 2)
  Wall: ground.shapes has entries 'wall'
  call ground.shapes["wall"].recurse("wall", 4)
    reached "bottom" object Curve
call vehicle.shapes["vehicle"].recurse("vehicle", 2)
  Composition: vehicle.shapes has entries 'body', 'wheels'
  call vehicle.shapes["body"].recurse("body", 4)
    Composition: body.shapes has entries 'over', 'under'
    call body.shapes["over"].recurse("over", 6)
      Rectangle: over.shapes has entries 'rectangle'
      call over.shapes["rectangle"].recurse("rectangle", 8)
        reached "bottom" object Curve
    call body.shapes["under"].recurse("under", 6)
      Rectangle: under.shapes has entries 'rectangle'
      call under.shapes["rectangle"].recurse("rectangle", 8)
        reached "bottom" object Curve
  ...
```

This example should clearly demonstrate the principle that we can start at any object
in the tree and do a recursive set of calls with that object as root.

9.4.4   Scaling, Translating, and Rotating a Figure

With recursion, as explained in the previous section, we can within minutes equip *all* classes in the Shape hierarchy, both present and future ones, with the ability to scale the figure, translate it, or rotate it. This added functionality requires only a few lines of code.

Scaling   We start with the simplest of the three geometric transformations, namely scaling. For a Curve instance containing a set of n coordinates (x_i, y_i) that make up a curve, scaling by a factor a means that we multiply all the x and y coordinates by a:

$$x_i \leftarrow ax_i, \quad y_i \leftarrow ay_i, \quad i = 0, \ldots, n-1.$$

Here we apply the arrow as an assignment operator. The corresponding Python implementation in class Curve reads

```
class Curve:
    ...
    def scale(self, factor):
        self.x = factor*self.x
        self.y = factor*self.y
```

Note here that self.x and self.y are Numerical Python arrays, so that multiplication by a scalar number factor is a vectorized operation.

An even more efficient implementation is to make use of in-place multiplication in the arrays,

```
class Curve:
    ...
    def scale(self, factor):
        self.x *= factor
        self.y *= factor
```

as this saves the creation of temporary arrays like factor*self.x.

In an instance of a subclass of Shape, the meaning of a method scale is to run through all objects in the dictionary shapes and ask each object to scale itself. This is the same delegation of actions to subclass instances as we do in the draw (or recurse) method. All objects, except Curve instances, can share the same implementation of the scale method. Therefore, we place the scale method in the superclass Shape such that all subclasses inherit the method. Since scale and draw are so similar, we can easily implement the scale method in class Shape by copying and editing the draw method:

```
class Shape:
    ...
    def scale(self, factor):
        for shape in self.shapes:
            self.shapes[shape].scale(factor)
```

This is all we have to do in order to equip all subclasses of `Shape` with scaling functionality! Any piece of the figure will scale itself, in the same manner as it can draw itself.

Translation A set of coordinates (x_i, y_i) can be translated v_0 units in the x direction and v_1 units in the y direction using the formulas

$$x_i \leftarrow x_i + v_0, \quad y_i \leftarrow y_i + v_1, \quad i = 0, \ldots, n - 1.$$

The natural specification of the translation is in terms of the vector $v = (v_0, v_1)$. The corresponding Python implementation in class `Curve` becomes

```
class Curve:
    ...
    def translate(self, v):
        self.x += v[0]
        self.y += v[1]
```

The translation operation for a shape object is very similar to the scaling and drawing operations. This means that we can implement a common method `translate` in the superclass `Shape`. The code is parallel to the `scale` method:

```
class Shape:
    ....
    def translate(self, v):
        for shape in self.shapes:
            self.shapes[shape].translate(v)
```

Rotation Rotating a figure is more complicated than scaling and translating. A counter clockwise rotation of θ degrees for a set of coordinates (x_i, y_i) is given by

$$\bar{x}_i \leftarrow x_i \cos\theta - y_i \sin\theta,$$
$$\bar{y}_i \leftarrow x_i \sin\theta + y_i \cos\theta.$$

This rotation is performed around the origin. If we want the figure to be rotated with respect to a general point (x, y), we need to extend the formulas above:

$$\bar{x}_i \leftarrow x + (x_i - x)\cos\theta - (y_i - y)\sin\theta,$$
$$\bar{y}_i \leftarrow y + (x_i - x)\sin\theta + (y_i - y)\cos\theta.$$

The Python implementation in class `Curve`, assuming that θ is given in degrees and not in radians, becomes

```
def rotate(self, angle, center):
    angle = radians(angle)
    x, y = center
    c = cos(angle);   s = sin(angle)
    xnew = x + (self.x - x)*c - (self.y - y)*s
    ynew = y + (self.x - x)*s + (self.y - y)*c
    self.x = xnew
    self.y = ynew
```

The `rotate` method in class `Shape` follows the principle of the `draw`, `scale`, and `translate` methods.

We have already seen the `rotate` method in action when animating the rolling wheel at the end of Sect. 9.4.1.

9.5   Classes for DNA Analysis

We shall here exemplify the use of classes for performing DNA analysis as explained in Sects. 3.3.1, 6.5.1, 6.5.2, 6.5.3, 6.5.4, 6.5.5, and 8.3.4. Basically, we create a class `Gene` to represent a DNA sequence (string) and a class `Region` to represent a subsequence (substring), typically an exon or intron.

9.5.1   Class for Regions

The class for representing a region of a DNA string is quite simple:

```
class Region(object):
    def __init__(self, dna, start, end):
        self._region = dna[start:end]

    def get_region(self):
        return self._region

    def __len__(self):
        return len(self._region)

    def __eq__(self, other):
        """Check if two Region instances are equal."""
        return self._region == other._region

    def __add__(self, other):
        """Add Region instances: self + other"""
        return self._region + other._region

    def __iadd__(self, other):
        """Increment Region instance: self += other"""
        self._region += other._region
        return self
```

Besides storing the substring and giving access to it through `get_region`, we have also included the possibility to

- say `len(r)` if r is a Region instance
- check if two Region instances are equal
- write `r1 + r2` for two instances r1 and r2 of type Region
- perform `r1 += r2`

The latter two operations are convenient for making one large string out of all exon or intron regions.

9.5.2 Class for Genes

The class for gene will be longer and more complex than class `Region`. We already have a bunch of functions performing various types of analysis. The idea of the `Gene` class is that these functions are methods in the class operating on the DNA string and the exon regions stored in the class. Rather than recoding all the functions as methods in the class we shall just let the class "wrap" the functions. That is, the class methods call up the functions we already have. This approach has two advantages: users can either choose the function-based or the class-based interface, and the programmer can reuse all the ready-made functions when implementing the class-based interface.

The selection of functions include

- `generate_string` for generating a random string from some alphabet
- `download` and `read_dnafile` (version `read_dnafile_v1`) for downloading data from the Internet and reading from file
- `read_exon_regions` (version `read_exon_regions_v2`) for reading exon regions from file
- `tofile_with_line_sep` (version `tofile_with_line_sep_v2`) for writing strings to file
- `read_genetic_code` (version `read_genetic_code_v2`) for loading the mapping from triplet codes to 1-letter symbols for amino acids
- `get_base_frequencies` (version `get_base_frequencies_v2`) for finding frequencies of each base
- `format_frequencies` for formatting base frequencies with two decimals
- `create_mRNA` for computing an mRNA string from DNA and exon regions
- `mutate` for mutating a base at a random position
- `create_markov_chain`, `transition`, and `mutate_via_markov_chain` for mutating a base at a random position according to randomly generated transition probabilities
- `create_protein_fixed` for proper creation of a protein sequence (string)

The set of plain functions for DNA analysis is found in the file `dna_functions.py`, while `dna_classes.py` contains the implementations of classes `Gene` and `Region`.

Basic features of class gene Class `Gene` is supposed to hold the DNA sequence and the associated exon regions. A simple constructor expects the exon regions to be specified as a list of (start, end) tuples indicating the start and end of each region:

```
class Gene(object):
    def __init__(self, dna, exon_regions):
        self._dna = dna

        self._exon_regions = exon_regions
        self._exons = []
        for start, end in exon_regions:
            self._exons.append(Region(dna, start, end))
```

```
    # Compute the introns (regions between the exons)
    self._introns = []
    prev_end = 0
    for start, end in exon_regions:
        self._introns.append(Region(dna, prev_end, start))
        prev_end = end
    self._introns.append(Region(dna, end, len(dna)))
```

The methods in class `Gene` are trivial to implement when we already have the functionality in stand-alone functions. Here are a few examples on methods:

```
from dna_functions import *

class Gene(object):
    ...

    def write(self, filename, chars_per_line=70):
        """Write DNA sequence to file with name filename."""
        tofile_with_line_sep(self._dna, filename, chars_per_line)

    def count(self, base):
        """Return no of occurrences of base in DNA."""
        return self._dna.count(base)

    def get_base_frequencies(self):
        """Return dict of base frequencies in DNA."""
        return get_base_frequencies(self._dna)

    def format_base_frequencies(self):
        """Return base frequencies formatted with two decimals."""
        return format_frequencies(self.get_base_frequencies())
```

Flexible constructor The constructor can be made more flexible. First, the exon regions may not be known so we should allow `None` as value and in fact use that as default value. Second, exon regions at the start and/or end of the DNA string will lead to empty intron `Region` objects so a proper test on nonzero length of the introns must be inserted. Third, the data for the DNA string and the exon regions can either be passed as arguments or downloaded and read from file. Two different initializations of `Gene` objects are therefore

```
g1 = Gene(dna, exon_regions)  # user has read data from file
g2 = Gene((urlbase, dna_file), (urlbase, exon_file))  # download
```

One can pass `None` for `urlbase` if the files are already at the computer. The flexible constructor has, not surprisingly, much longer code than the first version. The implementation illustrates well how the concept of overloaded constructors in other languages, like C++ and Java, are dealt with in Python (overloaded constructors take different types of arguments to initialize an instance):

```
class Gene(object):
    def __init__(self, dna, exon_regions):
        """
        dna: string or (urlbase,filename) tuple
        exon_regions: None, list of (start,end) tuples
                      or (urlbase,filename) tuple
        In case of (urlbase,filename) tuple the file
        is downloaded and read.
        """
        if isinstance(dna, (list,tuple)) and \
           len(dna) == 2 and isinstance(dna[0], str) and \
           isinstance(dna[1], str):
            download(urlbase=dna[0], filename=dna[1])
            dna = read_dnafile(dna[1])
        elif isinstance(dna, str):
            pass # ok type (the other possibility)
        else:
            raise TypeError(
                'dna=%s %s is not string or (urlbase,filename) '\
                'tuple' % (dna, type(dna)))

        self._dna = dna

        er = exon_regions
        if er is None:
            self._exons = None
            self._introns = None
        else:
            if isinstance(er, (list,tuple)) and \
               len(er) == 2 and isinstance(er[0], str) and \
               isinstance(er[1], str):
                download(urlbase=er[0], filename=er[1])
                exon_regions = read_exon_regions(er[1])
            elif isinstance(er, (list,tuple)) and \
                 isinstance(er[0], (list,tuple)) and \
                 isinstance(er[0][0], int) and \
                 isinstance(er[0][1], int):
                pass # ok type (the other possibility)
            else:
                raise TypeError(
                    'exon_regions=%s %s is not list of (int,int) '
                    'or (urlbase,filename) tuple' % (er, type(era)))

            self._exon_regions = exon_regions
            self._exons = []
            for start, end in exon_regions:
                self._exons.append(Region(dna, start, end))

            # Compute the introns (regions between the exons)
            self._introns = []
            prev_end = 0
            for start, end in exon_regions:
                if start - prev_end > 0:
                    self._introns.append(
                        Region(dna, prev_end, start))
                prev_end = end
            if len(dna) - end > 0:
                self._introns.append(Region(dna, end, len(dna)))
```

Note that we perform quite detailed testing of the object type of the data structures supplied as the dna and exon_regions arguments. This can well be done to ensure safe use also when there is only one allowed type per argument.

Other methods A create_mRNA method, returning the mRNA as a string, can be coded as

```
def create_mRNA(self):
    """Return string for mRNA."""
    if self._exons is not None:
        return create_mRNA(self._dna, self._exon_regions)
    else:
        raise ValueError(
            'Cannot create mRNA for gene with no exon regions')
```

Also here we rely on calling an already implemented function, but include some testing whether asking for mRNA is appropriate.

Methods for creating a mutated gene are also included:

```
def mutate_pos(self, pos, base):
    """Return Gene with a mutation to base at position pos."""
    dna = self._dna[:pos] + base + self._dna[pos+1:]
    return Gene(dna, self._exon_regions)

def mutate_random(self, n=1):
    """
    Return Gene with n mutations at a random position.
    All mutations are equally probable.
    """
    mutated_dna = self._dna
    for i in range(n):
        mutated_dna = mutate(mutated_dna)
    return Gene(mutated_dna, self._exon_regions)

def mutate_via_markov_chain(markov_chain):
    """
    Return Gene with a mutation at a random position.
    Mutation into new base based on transition
    probabilities in the markov_chain dict of dicts.
    """
    mutated_dna = mutate_via_markov_chain(
        self._dna, markov_chain)
    return Gene(mutated_dna, self._exon_regions)
```

Some "get" methods that give access to the fundamental attributes of the class can be included:

```
def get_dna(self):
    return self._dna

def get_exons(self):
    return self._exons
```

```
    def get_introns(self):
        return self._introns
```

Alternatively, one could access the attributes directly: `gene._dna`, `gene._exons`, etc. In that case we should remove the leading underscore as this underscore signals that these attributes are considered "protected", i.e., not to be directly accessed by the user. The "protection" in "get" functions is more mental than actual since we anyway give the data structures in the hands of the user and she can do whatever she wants (even delete them).

Special methods for the length of a gene, adding genes, checking if two genes are identical, and printing of compact gene information are relevant to add:

```
    def __len__(self):
        return len(self._dna)

    def __add__(self, other):
        """self + other: append other to self (DNA string)."""
        if self._exons is None and other._exons is None:
            return Gene(self._dna + other._dna, None)
        else:
            raise ValueError(
                'cannot do Gene + Gene with exon regions')

    def __iadd__(self, other):
        """self += other: append other to self (DNA string)."""
        if self._exons is None and other._exons is None:
            self._dna += other._dna
            return self
        else:
            raise ValueError(
                'cannot do Gene += Gene with exon regions')

    def __eq__(self, other):
        """Check if two Gene instances are equal."""
        return self._dna == other._dna and \
               self._exons == other._exons

    def __str__(self):
        """Pretty print (condensed info)."""
        s = 'Gene: ' + self._dna[:6] + '...' + self._dna[-6:] + \
            ', length=%d' % len(self._dna)
        if self._exons is not None:
            s += ', %d exon regions' % len(self._exons)
        return s
```

Here is an interactive session demonstrating how we can work with class `Gene` objects:

```
>>> from dna_classes import Gene
>>> g1 = Gene('ATCCGTAATTGCGCA', [(2,4), (6,9)])
>>> print g1
Gene: ATCCGT...TGCGCA, length=15, 2 exon regions
>>> g2 = g1.mutate_random(10)
>>> print g2
```

```
Gene: ATCCGT...TGTGCT, length=15, 2 exon regions
>>> g1 == g2
False
>>> g1 += g2   # expect exception
Traceback (most recent call last):
...
ValueError: cannot do Gene += Gene with exon regions
>>> g1b = Gene(g1.get_dna(), None)
>>> g2b = Gene(g2.get_dna(), None)
>>> print g1b
Gene: ATCCGT...TGCGCA, length=15
>>> g3 = g1b + g2b
>>> g3.format_base_frequencies()
'A: 0.17, C: 0.23, T: 0.33, G: 0.27'
```

9.5.3   Subclasses

There are two fundamental types of genes: the most common type that codes for proteins (indirectly via mRNA) and the type that only codes for RNA (without being further processed to proteins). The product of a gene, mRNA or protein, depends on the type of gene we have. It is then natural to create two subclasses for the two types of gene and have a method `get_product` which returns the product of that type of gene.

The `get_product` method can be declared in class `Gene`:

```
def get_product(self):
    raise NotImplementedError(
        'Subclass %s must implement get_product' % \
        self.__class__.__name__)
```

The exception here will be triggered by an instance (`self`) of any subclass that just inherits `get_product` from class `Gene` without implementing a subclass version of this method.

The two subclasses of `Gene` may take this simple form:

```
class RNACodingGene(Gene):
    def get_product(self):
        return self.create_mRNA()

class ProteinCodingGene(Gene):
    def __init__(self, dna, exon_positions):
        Gene.__init__(self, dna, exon_positions)
        urlbase = 'http://hplgit.github.com/bioinf-py/data/'
        genetic_code_file = 'genetic_code.tsv'
        download(urlbase, genetic_code_file)
        code = read_genetic_code(genetic_code_file)
        self.genetic_code = code

    def get_product(self):
        return create_protein_fixed(self.create_mRNA(),
                                    self.genetic_code)
```

A demonstration of how to load the lactase gene and create the lactase protein is done with

```
def test_lactase_gene():
    urlbase = 'http://hplgit.github.com/bioinf-py/data/'
    lactase_gene_file = 'lactase_gene.txt'
    lactase_exon_file = 'lactase_exon.tsv'
    lactase_gene = ProteinCodingGene(
        (urlbase, lactase_gene_file),
        (urlbase, lactase_exon_file))

    protein = lactase_gene.get_product()
    tofile_with_line_sep(protein, 'output', 'lactase_protein.txt')
```

Now, envision that the Lactase gene would instead have been an RNA-coding gene. The only necessary changes would have been to exchange `ProteinCoding Gene` by `RNACodingGene` in the assignment to `lactase_gene`, and one would get out a final RNA product instead of a protein.

9.6 Summary

9.6.1 Chapter Topics

A subclass inherits everything from its superclass, in particular all data attributes and methods. The subclass can add new data attributes, overload methods, and thereby enrich or restrict functionality of the superclass.

Subclass example Consider class `Gravity` from Sect. 7.7.1 for representing the gravity force GMm/r^2 between two masses m and M being a distance r apart. Suppose we want to make a class for the electric force between two charges q_1 and q_2, being a distance r apart in a medium with permittivity ϵ_0 is Gq_1q_2/r^2, where $G^{-1} = 4\pi\epsilon_0$. We use the approximate value $G = 8.99 \cdot 10^9 \, \text{Nm}^2/\text{C}^2$ (C is the Coulomb unit used to measure electric charges such as q_1 and q_2). Since the electric force is similar to the gravity force, we can easily implement the electric force as a subclass of `Gravity`. The implementation just needs to redefine the value of G!

```
class CoulombsLaw(Gravity):
    def __init__(self, q1, q2):
        Gravity.__init__(self, q1, q2)
        self.G = 8.99E9
```

We can now call the inherited `force(r)` method to compute the electric force and the `visualize` method to make a plot of the force:

```
c = CoulombsLaw(1E-6, -2E-6)
print 'Electric force:', c.force(0.1)
c.visualize(0.01, 0.2)
```

However, the `plot` method inherited from class `Gravity` has an inappropriate title referring to "Gravity force" and the masses m and M. An easy fix could be to have the plot title as a data attribute set in the constructor. The subclass can then override the contents of this attribute, as it overrides self.G. It is quite common to discover that a class needs adjustments if it is to be used as superclass.

Subclassing in general The typical sketch of creating a subclass goes as follows:

```
class SuperClass(object):
    def __init__(self, p, q):
        self.p, self.q = p, q

    def where(self):
        print 'In superclass', self.__class__.__name__

    def compute(self, x):
        self.where()
        return self.p*x + self.q


class SubClass(SuperClass):
    def __init__(self, p, q, a):
        SuperClass.__init__(self, p, q)
        self.a = a

    def where(self):
        print 'In subclass', self.__class__.__name__

    def compute(self, x):
        self.where()
        return SuperClass.compute(self, x) + self.a*x**2
```

This example shows how a subclass extends a superclass with one data attribute (a). The subclass' compute method calls the corresponding superclass method, as well as the overloaded method where. Let us invoke the compute method through superclass and subclass instances:

```
>>> super = SuperClass(1, 2)
>>> sub = SubClass(1, 2, 3)
>>> v1 = super.compute(0)
In superclass SuperClass
>>> v2 = sub.compute(0)
In subclass SubClass
In subclass SubClass
```

Observe that in the subclass sub, method compute calls self.where, which translates to the where method in SubClass. Then the compute method in SuperClass is invoked, and this method also makes a self.where call, which is a call to SubClass' where method (think of what self is here, it is sub, so it is natural that we get where in the subclass (sub.where) and not where in the superclass part of sub).

In this example, classes SuperClass and SubClass constitute a class hierarchy. Class SubClass inherits the attributes p and q from its superclass, and overrides the methods where and compute.

Terminology The important computer science topics in this chapter are

- superclass
- subclass
- inheritance
- class hierarchies
- tree structures
- recursion

9.6.2 Example: Input Data Reader

The summarizing example of this chapter concerns a class hierarchy for simplifying reading input data into programs. Input data may come from several different sources: the command line, a file, or from a dialog with the user, either of input form or in a graphical user interface (GUI). Therefore it makes sense to create a class hierarchy where subclasses are specialized to read from different sources and where the common code is placed in a superclass. The resulting tool will make it easy for you to let your programs read from many different input sources by adding just a few lines.

Problem Let us motivate the problem by a case where we want to write a program for dumping n function values of $f(x)$ to a file for $x \in [a, b]$. The core part of the program typically reads

```
import numpy as np
with open(filename, 'w') as outfile:
    for x in np.linspace(a, b, n):
        outfile.write('%12g  %12g\n' % (x, f(x)))
```

Our purpose is to read data into the variables a, b, n, filename, and f. For the latter we want to specify a formula and use the StringFunction tool (Sect. 4.3.3) to make the function f:

```
from scitools.StringFunction import StringFunction
f = StringFunction(formula)
```

How can we read a, b, n, formula, and filename conveniently into the program?

The basic idea is that we place the input data in a dictionary, and create a tool that can update this dictionary from sources like the command line, a file, a GUI, etc. Our dictionary is then

```
p = dict(formula='x+1', a=0, b=1, n=2, filename='tmp.dat')
```

This dictionary specifies the names of the input parameters to the program and the default values of these parameters.

Using the tool is a matter of feeding p into the constructor of a subclass in the tools' class hierarchy and extract the parameters into, for example, distinct variables:

```
inp = Subclassname(p)
a, b, filename, formula, n = inp.get_all()
```

Depending on what we write as Subclassname, the five variables can be read from
the command line, the terminal window, a file, or a GUI. The task now is to implement a class hierarchy to facilitate the described flexible reading of input data.

Solution We first create a very simple superclass ReadInput. Its main purpose
is to store the parameter dictionary as a data attribute, provide a method get to
extract single values, and a method get_all to extract all parameters into distinct
variables:

```
class ReadInput(object):
    def __init__(self, parameters):
        self.p = parameters

    def get(self, parameter_name):
        return self.p[parameter_name]

    def get_all(self):
        return [self.p[name] for name in sorted(self.p)]

    def __str__(self):
        import pprint
        return pprint.pformat(self.p)
```

Note that we in the get_all method must sort the keys in self.p such that the
list of returned variables is well defined. In the calling program we can then list
variables in the same order as the alphabetic order of the parameter names, for
example:

```
a, b, filename, formula, n = inp.get_all()
```

The __str__ method applies the pprint module to get a pretty print of all the
parameter names and their values.

Class ReadInput cannot read from any source – subclasses are supposed to do
this. The forthcoming text describes various types of subclasses for various types
of reading input.

Prompting the user The perhaps simplest way of getting data into a program is
to use raw_input. We then prompt the user with a text Give name: and get
an appropriate object back (recall that strings must be enclosed in quotes). The
subclass PromptUser for doing this then reads

```
class PromptUser(ReadInput):
    def __init__(self, parameters):
        ReadInput.__init__(self, parameters)
        self._prompt_user()
```

```
def _prompt_user(self):
    for name in self.p:
        self.p[name] = eval(raw_input("Give " + name + ": "))
```

Note the underscore in `_prompt_user`: the underscore signifies that this is a "private" method in the `PromptUser` class, not intended to be called by users of the class.

There is a major difficulty with using `eval` on the input from the user. When the input is intended to be a string object, such as a filename, say `tmp.inp`, the program will perform the operation `eval(tmp.inp)`, which leads to an exception because `tmp.inp` is treated as a variable `inp` in a module `tmp` and not as the string `'tmp.inp'`. To solve this problem, we use the `str2obj` function from the `scitools.misc` module. This function will return the right Python object also in the case where the argument should result in a string object (see Sect. 4.11.1 for some information about `str2obj`). The bottom line is that `str2obj` acts as a safer `eval(raw_input(...))` call. The key assignment in class `PromptUser` is then changed to

```
self.p[name] = str2obj(raw_input("Give " + name + ": "))
```

Reading from file We can also place `name = value` commands in a file and load this information into the dictionary `self.p`. An example of a file can be

```
formula    = sin(x) + cos(x)
filename   = tmp.dat
a          = 0
b          = 1
```

In this example we have omitted `n`, so we rely on its default value.

A problem is how to give the filename. The easy way out of this problem is to read from standard input, and just redirect standard input from a file when we run the program. For example, if the filename is `tmp.inp`, we run the program as follows in a terminal window

---
Terminal

```
Terminal> python myprog.py < tmp.inp
```

(The redirection of standard input from a file does not work in IPython so we are in this case forced to run the program in a terminal window.)

To interpret the contents of the file, we read line by line, split each line with respect to `=`, use the left-hand side as the parameter name and the right-hand side as the corresponding value. It is important to strip away unnecessary blanks in the name and value. The complete class now reads

```
class ReadInputFile(ReadInput):
    def __init__(self, parameters):
        ReadInput.__init__(self, parameters)
        self._read_file()
```

```
def _read_file(self, infile=sys.stdin):
    for line in infile:
        if "=" in line:
            name, value = line.split("=")
            self.p[name.strip()] = str2obj(value.strip())
```

A nice feature with reading from standard input is that if we do not redirect standard input to a file, the program will prompt the user in the terminal window, where the user can give commands of the type name = value for setting selected input data. A Ctrl+d is needed to terminate the interactive session in the terminal window and continue execution of the program.

Reading from the command line   For input from the command line we assume that parameters and values are given as option-value pairs, e.g., as in

```
--a 1 --b 10 --n 101 --formula "sin(x) + cos(x)"
```

We apply the argparse module (see Sect. 4.4) to parse the command-line arguments. The list of legal option names must be constructed from the list of keys in the self.p dictionary. The complete class takes the form

```
class ReadCommandLine(ReadInput):
    def __init__(self, parameters):
        self.sys_argv = sys.argv[1:]   # copy
        ReadInput.__init__(self, parameters)
        self._read_command_line()

    def _read_command_line(self):
        parser = argparse.ArgumentParser()
        # Make argparse list of options
        for name in self.p:
            # Default type: str
            parser.add_argument('--'+name, default=self.p[name])

        args = parser.parse_args()
        for name in self.p:
            self.p[name] = str2obj(getattr(args, name))

import Tkinter
try:
```

We could specify the type of a parameter as type(self.p[name]) or self.p[name].__class__, but if a float parameter has been given an integer default value, the type will be int and argparse will not accept a decimal number as input. Our more general strategy is to drop specifying the type, which implies that all parameters in the args object become strings. We then use the str2obj function to convert to the right type, a technique that is used throughout the ReadInput module.

Reading from a gui   We can with a little extra effort also make a graphical user interface (GUI) for reading the input data. An example of a user interface is displayed

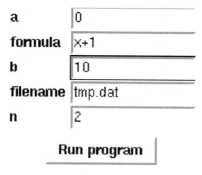

Fig. 9.13 Screen dump of a graphical user interface to read input data into a program (class GUI in the ReadInput hierarchy)

in Fig. 9.13. Since the technicalities of the implementation is beyond the scope of this book, we do not show the subclass GUI that creates the GUI and loads the user input into the self.p dictionary.

More flexibility in the superclass Some extra flexibility can easily be added to the get method in the superclass. Say we want to extract a variable number of parameters:

```
a, b, n = inp.get('a', 'b', 'n')   # 3 variables
n = inp.get('n')                   # 1 variable
```

The key to this extension is to use a variable number of arguments as explained in Sect. H.7.1:

```
class ReadInput(object):
    ...
    def get(self, *parameter_names):
        if len(parameter_names) == 1:
            return self.p[parameter_names[0]]
        else:
            return [self.p[name] for name in parameter_names]
```

Demonstrating the tool Let us show how we can use the classes in the ReadInput hierarchy. We apply the motivating example described earlier. The name of the program is demo_ReadInput.py. As first command-line argument it takes the name of the input source, given as the name of a subclass in the ReadInput hierarchy. The code for loading input data from any of the sources supported by the ReadInput hierarchy goes as follows:

```
p = dict(formula='x+1', a=0, b=1, n=2, filename='tmp.dat')
from ReadInput import *
input_reader = eval(sys.argv[1])  # PromptUser, ReadInputFile, ...
del sys.argv[1]   # otherwise argparse don't like our extra option
inp = input_reader(p)
a, b, filename, formula, n = inp.get_all()
print inp
```

Note how convenient `eval` is to automatically create the right subclass for reading input data.

Our first try on running this program applies the `PromptUser` class:

```
─────────────────────────────── Terminal ───────────────────────────────
demo_ReadInput.py PromptUser
Give a: 0
Give formula: sin(x) + cos(x)
Give b: 10
Give filename: function_data
Give n: 101
{'a': 0,
 'b': 10,
 'filename': 'function_data',
 'formula': 'sin(x) + cos(x)',
 'n': 101}
```

The next example reads data from a file `tmp.inp` with the same contents as shown in paragraph above about reading from file.

```
─────────────────────────────── Terminal ───────────────────────────────
Terminal> demo_ReadInput.py ReadFileInput < tmp.inp
{'a': 0, 'b': 1, 'filename': 'tmp.dat',
 'formula': 'sin(x) + cos(x)', 'n': 2}
```

We can also drop the redirection of standard input to a file, and instead run an interactive session in IPython or the terminal window:

```
─────────────────────────────── Terminal ───────────────────────────────
demo_ReadInput.py ReadFileInput
n = 101
filename = myfunction_data_file.dat
^D
{'a': 0,
 'b': 1,
 'filename': 'myfunction_data_file.dat',
 'formula': 'x+1',
 'n': 101}
```

Note that `Ctrl+d` is needed to end the interactive session with the user and continue program execution.

Command-line arguments can also be specified:

```
─────────────────────────────── Terminal ───────────────────────────────
demo_ReadInput.py ReadCommandLine \
        --a -1 --b 1 --formula "sin(x) + cos(x)"
{'a': -1, 'b': 1, 'filename': 'tmp.dat',
 'formula': 'sin(x) + cos(x)', 'n': 2}
```

Finally, we can run the program with a GUI,

```
                                    Terminal
demo_ReadInput.py GUI
{'a': -1, 'b': 10, 'filename': 'tmp.dat',
 'formula': 'x+1', 'n': 2}
```

The GUI is shown in Fig. 9.13.

Fortunately, it is now quite obvious how to apply the ReadInput hierarchy of classes in your own programs to simplify input. Especially in applications with a large number of parameters one can initially define these in a dictionary and then automatically create quite comprehensive user interfaces where the user can specify only some subset of the parameters (if the default values for the rest of the parameters are suitable).

9.7 Exercises

Exercise 9.1: Demonstrate the magic of inheritance
Consider class Line from Sect. 9.1.1 and a subclass Parabola0 defined as

```
class Parabola0(Line):
    pass
```

That is, class Parabola0 does not have any own code, but it inherits from class Line. Demonstrate in a program or interactive session, using dir and looking at the __dict__ object, (see Sect. 7.5.6) that an instance of class Parabola0 contains everything (i.e., all attributes) that an instance of class Line contains.
Filename: dir_subclass.

Exercise 9.2: Make polynomial subclasses of parabolas
The task in this exercise is to make a class Cubic for cubic functions

$$c_3 x^3 + c_2 x^2 + c_1 x + c_0$$

with a call operator and a table method as in classes Line and Parabola from Sect. 9.1. Implement class Cubic by inheriting from class Parabola, and call up functionality in class Parabola in the same way as class Parabola calls up functionality in class Line.

Make a similar class Poly4 for 4-th degree polynomials

$$c_4 x^4 + c_3 x^3 + c_2 x^2 + c_1 x + c_0$$

by inheriting from class Cubic. Insert print statements in all the __call__ methods such that you can easily watch the program flow and see when __call__ in the different classes is called.

Evaluate cubic and a 4-th degree polynomial at a point, and observe the printouts from all the superclasses.
Filename: Cubic_Poly4.

Remarks This exercise follows the idea from Sect. 9.1 where more complex polynomials are subclasses of simpler ones. Conceptually, a cubic polynomial *is not* a parabola, so many programmers will not accept class Cubic as a subclass of Parabola; it should be the other way around, and Exercise 9.2 follows that approach. Nevertheless, one can use inheritance solely for sharing code and not for expressing that a subclass is a kind of the superclass. For code sharing it is natural to start with the simplest polynomial as superclass and add terms to the inherited data structure as we make subclasses for higher degree polynomials.

Exercise 9.3: Implement a class for a function as a subclass

Implement a class for the function $f(x) = A\sin(wx) + ax^2 + bx + c$. The class should have a call operator for evaluating the function for some argument x, and a constructor that takes the function parameters A, w, a, b, and c as arguments. Also a table method as in classes Line and Parabola should be present. Implement the class by deriving it from class Parabola and call up functionality already implemented in class Parabola whenever possible.
Filename: sin_plus_quadratic.

Exercise 9.4: Create an alternative class hierarchy for polynomials

Let class Polynomial from Sect. 7.3.7 be a superclass and implement class Parabola as a subclass. The constructor in class Parabola should take the three coefficients in the parabola as separate arguments. Try to reuse as much code as possible from the superclass in the subclass. Implement class Line as a subclass specialization of class Parabola.

Which class design do you prefer, class Line as a subclass of Parabola and Polynomial, or Line as a superclass with extensions in subclasses? (See also remark in Exercise 9.2.)
Filename: Polynomial_hier.

Exercise 9.5: Make circle a subclass of an ellipse

Section 7.2.3 presents class Circle. Make a similar class Ellipse for representing an ellipse. Then create a new class Circle that is a subclass of Ellipse.
Filename: Ellipse_Circle.

Exercise 9.6: Make super- and subclass for a point

A point (x, y) in the plane can be represented by a class:

```
class Point(object):
    def __init__(self, x, y):
        self.x, self.y = x, y

    def __str__(self):
        return '(%g, %g)' % (self.x, self.y)
```

We can extend the Point class to also contain the representation of the point in polar coordinates. To this end, create a subclass PolarPoint whose constructor takes the polar representation of a point, (r, θ), as arguments. Store r and θ as data attributes and call the superclass constructor with the corresponding x and y

values (recall the relations $x = r\cos\theta$ and $y = r\sin\theta$ between Cartesian and polar coordinates). Add a `__str__` method in class `PolarPoint` which prints out r, θ, x, and y. Write a test function that creates two `PolarPoint` instances and compares the four data attributes x, y, r, and theta with the expected values. Filename: `PolarPoint`.

Exercise 9.7: Modify a function class by subclassing
Consider a class F implementing the function $f(t;a,b) = e^{-at}\sin(bt)$:

```
class F(object):
    def __init__(self, a, b):
        self.a, self.b = a, b
    def __call__(self, t):
        return exp(-self.a*t)*sin(self.b*t)
```

We now want to study how the function $f(t;a,b)$ varies with the parameter b, given t and a. Mathematically, this means that we want to compute $g(b;t,a) = f(t;a,b)$. Write a subclass Fb of F with a new `__call__` method for evaluating $g(b;t,a)$. Do not reimplement the formula, but call the `__call__` method in the superclass to evaluate $f(t;a,b)$. The Fs should work as follows:

```
f = Fs(t=2, a=4.5)
print f(3)   # b=3
```

Hint Before calling `__call__` in the superclass, the data attribute b in the superclass must be set to the right value.
Filename: `Fb`.

Exercise 9.8: Explore the accuracy of difference formulas
The purpose of this exercise is to investigate the accuracy of the `Backward1`, `Forward1`, `Forward3`, `Central2`, `Central4`, `Central6` methods for the function

$$v(x) = \frac{1 - e^{x/\mu}}{1 - e^{1/\mu}}.$$

Compute the errors in the approximations for $x = 0, 0.9$ and $\mu = 1, 0.01$. Illustrate in a plot how the $v(x)$ function looks like for these two μ values.

Hint Modify the `src/oo/Diff2_examples.py` program which produces tables of errors of difference approximations as discussed at the end of Sect. 9.2.4.
Filename: `boundary_layer_derivative`.

Exercise 9.9: Implement a subclass
Make a subclass Sine1 of class FuncWithDerivatives from Sect. 9.1.6 for the $\sin x$ function. Implement the function only, and rely on the inherited df and ddf methods for computing the derivatives. Make another subclass Sine2 for $\sin x$ where you also implement the df and ddf methods using analytical expressions for the derivatives. Compare Sine1 and Sine2 for computing the first- and second-order derivatives of $\sin x$ at two x points.
Filename: `Sine12`.

Exercise 9.10: Make classes for numerical differentiation
Carry out Exercise 7.16. Find the common code in the classes `Derivative`, `Backward`, and `Central`. Move this code to a superclass, and let the three mentioned classes be subclasses of this superclass. Compare the resulting code with the hierarchy shown in Sect. 9.2.1.
Filename: `numdiff_classes`.

Exercise 9.11: Implement a new subclass for differentiation
A one-sided, three-point, second-order accurate formula for differentiating a function $f(x)$ has the form

$$f'(x) \approx \frac{f(x-2h) - 4f(x-h) + 3f(x)}{2h}. \tag{9.17}$$

Implement this formula in a subclass `Backward2` of class `Diff` from Sect. 9.2. Compare `Backward2` with `Backward1` for $g(t) = e^{-t}$ for $t = 0$ and $h = 2^{-k}$ for $k = 0, 1, \ldots, 14$ (write out the errors in $g'(t)$).
Filename: `Backward2`.

Exercise 9.12: Understand if a class can be used recursively
Suppose you want to compute $f''(x)$ of some mathematical function $f(x)$, and that you apply some class from Sect. 9.2 twice, e.g.,

```
ddf = Central2(Central2(f))
```

Will this work?

Hint Follow the program flow, and find out what the resulting formula will be. Then see if this formula coincides with a formula you know for approximating $f''(x)$ (actually, to recover the well-known formula with an h parameter, you would use $h/2$ in the nested calls to `Central2`).

Exercise 9.13: Represent people by a class hierarchy
Classes are often used to model objects in the real world. We may represent the data about a person in a program by a class `Person`, containing the person's name, address, phone number, date of birth, and nationality. A method `__str__` may print the person's data. Implement such a class `Person`.

A worker is a person with a job. In a program, a worker is naturally represented as class `Worker` derived from class `Person`, because a worker *is* a person, i.e., we have an is-a relationship. Class `Worker` extends class `Person` with additional data, say name of company, company address, and job phone number. The print functionality must be modified accordingly. Implement this `Worker` class.

A scientist is a special kind of a worker. Class `Scientist` may therefore be derived from class `Worker`. Add data about the scientific discipline (physics, chemistry, mathematics, computer science, ...). One may also add the type of scientist: theoretical, experimental, or computational. The value of such a type attribute should not be restricted to just one category, since a scientist may be classified as, e.g., both experimental and computational (i.e., you can represent the value as a list or tuple). Implement class `Scientist`.

Researcher, postdoc, and professor are special cases of a scientist. One can either create classes for these job positions, or one may add an attribute (`position`) for this information in class `Scientist`. We adopt the former strategy. When, e.g., a researcher is represented by a class `Researcher`, no extra data or methods are needed. In Python we can create such an empty class by writing `pass` (the empty statement) as the class body:

```
class Researcher(Scientist):
    pass
```

Finally, make a demo program where you create and print instances of classes `Person`, `Worker`, `Scientist`, `Researcher`, `Postdoc`, and `Professor`. Print out the attribute contents of each instance (use the `dir` function).

Remark An alternative design is to introduce a class `Teacher` as a special case of `Worker` and let `Professor` be both a `Teacher` and `Scientist`, which is natural. This implies that class `Professor` has two superclasses, `Teacher` and `Scientist`, or equivalently, class `Professor` inherits from two superclasses. This is known as *multiple inheritance* and technically achieved as follows in Python:

```
class Professor(Teacher, Scientist):
    pass
```

It is a continuous debate in computer science whether multiple inheritance is a good idea or not. One obvious problem in the present example is that class `Professor` inherits two names, one via `Teacher` and one via `Scientist` (both these classes inherit from `Person`).
Filename: `Person`.

Exercise 9.14: Add a new class in a class hierarchy

a) Add the Monte Carlo integration method from Sect. 8.5.2 as a subclass `MCint` in the `Integrator` hierarchy explained in Sect. 9.3. Import the superclass `Integrator` from the `integrate` module in the file with the new integration class.

b) Make a test function for class `MCint` where you fix the seed of the random number generator, use three function evaluations only, and compare the result of this Monte Carlo integration with results calculated by hand using the same three random numbers.

c) Run the Monte Carlo integration class in a case with known analytical solution and see how the error in the integral changes with $n = 10^k$ function evaluations, $k = 3, 4, 5, 6$.

Filename: `MCint_class`.

Exercise 9.15: Compute convergence rates of numerical integration methods
Numerical integration methods can compute "any" integral $\int_a^b f(x)dx$, but the result is not exact. The methods have a parameter n, closely related to the number of evaluations of the function f, that can be increased to achieve more accurate results. In this exercise we want to explore the relation between the error E in the numerical approximation to the integral and n. Different numerical methods have different relations.

The relations are of the form

$$E = Cn^r,$$

where and C and $r < 0$ are constants to be determined. That is, r is the most important of these parameters, because if Simpson's method has a more negative r than the Trapezoidal method, it means that increasing n in Simpson's method reduces the error more effectively than increasing n in the Trapezoidal method.

One can estimate r from numerical experiments. For a chosen $f(x)$, where the exact value of $\int_a^b f(x)dx$ is available, one computes the numerical approximation for $N + 1$ values of n: $n_0 < n_1 < \cdots < n_N$ and finds the corresponding errors $E_0, E_1, \ldots, E_N$ (the difference between the exact value and the value produced by the numerical method).

One way to estimate r goes as follows. For two successive experiments we have

$$E_i = Cn_i^r.$$

and

$$E_{i+1} = Cn_{i+1}^r$$

Divide the first equation by the second to eliminate C, and then take the logarithm to solve for r:

$$r = \frac{\ln(E_i/E_{i+1})}{\ln(n_i/n_{i+1})}.$$

We can compute r for all pairs of two successive experiments. Say r_i is the r value found from experiment i and $i + 1$,

$$r_i = \frac{\ln(E_i/E_{i+1})}{\ln(n_i/n_{i+1})}, \quad i = 0, 1, \ldots, N - 1.$$

Usually, the last value, r_{N-1}, is the best approximation to the true r value. Knowing r, we can compute C as $E_i n_i^{-r}$ for any i.

Use the method above to estimate r and C for the Midpoint method, the Trapezoidal method, and Simpson's method. Make your own choice of integral problem: $f(x)$, a, and b. Let the parameter n be the number of function evaluations in each method, and run the experiments with $n = 2^k + 1$ for $k = 2, \ldots, 11$. The `Integrator` hierarchy from Sect. 9.3 has all the requested methods implemented. Filename: `integrators_convergence`.

Exercise 9.16: Add common functionality in a class hierarchy
Suppose you want to use classes in the `Integrator` hierarchy from Sect. 9.3. to calculate integrals of the form

$$F(x) = \int_a^x f(t)dt \; .$$

Such functions $F(x)$ can be efficiently computed by the method from Exercise 7.22. Implement this computation of $F(x)$ in an additional method in the superclass `Integrator`. Test that the implementation is correct for $f(x) = 2x - 3$ for all the implemented integration methods (the Midpoint, Trapezoidal and Gauss-Legendre methods, as well as Simpson's rule, integrate a linear function exactly).
Filename: `integrate_efficient`.

Exercise 9.17: Make a class hierarchy for root finding
Given a general nonlinear equation $f(x) = 0$, we want to implement classes for solving such an equation, and organize the classes in a class hierarchy. Make classes for three methods: Newton's method (in Sect. A.1.10), the Bisection method (in Sect. 4.11.2), and the Secant method (in Exercise A.10).

It is not obvious how such a hierarchy should be organized. One idea is to let the superclass store the $f(x)$ function and its derivative $f'(x)$ (if provided – if not, use a finite difference approximation for $f'(x)$). A method

```
def solve(start_values=[0], max_iter=100, tolerance=1E-6):
    ...
```

in the superclass can implement a general iteration loop. The `start_values` argument is a list of starting values for the algorithm in question: one point for Newton, two for Secant, and an interval $[a,b]$ containing a root for Bisection. Let `solve` define a list `self.x` holding all the computed approximations. The initial value of `self.x` is simply `start_values`. For the Bisection method, one can use the convention $a, b, c = $ `self.x[-3:]`, where $[a, b]$ represents the most recently computed interval and c is its midpoint. The `solve` method can return an approximate root x, the corresponding $f(x)$ value, a boolean indicator that is `True` if $|f(x)|$ is less than the `tolerance` parameter, and a list of all the approximations and their f values (i.e., a list of $(x, f(x))$ tuples).

Do Exercise A.11 using the new class hierarchy.
Filename: `Rootfinders`.

Exercise 9.18: Make a calculus calculator class
Given a function $f(x)$ defined on a domain $[a, b]$, the purpose of many mathematical exercises is to sketch the function curve $y = f(x)$, compute the derivative $f'(x)$, find local and global extreme points, and compute the integral $\int_a^b f(x)dx$. Make a class `CalculusCalculator` which can perform all these actions for any function $f(x)$ using numerical differentiation and integration, and the method explained in Exercise 7.34. for finding extrema.

Here is an interactive session with the class where we analyze $f(x) = x^2 e^{-0.2x} \sin(2\pi x)$ on $[0, 6]$ with a grid (set of x coordinates) of 700 points:

```
>>> from CalculusCalculator import *
>>> def f(x):
...     return x**2*exp(-0.2*x)*sin(2*pi*x)
...
>>> c = CalculusCalculator(f, 0, 6, resolution=700)
>>> c.plot()                 # plot f
>>> c.plot_derivative()      # plot f'
>>> c.extreme_points()

All minima: 0.8052, 1.7736, 2.7636, 3.7584, 4.7556, 5.754, 0
All maxima: 0.3624, 1.284, 2.2668, 3.2604, 4.2564, 5.2548, 6
Global minimum: 5.754
Global maximum: 5.2548

>>> c.integral
-1.7353776102348935
>>> c.df(2.51)       # c.df(x) is the derivative of f
-24.056988888465636
>>> c.set_differentiation_method(Central4)
>>> c.df(2.51)
-24.056988832723189
>>> c.set_integration_method(Simpson)   # more accurate integration
>>> c.integral
-1.7353857856973565
```

Design the class such that the above session can be carried out.

Hint Use classes from the `Diff` and `Integrator` hierarchies (Sects. 9.2 and 9.3) for numerical differentiation and integration (with, e.g., `Central2` and `Trapezoidal` as default methods for differentiation and integration). The method `set_differentiation_method` takes a subclass name in the `Diff` hierarchy as argument, and makes a data attribute `df` that holds a subclass instance for computing derivatives. With `set_integration_method` we can similarly set the integration method as a subclass name in the `Integrator` hierarchy, and then compute the integral $\int_a^b f(x)dx$ and store the value in the attribute `integral`. The `extreme_points` method performs a `print` on a `MinMax` instance, which is stored as an attribute in the calculator class.
Filename: `CalculusCalculator`.

Exercise 9.19: Compute inverse functions
Extend class `CalculusCalculator` from Exercise 9.18 to offer computations of inverse functions.

Hint A numerical way of computing inverse functions is explained in Sect. A.1.11. Other, perhaps more attractive methods are described in Exercises E.17–E.20.
Filename: `CalculusCalculator2`.

Exercise 9.20: Make line drawing of a person; program

A very simple sketch of a human being can be made of a circle for the head, two lines for the arms, one vertical line, a triangle, or a rectangle for the torso, and two lines for the legs. Make such a drawing in a program, utilizing appropriate classes in the Shape hierarchy.

Filename: `draw_person`.

Exercise 9.21: Make line drawing of a person; class

Use the code from Exercise 9.20 to make a subclass of Shape that draws a person. Supply the following arguments to the constructor: the center point of the head and the radius R of the head. Let the arms and the torso be of length $4R$, and the legs of length $6R$. The angle between the legs can be fixed (say 30 degrees), while the angle of the arms relative to the torso can be an argument to the constructor with a suitable default value.

Filename: `Person`.

Exercise 9.22: Animate a person with waving hands

Make a subclass of the class from Exercise 9.21 where the constructor can take an argument describing the angle between the arms and the torso. Use this new class to animate a person who waves her/his hands.

Filename: `waving_person`.

Sequences and Difference Equations

A

From mathematics you probably know the concept of a *sequence*, which is nothing but a collection of numbers with a specific order. A general sequence is written as

$$x_0, \; x_1, \; x_2, \; \ldots, \; x_n, \ldots,$$

One example is the sequence of all odd numbers:

$$1, 3, 5, 7, \ldots, 2n + 1, \ldots$$

For this sequence we have an explicit formula for the n-th term: $2n + 1$, and n takes on the values 0, 1, 2, …. We can write this sequence more compactly as $(x_n)_{n=0}^{\infty}$ with $x_n = 2n + 1$. Other examples of infinite sequences from mathematics are

$$1, \; 4, \; 9, \; 16, \; 25, \; \ldots \quad (x_n)_{n=0}^{\infty}, \; x_n = (n+1)^2, \tag{A.1}$$

$$1, \; \frac{1}{2}, \; \frac{1}{3}, \; \frac{1}{4}, \; \ldots \quad (x_n)_{n=0}^{\infty}, \; x_n = \frac{1}{n+1}. \tag{A.2}$$

The former sequences are infinite, because they are generated from all integers ≥ 0 and there are infinitely many such integers. Nevertheless, most sequences from real life applications are finite. If you put an amount x_0 of money in a bank, you will get an interest rate and therefore have an amount x_1 after one year, x_2 after two years, and x_N after N years. This process results in a finite sequence of amounts

$$x_0, x_1, x_2, \ldots, x_N, \quad (x_n)_{n=0}^{N}.$$

Usually we are interested in quite small N values (typically $N \leq 20-30$). Anyway, the life of the bank is finite, so the sequence definitely has an end.

For some sequences it is not so easy to set up a general formula for the n-th term. Instead, it is easier to express a relation between two or more consecutive elements. One example where we can do both things is the sequence of odd numbers. This sequence can alternatively be generated by the formula

$$x_{n+1} = x_n + 2. \tag{A.3}$$

© Springer-Verlag Berlin Heidelberg 2016
H.P. Langtangen, *A Primer on Scientific Programming with Python*,
Texts in Computational Science and Engineering 6, DOI 10.1007/978-3-662-49887-3

To start the sequence, we need an *initial condition* where the value of the first element is specified:

$$x_0 = 1 \, .$$

Relations like (A.3) between consecutive elements in a sequence is called recurrence relations or *difference equations*. Solving a difference equation can be quite challenging in mathematics, but it is almost trivial to solve it on a computer. That is why difference equations are so well suited for computer programming, and the present appendix is devoted to this topic. Necessary background knowledge is programming with loops, arrays, and command-line arguments and visualization of a function of one variable.

The program examples regarding difference equations are found in the folder `src/diffeq`[1].

A.1 Mathematical Models Based on Difference Equations

The objective of science is to understand complex phenomena. The phenomenon under consideration may be a part of nature, a group of social individuals, the traffic situation in Los Angeles, and so forth. The reason for addressing something in a scientific manner is that it appears to be complex and hard to comprehend. A common scientific approach to gain understanding is to create a model of the phenomenon, and discuss the properties of the model instead of the phenomenon. The basic idea is that the model is easier to understand, but still complex enough to preserve the basic features of the problem at hand.

> *Essentially, all models are wrong, but some are useful.* George E. P. Box, statistician, 1919–2013.

Modeling is, indeed, a general idea with applications far beyond science. Suppose, for instance, that you want to invite a friend to your home for the first time. To assist your friend, you may send a map of your neighborhood. Such a map is a model: it exposes the most important landmarks and leaves out billions of details that your friend can do very well without. This is the essence of modeling: a good model should be as simple as possible, but still rich enough to include the important structures you are looking for.

> *Everything should be made as simple as possible, but not simpler.*
> Paraphrased quote attributed to Albert Einstein, physicist, 1879–1955.

Certainly, the tools we apply to model a certain phenomenon differ a lot in various scientific disciplines. In the natural sciences, mathematics has gained a unique position as the key tool for formulating models. To establish a model, you need to understand the problem at hand and describe it with mathematics. Usually, this process results in a set of equations, i.e., the model consists of equations that must be solved in order to see how realistically the model describes a phenomenon. Difference equations represent one of the simplest yet most effective type of equations

---

[1] http://tinyurl.com/pwyasaa/diffeq

arising in mathematical models. The mathematics is simple and the programming is simple, thereby allowing us to focus more on the modeling part. Below we will derive and solve difference equations for diverse applications.

A.1.1   Interest Rates

Our first difference equation model concerns how much money an initial amount x_0 will grow to after n years in a bank with annual interest rate p. You learned in school the formula

$$x_n = x_0 \left(1 + \frac{p}{100}\right)^n .$$ (A.4)

Unfortunately, this formula arises after some limiting assumptions, like that of a constant interest rate over all the n years. Moreover, the formula only gives us the amount after each year, not after some months or days. It is much easier to compute with interest rates if we set up a more fundamental model in terms of a difference equation and then solve this equation on a computer.

The fundamental model for interest rates is that an amount x_{n-1} at some point of time t_{n-1} increases its value with p percent to an amount x_n at a new point of time t_n:

$$x_n = x_{n-1} + \frac{p}{100} x_{n-1} .$$ (A.5)

If n counts years, p is the annual interest rate, and if p is constant, we can with some arithmetics derive the following solution to (A.5):

$$x_n = \left(1 + \frac{p}{100}\right) x_{n-1} = \left(1 + \frac{p}{100}\right)^2 x_{n-2} = \ldots = \left(1 + \frac{p}{100}\right)^n x_0 .$$

Instead of first deriving a formula for x_n and then program this formula, we may attack the fundamental model (A.5) in a program (`growth_years.py`) and compute x_1, x_2, and so on in a loop:

```
from scitools.std import *
x0 = 100                        # initial amount
p = 5                           # interest rate
N = 4                           # number of years
index_set = range(N+1)
x = zeros(len(index_set))

# Compute solution
x[0] = x0
for n in index_set[1:]:
    x[n] = x[n-1] + (p/100.0)*x[n-1]
print x
plot(index_set, x, 'ro', xlabel='years', ylabel='amount')
```

The output of x is

```
[ 100.        105.        110.25      115.7625    121.550625]
```

Programmers of mathematical software who are trained in making programs more efficient, will notice that it is not necessary to store all the x_n values in an array or use a list with all the indices $0, 1, \ldots, N$. Just one integer for the index and two floats for x_n and x_{n-1} are strictly necessary. This can save quite some memory for large values of N. Exercise A.3 asks you to develop such a memory-efficient program.

Suppose now that we are interested in computing the growth of money after N days instead. The interest rate per day is taken as $r = p/D$ if p is the annual interest rate and D is the number of days in a year. The fundamental model is the same, but now n counts days and p is replaced by r:

$$x_n = x_{n-1} + \frac{r}{100}x_{n-1}. \tag{A.6}$$

A common method in international business is to choose $D = 360$, yet let n count the exact number of days between two dates (see the Wikipedia entry Day count convention[2] for an explanation). Python has a module datetime for convenient calculations with dates and times. To find the number of days between two dates, we perform the following operations:

```
>>> import datetime
>>> date1 = datetime.date(2007, 8, 3)   # Aug 3, 2007
>>> date2 = datetime.date(2008, 8, 4)   # Aug 4, 2008
>>> diff = date2 - date1
>>> print diff.days
367
```

We can modify the previous program to compute with days instead of years:

```
from scitools.std import *
x0 = 100                    # initial amount
p = 5                       # annual interest rate
r = p/360.0                 # daily interest rate
import datetime
date1 = datetime.date(2007, 8, 3)
date2 = datetime.date(2011, 8, 3)
diff = date2 - date1
N = diff.days
index_set = range(N+1)
x = zeros(len(index_set))

# Compute solution
x[0] = x0
for n in index_set[1:]:
    x[n] = x[n-1] + (r/100.0)*x[n-1]
print x
plot(index_set, x, 'ro', xlabel='days', ylabel='amount')
```

Running this program, called growth_days.py, prints out 122.5 as the final amount.

[2] http://en.wikipedia.org/wiki/Day_count_convention

It is quite easy to adjust the formula (A.4) to the case where the interest is added every day instead of every year. However, the strength of the model (A.6) and the associated program `growth_days.py` becomes apparent when r varies in time – and this is what happens in real life. In the model we can just write $r(n)$ to explicitly indicate the dependence upon time. The corresponding time-dependent annual interest rate is what is normally specified, and $p(n)$ is usually a piecewise constant function (the interest rate is changed at some specific dates and remains constant between these days). The construction of a corresponding array p in a program, given the dates when p changes, can be a bit tricky since we need to compute the number of days between the dates of changes and index p properly. We do not dive into these details now, but readers who want to compute p and who is ready for some extra brain training and index puzzling can attack Exercise A.8. For now we assume that an array p holds the time-dependent annual interest rates for each day in the total time period of interest. The `growth_days.py` program then needs a slight modification, typically,

```
p = zeros(len(index_set))
# set up p (might be challenging!)
r = p/360.0                          # daily interest rate
...
for n in index_set[1:]:
    x[n] = x[n-1] + (r[n-1]/100.0)*x[n-1]
```

For the very simple (and not-so-relevant) case where p grows linearly (i.e., daily changes) from 4 to 6 percent over the period of interest, we have made a complete program in the file `growth_days_timedep.py`. You can compare a simulation with linearly varying p between 4 and 6 and a simulation using the average p value 5 throughout the whole time interval.

A difference equation with $r(n)$ is quite difficult to solve mathematically, but the n-dependence in r is easy to deal with in the computerized solution approach.

A.1.2   The Factorial as a Difference Equation

The difference equation

$$x_n = n x_{n-1}, \quad x_0 = 1 \tag{A.7}$$

can quickly be solved recursively:

$$\begin{aligned}
x_n &= n x_{n-1} \\
&= n(n-1)x_{n-2} \\
&= n(n-1)(n-2)x_{n-3} \\
&= n(n-1)(n-2)\cdots 1 .
\end{aligned}$$

The result x_n is nothing but the factorial of n, denoted as $n!$. Equation (A.7) then gives a standard recipe to compute $n!$.

A.1.3 Fibonacci Numbers

Every textbook with some material on sequences usually presents a difference equation for generating the famous Fibonacci numbers[3]:

$$x_n = x_{n-1} + x_{n-2}, \quad x_0 = 1, \; x_1 = 1, \; n = 2, 3, \ldots \quad (A.8)$$

This equation has a relation between three elements in the sequence, not only two as in the other examples we have seen. We say that this is a difference equation of second order, while the previous examples involving two n levels are said to be difference equations of first order. The precise characterization of (A.8) is a homogeneous difference equation of second order. Such classification is not important when computing the solution in a program, but for mathematical solution methods by pen and paper, the classification helps determine the most suitable mathematical technique for solving the problem.

A straightforward program for generating Fibonacci numbers takes the form (fibonacci1.py):

```
import sys
import numpy as np
N = int(sys.argv[1])
x = np.zeros(N+1, int)
x[0] = 1
x[1] = 1
for n in range(2, N+1):
    x[n] = x[n-1] + x[n-2]
    print n, x[n]
```

Since x_n is an infinite sequence we could try to run the program for very large N. This causes two problems: the storage requirements of the x array may become too large for the computer, but long before this happens, x_n grows in size far beyond the largest integer that can be represented by int elements in arrays (the problem appears already for $N = 50$). A possibility is to use array elements of type int64, which allows computation of twice as many numbers as with standard int elements (see the program fibonacci1_int64.py). A better solution is to use float elements in the x array, despite the fact that the numbers x_n are integers. With float96 elements we can compute up to $N = 23600$ (see the program fibinacci1_float.py).

The best solution goes as follows. We observe, as mentioned after the growth_years.py program and also explained in Exercise A.3, that we need only three variables to generate the sequence. We can therefore work with just three standard int variables in Python:

```
import sys
N = int(sys.argv[1])
xnm1 = 1
xnm2 = 1
```

[3] http://en.wikipedia.org/wiki/Fibonacci_number

```
n = 2
while n <= N:
    xn = xnm1 + xnm2
    print 'x_%d = %d' % (n, xn)
    xnm2 = xnm1
    xnm1 = xn
    n += 1
```

Here xnm1 denotes x_{n-1} and xnm2 denotes x_{n-2}. To prepare for the next pass in the loop, we must shuffle the xnm1 down to xnm2 and store the new x_n value in xnm1. The nice thing with integers in Python (contrary to int elements in NumPy arrays) is that they can hold integers of arbitrary size. More precisely, when the integer is too large for the ordinary int object, xn becomes a long object that can hold integers as big as the computer's memory allows. We may try a run with N set to 250:

```
x_2 = 2
x_3 = 3
x_4 = 5
x_5 = 8
x_6 = 13
x_7 = 21
x_8 = 34
x_9 = 55
x_10 = 89
x_11 = 144
x_12 = 233
x_13 = 377
x_14 = 610
x_15 = 987
x_16 = 1597
...
x_249 = 7896325826131730509282738943634332893686268675876375
x_250 = 12776523572924732586037033894655031898659556447352249
```

In mathematics courses you learn how to derive a formula for the n-th term in a Fibonacci sequence. This derivation is much more complicated than writing a simple program to generate the sequence, but there is a lot of interesting mathematics both in the derivation and the resulting formula!

A.1.4   Growth of a Population

Let x_{n-1} be the number of individuals in a population at time t_{n-1}. The population can consists of humans, animals, cells, or whatever objects where the number of births and deaths is proportional to the number of individuals. Between time levels t_{n-1} and t_n, bx_{n-1} individuals are born, and dx_{n-1} individuals die, where b and d are constants. The net growth of the population is then $(b - d)x_n$. Introducing $r = (b - d)100$ for the net growth factor measured in percent, the new number of individuals become

$$x_n = x_{n-1} + \frac{r}{100}x_{n-1} . \tag{A.9}$$

This is the same difference equation as (A.5). It models growth of populations quite well as long as there are optimal growing conditions for each individual. If not, one can adjust the model as explained in Sect. A.1.5.

To solve (A.9) we need to start out with a known size x_0 of the population. The b and d parameters depend on the time difference $t_n - t_{n-1}$, i.e., the values of b and d are smaller if n counts years than if n counts generations.

A.1.5   Logistic Growth

The model (A.9) for the growth of a population leads to exponential increase in the number of individuals as implied by the solution (A.4). The size of the population increases faster and faster as time n increases, and $x_n \to \infty$ when $n \to \infty$. In real life, however, there is an upper limit M of the number of individuals that can exist in the environment at the same time. Lack of space and food, competition between individuals, predators, and spreading of contagious diseases are examples on factors that limit the growth. The number M is usually called the *carrying capacity* of the environment, the maximum population which is sustainable over time. With limited growth, the growth factor r must depend on time:

$$x_n = x_{n-1} + \frac{r(n-1)}{100} x_{n-1} . \tag{A.10}$$

In the beginning of the growth process, there is enough resources and the growth is exponential, but as x_n approaches M, the growth stops and r must tend to zero. A simple function $r(n)$ with these properties is

$$r(n) = \varrho \left(1 - \frac{x_n}{M} \right) . \tag{A.11}$$

For small n, $x_n \ll M$ and $r(n) \approx \varrho$, which is the growth rate with unlimited resources. As $n \to M$, $r(n) \to 0$ as we want. The model (A.11) is used for *logistic growth*. The corresponding *logistic difference equation* becomes

$$x_n = x_{n-1} + \frac{\varrho}{100} x_{n-1} \left(1 - \frac{x_{n-1}}{M} \right) . \tag{A.12}$$

Below is a program (`growth_logistic.py`) for simulating $N = 200$ time intervals in a case where we start with $x_0 = 100$ individuals, a carrying capacity of $M = 500$, and initial growth of $\varrho = 4$ percent in each time interval:

```
from scitools.std import *
x0 = 100              # initial amount of individuals
M = 500               # carrying capacity
rho = 4               # initial growth rate in percent
N = 200               # number of time intervals
index_set = range(N+1)
x = zeros(len(index_set))
```

```
# Compute solution
x[0] = x0
for n in index_set[1:]:
    x[n] = x[n-1] + (rho/100.0)*x[n-1]*(1 - x[n-1]/float(M))
print x
plot(index_set, x, 'r', xlabel='time units',
     ylabel='no of individuals', hardcopy='tmp.pdf')
```

Figure A.1 shows how the population stabilizes, i.e., that x_n approaches M as N becomes large (of the same magnitude as M).

If the equation stabilizes as $n \to \infty$, it means that $x_n = x_{n-1}$ in this limit. The equation then reduces to

$$x_n = x_n + \frac{\varrho}{100} x_n \left(1 - \frac{x_n}{M}\right) .$$

By inserting $x_n = M$ we see that this solution fulfills the equation. The same solution technique (i.e., setting $x_n = x_{n-1}$) can be used to check if x_n in a difference equation approaches a limit or not.

Mathematical models like (A.12) are often easier to work with if we *scale* the variables. Basically, this means that we divide each variable by a characteristic size of that variable such that the value of the new variable is typically 1. In the present case we can scale x_n by M and introduce a new variable,

$$y_n = \frac{x_n}{M} .$$

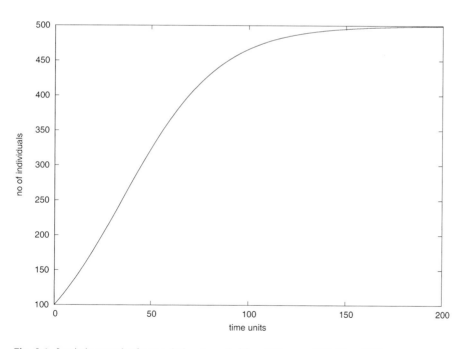

Fig. A.1 Logistic growth of a population ($\varrho = 4$, $M = 500$, $x_0 = 100$, $N = 200$).

Similarly, x_0 is replaced by $y_0 = x_0/M$. Inserting $x_n = M y_n$ in (A.12) and dividing by M gives

$$y_n = y_{n-1} + q y_{n-1} (1 - y_{n-1}), \tag{A.13}$$

where $q = \varrho/100$ is introduced to save typing. Equation (A.13) is simpler than (A.12) in that the solution lies approximately between y_0 and 1 (values larger than 1 can occur, see Exercise A.19), and there are only two dimensionless input parameters to care about: q and y_0. To solve (A.12) we need knowledge of three parameters: x_0, ϱ, and M.

A.1.6   Payback of a Loan

A loan L is to be paid back over N months. The payback in a month consists of the fraction L/N plus the interest increase of the loan. Let the annual interest rate for the loan be p percent. The monthly interest rate is then $\frac{p}{12}$. The value of the loan after month n is x_n, and the change from x_{n-1} can be modeled as

$$x_n = x_{n-1} + \frac{p}{12 \cdot 100} x_{n-1} - \left(\frac{p}{12 \cdot 100} x_{n-1} + \frac{L}{N} \right), \tag{A.14}$$

$$= x_{n-1} - \frac{L}{N}, \tag{A.15}$$

for $n = 1, \ldots, N$. The initial condition is $x_0 = L$. A major difference between (A.15) and (A.6) is that all terms in the latter are proportional to x_n or x_{n-1}, while (A.15) also contains a constant term (L/N). We say that (A.6) is homogeneous and linear, while (A.15) is inhomogeneous (because of the constant term) and linear. The mathematical solution of inhomogeneous equations are more difficult to find than the solution of homogeneous equations, but in a program there is no big difference: we just add the extra term $-L/N$ in the formula for the difference equation.

The solution of (A.15) is not particularly exciting (just use (A.15) repeatedly to derive the solution $x_n = L - nL/N$). What is more interesting, is what we pay each month, y_n. We can keep track of both y_n and x_n in a variant of the previous model:

$$y_n = \frac{p}{12 \cdot 100} x_{n-1} + \frac{L}{N}, \tag{A.16}$$

$$x_n = x_{n-1} + \frac{p}{12 \cdot 100} x_{n-1} - y_n . \tag{A.17}$$

Equations (A.16)–(A.17) is a system of difference equations. In a computer code, we simply update y_n first, and then we update x_n, inside a loop over n. Exercise A.4 asks you to do this.

A.1.7   The Integral as a Difference Equation

Suppose a function $f(x)$ is defined as the integral

$$f(x) = \int_a^x g(t)dt .\tag{A.18}$$

Our aim is to evaluate $f(x)$ at a set of points $x_0 = a < x_1 < \cdots < x_N$. The value $f(x_n)$ for any $0 \le n \le N$ can be obtained by using the Trapezoidal rule for integration:

$$f(x_n) = \sum_{k=0}^{n-1} \frac{1}{2}(x_{k+1} - x_k)(g(x_k) + g(x_{k+1})),\tag{A.19}$$

which is nothing but the sum of the areas of the trapezoids up to the point x_n (the plot to the right in Fig. 5.22 illustrates the idea.) We realize that $f(x_{n+1})$ is the sum above plus the area of the next trapezoid:

$$f(x_{n+1}) = f(x_n) + \frac{1}{2}(x_{n+1} - x_n)(g(x_n) + g(x_{n+1})) .\tag{A.20}$$

This is a much more efficient formula than using (A.19) with n replaced by $n + 1$, since we do not need to recompute the areas of the first n trapezoids.

Formula (A.20) gives the idea of computing all the $f(x_n)$ values through a difference equation. Define f_n as $f(x_n)$ and consider $x_0 = a$, and $x_1, \ldots, x_N$ as given. We know that $f_0 = 0$. Then

$$f_n = f_{n-1} + \frac{1}{2}(x_n - x_{n-1})(g(x_{n-1}) + g(x_n)),\tag{A.21}$$

for $n = 1, 2, \ldots, N$. By introducing g_n for $g(x_n)$ as an extra variable in the difference equation, we can avoid recomputing $g(x_n)$ when we compute f_{n+1}:

$$g_n = g(x_n),\tag{A.22}$$

$$f_n = f_{n-1} + \frac{1}{2}(x_n - x_{n-1})(g_{n-1} + g_n),\tag{A.23}$$

with initial conditions $f_0 = 0$ and $g_0 = g(a)$.

A function can take g, a, x, and N as input and return arrays x and f for $x_0, \ldots, x_N$ and the corresponding integral values $f_0, \ldots, f_N$:

```
def integral(g, a, x, N=20):
    index_set = range(N+1)
    x = np.linspace(a, x, N+1)
    g_ = np.zeros_like(x)
    f = np.zeros_like(x)
    g_[0] = g(x[0])
    f[0] = 0

    for n in index_set[1:]:
        g_[n] = g(x[n])
        f[n] = f[n-1] + 0.5*(x[n] - x[n-1])*(g_[n-1] + g_[n])
    return x, f
```

Note that g is used for the integrand function to call so we introduce g_ to be the array holding sequence of g(x[n]) values.

Our first task, after having implemented a mathematical calculation, is to verify the result. Here we can make use of the nice fact that the Trapezoidal rule is exact for linear functions $g(t)$:

```
def test_integral():
    def g_test(t):
        """Linear integrand."""
        return 2*t + 1

    def f_test(x, a):
        """Exact integral of g_test."""
        return x**2 + x - (a**2 + a)

    a = 2
    x, f = integral(g_test, a, x=10)
    f_exact = f_test(x, a)
    assert np.allclose(f_exact, f)
```

A realistic application is to apply the `integral` function to some $g(t)$ where there is no formula for the analytical integral, e.g.,

$$g(t) = \frac{1}{\sqrt{2\pi}} \exp\left(-t^2\right).$$

The code may look like

```
def demo():
    """Integrate the Gaussian function."""
    from numpy import sqrt, pi, exp

    def g(t):
        return 1./sqrt(2*pi)*exp(-t**2)

    x, f = integral(g, a=-3, x=3, N=200)
    integrand = g(x)
    from scitools.std import plot
    plot(x, f, 'r-',
         x, integrand, 'y-',
         legend=('f', 'g'),
         legend_loc='upper left',
         savefig='tmp.pdf')
```

Figure A.2 displays the integrand and the integral. All the code is available in the file `integral.py`.

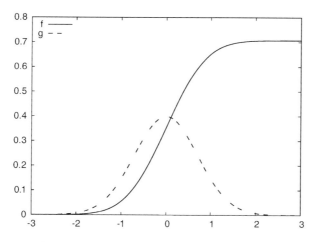

Fig. A.2 Integral of $\frac{1}{\sqrt{2\pi}} \exp\left(-t^2\right)$ from -3 to x.

A.1.8   Taylor Series as a Difference Equation

Consider the following system of two difference equations

$$e_n = e_{n-1} + a_{n-1}, \tag{A.24}$$

$$a_n = \frac{x}{n} a_{n-1}, \tag{A.25}$$

with initial conditions $e_0 = 0$ and $a_0 = 1$. We can start to nest the solution:

$$
\begin{aligned}
e_1 &= 0 + a_0 = 0 + 1 = 1, \\
a_1 &= x, \\
e_2 &= e_1 + a_1 = 1 + x, \\
a_2 &= \frac{x}{2} a_1 = \frac{x^2}{2}, \\
e_3 &= e_2 + a_2 = 1 + x + \frac{x^2}{2}, \\
e_4 &= 1 + x + \frac{x^2}{2} + \frac{x^3}{3 \cdot 2}, \\
e_5 &= 1 + x + \frac{x^2}{2} + \frac{x^3}{3 \cdot 2} + \frac{x^4}{4 \cdot 3 \cdot 2}
\end{aligned}
$$

The observant reader who has heard about Taylor series (see Sect. B.4) will recognize this as the Taylor series of e^x:

$$e^x = \sum_{n=0}^{\infty} \frac{x^n}{n!}. \tag{A.26}$$

How do we derive a system like (A.24)–(A.25) for computing the Taylor polynomial approximation to e^x? The starting point is the sum $\sum_{n=0}^{\infty} \frac{x^n}{n!}$. This sum is coded by adding new terms to an accumulation variable in a loop. The mathematical counterpart to this code is a difference equation

$$e_{n+1} = e_n + \frac{x^n}{n!}, \quad e_0 = 0, \; n = 0, 1, 2, \ldots. \tag{A.27}$$

or equivalently (just replace n by $n - 1$):

$$e_n = e_{n-1} + \frac{x^{n-1}}{n-1!}, \quad e_0 = 0, \; n = 1, 2, 3, \ldots. \tag{A.28}$$

Now comes the important observation: the term $x^n/n!$ contains many of the computations we already performed for the previous term $x^{n-1}/(n-1)!$ because

$$\frac{x^n}{n!} = \frac{x \cdot x \cdots x}{n(n-1)(n-2) \cdots 1}, \quad \frac{x^{n-1}}{(n-1)!} = \frac{x \cdot x \cdots x}{(n-1)(n-2)(n-3) \cdots 1}.$$

Let $a_n = x^n/n!$. We see that we can go from a_{n-1} to a_n by multiplying a_{n-1} by x/n:

$$\frac{x}{n} a_{n-1} = \frac{x}{n} \frac{x^{n-1}}{(n-1)!} = \frac{x^n}{n!} = a_n, \tag{A.29}$$

which is nothing but (A.25). We also realize that $a_0 = 1$ is the initial condition for this difference equation. In other words, (A.24) sums the Taylor polynomial, and (A.25) updates each term in the sum.

The system (A.24)–(A.25) is very easy to implement in a program and constitutes an efficient way to compute (A.26). The function `exp_diffeq` does the work:

```
def exp_diffeq(x, N):
    n = 1
    an_prev = 1.0   # a_0
    en_prev = 0.0   # e_0
    while n <= N:
        en = en_prev + an_prev
        an = x/n*an_prev
        en_prev = en
        an_prev = an
        n += 1
    return en
```

Observe that we do not store the sequences in arrays, but make use of the fact that only the most recent sequence element is needed to calculate a new element. The above function along with a direct evaluation of the Taylor series for e^x and a comparison with the exact result for various N values can be found in the file `exp_Taylor_series_diffeq.py`.

A.1.9   Making a Living from a Fortune

Suppose you want to live on a fortune F. You have invested the money in a safe way that gives an annual interest of p percent. Every year you plan to consume an amount c_n, where n counts years. The development of your fortune x_n from one year to the other can then be modeled by

$$x_n = x_{n-1} + \frac{p}{100}x_{n-1} - c_{n-1}, \quad x_0 = F . \tag{A.30}$$

A simple example is to keep c constant, say q percent of the interest the first year:

$$x_n = x_{n-1} + \frac{p}{100}x_{n-1} - \frac{pq}{10^4}F, \quad x_0 = F . \tag{A.31}$$

A more realistic model is to assume some inflation of I percent per year. You will then like to increase c_n by the inflation. We can extend the model in two ways. The simplest and clearest way, in the author's opinion, is to track the evolution of two sequences x_n and c_n:

$$x_n = x_{n-1} + \frac{p}{100}x_{n-1} - c_{n-1}, \quad x_0 = F, \ c_0 = \frac{pq}{10^4}F, \tag{A.32}$$

$$c_n = c_{n-1} + \frac{I}{100}c_{n-1} . \tag{A.33}$$

This is a system of two difference equations with two unknowns. The solution method is, nevertheless, not much more complicated than the method for a difference equation in one unknown, since we can first compute x_n from (A.32) and then update the c_n value from (A.33). You are encouraged to write the program (see Exercise A.5).

Another way of making a difference equation for the case with inflation, is to use an explicit formula for c_{n-1}, i.e., solve (A.32) and end up with a formula like (A.4). Then we can insert the explicit formula

$$c_{n-1} = \left(1 + \frac{I}{100}\right)^{n-1}\frac{pq}{10^4}F$$

in (A.30), resulting in only one difference equation to solve.

A.1.10 Newton's Method

The difference equation

$$x_n = x_{n-1} - \frac{f(x_{n-1})}{f'(x_{n-1})}, \quad x_0 \text{ given}, \tag{A.34}$$

generates a sequence x_n where, if the sequence converges (i.e., if $x_n - x_{n-1} \to 0$), x_n approaches a root of $f(x)$. That is, $x_n \to x$, where x solves the equation $f(x) = 0$. Equation (A.34) is the famous Newton's method for solving nonlinear algebraic equations $f(x) = 0$. When $f(x)$ is not linear, i.e., $f(x)$ is not on the form $ax + b$ with constant a and b, (A.34) becomes a *nonlinear difference equation*. This complicates analytical treatment of difference equations, but poses no extra difficulties for numerical solution.

We can quickly sketch the derivation of (A.34). Suppose we want to solve the equation

$$f(x) = 0$$

and that we already have an approximate solution x_{n-1}. If $f(x)$ were linear, $f(x) = ax + b$, it would be very easy to solve $f(x) = 0$: $x = -b/a$. The idea is therefore to approximate $f(x)$ in the vicinity of $x = x_{n-1}$ by a linear function, i.e., a straight line $f(x) \approx \tilde{f}(x) = ax + b$. This line should have the same slope as $f(x)$, i.e., $a = f'(x_{n-1})$, and both the line and f should have the same value at $x = x_{n-1}$. From this condition one can find $b = f(x_{n-1}) - x_{n-1}f'(x_{n-1})$. The approximate function (line) is then

$$\tilde{f}(x) = f(x_{n-1}) + f'(x_{n-1})(x - x_{n-1}) \,. \tag{A.35}$$

This expression is just the two first terms of a Taylor series approximation to $f(x)$ at $x = x_{n-1}$. It is now easy to solve $\tilde{f}(x) = 0$ with respect to x, and we get

$$x = x_{n-1} - \frac{f(x_{n-1})}{f'(x_{n-1})} \,. \tag{A.36}$$

Since $\tilde{f}$ is only an approximation to f, x in (A.36) is only an approximation to a root of $f(x) = 0$. Hopefully, the approximation is better than x_{n-1} so we set $x_n = x$ as the next term in a sequence that we hope converges to the correct root. However, convergence depends highly on the shape of $f(x)$, and there is no guarantee that the method will work.

The previous programs for solving difference equations have typically calculated a sequence x_n up to $n = N$, where N is given. When using (A.34) to find roots of nonlinear equations, we do not know a suitable N in advance that leads to an x_n where $f(x_n)$ is sufficiently close to zero. We therefore have to keep on increasing n until $f(x_n) < \epsilon$ for some small ϵ. Of course, the sequence diverges, we will keep on forever, so there must be some maximum allowable limit on n, which we may take as N.

It can be convenient to have the solution of (A.34) as a function for easy reuse. Here is a first rough implementation:

```
def Newton(f, x, dfdx, epsilon=1.0E-7, N=100):
    n = 0
    while abs(f(x)) > epsilon and n <= N:
        x = x - f(x)/dfdx(x)
        n += 1
    return x, n, f(x)
```

This function might well work, but `f(x)/dfdx(x)` can imply integer division, so we should ensure that the numerator or denumerator is of `float` type. There are also two function evaluations of `f(x)` in every pass in the loop (one in the loop body and one in the `while` condition). We can get away with only one evaluation if we store the `f(x)` in a local variable. In the small examples with $f(x)$ in the present course, twice as many function evaluations of f as necessary does not matter, but the same `Newton` function can in fact be used for much more complicated functions, and in those cases twice as much work can be noticeable. As a programmer, you should therefore learn to optimize the code by removing unnecessary computations.

Another, more serious, problem is the possibility dividing by zero. Almost as serious, is dividing by a very small number that creates a large value, which might cause Newton's method to diverge. Therefore, we should test for small values of $f'(x)$ and write a warning or raise an exception.

Another improvement is to add a boolean argument `store` to indicate whether we want the $(x, f(x))$ values during the iterations to be stored in a list or not. These intermediate values can be handy if we want to print out or plot the convergence behavior of Newton's method.

An improved `Newton` function can now be coded as

```
def Newton(f, x, dfdx, epsilon=1.0E-7, N=100, store=False):
    f_value = f(x)
    n = 0
    if store: info = [(x, f_value)]
    while abs(f_value) > epsilon and n <= N:
        dfdx_value = float(dfdx(x))
        if abs(dfdx_value) < 1E-14:
            raise ValueError("Newton: f'(%g)=%g" % (x, dfdx_value))

        x = x - f_value/dfdx_value

        n += 1
        f_value = f(x)
        if store: info.append((x, f_value))
    if store:
        return x, info
    else:
        return x, n, f_value
```

Note that to use the `Newton` function, we need to calculate the derivative $f'(x)$ and implement it as a Python function and provide it as the `dfdx` argument. Also note that what we return depends on whether we store $(x, f(x))$ information during the iterations or not.

It is quite common to test if `dfdx(x)` is zero in an implementation of Newton's method, but this is not strictly necessary in Python since an exception `ZeroDivisionError` is always raised when dividing by zero.

We can apply the `Newton` function to solve the equation $e^{-0.1x^2} \sin(\frac{\pi}{2}x) = 0$:

```
from math import sin, cos, exp, pi
import sys
from Newton import Newton

def g(x):
    return exp(-0.1*x**2)*sin(pi/2*x)

def dg(x):
    return -2*0.1*x*exp(-0.1*x**2)*sin(pi/2*x) + \
           pi/2*exp(-0.1*x**2)*cos(pi/2*x)
```

```
x0 = float(sys.argv[1])
x, info = Newton(g, x0, dg, store=True)
print 'root:', x
for i in range(len(info)):
    print 'Iteration %3d: f(%g)=%g' % \
          (i, info[i][0], info[i][1])
```

The `Newton` function and this program can be found in the file `Newton.py`. Running this program with an initial x value of 1.7 results in the output

```
root: 1.999999999768449
Iteration  0: f(1.7)=0.340044
Iteration  1: f(1.99215)=0.00828786
Iteration  2: f(1.99998)=2.53347e-05
Iteration  3: f(2)=2.43808e-10
```

Fortunately you realize that the exponential function can never be zero, so the solutions of the equation must be the zeros of the sine function, i.e., $\frac{\pi}{2}x = i\pi$ for all integers $i = \ldots, -2, 1, 0, 1, 2, \ldots$. This gives $x = 2i$ as the solutions. We see from the output that the convergence is fast towards the solution $x = 2$. The error is of the order 10^{-10} even though we stop the iterations when $f(x) \leq 10^{-7}$.

Trying a start value of 3, we would expect the method to find the root $x = 2$ or $x = 4$, but now we get

```
root: 42.49723316011362
Iteration  0: f(3)=-0.40657
Iteration  1: f(4.66667)=0.0981146
Iteration  2: f(42.4972)=-2.59037e-79
```

We have definitely solved $f(x) = 0$ in the sense that $|f(x)| \leq \epsilon$, where ϵ is a small value (here $\epsilon \sim 10^{-79}$). However, the solution $x \approx 42.5$ is *not* close to the correct solution ($x = 42$ and $x = 44$ are the solutions closest to the computed x). Can you use your knowledge of how the Newton method works and figure out why we get such strange behavior?

The demo program `Newton_movie.py` can be used to investigate the strange behavior. This program takes five command-line arguments: a formula for $f(x)$, a formula for $f'(x)$ (or the word `numeric`, which indicates a numerical approximation of $f'(x)$), a guess at the root, and the minimum and maximum x values in the plots. We try the following case with the program:

Terminal

```
Newton_movie.py 'exp(-0.1*x**2)*sin(pi/2*x)' numeric 3 -3 43
```

As seen, we start with $x = 3$ as the initial guess. In the first step of the method, we compute a new value of the root, now $x = 4.66667$. As we see in Fig. A.3, this root is near an extreme point of $f(x)$ so that the derivative is small, and the resulting straight line approximation to $f(x)$ at this root becomes quite flat. The result is a new guess at the root: $x42.5$. This root is far away from the last root,

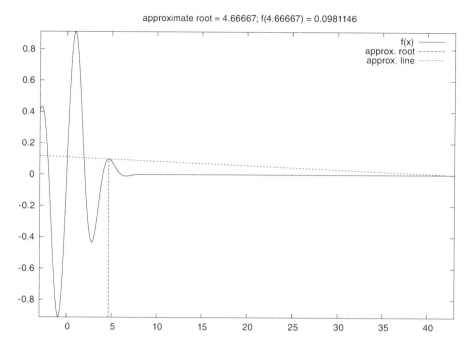

Fig. A.3 Failure of Newton's method to solve $e^{-0.1x^2} \sin(\frac{\pi}{2}x) = 0$. The plot corresponds to the second root found (starting with $x = 3$).

but the second problem is that $f(x)$ is quickly damped as we move to increasing x values, and at $x = 42.5$ f is small enough to fulfill the convergence criterion. Any guess at the root out in this region would satisfy that criterion.

You can run the `Newton_movie.py` program with other values of the initial root and observe that the method usually finds the nearest roots.

A.1.11   The Inverse of a Function

Given a function $f(x)$, the inverse function of f, say we call it $g(x)$, has the property that if we apply g to the value $f(x)$, we get x back:

$$g(f(x)) = x .$$

Similarly, if we apply f to the value $g(x)$, we get x:

$$f(g(x)) = x . \tag{A.37}$$

By hand, you substitute $g(x)$ by (say) y in (A.37) and solve (A.37) with respect to y to find some x expression for the inverse function. For example, given $f(x) = x^2 - 1$, we must solve $y^2 - 1 = x$ with respect to y. To ensure a unique solution for y, the x values have to be limited to an interval where $f(x)$ is monotone, say $x \in [0, 1]$ in the present example. Solving for y gives $y = \sqrt{1 + x}$, therefore $g(x) = \sqrt{1 + x}$. It is easy to check that $f(g(x)) = (\sqrt{1 + x})^2 - 1 = x$.

Numerically, we can use the "definition" (A.37) of the inverse function g at one point at a time. Suppose we have a sequence of points $x_0 < x_1 < \cdots < x_N$ along the x axis such that f is monotone in $[x_0, x_N]$: $f(x_0) > f(x_1) > \cdots > f(x_N)$ or $f(x_0) < f(x_1) < \cdots < f(x_N)$. For each point x_i, we have

$$f(g(x_i)) = x_i .$$

The value $g(x_i)$ is unknown, so let us call it γ. The equation

$$f(\gamma) = x_i \tag{A.38}$$

can be solved be respect γ. However, (A.38) is in general nonlinear if f is a nonlinear function of x. We must then use, e.g., Newton's method to solve (A.38). Newton's method works for an equation phrased as $f(x) = 0$, which in our case is $f(\gamma) - x_i = 0$, i.e., we seek the roots of the function $F(\gamma) \equiv f(\gamma) - x_i$. Also the derivative $F'(\gamma)$ is needed in Newton's method. For simplicity we may use an approximate finite difference:

$$\frac{dF}{d\gamma} \approx \frac{F(\gamma + h) - F(\gamma - h)}{2h} .$$

As start value γ_0, we can use the previously computed g value: g_{i-1}. We introduce the short notation $\gamma = \mathrm{Newton}(F, \gamma_0)$ to indicate the solution of $F(\gamma) = 0$ with initial guess γ_0.

The computation of all the $g_0, \ldots, g_N$ values can now be expressed by

$$g_i = \mathrm{Newton}(F, g_{i-1}), \quad i = 1, \ldots, N, \tag{A.39}$$

and for the first point we may use x_0 as start value (for instance):

$$g_0 = \mathrm{Newton}(F, x_0) . \tag{A.40}$$

Equations (A.39)–(A.40) constitute a difference equation for g_i, since given g_{i-1}, we can compute the next element of the sequence by (A.39). Because (A.39) is a nonlinear equation in the new value g_i, and (A.39) is therefore an example of a *nonlinear difference equation*.

The following program computes the inverse function $g(x)$ of $f(x)$ at some discrete points $x_0, \ldots, x_N$. Our sample function is $f(x) = x^2 - 1$:

```
from Newton import Newton
from scitools.std import *

def f(x):
    return x**2 - 1

def F(gamma):
    return f(gamma) - xi
```

```
def dFdx(gamma):
    return (F(gamma+h) - F(gamma-h))/(2*h)

h = 1E-6
x = linspace(0.01, 3, 21)
g = zeros(len(x))

for i in range(len(x)):
    xi = x[i]

    # Compute start value (use last g[i-1] if possible)
    if i == 0:
        gamma0 = x[0]
    else:
        gamma0 = g[i-1]

    gamma, n, F_value = Newton(F, gamma0, dFdx)
    g[i] = gamma

plot(x, f(x), 'r-', x, g, 'b-',
    title='f1', legend=('original', 'inverse'))
```

Note that with $f(x) = x^2 - 1$, $f'(0) = 0$, so Newton's method divides by zero and breaks down unless with let $x_0 > 0$, so here we set $x_0 = 0.01$. The f function can easily be edited to let the program compute the inverse of another function. The F function can remain the same since it applies a general finite difference to approximate the derivative of the f(x) function. The complete program is found in the file inverse_function.py.

A.2   Programming with Sound

Sound on a computer is nothing but a sequence of numbers. As an example, consider the famous A tone at 440 Hz. Physically, this is an oscillation of a tuning fork, loudspeaker, string or another mechanical medium that makes the surrounding air also oscillate and transport the sound as a compression wave. This wave may hit our ears and through complicated physiological processes be transformed to an electrical signal that the brain can recognize as sound. Mathematically, the oscillations are described by a sine function of time:

$$s(t) = A \sin (2\pi f t), \tag{A.41}$$

where A is the amplitude or strength of the sound and f is the frequency (440 Hz for the A in our example). In a computer, $s(t)$ is represented at discrete points of time. CD quality means 44100 samples per second. Other sample rates are also possible, so we introduce r as the sample rate. An f Hz tone lasting for m seconds with sample rate r can then be computed as the sequence

$$s_n = A \sin \left(2\pi f \frac{n}{r}\right), \quad n = 0, 1, \ldots, m \cdot r. \tag{A.42}$$

With Numerical Python this computation is straightforward and very efficient. Introducing some more explanatory variable names than r, A, and m, we can write a function for generating a note:

```
import numpy as np

def note(frequency, length, amplitude=1, sample_rate=44100):
    time_points = np.linspace(0, length, length*sample_rate)
    data = np.sin(2*np.pi*frequency*time_points)
    data = amplitude*data
    return data
```

A.2.1   Writing Sound to File

The note function above generates an array of float data representing a note. The sound card in the computer cannot play these data, because the card assumes that the information about the oscillations appears as a sequence of two-byte integers. With an array's astype method we can easily convert our data to two-byte integers instead of floats:

```
data = data.astype(numpy.int16)
```

That is, the name of the two-byte integer data type in numpy is int16 (two bytes are 16 bits). The maximum value of a two-byte integer is $2^{15} - 1$, so this is also the maximum amplitude. Assuming that amplitude in the note function is a relative measure of intensity, such that the value lies between 0 and 1, we must adjust this amplitude to the scale of two-byte integers:

```
max_amplitude = 2**15 - 1
data = max_amplitude*data
```

The data array of int16 numbers can be written to a file and played as an ordinary file in CD quality. Such a file is known as a wave file or simply a WAV file since the extension is .wav. Python has a module wave for creating such files. Given an array of sound, data, we have in SciTools a module sound with a function write for writing the data to a WAV file (using functionality from the wave module):

```
import scitools.sound
scitools.sound.write(data, 'Atone.wav')
```

You can now use your favorite music player to play the Atone.wav file, or you can play it from within a Python program using

```
scitools.sound.play('Atone.wav')
```

The write function can take more arguments and write, e.g., a stereo file with two channels, but we do not dive into these details here.

A.2.2   Reading Sound from File

Given a sound signal in a WAV file, we can easily read this signal into an array and mathematically manipulate the data in the array to change the flavor of the sound, e.g., add echo, treble, or bass. The recipe for reading a WAV file with name `filename` is

```
data = scitools.sound.read(filename)
```

The `data` array has elements of type `int16`. Often we want to compute with this array, and then we need elements of `float` type, obtained by the conversion

```
data = data.astype(float)
```

The `write` function automatically transforms the element type back to `int16` if we have not done this explicitly.

One operation that we can easily do is adding an echo. Mathematically this means that we add a damped delayed sound, where the original sound has weight β and the delayed part has weight $1 - \beta$, such that the overall amplitude is not altered. Let d be the delay in seconds. With a sampling rate r the number of indices in the delay becomes dr, which we denote by b. Given an original sound sequence s_n, the sound with echo is the sequence

$$e_n = \beta s_n + (1 - \beta)s_{n-b} . \tag{A.43}$$

We cannot start n at 0 since $e_0 = s_{0-b} = s_{-b}$ which is a value outside the sound data. Therefore we define $e_n = s_n$ for $n = 0, 1, \ldots, b$, and add the echo thereafter. A simple loop can do this (again we use descriptive variable names instead of the mathematical symbols introduced):

```
def add_echo(data, beta=0.8, delay=0.002, sample_rate=44100):
    newdata = data.copy()
    shift = int(delay*sample_rate)   # b (math symbol)
    for i in range(shift, len(data)):
        newdata[i] = beta*data[i] + (1-beta)*data[i-shift]
    return newdata
```

The problem with this function is that it runs slowly, especially when we have sound clips lasting several seconds (recall that for CD quality we need 44100 numbers per second). It is therefore necessary to vectorize the implementation of the difference equation for adding echo. The update is then based on adding slices:

```
newdata[shift:] = beta*data[shift:] + \
                  (1-beta)*data[:len(data)-shift]
```

A.2.3   Playing Many Notes

How do we generate a melody mathematically in a computer program? With the `note` function we can generate a note with a certain amplitude, frequency, and duration. The note is represented as an array. Putting sound arrays for different notes after each other will make up a melody. If we have several sound arrays data1, data2, data3, ..., we can make a new array consisting of the elements in the first array followed by the elements of the next array followed by the elements in the next array and so forth:

```
data = numpy.concatenate((data1, data2, data3, ...))
```

The frequency of a note[4] that is h half tones up from a base frequency f is given by $f\, 2^{h/12}$. With the tone A at $440\,\mathrm{Hz}$, we can define notes and the corresponding frequencies as

```
base_freq = 440.0
notes = ['A', 'A#', 'B', 'C', 'C#', 'D', 'D#', 'E',
         'F', 'F#', 'G', 'G#']
notes2freq = {notes[i]: base_freq*2**(i/12.0)
              for i in range(len(notes))}
```

With the notes to frequency mapping a melody can be made as a series of notes with specified duration:

```
l = .2  # basic duration unit
tones = [('E', 3*l), ('D', l), ('C#', 2*l), ('B', 2*l), ('A', 2*l),
         ('B', 2*l), ('C#', 2*l), ('D', 2*l), ('E', 3*l),
         ('F#', l), ('E', 2*l), ('D', 2*l), ('C#', 4*l)]

samples = []
for tone, duration in tones :
    s = note(notes2freq[tone], duration)
    samples.append(s)

data = np.concatenate(samples)
data *= 2**15-1
scitools.sound.write(data, "melody.wav")
```

Playing the resulting file melody.wav reveals that this is the opening of the most-played tune during international cross country skiing competitions.

All the notes had the same amplitude in this example, but more dynamics can easily be added by letting the elements in tones be triplets with tone, duration, and amplitude. The basic code above is found in the file melody.py.

A.2.4   Music of a Sequence

Problem The purpose of this example is to listen to the sound generated by two mathematical sequences. The first one is given by an explicit formula, constructed to oscillate around 0 with decreasing amplitude:

$$x_n = e^{-4n/N} \sin(8\pi n/N)\,. \tag{A.44}$$

---

[4] http://en.wikipedia.org/wiki/Note

The other sequence is generated by the difference equation (A.13) for logistic growth, repeated here for convenience:

$$x_n = x_{n-1} + qx_{n-1}(1 - x_{n-1}), \quad x = x_0. \tag{A.45}$$

We let $x_0 = 0.01$ and $q = 2$. This leads to fast initial growth toward the limit 1, and then oscillations around this limit (this problem is studied in Exercise A.19).

The absolute value of the sequence elements x_n are of size between 0 and 1, approximately. We want to transform these sequence elements to tones, using the techniques of Sect. A.2. First we convert x_n to a frequency the human ear can hear. The transformation

$$y_n = 440 + 200x_n \tag{A.46}$$

will make a standard A reference tone out of $x_n = 0$, and for the maximum value of x_n around 1 we get a tone of 640 Hz. Elements of the sequence generated by (A.44) lie between -1 and 1, so the corresponding frequencies lie between 240 Hz and 640 Hz. The task now is to make a program that can generate and play the sounds.

Solution   Tones can be generated by the `note` function from the `scitools.sound` module. We collect all tones corresponding to all the y_n frequencies in a list `tones`. Letting N denote the number of sequence elements, the relevant code segment reads

```
from scitools.sound import *
freqs = 440 + x*200
tones = []
duration = 30.0/N      # 30 sec sound in total
for n in range(N+1):
    tones.append(max_amplitude*note(freqs[n], duration, 1))
data = concatenate(tones)
write(data, filename)
data = read(filename)
play(filename)
```

It is illustrating to plot the sequences too,

```
plot(range(N+1), freqs, 'ro')
```

To generate the sequences (A.44) and (A.45), we make two functions, `oscillations` and `logistic`, respectively. These functions take the number of sequence elements (N) as input and return the sequence stored in an array.

In another function `make_sound` we compute the sequence, transform the elements to frequencies, generate tones, write the tones to file, and play the sound file.

As always, we collect the functions in a module and include a test block where we can read the choice of sequence and the sequence length from the command line. The complete module file looks as follows:

```
from scitools.sound import *
from scitools.std import *

def oscillations(N):
    x = zeros(N+1)
    for n in range(N+1):
        x[n] = exp(-4*n/float(N))*sin(8*pi*n/float(N))
    return x

def logistic(N):
    x = zeros(N+1)
    x[0] = 0.01
    q = 2
    for n in range(1, N+1):
        x[n] = x[n-1] + q*x[n-1]*(1 - x[n-1])
    return x

def make_sound(N, seqtype):
    filename = 'tmp.wav'
    x = eval(seqtype)(N)
    # Convert x values to frequences around 440
    freqs = 440 + x*200
    plot(range(N+1), freqs, 'ro')
    # Generate tones
    tones = []
    duration = 30.0/N      # 30 sec sound in total
    for n in range(N+1):
        tones.append(max_amplitude*note(freqs[n], duration, 1))
    data = concatenate(tones)
    write(data, filename)
    data = read(filename)
    play(filename)
```

This code should be quite easy to read at the present stage in the book. However, there is one statement that deserves a comment:

```
x = eval(seqtype)(N)
```

The seqtype argument reflects the type of sequence and is a string that the user provides on the command line. The values of the string equal the function names oscillations and logistic. With eval(seqtype) we turn the string into a function name. For example, if seqtype is 'logistic', performing an eval(seqtype)(N) is the same as if we had written logistic(N). This technique allows the user of the program to choose a function call inside the code. Without eval we would need to explicitly test on values:

```
if seqtype == 'logistic':
    x = logistic(N)
elif seqtype == 'oscillations':
    x = oscillations(N)
```

This is not much extra code to write in the present example, but if we have a large number of functions generating sequences, we can save a lot of boring if-else code by using the eval construction.

The next step, as a reader who have understood the problem and the implementation above, is to run the program for two cases: the `oscillations` sequence with $N = 40$ and the `logistic` sequence with $N = 100$. By altering the q parameter to lower values, you get other sounds, typically quite boring sounds for non-oscillating logistic growth ($q < 1$). You can also experiment with other transformations of the form (A.46), e.g., increasing the frequency variation from 200 to 400.

A.3 Exercises

Exercise A.1: Determine the limit of a sequence

a) Write a Python function for computing and returning the sequence

$$a_n = \frac{7 + 1/(n + 1)}{3 - 1/(n + 1)^2}, \quad n = 0, 2, \ldots, N .$$

Write out the sequence for $N = 100$. Find the exact limit as $N \to \infty$ and compare with a_N.

b) Write a Python function for computing and returning the sequence

$$D_n = \frac{\sin(2^{-n})}{2^{-n}}, \quad n = 0, \ldots, N .$$

Determine the limit of this sequence for large N.

c) Given the sequence

$$D_n = \frac{f(x + h) - f(x)}{h}, \quad h = 2^{-n}, \tag{A.47}$$

make a function `D(f, x, N)` that takes a function $f(x)$, a value x, and the number N of terms in the sequence as arguments, and returns the sequence D_n for $n = 0, 1, \ldots, N$. Make a call to the D function with $f(x) = \sin x$, $x = 0$, and $N = 80$. Plot the evolution of the computed D_n values, using small circles for the data points.

d) Make another call to D where $x = \pi$ and plot this sequence in a separate figure. What would be your expected limit?

e) Explain why the computations for $x = \pi$ go wrong for large N.

Hint Print out the numerator and denominator in D_n.
Filename: `sequence_limits`.

Exercise A.2: Compute π via sequences

The following sequences all converge to π:

$$(a_n)_{n=1}^{\infty}, \quad a_n = 4 \sum_{k=1}^{n} \frac{(-1)^{k+1}}{2k-1},$$

$$(b_n)_{n=1}^{\infty}, \quad b_n = \left(6 \sum_{k=1}^{n} k^{-2}\right)^{1/2},$$

$$(c_n)_{n=1}^{\infty}, \quad c_n = \left(90 \sum_{k=1}^{n} k^{-4}\right)^{1/4},$$

$$(d_n)_{n=1}^{\infty}, \quad d_n = \frac{6}{\sqrt{3}} \sum_{k=0}^{n} \frac{(-1)^k}{3^k(2k+1)},$$

$$(e_n)_{n=1}^{\infty}, \quad e_n = 16 \sum_{k=0}^{n} \frac{(-1)^k}{5^{2k+1}(2k+1)} - 4 \sum_{k=0}^{n} \frac{(-1)^k}{239^{2k+1}(2k+1)}.$$

Make a function for each sequence that returns an array with the elements in the sequence. Plot all the sequences, and find the one that converges fastest toward the limit π.
Filename: `pi_sequences`.

Exercise A.3: Reduce memory usage of difference equations

Consider the program `growth_years.py` from Sect. A.1.1. Since x_n depends on x_{n-1} only, we do not need to store all the $N+1$ x_n values. We actually only need to store x_n and its previous value x_{n-1}. Modify the program to use two variables and not an array for the entire sequence. Also avoid the `index_set` list and use an integer counter for n and a `while` loop instead. Write the sequence to file such that it can be visualized later.
Filename: `growth_years_efficient`.

Exercise A.4: Compute the development of a loan

Solve (A.16)–(A.17) in a Python function.
Filename: `loan`.

Exercise A.5: Solve a system of difference equations

Solve (A.32)–(A.33) in a Python function and plot the x_n sequence.
Filename: `fortune_and_inflation1`.

Exercise A.6: Modify a model for fortune development

In the model (A.32)–(A.33) the new fortune is the old one, plus the interest, minus the consumption. During year n, x_n is normally also reduced with t percent tax on the earnings $x_{n-1} - x_{n-2}$ in year $n-1$.

a) Extend the model with an appropriate tax term, implement the model, and demonstrate in a plot the effect of tax ($t = 27$) versus no tax ($t = 0$).

b) Suppose you expect to live for N years and can accept that the fortune x_n vanishes after N years. Choose some appropriate values for p, q, I, and t, and experiment with the program to find how large the initial c_0 can be in this case.

Filename: `fortune_and_inflation2`.

Exercise A.7: Change index in a difference equation

A mathematically equivalent equation to (A.5) is

$$x_{i+1} = x_i + \frac{p}{100}x_i, \tag{A.48}$$

since the name of the index can be chosen arbitrarily. Suppose someone has made the following program for solving (A.48):

```
from scitools.std import *
x0 = 100              # initial amount
p = 5                 # interest rate
N = 4                 # number of years
index_set = range(N+1)
x = zeros(len(index_set))

# Compute solution
x[0] = x0
for i in index_set[1:]:
    x[i+1] = x[i] + (p/100.0)*x[i]
print x
plot(index_set, x, 'ro', xlabel='years', ylabel='amount')
```

This program does not work. Make a correct version, but keep the difference equations in its present form with the indices i+1 and i.
Filename: `growth1_index_ip1`.

Exercise A.8: Construct time points from dates

A certain quantity p (which may be an interest rate) is piecewise constant and undergoes changes at some specific dates, e.g.,

$$p \text{ changes to} \begin{cases} 4.5 & \text{on Jan 4, 2019} \\ 4.75 & \text{on March 21, 2019} \\ 6.0 & \text{on April 1, 2019} \\ 5.0 & \text{on June 30, 2019} \\ 4.5 & \text{on Nov 1, 2019} \\ 2.0 & \text{on April 1, 2020} \end{cases} \tag{A.49}$$

Given a start date d_1 and an end date d_2, fill an array p with the right p values, where the array index counts days. Use the `datetime` module to compute the number of days between dates.
Filename: `dates2days`.

Exercise A.9: Visualize the convergence of Newton's method

Let $x_0, x_1, \ldots, x_N$ be the sequence of roots generated by Newton's method applied to a nonlinear algebraic equation $f(x) = 0$ (see Sect. A.1.10). In this exercise, the purpose is to plot the sequences $(x_n)_{n=0}^N$ and $(|f(x_n)|)_{n=0}^N$ such that we can understand how Newton's method converges or diverges.

a) Make a general function

```
Newton_plot(f, x, dfdx, xmin, xmax, epsilon=1E-7)
```

for this purpose. The arguments `f` and `dfdx` are Python functions representing the $f(x)$ function in the equation and its derivative $f'(x)$, respectively. Newton's method is run until $|f(x_N)| \leq \epsilon$, and the ϵ value is available as the `epsilon` argument. The `Newton_plot` function should make three separate plots of $f(x)$, $(x_n)_{n=0}^N$, and $(|f(x_n)|)_{n=0}^N$ on the screen and also save these plots to PNG files. The relevant x interval for plotting of $f(x)$ is given by the arguments `xmin` and `xmax`. Because of the potentially wide scale of values that $|f(x_n)|$ may exhibit, it may be wise to use a logarithmic scale on the y axis.

Hint You can save quite some coding by calling the improved `Newton` function from Sect. A.1.10, which is available in the module file `Newton.py`.

b) Demonstrate the function on the equation $x^6 \sin \pi x = 0$, with $\epsilon = 10^{-13}$. Try different starting values for Newton's method: $x_0 = -2.6, -1.2, 1.5, 1.7, 0.6$. Compare the results with the exact solutions $x = \ldots, -2-1, 0, 1, 2, \ldots$.

c) Use the `Newton_plot` function to explore the impact of the starting point x_0 when solving the following nonlinear algebraic equations:

$$\sin x = 0, \tag{A.50}$$

$$x = \sin x, \tag{A.51}$$

$$x^5 = \sin x, \tag{A.52}$$

$$x^4 \sin x = 0, \tag{A.53}$$

$$x^4 = 16, \tag{A.54}$$

$$x^{10} = 1, \tag{A.55}$$

$$\tanh x = 0, \tag{A.56}$$

$$\tanh x = x^{10}. \tag{A.57}$$

Hint Such an experimental investigation is conveniently recorded in an IPython notebook. See Sect. H.4 for a quick introduction to notebooks.

Filename: `Newton2`.

Exercise A.10: Implement the secant method

Newton's method (A.34) for solving $f(x) = 0$ requires the derivative of the function $f(x)$. Sometimes this is difficult or inconvenient. The derivative can be approximated using the last two approximations to the root, x_{n-2} and x_{n-1}:

$$f'(x_{n-1}) \approx \frac{f(x_{n-1}) - f(x_{n-2})}{x_{n-1} - x_{n-2}}.$$

Using this approximation in (A.34) leads to the Secant method:

$$x_n = x_{n-1} - \frac{f(x_{n-1})(x_{n-1} - x_{n-2})}{f(x_{n-1}) - f(x_{n-2})}, \quad x_0, x_1 \text{ given}. \tag{A.58}$$

Here $n = 2, 3, \ldots$. Make a program that applies the Secant method to solve $x^5 = \sin x$.

Filename: `Secant`.

Exercise A.11: Test different methods for root finding

Make a program for solving $f(x) = 0$ by Newton's method (Sect. A.1.10), the Bisection method (Sect. 4.11.2), and the Secant method (Exercise A.10). For each method, the sequence of root approximations should be written out (nicely formatted) on the screen. Read $f(x)$, $f'(x)$, a, b, x_0, and x_1 from the command line. Newton's method starts with x_0, the Bisection method starts with the interval $[a, b]$, whereas the Secant method starts with x_0 and x_1.

Run the program for each of the equations listed in Exercise A.9d. You should first plot the $f(x)$ functions so you know how to choose x_0, x_1, a, and b in each case.

Filename: `root_finder_examples`.

Exercise A.12: Make difference equations for the Midpoint rule

Use the ideas of Sect. A.1.7 to make a similar system of difference equations and corresponding implementation for the Midpoint integration rule:

$$\int_a^b f(x)dx \approx h \sum_{i=0}^{n-1} f(a - \frac{1}{2}h + ih),$$

where $h = (b - a)/n$ and n counts the number of function evaluations (i.e., rectangles that approximate the area under the curve).

Filename: `diffeq_midpoint`.

Exercise A.13: Compute the arc length of a curve

Sometimes one wants to measure the length of a curve $y = f(x)$ for $x \in [a, b]$. The arc length from $f(a)$ to some point $f(x)$ is denoted by $s(x)$ and defined through an integral

$$s(x) = \int_a^x \sqrt{1 + [f'(\xi)]^2}d\xi. \tag{A.59}$$

We can compute $s(x)$ via difference equations as explained in Sect. A.1.7.

a) Make a Python function `arclength(f, a, b, n)` that returns an array s with $s(x)$ values for n uniformly spaced coordinates x in $[a, b]$. Here `f(x)` is the Python implementation of the function that defines the curve we want to compute the arc length of.

b) How can you verify that the `arclength` function works correctly? Construct test case(s) and write corresponding test functions for automating the tests.

Hint Check the implementation for curves with known arc length, e.g., a semi-circle and a straight line.

c) Apply the function to

$$f(x) = \int_{-2}^{x} = \frac{1}{\sqrt{2\pi}} e^{-4t^2} dt, \quad x \in [-2, 2].$$

Compute $s(x)$ and plot it together with $f(x)$.

Filename: `arclength`.

Exercise A.14: Find difference equations for computing sin x
The purpose of this exercise is to derive and implement difference equations for computing a Taylor polynomial approximation to $\sin x$:

$$\sin x \approx S(x;n) = \sum_{j=0}^{n} (-1)^j \frac{x^{2j+1}}{(2j+1)!}. \tag{A.60}$$

To compute $S(x;n)$ efficiently, write the sum as $S(x;n) = \sum_{j=0}^{n} a_j$, and derive a relation between two consecutive terms in the series:

$$a_j = -\frac{x^2}{(2j+1)2j} a_{j-1}. \tag{A.61}$$

Introduce $s_j = S(x; j-1)$ and a_j as the two sequences to compute. We have $s_0 = 0$ and $a_0 = x$.

a) Formulate the two difference equations for s_j and a_j.

Hint Section A.1.8 explains how this task and the associated programming can be solved for the Taylor polynomial approximation of e^x.

b) Implement the system of difference equations in a function `sin_Taylor(x, n)`, which returns s_{n+1} and $|a_{n+1}|$. The latter is the first neglected term in the sum (since $s_{n+1} = \sum_{j=0}^{n} a_j$) and may act as a rough measure of the size of the error in the Taylor polynomial approximation.

c) Verify the implementation by computing the difference equations for $n = 2$ by hand (or in a separate program) and comparing with the output from the `sin_Taylor` function. Automate this comparison in a test function.

d) Make a table or plot of s_n for various x and n values to illustrate that the accuracy of a Taylor polynomial (around $x = 0$) improves as n increases and x decreases.

Hint Be aware of the fact that `sin_Taylor(x, n)` can give extremely inaccurate approximations to $\sin x$ if x is not sufficiently small and n sufficiently large. In a plot you must therefore define the axis appropriately.
Filename: `sin_Taylor_series_diffeq`.

Exercise A.15: Find difference equations for computing cos x

Solve Exercise A.14 for the Taylor polynomial approximation to $\cos x$. (The relevant expression for the Taylor series is easily found in a mathematics textbook or by searching on the Internet.)

Filename: `cos_Taylor_series_diffeq`.

Exercise A.16: Make a guitar-like sound

Given start values $x_0, x_1, \ldots, x_p$, the following difference equation is known to create guitar-like sound:

$$x_n = \frac{1}{2}(x_{n-p} + x_{n-p-1}), \quad n = p+1, \ldots, N . \tag{A.62}$$

With a sampling rate r, the frequency of this sound is given by r/p. Make a program with a function `solve(x, p)` which returns the solution array x of (A.62). To initialize the array `x[0:p+1]` we look at two methods, which can be implemented in two alternative functions:

- $x_0 = 1, x_1 = x_2 = \cdots = x_p = 0$
- $x_0, \ldots, x_p$ are uniformly distributed random numbers in $[-1, 1]$

Import `max_amplitude`, `write`, and `play` from the `scitools.sound` module. Choose a sampling rate r and set $p = r/440$ to create a 440 Hz tone (A). Create an array x1 of zeros with length $3r$ such that the tone will last for 3 seconds. Initialize x1 according to method 1 above and solve (A.62). Multiply the x1 array by `max_amplitude`. Repeat this process for an array x2 of length $2r$, but use method 2 for the initial values and choose p such that the tone is 392 Hz (G). Concatenate x1 and x2, call `write` and then `play` to play the sound. As you will experience, this sound is amazingly similar to the sound of a guitar string, first playing A for 3 seconds and then playing G for 2 seconds.

The method (A.62) is called the Karplus-Strong algorithm and was discovered in 1979 by a researcher, Kevin Karplus, and his student Alexander Strong, at Stanford University.

Filename: `guitar_sound`.

Exercise A.17: Damp the bass in a sound file

Given a sequence $x_0, \ldots, x_{N-1}$, the following *filter* transforms the sequence to a new sequence $y_0, \ldots, y_{N-1}$:

$$y_n = \begin{cases} x_n, & n = 0 \\ -\frac{1}{4}(x_{n-1} - 2x_n + x_{n+1}), & 1 \leq n \leq N-2 \\ x_n, & n = N-1 \end{cases} \tag{A.63}$$

If x_n represents sound, y_n is the same sound but with the bass damped. Load some sound file, e.g.,

```
x = scitools.sound.Nothing_Else_Matters()
# or
x = scitools.sound.Ja_vi_elsker()
```

to get a sound sequence. Apply the filter (A.63) and play the resulting sound. Plot the first 300 values in the x_n and y_n signals to see graphically what the filter does with the signal.
Filename: `damp_bass`.

Exercise A.18: Damp the treble in a sound file
Solve Exercise A.17 to get some experience with coding a filter and trying it out on a sound. The purpose of this exercise is to explore some other filters that reduce the treble instead of the bass. Smoothing the sound signal will in general damp the treble, and smoothing is typically obtained by letting the values in the new filtered sound sequence be an average of the neighboring values in the original sequence.

The simplest smoothing filter can apply a standard average of three neighboring values:

$$y_n = \begin{cases} x_n, & n = 0 \\ \frac{1}{3}(x_{n-1} + x_n + x_{n+1}), & 1 \le n \le N - 2 \\ x_n, & n = N - 1 \end{cases} \tag{A.64}$$

Two other filters put less emphasis on the surrounding values:

$$y_n = \begin{cases} x_n, & n = 0 \\ \frac{1}{4}(x_{n-1} + 2x_n + x_{n+1}), & 1 \le n \le N - 2 \\ x_n, & n = N - 1 \end{cases} \tag{A.65}$$

$$y_n = \begin{cases} x_n, & n = 0, 1 \\ \frac{1}{16}(x_{n-2} + 4x_{n-1} + 6x_n + 4x_{n+1} + x_{n+2}), & 2 \le n \le N - 3 \\ x_n, & n = N - 2, N - 1 \end{cases} \tag{A.66}$$

Apply all these three filters to a sound file and listen to the result. Plot the first 300 values in the x_n and y_n signals for each of the three filters to see graphically what the filter does with the signal.
Filename: `damp_treble`.

Exercise A.19: Demonstrate oscillatory solutions of the logistic equation

a) Write a program to solve the difference equation (A.13):

$$y_n = y_{n-1} + q y_{n-1} (1 - y_{n-1}), \quad n = 0, \ldots, N .$$

Read the input parameters y_0, q, and N from the command line. The variables and the equation are explained in Sect. A.1.5.
b) Equation (A.13) has the solution $y_n = 1$ as $n \to \infty$. Demonstrate, by running the program, that this is the case when $y_0 = 0.3$, $q = 1$, and $N = 50$.
c) For larger q values, y_n does not approach a constant limit, but y_n oscillates instead around the limiting value. Such oscillations are sometimes observed in wildlife populations. Demonstrate oscillatory solutions when q is changed to 2 and 3.

d) It could happen that y_n stabilizes at a constant level for larger N. Demonstrate that this is not the case by running the program with $N = 1000$.

Filename: `growth_logistic2`.

Exercise A.20: Automate computer experiments
It is tedious to run a program like the one from Exercise A.19 repeatedly for a wide range of input parameters. A better approach is to let the computer do the manual work. Modify the program from Exercise A.19 such that the computation of y_n and the plot is made in a function. Let the title in the plot contain the parameters y_0 and q (N is easily visible from the x axis). Also let the name of the plot file reflect the values of y_0, q, and N. Then make loops over y_0 and q to perform the following more comprehensive set of experiments:

- $y_0 = 0.01, 0.3$
- $q = 0.1, 1, 1.5, 1.8, 2, 2.5, 3$
- $N = 50$

How does the initial condition (the value y_0) seem to influence the solution?

Hint If you do no want to get a lot of plots on the screen, which must be killed, drop the call to `show()` in Matplotlib or use `show=False` as argument to `plot` in SciTools.
Filename: `growth_logistic3`.

Exercise A.21: Generate an HTML report
Extend the program made in Exercise A.20 with a report containing all the plots. The report can be written in HTML and displayed by a web browser. The plots must then be generated in PNG format. The source of the HTML file will typically look as follows:

```
<html>
<body>
<p><img src="tmp_y0_0.01_q_0.1_N_50.png">
<p><img src="tmp_y0_0.01_q_1_N_50.png">
<p><img src="tmp_y0_0.01_q_1.5_N_50.png">
<p><img src="tmp_y0_0.01_q_1.8_N_50.png">
...
<p><img src="tmp_y0_0.01_q_3_N_1000.png">
</html>
</body>
```

Let the program write out the HTML text to a file. You may let the function making the plots return the name of the plot file such that this string can be inserted in the HTML file.
Filename: `growth_logistic4`.

Exercise A.22: Use a class to archive and report experiments

The purpose of this exercise is to make the program from Exercise A.21 more flexible by creating a Python class that runs and archives all the experiments (provided you know how to program with Python classes). Here is a sketch of the class:

```python
class GrowthLogistic(object):
    def __init__(self, show_plot_on_screen=False):
        self.experiments = []
        self.show_plot_on_screen = show_plot_on_screen
        self.remove_plot_files()

    def run_one(self, y0, q, N):
        """Run one experiment."""
        # Compute y[n] in a loop...
        plotfile = 'tmp_y0_%g_q_%g_N_%d.png' % (y0, q, N)
        self.experiments.append({'y0': y0, 'q': q, 'N': N,
                                 'mean': mean(y[20:]),
                                 'y': y, 'plotfile': plotfile})
        # Make plot...

    def run_many(self, y0_list, q_list, N):
        """Run many experiments."""
        for q in q_list:
            for y0 in y0_list:
                self.run_one(y0, q, N)

    def remove_plot_files(self):
        """Remove plot files with names tmp_y0*.png."""
        import os, glob
        for plotfile in glob.glob('tmp_y0*.png'):
            os.remove(plotfile)

    def report(self, filename='tmp.html'):
        """
        Generate an HTML report with plots of all
        experiments generated so far.
        """
        # Open file and write HTML header...
        for e in self.experiments:
            html.write('<p><img src="%s">\n' % e['plotfile'])
        # Write HTML footer and close file...
```

Each time the `run_one` method is called, data about the current experiment is stored in the `experiments` list. Note that `experiments` contains a list of dictionaries. When desired, we can call the `report` method to collect all the plots made so far in an HTML report. A typical use of the class goes as follows:

```python
N = 50
g = GrowthLogistic()
g.run_many(y0_list=[0.01, 0.3],
           q_list=[0.1, 1, 1.5, 1.8] + [2, 2.5, 3], N=N)
g.run_one(y0=0.01, q=3, N=1000)
g.report()
```

Make a complete implementation of class GrowthLogistic and test it with the small program above. The program file should be constructed as a module.
Filename: growth_logistic5.

Exercise A.23: Explore logistic growth interactively

Class GrowthLogistic from Exercise A.22 is very well suited for interactive exploration. Here is a possible sample session for illustration:

```
>>> from growth_logistic5 import GrowthLogistic
>>> g = GrowthLogistic(show_plot_on_screen=True)
>>> q = 3
>>> g.run_one(0.01, q, 100)
>>> y = g.experiments[-1]['y']
>>> max(y)
1.3326056469620293
>>> min(y)
0.0029091569028512065
```

Extend this session with an investigation of the oscillations in the solution y_n. For this purpose, make a function for computing the local maximum values y_n and the corresponding indices where these local maximum values occur. We can say that y_i is a local maximum value if

$$y_{i-1} < y_i > y_{i+1}.$$

Plot the sequence of local maximum values in a new plot. If $I_0, I_1, I_2, \ldots$ constitute the set of increasing indices corresponding to the local maximum values, we can define the periods of the oscillations as $I_1 - I_0, I_2 - I_1$, and so forth. Plot the length of the periods in a separate plot. Repeat this investigation for $q = 2.5$.
Filename: GrowthLogistic_interactive.

Exercise A.24: Simulate the price of wheat

The demand for wheat in year t is given by

$$D_t = ap_t + b,$$

where $a < 0, b > 0$, and p_t is the price of wheat. Let the supply of wheat be

$$S_t = Ap_{t-1} + B + \ln(1 + p_{t-1}),$$

where A and B are given constants. We assume that the price p_t adjusts such that all the produced wheat is sold. That is, $D_t = S_t$.

a) For $A = 1, a = -3, b = 5, B = 0$, find from numerical computations, a stable price such that the production of wheat from year to year is constant. That is, find p such that $ap + b = Ap + B + \ln(1 + p)$.

b) Assume that in a very dry year the production of wheat is much less than planned. Given that price this year, p_0, is 4.5 and $D_t = S_t$, compute in a program how the prices $p_1, p_2, \ldots, p_N$ develop. This implies solving the difference equation

$$ap_t + b = Ap_{t-1} + B + \ln(1 + p_{t-1}).$$

From the p_t values, compute S_t and plot the points (p_t, S_t) for $t = 0, 1, 2, \ldots, N$. How do the prices move when $N \to \infty$?

Filename: `wheat`.

Introduction to Discrete Calculus

B

This appendix is authored by Aslak Tveito

In this chapter we will discuss how to differentiate and integrate functions on a computer. To do that, we have to care about how to treat mathematical functions on a computer. Handling mathematical functions on computers is not entirely straightforward: a function $f(x)$ contains an infinite amount of information (function values at an infinite number of x values on an interval), while the computer can only store a finite amount of data. Think about the $\cos x$ function. There are typically two ways we can work with this function on a computer. One way is to run an algorithm to compute its value, like that in Exercise 3.37, or we simply call `math.cos(x)` (which runs a similar type of algorithm), to compute an approximation to $\cos x$ for a given x, using a finite number of calculations. The other way is to store $\cos x$ values in a table for a finite number of x values (of course, we need to run an algorithm to populate the table with $\cos x$ numbers). and use the table in a smart way to compute $\cos x$ values. This latter way, known as a *discrete* representation of a function, is in focus in the present chapter. With a discrete function representation, we can easily integrate and differentiate the function too. Read on to see how we can do that.

The folder `src/discalc`[1] contains all the program example files referred to in this chapter.

B.1 Discrete Functions

Physical quantities, such as temperature, density, and velocity, are usually defined as continuous functions of space and time. However, as mentioned in above, discrete versions of the functions are more convenient on computers. We will illustrate the concept of discrete functions through some introductory examples. In fact, we use discrete functions when plotting curves on a computer: we define a finite set of coordinates x and store the corresponding function values f(x) in an array. A plotting program will then draw straight lines between the function values. A discrete representation of a continuous function is, from a programming point of view, nothing

---

[1] http://tinyurl.com/pwyasaa/discalc

but storing a finite set of coordinates and function values in an array. Nevertheless, we will in this chapter be more formal and describe discrete functions by precise mathematical terms.

B.1.1   The Sine Function

Suppose we want to generate a plot of the sine function for values of x between 0 and π. To this end, we define a set of x-values and an associated set of values of the sine function. More precisely, we define $n + 1$ points by

$$x_i = ih \quad \text{for } i = 0, 1, \ldots, n, \tag{B.1}$$

where $h = \pi/n$ and $n \geq 1$ is an integer. The associated function values are defined as

$$s_i = \sin(x_i) \quad \text{for } i = 0, 1, \ldots, n. \tag{B.2}$$

Mathematically, we have a sequence of coordinates $(x_i)_{i=0}^n$ and of function values $(s_i)_{i=0}^n$. (Here we have used the sequence notation $(x_i)_{i=0}^n = x_0, x_1, \ldots, x_n$.) Often we "merge" the two sequences to one sequence of points: $(x_i, s_i)_{i=0}^n$. Sometimes we also use a shorter notation, just x_i, s_i, or (x_i, s_i) if the exact limits are not of importance. The set of coordinates $(x_i)_{i=0}^n$ constitutes a *mesh* or a *grid*. The individual coordinates x_i are known as *nodes* in the mesh (or grid). The discrete representation of the sine function on $[0, \pi]$ consists of the mesh and the corresponding sequence of function values $(s_i)_{i=0}^n$ at the nodes. The parameter n is often referred to as the *mesh resolution*.

In a program, we represent the mesh by a coordinate array, say x, and the function values by another array, say s. To plot the sine function we can simply write

```
from scitools.std import *

n = int(sys.argv[1])
x = linspace(0, pi, n+1)
s = sin(x)
plot(x, s, legend='sin(x), n=%d' % n, savefig='tmp.pdf')
```

Figure B.1 shows the resulting plot for $n = 5, 10, 20$ and 100. Because the plotting program draws straight lines between the points in x and s, the curve looks smoother the more points we use, and since $\sin(x)$ is a smooth function, the plots in Fig. B.1 do not look sufficiently good. However, we can with our eyes hardly distinguish the plot with 100 points from the one with 20 points, so 20 points seem sufficient in this example.

There are no tests on the validity of the input data (n) in the previous program. A program including these tests reads

```
from scitools.std import *

n = int(sys.argv[1])
x = linspace(0, pi, n+1)
```

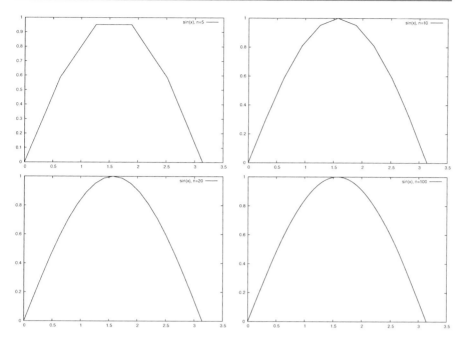

Fig. B.1 Plots of $\sin(x)$ with various n

```
s = sin(x)
plot(x, s, legend='sin(x), n=%d' % n, savefig='tmp.pdf')
```

Such tests are important parts of a good programming philosophy. However, for the programs displayed in this text, we skip such tests in order to make the programs more compact and readable and to strengthen the focus on the mathematics in the programs. In the versions of these programs in the files that can be downloaded you will, hopefully, always find a test on input data.

B.1.2   Interpolation

Suppose we have a discrete representation of the sine function: $(x_i, s_i)_{i=0}^n$. At the nodes we have the exact sine values s_i, but what about the points in between these nodes? Finding function values between the nodes is called *interpolation*, or we can say that we *interpolate* a discrete function.

A graphical interpolation procedure could be to look at one of the plots in Fig. B.1 to find the function value corresponding to a point x between the nodes. Since the plot is a straight line from node value to node value, this means that a function value between two nodes is found from a straight line approximation to the underlying continuous function. (Strictly speaking, we also assume that the function to be interpolated is rather smooth. It is easy to see that if the function is very wild, i.e., the values of the function change very rapidly, this procedure may fail even for very large values of n. Section 5.4.2 provides an example.) We formulate this procedure precisely in terms of mathematics in the next paragraph.

Assume that we know that a given x^* lies in the interval from $x = x_k$ to x_{k+1}, where the integer k is given. In the interval $x_k \leq x < x_{k+1}$, we define the linear function that passes through (x_k, s_k) and (x_{k+1}, s_{k+1}):

$$S_k(x) = s_k + \frac{s_{k+1} - s_k}{x_{k+1} - x_k}(x - x_k). \tag{B.3}$$

That is, $S_k(x)$ coincides with $\sin(x)$ at x_k and x_{k+1}, and between these nodes, $S_k(x)$ is linear. We say that $S_k(x)$ interpolates the discrete function $(x_i, s_i)_{i=0}^n$ on the interval $[x_k, x_{k+1}]$.

B.1.3   Evaluating the Approximation

Given the values $(x_i, s_i)_{i=0}^n$ and the formula (B.3), we want to compute an approximation of the sine function for any x in the interval from $x = 0$ to $x = \pi$. In order to do that, we have to compute k for a given value of x. More precisely, for a given x we have to find k such that $x_k \leq x \leq x_{k+1}$. We can do that by defining

$$k = \lfloor x/h \rfloor$$

where the function $\lfloor z \rfloor$ denotes the largest integer that is smaller than z. In Python, $\lfloor z \rfloor$ is computed by `int(z)`. The program below takes x and n as input and computes the approximation of $\sin(x)$. The program prints the approximation $S(x)$ and the exact value of $\sin(x)$ so we can look at the development of the error when n is increased. (The value is not really exact, it is the value of $\sin(x)$ provided by the computer, `math.sin(x)`, and this value is calculated from an algorithm that only yields an approximation to $\sin(x)$. Exercise 3.37 provides an example of the type of algorithm in question.)

```
from numpy import *
import sys

xp = eval(sys.argv[1])
n = int(sys.argv[2])

def S_k(k):
    return s[k] + \
            ((s[k+1] - s[k])/(x[k+1] - x[k]))*(xp - x[k])
h = pi/n
x = linspace(0, pi, n+1)
s = sin(x)
k = int(xp/h)

print 'Approximation of sin(%s):   ' % xp, S_k(k)
print 'Exact value of sin(%s):   ' % xp, sin(xp)
print 'Eror in approximation:      ', sin(xp) - S_k(k)
```

To study the approximation, we put $x = \sqrt{2}$ and use the program `eval_sine.py` for $n = 5, 10$ and 20.

```
Terminal
Terminal> python src-discalc/eval_sine.py 'sqrt(2)' 5
Approximation of sin(1.41421356237):   0.951056516295
Exact value of sin(1.41421356237):     0.987765945993
Eror in approximation:      0.0367094296976
```

```
Terminal
Terminal> python src-discalc/eval_sine.py 'sqrt(2)' 10
Approximation of sin(1.41421356237):   0.975605666221
Exact value of sin(1.41421356237):     0.987765945993
Eror in approximation:      0.0121602797718
```

```
Terminal
Terminal> python src-discalc/eval_sine.py 'sqrt(2)' 20
Approximation of sin(1.41421356237):   0.987727284363
Exact value of sin(1.41421356237):     0.987765945993
Eror in approximation:      3.86616296923e-05
```

Note that the error is reduced as the n increases.

B.1.4 Generalization

In general, we can create a discrete version of a continuous function as follows. Suppose a continuous function $f(x)$ is defined on an interval ranging from $x = a$ to $x = b$, and let $n \geq 1$, be a given integer. Define the distance between nodes,

$$h = \frac{b - a}{n},$$

and the nodes

$$x_i = a + ih \quad \text{for } i = 0, 1, \ldots, n. \tag{B.4}$$

The discrete function values are given by

$$y_i = f(x_i) \quad \text{for } i = 0, 1, \ldots, n. \tag{B.5}$$

Now, $(x_i, y_i)_{i=0}^{n}$ is the discrete version of the continuous function $f(x)$. The program `discrete_func.py` takes f, a, b and n as input, computes the discrete version of f, and then applies the discrete version to make a plot of f.

```
def discrete_func(f, a, b, n):
    x = linspace(a, b, n+1)
```

```
    y = zeros(len(x))
    for i in xrange(len(x)):
        y[i] = func(x[i])
    return x, y

from scitools.std import *

f_formula = sys.argv[1]
a = eval(sys.argv[2])
b = eval(sys.argv[3])
n = int(sys.argv[4])
f = StringFunction(f_formula)

x, y = discrete_func(f, a, b, n)
plot(x, y)
```

We can equally well make a vectorized version of the `discrete_func` function:

```
def discrete_func(f, a, b, n):
    x = linspace(a, b, n+1)
    y = f(x)
    return x, y
```

However, for the `StringFunction` tool to work properly in vectorized mode, we need to follow the recipe in Sect. 5.5.1:

```
f = StringFunction(f_formula)
f.vectorize(globals())
```

The corresponding vectorized program is found in the file `discrete_func_vec.py`.

B.2   Differentiation Becomes Finite Differences

You have heard about derivatives. Probably, the following formulas are well known to you:

$$\frac{d}{dx}\sin(x) = \cos(x),$$
$$\frac{d}{dx}\ln(x) = \frac{1}{x},$$
$$\frac{d}{dx}x^m = mx^{m-1},$$

But why is differentiation so important? The reason is quite simple: the derivative is a mathematical expression of change. And change is, of course, essential in modeling various phenomena. If we know the state of a system, and we know the laws of change, then we can, in principle, compute the future of that system. Appendix C treats this topic in detail. Appendix A also computes the future of systems, based on modeling changes, but without using differentiation. In Appendix C you will see that reducing the step size in the difference equations results in derivatives instead of pure differences. However, differentiation of continuous functions is somewhat hard on a computer, so we often end up replacing the derivatives by differences. This idea is quite general, and every time we use a discrete representation of a function, differentiation becomes differences, or *finite differences* as we usually say.

The mathematical definition of differentiation reads

$$f'(x) = \lim_{\varepsilon \to 0} \frac{f(x + \varepsilon) - f(x)}{\varepsilon}.$$

You have probably seen this definition many times, but have you understood what it means and do you think the formula has a great practical value? Although the definition requires that we pass to the limit, we obtain quite good approximations of the derivative by using a fixed positive value of ε. More precisely, for a small $\varepsilon > 0$, we have

$$f'(x) \approx \frac{f(x + \varepsilon) - f(x)}{\varepsilon}.$$

The fraction on the right-hand side is a finite difference approximation to the derivative of f at the point x. Instead of using ε it is more common to introduce $h = \varepsilon$ in finite differences, i.e., we like to write

$$f'(x) \approx \frac{f(x + h) - f(x)}{h}. \tag{B.6}$$

B.2.1   Differentiating the Sine Function

In order to get a feeling for how good the approximation (B.6) to the derivative really is, we explore an example. Consider $f(x) = \sin(x)$ and the associated derivative $f'(x) = \cos(x)$. If we put $x = 1$, we have

$$f'(1) = \cos(1) \approx 0.540,$$

and by putting $h = 1/100$ in (B.6) we get

$$f'(1) \approx \frac{f(1 + 1/100) - f(1)}{1/100} = \frac{\sin(1.01) - \sin(1)}{0.01} \approx 0.536.$$

The program `forward_diff.py`, shown below, computes the derivative of $f(x)$ using the approximation (B.6), where x and h are input parameters.

```
def diff(f, x, h):
    return (f(x+h) - f(x))/float(h)

from math import *
import sys

x = eval(sys.argv[1])
h = eval(sys.argv[2])
```

```
approx_deriv = diff(sin, x, h)
exact = cos(x)
print 'The approximated value is: ', approx_deriv
print 'The correct value is:      ', exact
print 'The error is:              ', exact - approx_deriv
```

Running the program for $x = 1$ and $h = 1/1000$ gives

```
                          Terminal
Terminal> python src-discalc/forward_diff.py 1 0.001
The approximated value is:  0.53988148036
The correct value is:       0.540302305868
The error is:               0.000420825507813
```

B.2.2 Differences on a Mesh

Frequently, we will need finite difference approximations to a discrete function defined on a mesh. Suppose we have a discrete representation of the sine function: $(x_i, s_i)_{i=0}^n$, as introduced in Sect. B.1.1. We want to use (B.6) to compute approximations to the derivative of the sine function at the nodes in the mesh. Since we only have function values at the nodes, the h in (B.6) must be the difference between nodes, i.e., $h = x_{i+1} - x_i$. At node x_i we then have the following approximation of the derivative:

$$z_i = \frac{s_{i+1} - s_i}{h}, \tag{B.7}$$

for $i = 0, 1, \ldots, n - 1$. Note that we have not defined an approximate derivative at the end point $x = x_n$. We cannot apply (B.7) directly since s_{n+1} is undefined (outside the mesh). However, the derivative of a function can also be defined as

$$f'(x) = \lim_{\varepsilon \to 0} \frac{f(x) - f(x - \varepsilon)}{\varepsilon},$$

which motivates the following approximation for a given $h > 0$,

$$f'(x) \approx \frac{f(x) - f(x - h)}{h}. \tag{B.8}$$

This alternative approximation to the derivative is referred to as a *backward difference* formula, whereas the expression (B.6) is known as a *forward difference* formula. The names are natural: the forward formula goes forward, i.e., in the direction of increasing x and i to collect information about the change of the function, while the backward formula goes backwards, i.e., toward smaller x and i value to fetch function information.

At the end point we can apply the backward formula and thus define

$$z_n = \frac{s_n - s_{n-1}}{h}. \tag{B.9}$$

We now have an approximation to the derivative at all the nodes. A plain specialized program for computing the derivative of the sine function on a mesh and comparing this discrete derivative with the exact derivative is displayed below (the name of the file is diff_sine_plot1.py).

```
from scitools.std import *

n = int(sys.argv[1])
h = pi/n
x = linspace(0, pi, n+1)
s = sin(x)
z = zeros(len(s))
for i in xrange(len(z)-1):
    z[i] = (s[i+1] - s[i])/h
# Special formula for end point_
z[-1] = (s[-1] - s[-2])/h
plot(x, z)

xfine = linspace(0, pi, 1001) # for more accurate plot
exact = cos(xfine)
hold()
plot(xfine, exact)
legend('Approximate function', 'Correct function')
title('Approximate and discrete functions, n=%d' % n)
```

In Fig. B.2 we see the resulting graphs for $n = 5, 10, 20$ and 100. Again, we note that the error is reduced as n increases.

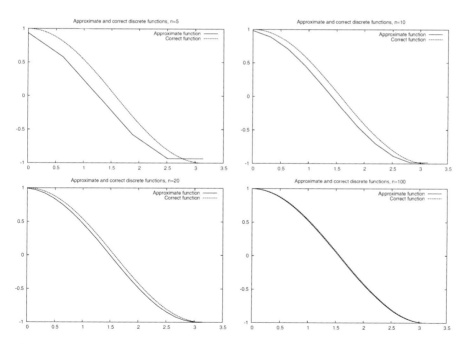

Fig. B.2 Plots for exact and approximate derivatives of $\sin(x)$ with varying values of the resolution n

B.2.3   Generalization

The discrete version of a continuous function $f(x)$ defined on an interval $[a, b]$ is given by $(x_i, y_i)_{i=0}^n$ where

$$x_i = a + ih,$$

and

$$y_i = f(x_i)$$

for $i = 0, 1, \ldots, n$. Here, $n \geq 1$ is a given integer, and the spacing between the nodes is given by

$$h = \frac{b - a}{n}.$$

A discrete approximation of the derivative of f is given by $(x_i, z_i)_{i=0}^n$ where

$$z_i = \frac{y_{i+1} - y_i}{h}$$

$i = 0, 1, \ldots, n - 1$, and

$$z_n = \frac{y_n - y_{n-1}}{h}.$$

The collection $(x_i, z_i)_{i=0}^n$ is the discrete derivative of the discrete version $(x_i, f_i)_{i=0}^n$ of the continuous function $f(x)$. The program below, found in the file `diff_func.py`, takes f, a, b, and n as input and computes the discrete derivative of f on the mesh implied by a, b, and h, and then a plot of f and the discrete derivative is made.

```
def diff(f, a, b, n):
    x = linspace(a, b, n+1)
    y = zeros(len(x))
    z = zeros(len(x))
    h = (b-a)/float(n)
    for i in xrange(len(x)):
        y[i] = func(x[i])
    for i in xrange(len(x)-1):
        z[i] = (y[i+1] - y[i])/h
    z[n] = (y[n] - y[n-1])/h
    return y, z

from scitools.std import *

f_formula = sys.argv[1]
a = eval(sys.argv[2])
b = eval(sys.argv[3])
n = int(sys.argv[4])
f = StringFunction(f_formula)
y, z = diff(f, a, b, n)
plot(x, y, 'r-', x, z, 'b-',
     legend=('function', 'derivative'))
```

B.3 Integration Becomes Summation

Some functions can be integrated analytically. You may remember the following cases,

$$\int x^m dx = \frac{1}{m+1} x^{m+1} \quad \text{for } m \neq -1,$$

$$\int \sin(x) dx = -\cos(x),$$

$$\int \frac{x}{1+x^2} dx = \frac{1}{2} \ln\left(x^2 + 1\right).$$

These are examples of so-called indefinite integrals. If the function can be integrated analytically, it is straightforward to evaluate an associated definite integral. Recall, in general, that

$$[f(x)]_a^b = f(b) - f(a).$$

Some particular examples are

$$\int_0^1 x^m dx = \left[\frac{1}{m+1} x^{m+1}\right]_0^1 = \frac{1}{m+1},$$

$$\int_0^\pi \sin(x) dx = [-\cos(x)]_0^\pi = 2,$$

$$\int_0^1 \frac{x}{1+x^2} dx = \left[\frac{1}{2} \ln\left(x^2 + 1\right)\right]_0^1 = \frac{1}{2} \ln 2.$$

But lots of functions cannot be integrated analytically and therefore definite integrals must be computed using some sort of numerical approximation. Above, we introduced the discrete version of a function, and we will now use this construction to compute an approximation of a definite integral.

B.3.1 Dividing into Subintervals

Let us start by considering the problem of computing the integral of $\sin(x)$ from $x = 0$ to $x = \pi$. This is not the most exciting or challenging mathematical problem you can think of, but it is good practice to start with a problem you know well when you want to learn a new method. In Sect. B.1.1 we introduce a discrete function $(x_i, s_i)_{i=0}^n$ where $h = \pi/n$, $s_i = \sin(x_i)$ and $x_i = ih$ for $i = 0, 1, \ldots, n$. Furthermore, in the interval $x_k \leq x < x_{k+1}$, we defined the linear function

$$S_k(x) = s_k + \frac{s_{k+1} - s_k}{x_{k+1} - x_k}(x - x_k).$$

We want to compute an approximation of the integral of the function $\sin(x)$ from $x = 0$ to $x = \pi$. The integral

$$\int_0^\pi \sin(x)dx$$

can be divided into subintegrals defined on the intervals $x_k \le x < x_{k+1}$, leading to the following sum of integrals:

$$\int_0^\pi \sin(x)dx = \sum_{k=0}^{n-1} \int_{x_k}^{x_{k+1}} \sin(x)dx.$$

To get a feeling for this split of the integral, let us spell the sum out in the case of only four subintervals. Then $n = 4$, $h = \pi/4$,

$$x_0 = 0,$$
$$x_1 = \pi/4,$$
$$x_2 = \pi/2,$$
$$x_3 = 3\pi/4$$
$$x_4 = \pi.$$

The interval from 0 to π is divided into four intervals of equal length, and we can divide the integral similarly,

$$\int_0^\pi \sin(x)dx = \int_{x_0}^{x_1} \sin(x)dx + \int_{x_1}^{x_2} \sin(x)dx +$$
$$\int_{x_2}^{x_3} \sin(x)dx + \int_{x_3}^{x_4} \sin(x)dx. \qquad \text{(B.10)}$$

So far we have changed nothing – the integral can be split in this way - with no approximation at all. But we have reduced the problem of approximating the integral

$$\int_0^\pi \sin(x)dx$$

down to approximating integrals on the subintervals, i.e. we need approximations of all the following integrals

$$\int_{x_0}^{x_1} \sin(x)dx, \int_{x_1}^{x_2} \sin(x)dx, \int_{x_2}^{x_3} \sin(x)dx, \int_{x_3}^{x_4} \sin(x)dx.$$

The idea is that the function to be integrated changes less over the subintervals than over the whole domain $[0, \pi]$ and it might be reasonable to approximate the sine by a straight line, $S_k(x)$, over each subinterval. The integration over a subinterval will then be very easy.

B.3.2   Integration on Subintervals

The task now is to approximate integrals on the form

$$\int\limits_{x_k}^{x_{k+1}} \sin(x)dx.$$

Since

$$\sin(x) \approx S_k(x)$$

on the interval (x_k, x_{k+1}), we have

$$\int\limits_{x_k}^{x_{k+1}} \sin(x)dx \approx \int\limits_{x_k}^{x_{k+1}} S_k(x)dx.$$

In Fig. B.3 we have graphed $S_k(x)$ and $\sin(x)$ on the interval (x_k, x_{k+1}) for $k = 1$ in the case of $n = 4$. We note that the integral of $S_1(x)$ on this interval equals the area of a trapezoid, and thus we have

$$\int\limits_{x_1}^{x_2} S_1(x)dx = \frac{1}{2} \left(S_1(x_2) + S_1(x_1) \right) (x_2 - x_1),$$

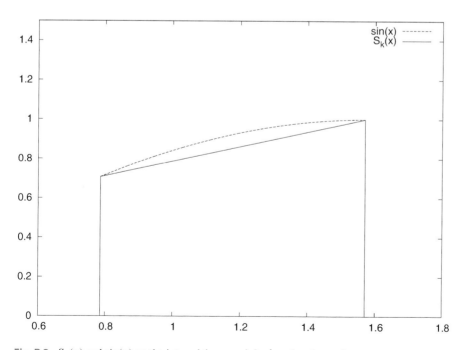

Fig. B.3 $S_k(x)$ and $\sin(x)$ on the interval (x_k, x_{k+1}) for $k = 1$ and $n = 4$

so

$$\int_{x_1}^{x_2} S_1(x)dx = \frac{h}{2}(s_2 + s_1),$$

and in general we have

$$\int_{x_k}^{x_{k+1}} \sin(x)dx \approx \frac{1}{2}(s_{k+1} + s_k)(x_{k+1} - x_k)$$

$$= \frac{h}{2}(s_{k+1} + s_k).$$

B.3.3   Adding the Subintervals

By adding the contributions from each subinterval, we get

$$\int_0^{\pi} \sin(x)dx = \sum_{k=0}^{n-1} \int_{x_k}^{x_{k+1}} \sin(x)dx$$

$$\approx \sum_{k=0}^{n-1} \frac{h}{2}(s_{k+1} + s_k),$$

so

$$\int_0^{\pi} \sin(x)dx \approx \frac{h}{2}\sum_{k=0}^{n-1}(s_{k+1} + s_k). \tag{B.11}$$

In the case of $n = 4$, we have

$$\int_0^{\pi} \sin(x)dx \approx \frac{h}{2}[(s_1 + s_0) + (s_2 + s_1) + (s_3 + s_2) + (s_4 + s_3)]$$

$$= \frac{h}{2}[s_0 + 2(s_1 + s_2 + s_3) + s_4].$$

One can show that (B.11) can be alternatively expressed as

$$\int_0^{\pi} \sin(x)dx \approx \frac{h}{2}\left[s_0 + 2\sum_{k=1}^{n-1} s_k + s_n\right]. \tag{B.12}$$

This approximation formula is referred to as the Trapezoidal rule of numerical integration. Using the more general program trapezoidal.py, presented in the next section, on integrating $\int_0^{\pi} \sin(x)dx$ with $n = 5, 10, 20$ and 100 yields the numbers 1.5644, 1.8864, 1.9713, and 1.9998 respectively. These numbers are to be compared to the exact value 2. As usual, the approximation becomes better the more points (n) we use.

B.3.4   Generalization

An approximation of the integral

$$\int_a^b f(x)dx$$

can be computed using the discrete version of a continuous function $f(x)$ defined on an interval $[a, b]$. We recall that the discrete version of f is given by $(x_i, y_i)_{i=0}^n$ where

$$x_i = a + ih, \text{ and } y_i = f(x_i)$$

for $i = 0, 1, \ldots, n$. Here, $n \geq 1$ is a given integer and $h = (b - a)/n$. The Trapezoidal rule can now be written as

$$\int_a^b f(x)dx \approx \frac{h}{2}\left[y_0 + 2\sum_{k=1}^{n-1} y_k + y_n \right].$$

The program `trapezoidal.py` implements the Trapezoidal rule for a general function f.

```
def trapezoidal(f, a, b, n):
    h = (b-a)/float(n)
    I = f(a) + f(b)
    for k in xrange(1, n, 1):
        x = a + k*h
        I += 2*f(x)
    I *= h/2
    return I


from math import *
from scitools.StringFunction import StringFunction
import sys

def test(argv=sys.argv):
    f_formula = argv[1]
    a = eval(argv[2])
    b = eval(argv[3])
    n = int(argv[4])
    f = StringFunction(f_formula)
    I = trapezoidal(f, a, b, n)
    print 'Approximation of the integral: ', I

if __name__ == '__main__':
    test()
```

We have made the file as a module such that you can easily import the `trapezoidal` function in another program. Let us do that: we make a table of how the approximation and the associated error of an integral are reduced as n is increased. For this purpose, we want to integrate $\int_{t_1}^{t_2} g(t)dt$, where

$$g(t) = -ae^{-at}\sin(\pi wt) + \pi we^{-at}\cos(\pi wt).$$

The exact integral $G(t) = \int g(t)dt$ equals

$$G(t) = e^{-at}\sin(\pi wt).$$

Here, a and w are real numbers that we set to 1/2 and 1, respectively, in the program. The integration limits are chosen as $t_1 = 0$ and $t_2 = 4$. The integral then equals zero. The program and its output appear below.

```
from trapezoidal import trapezoidal
from math import exp, sin, cos, pi

def g(t):
    return -a*exp(-a*t)*sin(pi*w*t) + pi*w*exp(-a*t)*cos(pi*w*t)

def G(t):  # integral of g(t)
    return exp(-a*t)*sin(pi*w*t)

a = 0.5
w = 1.0
t1 = 0
t2 = 4
exact = G(t2) - G(t1)
for n in 2, 4, 8, 16, 32, 64, 128, 256, 512:
    approx = trapezoidal(g, t1, t2, n)
    print 'n=%3d approximation=%12.5e  error=%12.5e' % \
        (n, approx, exact-approx)
```

```
n=  2 approximation= 5.87822e+00  error=-5.87822e+00
n=  4 approximation= 3.32652e-01  error=-3.32652e-01
n=  8 approximation= 6.15345e-02  error=-6.15345e-02
n= 16 approximation= 1.44376e-02  error=-1.44376e-02
n= 32 approximation= 3.55482e-03  error=-3.55482e-03
n= 64 approximation= 8.85362e-04  error=-8.85362e-04
n=128 approximation= 2.21132e-04  error=-2.21132e-04
n=256 approximation= 5.52701e-05  error=-5.52701e-05
n=512 approximation= 1.38167e-05  error=-1.38167e-05
```

We see that the error is reduced as we increase n. In fact, as n is doubled we realize that the error is roughly reduced by a factor of 4, at least when $n > 8$. This is an important property of the Trapezoidal rule, and checking that a program reproduces this property is an important check of the validity of the implementation.

B.4   Taylor Series

The single most important mathematical tool in computational science is the Taylor series. It is used to derive new methods and also for the analysis of the accuracy of approximations. We will use the series many times in this text. Right here, we just introduce it and present a few applications.

B.4.1   Approximating Functions Close to One Point

Suppose you know the value of a function f at some point x_0, and you are interested in the value of f close to x. More precisely, suppose we know $f(x_0)$ and we want an approximation of $f(x_0 + h)$ where h is a small number. If the function is smooth and h is really small, our first approximation reads

$$f(x_0 + h) \approx f(x_0). \tag{B.13}$$

That approximation is, of course, not very accurate. In order to derive a more accurate approximation, we have to know more about f at x_0. Suppose that we know the value of $f(x_0)$ and $f'(x_0)$, then we can find a better approximation of $f(x_0 + h)$ by recalling that

$$f'(x_0) \approx \frac{f(x_0 + h) - f(x_0)}{h}.$$

Hence, we have

$$f(x_0 + h) \approx f(x_0) + h f'(x_0). \tag{B.14}$$

B.4.2   Approximating the Exponential Function

Let us be a bit more specific and consider the case of

$$f(x) = e^x$$

around

$$x_0 = 0.$$

Since $f'(x) = e^x$, we have $f'(0) = 1$, and then it follows from (B.14) that

$$e^h \approx 1 + h.$$

The little program below (found in `Taylor1.py`) prints e^h and $1 + h$ for a range of h values.

```
from math import exp
for h in 1, 0.5, 1/20.0, 1/100.0, 1/1000.0:
    print 'h=%8.6f exp(h)=%11.5e  1+h=%g' % (h, exp(h), 1+h)
```

```
h=1.000000 exp(h)=2.71828e+00  1+h=2
h=0.500000 exp(h)=1.64872e+00  1+h=1.5
h=0.050000 exp(h)=1.05127e+00  1+h=1.05
h=0.010000 exp(h)=1.01005e+00  1+h=1.01
h=0.001000 exp(h)=1.00100e+00  1+h=1.001
```

As expected, $1 + h$ is a good approximation to e^h the smaller h is.

B.4.3 More Accurate Expansions

The approximations given by (B.13) and (B.14) are referred to as Taylor series. You can read much more about Taylor series in any Calculus book. More specifically, (B.13) and (B.14) are known as the zeroth- and first-order Taylor series, respectively. The second-order Taylor series is given by

$$f(x_0 + h) \approx f(x_0) + hf'(x_0) + \frac{h^2}{2}f''(x_0), \tag{B.15}$$

the third-order series is given by

$$f(x_0 + h) \approx f(x_0) + hf'(x_0) + \frac{h^2}{2}f''(x_0) + \frac{h^3}{6}f'''(x_0), \tag{B.16}$$

and the fourth-order series reads

$$f(x_0 + h) \approx f(x_0) + hf'(x_0) + \frac{h^2}{2}f''(x_0) + \frac{h^3}{6}f'''(x_0) + \frac{h^4}{24}f''''(x_0). \tag{B.17}$$

In general, the n-th order Taylor series is given by

$$f(x_0 + h) \approx \sum_{k=0}^{n} \frac{h^k}{k!} f^{(k)}(x_0), \tag{B.18}$$

where we recall that $f^{(k)}$ denotes the $k-th$ derivative of f, and

$$k! = 1 \cdot 2 \cdot 3 \cdot 4 \cdots (k-1) \cdot k$$

is the factorial. By again considering $f(x) = e^x$ and $x_0 = 0$, we have

$$f(x_0) = f'(x_0) = f''(x_0) = f'''(x_0) = f''''(x_0) = 1$$

which gives the following Taylor series:

$$e^h \approx 1 + h + \frac{1}{2}h^2 + \frac{1}{6}h^3 + \frac{1}{24}h^4.$$

The program below, called `Taylor2.py`, prints the error of these approximations for a given value of h (note that we can easily build up a Taylor series in a list by adding a new term to the last computed term in the list).

```
from math import exp
import sys

h = float(sys.argv[1])
Taylor_series = []
Taylor_series.append(1)
Taylor_series.append(Taylor_series[-1] + h)
Taylor_series.append(Taylor_series[-1] + (1/2.0)*h**2)
Taylor_series.append(Taylor_series[-1] + (1/6.0)*h**3)
Taylor_series.append(Taylor_series[-1] + (1/24.0)*h**4)

print 'h =', h
for order in range(len(Taylor_series)):
    print 'order=%d, error=%g' % \
            (order, exp(h) - Taylor_series[order])
```

By running the program with $h = 0.2$, we have the following output:

```
h = 0.2
order=0, error=0.221403
order=1, error=0.0214028
order=2, error=0.00140276
order=3, error=6.94248e-05
order=4, error=2.75816e-06
```

We see how much the approximation is improved by adding more terms. For $h = 3$ all these approximations are useless:

```
h = 3.0
order=0, error=19.0855
order=1, error=16.0855
order=2, error=11.5855
order=3, error=7.08554
order=4, error=3.71054
```

However, by adding more terms we can get accurate results for any h. The method from Sect. A.1.8 computes the Taylor series for e^x with n terms in general. Running the associated program exp_Taylor_series_diffeq.py for various values of h shows how much is gained by adding more terms to the Taylor series. For $h = 3$, $e^3 = 20.086$ and we have

$n + 1$	Taylor series
2	4
4	13
8	19.846
16	20.086

For $h = 50$, $e^{50} = 5.1847 \cdot 10^{21}$ and we have

$n+1$	Taylor series
2	51
4	$2.2134 \cdot 10^4$
8	$1.7960 \cdot 10^8$
16	$3.2964 \cdot 10^{13}$
32	$1.3928 \cdot 10^{19}$
64	$5.0196 \cdot 10^{21}$
128	$5.1847 \cdot 10^{21}$

Here, the evolution of the series as more terms are added is quite dramatic (and impressive!).

B.4.4   Accuracy of the Approximation

Recall that the Taylor series is given by

$$f(x_0 + h) \approx \sum_{k=0}^{n} \frac{h^k}{k!} f^{(k)}(x_0).$$ (B.19)

This can be rewritten as an equality by introducing an error term,

$$f(x_0 + h) = \sum_{k=0}^{n} \frac{h^k}{k!} f^{(k)}(x_0) + O(h^{n+1}).$$ (B.20)

Let's look a bit closer at this for $f(x) = e^x$. In the case of $n = 1$, we have

$$e^h = 1 + h + O(h^2).$$ (B.21)

This means that there is a constant c that does not depend on h such that

$$\left| e^h - (1 + h) \right| \leq ch^2,$$ (B.22)

so the error is reduced quadratically in h. This means that if we compute the fraction

$$q_h^1 = \frac{\left| e^h - (1 + h) \right|}{h^2},$$

we expect it to be bounded as h is reduced. The program `Taylor_err1.py` prints q_h^1 for $h = 1/10, 1/20, 1/100$ and $1/1000$.

```
from numpy import exp, abs

def q_h(h):
    return abs(exp(h) - (1+h))/h**2

print " h     q_h"
for h in 0.1, 0.05, 0.01, 0.001:
    print "%5.3f %f" %(h, q_h(h))
```

We can run the program and watch the output:

```
Terminal
```

```
Terminal> python src-discalc/Taylor_err1.py
  h      q_h
0.100 0.517092
0.050 0.508439
0.010 0.501671
0.001 0.500167
```

We observe that $q_h \approx 1/2$ and it is definitely bounded independent of h. The program `Taylor_err2.py` prints

$$q_h^0 = \frac{\left|e^h - 1\right|}{h},$$

$$q_h^1 = \frac{\left|e^h - (1 + h)\right|}{h^2},$$

$$q_h^2 = \frac{\left|e^h - \left(1 + h + \frac{h^2}{2}\right)\right|}{h^3},$$

$$q_h^3 = \frac{\left|e^h - \left(1 + h + \frac{h^2}{2} + \frac{h^3}{6}\right)\right|}{h^4},$$

$$q_h^4 = \frac{\left|e^h - \left(1 + h + \frac{h^2}{2} + \frac{h^3}{6} + \frac{h^4}{24}\right)\right|}{h^5},$$

for $h = 1/5, 1/10, 1/20$ and $1/100$.

```python
from numpy import exp, abs

def q_0(h):
    return abs(exp(h) - 1) / h
def q_1(h):
    return abs(exp(h) - (1 + h)) / h**2
def q_2(h):
    return abs(exp(h) - (1 + h + (1/2.0)*h**2)) / h**3
def q_3(h):
    return abs(exp(h) - (1 + h + (1/2.0)*h**2 + \
                         (1/6.0)*h**3)) / h**4
def q_4(h):
    return abs(exp(h) - (1 + h + (1/2.0)*h**2 + (1/6.0)*h**3 + \
                         (1/24.0)*h**4)) / h**5
hlist = [0.2, 0.1, 0.05, 0.01]
print "%-05s %-09s %-09s %-09s %-09s %-09s" \
      %("h", "q_0", "q_1", "q_2", "q_3", "q_4")
for h in hlist:
    print "%.02f  %04f  %04f  %04f  %04f  %04f" \
          %(h, q_0(h), q_1(h), q_2(h), q_3(h), q_4(h))
```

By using the program, we get the following table:

```
h      q_0        q_1        q_2        q_3        q_4
0.20   1.107014   0.535069   0.175345   0.043391   0.008619
0.10   1.051709   0.517092   0.170918   0.042514   0.008474
0.05   1.025422   0.508439   0.168771   0.042087   0.008403
0.01   1.005017   0.501671   0.167084   0.041750   0.008344
```

Again we observe that the error of the approximation behaves as indicated in (B.20).

B.4.5   Derivatives Revisited

We observed above that

$$f'(x) \approx \frac{f(x+h) - f(x)}{h}.$$

By using the Taylor series, we can obtain this approximation directly, and also get an indication of the error of the approximation. From (B.20) it follows that

$$f(x+h) = f(x) + hf'(x) + O(h^2),$$

and thus

$$f'(x) = \frac{f(x+h) - f(x)}{h} + O(h), \tag{B.23}$$

so the error is proportional to h. We can investigate if this is the case through some computer experiments. Take $f(x) = \ln(x)$, so that $f'(x) = 1/x$. The program diff_ln_err.py prints h and

$$\frac{1}{h}\left| f'(x) - \frac{f(x+h) - f(x)}{h} \right| \tag{B.24}$$

at $x = 10$ for a range of h values.

```python
def error(h):
    return (1.0/h)*abs(df(x) - (f(x+h)-f(x))/h)

from math import log as ln

def f(x):
    return ln(x)

def df(x):
    return 1.0/x

x = 10
hlist = []
for h in 0.2, 0.1, 0.05, 0.01, 0.001:
    print "%.4f    %4f" % (h, error(h))
```

From the output

```
0.2000    0.004934
0.1000    0.004967
0.0500    0.004983
0.0100    0.004997
0.0010    0.005000
```

we observe that the quantity in (B.24) is constant (≈ 0.5) independent of h, which indicates that the error is proportional to h.

B.4.6   More Accurate Difference Approximations

We can also use the Taylor series to derive more accurate approximations of the derivatives. From (B.20), we have

$$f(x + h) \approx f(x) + hf'(x) + \frac{h^2}{2}f''(x) + O(h^3). \tag{B.25}$$

By using $-h$ instead of h, we get

$$f(x - h) \approx f(x) - hf'(x) + \frac{h^2}{2}f''(x) + O(h^3). \tag{B.26}$$

By subtracting (B.26) from (B.25), we have

$$f(x + h) - f(x - h) = 2hf'(x) + O(h^3),$$

and consequently

$$f'(x) = \frac{f(x + h) - f(x - h)}{2h} + O(h^2). \tag{B.27}$$

Note that the error is now $O(h^2)$ whereas the error term of (B.23) is $O(h)$. In order to see if the error is actually reduced, let us compare the following two approximations

$$f'(x) \approx \frac{f(x + h) - f(x)}{h} \text{ and } f'(x) \approx \frac{f(x + h) - f(x - h)}{2h}$$

by applying them to the discrete version of $\sin(x)$ on the interval $(0, \pi)$. As usual, we let $n \geq 1$ be a given integer, and define the mesh

$$x_i = ih \quad \text{for } i = 0, 1, \ldots, n,$$

where $h = \pi/n$. At the nodes, we have the functional values

$$s_i = \sin(x_i) \quad \text{for } i = 0, 1, \ldots, n,$$

and at the inner nodes we define the first (F) and second (S) order approximations of the derivatives given by

$$d_i^F = \frac{s_{i+1} - s_i}{h},$$

and

$$d_i^S = \frac{s_{i+1} - s_{i-1}}{2h},$$

respectively for $i = 1, 2, \ldots, n - 1$. These values should be compared to the exact derivative given by

$$d_i = \cos(x_i) \quad \text{for } i = 1, 2, \ldots, n - 1.$$

The following program, found in diff_1st2nd_order.py, plots the discrete functions $(x_i, d_i)_{i=1}^{n-1}$, $(x_i, d_i^F)_{i=1}^{n-1}$, and $(x_i, d_i^S)_{i=1}^{n-1}$ for a given n. Note that the first three functions in this program are completely general in that they can be used for any $f(x)$ on any mesh. The special case of $f(x) = \sin(x)$ and comparing first- and second-order formulas is implemented in the example function. This latter function is called in the test block of the file. That is, the file is a module and we can reuse the first three functions in other programs (in particular, we can use the third function in the next example).

```
def first_order(f, x, h):
    return (f(x+h) - f(x))/h

def second_order(f, x, h):
    return (f(x+h) - f(x-h))/(2*h)

def derivative_on_mesh(formula, f, a, b, n):
    """
    Differentiate f(x) at all internal points in a mesh
    on [a,b] with n+1 equally spaced points.
    The differentiation formula is given by formula(f, x, h).
    """
    h = (b-a)/float(n)
    x = linspace(a, b, n+1)
    df = zeros(len(x))
    for i in xrange(1, len(x)-1):
        df[i] = formula(f, x[i], h)
    # Return x and values at internal points only
    return x[1:-1], df[1:-1]

def example(n):
    a = 0; b = pi;
    x, dF = derivative_on_mesh(first_order,  sin, a, b, n)
    x, dS = derivative_on_mesh(second_order, sin, a, b, n)
    # Accurate plot of the exact derivative at internal points
    h = (b-a)/float(n)
    xfine = linspace(a+h, b-h, 1001)
    exact = cos(xfine)
    plot(x, dF, 'r-', x, dS, 'b-', xfine, exact, 'y-',
```

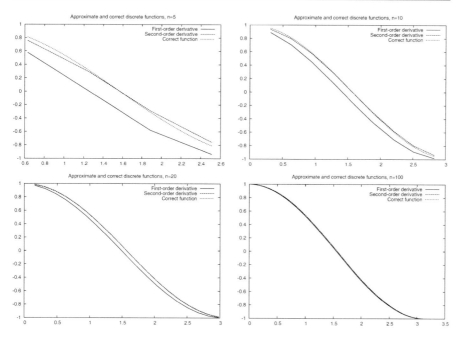

Fig. B.4 Plots of exact and approximate derivatives with various number of mesh points n

```
        legend=('First-order derivative',
                'Second-order derivative',
                'Correct function'),
        title='Approximate and correct discrete '\
              'functions, n=%d' % n)

# Main program
from scitools.std import *
n = int(sys.argv[1])
example(n)
```

The result of running the program with four different n values is presented in Fig. B.4. Observe that d_i^S is a better approximation to d_i than d_i^F, and note that both approximations become very good as n is getting large.

B.4.7 Second-Order Derivatives

We have seen that the Taylor series can be used to derive approximations of the derivative. But what about higher order derivatives? Next we shall look at second order derivatives. From (B.20) we have

$$f(x_0 + h) = f(x_0) + hf'(x_0) + \frac{h^2}{2} f''(x_0) + \frac{h^3}{6} f'''(x_0) + O(h^4),$$

and by using $-h$, we have

$$f(x_0 - h) = f(x_0) - hf'(x_0) + \frac{h^2}{2}f''(x_0) - \frac{h^3}{6}f'''(x_0) + O(h^4)$$

By adding these equations, we have

$$f(x_0 + h) + f(x_0 - h) = 2f(x_0) + h^2 f''(x_0) + O(h^4),$$

and thus

$$f''(x_0) = \frac{f(x_0 - h) - 2f(x_0) + f(x_0 + h)}{h^2} + O(h^2). \qquad \text{(B.28)}$$

For a discrete function $(x_i, y_i)_{i=0}^n$, $y_i = f(x_i)$, we can define the following approximation of the second derivative,

$$d_i = \frac{y_{i-1} - 2y_i + y_{i+1}}{h^2}. \qquad \text{(B.29)}$$

We can make a function, found in the file `diff2nd.py`, that evaluates (B.29) on a mesh. As an example, we apply the function to

$$f(x) = \sin(e^x),$$

where the exact second-order derivative is given by

$$f''(x) = e^x \cos(e^x) - (\sin(e^x)) e^{2x}.$$

```
from diff_1st2nd_order import derivative_on_mesh
from scitools.std import *

def diff2nd(f, x, h):
    return (f(x+h) - 2*f(x) + f(x-h))/(h**2)

def example(n):
    a = 0;   b = pi

    def f(x):
        return sin(exp(x))

    def exact_d2f(x):
        e_x = exp(x)
        return e_x*cos(e_x) - sin(e_x)*exp(2*x)

    x, d2f = derivative_on_mesh(diff2nd, f, a, b, n)
    h = (b-a)/float(n)
    xfine = linspace(a+h, b-h, 1001)  # fine mesh for comparison
```

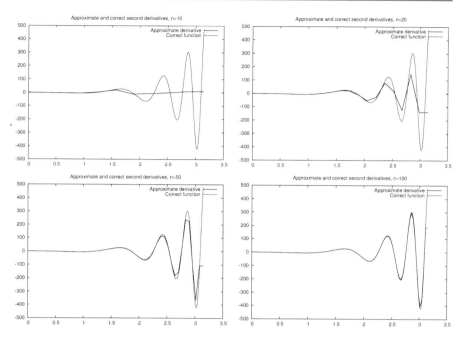

Fig. B.5 Plots of exact and approximate second-order derivatives with various mesh resolution n

```
exact = exact_d2f(xfine)
plot(x, d2f, 'r-', xfine, exact, 'b-',
    legend=('Approximate derivative',
            'Correct function'),
    title='Approximate and correct second order '\
          'derivatives, n=%d' % n,
    savefig='tmp.pdf')

try:
    n = int(sys.argv[1])
except:
    print "usage: %s n" % sys.argv[0]; sys.exit(1)

example(n)
```

In Fig. B.5 we compare the exact and the approximate derivatives for $n =$ 10, 20, 50, and 100. As usual, the error decreases when n becomes larger, but note here that the error is very large for small values of n.

B.5 Exercises

Exercise B.1: Interpolate a discrete function
In a Python function, represent the mathematical function

$$f(x) = \exp\left(-x^2\right)\cos(2\pi x)$$

on a mesh consisting of $q + 1$ equally spaced points on $[-1, 1]$, and return 1) the interpolated function value at $x = -0.45$ and 2) the error in the interpolated value. Call the function and write out the error for $q = 2, 4, 8, 16$.
Filename: `interpolate_exp_cos`.

Exercise B.2: Study a function for different parameter values
Develop a program that creates a plot of the function $f(x) = \sin(\frac{1}{x+\varepsilon})$ for x in the unit interval, where $\varepsilon > 0$ is a given input parameter. Use $n + 1$ nodes in the plot.

a) Test the program using $n = 10$ and $\varepsilon = 1/5$.
b) Refine the program such that it plots the function for two values of n; say n and $n + 10$.
c) How large do you have to choose n in order for the difference between these two functions to be less than 0.1?

Hint Each function gives an array. Create a `while` loop and use the `max` function of the arrays to retrieve the maximum value and compare these.

d) Let $\varepsilon = 1/10$ and recalculate.
e) Let $\varepsilon = 1/20$ and recalculate.
f) Try to find a formula for how large n needs to be for a given value of ε such that increasing n further does not change the plot so much that it is visible on the screen. Note that there is no exact answer to this question.

Filename: `plot_sin_eps`.

Exercise B.3: Study a function and its derivative
Consider the function

$$f(x) = \sin\left(\frac{1}{x + \varepsilon}\right)$$

for x ranging from 0 to 1, and the derivative

$$f'(x) = \frac{-\cos\left(\frac{1}{x+\varepsilon}\right)}{(x + \varepsilon)^2}.$$

Here, ε is a given input parameter.

a) Develop a program that creates a plot of the derivative of $f = f(x)$ based on a finite difference approximation using n computational nodes. The program should also graph the exact derivative given by $f' = f'(x)$ above.
b) Test the program using $n = 10$ and $\varepsilon = 1/5$.
c) How large do you have to choose n in order for the difference between these two functions to be less than 0.1?

Hint Each function gives an array. Create a `while` loop and use the `max` function of the arrays to retrieve the maximum value and compare these.

d) Let $\varepsilon = 1/10$ and recalculate.

e) Let $\varepsilon = 1/20$ and recalculate.

f) Try to determine experimentally how large n needs to be for a given value of ε such that increasing n further does not change the plot so much that you can view it on the screen. Note, again, that there is no exact solution to this problem.

Filename: `sin_deriv`.

Exercise B.4: Use the Trapezoidal method

The purpose of this exercise is to test the program `trapezoidal.py`.

a) Let

$$\bar{a} = \int_0^1 e^{4x}\,dx = \frac{1}{4}e^4 - \frac{1}{4}.$$

Compute the integral using the program `trapezoidal.py` and, for a given n, let $a(n)$ denote the result. Try to find, experimentally, how large you have to choose n in order for

$$|\bar{a} - a(n)| \le \varepsilon$$

where $\varepsilon = 1/100$.

b) Recalculate with $\varepsilon = 1/1000$.

c) Recalculate with $\varepsilon = 1/10{,}000$.

d) Try to figure out, in general, how large n has to be such that

$$|\bar{a} - a(n)| \le \varepsilon$$

for a given value of ε.

Filename: `trapezoidal_test_exp`.

Exercise B.5: Compute a sequence of integrals

a) Let

$$\bar{b}_k = \int_0^1 x^k\,dx = \frac{1}{k+1},$$

and let $b_k(n)$ denote the result of using the program `trapezoidal.py` to compute $\int_0^1 x^k\,dx$. For $k = 4, 6$ and 8, try to figure out, by doing numerical experiments, how large n needs to be in order for $b_k(n)$ to satisfy

$$\left|\bar{b}_k - b_k(n)\right| \le 0.0001.$$

Note that n will depend on k.

Hint Run the program for each k, look at the output, and calculate $\left|\bar{b}_k - b_k(n)\right|$ manually.

b) Try to generalize the result in the previous point to arbitrary $k \geq 2$.
c) Generate a plot of x^k on the unit interval for $k = 2, 4, 6, 8$, and 10, and try to figure out if the results obtained in (a) and (b) are reasonable taking into account that the program `trapezoidal.py` was developed using a piecewise linear approximation of the function.

Filename: `trapezoidal_test_power`.

Exercise B.6: Use the Trapezoidal method
The purpose of this exercise is to compute an approximation of the integral

$$I = \int_{-\infty}^{\infty} e^{-x^2} dx$$

using the Trapezoidal method.

a) Plot the function e^{-x^2} for x ranging from -10 to 10 and use the plot to argue that

$$\int_{-\infty}^{\infty} e^{-x^2} dx = 2 \int_{0}^{\infty} e^{-x^2} dx.$$

b) Let $T(n, L)$ be the approximation of the integral

$$2 \int_{0}^{L} e^{-x^2} dx$$

computed by the Trapezoidal method using n subintervals. Develop a program that computes the value of T for a given n and L.
c) Extend the program to write out values of $T(n, L)$ in a table with rows corresponding to $n = 100, 200, \ldots, 500$ and columns corresponding to $L = 2, 4, 6, 8, 10$.
d) Extend the program to also print a table of the errors in $T(n, L)$ for the same n and L values as in (c). The exact value of the integral is $\sqrt{\pi}$.

Filename: `integrate_exp`.

Remarks Numerical integration of integrals with finite limits requires a choice of n, while with infinite limits we also need to truncate the domain, i.e., choose L in the present example. The accuracy depends on both n and L.

Exercise B.7: Compute trigonometric integrals
The purpose of this exercise is to demonstrate a property of trigonometric functions that you will meet in later courses. In this exercise, you may compute the integrals using the program `trapezoidal.py` with $n = 100$.

a) Consider the integrals

$$I_{p,q} = 2 \int_0^1 \sin(p\pi x)\sin(q\pi x)dx$$

and fill in values of the integral $I_{p,q}$ in a table with rows corresponding to $q = 0, 1, \ldots, 4$ and columns corresponding to $p = 0, 1, \ldots, 4$.

b) Repeat the calculations for the integrals

$$I_{p,q} = 2 \int_0^1 \cos(p\pi x)\cos(q\pi x)dx.$$

c) Repeat the calculations for the integrals

$$I_{p,q} = 2 \int_0^1 \cos(p\pi x)\sin(q\pi x)dx.$$

Filename: `ortho_trig_funcs`.

Exercise B.8: Plot functions and their derivatives

a) Use the program `diff_func.py` to plot approximations of the derivative for the following functions defined on the interval ranging from $x = 1/1000$ to $x = 1$:

$$f(x) = \ln\left(x + \frac{1}{100}\right),$$
$$g(x) = \cos(e^{10x}),$$
$$h(x) = x^x.$$

b) Extend the program such that both the discrete approximation and the correct (analytical) derivative can be plotted. The analytical derivative should be evaluated in the same computational points as the numerical approximation. Test the program by comparing the discrete and analytical derivative of x^3.

c) Use the program to compare the analytical and discrete derivatives of the functions f, g, and h. How large do you have to choose n in each case in order for the plots to become indistinguishable on your screen. Note that the analytical derivatives are given by:

$$f'(x) = \frac{1}{x + \frac{1}{100}},$$
$$g'(x) = -10e^{10x}\sin\left(e^{10x}\right)$$
$$h'(x) = (\ln x)x^x + xx^{x-1}$$

Filename: `diff_functions`.

Exercise B.9: Use the Trapezoidal method

Develop an efficient program that creates a plot of the function

$$f(x) = \frac{1}{2} + \frac{1}{\sqrt{\pi}} \int_0^x e^{-t^2} dt$$

for $x \in [0, 10]$. The integral should be approximated using the Trapezoidal method and use as few function evaluations of e^{-t^2} as possible.

Filename: `plot_integral`.

Introduction to differential equations

C

This appendix is authored by Aslak Tveito

Differential equations have proven to be an immensely successful instrument for modeling phenomena in science and technology. It is hardly an exaggeration to say that differential equations are used to define mathematical models in virtually all parts of the natural sciences. In this chapter, we will take the first steps towards learning how to deal with differential equations on a computer. This is a core issue in Computational Science and reaches far beyond what we can cover in this text. However, the ideas you will see here are reused in lots of advanced applications, so this chapter will hopefully provide useful introduction to a topic that you will probably encounter many times later.

We will show you how to build programs for solving differential equations. More precisely, we will show how a differential equation can be formulated in a discrete manner suitable for analysis on a computer, and how to implement programs to compute the discrete solutions. The simplest differential equations can be solved analytically in the sense that you can write down an explicit formula for the solutions. However, differential equations arising in practical applications are usually rather complicated and thus have to be solved numerically on a computer. Therefore we focus on implementing numerical methods to solve the equations. Appendix E describes more advanced implementation techniques aimed at making an easy-to-use toolbox for solving differential equations. Exercises in the present appendix and Appendix E aim at solving a variety of differential equations arising in various disciplines of science.

As with all the other chapters, the source code can be found in `src`, in this case in the subdirectory `ode1`. The short form ODE (plural: ODEs) is commonly used as abbreviation for *ordinary differential equation*, which is the type of differential equation that we address in this appendix. Actually, differential equations are divided into two groups: ordinary differential equations and partial differential equations. Ordinary differential equations contain derivatives with respect to one variable (usually t in our examples), whereas partial differential equations contain derivatives with respect to more than one variable, typically with respect to space and time. A typical ordinary differential equation is

$$u'(t) = u(t),$$

and a typical partial differential equation is

$$\frac{\partial u}{\partial t} = \frac{\partial^2 u}{\partial x^2} + \frac{\partial^2 u}{\partial y^2},$$

The latter is known as the heat or diffusion equation.

C.1   The simplest case

Consider the problem of solving the following equation

$$u'(t) = t^3 . \tag{C.1}$$

The solution can be computed directly by integrating (C.1), which gives

$$u(t) = \frac{1}{4}t^4 + C,$$

where C is an arbitrary constant. To obtain a unique solution, we need an extra condition to determine C. Specifying $u(t_1)$ for some time point t_1 represents a possible extra condition. It is common to view (C.1) as an equation for the function $u(t)$ for $t \in [0, T]$, and the extra condition is usually that the start value $u(0)$ is known. This is called the *initial condition*. Say

$$u(0) = 1 . \tag{C.2}$$

In general, the solution of the differential equation (C.1) subject to the initial condition (C.2) is

$$u(t) = u(0) + \int_0^t u'(\tau)d\tau,$$

$$= 1 + \int_0^t \tau^3 d\tau$$

$$= 1 + \frac{1}{4}t^4 .$$

If you are confused by the use of t and τ, don't get too upset:

> *In mathematics you don't understand things. You just get used to them.* John von Neumann, mathematician, 1903–1957.

Let us go back and check the solution derived above. Does $u(t) = 1 + \frac{1}{4}t^4$ really satisfy the two requirements listed in (C.1) and (C.2)? Obviously, $u(0) = 1$, and $u'(t) = t^3$, so the solution is correct.

More generally, we consider the equation

$$u'(t) = f(t) \tag{C.3}$$

together with the initial condition

$$u(0) = u_0 .$$ (C.4)

Here we assume that $f(t)$ is a given function, and that u_0 is a given number. Then, by reasoning as above, we have

$$u(t) = u_0 + \int_0^T f(\tau)d\tau .$$ (C.5)

By using the methods introduced in Appendix B, we can find a discrete version of u by approximating the integral. Generally, an approximation of the integral

$$\int_0^T f(\tau)d\tau$$

can be computed using the discrete version of a continuous function $f(\tau)$ defined on an interval $[0, t]$. The discrete version of f is given by $(\tau_i, y_i)_{i=0}^n$ where

$$\tau_i = ih, \text{ and } y_i = f(\tau_i)$$

for $i = 0, 1, \ldots, n$. Here $n \geq 1$ is a given integer and $h = T/n$. The Trapezoidal rule can now be written as

$$\int_0^T f(\tau)d\tau \approx \frac{h}{2}\left[y_0 + 2\sum_{k=1}^{n-1} y_k + y_n \right].$$ (C.6)

By using this approximation, we find that an approximate solution of (C.3)–(C.4) is given by

$$u(t) \approx u_0 + \frac{h}{2}\left[y_0 + 2\sum_{k=1}^{n-1} y_k + y_n \right].$$

The program `integrate_ode.py` computes a numerical solution of (C.3)–(C.4), where the function f, the time t, the initial condition u_0, and the number of time-steps n are inputs to the program.

```
#!/usr/bin/env python

def integrate(T, n, u0):
    h = T/float(n)
    t = linspace(0, T, n+1)
    I = f(t[0])
    for k in iseq(1, n-1, 1):
        I += 2*f(t[k])
    I += f(t[-1])
    I *= (h/2)
```

```
    I += u0
    return I

from scitools.std import *

f_formula = sys.argv[1]
T  = eval(sys.argv[2])
u0 = eval(sys.argv[3])
n  = int(sys.argv[4])
f = StringFunction(f_formula, independent_variables='t')
print "Numerical solution of u'(t)=%s: %.4f" % \
      (f_formula, integrate(T, n, u0))
```

We apply the program for computing the solution of

$$u'(t) = te^{t^2},$$
$$u(0) = 0,$$

at time $T = 2$ using $n = 10, 20, 50$ and 100:

Terminal

```
Terminal> python src-ode1/integrate_ode.py 't*exp(t**2)' 2 0 10
Numerical solution of u'(t)=t*exp(t**2): 28.4066
```

Terminal

```
Terminal> python src-ode1/integrate_ode.py 't*exp(t**2)' 2 0 20
Numerical solution of u'(t)=t*exp(t**2): 27.2060
```

Terminal

```
Terminal> python src-ode1/integrate_ode.py 't*exp(t**2)' 2 0 50
Numerical solution of u'(t)=t*exp(t**2): 26.8644
```

Terminal

```
Terminal> python src-ode1/integrate_ode.py 't*exp(t**2)' 2 0 100
Numerical solution of u'(t)=t*exp(t**2): 26.8154
```

The exact solution is given by $\frac{1}{2}e^{2^2} - \frac{1}{2} \approx 26.799$, so we see that the approximate solution becomes better as n is increased, as expected.

C.2   Exponential Growth

The example above was really not much of a differential equation, because the solution was obtained by straightforward integration. Equations of the form

$$u'(t) = f(t) \tag{C.7}$$

arise in situations where we can explicitly specify the derivative of the unknown function u. Usually, the derivative is specified in terms of the solution itself. Consider, for instance, population growth under idealized conditions as modeled in Sect. A.1.4. We introduce the symbol v_i for the number of individuals at time τ_i (v_i corresponds to x_n in Sect. A.1.4). The basic model for the evolution of v_i is (A.9):

$$v_i = (1 + r)v_{i-1}, \quad i = 1, 2, \ldots, \text{ and } v_0 \text{ known}. \tag{C.8}$$

As mentioned in Sect. A.1.4, r depends on the time difference $\Delta\tau = \tau_i - \tau_{i-1}$: the larger $\Delta\tau$ is, the larger r is. It is therefore natural to introduce a growth rate α that is independent of $\Delta\tau$: $\alpha = r/\Delta\tau$. The number α is then fixed regardless of how long jumps in time we take in the difference equation for v_i. In fact, α equals the growth in percent, divided by 100, over a time interval of unit length.

The difference equation now reads

$$v_i = v_{i-1} + \alpha\Delta\tau\, v_{i-1}\,.$$

Rearranging this equation we get

$$\frac{v_i - v_{i-1}}{\Delta\tau} = \alpha v_{i-1}\,. \tag{C.9}$$

Assume now that we shrink the time step $\Delta\tau$ to a small value. The left-hand side of (C.9) is then an approximation to the time-derivative of a function $v(\tau)$ expressing the number of individuals in the population at time τ. In the limit $\Delta\tau \to 0$, the left-hand side becomes the derivative exactly, and the equation reads

$$v'(\tau) = \alpha v(\tau)\,. \tag{C.10}$$

As for the underlying difference equation, we need a start value $v(0) = v_0$. We have seen that reducing the time step in a difference equation to zero, we get a differential equation.

Many like to scale an equation like (C.10) such that all variables are without physical dimensions and their maximum absolute value is typically of the order of unity. In the present model, this means that we introduce new dimensionless variables

$$u = \frac{v}{v_0}, \quad t = \frac{\tau}{\alpha}$$

and derive an equation for $u(t)$. Inserting $v = v_0 u$ and $\tau = \alpha t$ in (C.10) gives the prototype equation for population growth:

$$u'(t) = u(t) \tag{C.11}$$

with the initial condition

$$u(0) = 1\,. \tag{C.12}$$

When we have computed the dimensionless $u(t)$, we can find the function $v(\tau)$ as

$$v(\tau) = v_0 u(\tau/\alpha)\,.$$

We shall consider practical applications of population growth equations later, but let's start by looking at the idealized case (C.11).

Analytical solution Our differential equation can be written in the form

$$\frac{du}{dt} = u,$$

which can be rewritten as

$$\frac{du}{u} = dt,$$

and then integration on both sides yields

$$\ln(u) = t + c,$$

where c is a constant that has to be determined by using the initial condition. Putting $t = 0$, we have

$$\ln(u(0)) = c,$$

hence

$$c = \ln(1) = 0,$$

and then

$$\ln(u) = t,$$

so we have the solution

$$u(t) = e^t . \tag{C.13}$$

Let us now check that this function really solves (C.7)–(C.11). Obviously, $u(0) = e^0 = 1$, so (C.11) is fine. Furthermore

$$u'(t) = e^t = u(t),$$

thus (C.7) also holds.

Numerical solution We have seen that we can find a formula for the solution of the equation of exponential growth. So the problem is solved, and it is trivial to write a program to graph the solution. We will, however, go one step further and develop a numerical solution strategy for this problem. We don't really need such a method for this problem since the solution is available in terms of a formula, but as mentioned earlier, it is good practice to develop methods for problems where we know the solution; then we are more confident when we are confronted with more challenging problems.

Suppose we want to compute a numerical approximation of the solution of

$$u'(t) = u(t) \tag{C.14}$$

equipped with the initial condition

$$u(0) = 1 . \tag{C.15}$$

We want to compute approximations from time $t = 0$ to time $t = 1$. Let $n \geq 1$ be a given integer, and define

$$\Delta t = 1/n . \tag{C.16}$$

Furthermore, let u_k denote an approximation of $u(t_k)$ where

$$t_k = k \Delta t \qquad (C.17)$$

for $k = 0, 1, \ldots, n$. The key step in developing a numerical method for this differential equation is to invoke the Taylor series as applied to the exact solution,

$$u(t_{k+1}) = u(t_k) + \Delta t u'(t_k) + O(\Delta t^2), \qquad (C.18)$$

which implies that

$$u'(t_k) \approx \frac{u(t_{k+1}) - u(t_k)}{\Delta t}. \qquad (C.19)$$

By using (C.14), we get

$$\frac{u(t_{k+1}) - u(t_k)}{\Delta t} \approx u(t_k). \qquad (C.20)$$

Recall now that $u(t_k)$ is the exact solution at time t_k, and that u_k is the approximate solution at the same point in time. We now want to determine u_k for all $k \geq 0$. Obviously, we start by defining

$$u_0 = u(0) = 1.$$

Since we want $u_k \approx u(t_k)$, we require that u_k satisfy the following equality

$$\frac{u_{k+1} - u_k}{\Delta t} = u_k \qquad (C.21)$$

motivated by (C.20). It follows that

$$u_{k+1} = (1 + \Delta t) u_k. \qquad (C.22)$$

Since u_0 is known, we can compute u_1, u_2 and so on by using the formula above. The formula is implemented in the program exp_growth.py. Actually, we do not need the method and we do not need the program. It follows from (C.22) that

$$u_k = (1 + \Delta t)^k u_0$$

for $k = 0, 1, \ldots, n$ which can be evaluated on a pocket calculator or even on your cellular phone. But again, we show examples where everything is as simple as possible (but not simpler!) in order to prepare your mind for more complex matters ahead.

```
#!/usr/bin/env python

def compute_u(u0, T, n):
    """Solve u'(t)=u(t), u(0)=u0 for t in [0,T] with n steps."""
    u = u0
    dt = T/float(n)
```

```
    for k in range(0, n, 1):
        u = (1+dt)*u
    return u   # u(T)

import sys

n = int(sys.argv[1])
# Special test case: u'(t)=u, u(0)=1, t in [0,1]
T = 1; u0 = 1
print 'u(1) =', compute_u(u0, T, n)
```

Observe that we do not store the u values: we just overwrite a `float` object u by its new value as this saves a lot of storage if n is large.

Running the program for $n = 5, 10, 20$ and 100, we get the approximations 2.4883, 2.5937, 2.6533, and 2.7048. The exact solution at time $t = 1$ is given by $u(1) = e^1 \approx 2.7183$, so again the approximations become better as n is increased.

An alternative program, where we plot $u(t)$ and therefore store all the u_k and $t_k = k\,\Delta t$ values, is shown below.

```
#!/usr/bin/env python

def compute_u(u0, T, n):
    """Solve u'(t)=u(t), u(0)=u0 for t in [0,T] with n steps."""
    t = linspace(0, T, n+1)
    t[0] = 0
    u = zeros(n+1)
    u[0] = u0
    dt = T/float(n)
    for k in range(0, n, 1):
        u[k+1] = (1+dt)*u[k]
        t[k+1] = t[k] + dt
    return u, t

from scitools.std import *

n = int(sys.argv[1])
# Special test case: u'(t)=u, u(0)=1, t in [0,1]
T = 1; u0 = 1
u, t = compute_u(u0, T, n)
plot(t, u)
tfine = linspace(0, T, 1001) # for accurate plot
v = exp(tfine)               # correct solution
hold('on')
plot(tfine, v)
legend(['Approximate solution', 'Correct function'])
title('Approximate and correct discrete functions, n=%d' % n)
savefig('tmp.pdf')
```

Using the program for $n = 5, 10, 20$, and 100, results in the plots in Fig. C.1. The convergence towards the exponential function is evident from these plots.

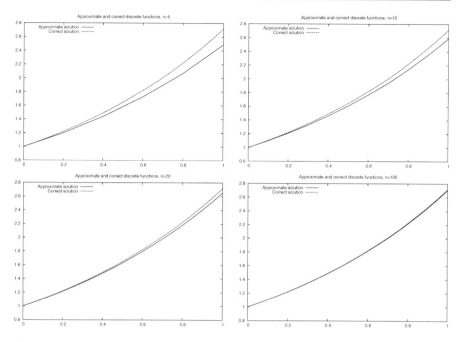

Fig. C.1 Plots of exact and approximate solutions of $u'(t) = u(t)$ with varying number of time steps in $[0, 1]$.

C.3 Logistic Growth

Exponential growth can be modelled by the following equation

$$u'(t) = \alpha u(t)$$

where $a > 0$ is a given constant. If the initial condition is given by

$$u(0) = u_0$$

the solution is given by

$$u(t) = u_0 e^{\alpha t}.$$

Since $a > 0$, the solution becomes very large as t increases. For a short time, such growth of a population may be realistic, but over a longer time, the growth of a population is restricted due to limitations of the environment, as discussed in Sect. A.1.5. Introducing a logistic growth term as in (A.12) we get the differential equation

$$u'(t) = \alpha u(t) \left(1 - \frac{u(t)}{R} \right), \tag{C.23}$$

where α is the growth-rate, and R is the carrying capacity (which corresponds to M in Sect. A.1.5). Note that R is typically very large, so if $u(0)$ is small, we have

$$\frac{u(t)}{R} \approx 0$$

for small values of t, and thus we have exponential growth for small t:

$$u'(t) \approx a u(t).$$

But as t increases, and u grows, the term $u(t)/R$ will become important and limit the growth.

A numerical scheme for the logistic equation (C.23) is given by

$$\frac{u_{k+1} - u_k}{\Delta t} = \alpha u_k \left(1 - \frac{u_k}{R} \right),$$

which we can solve with respect to the unknown u_{k+1}:

$$u_{k+1} = u_k + \Delta t \alpha u_k \left(1 - \frac{u_k}{R} \right).$$

This is the form of the equation that is suited for implementation.

C.4   A Simple Pendulum

So far we have considered scalar ordinary differential equations, i.e., equations with one single function $u(t)$ as unknown. Now we shall deal with systems of ordinary differential equations, where in general n unknown functions are coupled in a system of n equations. Our introductory example will be a system of two equations having two unknown functions $u(t)$ and $v(t)$. The example concerns the motion of a pendulum, see Fig. C.2. A sphere with mass m is attached to a massless rod of length L and oscillates back and forth due to gravity. Newton's second law of motion applied to this physical system gives rise the differential equation

$$\theta''(t) + \alpha \sin(\theta) = 0, \tag{C.24}$$

where $\theta = \theta(t)$ is the angle the rod makes with the vertical, measured in radians, and $\alpha = g/L$ (g is the acceleration of gravity). The unknown function to solve for is θ, and knowing θ, we can quite easily compute the position of the sphere, its velocity, and its acceleration, as well as the tension force in the rod. Since the highest derivative in (C.24) is of second order, we refer to (C.24) as a *second-order differential equations*. Our previous examples in this chapter involved only first-order derivatives, and therefore they are known as *first-order differential equations*.

Equation (C.24) can be solved by the same numerical method as we use in Sect. D.1.2, because (C.24) is very similar to Equation (D.8), which is the topic of Appendix D. The only difference is that (D.8) has extra terms, which can be skipped, while the kS term in (D.8) must be extended to $\alpha \sin(S)$ to make (D.8) identical to (C.24). This extension is easily performed. However, here we shall not solve the second-order equation (C.24) as it stands. We shall instead rewrite it as a system of two first-order equations so that we can use numerical methods for first-order equations to solve it.

To transform a second-order equation to a system of two first-order equations, we introduce a new variable for the first-order derivative (the angular velocity of the sphere): $v(t) = \theta'(t)$. Using v and θ in (C.24) yields

$$v'(t) + \alpha \sin(\theta) = 0.$$

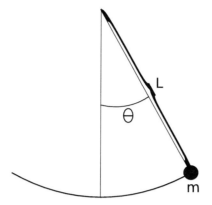

Fig. C.2 A pendulum with m = mass, L = length of massless rod and $\theta = \theta(t)$ = angle.

In addition, we have the relation

$$v = \theta'(t)$$

between v and θ. This means that (C.24) is equivalent to the following system of two coupled first-order differential equations:

$$\theta'(t) = v(t), \tag{C.25}$$
$$v'(t) = -\alpha \sin(\theta). \tag{C.26}$$

As for scalar differential equations, we need initial conditions, now two conditions because we have two unknown functions:

$$\theta(0) = \theta_0,$$
$$v(0) = v_0,$$

Here we assume the initial angle θ_0 and the initial angular velocity v_0 to be given.

It is common to group the unknowns and the initial conditions in 2-vectors: $(\theta(t), v(t))$ and (θ_0, v_0). One may then view (C.25)–(C.26) as a *vector equation*, whose first component equation is (C.25), and the second component equation is (C.26). In Python software, this vector notation makes solution methods for scalar equations (almost) immediately available for vector equations, i.e., systems of ordinary differential equations.

In order to derive a numerical method for the system (C.25)–(C.26), we proceed as we did above for one equation with one unknown function. Say we want to compute the solution from $t = 0$ to $t = T$ where $T > 0$ is given. Let $n \geq 1$ be a given integer and define the time step

$$\Delta t = T/n.$$

Furthermore, we let (θ_k, v_k) denote approximations of the exact solution $(\theta(t_k), v(t_k))$ for $k = 0, 1, \ldots, n$. A Forward Euler type of method will now read

$$\frac{\theta_{k+1} - \theta_k}{\Delta t} = v_k, \tag{C.27}$$

$$\frac{v_{k+1} - v_k}{\Delta t} = -\alpha \sin(\theta_k). \tag{C.28}$$

This scheme can be rewritten in a form more suitable for implementation:

$$\theta_{k+1} = \theta_k + \Delta t \, v_k, \tag{C.29}$$

$$v_{k+1} = v_k - \alpha \Delta t \sin(\theta_k). \tag{C.30}$$

The next program, pendulum.py, implements this method in the function pendulum. The input parameters to the model, θ_0, v_0,, the final time T, and the number of time-steps n, must be given on the command line.

```python
#!/usr/bin/env python

def pendulum(T, n, theta0, v0, alpha):
    """Return the motion (theta, v, t) of a pendulum."""
    dt = T/float(n)
    t = linspace(0, T, n+1)
    v = zeros(n+1)
    theta = zeros(n+1)
    v[0] = v0
    theta[0] = theta0
    for k in range(n):
        theta[k+1] = theta[k] + dt*v[k]
        v[k+1] = v[k] - alpha*dt*sin(theta[k+1])
    return theta, v, t

from scitools.std import *

n = int(sys.argv[1])
T = eval(sys.argv[2])
v0 = eval(sys.argv[3])
theta0 = eval(sys.argv[4])
alpha = eval(sys.argv[5])

theta, v, t = pendulum(T, n, theta0, v0)
plot(t, v, xlabel='t', ylabel='velocity')
figure()
plot(t, theta, xlabel='t', ylabel='velocity')
```

By running the program with the input data $\theta_0 = \pi/6$, $v_0 = 0$, $\alpha = 5$, $T = 10$ and $n = 1000$, we get the results shown in Fig. C.3. The angle $\theta = \theta(t)$ is displayed in the left panel and the velocity is given in the right panel.

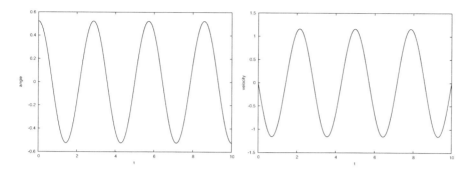

Fig. C.3 Plot of the motion of a pendulum. Left: the angle $\theta(t)$. Right: angular velocity θ'.

C.5   A Model for the Spreading of a Disease

Mathematical models are used intensively to analyze the spread of infectious diseases. In the simplest case, we may consider a population, that is supposed to be constant, consisting of two groups; the susceptibles (S) who can catch the disease, and the infectives (I) who have the disease and are able to transmit it. A system of differential equations modelling the evolution of S and I is given by

$$S' = -rSI, \tag{C.31}$$
$$I' = rSI - aI . \tag{C.32}$$

Here r and a are given constants reflecting the characteristics of the epidemic. The initial conditions are given by

$$S(0) = S_0,$$
$$I(0) = I_0,$$

where the initial state (S_0, I_0) is assumed to be known.

Suppose we want to compute numerical solutions of this system from time $t = 0$ to $t = T$. Then, by reasoning as above, we introduce the time step

$$\Delta t = T/n$$

and the approximations (S_k, I_k) of the solution $(S(t_k), I(t_k))$. An explicit Forward Euler method for the system takes the following form,

$$\frac{S_{k+1} - S_k}{\Delta t} = -rS_k I_k,$$
$$\frac{I_{k+1} - I_k}{\Delta t} = rS_k I_k - aI_k,$$

which can be rewritten on computational form

$$S_{k+1} = S_k - \Delta t \, r S_k I_k,$$
$$I_{k+1} = I_k + \Delta t \, (r S_k I_k - a I_k) \; .$$

This scheme is implemented in the program `exp_epidemic.py` where r, a, S_0, I_0, n and T are input data given on the command line. The function `epidemic` computes the solution (S, I) to the differential equation system. This pair of time-dependent functions is then plotted in two separate plots.

```
#!/usr/bin/env python

def epidemic(T, n, S0, I0, r, a):
    dt = T/float(n)
    t = linspace(0, T, n+1)
    S = zeros(n+1)
    I = zeros(n+1)
    S[0] = S0
    I[0] = I0
    for k in range(n):
        S[k+1] = S[k] - dt*r*S[k]*I[k]
        I[k+1] = I[k] + dt*(r*S[k]*I[k] - a*I[k])
    return S, I, t

from scitools.std import *

n = int(sys.argv[1])
T = eval(sys.argv[2])
S0 = eval(sys.argv[3])
I0 = eval(sys.argv[4])
r = eval(sys.argv[5])
a = eval(sys.argv[6])

S, I, t = epidemic(T, n, S0, I0, r, a)
plot(t, S, xlabel='t', ylabel='Susceptibles')
plot(t, I, xlabel='t', ylabel='Infectives')
```

We want to apply the program to a specific case where an influenza epidemic hit a British boarding school with a total of 763 boys. (The data are taken from Murray [18], and Murray found the data in the British Medical Journal, March 4, 1978.) The epidemic lasted from 21st January to 4th February in 1978. We let $t = 0$ denote 21st of January and we define $T = 14$ days. We put $S_0 = 762$ and $I_0 = 1$ which means that one person was ill at $t = 0$. In the Fig. C.4 we see the numerical results using $r = 2.18 \times 10^{-3}, a = 0.44, n = 1000$. Also, we have plotted actual the measurements, and we note that the simulations fit the real data quite well.

Reader interested in mathematical models for the spreading of infectious diseases may consult the excellent book [18] on Mathematical Biology by J.D. Murray.

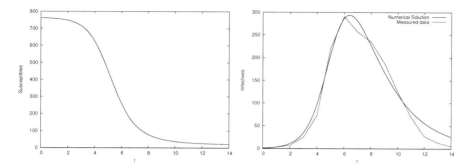

Fig. C.4 Graphs of susceptibles (left) and infectives (right) for an influenza in a British boarding school in 1978.

C.6   Exercises

Exercise C.1: Solve a nonhomogeneous linear ODE
Solve the ODE problem

$$u' = 2u - 1, \quad u(0) = 2, \quad t \in [0, 6]$$

using the Forward Euler method. Choose $\Delta t = 0.25$. Plot the numerical solution together with the exact solution $u(t) = \frac{1}{2} + \frac{3}{2}e^{2t}$.
Filename: `nonhomogeneous_linear_ODE`.

Exercise C.2: Solve a nonlinear ODE
Solve the ODE problem

$$u' = u^q, \quad u(0) = 1, \quad t \in [0, T]$$

using the Forward Euler method. The exact solution reads $u(t) = e^t$ for $q = 1$ and $u(t) = (t(1 - q) + 1)^{1/(1-q)}$ for $q > 1$ and $t(1 - q) + 1 > 0$. Read q, Δt, and T from the command line, solve the ODE, and plot the numerical and exact solution. Run the program for different cases: $q = 2$ and $q = 3$, with $\Delta t = 0.01$ and $\Delta t = 0.1$. Set $T = 6$ if $q = 1$ and $T = 1/(q - 1) - 0.1$ otherwise.
Filename: `nonlinear_ODE`.

Exercise C.3: Solve an ODE for $y(x)$
We have given the following ODE problem:

$$\frac{dy}{dx} = \frac{1}{2(y - 1)}, \quad y(0) = 1 + \sqrt{\epsilon}, \quad x \in [0, 4], \tag{C.33}$$

where $\epsilon > 0$ is a small number. Formulate a Forward Euler method for this ODE problem and compute the solution for varying step size in x: $\Delta x = 1$, $\Delta x = 0.25$, $\Delta x = 0.01$. Plot the numerical solutions together with the exact solution $y(x) = 1 + \sqrt{x + \epsilon}$, using 1001 x coordinates for accurate resolution of the latter. Set ϵ to 10^{-3}. Study the numerical solution with $\Delta x = 1$, and use that insight to explain why this problem is hard to solve numerically.
Filename: `yx_ODE`.

Exercise C.4: Experience instability of an ODE

Consider the ODE problem

$$u' = \alpha u, \quad u(0) = u_0,$$

solved by the Forward Euler method. Show by repeatedly applying the scheme that

$$u_k = (1 + \alpha \Delta t)^k u_0.$$

We now turn to the case $\alpha < 0$. Show that the numerical solution will oscillate if $\Delta t > -1/\alpha$. Make a program for computing u_k, set $\alpha = -1$, and demonstrate oscillatory solutions for $\Delta t = 1.1, 1.5, 1.9$. Recall that the exact solution, $u(t) = e^{\alpha t}$, never oscillates.

What happens if $\Delta t > -2/\alpha$? Try it out in the program and explain why we do not experience that $u_k \to 0$ as $k \to \infty$.

Filename: `unstable_ODE`.

Exercise C.5: Solve an ODE with time-varying growth

Consider the ODE for exponential growth,

$$u' = \alpha u, \quad u(0) = 1, \quad t \in [0, T].$$

Now we introduce a time-dependent α such that the growth decreases with time: $\alpha(t) = a - bt$. Solve the problem for $a = 1$, $b = 0.1$, and $T = 10$. Plot the solution and compare with the corresponding exponential growth using the mean value of $\alpha(t)$ as growth factor: $e^{(a - bT/2)t}$.

Filename: `time_dep_growth`.

A Complete Differential Equation Project

The examples in the ordinary chapters of this book are quite compact and composed to convey programming constructs in a gentle pedagogical way. In this appendix the idea is to solve a more comprehensive real-world problem by programming. The problem solving process gets quite advanced because we bring together elements from physics, mathematics, and programming, in a way that a scientific programmer must master. Each individual element is quite straightforward in the sense that you have probably met the element already, either in high school physics or mathematics, or in this book. The challenge is to understand the problem, and analyze it by breaking it into a set of simpler elements. It is not necessary to understand this problem solving process in detail. As a computer programmer, all you need to understand is how you translate the given algorithm into a working program and how to test the program. We anticipate that this task should be doable without a thorough understanding of the physics and mathematics of the problem.

Sections D.1 and D.2 require basic knowledge of loops, lists, functions, and command-line parsing via the `argparse` module, while Sect. D.3 also requires knowledge about curve plotting.

All Python files associated with this appendix are found in `src/boxspring`[1].

D.1 About the Problem: Motion and Forces in Physics

D.1.1 The Physical Problem

We shall study a simple device for modeling oscillating systems. A box with mass m and height b is attached to a spring of length L as shown in Fig. D.1. The end of the spring is attached to a plate, which we can move up and down with a displacement $w(t)$, where t denotes time. There are two ways the box can be set in motion: we can either stretch or compress the string initially by moving the box up or down, or we can move the plate. If $w = 0$ the box oscillates freely, otherwise we have what is called driven oscillations.

Why will such a system oscillate? When the box moves downward, the spring is stretched, which results in a force that tries to move the box upward. The more

---

[1] http://tinyurl.com/pwyasaa/boxspring

Fig. D.1 An oscillating system with a box attached to a spring

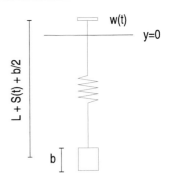

we stretch the spring, the bigger the force against the movement becomes. The box eventually stops and starts moving upward with an upward acceleration. At some point the spring is not stretched anymore and there is no spring force on the box, but because of inertia, the box continues its motion upward. This causes the spring to get compressed, causing a force from the spring on the box that acts downward, against the upward movement. The downward force increases in intensity and manages to stop the upward motion. The process repeats itself and results in an oscillatory motion of the box. Since the spring tries to restore the position of the box, we refer to the spring force as a *restoring force*.

You have probably experienced that oscillations in such springs tend to die out with time. There is always a *damping force* that works against the motion. This damping force may be due to a not perfectly elastic string, and the force can be quite small, but we can also explicitly attach the spring to a damping mechanism to obtain a stronger, controllable damping of the oscillations (as one wants in a car or a mountain bike). We will assume that there is some damping force present in our system, and this can well be a damping mechanism although this is not explicitly included in Fig. D.1.

Oscillating systems of the type depicted in Fig. D.1 have a huge number of applications throughout science and technology. One simple example is the spring system in a car or bicycle, which you have probably experienced on a bumpy road (the bumps lead to a $w(t)$ function). When your washing machine jumps up and down, it acts as a highly damped oscillating system (and the $w(t)$ function is related to uneven distribution of the mass of the clothes). The pendulum in a wall clock is another oscillating system, not with a spring, but physically the system can (for small oscillations) be modeled as a box attached to a spring because gravity makes a spring-like force on a pendulum (in this case, $w(t) = 0$). Other examples on oscillating systems where this type of equation arise are briefly mentioned in Exercise E.50. The bottom line is that understanding the dynamics of Fig. D.1 is the starting point for understanding the behavior of a wide range of oscillating phenomena in nature and technical devices.

Goal of the computations Our aim is to compute the position of the box as a function of time. If we know the position, we can compute the velocity, the acceleration, the spring force, and the damping force. The mathematically difficult thing is to calculate the position – everything else is much easier. More precisely, to compute the position we must solve a differential equation while the other quantities can be computed by differentiation and simple arithmetics. Solving differential equations is historically considered very difficult, but computers have simplified this task dramatically.

We assume that the box moves in the vertical direction only, so we introduce $Y(t)$ as the vertical position of the center point of the box. We shall derive a mathematical equation that has $Y(t)$ as solution. This equation can be solved by an algorithm, which can be implemented in a program. Our focus is on the implementation, since this is a book about programming, but for the reader interested in how computers play together with physics and mathematics in science and technology, we also outline how the equation and algorithm arise.

The key quantities   Let S be the stretch of the spring, where $S > 0$ means stretch and $S < 0$ implies compression. The length of the spring when it is unstretched is L, so at a given point of time t the actual length is $L + S(t)$. Given the position of the plate, $w(t)$, the length of the spring, $L + S(t)$, and the height of the box, b, the position $Y(t)$ is then, according to Fig. D.1,

$$Y(t) = w(t) - (L + S(t)) - \frac{b}{2}. \tag{D.1}$$

You can think as follows: we first "go up" to the plate at $y = w(t)$, then down $L + S(t)$ along the spring and then down $b/2$ to the center of the box. While L, w, and b must be known as input data, $S(t)$ is unknown and will be output data from the program.

D.1.2   The Computational Algorithm

Let us now go straight to the programming task and present the recipe for computing $Y(t)$. The algorithm below actually computes $S(t)$, but at any point of time we can easily find $Y(t)$ from (D.1) if we know $S(t)$. The $S(t)$ function is computed at discrete points of time, $t = t_i = i\Delta t$, for $i = 0, 1, \ldots, N$. We introduce the notation S_i for $S(t_i)$. The S_i values can be computed by the following algorithm.

Set the initial stretch S_0 from input data and compute S_1 by

$$S_{i+1} = \frac{1}{2m}\left(2m S_i - \Delta t^2 k S_i + m\left(w_{i+1} - 2w_i + w_{i-1}\right) + \Delta t^2 mg\right), \tag{D.2}$$

with $i = 0$. Then for $i = 1, 2, \ldots, N - 1$, compute S_{i+1} by

$$S_{i+1} = (m + \gamma)^{-1}\left(2m S_i - m S_{i-1} + \gamma\Delta t\ S_{i-1} - \Delta t^2 k S_i + \right.$$
$$\left. m(w_{i+1} - 2w_i + w_{i-1}) + \Delta t^2 mg\right). \tag{D.3}$$

The parameter γ equals $\frac{1}{2}\beta\Delta t$. The input data to the algorithm are the mass of the box m, a coefficient k characterizing the spring, a coefficient β characterizing the amount of damping in the system, the acceleration of gravity g, the movement of the plate $w(t)$, the initial stretch of the spring S_0, the number of time steps N, and the time Δt between each computation of S values. The smaller we choose Δt, the more accurate the computations become.

Now you have two options, either read the derivation of this algorithm in Sects. D.1.3 and D.1.4 or jump right to implementation in Sect. D.2.

D.1.3   Derivation of the Mathematical Model

To derive the algorithm we need to make a mathematical model of the oscillating system. This model is based on physical laws. The most important physical law for a moving body is Newton's second law of motion:

$$F = ma, \tag{D.4}$$

where F is the sum of all forces on the body, m is the mass of the body, and a is the acceleration of the body. The body here is our box.

Let us first find all the forces on the box. Gravity acts downward with magnitude mg. We introduce $F_g = -mg$ as the gravity force, with a minus sign because a negative force acts downward, in negative y direction.

The spring force on the box acts upward if the spring is stretched, i.e., if $S > 0$ we have a positive spring force F_s. The size of the force is proportional to the amount of stretching, so we write $F_s = kS$, where k is commonly known as the *spring constant*. (Spring forces are often written in the canonical form $F = -kx$, where x is the stretch. The reason that we have no minus sign is that our stretch S is positive in the downward, negative direction.) We also assume that we have a damping force that is always directed toward the motion and proportional with the *velocity of the stretch*, $-dS/dt$. Naming the proportionality constant β, we can write the damping force as $F_d = \beta dS/dt$. Note that when $dS/dt > 0$, S increases in time and the box moves downward, the F_d force then acts upward, against the motion, and must be positive. This is the way we can check that the damping force expression has the right sign.

The sum of all forces is now

$$F = F_g + F_s + F_d,$$
$$= -mg + kS + \beta \frac{dS}{dt}. \tag{D.5}$$

We now know the left-hand side of (D.4), but S is unknown to us. The acceleration a on the right-hand side of (D.4) is also unknown. However, acceleration is related to movement and the S quantity, and through this relation we can eliminate a as a second unknown. From physics, it is known that the acceleration of a body is the second derivative in time of the position of the body, so in our case,

$$a = \frac{d^2Y}{dt^2},$$
$$= \frac{d^2w}{dt^2} - \frac{d^2S}{dt^2}, \tag{D.6}$$

(remember that L and b are constant).

Equation (D.4) now reads

$$-mg + kS + \beta\frac{dS}{dt} = m\left(\frac{d^2w}{dt^2} - \frac{d^2S}{dt^2}\right). \tag{D.7}$$

It is common to collect the unknown terms on the left-hand side and the known quantities on the right-hand side, and let higher-order derivatives appear before lower-order derivatives. With such a reordering of terms we get

$$m\frac{d^2S}{dt^2} + \beta\frac{dS}{dt} + kS = m\frac{d^2w}{dt^2} + mg. \tag{D.8}$$

This is the equation governing our physical system. If we solve the equation for $S(t)$, we have the position of the box according to (D.1), the velocity v as

$$v(t) = \frac{dY}{dt} = \frac{dw}{dt} - \frac{dS}{dt}, \tag{D.9}$$

the acceleration as (D.6), and the various forces can be easily obtained from the formulas in (D.5).

A key question is if we can solve (D.8). If $w = 0$, there is in fact a well-known solution, which can be written

$$S(t) = \frac{m}{k}g + \begin{cases} e^{-\zeta t}\left(c_1 e^{t\sqrt{\beta^2-1}} + c_2 e^{-t\sqrt{\zeta^2-1}}\right), & \zeta > 1, \\ e^{-\zeta t}(c_1 + c_2 t), & \zeta = 1, \\ e^{-\zeta t}\left[c_1\cos\left(\sqrt{1-\zeta^2}t\right) + c_2\sin\left(\sqrt{1-\zeta^2}t\right)\right], & \zeta < 1. \end{cases}$$
$$\tag{D.10}$$

Here, ζ is a short form for $\beta/2$, and c_1 and c_2 are arbitrary constants. That is, the solution (D.10) is not unique.

To make the solution unique, we must determine c_1 and c_2. This is done by specifying the state of the system at some point of time, say $t = 0$. In the present type of mathematical problem we must specify S and dS/dt. We allow the spring to be stretched an amount S_0 at $t = 0$. Moreover, we assume that there is no ongoing increase or decrease in the stretch at $t = 0$, which means that $dS/dt = 0$. In view of (D.9), this condition implies that the velocity of the box is that of the plate, and if the latter is at rest, the box is also at rest initially. The conditions at $t = 0$ are called *initial conditions*:

$$S(0) = S_0, \quad \frac{dS}{dt}(0) = 0. \tag{D.11}$$

These two conditions provide two equations for the two unknown constants c_1 and c_2. Without the initial conditions two things happen: (i) there are infinitely many solutions to the problem, and (ii) the computational algorithm in a program cannot start.

Also when $w \neq 0$ one can find solutions $S(t)$ of (D.8) in terms of mathematical expressions, but only for some very specific choices of $w(t)$ functions. With a program we can compute the solution $S(t)$ for any "reasonable" $w(t)$ by a quite simple method. The method gives only an approximate solution, but the approximation can usually be made as good as desired. This powerful solution method is described below.

D.1.4   Derivation of the Algorithm

To solve (D.8) on a computer, we do two things:

- We calculate the solution at some discrete time points $t = t_i = i\Delta t$, $i = 0, 1, 2, \ldots, N$.
- We replace the derivatives by finite differences, which are approximate expressions for the derivatives.

The first and second derivatives can be approximated by the following finite differences

$$\frac{dS}{dt}(t_i) \approx \frac{S(t_{i+1}) - S(t_{i-1})}{2\Delta t}, \tag{D.12}$$

$$\frac{d^2 S}{dt^2}(t_i) \approx \frac{S(t_{i+1}) - 2S(t_i) + S(t_{i-1})}{\Delta t^2}. \tag{D.13}$$

Derivations of such formulas can be found in Appendices B and C. It is common to save some writing by introducing S_i as a short form for $S(t_i)$. The formulas then read

$$\frac{dS}{dt}(t_i) \approx \frac{S_{i+1} - S_{i-1}}{2\Delta t}, \tag{D.14}$$

$$\frac{d^2 S}{dt^2}(t_i) \approx \frac{S_{i+1} - 2S_i + S_{i-1}}{\Delta t^2}. \tag{D.15}$$

Let (D.8) be valid at a point of time t_i:

$$m\frac{d^2 S}{dt^2}(t_i) + \beta\frac{dS}{dt}(t_i) + kS(t_i) = m\frac{d^2 w}{dt^2}(t_i) + mg. \tag{D.16}$$

We now insert (D.14) and (D.15) in (D.16) (observe that we can approximate $d^2 w/dt^2$ in the same way as we approximate $d^2 S/dt^2$):

$$m\frac{S_{i+1} - 2S_i + S_{i-1}}{\Delta t^2} + \beta\frac{S_{i+1} - S_{i-1}}{2\Delta t} + kS_i = m\frac{w_{i+1} - 2w_i + w_{i-1}}{\Delta t^2} + mg. \tag{D.17}$$

The computational algorithm starts with knowing S_0, then S_1 is computed, then S_2, and so on. Therefore, in (D.17) we can assume that S_i and S_{i-1} are already computed, and that S_{i+1} is the new unknown to calculate. Let us as usual put the unknown terms on the left-hand side (and multiply by Δt^2):

$$mS_{i+1} + \gamma S_{i+1} = 2mS_i - mS_{i-1} + \gamma S_{i-1} - \Delta t^2 kS_i +$$
$$m(w_{i+1} - 2w_i + w_{i-1}) + \Delta t^2 mg, \tag{D.18}$$

where we have introduced the short form $\gamma = \frac{1}{2}\beta\Delta t$ to save space. Equation (D.18) can easily be solved for S_{i+1}:

$$S_{i+1} = (m + \gamma)^{-1}\big(2mS_i - mS_{i-1} + \gamma S_{i-1} - \Delta t^2 kS_i +$$
$$m(w_{i+1} - 2w_i + w_{i-1}) + \Delta t^2 mg\big), \tag{D.19}$$

One fundamental problem arises when we try to start the computations. We know S_0 and want to apply (D.19) for $i = 0$ to calculate S_1. However, (D.19) involves S_{i-1}, that is, S_{-1}, which is an unknown value at a point of time *before* we compute the motion. The initial conditions come to rescue here. Since $dS/dt = 0$ at $t = 0$ (or $i = 0$), we can approximate this condition as

$$\frac{S_1 - S_{-1}}{2\Delta t} = 0 \quad \Rightarrow \quad S_{-1} = S_1 . \tag{D.20}$$

Inserting this relation in (D.19) when $i = 0$ gives a special formula for S_1 (or S_{i+1} with $i = 0$, if we want):

$$S_{i+1} = \frac{1}{2m} \left(2m S_i - \Delta t^2 k S_i + m \left(w_{i+1} - 2w_i + w_{i-1} \right) + \Delta t^2 mg \right) . \tag{D.21}$$

Remember that $i = 0$ in this formula. The overall algorithm is summarized below.

1. Initialize S_0 from initial condition
2. Use (D.21) to compute S_{i+1} for $i = 0$
3. For $i = 0, 1, 2, \ldots, N - 1$, use (D.19) to compute S_{i+1}

D.2   Program Development and Testing

D.2.1   Implementation

The aim now is to implement the algorithm from Sect. D.1.2 in a Python program. There are naturally two parts of the program, one where we read input data such as L, m, and $w(t)$, and one part where we run the computational algorithm. Let us write a function for each part.

The set of input data to the program consists of the mathematical symbols

- m (the mass of the box)
- b (the height of the box)
- L (the length of the unstretched spring)
- β (coefficient for the damping force)
- k (coefficient for the spring force)
- Δt (the time step between each S_i calculation)
- N (the number of computed time steps)
- S_0 (the initial stretch of the spring)
- $w(t)$ (the vertical displacement of the plate)
- g (acceleration of gravity)

We make a function `init_prms` for initializing these input parameters from option-value pairs on the command line. That is, the user provides pairs like `-m 2` and `-dt` `0.1` (for Δt). The `argparse` module from Sect. 4.4 can be used for this purpose. We supply default values for all parameters as arguments to the `init_prms` function. The function returns all these parameters with the changes that the user has specified on the command line. The w parameter is given as a string expression

(called `w_formula` below), and the `StringFunction` tool from Sect. 4.3.3 can be
used to turn the formula into a working Python function. An algorithmic sketch
of the tasks in the `init_prms` function can be expressed by some pseudo Python
code:

```
def init_prms(m, b, L, k, beta, S0, dt, g, w_formula, N):
    import argparse
    parser = argparse.ArgumentParser()
    parser.add_argument('--m', '--mass',
                        type=float, default=m)
    parser.add_argument('--b', '--boxheight',
                        type=float, default=b)
    ...
    args = parser.parse_args()

    from scitools.StringFunction import StringFunction
    w = StringFunction(args.w, independent_variables='t')
    return args.m, args.b, args.L, args.k, args.beta, \
           args.S0, args.dt, args.g, w, args.N
```

With such a sketch as a start, we can complete the indicated code and arrive at
a working function for specifying input parameters to the mathematical model:

```
def init_prms(m, b, L, k, beta, S0, dt, g, w_formula, N):
    import argparse
    parser = argparse.ArgumentParser()
    parser.add_argument('--m', '--mass',
                        type=float, default=m)
    parser.add_argument('--b', '--boxheight',
                        type=float, default=b)
    parser.add_argument('--L', '--spring-length',
                        type=float, default=L)
    parser.add_argument('--k', '--spring-stiffness',
                        type=float, default=k)
    parser.add_argument('--beta', '--spring-damping',
                        type=float, default=beta)
    parser.add_argument('--S0', '--initial-position',
                        type=float, default=S0)
    parser.add_argument('--dt','--timestep',
                        type=float, default=dt)
    parser.add_argument('--g', '--gravity',
                        type=float, default=g)
    parser.add_argument('--w', type=str, default=w_formula)
    parser.add_argument('--N', type=int, default=N)
    args = parser.parse_args()

    from scitools.StringFunction import StringFunction
    w = StringFunction(args.w, independent_variables='t')
    return args.m, args.b, args.L, args.k, args.beta, \
           args.S0, args.dt, args.g, w, args.N
```

You may wonder why we specify g (gravity) since this is a known constant, but
it is useful to turn off the gravity force to test the program. Just imagine the oscillations take place in the horizontal direction – the mathematical model is the same,

but $F_g = 0$, which we can obtain in our program by setting the input parameter g to zero.

The computational algorithm is quite easy to implement, as there is a quite direct translation of the mathematical algorithm in Sect. D.1.2 to valid Python code. The S_i values can be stored in a list or array with indices going from 0 to N. We use lists and instead of arrays here to allow readers not familiar with the latter concept to follow the code.

The function for computing S_i reads

```
def solve(m, k, beta, S0, dt, g, w, N):
    S = [0.0]*(N+1)        # output list
    gamma = beta*dt/2.0  # short form
    t = 0
    S[0] = S0
    # Special formula for first time step
    i = 0
    S[i+1] = (1/(2.0*m))*(2*m*S[i] - dt**2*k*S[i] +
            m*(w(t+dt) - 2*w(t) + w(t-dt)) + dt**2*m*g)
    t = dt

    for i in range(1,N):
        S[i+1] = (1/(m + gamma))*(2*m*S[i] - m*S[i-1] +
                                gamma*dt*S[i-1] - dt**2*k*S[i] +
                                m*(w(t+dt) - 2*w(t) + w(t-dt))
                                + dt**2*m*g)
        t += dt
    return S
```

The primary challenge in coding the algorithm is to set the index i and the time t right. Recall that in the updating formula for S[i+1] at time t+dt, the time on the right-hand side shall be the time at time step i, so the t+=dt update must come after S[i+1] is computed. The same is important in the special formula for the first time step as well.

A main program will typically first set some default values of the 10 input parameters, then call init_prms to let the user adjust the default values, and then call solve to compute the S_i values:

```
# Default values
from math import pi
m = 1; b = 2; L = 10; k = 1; beta = 0; S0 = 1;
dt = 2*pi/40; g = 9.81; w_formula = '0'; N = 80;

m, b, L, k, beta, S0, dt, g, w, N = \
    init_prms(m, b, L, k, beta, S0, dt, g, w_formula, N)
S = solve(m, k, beta, S0, dt, g, w, N)
```

So, what shall we do with the solution S? We can write out the values of this list, but the numbers do not give an immediate feeling for how the box moves. It will be better to graphically illustrate the $S(t)$ function, or even better, the $Y(t)$ function. This is straightforward and treated in Sect. D.3. In Sect. 9.4, we develop a drawing tool for drawing figures like Fig. D.1. By drawing the box, string, and plate at every time level we compute S_i, we can use this tool to make a moving figure that

illustrates the dynamics of the oscillations. Already now you can play around with
a program doing that (boxspring_figure_anim.py).

D.2.2   Callback Functionality

It would be nice to make some graphics of the system while the computations take
place, not only after the S list is ready. The user must then put some relevant statements in between the statements in the algorithm. However, such modifications
will depend on what type of analysis the user wants to do. It is a bad idea to mix
user-specific statements with general statements in a general algorithm. We therefore let the user provide a function that the algorithm can call after each S_i value is
computed. This is commonly called a *callback* function (because a general function
calls back to the user's program to do a user-specific task). To this callback function we send three key quantities: the S list, the point of time (t), and the time step
number ($i + 1$), so that the user's code gets access to these important data.

If we just want to print the solution to the screen, the callback function can be as
simple as

```
def print_S(S, t, step):
    print 't=%.2f  S[%d]=%+g' % (t, step, S[step])
```

In the solve function we take the callback function as a keyword argument
user_action. The default value can be an empty function, which we can define separately:

```
def empty_func(S, time, time_step_no):
    return None

def solve(m, k, beta, S0, dt, g, w, N,
          user_action=empty_func):
    ...
```

However, it is quicker to just use a lambda function (see Sect. 3.1.14):

```
def solve(m, k, beta, S0, dt, g, w, N,
          user_action=lambda S, time, time_step_no: None):
```

The new solve function has a call to user_action each time a new S value has
been computed:

```
def solve(m, k, beta, S0, dt, g, w, N,
          user_action=lambda S, time, time_step_no: None):
    """Calculate N steps forward. Return list S."""
    S = [0.0]*(N+1)       # output list
    gamma = beta*dt/2.0  # short form
    t = 0
    S[0] = S0
    user_action(S, t, 0)
    # Special formula for first time step
    i = 0
```

```
        S[i+1] = (1/(2.0*m))*(2*m*S[i] - dt**2*k*S[i] +
                m*(w(t+dt) - 2*w(t) + w(t-dt)) + dt**2*m*g)
        t = dt
        user_action(S, t, i+1)

        # Time loop
        for i in range(1,N):
            S[i+1] = (1/(m + gamma))*(2*m*S[i] - m*S[i-1] +
                                      gamma*dt*S[i-1] - dt**2*k*S[i] +
                                      m*(w(t+dt) - 2*w(t) + w(t-dt))
                                      + dt**2*m*g)
            t += dt
            user_action(S, t, i+1)

    return S

def test_constant():
    """Test constant solution."""
    from math import pi
    m = 10.0; k = 5.0; g = 9.81;
    S0 = m/k*g
    m, b, L, k, beta, S0, dt, g, w, N = \
        init_prms(m=10, b=0, L=5, k=5, beta=0, S0=S0,
                  dt=2*pi/40, g=g, w_formula='0', N=40)
    S = solve(m, k, beta, S0, dt, g, w, N)
    S_ref = S0    # S[i] = S0 is the reference value
    tol = 1E-13
    for S_ in S:
        assert abs(S_ref - S_) < tol

def test_general_solve():
    def print_S(S, t, step):
        print 't=%.2f  S[%d]=%+g' % (t, step, S[step])

    # Default values
    from math import pi
    m = 1; b = 2; L = 10; k = 1; beta = 0; S0 = 1;
    dt = 2*pi/40; g = 9.81; w_formula = '0'; N = 80;

    m, b, L, k, beta, S0, dt, g, w, N = \
        init_prms(m, b, L, k, beta, S0, dt, g, w_formula, N)
    S = solve(m, k, beta, S0, dt, g, w, N,
              user_action=print_S)
    S_reference = [
        1, 1.1086890184669964, 1.432074279830456, 1.9621765725933782,
        2.685916146951562, 3.5854354446841863]
    for S_new, S_ref in zip(S, S_reference):
        assert abs(S_ref - S_new) < 1E-14

def demo():
    def print_S(S, t, step):
        """Callback function: user_action."""
        print 't=%.2f  S[%d]=%+g' % (t, step, S[step])

    from math import pi
```

```
    m, b, L, k, beta, S0, dt, g, w, N = \
        init_prms(m=1, b=2, L=10, k=1, beta=0, S0=0,
                  dt=2*pi/40, g=9.81, w_formula='0', N=80)

    S = solve(m, k, beta, S0, dt, g, w, N,
              user_action=print_S)

    import matplotlib.pyplot as plt
    plt.plot(S)
    plt.show()

if __name__ == '__main__':
    #demo()
    test_constant()
```

The two last arguments to `user_action` must be carefully set: these should be
time value and index for the most recently computed S value.

D.2.3   Making a Module

The `init_prms` and `solve` functions can now be combined with many different
types of main programs and `user_action` functions. It is therefore preferable to
have the general `init_prms` and `solve` functions in a module boxspring and
import these functions in more user-specific programs. Making a module out of
`init_prms` and `solve` is, according to Sect. 4.9, quite trivial as we just need to put
the functions in a file `boxspring.py`.

It is a good habit to include a demo function that quickly tells users how to
operate the key functions in the module. The demo function is often called from the
test block. Here is an example:

```
def demo():
    def print_S(S, t, step):
        """Callback function: user_action."""
        print 't=%.2f  S[%d]=%+g' % (t, step, S[step])

    from math import pi

    m, b, L, k, beta, S0, dt, g, w, N = \
        init_prms(m=1, b=2, L=10, k=1, beta=0, S0=0,
                  dt=2*pi/40, g=9.81, w_formula='0', N=80)

    S = solve(m, k, beta, S0, dt, g, w, N,
              user_action=print_S)

    import matplotlib.pyplot as plt
    plt.plot(S)
    plt.show()

if __name__ == '__main__':
    #demo()
    test_constant()
```

D.2.4   Verification

To check that the program works correctly, we need a series of problems where the solution is known. These test cases must be specified by someone with a good physical and mathematical understanding of the problem being solved. We already have a solution formula (D.10) that we can compare the computations with, and more tests can be made in the case $w \neq 0$ as well.

However, before we even think of checking the program against the formula (D.10), we should perform some much simpler tests. The simplest test is to see what happens if we do nothing with the system. This solution is of course not very exciting – the box is at rest, but it is in fact exciting to see if our program reproduces the boring solution. Many bugs in the program can be found this way! So, let us run the program `boxspring.py` with `-S0 0` as the only command-line argument. The output reads

```
t=0.00   S[0]=+0
t=0.16   S[1]=+0.121026
t=0.31   S[2]=+0.481118
t=0.47   S[3]=+1.07139
t=0.63   S[4]=+1.87728
t=0.79   S[5]=+2.8789
t=0.94   S[6]=+4.05154
t=1.10   S[7]=+5.36626
...
```

Some motion takes place! All `S[1]`, `S[2]`, and so forth should be zero. What is the error?

There are two directions to follow now: we can either visualize the solution to understand more of what the computed $S(t)$ function looks like (perhaps this explains what is wrong), or we can dive into the algorithm and compute `S[1]` by hand to understand why it does not become zero. Let us follow both paths.

First we print out all terms on the right-hand side of the statement that computes `S[1]`. All terms except the last one ($\Delta t^2 mg$) are zero. The gravity term causes the spring to be stretched downward, which causes oscillations. We can see this from the governing equation (D.8) too: if there is no motion, $S(t) = 0$, the derivatives are zero (and $w = 0$ is default in the program), and then we are left with

$$ kS = mg \quad \Rightarrow \quad S = \frac{m}{k}g . \tag{D.22} $$

This result means that if the box is at rest, the spring is stretched (which is reasonable!). Either we have to start with $S(0) = \frac{m}{k}g$ in the equilibrium position, or we have to turn off the gravity force by setting `-g 0` on the command line. Setting either `-S0 0 -g 0` or `-S0 9.81` shows that the whole S list contains either zeros or 9.81 values (recall that $m = k = 1$ so $S_0 = g$). This constant solution is correct, and the coding looks promising.

Our first verification case should be automated in a test function that follows the conventions for the nose testing framework: the function name starts with `test_`, the function takes no arguments, and the tests raise an `AssertionError` if a test fails (e.g., by using `assert success`, where `success` is a boolean variable reflecting whether the test is successful or not).

A relevant test function sets the special input corresponding to the equilibrium position and must test that every `S[i]` equals the constant reference value within a tolerance:

```
def test_constant():
    """Test constant solution."""
    from math import pi
    m = 10.0; k = 5.0; g = 9.81;
    S0 = m/k*g
    m, b, L, k, beta, S0, dt, g, w, N = \
        init_prms(m=10, b=0, L=5, k=5, beta=0, S0=S0,
                  dt=2*pi/40, g=g, w_formula='0', N=40)



    S = solve(m, k, beta, S0, dt, g, w, N)
    S_ref = S0    # S[i] = S0 is the reference value
    tol = 1E-13
    for S_ in S:
        assert abs(S_ref - S_) < tol
```

For the first test where $S_0 \neq mg/k$, we can plot the solution to see what it looks like, using the program `boxspring_plot`:

Terminal

```
boxspring_plot.py --S0 0 --N 200
```

Figure D.2 shows the function $Y(t)$ for this case where the initial stretch is zero, but gravity is causing a motion. With some mathematical analysis of this problem we can establish that the solution is correct. We have that $m = k = 1$ and $w = \beta = 0$, which implies that the governing equation is

$$\frac{d^2 S}{dt^2} + S = g, \quad S(0) = 0, \, dS/dt(0) = 0.$$

Without the g term this equation is simple enough to be solved by basic techniques you can find in most introductory books on differential equations. Let us therefore get rid of the g term by a little trick: we introduce a new variable $T = S - g$, and by inserting $S = T + g$ in the equation, the g is gone:

$$\frac{d^2 T}{dt^2} + T = 0, \quad T(0) = -g, \, \frac{dT}{dt}(0) = 0. \tag{D.23}$$

This equation is of a very well-known type and the solution reads $T(t) = -g \cos t$, which means that $S(t) = g(1 - \cos t)$ and

$$Y(t) = -L - g(1 - \cos t) - \frac{b}{2}.$$

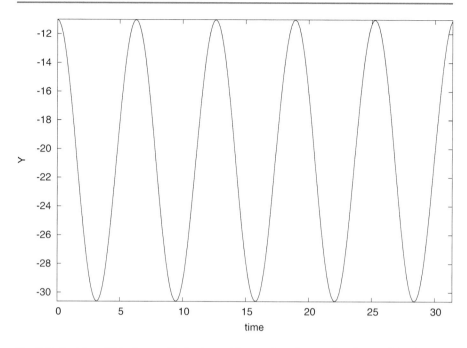

Fig. D.2 Positions $Y(t)$ of an oscillating box with $m = k = 1$, $w = \beta = 0$, $g = 9.81$, $L = 10$, and $b = 2$

With $L = 10$, $g \approx 10$, and $b = 2$ we get oscillations around $y \approx 21$ with a period of 2π and a start value $Y(0) = -L - b/2 = 11$. A rough visual inspection of the plot shows that this looks right. A more thorough analysis would be to make a test of the numerical values in a new callback function (the program is found in boxspring_test1.py):

```
from boxspring import init_prms, solve
from math import cos

def exact_S_solution(t):
    return g*(1 - cos(t))

def check_S(S, t, step):
    error = exact_S_solution(t) - S[step]
    print 't=%.2f  S[%d]=%+g error=%g' % (t, step, S[step], error)

# Fixed values for a test
from math import pi
m = 1; b = 2; L = 10; k = 1; beta = 0; S0 = 0
dt = 2*pi/40; g = 9.81; N = 200

def w(t):
    return 0

S = solve(m, k, beta, S0, dt, g, w, N, user_action=check_S)
```

The output from this program shows increasing errors with time, up as large values as 0.3. The difficulty is to judge whether this is the error one must expect because the program computes an approximate solution, or if this error points to a bug in the program – or a wrong mathematical formula.

From these sessions on program testing you will probably realize that verification of mathematical software is challenging. In particular, the design of the test problems and the interpretation of the numerical output require quite some experience with the interplay between physics (or another application discipline), mathematics, and programming.

D.3   Visualization

The purpose of this section is to add graphics to the oscillating system application developed in Sect. D.2. Recall that the function `solve` solves the problem and returns a list S with indices from 0 to N. Our aim is to plot this list and various physical quantities computed from it.

D.3.1   Simultaneous Computation and Plotting

The `solve` function makes a call back to the user's code through a callback function (the `user_action` argument to `solve`) at each time level. The callback function has three arguments: S, the time, and the current time step number. Now we want the callback function to plot the position $Y(t)$ of the box during the computations. In principle this is easy, but S is longer than we want to plot, because S is allocated for the whole time simulation while the `user_action` function is called at time levels where only the indices in S up to the current time level have been computed (the rest of the elements in S are zero). We must therefore use a sublist of S, from time zero and up to the current time. The callback function we send to `solve` as the `user_action` argument can then be written like this:

```
def plot_S(S, t, step):
    if step == 0:          # nothing to plot yet
        return None

    tcoor = linspace(0, t, step+1)
    S = array(S[:len(tcoor)])
    Y = w(tcoor) - L - S - b/2.
    plot(tcoor, Y)
```

Note that L, dt, b, and w must be global variables in the user's main program.

The major problem with the `plot_S` function shown is that the `w(tcoor)` evaluation does not work. The reason is that w is a StringFunction object, and according to Sect. 5.5.1, StringFunction objects do not work with array arguments unless we call their `vectorize` function once. We therefore need to do a

```
w.vectorize(globals())
```

before calling `solve` (which calls `plot_S` repeatedly). Here is the main program with this important statement:

```
from boxspring import init_prms, solve
from scitools.std import *

# Default values
m = 1; b = 2; L = 10; k = 1; beta = 0; S0 = 1;
dt = 2*pi/40; g = 9.81; w_formula = '0'; N = 200;

m, b, L, k, beta, S0, dt, g, w, N = \
    init_prms(m, b, L, k, beta, S0, dt, g, w_formula, N)

w.vectorize(globals())

S = solve(m, k, beta, S0, dt, g, w, N, user_action=plot_S)
```

Now the `plot_S` function works fine. You can try the program out by running

```
Terminal
boxspring_plot_v1.py
```

Fixing axes Both the t and the y axes adapt to the solution array in every plot. The adaptation of the y is okay since it is difficult to predict the future minimum and maximum values of the solution, and hence it is most natural to just adapt the y axis to the computed Y points so far in the simulation. However, the t axis should be fixed throughout the simulation, and this is easy since we know the start and end times. The end time is $T = N\Delta t$ and a relevant `plot` command becomes

```
plot(tcoor, Y,
     axis=[0, N*dt, min(Y), max(Y)],
     xlabel='time', ylabel='Y')
```

At the end of the simulation it can be nice to make a hardcopy of the last `plot` command performed in the `plot_S` function. We then just call

```
savefig('tmp_Y.pdf')
```

after the `solve` function is finished.

In the beginning of the simulation it is convenient to skip plotting for a number of steps until there are some interesting points to visualize and to use for computing the axis extent. We also suggest to apply the recipe at the end of Sect. 5.5.1 to vectorize w. More precisely, we use `w.vectorize` in general, but turn to numpy's `vectorize` feature only if the string formula contains an inline `if-else` test (to avoid requiring users to use `where` to vectorize the string expressions). One reason for paying attention to `if-else` tests in the w formula is that sudden movements of the plate are of interest, and this gives rise to step functions and strings like '1 if t>0 else 0'. A main program with all these features is listed next.

```
from boxspring import init_prms, solve
from scitools.std import *

def plot_S(S, t, step):
    first_plot_step = 10             # skip the first steps
    if step < first_plot_step:
        return

    tcoor = linspace(0, t, step+1)   # t = dt*step
    S = array(S[:len(tcoor)])
    Y = w(tcoor) - L - S - b/2.0      # (w, L, b are global vars.)

    plot(tcoor, Y,
         axis=[0, N*dt, min(Y), max(Y)],
         xlabel='time', ylabel='Y')


# Default values
m = 1; b = 2; L = 10; k = 1; beta = 0; S0 = 1
dt = 2*pi/40; g = 9.81; w_formula = '0'; N = 200

m, b, L, k, beta, S0, dt, g, w, N = \
    init_prms(m, b, L, k, beta, S0, dt, g, w_formula, N)

# Vectorize the StringFunction w
w_formula = str(w)   # keep this to see if w=0 later
if ' else ' in w_formula:
    w = vectorize(w)         # general vectorization
else:
    w.vectorize(globals())  # more efficient (when no if)

S = solve(m, k, beta, S0, dt, g, w, N, user_action=plot_S)

# First make a hardcopy of the the last plot of Y
savefig('tmp_Y.pdf')
```

D.3.2   Some Applications

What if we suddenly, right after $t = 0$, move the plate upward from $y = 0$ to $y = 1$? This will set the system in motion, and the task is to find out what the motion looks like.

There is no initial stretch in the spring, so the initial condition becomes $S_0 = 0$. We turn off gravity for simplicity and try a $w = 1$ function since the plate has the position $y = w = 1$ for $t > 0$:

| Terminal |
```
boxspring_plot.py --w '1' --S 0 --g 0
```

Nothing happens. The reason is that we specify $w(t) = 1$, but in the equation only d^2w/dt^2 has an effect and this quantity is zero. What we need to specify is a step function: $w = 0$ for $t \le 0$ and $w = 1$ for $t > 0$. In Python such a function can

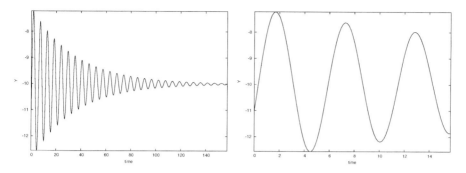

Fig. D.3 Plot of the position of an oscillating box where the end point of the spring ($w(t)$) is given a sudden movement at $t = 0$. Other parameters are $m = k = 1$, $\beta = 0.1$, $g = 0$, $S_0 = 0$. *Left*: 1000 time steps. *Right*: 100 steps for magnifying the first oscillation cycles

be specified as a string expression `'1 if t>0 else 0'`. With a step function we obtain the right initial jump of the plate:

```
                                    Terminal
boxspring_plot.py --w '1 if t > 0 else 0' \
         --S0 0 --g 0 --N 1000 --beta 0.1
```

Figure D.3 displays the solution. We see that the damping parameter has the effect of reducing the amplitude of $Y(t)$, and the reduction looks exponential, which is in accordance with the exact solution (D.10) (although this formula is not valid in the present case because $w \neq 0$ – but one gets the same exponential reduction even in this case). The box is initially located at $Y = 0 - (10 + 0) - 2/2 = -11$. During the first time step we get a stretch $S = 0.5$ and the plate jumps up to $y = 1$ so the box jumps to $Y = 1 - (10 + 0.5) - 2/2 = -10.5$. In Fig. D.3 (right) we see that the box starts correctly out and jumps upwards, as expected.

More exciting motions of the box can be obtained by moving the plate back and forth in time, see for instance Fig. D.4.

D.3.3 Remark on Choosing Δt

If you run the `boxspring_plot.py` program with a large -dt argument (for Δt), strange things may happen. Try -dt 2 -N 20 as command-line arguments and observe that Y jumps up and down in a saw tooth fashion so we clearly have too large time steps. Then try -dt 2.1 -N 20 and observe that Y takes on very large values (10^5). This highly non-physical result points to an error in the program. However, the problem is not in the program, but in the numerical method used to solve (D.8). This method becomes unstable and hence useless if Δt is larger than a critical value. We shall not dig further into such problems, but just notice that mathematical models on a computer must be used with care, and that a serious user of simulation programs must understand how the mathematical methods work in detail and what their limitations are.

D.3.4 Comparing Several Quantities in Subplots

So far we have plotted Y, but there are other interesting quantities to look at, e.g., S, w, the spring force, and the damping force. The spring force and S are proportional, so only one of these is necessary to plot. Also, the damping force is relevant only if $\beta \neq 0$, and w is only relevant if the string formula is different from the default value '0'.

All the mentioned additional plots can be placed in the same figure for comparison. To this end, we apply the subfigure command in Easyviz and create a row of individual plots. How many plots we have depends on the values of str(w) and beta. The relevant code snippet for creating the additional plots is given below and appears after the part of the main program shown above.

```
# Make plots of several additional interesting quantities
tcoor = linspace(0, N*dt, N+1)
S = array(S)

plots = 2          # number of rows of plots
if beta != 0:
    plots += 1
if w_formula != '0':
    plots += 1

# Position Y(t)
plot_row = 1
subplot(plots, 1, plot_row)
Y = w(tcoor) - L - S - b/2.0
plot(tcoor, Y, xlabel='time', ylabel='Y')

# Spring force (and S)
plot_row += 1
subplot(plots, 1, plot_row)
Fs = k*S
plot(tcoor, Fs, xlabel='time', ylabel='spring force')

# Friction force
if beta != 0:
    plot_row += 1
    subplot(plots, 1, plot_row)
    Fd = beta*diff(S)  # diff is in numpy
    # len(diff(S)) = len(S)-1 so we use tcoor[:-1]:
    plot(tcoor[:-1], Fd, xlabel='time', ylabel='damping force')

# Excitation
if w_formula != '0':
    plot_row += 1
    subplot(plots, 1, plot_row)
    w_array = w(tcoor)
    plot(tcoor, w_array, xlabel='time', ylabel='w(t)')

savefig('tmp.pdf')  # save this multi-axis plot in a file
```

Figure D.4 displays what the resulting plot looks like for a test case with an oscillating plate (w). The command for this run is

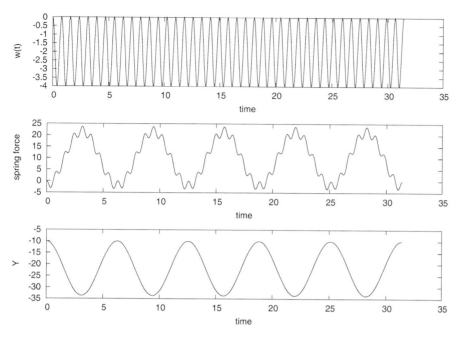

Fig. D.4 Plot of the plate position $w(t)$, the spring force (proportional to $S(t)$)), and the position $Y(t)$ for a test problem where $w(t) = 2(\cos(8t) - 1)$, $\beta = g = 0$, $m = k = 1$, $S_0 = 0$, $\Delta t = 0.5236$, and $N = 600$

```
Terminal
boxspring_plot.py  --S0 0  --w '2*(cos(8*t)-1)'  \
                   --N 600  --dt 0.05236
```

The rapid oscillations of the plate require us to use a smaller Δt and more steps (larger N).

D.3.5   Comparing Approximate and Exact Solutions

To illustrate multiple curves in the same plot and animations we turn to a slightly different program. The task now is to visually investigate how the accuracy of the computations depends on the Δt parameter. The smaller Δt is, the more accurate the solution S is. To look into this topic, we need a test problem with known solution. Setting $m = k = 1$ and $w = 0 = \beta = 0$ implies the exact solution $S(t) = g(1 - \cos t)$ (see Sect. D.2.4). The `boxspring_test1.py` program from Sect. D.2.4 can easily be extended to plot the calculated solution together with the exact solution. We drop the `user_action` callback function and just make the plot after having the complete solution S returned from the `solve` function:

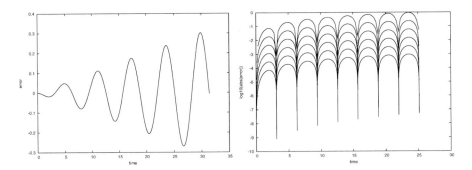

Fig. D.5 Error plots for a test problem involving an oscillating system: *Left*: the error as a function of time. *Right*: the logarithm of the absolute value of the error as a function of time, where Δt is reduced by one half from one curve to the next one below

```
tcoor = linspace(0, N*dt, len(S))
exact = exact_S_solution(tcoor)
plot(tcoor, S, 'r', tcoor, exact, 'b',
    xlabel='time', ylabel='S',
    legend=('computed S(t)', 'exact S(t)'),
    savefig='tmp_S.pdf')
```

The two curves tend to lie upon each other, so to get some more insight into the details of the error, we plot the error itself, in a separate plot window:

```
figure()      # new plot window
S = array(S)  # turn list into NumPy array for computations
error = exact - S
plot(tcoor, error, xlabel='time', ylabel='error',
    savefig='tmp_error.pdf')
```

The error increases in time as the plot in Fig. D.5 (left) clearly shows.

D.3.6   Evolution of the Error as Δt Decreases

Finally, we want to investigate how the error curve evolves as the time step Δt decreases. In a loop we halve Δt in each pass, solve the problem, compute the error, and plot the error curve. From the finite difference formulas involved in the computational algorithm, we can expect that the error is of order Δt^2. That is, if Δt is halved, the error should be reduced by 1/4.

The resulting plot of error curves is not very informative because the error reduces too quickly (by several orders of magnitude). A better plot is obtained by taking the logarithm of the error. Since an error curve may contain positive and negative elements, we take the absolute value of the error before taking the logarithm. We also note that S_0 is always correct, so it is necessary to leave out the initial value of the error array to avoid the logarithm of zero.

The ideas of the previous two paragraphs can be summarized in a Python code snippet:

```
figure()       # new plot window
dt = 2*pi/10
tstop = 8*pi   # 4 periods
N = int(tstop/dt)
for i in range(6):
    dt /= 2.0
    N *= 2
    S = solve(m, k, beta, S0, dt, g, w, N)
    S = array(S)
    tcoor = linspace(0, tstop, len(S))
    exact = exact_S_solution(tcoor)
    abserror = abs(exact - S)
    # Drop abserror[0] since it is always zero and causes
    # problems for the log function:
    logerror = log10(abserror[1:])

    plot(tcoor[1:], logerror, 'r', xlabel='time',
         ylabel='log10(abs(error))')
    hold('on')
savefig('tmp_errors.pdf')
```

The resulting plot is shown in Fig. D.5 (right).

Visually, it seems to be a constant distance between the curves in Fig. D.5 (right). Let d denote this difference and let E_i be the absolute error curve associated with Δt in the i-th pass in the loop. What we plot is $\log_{10} E_i$. The difference between two curves is then $D_{i+1} = \log_{10} E_i - \log_{10} E_{i+1} = \log_{10}(E_i / E_{i+1})$. If this difference is roughly 0.5 as we see from Fig. D.5 (right), we have

$$\log_{10} \frac{E_i}{E_{i+1}} = d = 0.5 \quad \Rightarrow \quad E_{i+1} = \frac{1}{3.16} E_i .$$

That is, the error is reduced, but not by the theoretically expected factor 4. Let us investigate this topic in more detail by plotting D_{i+1}.

We make a loop as in the last code snippet, but store the `logerror` array from the previous pass in the loop (E_i) in a variable `logerror_prev` such that we can compute the difference D_{i+1} as

```
logerror_diff = logerror_prev - logerror
```

There are two problems to be aware of now in this array subtraction: (i) the `logerror_prev` array is not defined before the second pass in the loop (when i is one or greater), and (ii) `logerror_prev` and `logerror` have different lengths since `logerror` has twice as many time intervals as `logerror_prev`. Numerical Python does not know how to compute this difference unless the arrays have the same length. We therefore need to use every two elements in `logerror`:

```
logerror_diff = logerror_prev - logerror[::2]
```

An additional problem now arises because the set of time coordinates, `tcoor`, in the current pass of the loop also has twice as many intervals so we need to plot `logerror_diff` against `tcoor[::2]`.

The complete code snippet for plotting differences between the logarithm of the absolute value of the errors now becomes

```
figure()
dt = 2*pi/10
tstop = 8*pi   # 4 periods
N = int(tstop/dt)
for i in range(6):
    dt /= 2.0
    N *= 2
    S = solve(m, k, beta, S0, dt, g, w, N)
    S = array(S)
    tcoor = linspace(0, tstop, len(S))
    exact = exact_S_solution(tcoor)
    abserror = abs(exact - S)
    logerror = log10(abserror[1:])
    if i > 0:
        logerror_diff = logerror_prev - logerror[::2]
        plot(tcoor[1::2], logerror_diff, 'r', xlabel='time',
            ylabel='difference in log10(abs(error))')
        hold('on')
        meandiff = mean(logerror_diff)
        print 'average log10(abs(error)) difference:', meandiff
    logerror_prev = logerror
savefig('tmp_errors_diff.pdf')
```

Figure D.6 shows the result. We clearly see that the differences between the curves in Fig. D.5 (right) are almost the same even if Δt is reduced by several orders of magnitude.

In the loop we also print out the average value of the difference curves in Fig. D.6:

```
average log10(abs(error)) difference:  0.558702094666
average log10(abs(error)) difference:  0.56541814902
average log10(abs(error)) difference:  0.576489014172
average log10(abs(error)) difference:  0.585704362507
average log10(abs(error)) difference:  0.592109360025
```

These values are almost constant. Let us use 0.57 as an representative value and see what it implies. Roughly speaking, we can then say that

$$\log_{10} E_i - \log_{10} E_{i+1} = 0.57 \,.$$

Collecting the two first terms and applying the exponential function 10^x on both sides we get that

$$E_{i+1} = \frac{1}{3.7} E_i \,.$$

This error reduction when Δt is decreased is not quite as good as we would theoretically expect (1/4), but it is close. The purpose of this brief analysis is primarily to show how errors can be explored by plotting, and how we can take advantage of array computing to produce various quantities of interest in a problem. A more

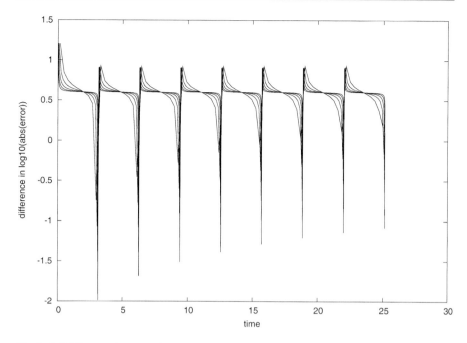

Fig. D.6 Differences between the curves in Fig. D.5 (right)

thorough investigation of how the error depends on Δt would use time integrals of the error instead of the complete error curves.

Again we mention that the complete problem analyzed in this appendix is challenging to understand because of its mix of physics, mathematics, and programming. In real life, however, problem solving in science and industry involve multidisciplinary projects where people with different competence work together. As a scientific programmer you must then be able to fully understand what to program and how to verify the results. This is a requirement in the current summarizing example too. You have to accept that your programming problem is buried in a lot of physical and mathematical details.

Having said this, we expect that most readers of this book also gain a background in physics and mathematics so that the present summarizing example can be understood in complete detail, at least at some later stage.

D.4 Exercises

Exercise D.1: Model sudden movements of the plate
Set up a problem with the `boxspring_plot.py` program where the initial stretch in the spring is 1 and there is no gravity force. Between $t = 20$ and $t = 30$ we move the plate suddenly from 0 to 2 and back again:

$$w(t) = \begin{cases} 2, & 20 < t < 30, \\ 0, & \text{otherwise} \end{cases}$$

Run this problem and view the solution.

Exercise D.2: Write a callback function
Doing Exercise D.1 shows that the Y position increases significantly in magnitude when the plate jumps upward and back again at $t = 20$ and $t = 30$, respectively. Make a program where you import from the `boxspring` module and provide a callback function that checks if $Y < 9$ and then aborts the program.
Filename: `boxspring_Ycrit`.

Exercise D.3: Improve input to the simulation program
The oscillating system in Sect. D.1 has an equilibrium position $S = mg/k$, see (D.22). A natural case is to let the box start at rest in this position and move the plate to induce oscillations. We must then prescribe $S_0 = mg/k$ on the command line, but the numerical value depends on the values of m and g that we might also give in the command line. However, it is possible to specify `-S0 m*g/k` on the command line if we in the `init_prms` function first let S0 be a string in the `elif` test and then, after the `for` loop, execute `S0 = eval(S0)`. At that point, m and k are read from the command line so that `eval` will work on `'m*g/k'`, or any other expression involving data from the command. Implement this idea.

A first test problem is to start from rest in the equilibrium position $S(0) = mg/k$ and give the plate a sudden upward change in position from $y = 0$ to $y = 1$. That is,

$$w(t) = \begin{cases} 0, & t \le 0, \\ 1, & t > 0 \end{cases}$$

You should get oscillations around the displaced equilibrium position $Y = w - L - S_0 = -9 - 2g$.
Filename: `boxspring2`.

Programming of Differential Equations

E

Appendices C and D give a brief introduction to differential equations, with a focus on a few specific equations and programs that are tailored to these equations. The present appendix views differential equations from a more abstract point of view, which allows us to formulate numerical methods and create general software that are applicable to a large family of widely different differential equation problems from mathematics, physics, biology, and finance. More specifically, the abstract view is motivated by the slogan *implement once, apply anywhere*. We have in fact met this principle several places in the book: differentiation ($f''(x)$) in Sects. 3.1.12 and 7.3.2, integration ($\int_a^b f(x)dx$) in Sects. 3.4.2 and 7.3.3, and root finding ($f(x) = 0$) in Sect. 4.11.2 and Sect. A.1.10. In all of the referred implementations, we work with a *general function* $f(x)$ so that any problem can be solved by the same piece of code as long as we can define the problem in terms of a function $f(x)$. This is an excellent demonstration of the power of mathematics, and this abstract view of problems in terms of some $f(x)$ is especially powerful in numerical methods and programming. Now we shall formulate differential equations on the abstract form $u' = f(u, t)$ and create software that can solve any equation for which the $f(u, t)$ is given.

Before studying the present appendix, the reader should have some familiarity with differential equations at the level of Appendix C. Appendix D can also be advantageous to read, although this is not strictly required. Only fundamental programming skills regarding loops, lists, arrays, functions, if tests, command-line arguments, and curve plotting are required for the basic function-based material in this appendix. However, Sects. E.1.7 and E.2.4, as well as many exercises, use classes to a large extent and hence demand familiarity with the class concept from Chap. 7. The material on object-oriented programming in Sect. E.3 requires good knowledge of class hierarchies and inheritance from Chap. 9.

All computer codes associated with this appendix is found in `src/ode2`[1].

---

[1] http://tinyurl.com/pwyasaa/ode2

E.1   Scalar Ordinary Differential Equations

We shall in this appendix work with ordinary differential equations (ODEs) written on the abstract form

$$u'(t) = f(u(t), t) . \tag{E.1}$$

There is an infinite number of solutions to such an equation, so to make the solution $u(t)$ unique, we must also specify an initial condition

$$u(0) = U_0 . \tag{E.2}$$

Given $f(u, t)$ and U_0, our task is to compute $u(t)$.

At first sight, (E.1) is only a first-order differential equation, since only u' and not higher-order derivatives like u' are present in the equation. However, equations with higher-order derivatives can also be written on the abstract form (E.1) by introducing auxiliary variables and interpreting u and f as vector functions. This rewrite of the original equation leads to a system of first-order differential equations and will be treated in Sect. E.2. The bottom line is that a very large family of differential equations can be written as (E.1). Forthcoming examples will provide evidence.

We shall first assume that $u(t)$ is a *scalar function*, meaning that it has one number as value, which can be represented as a `float` object in Python. We then refer to (E.1) as a *scalar differential equation*. The counterpart *vector function* means that u is a vector of scalar functions and the equation is known as a *system of ODEs* (also known as a *vector ODE*). The value of a vector function is a list or array in a program. Systems of ODEs are treated in Sect. E.2.

E.1.1   Examples on Right-Hand-Side Functions

To write a specific differential equation on the form (E.1) we need to identify what the f function is. Say the equation reads

$$y^2 y' = x, \quad y(0) = Y,$$

with $y(x)$ as the unknown function. First, we need to introduce u and t as new symbols: $u = y, t = x$. This gives the equivalent equation $u^2 u' = t$ and the initial condition $u(0) = Y$. Second, the quantity u' must be isolated on the left-hand side of the equation in order to bring the equation on the form (E.1). Dividing by u^2 gives

$$u' = t u^{-2} .$$

This fits the form (E.1), and the $f(u, t)$ function is simply the formula involving u and t on the right-hand side:

$$f(u, t) = t u^{-2} .$$

The t parameter is very often absent on the right-hand side such that f involves u only.

Let us list some common scalar differential equations and their corresponding f functions. Exponential growth of money or populations is governed by

$$u' = \alpha u, \tag{E.3}$$

where $\alpha > 0$ is a given constant expressing the growth rate of u. In this case,

$$f(u,t) = \alpha u. \tag{E.4}$$

A related model is the logistic ODE for growth of a population under limited resources:

$$u' = \alpha u \left(1 - \frac{u}{R}\right), \tag{E.5}$$

where $\alpha > 0$ is the initial growth rate and R is the maximum possible value of u. The corresponding f is

$$f(u,t) = \alpha u \left(1 - \frac{u}{R}\right). \tag{E.6}$$

Radioactive decay of a substance has the model

$$u' = -au, \tag{E.7}$$

where $a > 0$ is the rate of decay of u. Here,

$$f(u,t) = -au. \tag{E.8}$$

A body falling in a fluid can be modeled by

$$u' + b|u|u = g, \tag{E.9}$$

where $b > 0$ models the fluid resistance, g is the acceleration of gravity, and u is the body's velocity (see Exercise E.8). By solving for u' we find

$$f(u,t) = -b|u|u + g. \tag{E.10}$$

Finally, Newton's law of cooling is the ODE

$$u' = -h(u - s), \tag{E.11}$$

where u is the temperature of a body, $h > 0$ is a proportionality constant, normally to be estimated from experiments, and s is the temperature of the surroundings. Obviously,

$$f(u,t) = -h(u - s). \tag{E.12}$$

E.1.2   The Forward Euler Scheme

Our task now is to define numerical methods for solving equations of the form (E.1). The simplest such method is the Forward Euler scheme. Equation (E.1) is to be solved for $t \in (0, T]$, and we seek the solution u at discrete time points

$t_i = i\Delta t$, $i = 1, 2, \ldots, n$. Clearly, $t_n = n\Delta t = T$, determining the number of points n as $T/\Delta t$. The corresponding values $u(t_i)$ are often abbreviated as u_i, just for notational simplicity.

Equation (E.1) is to be fulfilled at all time points $t \in (0, T]$. However, when we solve (E.1) numerically, we only require the equation to be satisfied at the discrete time points $t_1, t_2, \ldots, t_n$. That is,

$$u'(t_k) = f(u(t_k), t_k),$$

for $k = 1, \ldots, n$. The fundamental idea of the Forward Euler scheme is to approximate $u'(t_k)$ by a one-sided, forward difference:

$$u'(t_k) \approx \frac{u(t_{k+1}) - u(t_k)}{\Delta t} = \frac{u_{k+1} - u_k}{\Delta t}.$$

This removes the derivative and leaves us with the equation

$$\frac{u_{k+1} - u_k}{\Delta t} = f(u_k, t_k).$$

We assume that u_k is already computed, so that the only unknown in this equation is u_{k+1}, which we can solve for:

$$u_{k+1} = u_k + \Delta t f(u_k, t_k). \tag{E.13}$$

This is the Forward Euler scheme for a scalar first-order differential equation $u' = f(u, t)$.

Equation (E.13) has a recursive nature. We start with the initial condition, $u_0 = U_0$, and compute u_1 as

$$u_1 = u_0 + \Delta t f(u_0, t_0).$$

Then we can continue with

$$u_2 = u_1 + \Delta t f(u_1, t_1),$$

and then with u_3 and so forth. This recursive nature of the method also demonstrates that we *must* have an initial condition – otherwise the method cannot start.

E.1.3   Function Implementation

The next task is to write a general piece of code that implements the Forward Euler scheme (E.13). The complete original (continuous) mathematical problem is stated as

$$u' = f(u, t), \; t \in (0, T], \quad u(0) = U_0, \tag{E.14}$$

while the discrete numerical problem reads

$$u_0 = U_0, \tag{E.15}$$

$$u_{k+1} = u_k + \Delta t f(u_k, t_k), \; t_k = k\Delta t, k = 1, \ldots, n, \; n = T/\Delta t. \tag{E.16}$$

We see that the input data to the numerical problem consist of f, U_0, T, and Δt or n. The output consists of $u_1, u_2, \ldots, u_n$ and the corresponding set of time points $t_1, t_2, \ldots, t_n$.

Let us implement the Forward Euler method in a function `ForwardEuler` that takes f, U_0, T, and n as input, and that returns $u_0, \ldots, u_n$ and $t_0, \ldots, t_n$:

```python
import numpy as np

def ForwardEuler(f, U0, T, n):
    """Solve u'=f(u,t), u(0)=U0, with n steps until t=T."""
    t = np.zeros(n+1)
    u = np.zeros(n+1)  # u[k] is the solution at time t[k]
    u[0] = U0
    t[0] = 0
    dt = T/float(n)
    for k in range(n):
        t[k+1] = t[k] + dt
        u[k+1] = u[k] + dt*f(u[k], t[k])
    return u, t
```

Note the close correspondence between the implementation and the mathematical specification of the problem to be solved. The argument f to the `ForwardEuler` function must be a Python function `f(u, t)` implementing the $f(u, t)$ function in the differential equation. In fact, f is the definition of the equation to be solved. For example, we may solve $u' = u$ for $t \in (0, 3)$, with $u(0) = 1$, and $\Delta t = 0.1$ by the following code utilizing the `ForwardEuler` function:

```python
def f(u, t):
    return u

u, t = ForwardEuler(f, U0=1, T=4, n=20)
```

With the u and t arrays we can easily plot the solution or perform data analysis on the numbers.

E.1.4   Verifying the Implementation

Visual comparison Many computational scientists and engineers look at a plot to see if a numerical and exact solution are sufficiently close, and if so, they conclude that the program works. This is, however, not a very reliable test. Consider a first try at running `ForwardEuler(f, U0=1, T=4, n=10)`, which gives the plot to the left in Fig. E.1. The discrepancy between the solutions is large, and the viewer may be uncertain whether the program works correctly or not. Running n=20 should give a better solution, depicted to the right in Fig. E.1, but is the improvement good enough? Increasing n drives the numerical curve closer to the exact one. This brings evidence that the program is correct, but there could potentially be errors in the code that makes the curves further apart than what is implied by the numerical approximations alone. We cannot know if such a problem exists.

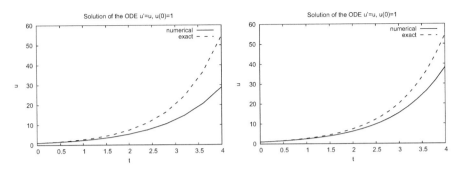

Fig. E.1 Comparison of numerical exact solution for 10 intervals (*left*) and 20 (intervals)

Comparing with hand calculations A more rigorous way of verifying the implementation builds on a simple principle: we run the algorithm by hand a few times and compare the results with those in the program. For most practical purposes, it suffices to compute u_1 and u_2 by hand:

$$u_1 = 1 + 0.1 \cdot 1 = 1.1, \quad u_2 = 1.1 + 0.1 \cdot 1.1 = 1.21 \,.$$

These values are to be compared with the numbers produced by the code. A correct program will lead to deviations that are zero (to machine precision). Any such test should be wrapped in a proper test function such that it can easily be repeated later. Here, it means we make a function

```
def test_ForwardEuler_against_hand_calculations():
    """Verify ForwardEuler against hand calc. for 3 time steps."""
    u, t = ForwardEuler(f, U0=1, T=0.2, n=2)
    exact = np.array([1, 1.1, 1.21])  # hand calculations
    error = np.abs(exact - u).max()
    success = error < 1E-14
    assert success, '|exact - u| = %g != 0' % error
```

The test function is written in a way that makes it trivial to integrate it in the nose testing framework, see Sect. H.9 and the brief examples in Sects. 3.3.3, 3.4.2, and 4.9.4. This means that the name starts with `test_`, there are no function arguments, and the check for passing the test is done with `assert success`. The test fails if the boolean variable `success` is `False`. The string after `assert success` is a message that will be written out if the test fails. The error measure is most conveniently a scalar number, which here is taken as the absolute value of the largest deviation between the exact and the numerical solution. Although we expect the error measure to be zero, we are prepared for rounding errors and must use a tolerance when testing if the test has passed.

Comparing with an exact numerical solution Another effective way to verify the code, is to find a problem that can be solved exactly by the numerical method we use. That is, we seek a problem where we do not have to deal with numerical approximation errors when comparing the exact solution with the one produced by the program. It turns out that if the solution $u(t)$ is linear in t, the Forward Euler method will reproduce this solution exactly. Therefore, we choose $u(t) = at + U_0$, with (e.g.) $a = 0.2$ and $U_0 = 3$. The corresponding f is the derivative of u, i.e., $f(u,t) = a$. This is obviously a very simple right-hand side without any u or t. However, we can make f more complicated by adding something that is zero, e.g., some expression with $u - (at + U_0)$, say $(u - (at + U_0))^4$, so that

$$f(u,t) = a + (u - (at + U_0))^4. \tag{E.17}$$

We implement our special f and the exact solution in two functions f and u_exact, and compute a scalar measure of the error. As a above, we place the test inside a test function and make an assertion that the error is sufficiently close to zero:

```python
def test_ForwardEuler_against_linear_solution():
    """Use knowledge of an exact numerical solution for testing."""
    def f(u, t):
        return 0.2 + (u - u_exact(t))**4

    def u_exact(t):
        return 0.2*t + 3

    u, t = ForwardEuler(f, U0=u_exact(0), T=3, n=5)
    u_e = u_exact(t)
    error = np.abs(u_e - u).max()
    success = error < 1E-14
    assert success, '|exact - u| = %g != 0' % error
```

A "silent" execution of the function indicates that the test works.

The shown functions are collected in a file `ForwardEuler_func.py`.

E.1.5   From Discrete to Continuous Solution

The numerical solution of an ODE is a *discrete function* in the sense that we only know the function values $u_0, u_1, ldots, u_N$ at some discrete points $t_0, t_1, \ldots, t_N$ in time. What if we want to know u between two computed points? For example, what is u between t_i and t_{i+1}, say at the midpoint $t = t_i + \frac{1}{2}\Delta t$? One can use *interpolation* techniques to find this value u. The simplest interpolation technique is to assume that u varies linearly on each time interval. On the interval $[t_i, t_{i+1}]$ the linear variation of u becomes

$$u(t) = u_i + \frac{u_{i+1} - ui}{t_{i+1} - t_i}(t - t_i).$$

We can then evaluate, e.g., $u(t_i + \frac{1}{2}\Delta t)$ from this formula and show that it becomes $(u_i + u_{i+1})/2$.

The function `scitools.std.wrap2callable` can automatically convert a discrete function to a continuous function:

```python
from scitools.std import wrap2callable
u_cont = wrap2callable((t, u))
```

From the arrays t and u, wrap2callable constructs a continuous function based on linear interpolation. The result u_cont is a Python function that we can evaluate for any value of its argument t:

```
dt = t[i+1] - t[i]
t = t[i] + 0.5*dt
value = u_cont(t)
```

In general, the wrap2callable function is handy when you have computed some discrete function and you want to evaluate this discrete function at any point.

E.1.6   Switching Numerical Method

There are numerous alternative numerical methods for solving (E.13). One of the simplest is Heun's method:

$$u_* = u_k + \Delta t f(u_k, t_k), \tag{E.18}$$

$$u_{k+1} = u_k + \frac{1}{2}\Delta t f(u_k, t_k) + \frac{1}{2}\Delta t f(u_*, t_{k+1}). \tag{E.19}$$

This scheme is easily implemented in the ForwardEuler function by replacing the Forward Euler formula

```
u[k+1] = u[k] + dt*f(u[k], t[k])
```

by (E.18) and (E.19):

```
u_star = u[k] + dt*f(u[k], t[k])
u[k+1] = u[k] + 0.5*dt*f(u[k], t[k]) + 0.5*dt*f(u_star, t[k+1])
```

We can, especially if f is expensive to calculate, eliminate a call f(u[k], t[k]) by introducing an auxiliary variable:

```
f_k = f(u[k], t[k])
u_star = u[k] + dt*f_k
u[k+1] = u[k] + 0.5*dt*f_k + 0.5*dt*f(u_star, t[k+1])
```

E.1.7   Class Implementation

As an alternative to the general ForwardEuler function in Sect. E.1.3, we shall now implement the numerical method in a class. This requires, of course, familiarity with the class concept from Chap. 7.

Class wrapping of a function Let us start with simply wrapping the ForwardEuler function in a class ForwardEuler_v1 (the postfix _v1 indicates that this is the very first class version). That is, we take the code in the `ForwardEuler` function and distribute it among methods in a class.

The constructor can store the input data of the problem and initialize data structures, while a `solve` method can perform the time stepping procedure:

```python
import numpy as np

class ForwardEuler_v1(object):
    def __init__(self, f, U0, T, n):
        self.f, self.U0, self.T, self.n = f, dt, U0, T, n
        self.dt = T/float(n)
        self.u = np.zeros(n+1)
        self.t = np.zeros(n+1)

    def solve(self):
        """Compute solution for 0 <= t <= T."""
        self.u[0] = float(self.U0)
        self.t[0] = float(0)

        for k in range(self.n):
            self.k = k
            self.t[k+1] = self.t[k] + self.dt
            self.u[k+1] = self.advance()
        return self.u, self.t

    def advance(self):
        """Advance the solution one time step."""
        u, dt, f, k, t = \
            self.u, self.dt, self.f, self.k, self.t

        u_new = u[k] + dt*f(u[k], t[k])
        return u_new
```

Note that we have introduced a third class method, `advance`, which isolates the numerical scheme. The motivation is that, by observation, the constructor and the `solve` method are completely general as they remain unaltered if we change the numerical method (at least this is true for a wide class of numerical methods). The only difference between various numerical schemes is the updating formula. It is therefore a good programming habit to isolate the updating formula so that another scheme can be implemented by just replacing the `advance` method – without touching any other parts of the class.

Also note that we in the `advance` method "strip off" the `self` prefix by introducing local symbols with exactly the same names as in the mathematical specification of the numerical method. This is important if we want a visually one-to-one correspondence between the mathematics and the computer code.

Application of the class goes as follows, here for the model problem $u' = u$, $u(0) = 1$:

```python
def f(u, t):
    return u

solver = ForwardEuler_v1(f, U0=1, T=3, n=15)
u, t = solver.solve()
```

Switching numerical method Implementing, for example, Heun's method (E.18)–(E.19) is a matter of replacing the `advance` method by

```
def advance(self):
    """Advance the solution one time step."""
    u, dt, f, k, t = \
        self.u, self.dt, self.f, self.k, self.t

    u_star = u[k] + dt*f(u[k], t[k])
    u_new = u[k] + \
            0.5*dt*f(u[k], t[k]) + 0.5*dt*f(u_star, t[k+1])
    return u_new
```

Checking input data is always a good habit, and in the present class the constructor may test that the `f` argument is indeed an object that can be called as a function:

```
if not callable(f):
    raise TypeError('f is %s, not a function' % type(f))
```

Any function `f` or any instance of a class with a `__call__` method will make `callable(f)` evaluate to True.

A more flexible class Say we solve $u' = f(u,t)$ from $t = 0$ to $t = T_1$. We can continue the solution for $t > T_1$ simply by restarting the whole procedure with initial conditions at $t = T_1$. Hence, the implementation should allow several consequtive solve steps.

Another fact is that the time step Δt does not need to be constant. Allowing small Δt in regions where u changes rapidly and letting Δt be larger in areas where u is slowly varying, is an attractive solution strategy. The Forward Euler method can be reformulated for a variable time step size $t_{k+1} - t_k$:

$$u_{k+1} = u_k + (t_{k+1} - t_k)f(u_k, t_k). \tag{E.20}$$

Similarly, Heun's method and many other methods can be formulated with a variable step size simply by replacing Δt with $t_{k+1} - t_k$. It then makes sense for the user to provide a list or array with time points for which a solution is sought: $t_0, t_1, \ldots, t_n$. The `solve` method can accept such a set of points.

The mentioned extensions lead to a modified class:

```
class ForwardEuler(object):
    def __init__(self, f):
        if not callable(f):
            raise TypeError('f is %s, not a function' % type(f))
        self.f = f

    def set_initial_condition(self, U0):
        self.U0 = float(U0)
```

```
    def solve(self, time_points):
        """Compute u for t values in time_points list."""
        self.t = np.asarray(time_points)
        self.u = np.zeros(len(time_points))
        # Assume self.t[0] corresponds to self.U0
        self.u[0] = self.U0

        for k in range(len(self.t)-1):
            self.k = k
            self.u[k+1] = self.advance()
        return self.u, self.t

    def advance(self):
        """Advance the solution one time step."""
        u, f, k, t = self.u, self.f, self.k, self.t
        dt = t[k+1] - t[k]
        u_new = u[k] + dt*f(u[k], t[k])
        return u_new
```

Usage of the class We must instantiate an instance, call the `set_initial_condition` method, and then call the `solve` method with a list or array of the time points we want to compute u at:

```
def f(u, t):
    """Right-hand side function for the ODE u' = u."""
    return u

solver = ForwardEuler(f)
solver.set_initial_condition(2.5)
u, t = solver.solve(np.linspace(0, 4, 21))
```

A simple `plot(t, u)` command can visualize the solution.

Verification It is natural to perform the same verifications as we did for the `ForwardEuler` function in Sect. E.1.4. First, we test the numerical solution against hand calculations. The implementation makes use of the same test function, just the way of calling up the numerical solver is different:

```
def test_ForwardEuler_against_hand_calculations():
    """Verify ForwardEuler against hand calc. for 2 time steps."""
    solver = ForwardEuler(lambda u, t: u)
    solver.set_initial_condition(1)
    u, t = solver.solve([0, 0.1, 0.2])
    exact = np.array([1, 1,1, 1.21])  # hand calculations
    error = np.abs(exact - u).max()
    assert error < 1E-14, '|exact - u| = %g != 0' % error
```

We have put some efforts into making this test very compact, mainly to demonstrate how Python allows very short, but still readable code. With a lambda function we can define the right-hand side of the ODE directly in the constructor argument. The `solve` method accepts a list, tuple, or array of time points and turns the data into an array anyway. Instead of a separate boolean variable `success` we have inserted the test inequality directly in the `assert` statement.

The second verification method applies the fact that the Forward Euler scheme is exact for a u that is linear in t. We perform a slightly more complicated test than in Sect. E.1.4: now we first solve for the points $0, 0.4, 1, 1.2$, and then we continue the solution process for $t_1 = 1.4$ and $t_2 = 1.5$.

```python
def test_ForwardEuler_against_linear_solution():
    """Use knowledge of an exact numerical solution for testing."""
    u_exact = lambda t: 0.2*t + 3
    solver = ForwardEuler(lambda u, t: 0.2 + (u - u_exact(t))**4)

    # Solve for first time interval [0, 1.2]
    solver.set_initial_condition(u_exact(0))
    u1, t1 = solver.solve([0, 0.4, 1, 1.2])

    # Continue with a new time interval [1.2, 1.5]
    solver.set_initial_condition(u1[-1])
    u2, t2 = solver.solve([1.2, 1.4, 1.5])

    # Append u2 to u1 and t2 to t1
    u = np.concatenate((u1, u2))
    t = np.concatenate((t1, t2))

    u_e = u_exact(t)
    error = np.abs(u_e - u).max()
    assert error < 1E-14, '|exact - u| = %g != 0' % error
```

Making a module   It is a well-established programming habit to have class implementations in files that act as Python modules. This means that all code is collected within classes or functions, and that the main program is executed in a test block. Upon import, no test or demonstration code should be executed.

Everything we have made so far is in classes or functions, so the remaining task to make a module, is to construct the test block.

```python
if __name__ == '__main__':
    import sys
    if len(sys.argv) >= 2 and sys.argv[1] == 'test':
        test_ForwardEuler_v1_against_hand_calculations()
        test_ForwardEuler_against_hand_calculations()
        test_ForwardEuler_against_linear_solution()
```

The `ForwardEuler_func.py` file with functions from Sects. E.1.3 and E.1.4 is in theory a module, but not sufficiently cleaned up. Exercise E.15 encourages you to turn the file into a proper module.

Remark   We do not need to call the test functions from the test block, since we can let nose run the tests automatically, see Sect. H.9, by `nosetests -s ForwardEuler.py`.

E.1.8   Logistic Growth via a Function-Based Approach

A more exciting application than the verification problems above is to simulate logistic growth of a population. The relevant ODE reads

$$u'(t) = \alpha u(t) \left(1 - \frac{u(t)}{R} \right) .$$

The mathematical $f(u,t)$ function is simply the right-hand side of this ODE. The corresponding Python function is

```
def f(u, t):
    return alpha*u*(1 - u/R)
```

where `alpha` and `R` are global variables that correspond to α and R. These must be initialized before calling the `ForwardEuler` function (which will call the `f(u,t)` above):

```
alpha = 0.2
R = 1.0

from ForwardEuler_func2 import ForwardEuler
u, t = ForwardEuler(f, U0=0.1, T=40, n=400)
```

We have in this program assumed that Exercise E.15 has been carried out to clean up the `ForwardEuler_func.py` file such that it becomes a proper module file with the name `ForwardEuler_func2.py`.

With u and t computed we can proceed with visualizing the solution (see Fig. E.2):

```
from matplotlib.pyplot import *
plot(t, u)
xlabel('t'); ylabel('u')
title('Logistic growth: alpha=%s, R=%g, dt=%g' %
      (alpha, R, t[1]-t[0]))
savefig('tmp.pdf'); savefig('tmp.png')
show()
```

The complete code appears in the file `logistic_func.py`.

E.1.9   Logistic Growth via a Class-Based Approach

The task of this section is to redo the implementation of Sect. E.1.8 using a problem class to store the physical parameters and the $f(u,t)$ function, and using class `ForwardEuler` from Sect. E.1.7 to solve the ODE. Comparison with the code in Sect. E.1.8 will then exemplify what the difference between a function-based and a class-based implementation is. There will be two major differences. One is related to technical differences between programming with functions and programming

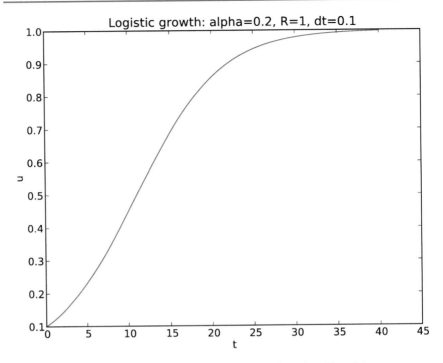

Fig. E.2 Plot of the solution of the ODE problem $u' = 0.2u(1 - u)$, $u(0) = 0.1$

with classes. The other is psychological: when doing class programming one often puts more efforts into making more functions, a complete module, a user interface, more testing, etc. A function-based approach, and in particular the present "flat" MATLAB-style program, tends to be more ad hoc and contain less general, reusable code. At least this is the author's experience over many years when observing students and professionals create different style of code with different type of programming techniques.

The style adopted for this class-based example have several important ingredients motivated by professional programming habits:

- Modules are imported as `import module` and calls to functions in the module are therefore prefixed with the module name such that we can easily see where different functionality comes from.
- All information about the original ODE problem is collected in a class.
- Physical and numerical parameters can be set on the command line.
- The main program is collected in a function.
- The implementation takes the form of a module such that other programs can reuse the class for representing data in a logistic problem.

The problem class Class Logistic holds the parameters of the ODE problem: U_0, α, R, and T as well as the $f(u, t)$ function. Whether T should be a member of class Logistic or not is a matter of taste, but the appropriate size of T is strongly linked to the other parameters so it is natural to specify them together. The number of time intervals, n, used in the numerical solution method is not a part of class Logistic since it influences the accuracy of the solution, but not the qualitative properties of the solution curve as the other parameters do.

The $f(u, t)$ function is naturally implemented as a `__call__` method such that the problem instance can act as both an instance and a callable function at the same time. In addition, we include a `__str__` for printing out the ODE problem. The complete code for the class looks like

```
class Logistic(object):
    """Problem class for a logistic ODE."""
    def __init__(self, alpha, R, U0, T):
        self.alpha, self.R, self.U0, self.T = alpha, float(R), U0, T

    def __call__(self, u, t):
        """Return f(u,t) for the logistic ODE."""
        return self.alpha*u*(1 - u/self.R)

    def __str__(self):
        """Return ODE and initial condition."""
        return "u'(t) = %g*u*(1 - u/%g), t in [0, %g]\nu(0)=%g" % \
               (self.alpha, self.R, self.T, self.U0)
```

Getting input from the command line We decide to specify α, R, U_0, T, and n, in that order, on the command line. A function for converting the command-line arguments into proper Python objects can be

```
def get_input():
    """Read alpha, R, U0, T, and n from the command line."""
    try:
        alpha = float(sys.argv[1])
        R = float(sys.argv[2])
        U0 = float(sys.argv[3])
        T = float(sys.argv[4])
        n = float(sys.argv[5])
    except IndexError:
        print 'Usage: %s alpha R U0 T n' % sys.argv[0]
        sys.exit(1)
    return alpha, R, U0, T, n
```

We have used a standard a `try-except` block to handle potential errors because of missing command-line arguments. A more user-friendly alternative would be to allow option-value pairs such that, e.g., T can be set by `-T 40` on the command line, but this requires more programming (with the `argparse` module).

Import statements The `import` statements necessary for the problem solving process are written as

```
import ForwardEuler
import numpy as np
import matplotlib.pyplot as plt
```

The two latter statements with their abbreviations have evolved as a standard in Python code for scientific computing.

Solving the problem The remaining necessary statements for solving a logistic problem are collected in a function

```
def logistic():
    alpha, R, U0, T, n = get_input()
    problem = Logistic(alpha=alpha, R=R, U0=U0)
    solver = ForwardEuler.ForwardEuler(problem)
    solver.set_initial_condition(problem.U0)
    time_points = np.linspace(0, T, n+1)
    u, t = solver.solve(time_points)

    plt.plot(t, u)
    plt.xlabel('t'); plt.ylabel('u')
    plt.title('Logistic growth: alpha=%s, R=%g, dt=%g'
              % (problem.alpha, problem.R, t[1]-t[0]))
    plt.savefig('tmp.pdf'); plt.savefig('tmp.png')
    plt.show()
```

Making a module Everything we have created is either a class or a function. The only remaining task to ensure that the file is a proper module is to place the call to the "main" function `logistic` in a test block:

```
if __name__ == '__main__':
    logistic()
```

The complete module is called `logistic_class.py`.

Pros and cons of the class-based approach If we quickly need to solve an ODE problem, it is tempting and efficient to go for the function-based code, because it is more direct and much shorter. A class-based module, with a user interface and often also test functions, usually gives more high-quality code that pays off when the software is expected to have a longer life time and will be extended to more complicated problems.

A pragmatic approach is to first make a quick function-based code, but refactor that code to a more reusable and extensible class version with test functions when you experience that the code frequently undergo changes. The present simple logistic ODE problem is, in my honest opinion, not complicated enough to defend a class version for practical purposes, but the primary goal here was to use a very simple mathematical problem for illustrating class programming.

E.2   Systems of Ordinary Differential Equations

The software developed so far in this appendix targets scalar ODEs of the form $u' = f(u, t)$ with initial condition $u(0) = U_0$. Our goal now is to build flexible software for solving scalar ODEs as well as systems of ODEs. That is, we want the same code to work both for systems and scalar equations.

E.2.1 Mathematical Problem

A scalar ODE involves the single equation

$$u'(t) = f(u(t), t)$$

with a single function $u(t)$ as unknown, while a system of ODEs involves n scalar ODEs and consequently n unknown functions. Let us denote the unknown functions in the system by $u^{(i)}(t)$, with i as a counter, $i = 0, \ldots, m - 1$. The system of n ODEs can then be written in the following abstract form:

$$\frac{du^{(0)}}{dt} = f^{(0)}(u^{(0)}, u^{(1)}, \ldots, u^{(m-1)}, t), \qquad (E.21)$$

$$\vdots$$

$$\frac{du^{(i)}}{dt} = f^{(i)}(u^{(0)}, u^{(1)}, \ldots, u^{(m-1)}, t), \qquad (E.22)$$

$$\vdots$$

$$\frac{du^{(m-1)}}{dt} = f^{(m-1)}(u^{(0)}, u^{(1)}, \ldots, u^{(m-1)}, t), \qquad (E.23)$$

In addition, we need n initial conditions for the n unknown functions:

$$u^{(i)}(0) = U_0^{(i)}, \quad i = 0, \ldots, m - 1. \qquad (E.24)$$

Instead of writing out each equation as in (E.21)–(E.23), mathematicians like to collect the individual functions $u^{(0)}, u^{(1)}, \ldots, u^{(m-1)}$ in a vector

$$u = (u^{(0)}, u^{(1)}, \ldots, u^{(m-1)}).$$

The different right-hand-side functions $f^{(0)}, f^{(1)}, \ldots, f^{(m-1)}$ in (E.21)–(E.23) can also be collected in a vector

$$f = (f^{(0)}, f^{(1)}, \ldots, f^{(m-1)}).$$

Similarly, we put the initial conditions also in a vector

$$U_0 = (U_0^{(0)}, U_0^{(1)}, \ldots, U_0^{(m-1)}).$$

With the vectors u, f, and U_0, we can write the ODE system (E.21)–(E.23) with initial conditions (E.24) as

$$u' = f(u, t), \quad u(0) = U_0. \qquad (E.25)$$

This is exactly the same notation as we used for a scalar ODE (!). The power of mathematics is that abstractions can be generalized so that new problems look like the familiar ones, and very often methods carry over to the new problems in the new notation without any changes. This is true for numerical methods for ODEs too.

Let us apply the Forward Euler scheme to each of the ODEs in the system (E.21)–(E.23):

$$u_{k+1}^{(0)} = u_k^{(0)} + \Delta t f^{(0)}(u_k^{(0)}, u_k^{(1)}, \ldots, u_k^{(m-1)}, t_k),   \tag{E.26}$$

$$\vdots$$

$$u_{k+1}^{(i)} = u_k^{(i)} + \Delta t f^{(i)}(u_k^{(0)}, u_k^{(1)}, \ldots, u_k^{(m-1)}, t_k),   \tag{E.27}$$

$$\vdots$$

$$u_{k+1}^{(m-1)} = u_k^{(m-1)} + \Delta t f^{(m-1)}(u_k^{(0)}, u_k^{(1)}, \ldots, u_k^{(m-1)}, t_k),   \tag{E.28}$$

Utilizing the vector notation, (E.26)–(E.28) can be compactly written as

$$u_{k+1} = u_k + \Delta t f(u_k, t_k),   \tag{E.29}$$

and this is again nothing but the formula we had for the Forward Euler scheme applied to a scalar ODE.

To summarize, the notation $u' = f(u, t)$, $u(0) = U_0$, is from now on used both for scalar ODEs and for systems of ODEs. In the former case, u and f are scalar functions, while in the latter case they are vectors. This great flexibility carries over to programming too: we can develop code for $u' = f(u, t)$ that works for scalar ODEs and systems of ODEs, the only difference being that u and f correspond to float objects for scalar ODEs and to arrays for systems of ODEs.

E.2.2   Example of a System of ODEs

An oscillating spring-mass system can be governed by a second-order ODE (see (D.8) in Appendix D for derivation):

$$mu'' + \beta u' + ku = F(t), \quad u(0) = U_0, \ u'(0) = 0.   \tag{E.30}$$

The parameters m, β, and k are known and $F(t)$ is a prescribed function. This second-order equation can be rewritten as two first-order equations by introducing two functions,

$$u^{(0)}(t) = u(t), \quad u^{(1)}(t) = u'(t).$$

The unknowns are now the position $u^{(0)}(t)$ and the velocity $u^{(1)}(t)$. We can then create equations where the derivative of the two new primary unknowns $u^{(0)}$ and $u^{(1)}$ appear alone on the left-hand side:

$$\frac{d}{dt} u^{(0)}(t) = u^{(1)}(t),   \tag{E.31}$$

$$\frac{d}{dt} u^{(1)}(t) = m^{-1}(F(t) - \beta u^{(1)} - ku^{(0)}).   \tag{E.32}$$

We write this system as $u'(t) = f(u, t)$ where now u and f are vectors, here of length two:

$$u(t) = (u^{(0)}(t), u^{(1)}(t))$$

$$f(t, u) = (u^{(1)}, m^{-1}(F(t) - \beta u^{(1)} - ku^{(0)})).   \tag{E.33}$$

Note that the vector $u(t)$ is different from the quantity u in (E.30)! There are, in fact, several interpretation of the symbol u, depending on the context: the exact solution u of (E.30), the numerical solution u of (E.30), the vector u in a rewrite of (E.30) as a first-order ODE system, and the array u in the software, holding the numerical approximation to $u(t) = (u^{(0)}(t), u^{(1)}(t))$.

E.2.3   Function Implementation

Let us have a look at how the software from Sects. E.1.3–E.1.6 changes if we try to apply it to systems of ODEs. We start with the ForwardEuler function listed in Sect. E.1.3 and the specific system from Sect. E.2.2. The right-hand-side function f(u, t) must now return the vector in (E.33), here as a NumPy array:

```
def f(u, t):
    return np.array([u[1], 1./m*(F(t) - beta*u[1] - k*u[0])])
```

Note that u is an array with two components, holding the values of the two unknown functions $u^{(0)}(t)$ and $u^{(1)}(t)$ at time t.

The initial conditions can also be specified as an array

```
U0 = np.array([0.1, 0])
```

What happens if we just send these f and U0 objects to the ForwardEuler function? To answer the question, we must examine each statement inside the function to see if the Python operations are still valid. But of greater importance, we must check that the right mathematics is carried out.

The first failure occurs with the statement

```
u = np.zeros(n+1)   # u[k] is the solution at time t[k]
```

Now, u should be an array of arrays, since the solution at each time level is an array. The length of U0 gives information on how many equations and unknowns there are in the system. An updated code is

```
if isinstance(U0, (float,int)):
    u = np.zeros(n+1)
else:
    neq = len(U0)
    u = np.zeros((n+1,neq))
```

Fortunately, the rest of the code now works regardless of whether u is a one- or two-dimensional array. In the former case, u[k+1] = u[k] + ... involves computations with float objects only, while in the latter case, u[k+1] picks out row $k + 1$ in the two-dimensional array u. This row is the array with the two unknown values at time t_{k+1}: $u^{(0)}(t_{k+1})$ and $u^{(1)}(t_{k+1})$. The statement u[k+1] = u[k] + ... then involves array arithmetics with arrays of length two in this specific example.

Allowing lists The specification of f and U0 using arrays is not as readable as
a plain list specification:

```
def f(u, t):
    return [u[1], 1./m*(F(t) - beta*u[1] - k*u[0])]

U0 = [0.1, 0]
```

Users would probably prefer the list syntax. With a little adjustment inside the
modified ForwardEuler function we can allow lists, tuples, or arrays for U0 and as
return objects from f. With U0 we just do

```
U0 = np.asarray(U0)
```

since np.asarray will just return U0 if it already is an array and otherwise copy
the data to an array.

The situation is a bit more demanding with the f function. The array operation
dt*f(u[k], t[k]) will not work unless f really returns an array (since lists or
tuples cannot be multiplied by a scalar dt). A trick is to wrap a function around the
user-provided right-hand-side function:

```
def ForwardEuler(f_user, dt, U0, T):
    def f(u, t):
        return np.asarray(f_user(u, t))
    ...
```

Now, dt*f(u[k], t[k]) will call f, which calls the user's f_user and turns
whatever is returned from that function into a NumPy array. A more compact syntax
arises from using a lambda function (see Sect. 3.1.14):

```
def ForwardEuler(f_user, dt, U0, T):
    f = lambda u, t: np.asarray(f_user(u, t))
    ...
```

The file ForwardEuler_sys_func.py contains the complete code. Verification of the implementation in terms of a test function is lacking, but Exercise E.24
encourages you to write such a function.

Let us apply the software to solve the equation $u'' + u = 0$, $u(0) = 0$, $u'(0) = 1$,
with solution $u(t) = \sin(t)$ and $u'(t) = \cos(t)$. The corresponding first-order ODE
system is derived in Sect. E.2.2. The right-hand side of the system is in the present
case $(u^{(1)}, -u^{(0)})$. The following Python function solves the problem

```
def demo(T=8*np.pi, n=200):
    def f(u, t):
        return [u[1], -u[0]]

    U0 = [0, 1]
    u, t = ForwardEuler(f, U0, T, n)
    u0 = u[:,0]
```

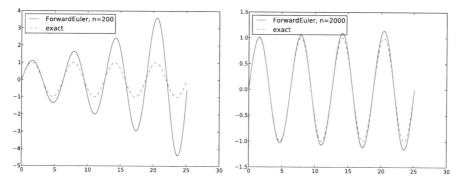

Fig. E.3 Comparison of large (*left*) and small (*right*) time step when solving $u'' + u = 0$ by the Forward Euler method

```
# Plot u0 vs t and compare with exact solution sin(t)
from matplotlib.pyplot import plot, show, savefig, legend
plot(t, u0, 'r-', t, np.sin(t), 'b--')
legend(['ForwardEuler, n=%d' % n, 'exact'], loc='upper left')
savefig('tmp.pdf')
show()
```

Storage of the solution of ODE systems

When solving systems of ODEs, the computed solution u is a two-dimensional array where u[k,i] holds the unknown function number i at time point number k: $u^{(i)}(t_k)$. Hence, to grab all the values associated with $u^{(0)}$, we fix i as 0 and let the k index take on all its legal values: u[:,0]. This slice of the u array refers to the piece of u where the discrete values $u^{(0)}(t_0), u^{(0)}(t_1), \ldots, u^{(0)}(t_n)$ are stored. The remaining part of u, u[:,1], holds all the discrete values of the computed $u'(t)$.

From visualizing the solution, see Fig. E.3, we realize that the Forward Euler method leads to a growing amplitude, while the exact solution has a constant amplitude. Fortunately, the amplification effect is reduced when Δt is reduced, but other methods, especially the 4th-order Runge-Kutta method, can solve this problem much more efficiently, see Sect. E.3.7.

To really understand how we generalized the code for scalar ODEs to systems of ODEs, it is recommended to do Exercise E.25.

E.2.4   Class Implementation

A class version of code in the previous section naturally starts with class ForwardEuler from Sect. E.1.7. The first task is to make similar adjustments of the code as we did for the ForwardEuler function: the trick with the lambda function for allowing the user's f to return a list is introduced in the constructor, and distinguishing between scalar and vector ODEs is required where self.U0 and self.u are created. The complete class looks as follows:

```python
class ForwardEuler(object):
    """
    Class for solving a scalar of vector ODE,

      du/dt = f(u, t)

    by the ForwardEuler solver.

    Class attributes:
    t: array of time values
    u: array of solution values (at time points t)
    k: step number of the most recently computed solution
    f: callable object implementing f(u, t)
    """
    def __init__(self, f):
        if not callable(f):
            raise TypeError('f is %s, not a function' % type(f))
        self.f = lambda u, t: np.asarray(f(u, t))

    def set_initial_condition(self, U0):
        if isinstance(U0, (float,int)):  # scalar ODE
            self.neq = 1
        else:                            # system of ODEs
            U0 = np.asarray(U0)
            self.neq = U0.size
        self.U0 = U0

    def solve(self, time_points):
        """Compute u for t values in time_points list."""
        self.t = np.asarray(time_points)
        n = self.t.size
        if self.neq == 1:  # scalar ODEs
            self.u = np.zeros(n)
        else:              # systems of ODEs
            self.u = np.zeros((n,self.neq))

        # Assume self.t[0] corresponds to self.U0
        self.u[0] = self.U0

        # Time loop
        for k in range(n-1):
            self.k = k
            self.u[k+1] = self.advance()
        return self.u, self.t

    def advance(self):
        """Advance the solution one time step."""
        u, f, k, t = self.u, self.f, self.k, self.t
        dt = t[k+1] - t[k]
        u_new = u[k] + dt*f(u[k], t[k])
        return u_new
```

You are strongly encouraged to do Exercise E.26 to understand class ForwardEuler listed above. This will also be an excellent preparation for the further generalizations of class ForwardEuler in Sect. E.3.

E.3   The ODESolver Class Hierarchy

This section takes class ForwardEuler from Sect. E.2.4 as a starting point for creating more flexible software where the user can switch problem and numerical method with very little coding. Also, the developer of the tool must be able to include a new numerical method with a minimum of coding. These requirements can be met by utilizing object-oriented programming. Recommended background material consists of Sects. 9.1–9.3.

E.3.1   Numerical Methods

Numerical methods for ODEs compute approximations u_k to the exact solution u at discrete time levels t_k, $k = 1, 2, 3, \ldots$. Some of the simplest, but also most widely used methods for ODEs are listed below.

The *Forward Euler method* has the formula

$$u_{k+1} = u_k + \Delta t \ f(u_k, t_k), \quad \Delta t = t_{k+1} - t_k \,. \tag{E.34}$$

The *Leapfrog method* (also called the Midpoint method) involves three time levels and is written as

$$u_{k+1} = u_{k-1} + 2\Delta t f(u_k, t_k), \quad 2\Delta t = t_{k+1} - t_{k-1} \tag{E.35}$$

for $k = 1, 2, \ldots$. The computation of u_1 requires u_{-1}, which is unknown, so for the first step we must use another method, for instance, (E.34). *Heun's method* is a two-step procedure,

$$u_* = u_k + \Delta t f(u_k, t_k), \tag{E.36}$$

$$u_{k+1} = u_k + \frac{1}{2}\Delta t f(u_k, t_k) + \frac{1}{2}\Delta t f(u_*, t_{k+1}), \tag{E.37}$$

with $\Delta t = t_{k+1} - t_k$. A closely related technique is the *2nd-order Runge-Kutta method*, commonly written as

$$u_{k+1} = u_k + K_2 \tag{E.38}$$

where

$$K_1 = \Delta t \ f(u_k, t_k), \tag{E.39}$$

$$K_2 = \Delta t \ f(u_k + \frac{1}{2}K_1, t_k + \frac{1}{2}\Delta t), \tag{E.40}$$

with $\Delta t = t_{k+1} - t_k$.

The perhaps most famous and most widely used method for solving ODEs is the *4th-order Runge-Kutta method*:

$$u_{k+1} = u_k + \frac{1}{6}\left(K_1 + 2K_2 + 2K_3 + K_4\right), \tag{E.41}$$

where

$$K_1 = \Delta t \ f(u_k, t_k), \tag{E.42}$$

$$K_2 = \Delta t \ f(u_k + \frac{1}{2}K_1, t_k + \frac{1}{2}\Delta t), \tag{E.43}$$

$$K_3 = \Delta t \ f(u_k + \frac{1}{2}K_2, t_k + \frac{1}{2}\Delta t), \tag{E.44}$$

$$K_4 = \Delta t \ f(u_k + K3, t_k + \Delta t), \tag{E.45}$$

and $\Delta t = t_{k+1} - t_k$. Another common technique is the *3rd-order Adams-Bashforth method*:

$$u_{k+1} = u_k + \frac{\Delta t}{12} \left(23 f(u_k, t_k) - 16 f(u_{k-1}, t_{k-1}) + 5 f(u_{k-2}, t_{k-2})\right), \tag{E.46}$$

with Δt constant. To start the scheme, one can apply a 2nd-order Runge-Kutta method or Heun's method to compute u_1 and u_2 before (E.46) is applied for $k \geq 2$. A more complicated solution procedure is the *Midpoint method with iterations*:

$$v_q = u_k + \frac{1}{2}\Delta t \left(f(v_{q-1}, t_{k+1}) + f(u_k, t_k)\right), \tag{E.47}$$

$$q = 1, \dots, N, \ v_0 = u_k$$

$$u_{k+1} = v_N . \tag{E.48}$$

At each time level, one runs the formula (E.47) N times, and the value v_N becomes u_{k+1}. Setting $N = 1$ recovers the Forward Euler scheme if f is independent of t, while $N = 2$ corresponds to Heun's method. We can either fix the value of N, or we can repeat (E.47) until the change in v_q is small, that is, until $|v_q - v_{q-1}| < \epsilon$, where ϵ is a small value.

Finally, we mention the *Backward Euler method*:

$$u_{k+1} = u_k + \Delta t \ f(u_{k+1}, t_{k+1}), \quad \Delta t = t_{k+1} - t_k . \tag{E.49}$$

If $f(u, t)$ is nonlinear in u, (E.49) constitutes a nonlinear equation in u_{k+1}, which must be solved by some method for nonlinear equations, say Newton's method.

All the methods listed above are valid both for scalar ODEs and for systems of ODEs. In the system case, the quantities $u, u_k, u_{k+1}, f, K_1, K_2$, etc., are vectors.

E.3.2   Construction of a Solver Hierarchy

Section E.2.4 presents a class `ForwardEuler` for implementing the Forward Euler scheme (E.34) both for scalar ODEs and systems. Only the advance method should be necessary to change in order to implement other numerical methods. Copying the `ForwardEuler` class and editing just the advance method is considered bad programming practice, because we get two copies the general parts of class `ForwardEuler`. As we implement more schemes, we end up with a lot of

copies of the same code. Correcting an error or improving the code in this general
part therefore requires identical edits in several almost identical classes.

A good programming practice is to collect all the common code in a superclass. Subclasses can implement the `advance` method, but share the constructor,
the `set_initial_condition` method, and the `solve` method with the superclass.

The superclass   We introduce class `ODESolver` as the superclass of various numerical methods for solving ODEs. Class `ODESolver` should provide all functionality
that is common to all numerical methods for ODEs:

- hold the solution $u(t)$ at discrete time points in an array `u`
- hold the corresponding time values `t`
- hold information about the $f(u, t)$ function, i.e., a callable Python object `f(u,
  t)`
- hold the current time step number k in a data attribute `k`
- hold the initial condition U_0
- implement the loop over all time steps

As already outlined in class `ForwardEuler` in Sects. E.1.7 and E.2.4, we implement
the last point as two methods: `solve` for performing the time loop and `advance` for
advancing the solution one time step. The latter method is empty in the superclass
since the method is to be implemented by various subclasses for various specific
numerical schemes.

A first version of class `ODESolver` follows directly from class `ForwardEuler`
in Sect. E.1.7, but letting `advance` be an empty method. However, there is one
more extension which will be handy in some problems, namely a possibility for
the user to terminate the simulation if the solution has certain properties. Throwing
a ball yields an example: the simulation should be stopped when the ball hits the
ground, instead of simulating an artificial movement down in the ground until the
final time T is reached. To implement the requested feature, the user can provide
a function `terminate(u, t, step_no)`, which returns `True` if the time loop is be
terminated. The arguments are the solution array `u`, the corresponding time points
`t`, and the current time step number `step_no`. For example, if we want to solve an
ODE until the solution is (close enough to) zero, we can supply the function

```
def terminate(u, t, step_no):
    eps = 1.0E-6                    # small number
    return abs(u[step_no]) < eps    # close enough to zero?
```

The `terminate` function is an optional argument to the `solve` method. By default,
a function that always returns `False` is used.

The suggested code for the superclass `ODESolver` takes the following form:

```
class ODESolver(object):
    def __init__(self, f):
        self.f = lambda u, t: np.asarray(f(u, t), float)

    def advance(self):
        """Advance solution one time step."""
        raise NotImplementedError
```

```
    def set_initial_condition(self, U0):
        if isinstance(U0, (float,int)):   # scalar ODE
            self.neq = 1
            U0 = float(U0)
        else:                             # system of ODEs
            U0 = np.asarray(U0)
            self.neq = U0.size
        self.U0 = U0

    def solve(self, time_points, terminate=None):
        if terminate is None:
            terminate = lambda u, t, step_no: False

        self.t = np.asarray(time_points)
        n = self.t.size
        if self.neq == 1:  # scalar ODEs
            self.u = np.zeros(n)
        else:              # systems of ODEs
            self.u = np.zeros((n,self.neq))

        # Assume that self.t[0] corresponds to self.U0
        self.u[0] = self.U0

        # Time loop
        for k in range(n-1):
            self.k = k
            self.u[k+1] = self.advance()
            if terminate(self.u, self.t, self.k+1):
                break  # terminate loop over k
        return self.u[:k+2], self.t[:k+2]
```

Note that we return just the parts of self.u and self.t that have been filled with values (the rest are zeroes): all elements up to the one with index k+1 are computed before terminate may return True. The corresponding slice of the array is then :k+2 since the upper limit is not included in the slice. If terminate never returns True we simply have that :k+1 is the entire array.

The Forward Euler method Subclasses implement specific numerical formulas for numerical solution of ODEs in the advance method. The Forward Euler scheme (E.34) is implemented by defining the subclass name and copying the advance method from the ForwardEuler class in Sect. E.1.7 or E.1.7:

```
class ForwardEuler(ODESolver):
    def advance(self):
        u, f, k, t = self.u, self.f, self.k, self.t
        dt = t[k+1] - t[k]
        u_new = u[k] + dt*f(u[k], t[k])
        return u_new
```

Remark on stripping off the self prefix

When we extract data attributes to local variables with short names, we should only use these local variables for reading values, not setting values. For example, if we do a k += 1 to update the time step counter, that increased value is not reflected in self.k (which is the "official" counter). On the other hand, changing a list *in-place*, say u[k+1] = ..., is reflected in self.u. Extracting data attributes in local variables is done for getting the code closer to the mathematics, but has a danger of introducing bugs that might be hard to track down.

The 4th-order Runge-Kutta method Below is an implementation of the 4th-order Runge-Kutta method (E.41):

```
class RungeKutta4(ODESolver):
    def advance(self):
        u, f, k, t = self.u, self.f, self.k, self.t
        dt = t[k+1] - t[k]
        dt2 = dt/2.0
        K1 = dt*f(u[k], t[k])
        K2 = dt*f(u[k] + 0.5*K1, t[k] + dt2)
        K3 = dt*f(u[k] + 0.5*K2, t[k] + dt2)
        K4 = dt*f(u[k] + K3, t[k] + dt)
        u_new = u[k] + (1/6.0)*(K1 + 2*K2 + 2*K3 + K4)
        return u_new
```

It is left as exercises to implement other numerical methods in the ODESolver class hierarchy. However, the Backward Euler method (E.49) requires a much more advanced implementation than the other methods so that particular method deserves its own section.

E.3.3 The Backward Euler Method

The Backward Euler scheme (E.49) leads in general to a *nonlinear* equation at a new time level, while all the other schemes listed in Sect. E.3.1 have a simple formula for the new u_{k+1} value. The nonlinear equation reads

$$u_{k+1} = u_k + \Delta t \, f(u_{k+1}, t_{k+1}).$$

For simplicity we assume that the ODE is scalar so the unknown u_{k+1} is a scalar. It might be easier to see that the equation for u_{k+1} is nonlinear if we rearrange the equation to

$$F(w) \equiv w - \Delta t f(w, t_{k+1}) - u_k = 0, \qquad (E.50)$$

where $w = u_{k+1}$. If now $f(u, t)$ is a nonlinear function of u, $F(w)$ will also be a nonlinear function of w.

To solve $F(w) = 0$ we can use the Bisection method from Sect. 4.11.2, Newton's method from Sect. A.1.10, or the Secant method from Exercise A.10. Here we apply Newton's method and the implementation given in src/diffeq/Newton.py. A disadvantage with Newton's method is that we need the derivative of F with respect to w, which requires the derivative $\partial f(w, t)/\partial w$. A quick solution is to use a numerical derivative, e.g., class Derivative from Sect. 7.3.2.

We make a subclass BackwardEuler. As we need to solve $F(w) = 0$ at every time step, we also need to implement the $F(w)$ function. This is conveniently done in a local function inside the advance method:

```
def advance(self):
    u, f, k, t = self.u, self.f, self.k, self.t

    def F(w):
        return w - dt*f(w, t[k+1]) - u[k]

    dFdw = Derivative(F)
    w_start = u[k] + dt*f(u[k], t[k])   # Forward Euler step
    u_new, n, F_value = self.Newton(F, w_start, dFdw, N=30)
    if n >= 30:
        print "Newton's failed to converge at t=%g "\
            "(%d iterations)" % (t, n)
    return u_new
```

The local variables in the advance function, such as dt and u, act as "global" variables for the F function. Hence, when F is sent away to some self.Newton function, F remembers the values of dt, f, t, and u (!). The derivative dF/dw is in our advance function computed numerically by a class Derivative, which is a slight modification of the similar class in Sect. 7.3.2, because we now want to use a more accurate, centered formula:

```
class Derivative(object):
    def __init__(self, f, h=1E-5):
        self.f = f
        self.h = float(h)

    def __call__(self, x):
        f, h = self.f, self.h
        return (f(x+h) - f(x-h))/(2*h)
```

This code is included in the ODESolver.py file.

The next step is to call Newton's method. For this purpose we need to import the Newton function from the Newton module. The Newton.py file must then reside in the same directory as ODESolver.py, or Newton.py must be in one of the directories listed in the sys.path list or the PYTHONPATH environment variable (see Sect. 4.9.7).

Having the Newton(f, x_start, dfdx, N) function from Sect. A.1.10 accessible in our ODESolver.py file, we can make a call and supply our F function as the argument f, a start value for the iteration, here called w_start, as the argument x, and the derivative dFdw for the argument dfdx. We rely on default values for the epsilon and store arguments, while the maximum number of iterations is set to N=30. The program is terminated if it happens that the number of iterations exceeds that value, because then the method is not considered to have converged (at least not quickly enough), and we have consequently not been able to compute the next u_{k+1} value.

The starting value for Newton's method must be chosen. As we expect the solution to not change much from one time level to the next, u_k could be a good initial guess. However, we can do better by using a simple Forward Euler step $u_k + \Delta t f(u_k, t_k)$, which is exactly what we do in the advance function above.

Since Newton's method always has the danger of converging slowly, it can be interesting to store the number of iterations at each time level as a data attribute in the BackwardEuler class. We can easily insert extra statement for this purpose:

```
def advance(self):
    ...
    u_new, n, F_value = Newton(F, w_start, dFdw, N=30)
    if k == 0:
        self.Newton_iter = []
    self.Newton_iter.append(n)
    ...
```

Note the need for creating an empty list (at the first call of advance) before we can append elements.

There is now one important question to ask: will the advance method work for systems of ODEs? In that case, $F(w)$ is a vector of functions. The implementation of F will work when w is a vector, because all the quantities involved in the formula are arrays or scalar variables. The dFdw instance will compute a numerical derivative of each component of the vector function dFdw.f (which is simply our F function). The call to the Newton function is more critical: It turns out that this function, as the algorithm behind it, works for scalar equations only. Newton's method can quite easily be extended to a system of nonlinear equations, but we do not consider that topic here. Instead we equip class BackwardEuler with a constructor that calls the f object and controls that the returned value is a float and not an array:

```
class BackwardEuler(ODESolver):
    def __init__(self, f):
        ODESolver.__init__(self, f)
        # Make a sample call to check that f is a scalar function:
        try:
            u = np.array([1]); t = 1
            value = f(u, t)
        except IndexError:  # index out of bounds for u
            raise ValueError('f(u,t) must return float/int')
```

Observe that we must explicitly call the superclass constructor and pass on the argument f to achieve the right storage and treatment of this argument.

Understanding class BackwardEuler implies a good understanding of classes in general; a good understanding of numerical methods for ODEs, for numerical differentiation, and for finding roots of functions; and a good understanding on how to combine different code segments from different parts of the book. Therefore, if you have digested class BackwardEuler, you have all reasons to believe that you have digested the key topics of this book.

E.3.4   Verification

The fundamental problem with testing approximate numerical methods is that we do not normally know what the output from the computer should be. In some special cases, however, we can find an exact solution of the discretized problem that the computer program solves. This exact solution should be reproduced to machine precision by the program. It turns out that most numerical methods for ordinary differential equations are able to exactly reproduce a linear solution. That is, if the solution of the differential equation is $u = at + b$, the numerical method will produce the same solution: $u_k = ak\Delta t + b$. We can use this knowledge to make a test function for verifying our implementations.

Let $u = at + b$ be the solution of the test problem. A corresponding ODE is obviously $u' = a$, with $u(0) = b$. A more demanding ODE arises from adding a term that is zero, e.g., $(u - (at + b))^5$. We therefore aim to solve

$$u' = a + (u - (at + b))^5, \quad u(0) = b.$$

Our test function loops over registered solvers in the `ODESolver` hierarchy, solves the test problem, and checks that the maximum deviation between the computed solution and the exact linear solution is within a tolerance:

```
registered_solver_classes = [
    ForwardEuler, RungeKutta4, BackwardEuler]

def test_exact_numerical_solution():
    a = 0.2; b = 3

    def f(u, t):
        return a + (u - u_exact(t))**5

    def u_exact(t):
        """Exact u(t) corresponding to f above."""
        return a*t + b

    U0 = u_exact(0)
    T = 8
    n = 10
    tol = 1E-15
    t_points = np.linspace(0, T, n)
    for solver_class in registered_solver_classes:
        solver = solver_class(f)
        solver.set_initial_condition(U0)
        u, t = solver.solve(t_points)
        u_e = u_exact(t)
        max_error = (u_e - u).max()
        msg = '%s failed with max_error=%g' % \
              (solver.__class__.__name__, max_error)
        assert max_error < tol, msg
```

Note how we can make a loop over class types (because the class is an ordinary object in Python). New subclasses can add their class type to the `registered_solver_classes` list and the test function will include such new classes in the test as well.

Remarks A more general testing technique is based on knowing how the error in a numerical method varies with the discretization parameter, here Δt. Say we know that a particular method has an error that decays as Δt^2. In a problem where the exact analytical solution is known, we can run the numerical method for several values of Δt and compute the corresponding numerical error in each case. If the computed errors decay like Δt^2, it brings quite strong evidence for a correct implementation. Such tests are called convergence tests and constitute the most general tool we have for verifying implementations of numerical algorithms. Exercise E.37 gives an introduction to the topic.

E.3.5   Example: Exponential Decay

Let us apply the classes in the ODESolver hierarchy to see how they solve the perhaps simplest of all ODEs: $u' = -u$, with initial condition $u(0) = 1$. The exact solution is $u(t) = e^{-t}$, which decays exponentially with time. Application of class ForwardEuler to solve this problem requires writing the following code:

```
import ODESolver

def f(u, t):
    return -u

solver = ODESolver.ForwardEuler(f)
solver.set_initial_condition(1.0)
t_points = linspace(0, 3, 31)
u, t = solver.solve(t_points)
plot(t, u)
```

We can run various values of Δt to see the effect on the accuracy:

```
# Test various dt values and plot
figure()
legends = []
T = 3
for dt in 2.0, 1.0, 0.5, 0.1:
    n = int(round(T/dt))
    solver = ODESolver.ForwardEuler(f)
    solver.set_initial_condition(1)
    u, t = solver.solve(linspace(0, T, n+1))
    plot(t, u)
    legends.append('dt=%g' % dt)
    hold('on')
plot(t, exp(-t), 'bo')
legends.append('exact')
legend(legends)
```

Figure E.4 shows alarming results. With $\Delta t = 2$ we get a completely wrong solution that becomes negative and then increasing. The value $\Delta t = 1$ gives a peculiar solution: $u_k = 0$ for $k \geq 1$! Qualitatively correct behavior appears with $\Delta t = 0.5$, and the results get quantitatively better as we decrease Δt. The solution corresponding to $\Delta = 0.1$ looks good from the graph.

Such strange results reveal that we most likely have programming errors in our implementation. Fortunately, we did some verification of the implementations in

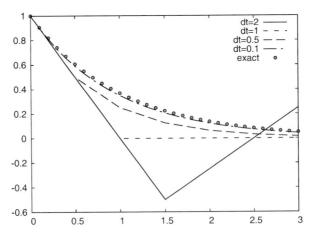

Fig. E.4 Solution of $u' = -u$ for $t \in [0, 3]$ by the Forward Euler method and $\Delta t \in \{2, 1, 0.5, 0.1\}$

Sect. E.3.4, so it might well happen that what we observe in the experiments are problems with the numerical method and not with the implementation.

We can in fact easily explain what we observe in Fig. E.4. For the equation in question, the Forward Euler method computes

$$u_1 = u_0 - \Delta t u_0 = (1 - \Delta t)u_0,$$

$$u_2 = u_1 - \Delta t u_1 = (1 - \Delta t)u_1 = (1 - \Delta t)^2 u_0,$$

$$\vdots$$

$$u_k = (1 - \Delta t)^k u_0.$$

With $\Delta t = 1$ we simply get $u_k = 0$ for $k \geq 1$. For $\Delta t > 1$, $1 - \Delta t < 0$, and $(1 - \Delta t)^k$ means raising a negative value to an integer power, which results in $u_k > 0$ for even k and $u_k < 0$ for odd k. Moreover, $|u_k|$ decreases with k. Such a growing, oscillating solution is of course qualitatively wrong when the exact solution is e^{-t} and monotonically decaying. The conclusion is that the Forward Euler method gives meaningless results for $\Delta t \geq 1$ in the present example.

A particular strength of the ODESolver hierarchy of classes is that we can trivially switch from one method to another. For example, we may demonstrate how superior the 4-th order Runge-Kutta method is for this equation: just replace ForwardEuler by RungeKutta4 in the previous code segment and re-run the program. It turns out that the 4-th order Runge-Kutta method gives a monotonically decaying numerical solution for all the tested Δt values. In particular, the solutions corresponding to $\Delta t = 0.5$ and $\Delta t = 0.1$ are visually very close to the exact solution. The conclusion is that the 4-th order Runge-Kutta method is a safer and more accurate method.

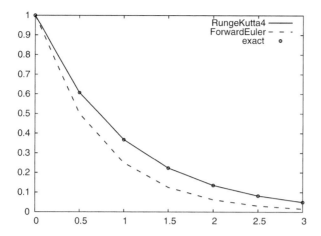

Fig. E.5 Comparison of the Forward Euler and the 4-th order Runge-Kutta method for solving $u' = -u$ for $t \in [0, 3]$ and a time step $\Delta t = 0.5$

Let us compare the two numerical methods in the case where $\Delta t = 0.5$:

```
# Test ForwardEuler vs RungeKutta4
figure()
legends = []
T = 3
dt = 0.5
n = int(round(T/dt))
t_points = linspace(0, T, n+1)
for solver_class in ODESolver.RungeKutta4, ODESolver.ForwardEuler:
    solver = solver_class(f)
    solver.set_initial_condition(1)
    u, t = solver.solve(t_points)
    plot(t, u)
    legends.append('%s' % solver_class.__name__)
    hold('on')
plot(t, exp(-t), 'bo')
legends.append('exact')
legend(legends)
```

Figure E.5 illustrates that differences in accuracy between the two methods. The complete program can be found in the file app1_decay.py.

E.3.6   Example: The Logistic Equation with Problem and Solver Classes

The logistic ODE (E.5) is copied here for convenience:

$$u'(t) = \alpha u(t) \left(1 - \frac{u(t)}{R} \right), \quad u(0) = U_0.$$

The right-hand side contains the parameters α and R. We know that $u \rightarrow R$ as $t \rightarrow \infty$, so at some point $\hat{t}$ in time we have approached the asymptotic value $u = R$ within a sufficiently small tolerance and should stop the simulation. This can be done by providing a function as the `tolerance` argument in the `solve` method.

Basic problem and solver classes Let us, as in Sect. E.1.9, implement the problem-dependent data in a class. This time we store all user-given physical data in the class:

```
import ODESolver
from scitools.std import plot, figure, savefig, title, show
#from matplotlib.pyplot import plot, figure, savefig, title, show
import numpy as np

class Problem(object):
    def __init__(self, alpha, R, U0, T):
        """
        alpha, R: parameters in the ODE.
        U0: initial condition.
        T: max length of time interval for integration;
        asympotic value R must be reached within 1%
        accuracy for some t <= T.
        """
        self.alpha, self.R, self.U0, self.T = alpha, R, U0, T

    def __call__(self, u, t):
        """Return f(u,t) for logistic ODE."""
        return self.alpha*u*(1 - u/self.R)

    def terminate(self, u, t, step_no):
        """Return True when asymptotic value R is reached."""
        tol = self.R*0.01
        return abs(u[step_no] - self.R) < tol

    def __str__(self):
        """Pretty print of physical parameters."""
        return 'alpha=%g, R=%g, U0=%g' % \
               (self.alpha, self.R, self.U0)
```

Note that the tolerance used in the `terminate` method is made dependent on the size of R: $|u - R|/R < 0.01$. For example, if $R = 1000$ we say the asymptotic value is reached when $u \geq 990$. Smaller tolerances will just lead to a solution curve where large parts of it show the boring behavior $u \approx R$.

The solution is obtained the usual way by short code:

```
solver = ODESolver.RungeKutta4(problem)
solver.set_initial_condition(problem.U0)
dt = 1.0
n = int(round(problem.T/dt))
t_points = np.linspace(0, T, n+1)
u, t = solver.solve(t_points, problem.terminate)
```

Let us pack these statements into a class `Solver`, which has two methods: `solve` and `plot`, and add some documentation and flexibility. The code may look like

```
class Solver(object):
    def __init__(self, problem, dt,
                    method=ODESolver.ForwardEuler):
        """
        problem: instance of class Problem.
        dt: time step.
        method: class in ODESolver hierarchy.
        """
        self.problem, self.dt = problem, dt
        self.solver = method

    def solve(self):
        solver = self.method(self.problem)
        solver.set_initial_condition(self.problem.U0)
        n = int(round(self.problem.T/self.dt))
        t_points = np.linspace(0, self.problem.T, n+1)
        self.u, self.t = solver.solve(t_points,
                                        self.problem.terminate)

        # The solution terminated if the limiting value was reached
        if solver.k+1 == n:  # no termination - we reached final T
            self.plot()
            raise ValueError(
                'termination criterion not reached, '\
                'give T > %g' % self.problem.T)

    def plot(self):
        filename = 'logistic_' + str(self.problem) + '.pdf'
        plot(self.t, self.u)
        title(str(self.problem) + ', dt=%g' % self.dt)
        savefig(filename)
        show()
```

Problem-dependent data related to the numerical quality of the solution, such as the time step here, go to the `Solver` class. That is, class `Problem` contains the physics and class `Solver` the numerics of the problem under investigation.

If the last computed time step, `solver.k+1`, equals the last possible index, `n`, `problem.terminate` never returned `True`, which means that the asymptotic limit was not reached. This is treated as an erroneous condition. To guide the user, we launch a plot before raising the exception with an instructive message. The complete code is found in the file `app2_logistic.py`.

Computing an appropriate Δt   Choosing an appropriate Δt is not always so easy. The impact of Δt can sometimes be dramatic, as demonstrated for the Forward Euler method in Sect. E.3.5. We could automate the process of finding a suitable Δt: start with a large Δt, and keep halving Δt until the difference between two solutions corresponding to two consequtive Δt values is small enough.

Say `solver` is a class `Solver` instance computed with time step Δt and `solver2` is the instance corresponding to a computation with $\Delta t/2$. Calculating the difference between `solver.u` and `solver2.u` is not trivial as one of the arrays has approximately twice as many elements as the other, and the last element in both arrays does not necessarily correspond to the same time value since the time stepping and the `terminate` function may lead to slightly different termination times.

A solution to these two problems is to turn each of the arrays `solver.u` and `solver2.u` into *continuous functions*, as explained in Sect. E.1.5, and then evaluate the difference at some selected time points up to the smallest value of `solver.t[-1]` and `solver2.t[-1]`. The code becomes

```
# Make continuous functions u(t) and u2(t)
u  = wrap2callable((solver. t, solver. u))
u2 = wrap2callable((solver2.t, solver2.u))
# Sample the difference in n points in [0, t_end]
n = 13
t_end = min(solver2.t[-1], solver.t[-1])
t = np.linspace(0, t_end, n)
u_diff = np.abs(u(t) - u2(t)).max()
```

The next step is to introduce a loop where we halve the time step in each iteration and solve the logistic ODE with the new time step and compute `u_diff` as shown above. A complete function takes the form

```
def find_dt(problem, method=ODESolver.ForwardEuler,
            tol=0.01, dt_min=1E-6):
    """
    Return a "solved" class Solver instance where the
    difference in the solution and one with a double
    time step is less than tol.

    problem: class Problem instance.
    method: class in ODESolver hierarchy.
    tol: tolerance (chosen relative to problem.R).
    dt_min: minimum allowed time step.
    """
    dt = problem.T/10  # start with 10 intervals
    solver = Solver(problem, dt, method)
    solver.solve()
    from scitools.std import wrap2callable

    good_approximation = False
    while not good_approximation:
        dt = dt/2.0
        if dt < dt_min:
            raise ValueError('dt=%g < %g - abort' % (dt, dt_min))

        solver2 = Solver(problem, dt, method)
        solver2.solve()

        # Make continuous functions u(t) and u2(t)
        u  = wrap2callable((solver. t, solver. u))
        u2 = wrap2callable((solver2.t, solver2.u))

        # Sample the difference in n points in [0, t_end]
        n = 13
        t_end = min(solver2.t[-1], solver.t[-1])
```

```
        t = np.linspace(0, t_end, n)
        u_diff = np.abs(u(t) - u2(t)).max()
        print u_diff, dt, tol
        if u_diff < tol:
            good_approximation = True
        else:
            solver = solver2
    return solver2
```

Setting the tolerance `tol` must be done with a view to the typical size of u, i.e., the size of R. With $R = 100$ and `tol=1`, the Forward Euler method meets the tolerance for $\Delta t = 0.25$. Switching to the 4-th order Runge-Kutta method makes $\Delta t = 1.625$ sufficient to meet the tolerance. Note that although the latter method can use a significantly larger time step, it also involves four times as many evaluations of the right-hand side function at each time step.

Finally, we show how to make a class that behaves as class `Solver`, but with automatic computation of the time step. If we do not provide a `dt` parameter to the constructor, the `find_dt` function just presented is used to compute `dt` and the solution, otherwise we use the standard `Solver.solve` code. This new class is conveniently realized as a subclass of `Solver` where we override the constructor and the `solve` method. The `plot` method can be inherited as is. The code becomes

```
class AutoSolver(Solver):
    def __init__(self, problem, dt=None,
                 method=ODESolver.ForwardEuler,
                 tol=0.01, dt_min=1E-6):
        Solver.__init__(self, problem, dt, method)
        if dt is None:
            solver = find_dt(self.problem, method,
                             tol, dt_min)
            self.dt = solver.dt
            self.u, self.t = solver.u, solver.t

    def solve(self, method=ODESolver.ForwardEuler):
        if hasattr(self, 'u'):
            # Solution was computed by find_dt in constructor
            pass
        else:
            Solver.solve(self)
```

The call `hasattr(self, 'u')` returns True if u is a data attribute in object `self`. Here this is used as an indicator that the solution was computed in the constructor by the `find_dt` function. A typical use is

```
problem = Problem(alpha=0.1, R=500, U0=2, T=130)
solver = AutoSolver(problem, tol=1)
solver.solve(method=ODESolver.RungeKutta4)
solver.plot()
```

Dealing with time-dependent coefficients   The carrying capacity of the environment, R, may vary with time, e.g., due to seasonal changes. Can we extend the previous code so that R can be specified either as a constant or as a function of time?

This is in fact easy if we in the implementation of the right-hand side function assume that R is a function of time. If R is given as a constant in the constructor of class Problem, we just wrap it as a function of time:

```
if isinstance(R, (float,int)):  # number?
    self.R = lambda t: R
elif callable(R):
    self.R = R
else:
    raise TypeError(
        'R is %s, has to be number of function' % type(R))
```

The terminate method is also affected as we need to base the tolerance on the R value at the present time level. Also the __str__ method must be changed since it is not meaningful to print a *function* self.R. That is, all methods in the generalized problem class, here called Problem2, must be altered. We have not chosen to make Problem2 a subclass of Problem, even though the interface is the same and the two classes are closely related. While Problem is clearly a special case of Problem2, as a constant R is a special case of a function (R), the opposite case is not true.

Class Problem2 becomes

```
class Problem2(Problem):
    def __init__(self, alpha, R, U0, T):
        """

        alpha, R: parameters in the ODE.
        U0: initial condition.
        T: max length of time interval for integration;
        asympotic value R must be reached within 1%
        accuracy for some t <= T.
        """
        self.alpha, self.U0, self.T = alpha, U0, T
        if isinstance(R, (float,int)):  # number?
            self.R = lambda t: R
        elif callable(R):
            self.R = R
        else:
            raise TypeError(
                'R is %s, has to be number of function' % type(R))

    def __call__(self, u, t):
        """Return f(u,t) for logistic ODE."""
        return self.alpha*u*(1 - u/self.R(t))

    def terminate(self, u, t, step_no):
        """Return True when asymptotic value R is reached."""
        tol = self.R(t[step_no])*0.01
        return abs(u[step_no] - self.R(t[step_no])) < tol

    def __str__(self):
        return 'alpha=%g, U0=%g' % (self.alpha, self.U0)
```

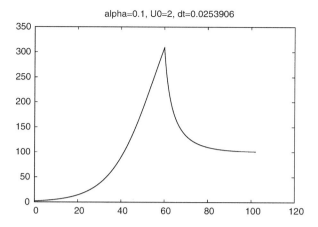

Fig. E.6 Solution of the logistic equation $u' = \alpha u \, (1 - u/R(t))$ when $R = 500$ for $t < 60$ and $R = 100$ for $t \geq 60$

We can compute the case where $R = 500$ for $t < 60$ and then reduced to $R = 100$ because of an environmental crisis (see Fig. E.6):

```
problem = Problem2(alpha=0.1, U0=2, T=130,
                   R=lambda t: 500 if t < 60 else 100)
solver = AutoSolver(problem, tol=1)
solver.solve(method=ODESolver.RungeKutta4)
solver.plot()
```

Note the use of a lambda function (Sect. 3.1.14) to save some typing when specifying R. The corresponding graph is made of two parts, basically exponential growth until the environment changes and then exponential reduction until u approaches the new R value and the change in u becomes small.

Reading input   Our final version of the problem class is equipped with functionality for reading data from the command line in addition to setting data explicitly in the program. We use the `argparse` module described in Sect. 4.4. The idea now is to have a constructor that just sets default values. Then we have a method for defining the command-line arguments and a method for transforming the `argparse` information to the attributes `alpha`, `U0`, `R`, and `T`. The R attribute is supposed to be a function, and we use the `StringFunction` tool to turn strings from the command-line into a Python function of time `t`.

The code of our new problem class is listed next.

```
class Problem3(Problem):
    def __init__(self):
        # Set default parameter values
        self.alpha = 1.
        self.R = StringFunction('1.0', independent_variable='t')
        self.U0 = 0.01
        self.T = 4.
```

```
def define_command_line_arguments(self, parser):
    """Add arguments to parser (argparse.ArgumentParser)."""

    def evalcmlarg(text):
        return eval(text)

    def toStringFunction(text):
        return StringFunction(text, independent_variable='t')

    parser.add_argument(
        '--alpha', dest='alpha', type=evalcmlarg,
        default=self.alpha,
        help='initial growth rate in logistic model')
    parser.add_argument(
        '--R', dest='R', type=toStringFunction, default=self.R,
        help='carrying capacity of the environment')
    parser.add_argument(
        '--U0', dest='U0', type=evalcmlarg, default=self.U0,
        help='initial condition')
    parser.add_argument(
        '--T', dest='T', type=evalcmlarg, default=self.T,
        help='integration in time interval [0,T]')
    return parser

def set(self, **kwargs):
    """
    Set parameters as keyword arguments alpha, R, U0, or T,
    or as args (object returned by parser.parse_args()).
    """
    for prm in ('alpha', 'U0', 'R', 'T'):
        if prm in kwargs:
            setattr(self, prm, kwargs[prm])
    if 'args' in kwargs:
        args = kwargs['args']
        for prm in ('alpha', 'U0', 'R', 'T'):
            if hasattr(args, prm):
                setattr(self, prm, getattr(args, prm))
            else:
                print 'Really strange', dir(args)

def __call__(self, u, t):
    """Return f(u,t) for logistic ODE."""
    return self.alpha*u*(1 - u/self.R(t))

def terminate(self, u, t, step_no):
    """Return True when asymptotic value R is reached."""
    tol = self.R(t[step_no])*0.01
    return abs(u[step_no] - self.R(t[step_no])) < tol

def __str__(self):
    s = 'alpha=%g, U0=%g' % (self.alpha, self.U0)
    if isinstance(self.R, StringFunction):
        s += ', R=%s' % str(self.R)
    return s
```

The calls to `parser.add_argument` are straightforward, but notice that we allow strings for α, U_0, and T to be interpreted by `eval`. The string for R is interpreted as a formula by `StringFunction`. The `set` method is flexible: it accepts any set of keyword arguments, and first checks if these are the names of the problem parameters, and thereafter if `args='` is given, the parameters are taken from the command line. The rest of the class is very similar to earlier versions.

The typical use of class `Problem3` is shown below. First we set parameters directly:

```
problem = Problem3()
problem.set(alpha=0.1, U0=2, T=130,
            R=lambda t: 500 if t < 60 else 100)
solver = AutoSolver(problem, tol=1)
solver.solve(method=ODESolver.RungeKutta4)
solver.plot()
```

Then we rely on reading parameters from the command line:

```
problem = Problem3()
import argparse
parser = argparse.ArgumentParser(
    description='Logistic ODE model')
parser = problem.define_command_line_arguments(parser)

# Try --alpha 0.11 --T 130 --U0 2 --R '500 if t < 60 else 300'
args = parser.parse_args()
problem.set(args=args)
solver = AutoSolver(problem, tol=1)
solver.solve(method=ODESolver.RungeKutta4)
solver.plot()
```

The last example using a problem class integrated with the command line is the most flexible way of implementing ODE models.

E.3.7   Example: An Oscillating System

The motion of a box attached to a spring, as described in detail in Appendix D, can be modeled by two first-order differential equations as listed in (E.33), here repeated with $F(t) = mw''(t)$, where the $w(t)$ function is the forced movement of the end of the spring.

$$\frac{du^{(0)}}{dt} = u^{(1)},$$
$$\frac{du^{(1)}}{dt} = w''(t) + g - m^{-1}\beta u^{(1)} - m^{-1}ku^{(0)}.$$

The code related to this example is found in `app3_osc.py`. Because our right-hand side f contains several parameters, we implement it as a class with the parameters as data attributes and a `__call__` method for returning the 2-vector f. We assume that the user of the class supplies the $w(t)$ function, so it is natural to compute $w''(t)$ by a finite difference formula.

```
class OscSystem:
    def __init__(self, m, beta, k, g, w):
        self.m, self.beta, self.k, self.g, self.w = \
                float(m), float(beta), float(k), float(g), w

    def __call__(self, u, t):
        u0, u1 = u
        m, beta, k, g, w = \
            self.m, self.beta, self.k, self.g, self.w
        # Use a finite difference for w''(t)
        h = 1E-5
        ddw = (w(t+h) - 2*w(t) + w(t-h))/(h**2)
        f = [u1, ddw  + g - beta/m*u1 - k/m*u0]
        return f
```

A simple test case arises if we set $m = k = 1$ and $\beta = g = w = 0$:

$$\frac{du^{(0)}}{dt} = u^{(1)},$$
$$\frac{du^{(1)}}{dt} = -u^{(0)}.$$

Suppose that $u^{(0)}(0) = 1$ and $u^{(1)}(0) = 0$. An exact solution is then

$$u^{(0)}(t) = \cos t, \quad u^{(1)}(t) = -\sin t.$$

We can use this case to check how the Forward Euler method compares with the 4-th order Runge-Kutta method:

```
import ODESolver
from scitools.std import *
#from matplotlib.pyplot import *
legends = []
f = OscSystem(1.0, 0.0, 1.0, 0.0, lambda t: 0)
u_init = [1, 0]      # initial condition
nperiods = 3.5       # no of oscillation periods
T = 2*pi*nperiods
for solver_class in ODESolver.ForwardEuler, ODESolver.RungeKutta4:
    if solver_class == ODESolver.ForwardEuler:
        npoints_per_period = 200
    elif solver_class == ODESolver.RungeKutta4:
        npoints_per_period = 20
    n = npoints_per_period*nperiods
    t_points = linspace(0, T, n+1)
    solver = solver_class(f)
    solver.set_initial_condition(u_init)
    u, t = solver.solve(t_points)

    # u is an array of [u0,u1] pairs for each time level,
    # get the u0 values from u for plotting
    u0_values = u[:, 0]
    u1_values = u[:, 1]
    u0_exact = cos(t)
    u1_exact = -sin(t)
```

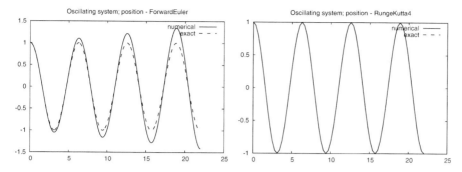

Fig. E.7 Solution of an oscillating system ($u'' + u = 0$ formulated as system of two ODEs) by the Forward Euler method with $\Delta t = 2\pi/200$ (*left*), and the 4-th order Runge-Kutta method with the same time step (*right*)

```
    figure()
    alg = solver_class.__name__    # (class) name of algorithm
    plot(t, u0_values, 'r-',
         t, u0_exact, 'b-')
    legend(['numerical', 'exact']),
    title('Oscillating system; position - %s' % alg)
    savefig('tmp_oscsystem_pos_%s.pdf' % alg)
    figure()
    plot(t, u1_values, 'r-',
         t, u1_exact, 'b-')
    legend(['numerical', 'exact'])
    title('Oscillating system; velocity - %s' % alg)
    savefig('tmp_oscsystem_vel_%s.pdf' % alg)
show()
```

For this particular application it turns out that the 4-th order Runge-Kutta is very accurate, even with few (20) time steps per oscillation. Unfortunately, the Forward Euler method leads to a solution with increasing amplitude in time. Figure E.7 shows a comparison between the two methods. Note that the Forward Euler method uses 10 times as many time steps as the 4-th order Runge-Kutta method and is still much less accurate. A very much smaller time step is needed to limit the growth of the Forward Euler scheme for oscillating systems.

E.3.8   Application 4: The Trajectory of a Ball

Exercise 1.13 derives the following two second-order differential equations for the motion of a ball (neglecting air resistance):

$$\frac{d^2x}{dt^2} = 0, \tag{E.51}$$

$$\frac{d^2y}{dt^2} = -g, \tag{E.52}$$

where (x, y) is the position of the ball (x is a horizontal measure and y is a vertical measure), and g is the acceleration of gravity. To use numerical methods for first-order equations, we must rewrite the system of two second-order equations as a system of four first-order equations. This is done by introducing to new unknowns, the velocities $v_x = dx/dt$ and $v_y = dy/dt$. We then have the first-order system of ODEs

$$\frac{dx}{dt} = v_x, \tag{E.53}$$

$$\frac{dv_x}{dt} = 0, \tag{E.54}$$

$$\frac{dy}{dt} = v_y, \tag{E.55}$$

$$\frac{dv_y}{dt} = -g. \tag{E.56}$$

The initial conditions are

$$x(0) = 0, \tag{E.57}$$

$$v_x(0) = v_0 \cos\theta, \tag{E.58}$$

$$y(0) = y_0, \tag{E.59}$$

$$v_y(0) = v_0 \sin\theta, \tag{E.60}$$

where v_0 is the initial magnitude of the velocity of the ball. The initial velocity has a direction that makes the angle θ with the horizontal.

The code related to this example is found in app4_ball.py. A function returning the right-hand side of our ODE system reads

```
def f(u, t):
    x, vx, y, vy = u
    g = 9.81
    return [vx, 0, vy, -g]
```

It makes sense to solve the ODE system as long as the ball as above the ground, i.e., as long as $y \geq 0$. We must therefore supply a terminate function as explained in Sect. E.3:

```
def terminate(u, t, step_no):
    y = u[:,2]                    # all the y coordinates
    return y[step_no] < 0
```

Observe that all the y values are given by u[:,2] and we want to test the value at the current step, which becomes u[step_no,2].

The main program for solving the ODEs can be set up as

```
v0 = 5
theta = 80*pi/180
U0 = [0, v0*cos(theta), 0, v0*sin(theta)]
T = 1.2; dt = 0.01; n = int(round(T/dt))
```

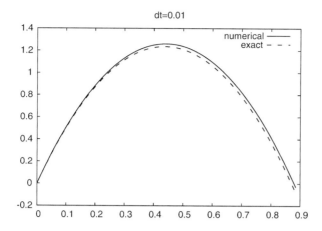

Fig. E.8 The trajectory of a ball solved as a system of four ODEs by the Forward Euler method

```
solver = ODESolver.ForwardEuler(f)
solver.set_initial_condition(U0)

def terminate(u, t, step_no):
    return False if u[step_no,2] >= 0 else True

u, t = solver.solve(linspace(0, T, n+1), terminate)
```

Now, `u[:,0]` represents all the $x(t)$ values, `u[:,1]` all the $v_x(t)$ values, `u[:,2]` all the $y(t)$ values, and `u[:,3]` all the $v_y(t)$ values. To plot the trajectory, y versus x, we write

```
x = u[:,0]
y = u[:,2]
plot(x, y)
```

The exact solution is given by (1.6), so we can easily assess the accuracy of the numerical solution. Figure E.8 shows a comparison of the numerical and the exact solution in this simple test problem. Note that even if we are just interested in y as a function of x, we first need to solve the complete ODE system for $x(t)$, $v_x(t)$, $y(t)$, and $v_y(t)$.

The real strength of the numerical approach is the ease with which we can add air resistance and lift to the system of ODEs. Insight in physics is necessary to derive what the additional terms are, but implementing the terms is trivial in our program above (do Exercise E.39).

E.3.9 Further Developments of ODESolver

The ODESolver hierarchy is a simplified prototype version of a more professional Python package for solving ODEs called Odespy. This package features a range of simple and sophisticated methods for solving scalar ODEs and systems of ODEs.

Some of the solvers are implemented in Python, while others call up well-known ODE software in Fortran. Like the ODESolver hierarchy, Odespy offers a unified interface to the different numerical methods, which means that the user can specify the ODE problem as a function f(u,t) and send this function to all solvers. This feature makes it easy to switch between solvers to test a wide collection of numerical methods for a problem.

Odespy can be downloaded from http://hplgit.github.com/odespy. It is installed by the usual python setup.py install command.

E.4   Exercises

Exercise E.1: Solve a simple ODE with function-based code

This exercise aims to solve the ODE problem $u - 10u' = 0$, $u(0) = 0.2$, for $t \in [0, 20]$.

a) Identify the mathematical function $f(u,t)$ in the generic ODE form $u' = f(u,t)$.
b) Implement the $f(u,t)$ function in a Python function.
c) Use the ForwardEuler function from Sect. E.1.3 to compute a numerical solution of the ODE problem. Use a time step $\Delta t = 5$.
d) Plot the numerical solution and the exact solution $u(t) = 0.2e^{0.1t}$.
e) Save the numerical solution to file. Decide upon a suitable file format.
f) Perform simulations for smaller Δt values and demonstrate visually that the numerical solution approaches the exact solution.

Filename: simple_ODE_func.

Exercise E.2: Solve a simple ODE with class-based code

Solve the same ODE problem as in Exercise E.1, but use the ForwardEuler class from Sect. E.1.7. Implement the right-hand side function f as a class too.
Filename: simple_ODE_class.

Exercise E.3: Solve a simple ODE with the ODEsolver hierarchy

Solve the same ODE problem as in Exercise E.1, but use the ForwardEuler class in the ODESolver hierarchy from Sect. E.3.
Filename: simple_ODE_class_ODESolver.

Exercise E.4: Solve an ODE specified on the command line

We want to make a program odesolver_cml.py which accepts an ODE problem to be specified on the command line. The command-line arguments are f u0 dt T, where f is the right-hand side $f(u,t)$ specified as a string formula, u0 is the initial condition, dt is the time step, and T is the final time of the simulation. A fifth optional argument can be given to specify the name of the numerical solution method (set any method of your choice as default value). A curve plot of the solution versus time should be produced and stored in a file plot.png.

Hint 1 Use `StringFunction` from `scitools.std` for convenient conversion of a formula on the command line to a Python function.

Hint 2 Use the `ODESolver` hierarchy to solve the ODE and let the fifth command-line argument be the class name in the `ODESolver` hierarchy.
Filename: `odesolver_cml`.

Exercise E.5: Implement a numerical method for ODEs
Implement the numerical method (E.36)–(E.37). How can you verify that the implementation is correct?
Filename: `Heuns_method`.

Exercise E.6: Solve an ODE for emptying a tank
A cylindrical tank of radius R is filled with water to a height h_0. By opening a valve of radius r at the bottom of the tank, water flows out, and the height of water, $h(t)$, decreases with time. We can derive an ODE that governs the height function $h(t)$.

Mass conservation of water requires that the reduction in height balances the outflow. In a time interval Δt, the height is reduced by Δh, which corresponds to a water volume of $\pi R^2 \Delta h$. The water leaving the tank in the same interval of time equals $\pi r^2 v \Delta t$, where v is the outflow velocity. It can be shown (from what is known as Bernoulli's equation) [15, 26] that

$$v(t) = \sqrt{2gh(t) + h'(t)^2},$$

where g is the acceleration of gravity. Note that $\Delta h > 0$ implies an increase in h, which means that $-\pi R^2 \Delta h$ is the corresponding decrease in volume that must balance the outflow loss of volume $\pi r^2 v \Delta t$. Elimination of v and taking the limit $\Delta t \to 0$ lead to the ODE

$$\frac{dh}{dt} = -\left(\frac{r}{R}\right)^2 \left(1 - \left(\frac{r}{R}\right)^4\right)^{-1/2} \sqrt{2gh}\,.$$

For practical applications $r \ll R$ so that $1 - (r/R)^4 \approx 1$ is a reasonable approximation, since other approximations are done as well: friction is neglected in the derivation, and we are going to solve the ODE by approximate numerical methods. The final ODE then becomes

$$\frac{dh}{dt} = -\left(\frac{r}{R}\right)^2 \sqrt{2gh}\,. \tag{E.61}$$

The initial condition follows from the initial height of water, h_0, in the tank: $h(0) = h_0$.

Solve (E.61) by a numerical method of your choice in a program. Set $r = 1$ cm, $R = 20$ cm, $g = 9.81$ m/s^2, and $h_0 = 1$ m. Use a time step of 10 seconds. Plot the solution, and experiment to see what a proper time interval for the simulation is. Make sure to test for $h < 0$ so that you do not apply the square root function to negative numbers. Can you find an analytical solution of the problem to compare the numerical solution with?
Filename: `tank_ODE`.

Exercise E.7: Solve an ODE for the arc length
Given a curve $y = f(x)$, the length of the curve from $x = x_0$ to some point x is
given by the function $s(x)$, which solves the problem

$$\frac{ds}{dx} = \sqrt{1 + [f'(x)]^2}, \quad s(x_0) = 0. \tag{E.62}$$

Since s does not enter the right-hand side, (E.62) can immediately be integrated
from x_0 to x (see Exercise A.13). However, we shall solve (E.62) as an ODE.
Use the Forward Euler method and compute the length of a straight line (for easy
verification) and a parabola: $f(x) = \frac{1}{2}x + 1$, $x \in [0, 2]$; $f(x) = x^2$, $x \in [0, 2]$.
Filename: `arclength_ODE`.

Exercise E.8: Simulate a falling or rising body in a fluid
A body moving vertically through a fluid (liquid or gas) is subject to three different
types of forces:

- the gravity force $F_g = -mg$, where m is the mass of the body and g is the
  acceleration of gravity;
- the drag force $F_d = -\frac{1}{2}C_D \varrho A|v|v$, where C_D is a dimensionless drag coefficient
  depending on the body's shape, ϱ is the density of the fluid, A is the cross-
  sectional area (produced by a cut plane, perpendicular to the motion, through the
  thickest part of the body), and v is the velocity;
- the uplift or buoyancy ("Archimedes") force $F_b = \varrho g V$, where V is the volume
  of the body.

(Roughly speaking, the F_d formula is suitable for medium to high velocities, while
for very small velocities, or very small bodies, F_d is proportional to the velocity,
not the velocity squared, see [26].)

Newton's second law applied to the body says that the sum of the listed forces
must equal the mass of the body times its acceleration a:

$$F_g + F_d + F_b = ma,$$

which gives

$$-mg - \frac{1}{2}C_D \varrho A|v|v + \varrho g V = ma.$$

The unknowns here are v and a, i.e., we have two unknowns but only one equation.
From kinematics in physics we know that the acceleration is the time derivative of
the velocity: $a = dv/dt$. This is our second equation. We can easily eliminate a
and get a single differential equation for v:

$$-mg - \frac{1}{2}C_D \varrho A|v|v + \varrho g V = m\frac{dv}{dt}.$$

A small rewrite of this equation is handy: we express m as $\varrho_b V$, where ϱ_b is the
density of the body, and we isolate dv/dt on the left-hand side,

$$\frac{dv}{dt} = -g\left(1 - \frac{\varrho}{\varrho_b}\right) - \frac{1}{2}C_D\frac{\varrho A}{\varrho_b V}|v|v. \tag{E.63}$$

This differential equation must be accompanied by an initial condition: $v(0) = V_0$.

a) Make a program for solving (E.63) numerically, using any numerical method of your choice.

Hint It is not strictly necessary, but it is an elegant Python solution to implement the right-hand side of (E.63) in the `__call__` method of a class where the parameters $g, \varrho, \varrho_b, C_D, A$, and V are data attributes.

b) To verify the program, assume a heavy body in air such that the F_b force can be neglected, and assume a small velocity such that the air resistance F_d can also be neglected. Mathematically, setting $\varrho = 0$ removes both these terms from the equation. The solution is then $v(t) = y'(t) = v_0 - gt$. Observe through experiments that the linear solution is exactly reproduced by the numerical solution regardless of the value of Δt. (Note that if you use the Forward Euler method, the method can become unstable for large Δt, see Sect. E.3.5. and time steps above the critical limit for stability cannot be used to reproduce the linear solution.) Write a test function for automatically checking that the numerical solution is $u_k = v_0 - gk\Delta t$ in this test case.

c) Make a function for plotting the forces F_g, F_b, and F_d as functions of t. Seeing the relative importance of the forces as time develops gives an increased understanding of how the different forces contribute to changing the velocity.

d) Simulate a skydiver in free fall before the parachute opens. We set the density of the human body as $\varrho_b = 1003\ \text{kg/m}^3$ and the mass as $m = 80$ kg, implying $V = m/\varrho_b = 0.08\ \text{m}^3$. We can base the cross-sectional area A the assumption of a circular cross section of diameter 50 cm, giving $A = \pi R^2 = 0.9\ \text{m}^2$. The density of air decreases with height, and we here use the value $0.79\ \text{kg/m}^3$ which is relevant for about 5000 m height. The C_D coefficient can be set as 0.6. Start with $v_0 = 0$.

e) A ball with the size of a soccer ball is placed in deep water, and we seek to compute its motion upwards. Contrary to the former example, where the buoyancy force F_b is very small, F_b is now the driving force, and the gravity force F_g is small. Set $A = \pi a^2$ with $a = 11$ cm. The mass of the ball is 0.43 kg, the density of water is $1000\ \text{kg/m}^3$, and C_D can be taken as 0.4. Start with $v_0 = 0$.

Filename: `body_in_fluid`.

Exercise E.9: Verify the limit of a solution as time grows
The solution of (E.63) often tends to a constant velocity, called the terminal velocity. This happens when the sum of the forces, i.e., the right-hand side in (E.63), vanishes.

a) Compute the formula for the terminal velocity by hand.

b) Solve the ODE using class `ODESolver` and call the `solve` method with a `terminate` function that terminates the computations when a constant velocity is reached, that is, when $|v(t_n) - v(t_{n-1})| \leq \epsilon$, where ϵ is a small number.

c) Run a series of Δt values and make a graph of the terminal velocity as a function of Δt for the two cases in Exercise E.8 d) and e). Indicate the exact terminal velocity in the plot by a horizontal line.

Filename: `body_in_fluid_termvel`.

Exercise E.10: Scale the logistic equation

Consider the logistic model (E.5):

$$u'(t) = \alpha u(t)\left(1 - \frac{u(t)}{R}\right), \quad u(0) = U_0.$$

This problem involves three input parameters: U_0, R, and α. Learning how u varies with U_0, R, and α requires much experimentation where we vary all three parameters and observe the solution. A much more effective approach is to *scale* the problem. By this technique the solution depends only on one parameter: U_0/R. This exercise tells how the scaling is done.

The idea of scaling is to introduce *dimensionless* versions of the independent and dependent variables:

$$v = \frac{u}{u_c}, \quad \tau = \frac{t}{t_c},$$

where u_c and t_c are characteristic sizes of u and t, respectively, such that the dimensionless variables v and τ are of approximately unit size. Since we know that $u \to R$ as $t \to \infty$, R can be taken as the characteristic size of u.

Insert $u = Rv$ and $t = t_c\tau$ in the governing ODE and choose $t_c = 1/\alpha$. Show that the ODE for the new function $v(\tau)$ becomes

$$\frac{dv}{d\tau} = v(1-v), \quad v(0) = v_0. \tag{E.64}$$

We see that the three parameters U_0, R, and α have disappeared from the ODE problem, and only one parameter $v_0 = U_0/R$ is involved.

Show that if $v(\tau)$ is computed, one can recover $u(t)$ by

$$u(t) = Rv(\alpha t). \tag{E.65}$$

Geometrically, the transformation from v to u is just a stretching of the two axis in the coordinate system.

Make a program `logistic_scaled.py` where you compute $v(\tau)$, given $v_0 = 0.05$, and then you use (E.65) to plot $u(t)$ for $R = 100, 500, 1000$ and $\alpha = 1$ in one figure, and $u(t)$ for $\alpha = 1, 5, 10$ and $R = 1000$ in another figure. Note how effectively you can generate $u(t)$ without needing to solve an ODE problem, and also note how varying R and α impacts the graph of $u(t)$.
Filename: `logistic_scaled`.

Exercise E.11: Compute logistic growth with time-varying carrying capacity

Use classes `Problem2` and `AutoSolver` from Sect. E.3.6 to study logistic growth when the carrying capacity of the environment, R, changes periodically with time: $R = 500$ for $it_s \le t < (i+1)t_s$ and $R = 200$ for $(i+1)t_s \le t < (i+2)t_s$, with $i = 0, 2, 4, 6, \ldots$. Use the same data as in Sect. E.3.6, and find some relevant sizes of the period of variation, t_s, to experiment with.
Filename: `seasonal_logistic_growth`.

Exercise E.12: Solve an ODE until constant solution

Newton's law of cooling,

$$\frac{dT}{dt} = -h(T - T_s) \qquad (E.66)$$

can be used to see how the temperature T of an object changes because of heat exchange with the surroundings, which have a temperature T_s. The parameter h, with unit s^{-1} is an experimental constant (heat transfer coefficient) telling how efficient the heat exchange with the surroundings is. For example, (E.66) may model the cooling of a hot pizza taken out of the oven. The problem with applying (E.66) is that h must be measured. Suppose we have measured T at $t = 0$ and t_1. We can use a rough Forward Euler approximation of (E.66) with one time step of length t_1,

$$\frac{T(t_1) - T(0)}{t_1} = -h(T(0) - T_s),$$

to make the estimate

$$h = \frac{T(t_1) - T(0)}{t_1(T_s - T(0))}. \qquad (E.67)$$

a) The temperature of a pizza is $200\,C$ at $t = 0$, when it is taken out of the oven, and $180\,C$ after 50 seconds, in a room with temperature $20\,C$. Find an estimate of h from the formula above.

b) Solve (E.66) numerically by a method of your choice to find the evolution of the temperature of the pizza. Plot the solution.

Hint You may solve the ODE the way you like, but the `solve` method in the classes in the `ODESolver` hierarchy accepts an optional `terminate` function that can be used to terminate the solution process when T is sufficiently close to T_s. Reading the first part of Sect. E.3.6 may be useful.

Filename: `pizza_cooling1`.

Exercise E.13: Use a problem class to hold data about an ODE

We address the same physical problem as in Exercise E.12, but we will now provide a class-based implementation for the user's code.

a) Make a class `Problem` containing the parameters h, T_s, $T(0)$, and Δt as data attributes. Let these parameters be set in the constructor. The right-hand side of the ODE can be implemented in a `__call__` method. If you use a class from the `ODESolver` hierarchy to solve the ODE, include the `terminate` function as a method in class `Problem`.

Create a stand-alone function `estimate_h(t0, t1, T0, T1)` which applies (E.67) from Exercise E.12 to estimate the h parameter based on the initial temperature and one data point $(t_1, T(t_1))$. You can use this function to estimate a value for h prior to calling the constructor in the `Problem` class.

Hint You may want to read Sect. E.3.6 to see why and how a class `Problem` is constructed.

b) Implement a test function `test_Problem()` for testing that class `Problem` works. (It is up to you to define how to test the class.)
c) What are the advantages and disadvantages with class `Problem` compared to using plain functions (in your view)?
d) We now want to run experiments with different values of some parameters: $T_s = 15, 22, 30$ C and $T(0) = 250, 200$ C. For each $T(0)$, plot T for the three T_s values. The estimated value of h in Exercise E.12 can be reused here.

Hint The typical elegant Python way to solve such a problem goes as follows. Write a function `solve(problem)` that takes a `Problem` object with name `problem` as argument and performs what it takes to solve one case (i.e., `solve` must solve the ODE and plot the solution). A dictionary can for each T_0 value hold a list of the cases to be plotted together. Then we loop over the problems in the dictionary of lists and call `solve` for each problem:

```
# Given h and dt
problems = {T_0: [Problem(h, T_s, T_0, dt)
                  for T_s in 15, 22, 30] for T_0 in 250, 200}
for T_0 in problems:
    hold('off')
    for problem in problems[T_0]:
        solve(problem)
        hold('on')
    savefig('T0_%g'.pdf % T_0)
```

When you become familiar with such code, and appreciate it, you can call yourself a professional programmer – with a deep knowledge of how lists, dictionaries, and classes can play elegantly together to conduct scientific experiments. In the present case we perform only a few experiments that could also have been done by six separate calls to the solver functionality, but in large-scale scientific and engineering investigations with hundreds of parameter combinations, the above code is still the same, only the generation of the `Problem` instances becomes more involved.
Filename: `pizza_cooling2`.

Exercise E.14: Derive and solve a scaled ODE problem
Use the scaling approach outlined in Exercise E.10 to "scale away" the parameters in the ODE in Exercise E.12. That is, introduce a new unknown $u = (T - T_s)/(T(0) - T_s)$ and a new time scale $\tau = th$. Find the ODE and the initial condition that governs the $u(\tau)$ function. Make a program that computes $u(\tau)$ until $|u| < 0.001$. Store the discrete u and τ values in a file `u_tau.dat` if that file is not already present (you can use `os.path.isfile(f)` to test if a file with name f exists). Create a function `T(u, tau, h, T0, Ts)` that loads the u and τ data from the `u_tau.dat` file and returns two arrays with T and t values, corresponding to the computed arrays for u and τ. Plot T versus t. Give the parameters h, T_s, and $T(0)$ on the command line. Note that this program is supposed to solve the ODE once and then recover any $T(t)$ solution by a simple scaling of the single $u(\tau)$ solution.
Filename: `pizza_cooling3`.

Exercise E.15: Clean up a file to make it a module

The ForwardEuler_func.py file is not well organized to be used as a module, say for doing

```
>>> from ForwardEuler_func import ForwardEuler
>>> u, t = ForwardEuler(lambda u, t: -u**2, U0=1, T=5, n=30)
```

The reason is that this import statement will execute a main program in the ForwardEuler_func.py file, involving plotting of the solution in an example. Also, the verification tests are run, which in more complicated problems could take considerable time and thus make the import statement hang until the tests are done.

Go through the file and modify it such that it becomes a module where no code is executed unless the module file is run as a program.
Filename: ForwardEuler_func2.

Exercise E.16: Simulate radioactive decay

The equation $u' = -au$ is a relevant model for radioactive decay, where $u(t)$ is the fraction of particles that remains in the radioactive substance at time t. The parameter a is the inverse of the so-called mean lifetime of the substance. The initial condition is $u(0) = 1$.

a) Introduce a class Decay to hold information about the physical problem: the parameter a and a __call__ method for computing the right-hand side $-au$ of the ODE.

b) Initialize an instance of class Decay with $a = \ln(2)/5600$ 1/y. The unit 1/y means one divided by year, so time is here measured in years, and the particular value of a corresponds to the Carbon-14 radioactive isotope whose decay is used extensively in dating organic material that is tens of thousands of years old.

c) Solve $u' = -au$ with a time step of 500 y, and simulate the radioactive decay for $T = 20,000$ y. Plot the solution. Write out the final $u(T)$ value and compare it with the exact value e^{-aT}.

Filename: radioactive_decay.

Exercise E.17: Compute inverse functions by solving an ODE

The inverse function g of some function $f(x)$ takes the value of $f(x)$ back to x again: $g(f(x)) = x$. The common technique to compute inverse functions is to set $y = f(x)$ and solve with respect to x. The formula on the right-hand side is then the desired inverse function $g(y)$. Section A.1.11 makes use of such an approach, where $y - f(x) = 0$ is solved numerically with respect to x for different discrete values of y.

We can formulate a general procedure for computing inverse functions from an ODE problem. If we differentiate $y = f(x)$ with respect to y, we get $1 = f'(x)\frac{dx}{dy}$ by the chain rule. The inverse function we seek is $x(y)$, but this function then fulfills the ODE

$$x'(y) = \frac{1}{f'(x)}. \tag{E.68}$$

That y is the independent coordinate and x a function of y can be a somewhat confusing notation, so we might introduce u for x and t for y:

$$u'(t) = \frac{1}{f'(u)}.$$

The initial condition is $x(0) = x_r$ where x_r solves the equation $f(x_r) = 0$ ($x(0)$ implies $y = 0$ and then from $y = f(x)$ it follows that $f(x(0)) = 0$).

Make a program that can use the described method to compute the inverse function of $f(x)$, given x_r. Use any numerical method of your choice for solving the ODE problem. Verify the implementation for $f(x) = 2x$. Apply the method for $f(x) = \sqrt{x}$ and plot $f(x)$ together with its inverse function.
Filename: `inverse_ODE`.

Exercise E.18: Make a class for computing inverse functions
The method for computing inverse functions described in Exercise E.17 is very general. The purpose now is to use this general approach to make a more reusable utility, here called `Inverse`, for computing the inverse of some Python function `f(x)` on some interval `I=[a,b]`. The utility can be used as follows to calculate the inverse of $\sin x$ on $I = [0, \pi/2]$:

```
def f(x):
    return sin(x)

# Compute the inverse of f
inverse = Inverse(f, x0=0, I=[0, pi/2], resolution=100)
x, y = Inverse.compute()

plot(y, x, 'r-',
     x, f(x), 'b-',
     y, asin(y), 'go')
legend(['computed inverse', 'f(x)', 'exact inverse'])
```

Here, x0 is the value of x at 0, or in general at the left point of the interval: `I[0]`. The parameter `resolution` tells how many equally sized intervals Δy we use in the numerical integration of the ODE. A default choice of 1000 can be used if it is not given by the user.

Write class `Inverse` and put it in a module. Include a test function `test_Inverse()` in the module for verifying that class `Inverse` reproduces the exact solution in the test problem $f(x) = 2x$.
Filename: `Inverse1`.

Exercise E.19: Add functionality to a class
Extend the module in Exercise E.18 such that the value of $x(0)$ (x0 in class `Inverse`'s constructor) does not need to be provided by the user.

Hint Class `Inverse` can compute a value of $x(0)$ as the root of $f(x) = 0$. You may use the Bisection method from Sect. 4.11.2, Newton's method from Sect. A.1.10, or the Secant method from Exercise A.10 to solve $f(x) = 0$. Class `Inverse` should figure out a suitable initial interval for the Bisection method or start values for the

Newton or Secant methods. Computing $f(x)$ for x at many points and examining these may help in solving $f(x) = 0$ without any input from the user.
Filename: `Inverse2`.

Exercise E.20: Compute inverse functions by interpolation

Instead of solving an ODE for computing the inverse function $g(y)$ of some function $f(x)$, as explained in Exercise E.17, one may use a simpler approach based on ideas from Sect. E.1.5. Say we compute discrete values of x and $f(x)$, stored in the arrays x and y. Doing a `plot(x, y)` shows $y = f(x)$ as a function of x, and doing `plot(y, x)` shows x as a function of y, i.e., we can trivially plot the inverse function $g(y)$ (!).

However, if we want the inverse function of $f(x)$ as some Python function `g(y)` that we can call for any y, we can use the tool `wrap2callable` from Sect. E.1.5 to turn the discrete inverse function, described by the arrays y (independent coordinate) and x (dependent coordinate), into a continuous function `g(y)`:

```
from scitools.std import wrap2callable
g = wrap2callable((y, x))


y = 0.5
print g(y)
```

The `g(y)` function applies linear interpolation in each interval between the points in the y array.

Implement this method in a program. Verify the implementation for $f(x) = 2x$, $x \in [0, 4]$, and apply the method to $f(x) = \sin x$ for $x \in [0, \pi/2]$.
Filename: `inverse_wrap2callable`.

Exercise E.21: Code the 4th-order Runge-Kutta method; function

Use the file `ForwardEuler_func.py` from Sect. E.1.3 as starting point for implementing the famous and widely used 4th-order Runge-Kutta method (E.41)–(E.45). Use the test function involving a linear $u(t)$ for verifying the implementation. Exercise E.23 suggests an application of the code.
Filename: `RK4_func`.

Exercise E.22: Code the 4th-order Runge-Kutta method; class

Carry out the steps in Exercise E.21, but base the implementation on the file `ForwardEuler.py` from Sect. E.1.7.
Filename: `RK4_class`.

Exercise E.23: Compare ODE methods

Investigate the accuracy of the 4th-order Runge-Kutta method and the Forward Euler scheme for solving the (challenging) ODE problem

$$\frac{dy}{dx} = \frac{1}{2(y-1)}, \quad y(0) = 1 + \sqrt{\epsilon}, \quad x \in [0, 4], \tag{E.69}$$

where ϵ is a small number, say $\epsilon = 0.001$. Start with four steps in $[0, 4]$ and reduce the step size repeatedly by a factor of two until you find the solutions sufficiently

accurate. Make a plot of the numerical solutions along with the exact solution $y(x) = 1 + \sqrt{x + \epsilon}$ for each step size.
Filename: yx_ODE_FE_vs_RK4.

Exercise E.24: Code a test function for systems of ODEs

The ForwardEuler_func.py file from Sect. E.1.3 does not contain any test function for verifying the implementation. We can use the fact that linear functions of time will be exactly reproduced by most numerical methods for ODEs. A simple system of two ODEs with linear solutions $v(t) = 2 + 3t$ and $w(t) = 3 + 4t$ is

$$v' = 3 + (3 + 4t - w)^3, \tag{E.70}$$

$$w' = 4 + (2 + 3t - v)^4 \tag{E.71}$$

Write a test function test_ForwardEuler() for comparing the numerical solution of this system with the exact solution.
Filename: ForwardEuler_sys_func2.

Exercise E.25: Code Heun's method for ODE systems; function

Use the file ForwardEuler_sys_func.py from Sect. E.2.3 as starting point for implementing Heun's method (E.36)–(E.37) for systems of ODEs. Verify the solution using the test function suggested in Exercise E.24.
Filename: Heun_sys_func.

Exercise E.26: Code Heun's method for ODE systems; class

Carry out the steps in Exercise E.25, but make a class implementation based on the file ForwardEuler_sys.py from Sect. E.2.4.
Filename: Heun_sys_class.

Exercise E.27: Implement and test the Leapfrog method

a) Implement the Leapfrog method specified in formula (E.35) from Sect. E.3.1 in a subclass of ODESolver. Place the code in a separate module file Leapfrog.py.
b) Make a test function for verifying the implementation.

Hint Use the fact that the method will exactly produce a linear u, see Sect. E.3.4.

c) Make a movie that shows how the Leapfrog method, the Forward Euler method, and the 4th-order Runge-Kutta method converge to the exact solution as Δt is reduced. Use the model problem $u' = u$, $u(0) = 1$, $t \in [0, 8]$, with $n = 2^k$ intervals, $k = 1, 2 \ldots, 14$. Place the movie generation in a function.
d) Repeat c) for the model problem $u' = -u$, $u(0) = 1$, $t \in [0, 5]$, with $n = 2^k$ intervals, $k = 1, 2 \ldots, 14$. In the movie, start with the finest resolution and reduce n until $n = 2$. The lessons learned is that Leapfrog can give completely wrong, oscillating solutions if the time step is not small enough.

Filename: Leapfrog.

Exercise E.28: Implement and test an Adams-Bashforth method
Do Exercise E.27 with the 3rd-order Adams-Bashforth method (E.46).
Filename: `AdamBashforth3`.

Exercise E.29: Solve two coupled ODEs for radioactive decay
Consider two radioactive substances A and B. The nuclei in substance A decay to
form nuclei of type B with a mean lifetime τ_A, while substance B decay to form
type A nuclei with a mean lifetime τ_B. Letting u_A and u_B be the fractions of the
initial amount of material in substance A and B, respectively, the following system
of ODEs governs the evolution of $u_A(t)$ and $u_B(t)$:

$$u'_A = u_B/\tau_B - u_A/\tau_A, \tag{E.72}$$
$$u'_B = u_A/\tau_A - u_B/\tau_B, \tag{E.73}$$

with $u_A(0) = u_B(0) = 1$.

a) Introduce a problem class, which holds the parameters τ_A and τ_B and offers
   a `__call__` method to compute the right-hand side vector of the ODE system,
   i.e., $(u_B/\tau_B - u_A/\tau_A, u_A/\tau_A - u_B/\tau_B)$.
b) Solve for u_A and u_B using a subclass in the ODESolver hierarchy and the parameter choices $\tau_A = 8$ minutes, $\tau_B = 40$ minutes, and $\Delta t = 10$ seconds.
c) Plot u_A and u_B against time measured in minutes.
d) From the ODE system it follows that the ratio $u_A/u_B \to \tau_A/\tau_B$ as $t \to \infty$
   (assuming $u'_A = u'_B = 0$ in the limit $t \to \infty$). Extend the problem class
   with a test method for checking that two given solutions u_A and u_B fulfill this
   requirement. Verify that this is indeed the case with the computed solutions in
   b).

Filename: `radioactive_decay2`.

Exercise E.30: Implement a 2nd-order Runge-Kutta method; function
Implement the 2nd-order Runge-Kutta method specified in formula (E.38). Use
a plain function `RungeKutta2` of the type shown in Sect. E.1.2 for the Forward
Euler method. Construct a test problem where you know the analytical solution,
and plot the difference between the numerical and analytical solution. Demonstrate
that the numerical solution approaches the exact solution as Δt is reduced.
Filename: `RungeKutta2_func`.

Exercise E.31: Implement a 2nd-order Runge-Kutta method; class

a) Make a new subclass `RungeKutta2` in the ODESolver hierarchy from Sect. E.3
   for solving ordinary differential equations with the 2nd-order Runge-Kutta
   method specified in formula (E.38).
b) Construct a test problem where you can find an exact solution. Run different values of Δt and demonstrate in a plot that the numerical solution approaches the
   exact solution as Δt is decreased. Put the code that creates the plot in a function.

c) Make a test function `test_RungeKutta2_against_hand_calc()` where you do the computations of u_1 and u_2 i Python based on the mathematical formulas. Thereafter, run the `RungeKutta2` class for two time steps and check that the two solutions are equal (within a small tolerance). Use an ODE where the right-hand side depends on t as well as u such that you can test that `RungeKutta2` treats the t argument in $f(u, t)$ correctly.

d) Make a module out of the `RungeKutta2` class and the associated functions. Call the functions from a test block in the module file.

Filename: `RungeKutta2`.

Exercise E.32: Code the iterated midpoint method; function

a) Implement the numerical method (E.47)–(E.48) as a function

```
iterated_Midpoint_method(f, U0, T, n, N)
```

where `f` is a Python implementation of $f(u, t)$, `U0` is the initial condition $u(0) = U_0$, `T` is the final time of the simulation, `n` is the number of time steps, and `N` is the parameter N in the method (E.47). The `iterated_Midpoint_method` should return two arrays: $u_0, \dots, u_n$ and $t_0, \dots, t_n$.

Hint You may want to build the function on the software described in Sect. E.1.3.

b) To verify the implementation, calculate by hand u_1 and u_2 when $N = 2$ for the ODE $u' = -2u$, $u(0) = 1$, with $\Delta t = 1/4$. Compare your hand calculations with the results of the program. Make a test function `test_iterated_Midpoint_method()` for automatically comparing the hand calculations with the output of the function in a).

c) Consider the ODE problem $u' = -2(t - 4)u$, $u(0) = e^{-16}$, $t \in [0, 8]$, with exact solution $u = e^{-(t-4)^2}$. Write a function for comparing the numerical and exact solution in a plot. Enable setting of Δt and N from the command line and use the function to study the behavior of the numerical solution as you vary Δt and N. Start with $\Delta t = 0.5$ and $N = 1$. Continue with reducing Δt and increasing N.

Filename: `MidpointIter_func`.

Exercise E.33: Code the iterated midpoint method; class

The purpose of this exercise is to implement the numerical method (E.47)–(E.48) in a class `MidpointIter`, like the `ForwardEuler` class from Sect. E.1.7. Also make a test function `test_MidpointIter()` where you apply the verification technique from Exercise E.32b.

Filename: `MidpointIter_class`.

Exercise E.34: Make a subclass for the iterated midpoint method
Implement the numerical method (E.47)–(E.48) in a subclass in the `ODESolver` hierarchy. The code should reside in a separate file where the `ODESolver` class is imported. One can either fix N or introduce an ϵ and iterate until the change in $|v_q - v_{q-1}|$ is less than ϵ. Allow the constructor to take both N and ϵ as arguments. Compute a new v_q as long as $q \leq N$ and $|v_q - v_{q-1}| > \epsilon$. Let $N = 20$ and $\epsilon = 10^{-6}$ by default. Store N as an attribute such that the user's code can access what N was in the last computation. Also write a test function for verifying the implementation. Filename: `MidpointIter`.

Exercise E.35: Compare the accuracy of various methods for ODEs
We want to see how various numerical methods treat the following ODE problem:

$$u' = -2(t-4)u, \quad u(0) = e^{-16}, \quad t \in (0, 10].$$

The exact solution is a Gaussian function: $u(t) = e^{-(t-4)^2}$. Compare the Forward Euler method with other methods of your choice in the same plot. Relevant methods are the 4th-order Runge-Kutta method (found in the `ODESolver.py` hierarchy) and methods from Exercises E.5, E.21, E.22, E.25, E.26, E.27, E.28 E.31, or E.34. Put the value of Δt in the title of the plot. Perform experiments with $\Delta t = 0.3, 0.25, 0.1, 0.05, 0.01, 0.001$ and report how the various methods behave. Filename: `methods4gaussian`.

Exercise E.36: Animate how various methods for ODEs converge
Make a movie for illustrating how three selected numerical methods converge to the exact solution for the problem described in Exercise E.35 as Δt is reduced. Start with $\Delta t = 1$, fix the y axis in $[-0.1, 1.1]$, and reduce Δt by a quite small factor, say 1.5, between each frame in the movie. The movie must last until all methods have their curves visually on top of the exact solution. Filename: `animate_methods4gaussian`.

Exercise E.37: Study convergence of numerical methods for ODEs
The approximation error when solving an ODE numerically is usually of the form $C\Delta t^r$, where C and r are constants that can be estimated from numerical experiments. The constant r, called the *convergence rate*, is of particular interest. Halving Δt halves the error if $r = 1$, but if $r = 3$, halving Δt reduces the error by a factor of 8.

Exercise 9.15 describes a method for estimating r from two consecutive experiments. Make a function

```
ODE_convergence(f, U0, u_e, method, dt=[])
```

that returns a series of estimated r values corresponding to a series of Δt values given as the `dt` list. The argument `f` is a Python implementation of $f(u,t)$ in the ODE $u' = f(u,t)$. The initial condition is $u(0) = U_0$, where U_0 is given as the `U0` argument, `u_e` is the exact solution $u_e(t)$ of the ODE, and `method` is the name of a class in the `ODESolver` hierarchy. The error between the exact solution u_e and the computed solution $u_0, u_1, \ldots, u_n$ can be defined as

$$e = \left(\Delta t \sum_{i=0}^{n} (u_e(t_i) - u_i)^2 \right)^{1/2} .$$

Call the `ODE_convergence` function for some numerical methods and print the estimated r values for each method. Make your own choice of the ODE problem and the collection of numerical methods.

Filename: `ODE_convergence`.

Exercise E.38: Find a body's position along with its velocity

In Exercise E.8 we compute the velocity $v(t)$. The position of the body, $y(t)$, is related to the velocity by $y'(t) = v(t)$. Extend the program from Exercise E.8 to solve the system

$$\frac{dy}{dt} = v,$$

$$\frac{dv}{dt} = -g \left(1 - \frac{\varrho}{\varrho_b} \right) - \frac{1}{2} C_D \frac{\varrho A}{\varrho_b V} |v| v .$$

Filename: `body_in_fluid2`.

Exercise E.39: Add the effect of air resistance on a ball

The differential equations governing the horizontal and vertical motion of a ball subject to gravity and air resistance read

$$\frac{d^2 x}{dt^2} = -\frac{3}{8} C_D \bar{\varrho} a^{-1} \sqrt{\left(\frac{dx}{dt} \right)^2 + \left(\frac{dy}{dt} \right)^2} \frac{dx}{dt}, \qquad (\text{E.74})$$

$$\frac{d^2 y}{dt^2} = -g - \frac{3}{8} C_D \bar{\varrho} a^{-1} \sqrt{\left(\frac{dx}{dt} \right)^2 + \left(\frac{dy}{dt} \right)^2} \frac{dy}{dt}, \qquad (\text{E.75})$$

where (x, y) is the position of the ball (x is a horizontal measure and y is a vertical measure), g is the acceleration of gravity, $C_D = 0.2$ is a drag coefficient, $\bar{\varrho}$ is the ratio of the density of air and the ball, and a is the radius of the ball.

Let the initial condition be $x = y = 0$ (start position in the origin) and

$$dx/dt = v_0 \cos \theta, \quad dy/dt = v_0 \sin \theta,$$

where v_0 is the magnitude of the initial velocity and θ is the angle the velocity makes with the horizontal.

a) Express the two second-order equations above as a system of four first-order equations with four initial conditions.

b) Implement the right-hand side in a problem class where the physical parameters C_D, $\bar{\varrho}$, a, v_0, and θ are stored along with the initial conditions. You may also want to add a `terminate` method in this class for checking when the ball hits the ground and then terminate the solution process.

c) Simulate a hard football kick where $v_0 = 120$ km/h and θ is 30 degrees. Take the density of the ball as 0.017 hg/m^3 and the radius as 11 cm. Solve the ODE system for $C_D = 0$ (no air resistance) and $C_D = 0.2$, and plot y as a function of x in both cases to illustrate the effect of air resistance. Make sure you express all units in kg, m, s, and radians.

Filename: `kick2D`.

Exercise E.40: Solve an ODE system for an electric circuit
An electric circuit with a resistor, a capacitor, an inductor, and a voltage source can be described by the ODE

$$L\frac{dI}{dt} + RI + \frac{Q}{C} = E(t), \tag{E.76}$$

where $L\,dI/dt$ is the voltage drop across the inductor, I is the current (measured in amperes, A), L is the inductance (measured in henrys, H), R is the resistance (measured in ohms, Ω), Q is the charge on the capacitor (measured in coulombs, C), C is the capacitance (measured in farads, F), $E(t)$ is the time-variable voltage source (measured in volts, V), and t is time (measured in seconds, s). There is a relation between I and Q:

$$\frac{dQ}{dt} = I. \tag{E.77}$$

Equations (E.76)–(E.77) is a system two ODEs. Solve these for $L = 1$ H, $E(t) = 2\sin\omega t$ V, $\omega^2 = 3.5$ s^{-2}, $C = 0.25$ C, $R = 0.2\ \Omega$, $I(0) = 1$ A, and $Q(0) = 1C$. Use the Forward Euler scheme with $\Delta t = 2\pi/(60\omega)$. The solution will, after some time, oscillate with the same period as $E(t)$, a period of $2\pi/\omega$. Simulate 10 periods. Filename: `electric_circuit`.

Remarks It turns out that the Forward Euler scheme overestimates the amplitudes of the oscillations. The more accurate 4th-order Runge-Kutta method is much better for this type of differential equation model.

Exercise E.41: Simulate the spreading of a disease by a SIR model
We shall in this exercise model epidemiological diseases such as measles or swine flu. Suppose we have three categories of people: susceptibles (S) who can get the disease, infected (I) who have developed the disease and who can infect susceptibles, and recovered (R) who have recovered from the disease and become immune. Let $S(t)$, $I(t)$, and $R(t)$ be the number of people in category S, I, and R, respectively. We have that $S + I + R = N$, where N is the size of the population, assumed constant here for simplicity.

When people mix in the population there are SI possible pairs of susceptibles and infected, and a certain fraction βSI per time interval meets with the result that the infected "successfully" infect the susceptible. During a time interval Δt, $\beta SI\Delta t$ get infected and move from the S to the I category:

$$S(t + \Delta t) = S(t) - \beta SI\Delta t.$$

We divide by Δt and let $\Delta \to 0$ to get the differential equation

$$S'(t) = -\beta S I \,. \tag{E.78}$$

A fraction νI of the infected will per time unit recover from the disease. In a time Δt, $\nu I \Delta t$ recover and move from the I to the R category. The quantity $1/\nu$ typically reflects the duration of the disease. In the same time interval, $\beta S I \Delta t$ come from the S to the I category. The accounting for the I category therefore becomes

$$I(t + \Delta t) = I(t) + \beta S I \Delta t - \nu I \Delta t,$$

which in the limit $\Delta t \to 0$ becomes the differential equation

$$I'(t) = \beta S I - \nu I \,. \tag{E.79}$$

Finally, the R category gets contributions from the I category:

$$R(t + \Delta t) = R(t) + \nu I \Delta t \,.$$

The corresponding ODE for R reads

$$R'(t) = \nu I \,. \tag{E.80}$$

In case the recovered do not become immune, we do not need the recovered category, since the recovered go directly out of the I category to the S category again. This gives a contribution νI to the equation for S and we end up with the S-I system (C.31)–(C.32) from Sect. C.5.

The system (E.78)–(E.80) is known as a SIR model in epidemiology (which is the name of the scientific field studying the spreading of epidemic diseases).

Make a function for solving the differential equations in the SIR model by any numerical method of your choice. Make a separate function for visualizing $S(t)$, $I(t)$, and $R(t)$ in the same plot.

Adding the equations shows that $S' + I' + R' = 0$, which means that $S + I + R$ must be constant. Perform a test at each time level for checking that $S + I + R$ equals $S_0 + I_0 + R_0$ within some small tolerance. If a subclass of ODESolver is used to solve the ODE system, the test can be implemented as a user-specified terminate function that is called by the solve method a every time level (simply return True for termination if $S + I + R$ is not sufficiently constant).

A specific population has 1500 susceptibles and one infected. We are interested in how the disease develops. Set $S(0) = 1500$, $I(0) = 1$, and $R(0) = 0$. Choose $\nu = 0.1$, $\Delta t = 0.5$, and $t \in [0, 60]$. Time t here counts days. Visualize first how the disease develops when $\beta = 0.0005$. Certain precautions, like staying inside, will reduce β. Try $\beta = 0.0001$ and comment from the plot how a reduction in β influences $S(t)$. (Put the comment as a multi-line string in the bottom of the program file.)

Filename: SIR.

Exercise E.42: Introduce problem and solver classes in the SIR model

The parameters ν and β in the SIR model in Exercise E.41 can be constants or functions of time. Now we shall make an implementation of the $f(u, t)$ function specifying the ODE system such that ν and β can be given as either a constant or a Python function. Introduce a class for $f(u, t)$, with the following code sketch:

```python
class ProblemSIR(object):
    def __init__(self, nu, beta, S0, I0, R0, T):
        """
        nu, beta: parameters in the ODE system
        S0, I0, R0: initial values
        T: simulation for t in [0,T]
        """
        if isinstance(nu, (float,int)):   # number?
            self.nu = lambda t: nu        # wrap as function
        elif callable(nu):
            self.nu = nu

        # same for beta and self.beta
        ...

        # store the other parameters

    def __call__(self, u, t):
        """Right-hand side function of the ODE system."""
        S, I, R = u
        return [-self.beta(t)*S*I,     # S equation
                ...,                   # I equation
                self.nu(t)*I]          # R equation

# Example:
problem = ProblemSIR(beta=lambda t: 0.0005 if t <= 12 else 0.0001,
                     nu=0.1, S0=1500, I0=1, R0=0, T=60)
solver = ODESolver.ForwardEuler(problem)
```

Write the complete code for class `ProblemSIR` based on the sketch of ideas above. The ν parameter is usually not varying with time as $1/\nu$ is a characteristic size of the period a person is sick, but introduction of new medicine during the disease might change the picture such that time dependence becomes relevant.

We can also make a class `SolverSIR` for solving the problem (see Sect. E.3.6 for similar examples):

```python
class SolverSIR(object):
    def __init__(self, problem, dt):
        self.problem, self.dt = problem, dt

    def solve(self, method=ODESolver.RungeKutta4):
        self.solver = method(self.problem)
        ic = [self.problem.S0, self.problem.I0, self.problem.R0]
        self.solver.set_initial_condition(ic)
        n = int(round(self.problem.T/float(self.dt)))
        t = np.linspace(0, self.problem.T, n+1)
```

```
      u, self.t = self.solver.solve(t)
      self.S, self.I, self.R = u[:,0], u[:,1], u[:,2]

  def plot(self):
      # plot S(t), I(t), and R(t)
```

After the breakout of a disease, authorities often start campaigns for decreasing the spreading of the disease. Suppose a massive campaign telling people to wash their hands more frequently is launched, with the effect that β is significantly reduced after some days. For the specific case simulated in Exercise E.41, let

$$\beta(t) = \begin{cases} 0.0005, & 0 \le t \le 12, \\ 0.0001, & t > 12 \end{cases}$$

Simulate this scenario with the `Problem` and `Solver` classes. Report the maximum number of infected people and compare it to the case where $\beta(t) = 0.0005$.
Filename: `SIR_class`.

Exercise E.43: Introduce vaccination in a SIR model
We shall now extend the SIR model in Exercise E.41 with a vaccination[2] program. If a fraction p of the susceptibles per time unit is being vaccinated, and we say that the vaccination is 100 percent effective, $pS\Delta t$ individuals will be removed from the S category in a time interval Δt. We place the vaccinated people in a new category V. The equations for S and V becomes

$$S' = -\beta SI - pS, \tag{E.81}$$
$$V' = pS. \tag{E.82}$$

The equations for I and R are not affected. The initial condition for V can be taken as $V(0) = 0$. The resulting model is named SIRV.

Try the same parameters as in Exercise E.41 in combination with $p = 0.1$ and compute the evolution of $S(t)$, $I(t)$, $R(t)$, and $V(t)$. Comment on the effect of vaccination on the maximum number of infected.

Hint You can of course edit the code from Exercise E.42, but it is much better to avoid duplicating code and use object-oriented programming to implement the extensions in the present exercise as subclasses of the classes from Exercise E.42.
Filename: `SIRV`.

Exercise E.44: Introduce a vaccination campaign in a SIR model
Let the vaccination campaign in Exercise E.43 start 6 days after the outbreak of the disease and let it last for 10 days,

$$p(t) = \begin{cases} 0.1, & 6 \le t \le 15, \\ 0, & \text{otherwise} \end{cases}$$

---

[2] https://www.youtube.com/watch?v=s_6QW9sNPEY

Plot the corresponding solutions $S(t)$, $I(t)$, $R(t)$, and $V(t)$. (It is clearly advantageous to have the SIRV model implemented as an extension to the classes in Exercise E.42.)

Filename: `SIRV_varying_p`.

Exercise E.45: Find an optimal vaccination period

Let the vaccination campaign in Exercise E.44 last for V_T days:

$$p(t) = \begin{cases} 0.1, & 6 \le t \le 6 + V_T, \\ 0, & \text{otherwise} \end{cases}$$

Compute the maximum number of infected people, $\max_t I(t)$, as a function of $V_T \in [0, 31]$, by running the model for $V_T = 0, 1, 2 \ldots, 31$. Plot this function. Determine from the plot the optimal V_T, i.e., the smallest vaccination period V_T such that increasing V_T has negligible effect on the maximum number of infected people.

Filename: `SIRV_optimal_duration`.

Exercise E.46: Simulate human-zombie interaction

Suppose the human population is attacked by zombies. This is quite a common happening in movies, and the "zombification" of humans acts much like the spreading of a disease. Let us make a differential equation model, inspired by the SIR model from Exercise E.41, to simulate how humans and zombies interact.

We introduce four categories of individuals:

1. S: susceptible humans who can become zombies.
2. I: infected humans, being bitten by zombies.
3. Z: zombies.
4. R: removed individuals, either conquered zombies or dead humans.

The corresponding functions counting how many individuals we have in each category are named $S(t)$, $I(t)$, $Z(t)$, and $R(t)$, respectively.

The type of zombies considered here is inspired by the standard for modern zombies set by the classic movie *The Night of the Living Dead*, by George A. Romero from 1968. Only a small extension of the SIR model is necessary to model the effect of human-zombie interaction mathematically. A fraction of the human susceptibles is getting bitten by zombies and moves to the infected category. A fraction of the infected is then turned into zombies. On the other hand, humans can conquer zombies.

Now we shall precisely set up all the dynamic features of the human-zombie populations we aim to model. Changes in the S category are due to three effects:

1. Susceptibles are infected by zombies, modeled by a term $-\Delta t \beta S Z$, similar to the S-I interaction in the SIR model.
2. Susceptibles die naturally or get killed and therefore enter the removed category. If the probability that one susceptible dies during a unit time interval is δ_S, the total expected number of deaths in a time interval Δt becomes $\Delta t \delta_S S$.

3. We also allow new humans to enter the area with zombies, as this effect may be necessary to successfully run a war on zombies. The number of new individuals in the S category arriving per time unit is denoted by Σ, giving an increase in $S(t)$ by $\Delta t \Sigma$ during a time Δt.

We could also add newborns to the S category, but we simply skip this effect since it will not be significant over time scales of a few days.

The balance of the S category is then

$$S' = \Sigma - \beta S Z - \delta_S S,$$

in the limit $\Delta t \to 0$.

The infected category gets a contribution $\Delta t \beta S Z$ from the S category, but loses individuals to the Z and R category. That is, some infected are turned into zombies, while others die. Movies reveal that infected may commit suicide or that others (susceptibles) may kill them. Let δ_I be the probability of being killed in a unit time interval. During time Δt, a total of $\delta_I \Delta t I$ will die and hence be transferred to the removed category. The probability that a single infected is turned into a zombie during a unit time interval is denoted by ρ, so that a total of $\Delta t \rho I$ individuals are lost from the I to the Z category in time Δt. The accounting in the I category becomes

$$I' = \beta S Z - \rho I - \delta_I I.$$

The zombie category gains $-\Delta t \rho I$ individuals from the I category. We disregard the effect that any removed individual can turn into a zombie again, as we consider that effect as pure magic beyond reasonable behavior, at least according to what is observed in the Romero movie tradition. A fundamental feature in zombie movies is that humans can conquer zombies. Here we consider zombie killing in a "man-to-man" human-zombie fight. This interaction resembles the nature of zombification (or the susceptible-infective interaction in the SIR model) and can be modeled by a loss $-\alpha S Z$ for some parameter α with an interpretation similar to that of β. The equation for Z then becomes

$$Z' = \rho I - \alpha S Z.$$

The accounting in the R category consists of a gain δS of natural deaths from the S category, a gain δI from the I category, and a gain $\alpha S Z$ from defeated zombies:

$$R' = \delta_S S + \delta_I I + \alpha S Z.$$

The complete SIZR model for human-zombie interaction can be summarized as

$$S' = \Sigma - \beta S Z - \delta_S S, \tag{E.83}$$

$$I' = \beta S Z - \rho I - \delta_I I, \tag{E.84}$$

$$Z' = \rho I - \alpha S Z, \tag{E.85}$$

$$R' = \delta_S S + \delta_I I + \alpha S Z. \tag{E.86}$$

The interpretations of the parameters are as follows:

- Σ: the number of new humans brought into the zombified area per unit time.
- β: the probability that a theoretically possible human-zombie pair actually meets physically, during a unit time interval, with the result that the human is infected.
- δ_S: the probability that a susceptible human is killed or dies, in a unit time interval.
- δ_I: the probability that an infected human is killed or dies, in a unit time interval.
- ρ: the probability that an infected human is turned into a zombie, during a unit time interval.
- α: the probability that, during a unit time interval, a theoretically possible human-zombie pair fights and the human kills the zombie.

Note that probabilities per unit time do not necessarily lie in the interval $[0, 1]$. The real probability, lying between 0 and 1, arises after multiplication by the time interval of interest.

Implement the SIZR model with a `Problem` and `Solver` class as explained in Exercise E.42, allowing parameters to vary in time. The time variation is essential to make a realistic model that can mimic what happens in movies.

Test the implementation with the following data: $\beta = 0.0012$, $\alpha = 0.0016$, $\delta_I = 0.014$, $\Sigma = 2$, $\rho = 1$, $S(0) = 10$, $Z(0) = 100$, $I(0) = 0$, $R(0) = 0$, and simulation time $T = 24$ hours. Other parameters can be set to zero. These values are estimated from the hysterical phase of the movie *The Night of the Living Dead*. The time unit is hours. Plot the S, I, Z, and R quantities.
Filename: `SIZR`.

Exercise E.47: Simulate a zombie movie
The movie *The Night of the Living Dead* has three phases:

1. The initial phase, lasting for (say) 4 hours, where two humans meet one zombie and one of the humans get infected. A rough (and uncertain) estimation of parameters in this phase, taking into account dynamics not shown in the movie, yet necessary to establish a more realistic evolution of the S and Z categories later in the movie, is $\Sigma = 20$, $\beta = 0.03$, $\rho = 1$, $S(0) = 60$, and $Z(0) = 1$. All other parameters are taken as zero when not specified.
2. The hysterical phase, when the zombie treat is evident. This phase lasts for 24 hours, and relevant parameters can be taken as $\beta = 0.0012$, $\alpha = 0.0016$, $\delta_I = 0.014$, $\Sigma = 2$, $\rho = 1$.
3. The counter attack by humans, estimated to last for 5 hours, with parameters $\alpha = 0.006$, $\beta = 0$ (humans no longer get infected), $\delta_S = 0.0067$, $\rho = 1$.

Use the program from Exercise E.46 to simulate all three phases of the movie.

Hint It becomes necessary to work with piecewise constant functions in time. These can be hardcoded for each special case, our one can employ a ready-made tool for such functions (actually developed in Exercise 3.32):

```
from scitools.std import PiecewiseConstant

# Define f(t) as 1.5 in [0,3], 0.1 in [3,4] and 1 in [4,7]
f = PiecewiseConstant(domain=[0, 7],
                      data=[(0, 1.5), (3, 0.1), (4, 1)])
```

Filename: Night_of_the_Living_Dead.

Exercise E.48: Simulate a war on zombies

A war on zombies can be implemented through large-scale effective attacks. A possible model is to increase α in the SIZR model from Exercise E.46 by some additional amount $\omega(t)$, where $\omega(t)$ varies in time to model strong attacks at $m + 1$ distinct points of time $T_0 < T_1 < \cdots < T_m$. Around these t values we want ω to have a large value, while in between the attacks ω is small. One possible mathematical function with this behavior is a sum of Gaussian functions:

$$\omega(t) = a \sum_{i=0}^{m} \exp\left(-\frac{1}{2}\left(\frac{t - T_i}{\sigma}\right)^2\right), \tag{E.87}$$

where a measures the strength of the attacks (the maximum value of $\omega(t)$) and σ measures the length of the attacks, which should be much less than the time between the points of attack: typically, 4σ measures the length of an attack, and we must have $4\sigma \ll T_i - T_{i-1}$ for $i = 1, \ldots, m$. We should choose a significantly larger than α to make the attacks in the war on zombies much stronger than the usual "man-to-man" killing of zombies.

Modify the model and the implementation from Exercise E.46 to include a war on zombies. We start out with 50 humans and 3 zombies and $\beta = 0.03$. This leads to rapid zombification. Assume that there are some small resistances against zombies from the humans, $\alpha = 0.2\beta$, throughout the simulations. In addition, the humans implement three strong attacks, $a = 50\alpha$, at 5, 10, and 18 hours after the zombification starts. The attacks last for about 2 hours ($\sigma = 0.5$). Set $\delta_S = \Delta_I = \Sigma = 0$, $\beta = 0.03$, and $\rho = 1$, simulate for $T = 20$ hours, and see if the war on zombies modeled by the suggested $\omega(t)$ is sufficient to save mankind.
Filename: war_on_zombies.

Exercise E.49: Explore predator-prey population interactions

Suppose we have two species in an environment: a predator and a prey. How will the two populations interact and change with time? A system of ordinary differential equations can give insight into this question. Let $x(t)$ and $y(t)$ be the size of the prey and the predator populations, respectively. In the absence of a predator, the population of the prey will follow the ODE derived in Sect. C.2:

$$\frac{dx}{dt} = rx,$$

with $r > 0$, assuming there are enough resources for exponential growth. Similarly, in the absence of prey, the predator population will just experience a death rate $m > 0$:

$$\frac{dy}{dt} = -my.$$

In the presence of the predator, the prey population will experience a reduction in the growth proportional to xy. The number of interactions (meetings) between x and y numbers of animals is xy, and in a certain fraction of these interactions the predator eats the prey. The predator population will correspondingly experience a growth in the population because of the xy interactions with the prey population. The adjusted growth of both populations can now be expressed as

$$\frac{dx}{dt} = rx - axy, \tag{E.88}$$

$$\frac{dy}{dt} = -my + bxy, \tag{E.89}$$

for positive constants r, m, a, and b. Solve this system and plot $x(t)$ and $y(t)$ for $r = m = 1$, $a = 0.3$, $b = 0.2$, $x(0) = 1$, and $y(0) = 1$, $t \in [0, 20]$. Try to explain the dynamics of the population growth you observe. Experiment with other values of a and b.
Filename: `predator_prey`.

Exercise E.50: Formulate a 2nd-order ODE as a system
In this and subsequent exercises we shall deal with the following second-order ordinary differential equation with two initial conditions:

$$m\ddot{u} + f(\dot{u}) + s(u) = F(t), \quad t > 0, \quad u(0) = U_0, \ \dot{u}(0) = V_0. \tag{E.90}$$

The notation $\dot{u}$ and $\ddot{u}$ means $u'(t)$ and $u''(t)$, respectively. Write (E.90) as a system of two first-order differential equations. Also set up the initial condition for this system.

Physical applications Equation (E.90) has a wide range of applications throughout science and engineering. A primary application is damped spring systems in, e.g., cars and bicycles: u is the vertical displacement of the spring system attached to a wheel; $\dot{u}$ is then the corresponding velocity; $F(t)$ resembles a bumpy road; $s(u)$ represents the force from the spring; and $f(\dot{u})$ models the damping force (friction) in the spring system. For this particular application f and s will normally be linear functions of their arguments: $f(\dot{u}) = \beta\dot{u}$ and $s(u) = ku$, where k is a spring constant and β some parameter describing viscous damping.

Equation (E.90) can also be used to describe the motions of a moored ship or oil platform in waves: the moorings act as a nonlinear spring $s(u)$; $F(t)$ represents environmental excitation from waves, wind, and current; $f(\dot{u})$ models damping of the motion; and u is the one-dimensional displacement of the ship or platform.

Oscillations of a pendulum can be described by (E.90): u is the angle the pendulum makes with the vertical; $s(u) = (mg/L)\sin(u)$, where L is the length of the pendulum, m is the mass, and g is the acceleration of gravity; $f(\dot{u}) = \beta|\dot{u}|\dot{u}$ models air resistance (with β being some suitable constant, see Exercises 1.11 and E.54); and $F(t)$ might be some motion of the top point of the pendulum.

Another application is electric circuits with $u(t)$ as the charge, $m = L$ as the inductance, $f(\dot{u}) = R\dot{u}$ as the voltage drop across a resistor R, $s(u) = u/C$ as the voltage drop across a capacitor C, and $F(t)$ as an electromotive force (supplied by a battery or generator).

Furthermore, Equation (E.90) can act as a (very) simplified model of many other oscillating systems: aircraft wings, lasers, loudspeakers, microphones, tuning forks, guitar strings, ultrasound imaging, voice, tides, the El Ni no phenomenon, climate changes – to mention some.

We remark that (E.90) is a possibly nonlinear generalization of Equation (D.8) explained in Sect. D.1.3. The case in Appendix D corresponds to the special choice of $f(\dot{u})$ proportional to the velocity $\dot{u}$, $s(u)$ proportional to the displacement u, and $F(t)$ as the acceleration $\ddot{w}$ of the plate and the action of the gravity force.

Exercise E.51: Solve $\ddot{u} + u = 0$
Make a function

```
def rhs(u, t):
    ...
```

for returning a list with two elements with the two right-hand side expressions in the first-order differential equation system from Exercise E.50. As usual, the u argument is an array or list with the two solution components u[0] and u[1] at some time t. Inside rhs, assume that you have access to three global Python functions friction(dudt), spring(u), and external(t) for evaluating $f(\dot{u})$, $s(u)$, and $F(t)$, respectively.

Test the rhs function in combination with the functions $f(\dot{u}) = 0$, $F(t) = 0$, $s(u) = u$, and the choice $m = 1$. The differential equation then reads $\ddot{u} + u = 0$. With initial conditions $u(0) = 1$ and $\dot{u}(0) = 0$, one can show that the solution is given by $u(t) = \cos(t)$. Apply three numerical methods: the 4th-order Runge-Kutta method and the Forward Euler method from the ODESolver module developed in Sect. E.3, as well as the 2nd-order Runge-Kutta method developed in Exercise E.31. Use a time step $\Delta t = \pi/20$.

Plot $u(t)$ and $\dot{u}(t)$ versus t together with the exact solutions. Also make a plot of $\dot{u}$ versus u (plot(u[:,0], u[:,1]) if u is the array returned from the solver's solve method). In the latter case, the exact plot should be a circle because the points on the curve are $(\cos t, \sin t)$, which all lie on a circle as t is varied. Observe that the ForwardEuler method results in a spiral and investigate how the spiral develops as Δt is reduced.

The kinetic energy K of the motion is given by $\frac{1}{2}m\dot{u}^2$, and the potential energy P (stored in the spring) is given by the work done by the spring force: $P = \int_0^u s(v)dv = \frac{1}{2}u^2$. Make a plot with K and P as functions of time for both the 4th-order Runge-Kutta method and the Forward Euler method, for the same physical problem described above. In this test case, the sum of the kinetic and potential energy should be constant. Compute this constant analytically and plot it together with the sum $K + P$ as calculated by the 4th-order Runge-Kutta method and the Forward Euler method.
Filename: oscillator_v1.

Exercise E.52: Make a tool for analyzing oscillatory solutions
The solution $u(t)$ of the equation (E.90) often exhibits an oscillatory behavior (for the test problem in Exercise E.51 we have that $u(t) = \cos t$). It is then of interest to find the wavelength of the oscillations. The purpose of this exercise is to find and

visualize the distance between peaks in a numerical representation of a continuous function.

Given an array $(y_0, \ldots, y_{n-1})$ representing a function $y(t)$ sampled at various points $t_0, \ldots, t_{n-1}$, a local maximum of $y(t)$ occurs at $t = t_k$ if $y_{k-1} < y_k > y_{k+1}$. Similarly, a local minimum of $y(t)$ occurs at $t = t_k$ if $y_{k-1} > y_k < y_{k+1}$. By iterating over the $y_1, \ldots, y_{n-2}$ values and making the two tests, one can collect local maxima and minima as (t_k, y_k) pairs. Make a function `minmax(t, y)` which returns two lists, `minima` and `maxima`, where each list holds pairs (2-tuples) of t and y values of local minima or maxima. Ensure that the t value increases from one pair to the next. The arguments `t` and `y` in `minmax` hold the coordinates $t_0, \ldots, t_{n-1}$ and $y_0, \ldots, y_{n-1}$, respectively.

Make another function `wavelength(peaks)` which takes a list `peaks` of 2-tuples with t and y values for local minima or maxima as argument and returns an array of distances between consecutive t values, i.e., the distances between the peaks. These distances reflect the local wavelength of the computed y function. More precisely, the first element in the returned array is `peaks[1][0]-peaks[0][0]`, the next element is `peaks[2][0]-peaks[1][0]`, and so forth.

Test the `minmax` and `wavelength` functions on y values generated by $y = e^{t/4}\cos(2t)$ and $y = e^{-t/4}\cos(t^2/5)$ for $t \in [0, 4\pi]$. Plot the $y(t)$ curve in each case, and mark the local minima and maxima computed by `minmax` with circles and boxes, respectively. Make a separate plot with the array returned from the `wavelength` function (just plot the array against its indices – the point is to see if the wavelength varies or not). Plot only the wavelengths corresponding to maxima.

Make a module with the `minmax` and `wavelength` function, and let the test block perform the tests specified above.
Filename: `wavelength`.

Exercise E.53: Implement problem, solver, and visualizer classes

The user-chosen functions f, s, and F in Exercise E.51 must be coded with particular names. It is then difficult to have several functions for $s(u)$ and experiment with these. A much more flexible code arises if we adopt the ideas of a problem and a solver class as explained in Sect. E.3.6. Specifically, we shall here make use of class `Problem3` in Sect. E.3.6 to store information about $f(\dot{u})$, $s(u)$, $F(t)$, $u(0)$, $\dot{u}(0)$, m, T, and the exact solution (if available). The solver class can store parameters related to the numerical quality of the solution, i.e., Δt and the name of the solver class in the `ODESolver` hierarchy. In addition we will make a visualizer class for producing plots of various kinds.

We want all parameters to be set on the command line, but also have sensible default values. As in Sect. E.3.6, the `argparse` module is used to read data from the command line. Class `Problem` can be sketched as follows:

```
class Problem(object):
    def define_command_line_arguments(self, parser):
        """Add arguments to parser (argparse.ArgumentParser)."""

        parser.add_argument(
            '--friction', type=func_dudt, default='0',
            help='friction function f(dudt)',
            metavar='<function expression>')
```

```
        parser.add_argument(
            '--spring', type=func_u, default='u',
            help='spring function s(u)',
            metavar='<function expression>')
        parser.add_argument(
            '--external', type=func_t, default='0',
            help='external force function F(t)',
            metavar='<function expression>')
        parser.add_argument(
            '--u_exact', type=func_t_vec, default='0',
            help='exact solution u(t) (0 or None: now known)',
            metavar='<function expression>')
        parser.add_argument(
            '--m', type=evalcmlarg, default=1.0, help='mass',
            type=float, metavar='mass')
        ...
        return parser

    def set(self, args):
        """Initialize parameters from the command line."""
        self.friction = args.friction
        self.spring = args.spring
        self.m = args.m
        ...

    def __call__(self, u, t):
        """Define the right-hand side in the ODE system."""
        m, f, s, F = \
            self.m, self.friction, self.spring, self.external
        ...
```

Several functions are specified as the `type` argument to `parser.add_argument`
for turning strings into proper objects, in particular `StringFunction` objects with
different independent variables:

```
def evalcmlarg(text):
    return eval(text)

def func_dudt(text):
    return StringFunction(text, independent_variable='dudt')

def func_u(text):
    return StringFunction(text, independent_variable='u')

def func_t(text):
    return StringFunction(text, independent_variable='t')

def func_t_vec(text):
    if text == 'None' or text == '0':
        return None
    else:
        f = StringFunction(text, independent_variable='t')
        f.vectorize(globals())
        return f
```

The use of `evalcmlarg` is essential: this function runs the strings from the command line through `eval`, which means that we can use mathematical formulas like `-T '4*pi'`.

Class `Solver` is relatively much shorter than class `Problem`:

```
class Solver(object):
    def __init__(self, problem):
        self.problem = problem

    def define_command_line_arguments(self, parser):
        """Add arguments to parser (argparse.ArgumentParser)."""
        # add --dt and --method
        ...
        return parser

    def set(self, args):
        self.dt = args.dt
        self.n = int(round(self.problem.T/self.dt))
        self.solver = eval(args.method)

    def solve(self):
        self.solver = self.method(self.problem)
        ic = [self.problem.initial_u, self.problem.initial_dudt]
        self.solver.set_initial_condition(ic)
        time_points = linspace(0, self.problem.T, self.n+1)
        self.u, self.t = self.solver.solve(time_points)
```

The `Visualizer` class holds references to a `Problem` and `Solver` instance and creates plots. The user can specify plots in an interactive dialog in the terminal window. Inside a loop, the user is repeatedly asked to specify a plot until the user responds with `quit`. The specification of a plot can be one of the words u, dudt, dudt-u, K, and `wavelength` which means a plot of $u(t)$ versus t, $\dot{u}(t)$ versus t, $\dot{u}$ versus u, K $(= \frac{1}{2}m\dot{u}^2$, kinetic energy) versus t, and u's wavelength versus its indices, respectively. The wavelength can be computed from the local maxima of u as explained in Exercise E.52.

A sketch of class `Visualizer` is given next:

```
class Visualizer(object):
    def __init__(self, problem, solver):
        self.problem = problem
        self.solver = solver

    def visualize(self):
        t = self.solver.t    # short form
        u, dudt = self.solver.u[:,0], self.solver.u[:,1]

        # Tag all plots with numerical and physical input values
        title = 'solver=%s, dt=%g, m=%g' % \
                (self.solver.method, self.solver.dt, self.problem.m)
        # Can easily get the formula for friction, spring and force
        # if these are string formulas.
        if isinstance(self.problem.friction, StringFunction):
            title += ' f=%s' % str(self.problem.friction)
```

```
    if isinstance(self.problem.spring, StringFunction):
        title += ' s=%s' % str(self.problem.spring)
    if isinstance(self.problem.external, StringFunction):
        title += ' F=%s' % str(self.problem.external)

    # Let the user interactively specify what
    # to be plotted
    plot_type = ''
    while plot_type != 'quit':
        plot_type = raw_input('Specify a plot: ')
        figure()
        if plot_type == 'u':
            # Plot u vs t
            if self.problem.u_exact is not None:
                hold('on')
                # Plot self.problem.u_exact vs t
            show()
            savefig('tmp_u.pdf')
        elif plot_type == 'dudt':
        ...
        elif plot_type == 'dudt-u':
        ...
        elif plot_type == 'K':
        ...
        elif plot_type == 'wavelength':
        ...
```

Make a complete implementation of the three proposed classes. Also make a `main` function that (i) creates a problem, solver, and visualizer, (ii) calls the functions to define command-line arguments in the problem and solver classes, (iii) reads the command line, (iv) passes on the command-line parser object to the problem and solver classes, (v) calls the solver, and (vi) calls the visualizer's `visualize` method to create plots. Collect the classes and functions in a module `oscillator`, which has a call to `main` in the test block.

The first task from Exercises E.51 can now be run as

```
Terminal
oscillator.py --method ForwardEuler --u_exact "cos(t)" \
        --dt "pi/20" --T "5*pi"
```

The other tasks from Exercises E.51 can be tested similarly.

Explore some of the possibilities of specifying several functions on the command line:

```
Terminal
oscillator.py --method RungeKutta4 --friction "0.1*dudt" \
        --external "sin(0.5*t)" --dt "pi/80" \
        --T "40*pi" --m 10

oscillator.py --method RungeKutta4 --friction "0.8*dudt" \
        --external "sin(0.5*t)" --dt "pi/80" \
        --T "120*pi" --m 50
```

Filename: `oscillator`.

Exercise E.54: Use classes for flexible choices of models
Some typical choices of $f(\dot{u})$, $s(u)$, and $F(t)$ in (E.90) are listed below:

- Linear friction force (low velocities): $f(\dot{u}) = 6\pi\mu R\dot{u}$ (Stokes drag), where R is the radius of a spherical approximation to the body's geometry, and μ is the viscosity of the surrounding fluid.
- Quadratic friction force (high velocities): $f(\dot{u}) = \frac{1}{2}C_D\varrho A|\dot{u}|\dot{u}$. See Exercise 1.11 for explanation of the symbols.
- Linear spring force: $s(u) = ku$, where k is a spring constant.
- Sinusoidal spring force: $s(u) = k\sin u$, where k is a constant.
- Cubic spring force: $s(u) = k(u - \frac{1}{6}u^3)$, where k is a spring constant.
- Sinusoidal external force: $F(t) = F_0 + A\sin\omega t$, where F_0 is the mean value of the force, A is the amplitude, and ω is the frequency.
- Bump force: $F(t) = H(t - t_1)(1 - H(t - t_2))F_0$, where $H(t)$ is the Heaviside function ($H = 0$ for $x < 0$ and $H = 1$ for $x \geq 0$), t_1 and t_2 are two given time points, and F_0 is the size of the force. This $F(t)$ is zero for $t < t_1$ and $t > t_2$, and F_0 for $t \in [t_1, t_2]$.
- Random force 1: $F(t) = F_0 + A \cdot U(t; B)$, where F_0 and A are constants, and $U(t; B)$ denotes a function whose value at time t is random and uniformly distributed in the interval $[-B, B]$.
- Random force 2: $F(t) = F_0 + A \cdot N(t; \mu, \sigma)$, where F_0 and A are constants, and $N(t; \mu, \sigma)$ denotes a function whose value at time t is random, Gaussian distributed number with mean μ and standard deviation σ.

Make a module `functions` where each of the choices above are implemented as a class with a `__call__` special method. Also add a class `Zero` for a function whose value is always zero. It is natural that the parameters in a function are set as arguments to the constructor. The different classes for spring functions can all have a common base class holding the k parameter as data attribute.
Filename: `functions`.

Exercise E.55: Apply software for oscillating systems
The purpose of this exercise is to demonstrate the use of the classes from Exercise E.54 to solve problems described by (E.90).

With a lot of models for $f(\dot{u})$, $s(u)$, and $F(t)$ available as classes in `functions.py`, the initialization of `self.friction`, `self.spring`, etc., from the command line does not work, because we assume simple string formulas on the command line. Now we want to write things like `-spring 'LinearSpring(1.0)'`. There is a quite simple remedy: replace all the special conversion functions to `StringFunction` objects by `evalcmlarg` in the type specifications in the `parser.add_argument` calls. If a `from functions import *` is also performed in the `oscillator.py` file, a simple `eval` will turn strings like `'LinearSpring(1.0)'` into living objects.

However, we shall here follow a simpler approach, namely dropping initializing parameters on the command line and instead set them directly in the code. Here is an example:

```
problem = Problem()
problem.m = 1.0
k = 1.2
problem.spring = CubicSpring(k)
problem.friction = Zero()
problem.T = 8*pi/sqrt(k)
...
```

This is the simplest way of making use of the objects in the functions module.

Note that the set method in classes Solver and Visualizer is unaffected by the new objects from the functions module, so flexible initialization via command-line arguments works as before for -dt, -method, and plot. One may also dare to call the set method in the problem object to set parameters like m, initial_u, etc., or one can choose the safer approach of not calling set but initialize all data attributes explicitly in the user's code.

Make a new file say oscillator_test.py where you import class Problem, Solver, and Visualizer, plus all classes from the functions module. Provide a main1 function for solving the following problem: $m = 1$, $u(0) = 1$, $\dot{u}(0) = 0$, no friction (use class Zero), no external forcing (class Zero), a linear spring $s(u) = u$, $\Delta t = \pi/20$, $T = 8\pi$, and exact $u(t) = \cos(t)$. Use the Forward Euler method.

Then make another function main2 for the case with $m = 5$, $u(0) = 1$, $\dot{u}(0) = 0$, linear friction $f(\dot{u}) = 0.1\dot{u}$, $s(u) = u$, $F(t) = \sin(\frac{1}{2}t)$, $\Delta t = \pi/80$, $T = 60\pi$, and no knowledge of an exact solution. Use the 4-th order Runge-Kutta method.

Let a test block use the first command-line argument to indicate a call to main1 or main2.
Filename: oscillator_test.

Exercise E.56: Model the economy of fishing
A population of fish is governed by the differential equation

$$\frac{dx}{dt} = \frac{1}{10}x\left(1 - \frac{x}{100}\right) - h, \quad x(0) = 500, \tag{E.91}$$

where $x(t)$ is the size of the population at time t and h is the harvest.

a) Assume $h = 0$. Find an exact solution for $x(t)$. For which value of t is $\frac{dx}{dt}$ largest? For which value of t is $\frac{1}{x}\frac{dx}{dt}$ largest?
b) Solve the differential equation (E.91) by the Forward Euler method. Plot the numerical and exact solution in the same plot.
c) Suppose the harvest h depends on the fishers' efforts, E, in the following way: $h = qxE$, with q as a constant. Set $q = 0.1$ and assume E is constant. Show the effect of E on $x(t)$ by plotting several curves, corresponding to different E values, in the same figure.
d) The fishers' total revenue is given by $\pi = ph - \frac{c}{2}E^2$, where p is a constant. In the literature about the economy of fisheries, one is often interested in how a fishery will develop in the case the harvest is not regulated. Then new fishers will appear as long as there is money to earn ($\pi > 0$). It can (for simplicity) be reasonable to model the dependence of E on π as

$$\frac{dE}{dt} = \gamma\pi, \tag{E.92}$$

where γ is a constant. Solve the system of differential equations for $x(t)$ and $E(t)$ by the 4th-order Runge-Kutta method, and plot the curve with points $(x(t), E(t))$ in the two cases $\gamma = 1/2$ and $\gamma \to \infty$. Choose $c = 0.3$, $p = 10$, $E(0) = 0.5$, and $T = 1$.

Filename: `fishery`.

Debugging

F

Testing a program to find errors usually takes much more time than to write the code. This appendix is devoted to tools and good habits for effective debugging. Section F.1 describes the Python debugger, a key tool for examining the internal workings of a code, while Sect. F.2 explains how to solve problems and write software to simplify the debugging process.

F.1 Using a Debugger

A debugger is a program that can help you find out what is going on in a computer program. You can stop the execution at any prescribed line number, print out variables, continue execution, stop again, execute statements one by one, and repeat such actions until you have tracked down abnormal behavior and found bugs.

Here we shall use the debugger to demonstrate the program flow of the code `Simpson.py` (which can integrate functions of one variable with the famous Simpson's rule). This development of this code is explained in Sect. 3.4.2. You are strongly encouraged to carry out the steps below on your computer to get a glimpse of what a debugger can do.

Step 1 Go to the folder `src/funcif`[1] where the program `Simpson.py` resides.

Step 2 If you use the Spyder Integrated Development Environment, choose *Debug* on the *Run* pull-down menu. If you run your programs in a plain terminal window, start IPython:

```
Terminal

Terminal> ipython
```

Run the program `Simpson.py` with the debugger on (-d):

```
In [1]: run -d Simpson.py
```

---

[1] http://tinyurl.com/pwyasaa/funcif

835

We now enter the debugger and get a prompt

```
ipdb>
```

After this prompt we can issue various debugger commands. The most important ones will be described as we go along.

Step 3 Type continue or just c to go to the first line in the file. Now you can see a printout of where we are in the program:

```
1---> 1 def Simpson(f, a, b, n=500):
      2     """
      3     Return the approximation of the integral of f
```

Each program line is numbered and the arrow points to the next line to be executed. This is called the *current line*.

Step 4 You can set a *break point* where you want the program to stop so that you can examine variables and perhaps follow the execution closely. We start by setting a break point in the application function:

```
ipdb> break application
Breakpoint 2 at /home/.../src/funcif/Simpson.py:30
```

You can also say break X, where X is a line number in the file.

Step 5 Continue execution until the break point by writing continue or c. Now the program stops at line 31 in the application function:

```
ipdb> c
> /home/.../src/funcif/Simpson.py(31)application()
2    30 def application():
---> 31     from math import sin, pi
     32     print 'Integral of 1.5*sin^3 from 0 to pi:'
```

Step 6 Typing step or just s executes one statement at a time:

```
ipdb> s
> /home/.../src/funcif/Simpson.py(32)application()
     31     from math import sin, pi
---> 32     print 'Integral of 1.5*sin^3 from 0 to pi:'
     33     for n in 2, 6, 12, 100, 500:

ipdb> s
Integral of 1.5*sin^3 from 0 to pi:
> /home/.../src/funcif/Simpson.py(33)application()
     32     print 'Integral of 1.5*sin^3 from 0 to pi:'
---> 33     for n in 2, 6, 12, 100, 500:
     34         approx = Simpson(h, 0, pi, n)
```

Typing another s reaches the call to Simpson, and a new s steps *into* the function Simpson:

```
ipdb> s
--Call--
> /home/.../src/funcif/Simpson.py(1)Simpson()
1---> 1 def Simpson(f, a, b, n=500):
      2      """
      3      Return the approximation of the integral of f
```

Type a few more s to step ahead of the if tests.

Step 7 Examining the contents of variables is easy with the print (or p) command:

```
ipdb> print f, a, b, n
<function h at 0x898ef44> 0 3.14159265359 2
```

We can also check the type of the objects:

```
ipdb> whatis f
Function h
ipdb> whatis a
<type 'int'>
ipdb> whatis b
<type 'float'>
ipdb> whatis n
<type 'int'>
```

Step 8 Set a new break point in the application function so that we can jump directly there without having to go manually through all the statements in the Simpson function. To see line numbers and corresponding statements around some line with number X, type list X. For example,

```
ipdb> list 32
     27 def h(x):
     28      return (3./2)*sin(x)**3
     29
     30 from math import sin, pi
     31
 2   32 def application():
     33      print 'Integral of 1.5*sin^3 from 0 to pi:'
     34      for n in 2, 6, 12, 100, 500:
     35          approx = Simpson(h, 0, pi, n)
     36          print 'n=%3d, approx=%18.15f, error=%9.2E' % \
     37              (n, approx, 2-approx)
```

We set a line break at line 35:

```
ipdb> break 35
Breakpoint 3 at /home/.../src/funcif/Simpson.py:35
```

Typing c continues execution up to the next break point, line 35.

Step 9 The command `next` or `n` is like `step` or `s` in that the current line is executed, but the execution does not step into functions, instead the function calls are just performed and the program stops at the next line:

```
ipdb> n
> /home/.../src/funcif/Simpson.py(36)application()
3    35            approx = Simpson(h, 0, pi, n)
---> 36           print 'n=%3d, approx=%18.15f, error=%9.2E' % \
     37                 (n, approx, 2-approx)
ipdb> print approx, n
1.9891717005835792 6
```

Step 10 The command `disable X Y Z` disables break points with numbers X, Y, and Z, and so on. To remove our three break points and continue execution until the program naturally stops, we write

```
ipdb> disable 1 2 3
ipdb> c
n=100, approx= 1.999999902476350, error= 9.75E-08
n=500, approx= 1.999999999844138, error= 1.56E-10

In [2]:
```

At this point, I hope you realize that a debugger is a very handy tool for monitoring the program flow, checking variables, and thereby understanding why errors occur.

F.2 How to Debug

Most programmers will claim that writing code consumes a small portion of the time it takes to develop a program: the major portion of the work concerns testing the program and finding errors.

> *Debugging is twice as hard as writing the code in the first place. Therefore, if you write the code as cleverly as possible, you are, by definition, not smart enough to debug it.* Brian W. Kernighan, computer scientist, 1942-.

Newcomers to programming often panic when their program runs for the first time and aborts with a seemingly cryptic error message. How do you approach the art of debugging? This appendix summarizes some important working habits in this respect. Some of the tips are useful for problem solving in general, not only when writing and testing Python programs.

F.2.1 A Recipe for Program Writing and Debugging

1. Understand the problem Make sure that you really understand the task the program is supposed to solve. We can make a general claim: if you do not understand the problem and the solution method, you will never be able to make a correct program. It may be argued that this claim is not entirely true: sometimes students with limited understanding of the problem are able to grab a similar program and guess at a few modifications, and actually get a program that works. But this technique is based on luck and not on understanding. The famous Norwegian computer scientist Kristen Nygaard (1926–2002) phrased it precisely: *Programming is understanding*. It may be necessary to read a problem description or exercise many times and study relevant background material before starting on the programming part of the problem solving process.

2. Work out examples Start with sketching one or more examples on input and output of the program. Such examples are important for controlling the understanding of the purpose of the program, and for verifying the implementation.

3. Decide on a user interface Find out how you want to get data into the program. You may want to grab data from the command-line, a file, or a dialog with questions and answers.

4. Make algorithms Identify the key tasks to be done in the program and sketch rough algorithms for these. Some programmers prefer to do this on a piece of paper, others prefer to start directly in Python and write Python-like code with comments to sketch the program (this is easily developed into real Python code later).

5. Look up information Few programmers can write the whole program without consulting manuals, books, and the Internet. You need to know and understand the basic constructs in a language and some fundamental problem solving techniques, but technical details can be looked up.

The more program examples you have studied (in this book, for instance), the easier it is to adapt ideas from an existing example to solve a new problem.

6. Write the program Be extremely careful with what you write. In particular, compare all mathematical statements and algorithms with the original mathematical expressions.

In longer programs, do not wait until the program is complete before you start testing it, test parts while you write.

7. Run the program If the program aborts with an error message from Python, these messages are fortunately quite precise and helpful. First, locate the line number where the error occurs and read the statement, then carefully read the error message. The most common errors (exceptions) are listed below.

SyntaxError: Illegal Python code.

```
  File "somefile.py", line 5
    x = . 5
        ^
SyntaxError: invalid syntax
```

Often the error is precisely indicated, as above, but sometimes you have to search for the error on the previous line.

NameError: A name (variable, function, module) is not defined.

```
  File "somefile.py", line 20, in <module>
    table(10)
  File "somefile.py", line 16, in table
    value, next, error = L(x, n)
  File "somefile.py", line 8, in L
    exact_error = log(1+x) - value_of_sum
NameError: global name 'value_of_sum' is not defined
```

Look at the last of the lines starting with File to see where in the program the error occurs. The most common reasons for a NameError are

- a misspelled name,
- a variable that is not initialized,
- a function that you have forgotten to define,
- a module that is not imported.

TypeError: An object of wrong type is used in an operation.

```
  File "somefile.py", line 17, in table
    value, next, error = L(x, n)
  File "somefile.py", line 7, in L
    first_neglected_term = (1.0/(n+1))*(x/(1.0+x))**(n+1)
TypeError: unsupported operand type(s) for +: 'float' and 'str'
```

Print out objects and their types (here: print x, type(x), n, type(n)), and you will most likely get a surprise. The reason for a TypeError is often far away from the line where the TypeError occurs.

ValueError: An object has an illegal value.

```
  File "somefile.py", line 8, in L
    y = sqrt(x)
ValueError: math domain error
```

Print out the value of objects that can be involved in the error (here: print x).

IndexError: An index in a list, tuple, string, or array is too large.

```
  File "somefile.py", line 21
    n = sys.argv[i+1]
IndexError: list index out of range
```

Print out the length of the list, and the index if it involves a variable (here: print len(sys.argv), i).

8. Verify the results Assume now that we have a program that runs without error messages from Python. Before judging the results of the program, set precisely up a test case where you know the exact solution. This is in general quite difficult. In complicated mathematical problems it is an art to construct good test problems and procedures for providing evidence that the program works.

If your program produces wrong answers, start to *examine intermediate results.* Never forget that your own hand calculations that you use to test the program may be wrong!

9. Use a debugger If you end up inserting a lot of `print` statements in the program for checking intermediate results, you might benefit from using a debugger as explained in Sect. F.1.

Some may think that this list of nine points is very comprehensive. However, the recipe just contains the steps that you should always carry out when developing programs. Never forget that computer programming is a difficult task.

> *Program writing is substantially more demanding than book writing. Why is it so? I think the main reason is that a larger attention span is needed when working on a large computer program than when doing other intellectual tasks.* Donald Knuth [11, p. 18], computer scientist, 1938-.

F.2.2 Application of the Recipe

Let us illustrate the points above in a specific programming problem: implementation of the Midpoint rule for numerical integration. The Midpoint rule for approximating an integral $\int_a^b f(x)dx$ reads

$$I = h \sum_{i=1}^{n} f(a + (i - \frac{1}{2})h), \quad h = \frac{b - a}{n}. \tag{F.1}$$

We just follow the individual steps in the recipe to develop the code.

1. Understand the problem In this problem we must understand how to program the formula (F.1). Observe that we do not need to understand how the formula is derived, because we do not apply the derivation in the program. What is important, is to notice that the formula is an *approximation* of an integral. Comparing the result of the program with the exact value of the integral will in general show a discrepancy. Whether we have an approximation error or a programming error is always difficult to judge. We will meet this difficulty below.

2. Work out examples As a test case we choose to integrate

$$f(x) = \sin^{-1}(x). \tag{F.2}$$

between 0 and π. From a table of integrals we find that this integral equals

$$\left[x \sin^{-1}(x) + \sqrt{1 - x^2} \right]_0^{\pi}. \tag{F.3}$$

The formula (F.1) gives an approximation to this integral, so the program will (most likely) print out a result different from (F.3). It would therefore be very helpful to construct a calculation where there are no approximation errors. Numerical integration rules usually integrate some polynomial of low order exactly. For the Midpoint rule it is obvious, if you understand the derivation of this rule, that a constant function will be integrated exactly. We therefore also introduce a test problem where we integrate $g(x) = 1$ from 0 to 10. The answer should be exactly 10.

Input and output The input to the calculations is the function to integrate, the integration limits a and b, and the n parameter (number of intervals) in the formula (F.1). The output from the calculations is the approximation to the integral.

3. Decide on a user interface We find it easiest at this beginning stage to program the two functions $f(x)$ and $g(x)$ directly in the program. We also specify the corresponding integration limits a and b in the program, but we read a common n for both integrals from the command line. Note that this is not a flexible user interface, but it suffices as a start for creating a working program. A much better user interface is to read f, a, b, and n from the command line, which will be done later in a more complete solution to the present problem.

4. Make algorithms Like most mathematical programming problems, also this one has a generic part and an application part. The generic part is the formula (F.1), which is applicable to an arbitrary function $f(x)$. The implementation should reflect that we can specify any Python function `f(x)` and get it integrated. This principle calls for calculating (F.1) in a Python function where the input to the computation (f, a, b, n) are arguments. The function heading can look as `integrate(f, a, b, n)`, and the value of (F.1) is returned.

The test part of the program consists of defining the test functions $f(x)$ and $g(x)$ and writing out the calculated approximations to the corresponding integrals.

A first rough sketch of the program can then be

```
def integrate(f, a, b, n):
    # compute integral, store in I
    return I

def f(x):
...

def g(x):
...

# test/application part:
n = sys.argv[1]
I = integrate(g, 0, 10, n)
print "Integral of g equals %g" % I
I = integrate(f, 0, pi, n)
# calculate and print out the exact integral of f
```

The next step is to make a detailed implementation of the `integrate` function. Inside this function we need to compute the sum (F.1). In general, sums are computed by a `for` loop over the summation index, and inside the loop we calculate a term in the sum and add it to an accumulation variable. Here is the algorithm in Python code:

```
s = 0
for i in range(1, n+1):
    s = s + f(a + (i-0.5)*h)
I = s*h
```

5. Look up information Our test function $f(x) = \sin^{-1}(x)$ must be evaluated in the program. How can we do this? We know that many common mathematical functions are offered by the `math` module. It is therefore natural to check if this module has an inverse sine function. The best place to look for Python modules is the Python Standard Library[2] [3] documentation, which has a search facility. Typing *math* brings up a link to the `math` module, there we find `math.asin` as the function we need. Alternatively, one can use the command line utility `pydoc` and write `pydoc math` to look up all the functions in the module.

In this simple problem, we use very basic programming constructs and there is hardly any need for looking at similar examples to get started with the problem solving process. We need to know how to program a sum, though, via a `for` loop and an accumulation variable for the sum. Examples are found in Sects. 2.1.4 and 3.1.8.

6. Write the program Here is our first attempt to write the program. You can find the whole code in the file `integrate_v1.py`.

```
def integrate(f, a, b, n):
    s = 0
    for i in range(1, n):
        s += f(a + i*h)
    return s

def f(x):
return asin(x)

def g(x):
return 1

# Test/application part
n = sys.argv[1]
I = integrate(g, 0, 10,  n)
print "Integral of g equals %g" % I
I = integrate(f, 0, pi,  n)
I_exact = pi*asin(pi) - sqrt(1 - pi**2) - 1
print "Integral of f equals %g (exact value is %g)' % \
   (I, I_exact)
```

7. Run the program We try a first execution from IPython

```
In [1]: run integrate_v1.py
```

[2] http://docs.python.org/2/library/

Unfortunately, the program aborts with an error:

```
File "integrate_v1.py", line 8
  return asin(x)
      ^
IndentationError: expected an indented block
```

We go to line 8 and look at that line and the surrounding code:

```
def f(x):
return asin(x)
```

Python expects that the return line is indented, because the function body must always be indented. By the way, we realize that there is a similar error in the g(x) function as well. We correct these errors:

```
def f(x):
    return asin(x)

def g(x):
    return 1
```

Running the program again makes Python respond with

```
File "integrate_v1.py", line 24
  (I, I_exact)
        ^
SyntaxError: EOL while scanning single-quoted string
```

There is nothing wrong with line 24, but line 24 is a part of the statement starting on line 23:

```
print "Integral of f equals %g (exact value is %g)' % \
      (I, I_exact)
```

A SyntaxError implies that we have written illegal Python code. Inspecting line 23 reveals that the string to be printed starts with a double quote, but ends with a single quote. We must be consistent and use the same enclosing quotes in a string. Correcting the statement,

```
print "Integral of f equals %g (exact value is %g)" % \
      (I, I_exact)
```

and rerunning the program yields the output

```
Traceback (most recent call last):
  File "integrate_v1.py", line 18, in <module>
    n = sys.argv[1]
NameError: name 'sys' is not defined
```

Obviously, we need to import sys before using it. We add import sys and run again:

```
Traceback (most recent call last):
  File "integrate_v1.py", line 19, in <module>
    n = sys.argv[1]
IndexError: list index out of range
```

This is a very common error: we index the list sys.argv out of range because we have not provided enough command-line arguments. Let us use $n = 10$ in the test and provide that number on the command line:

```
In [5]: run integrate_v1.py 10
```

We still have problems:

```
Traceback (most recent call last):
  File "integrate_v1.py", line 20, in <module>
    I = integrate(g, 0, 10,  n)
  File "integrate_v1.py", line 7, in integrate
    for i in range(1, n):
TypeError: range() integer end argument expected, got str.
```

It is the final File line that counts (the previous ones describe the nested functions calls up to the point where the error occurred). The error message for line 7 is very precise: the end argument to range, n, should be an integer, but it is a string. We need to convert the string sys.argv[1] to int before sending it to the integrate function:

```
n = int(sys.argv[1])
```

After a new edit-and-run cycle we have other error messages waiting:

```
Traceback (most recent call last):
  File "integrate_v1.py", line 20, in <module>
    I = integrate(g, 0, 10,  n)
  File "integrate_v1.py", line 8, in integrate
    s += f(a + i*h)
NameError: global name 'h' is not defined
```

The h variable is used without being assigned a value. From the formula (F.1) we see that $h = (b - a)/n$, so we insert this assignment at the top of the integrate function:

```
def integrate(f, a, b, n):
    h = (b-a)/n
    ...
```

A new run results in a new error:

```
Integral of g equals 9
Traceback (most recent call last):
  File "integrate_v1.py", line 23, in <module>
    I = integrate(f, 0, pi,  n)
NameError: name 'pi' is not defined
```

Looking carefully at all output, we see that the program managed to call the
integrate function with g as input and write out the integral. However, in the
call to integrate with f as argument, we get a NameError, saying that pi is
undefined. When we wrote the program we took it for granted that pi was π,
but we need to import pi from math to get this variable defined, before we call
integrate:

```
from math import pi
I = integrate(f, 0, pi,  n)
```

The output of a new run is now

```
Integral of g equals 9
Traceback (most recent call last):
  File "integrate_v1.py", line 24, in <module>
    I = integrate(f, 0, pi,  n)
  File "integrate_v1.py", line 9, in integrate
    s += f(a + i*h)
  File "integrate_v1.py", line 13, in f
    return asin(x)
NameError: global name 'asin' is not defined
```

A similar error occurred: asin is not defined as a function, and we need to import
it from math. We can either do a

```
from math import pi, asin
```

or just do the rough

```
from math import *
```

to avoid any further errors with undefined names from the math module (we will
get one for the sqrt function later, so we simply use the last "import all" kind of
statement).

There are still more errors:

```
Integral of g equals 9
Traceback (most recent call last):
  File "integrate_v1.py", line 24, in <module>
    I = integrate(f, 0, pi,  n)
  File "integrate_v1.py", line 9, in integrate
    s += f(a + i*h)
  File "integrate_v1.py", line 13, in f
    return asin(x)
ValueError: math domain error
```

Now the error concerns a wrong x value in the f function. Let us print out x:

```
def f(x):
    print x
    return asin(x)
```

The output becomes

```
Integral of g equals 9
0.314159265359
0.628318530718
0.942477796077
1.25663706144
Traceback (most recent call last):
  File "integrate_v1.py", line 25, in <module>
    I = integrate(f, 0, pi,  n)
  File "integrate_v1.py", line 9, in integrate
    s += f(a + i*h)
  File "integrate_v1.py", line 14, in f
    return asin(x)
ValueError: math domain error
```

We see that all the asin(x) computations are successful up to and including $x = 0.942477796077$, but for $x = 1.25663706144$ we get an error. A math domain error may point to a wrong x value for $\sin^{-1}(x)$ (recall that the domain of a function specifies the legal x values for that function).

To proceed, we need to think about the mathematics of our problem: Since $\sin(x)$ is always between -1 and 1, the inverse sine function cannot take x values outside the interval $[-1, 1]$. The problem is that we try to integrate $\sin^{-1}(x)$ from 0 to π, but only integration limits within $[-1, 1]$ make sense (unless we allow for complex-valued trigonometric functions). Our test problem is hence wrong from a mathematical point of view. We need to adjust the limits, say 0 to 1 instead of 0 to π. The corresponding program modification reads

```
I = integrate(f, 0, 1,  n)
```

We run again and get

```
Integral of g equals 9
0
0
0
0
0
0
0
0
0
Traceback (most recent call last):
  File "integrate_v1.py", line 26, in <module>
    I_exact = pi*asin(pi) - sqrt(1 - pi**2) - 1
ValueError: math domain error
```

It is easy to go directly to the `ValueError` now, but one should always examine the output from top to bottom. If there is strange output before Python reports an error, there may be an error indicated by our `print` statements. This is not the case in the present example, but it is a good habit to start at the top of the output anyway. We see that all our `print x` statements inside the `f` function say that x is zero. This must be wrong – the idea of the integration rule is to pick *n* different points in the integration interval [0, 1].

Our `f(x)` function is called from the `integrate` function. The argument to `f`, `a + i*h`, is seemingly always 0. Why? We print out the argument and the values of the variables that make up the argument:

```
def integrate(f, a, b, n):
    h = (b-a)/n
    s = 0
    for i in range(1, n):
        print a, i, h, a+i*h
        s += f(a + i*h)
    return s
```

Running the program shows that h is zero and therefore a+i*h is zero.

Why is h zero? We need a new `print` statement in the computation of h:

```
def integrate(f, a, b, n):
    h = (b-a)/n
    print b, a, n, h
    ...
```

The output shows that a, b, and n are correct. Now we have encountered a very common error in Python version 2 and C-like programming languages: integer division (see Sect. 1.3.1). The formula $(1 - 0)/10 = 1/10$ is zero according to integer division. The reason is that a and b are specified as 0 and 1 in the call to `integrate`, and 0 and 1 imply `int` objects. Then b-a becomes an `int`, and n is an `int`, causing an `int/int` division. We must ensure that b-a is `float` to get the right mathematical division in the computation of h:

```
def integrate(f, a, b, n):
    h = float(b-a)/n
    ...
```

Thinking that the problem with wrong *x* values in the inverse sine function is resolved, we may remove all the `print` statements in the program, and run again.

The output now reads

```
Integral of g equals 9
Traceback (most recent call last):
  File "integrate_v1.py", line 25, in <module>
    I_exact = pi*asin(pi) - sqrt(1 - pi**2) - 1
ValueError: math domain error
```

That is, we are back to the `ValueError` we have seen before. The reason is that `asin(pi)` does not make sense, and the argument to `sqrt` is negative. The error is simply that we forgot to adjust the upper integration limit in the computation of the exact result. This is another very common error. The correct line is

```
I_exact = 1*asin(1) - sqrt(1 - 1**2) - 1
```

We could have avoided the error by introducing variables for the integration limits, and a function for $\int f(x)dx$ would make the code cleaner:

```
a = 0; b = 1
def int_f_exact(x):
    return x*asin(x) - sqrt(1 - x**2)
I_exact = int_f_exact(b) - int_f_exact(a)
```

Although this is more work than what we initially aimed at, it usually saves time in the debugging phase to do things this proper way.

Eventually, the program seems to work! The output is just the result of our two `print` statements:

```
Integral of g equals 9
Integral of f equals 5.0073 (exact value is 0.570796)
```

8. Verify the results Now it is time to check if the numerical results are correct. We start with the simple integral of 1 from 0 to 10: the answer should be 10, not 9. Recall that for this particular choice of integration function, there is no approximation error involved (but there could be a small rounding error). Hence, there must be a programming error.

To proceed, we need to calculate some intermediate mathematical results by hand and compare these with the corresponding statements in the program. We choose a very simple test problem with $n = 2$ and $h = (10 - 0)/2 = 5$. The formula (F.1) becomes

$$I = 5 \cdot (1 + 1) = 10 \,.$$

Running the program with $n = 2$ gives

```
Integral of g equals 1
```

We insert some `print` statements inside the `integrate` function:

```
def integrate(f, a, b, n):
    h = float(b-a)/n
    s = 0
    for i in range(1, n):
        print 'i=%d, a+i*h=%g' % (i, a+i*h)
        s += f(a + i*h)
    return s
```

Here is the output:

```
i=1, a+i*h=5
Integral of g equals 1
i=1, a+i*h=0.5
Integral of f equals 0.523599 (exact value is 0.570796)
```

There was only one pass in the i loop in integrate. According to the formula, there should be n passes, i.e., two in this test case. The limits of i must be wrong. The limits are produced by the call range(1,n). We recall that such a call results in integers going from 1 up to n, but *not* including n. We need to include n as value of i, so the right call to range is range(1,n+1).

We make this correction and rerun the program. The output is now

```
i=1, a+i*h=5
i=2, a+i*h=10
Integral of g equals 2
i=1, a+i*h=0.5
i=2, a+i*h=1
Integral of f equals 2.0944 (exact value is 0.570796)
```

The integral of 1 is still not correct. We need more intermediate results!

In our quick hand calculation we knew that $g(x) = 1$ so all the $f(a + (i - \frac{1}{2})h)$ evaluations were rapidly replaced by ones. Let us now compute all the x coordinates $a + (i - \frac{1}{2})h$ that are used in the formula:

$$i = 1 : a + \left(i - \frac{1}{2}\right) h = 2.5, \quad i = 2 : a + \left(i - \frac{1}{2}\right) h = 7.5 .$$

Looking at the output from the program, we see that the argument to g has a different value – and fortunately we realize that the formula we have coded is wrong. It should be a+(i-0.5)*h.

We correct this error and run the program:

```
i=1, a+(i-0.5)*h=2.5
i=2, a+(i-0.5)*h=7.5
Integral of g equals 2
...
```

Still the integral is wrong. At this point you may give up programming, but the more skills you pick up in debugging, the more fun it is to hunt for errors! Debugging is like reading an exciting criminal novel: the detective follows different ideas and tracks, but never gives up before the culprit is caught.

Now we read the code more carefully and compare expressions with those in the mathematical formula. We should, of course, have done this already when writing the program, but it is easy to get excited when writing code and hurry for the end. This ongoing story of debugging probably shows that reading the code carefully can save much debugging time. (Actually, being extremely careful with what you write, and comparing all formulas with the mathematics, may be the best way to get more spare time when taking a programming course!)

We clearly add up all the f evaluations correctly, but then this sum must be multiplied by h, and we forgot that in the code. The `return` statement in `integrate`
must therefore be modified to

```
    return s*h
```

Eventually, the output is

```
Integral of g equals 10
Integral of f equals 0.568484 (exact value is 0.570796)
```

and we have managed to integrate a constant function in our program! Even the
second integral looks promising!

To judge the result of integrating the inverse sine function, we need to run
several increasing n values and see that the approximation gets better. For $n =
2, 10, 100, 1000$ we get $0.550371, 0.568484, 0.570714, 0.570794$, to be compared
to the exact value 0.570796. (This is not the mathematically exact value, because it
involves computations of $\sin^{-1}(x)$, which is only approximately calculated by the
`asin` function in the `math` module. However, the approximation error is very small
($\sim 10^{-16}$).) The decreasing error provides evidence for a correct program, but it is
not a strong proof. We should try out more functions. In particular, linear functions
are integrated exactly by the Midpoint rule. We can also measure the speed of the
decrease of the error and check that the speed is consistent with the properties of
the Midpoint rule, but this is a mathematically more advanced topic.

The very important lesson learned from these debugging sessions is that you
should start with a simple test problem where all formulas can be computed by
hand. If you start out with $n = 100$ and try to integrate the inverse sine function,
you will have a much harder job with tracking down all the errors.

9. Use a debugger Another lesson learned from these sessions is that we needed
many `print` statements to see intermediate results. It is an open question if it
would be more efficient to run a debugger and stop the code at relevant lines. In an
edit-and-run cycle of the type we met here, we frequently need to examine many
numerical results, correct something, and look at all the intermediate results again.
Plain `print` statements are often better suited for this massive output than the pure
manual operation of a debugger, unless one writes a program to automate the interaction with the debugger.

The correct code for the implementation of the Midpoint rule is found in
`integrate_v2.py`. Some readers might be frightened by all the energy it took
to debug this code, but this is just the nature of programming. The experience of
developing programs that finally work is very awarding.

*People only become computer programmers if they're obsessive about details, crave power
over machines, and can bear to be told day after day exactly how stupid they are.* Gregory
J. E. Rawlins [24], computer scientist.

Refining the user interface We briefly mentioned that the chosen user interface,
where the user can only specify n, is not particularly user friendly. We should

allow f, a, b, and n to be specified on the command line. Since f is a function and the command line can only provide strings to the program, we may use the `StringFunction` object from `scitools.std` to convert a string expression for the function to be integrated to an ordinary Python function (see Sect. 4.3.3). The other parameters should be easy to retrieve from the command line if Sect. 4.2 is understood. As suggested in Sect. 4.7, we enclose the input statements in a `try-except` block, here with a specific exception type `IndexError` (because an index in `sys.argv` out of bounds is the only type of error we expect to handle):

```
try:
    f_formula = sys.argv[1]
    a = eval(sys.argv[2])
    b = eval(sys.argv[3])
    n = int(sys.argv[4])
except IndexError:
    print 'Usage: %s f-formula a b n' % sys.argv[0]
    sys.exit(1)
```

Note that the use of `eval` allows us to specify a and b as `pi` or `exp(5)` or another mathematical expression.

With the input above we can perform the general task of the program:

```
from scitools.std import StringFunction
f = StringFunction(f_formula)
I = integrate(f, a, b, n)
print I
```

Writing a test function Instead of having these test statements as a main program we follow the good habits of Sect. 4.9 and make a module with

- the `integrate` function,
- a `test_integrate` function for testing the `integrate` function's ability to exactly integrate linear functions,
- a `main` function for reading data from the command line and calling `integrate` for the user's problem at hand.

Any module should also have a test block, as well as doc strings for the module itself and all functions.

The `test_integrate` function can perform a loop over some specified n values and check that the Midpoint rule integrates a linear function exactly. As always, we must be prepared for rounding errors, so "exactly" means errors less than (say) 10^{-14}. The relevant code becomes

```
def test_integrate():
    """Check that linear functions are integrated exactly."""

    def g(x):
        return p*x + q    # general linear function

    def int_g_exact(x):   # integral of g(x)
        return 0.5*p*x**2 + q*x
```

```
a = -1.2; b = 2.8      # "arbitrary" integration limits
p = -2;    q = 10
success = True          # True if all tests below are passed
for n in 1, 10, 100:
    I = integrate(g, a, b, n)
    I_exact = int_g_exact(b) - int_g_exact(a)
    error = abs(I_exact - I)
    if error > 1E-14:
        success = False
assert success
```

We have followed the programming standard that will make this test function automatically work with the nose test framework:

1. the name of the function starts with `test_`,
2. the function has no arguments,
3. checks of whether a test is passed or not are done with `assert`.

The `assert success` statement raises an `AssertionError` exception if `success`
is false, otherwise nothing happens. The nose testing framework searches for
functions whose name start with `test_`, execute each function, and record if an
`AssertionError` is raised. It is overkill to use nose for small programs, but in
larger projects with many functions in many files, nose can run all tests with a short
command and write back a notification that all tests passed.

The `main` function is simply a wrapping of the main program given above. The
test block may call or `test_integrate` function or `main`, depending on whether
the user will test the module or use it:

```
if __name__ == '__main__':
    if sys.argv[1] == 'verify':
        verify()
    else:
        # Compute the integral specified on the command line
        main()
```

Here is a short demo computing $\int_0^{2\pi}(\cos(x) + \sin(x))dx$ with the aid of the
`integrate.py` file:

Terminal
```
integrate.py 'cos(x)+sin(x)' 0 2*pi 10
-3.48786849801e-16
```

F.2.3   Getting Help from a Code Analyzer

The tools PyLint[3] and Flake8[4] can analyze your code and point out errors and undesired coding styles. Before point 7 in the lists above, *Run the program*, it can

[3] http://www.pylint.org/
[4] https://flake8.readthedocs.org/en/2.0/

be wise to run PyLint and/or Flake8 to be informed about problems with the code. Flake8 complains in general a lot less than PyLint, and might be a more useful tool for mathematical software (see the examples below), but both tools are very useful during program development to improve your code and ease further debugging.

Consider the first version of the `integrate` code, `integrate_v1.py`. Running Flake8 gives

```Terminal
Terminal> flake8 integrate_v1.py
integrate_v1.py:7:1: E302 expected 2 blank lines, found 1
integrate_v1.py:8:1: E112 expected an indented block
integrate_v1.py:8:7: E901 IndentationError: expected an indented block
integrate_v1.py:10:1: E302 expected 2 blank lines, found 1
integrate_v1.py:11:1: E112 expected an indented block
```

Flake8 checks if the program obeys the official Style Guide for Python Code[5] (known as *PEP8*). One of the rules in this guide is to have two blank lines before functions and classes (a habit that is often dropped in this book to reduce the length of code snippets), and our program breaks the rule before the f and g functions. More serious and useful is the notification of `expected an indented block` at lines 8 and 11, but this error is quickly found by running the program too.

PyLint reports much less about `integrate_v1.py`:

```Terminal
Terminal> pylint integrate_v1.py
E:  8, 0: expected an indented block (syntax-error)
```

Running Flak8 on `integrate_v2.py` leads to only three problems: missing two blank lines before functions and doing `from math import *`. Applying PyLint to `integrate_v2.py` results in many more problems:

```Terminal
Terminal> pylint integrate_v2.py
C: 20, 0: Exactly one space required after comma
I = integrate(f, 0, 1,  n)
                      ^ (bad-whitespace)
W: 19, 0: Redefining built-in 'pow' (redefined-builtin)
C:  1, 0: Missing module docstring (missing-docstring)
W:  1,14: Redefining name 'f' from outer scope (line 8)
W:  1,23: Redefining name 'n' from outer scope (line 16)
C:  1, 0: Invalid argument name "f" (invalid-name)
C:  1, 0: Invalid argument name "a" (invalid-name)
```

There is much more output, but let us summarize what PyLint does not like about the code:

1. Extra whitespace (after comma in a call to `integrate`)
2. Missing doc string at the beginning of the file

---
[5] http://www.python.org/dev/peps/pep-0008/

3. Missing doc strings in the functions
4. Same name f used as local variable in `integrate` and global function name in the `f(x)` function
5. Too short variable names: a, b, n, etc.
6. "Star import" of the form `from math import *`

In short programs where the one-to-one mapping between mathematical notation and the variable names is very important to make the code self-explanatory, this author thinks that only points 1–3 qualify for attention. Nevertheless, for larger non-mathematical programs all the style violations pointed out are serious and lead to code that is easier to read, debug, maintain, and use.

Migrating Python to Compiled Code

G

Python is a very convenient language for implementing scientific computations as the code can be made very close to the mathematical algorithms. However, the execution speed of the code is significantly lower than what can be obtained by programming in languages such as Fortran, C, or C++. These languages *compile* the program to machine language, which enables the computing resources to be utilized with very high efficiency. Frequently, and this includes almost all examples in the present book, Python is fast enough. But in the cases where speed really matters, can we increase the efficiency without rewriting the whole program in Fortran, C, or C++? The answer is yes, which will be illustrated through a case study in the forthcoming text.

Fortunately, Python was initially designed for being integrated with C. This feature has spawned the development of several techniques and tools for calling compiled languages from Python, allowing us to relatively easily reuse fast and well-tested scientific libraries in Fortran, C, or C++ from Python, *or* migrate slow Python code to compiled languages. It often turns out that only smaller parts of the code, usually `for` loops doing heavy numerical computations, suffer from low speed and can benefit from being implemented in Fortran, C, or C++.

The primary technique to be advocated here is to use Cython. Cython can be viewed as an extension of the Python language where variables can be declared with a type and other information such that Cython is able to automatically generate special-purpose, fast C code from the Python code. We will show how to utilize Cython and what the computational gain might be.

The present case study starts with stating a computational problem involving statistical simulations, which are known to cause long execution times, especially if accurate results are desired.

G.1 Pure Python Code for Monte Carlo Simulation

A short, intuitive algorithm in Python is first developed. Then this code is vectorized using functionality of the Numerical Python package. Later sections migrate the algorithm to Cython code and also plain C code for comparison. At the end the various techniques are ranked according to their computational efficiency.

G.1.1   The Computational Problem

A die is thrown *m* times. What is the probability of getting six eyes *at least n* times? For example, if $m = 5$ and $n = 3$, this is the same as asking for the probability that three or more out of five dice show six eyes.

The probability can be estimated by Monte Carlo simulation. Chapter 8.3 provides a background for this technique: We simulate the process a large number of times, N, and count how many times, M, the experiment turned out successfully, i.e., when we got at least *n* out of *m* dice with six eyes in a throw.

Monte Carlo simulation has traditionally been viewed as a very costly computational method, normally requiring very sophisticated, fast computer implementations in compiled languages. An interesting question is how useful high-level languages like Python and associated tools are for Monte Carlo simulation. This will now be explored.

G.1.2   A Scalar Python Implementation

Let us introduce the more descriptive variables `ndice` for *m* and `nsix` for *n*. The Monte Carlo method is simply a loop, repeated N times, where the body of the loop may directly express the problem at hand. Here, we draw `ndice` random integers `r` in $[1, 6]$ inside the loop and count of many (`six`) that equal 6. If `six >= nsix`, the experiment is a success and we increase the counter `M` by one.

A Python function implementing this approach may look as follows:

```python
import random

def dice6_py(N, ndice, nsix):
    M = 0                       # no of successful events
    for i in range(N):          # repeat N experiments
        six = 0                 # how many dice with six eyes?
        for j in range(ndice):
            r = random.randint(1, 6)  # roll die no. j
            if r == 6:
                six += 1
        if six >= nsix:         # successful event?
            M += 1
    p = float(M)/N
    return p
```

The `float(M)` transformation is important since M/N will imply integer division when M and N both are integers in Python v2.x and many other languages.

We will refer to this implementation is the *plain Python* implementation. Timing the function can be done by:

```python
import time
t0 = time.clock()
p = dice6_py(N, ndice, nsix)
t1 = time.clock()
print 'CPU time for loops in Python:', t1-t0
```

The table to appear later shows the performance of this plain, pure Python code relative to other approaches. There is a factor of 30+ to be gained in computational efficiency by reading on.

The function above can be verified by studying the (somewhat simplified) case $m = n$ where the probability becomes 6^{-n}. The probability quickly becomes small with increasing n. For such small probabilities the number of successful events M is small, and M/N will not be a good approximation to the probability unless M is reasonably large, which requires a very large N. For example, with $n = 4$ and $N = 10^5$ the average probability in 25 full Monte Carlo experiments is 0.00078 while the exact answer is 0.00077. With $N = 10^6$ we get the two correct significant digits from the Monte Carlo simulation, but the extra digit costs a factor of 10 in computing resources since the CPU time scales linearly with N.

G.1.3   A Vectorized Python Implementation

A vectorized version of the previous program consists of replacing the explicit loops in Python by efficient operations on vectors or arrays, using functionality in the Numerical Python (numpy) package. Each array operation takes place in C or Fortran and is hence much more efficient than the corresponding loop version in Python.

First, we must generate all the random numbers to be used in one operation, which runs fast since all numbers are then calculated in efficient C code. This is accomplished using the numpy.random module. Second, the analysis of the large collection of random numbers must be done by appropriate vector/array operations such that no looping in Python is needed. The solution algorithm must therefore be expressed through a series of function calls to the numpy library. Vectorization requires knowledge of the library's functionality and how to assemble the relevant building blocks to an algorithm without operations on individual array elements.

Generation of ndice random number of eyes for N experiments is performed by

```
import numpy as np
eyes = np.random.random_integers(1, 6, size=(N, ndice))
```

Each row in the eyes array corresponds to one Monte Carlo experiment.

The next step is to count the number of successes in each experiment. This counting should not make use of any loop. Instead we can test eyes == 6 to get a boolean array where an element i,j is True if throw (or die) number j in Monte Carlo experiment number i gave six eyes. Summing up the rows in this boolean array (True is interpreted as 1 and False as 0), we are interested in the rows where the sum is equal to or greater than nsix, because the number of such rows equals the number of successful events. The vectorized algorithm can be expressed as

```
def dice6_vec1(N, ndice, nsix):
    eyes = np.random.random_integers(1, 6, size=(N, ndice))
    compare = eyes == 6
    throws_with_6 = np.sum(compare, axis=1)  # sum over columns
    nsuccesses = throws_with_6 >= nsix
    M = np.sum(nsuccesses)
    p = float(M)/N
    return p
```

The use of `np.sum` instead of Python's own `sum` function is essential for the speed of this function: using `M = sum(nsucccesses)` instead slows down the code by a factor of almost 10! We shall refer to the `dice6_vec1` function as the *vectorized Python, version1* implementation.

The criticism against the vectorized version is that the original problem description, which was almost literally turned into Python code in the `dice6_py` function, has now become much more complicated. We have to decode the calls to various numpy functionality to actually realize that `dice6_py` and `dice6_vec` correspond to the same mathematics.

Here is another possible vectorized algorithm, which is easier to understand, because we retain the Monte Carlo loop and vectorize only each individual experiment:

```python
def dice6_vec2(N, ndice, nsix):
    eyes = np.random.random_integers(1, 6, (N, ndice))
    six = [6 for i in range(ndice)]
    M = 0
    for i in range(N):
        # Check experiment no. i:
        compare = eyes[i,:] == six
        if np.sum(compare) >= nsix:
            M += 1
    p = float(M)/N
    return p
```

We refer to this implementation as *vectorized Python, version 2*. As will be shown later, this implementation is significantly slower than the *plain Python* implementation (!) and very much slower than the *vectorized Python, version 1* approach. A conclusion is that readable, partially vectorized code, may run slower than straightforward scalar code.

G.2 Migrating Scalar Python Code to Cython

G.2.1 A Plain Cython Implementation

A Cython program starts with the scalar Python implementation, but all variables are specified with their types, using Cython's variable declaration syntax, like `cdef int M = 0` where we in standard Python just write `M = 0`. Adding such variable declarations in the scalar Python implementation is straightforward:

```python
import random

def dice6_cy1(int N, int ndice, int nsix):
    cdef int M = 0                  # no of successful events
    cdef int six, r
    cdef double p
    for i in range(N):              # repeat N experiments
        six = 0                     # how many dice with six eyes?
        for j in range(ndice):
            r = random.randint(1, 6) # roll die no. j
```

```
            if r == 6:
                six += 1
        if six >= nsix:         # successful event?
            M += 1
    p = float(M)/N
    return p
```

This code must be put in a separate file with extension .pyx. Running Cython on this file translates the Cython code to C. Thereafter, the C code must be compiled and linked to form a shared library, which can be imported in Python as a module. All these tasks are normally automated by a setup.py script. Let the dice6_cy1 function above be stored in a file dice6.pyx. A proper setup.py script looks as follows:

```
from distutils.core import setup
from distutils.extension import Extension
from Cython.Distutils import build_ext

setup(
  name='Monte Carlo simulation',
  ext_modules=[Extension('_dice6_cy', ['dice6.pyx'],)],
  cmdclass={'build_ext': build_ext},
)
```

Running

```
Terminal
Terminal> python setup.py build_ext --inplace
```

generates the C code and creates a (shared library) file _dice6_cy.so (known as a *C extension module*) which can be loaded into Python as a module with name _dice6_cy:

```
from _dice6_cy import dice6_cy1
import time
t0 = time.clock()
p = dice6_cy1(N, ndice, nsix)
t1 = time.clock()
print t1 - t0
```

We refer to this implementation as *Cython random.randint*. Although most of the statements in the dice6_cy1 function are turned into plain and fast C code, the speed is not much improved compared with the original scalar Python code.

To investigate what takes time in this Cython implementation, we can perform a profiling. The template for profiling a Python function whose call syntax is stored in some string statement, reads

```
import cProfile, pstats
cProfile.runctx(statement, globals(), locals(), 'tmp_profile.dat')
s = pstats.Stats('tmp_profile.dat')
s.strip_dirs().sort_stats('time').print_stats(30)
```

Data from the profiling are here stored in the file `tmp_profile.dat`. Our interest now is the `dice6_cy1` function so we set

```
statement = 'dice6_cy1(N, ndice, nsix)'
```

In addition, a Cython file in which there are functions we want to profile must start with the line

```
# cython: profile=True
```

to turn on profiling when creating the extension module. The profiling output from the present example looks like

```
        5400004 function calls in 7.525 CPU seconds

  Ordered by: internal time

  ncalls  tottime  percall  cumtime  percall filename:lineno(function)
  1800000   4.511    0.000    4.863    0.000 random.py:160(randrange)
  1800000   1.525    0.000    6.388    0.000 random.py:224(randint)
        1   1.137    1.137    7.525    7.525 dice6.pyx:6(dice6_cy1)
  1800000   0.352    0.000    0.352    0.000 {method 'random' ...
        1   0.000    0.000    7.525    7.525 {dice6_cy.dice6_cy1}
```

We easily see that it is the call to `random.randint` that consumes almost all the time. The reason is that the generated C code must call a Python module (`random`), which implies a lot of overhead. The C code should only call plain C functions, or if Python functions *must* be called, they should involve so much computations that the overhead in calling Python from C is negligible.

Instead of profiling the code to uncover inefficient constructs we can generate a visual representation of how the Python code is translated to C. Running

---
Terminal

```
Terminal> cython -a dice6.pyx
```

---

creates a file `dice6.html` which can be loaded into a web browser to inspect what Cython has done with the Python code.

```
Raw output: roll_dice.c

 1: import numpy as np
 2: cimport numpy as np
 3: import random
 4:
 5: def roll_dice1(int N, int ndice, int nsix):
 6:     cdef int M = 0                    # no of successful events
 7:     cdef int six, r
 8:     cdef double p
 9:     for i in range(N):
10:         six = 0                       # how many dice with six eyes?
11:         for j in range(ndice):
12:             # Roll die no. j
13:             r = random.randint(1, 6)
14:             if r == 6:
15:                 six += 1
16:         if six >= nsix:   # Successful event?
17:             M += 1
18:     p = float(M)/N
19:     return p
--
```

White lines indicate that the Python code is translated into C code, while the yellow lines indicate that the generated C code must make calls back to Python (using the Python C API, which implies overhead). Here, the `random.randint` call is in yellow, so this call is not translated to efficient C code.

G.2.2 A Better Cython Implementation

To speed up the previous Cython code, we have to get rid of the `random.randint` call every time we need a random variable. Either we must call some C function for generating a random variable or we must create a bunch of random numbers simultaneously as we did in the vectorized functions shown above. We first try the latter well-known strategy and apply the `numpy.random` module to generate all the random numbers we need at once:

```
import  numpy  as  np
cimport numpy  as  np

@cython.boundscheck(False)  # turn off array bounds check
@cython.wraparound(False)   # turn off negative indices ([-1,-1])
def dice6_cy2(int N, int ndice, int nsix):
    # Use numpy to generate all random numbers
    ...
    cdef np.ndarray[np.int_t, ndim=2, mode='c'] eyes = \
        np.random.random_integers(1, 6, (N, ndice))
```

This code needs some explanation. The `cimport` statement imports a special version of `numpy` for Cython and is needed *after* the standard `numpy` import. The declaration of the array of random numbers could just go as

```
cdef np.ndarray eyes = np.random.random_integers(1, 6, (N, ndice))
```

However, the processing of the `eyes` array will then be slow because Cython does not have enough information about the array. To generate optimal C code, we must provide information on the element types in the array, the number of dimensions of the array, that the array is stored in contiguous memory, that we do not want the overhead of checking whether indices are within their bounds or not, and that we do not need negative indices (which slows down array indexing). The latter two properties are taken care of by the `@cython.boundscheck(False)` and the `@cython.wraparound(False)` statements (decorators) right before the function, respectively, while the rest of the information is specified within square brackets in the `cdef np.ndarray` declaration. Inside the brackets, `np.int_t` denotes integer array elements (`np.int` is the usual data type object, but `np.int_t` is a Cython precompiled version of this object), `ndim=2` tells that the array has two dimensions (indices), and `mode='c'` indicates contiguous storage of the array. With all this extra information, Cython can generate C code that works with `numpy` arrays as efficiently as native C arrays.

The rest of the code is a plain copy of the `dice6_py` function, but with the `random.randint` call replaced by an array look-up `eyes[i,j]` to retrieve the next random number. The two loops will now be as efficient as if they were coded directly in pure C.

The complete code for the efficient version of the `dice6_cy1` function looks as follows:

```
import   numpy as np
cimport numpy as np
import cython
@cython.boundscheck(False)  # turn off array bounds check
@cython.wraparound(False)   # turn off negative indices ([-1,-1])
def dice6_cy2(int N, int ndice, int nsix):
    # Use numpy to generate all random numbers
    cdef int M = 0            # no of successful events
    cdef int six, r
    cdef double p
    cdef np.ndarray[np.int_t, ndim=2, mode='c'] eyes = \
        np.random.random_integers(1, 6, (N, ndice))
    for i in range(N):
        six = 0               # how many dice with six eyes?
        for j in range(ndice):
            r = eyes[i,j]     # roll die no. j
            if r == 6:
                six += 1
        if six >= nsix:       # successful event?
            M += 1
    p = float(M)/N
    return p
```

This Cython implementation is named *Cython numpy.random.*

The disadvantage with the `dice6_cy2` function is that large simulations (large N) also require large amounts of memory, which usually limits the possibility for high accuracy much more than the CPU time. It would be advantageous to have a fast random number generator a la `random.randint` in C. The C library `stdlib` has a generator of random integers, `rand()`, generating numbers from 0 to up `RAND_MAX`. Both the `rand` function and the `RAND_MAX` integer are easy to access in a Cython program:

```
from libc.stdlib cimport rand, RAND_MAX

r = 1 + int(6.0*rand()/RAND_MAX) # random integer 1,...,6
```

Note that `rand()` returns an integer so we must avoid integer division by ensuring that the denominator is a real number. We also need to explicitly convert the resulting real fraction to `int` since r is declared as `int`.

With this way of generating random numbers we can create a version of `dice6_cy1` that is as fast as `dice6_cy`, but avoids all the memory demands and the somewhat complicated array declarations of the latter:

```
from libc.stdlib cimport rand, RAND_MAX
def dice6_cy3(int N, int ndice, int nsix):
    cdef int M = 0            # no of successful events
    cdef int six, r
```

```
    cdef double p
    for i in range(N):
        six = 0                    # how many dice with six eyes?
        for j in range(ndice):
            # Roll die no. j
            r = 1 + int(6.0*rand()/RAND_MAX)
            if r == 6:
                six += 1
        if six >= nsix:        # successful event?
            M += 1
    p = float(M)/N
    return p
```

This final Cython implementation will be referred to as *Cython stdlib.rand*.

G.3 Migrating Code to C

G.3.1 Writing a C Program

A natural next improvement would be to program the Monte Carlo simulation loops directly in a compiled programming language, which guarantees optimal speed. Here we choose the C programming language for this purpose. The C version of our dice6 function and an associated main program take the form

```c
#include <stdio.h>
#include <stdlib.h>

double dice6(int N, int ndice, int nsix)
{
  int M = 0;
  int six, r, i, j;
  double p;

  for (i = 0; i < N; i++) {
    six = 0;
    for (j = 0; j < ndice; j++) {
      r = 1 + rand()/(RAND_MAX*6.0); /* roll die no. j */
      if (r == 6)
    six += 1;
    }
    if (six >= nsix)
      M += 1;
  }
  p = ((double) M)/N;
  return p;
}

int main(int nargs, const char* argv[])
{
  int N = atoi(argv[1]);
  int ndice = 6;
  int nsix = 3;
```

```
      double p = dice6(N, ndice, nsix);
      printf("C code: N=%d, p=%.6f\n", N, p);
      return 0;
}
```

This code is placed in a file `dice6_c.c`. The file can typically be compiled and
run by

```
──────────────────────────────┤ Terminal ├──────────────────────────────
Terminal> gcc -O3 -o dice6.capp dice6_c.c
Terminal> ./dice6.capp 1000000
```

This solution is later referred to as *C program*.

G.3.2   Migrating Loops to C Code via F2PY

Instead of programming the whole application in C, we may consider migrating the
loops to the C function `dice6` shown above and then have the rest of the program
(essentially the calling main program) in Python. This is a convenient solution if
we were to do many other, less CPU time critical things for convenience in Python.

There are many alternative techniques for calling C functions from Python. Here
we shall explain two. The first applies the program `f2py` to generate the necessary
code that glues Python and C. The `f2py` program was actually made for gluing
Python and Fortran, but it can work with C too. We need a specification of the
C function to call in terms of a Fortran 90 module. Such a module can be written by hand, but `f2py` can also generate it. To this end, we make a Fortran file
`dice6_c_signature.f` with the signature of the C function written in Fortran 77
syntax with some annotations:

```
      real*8 function dice6(n, ndice, nsix)
Cf2py intent(c) dice6
      integer n, ndice, nsix
Cf2py intent(c) n, ndice, nsix
      return
      end
```

The annotations `intent(c)` are necessary to tell `f2py` that the Fortran variables
are to be treated as plain C variables and not as pointers (which is the default interpretation of variables in Fortran). The `C2fpy` are special comment lines that `f2py`
recognizes, and these lines are used to provide extra information to `f2py` which
have no meaning in plain Fortran 77.

We must run `f2py` to generate a `.pyf` file with a Fortran 90 module specification
of the C function to call:

```
──────────────────────────────┤ Terminal ├──────────────────────────────
Terminal> f2py -m _dice6_c1 -h dice6_c.pyf \
          dice6_c_signature.f
```

Here `_dice6_c1` is the name of the module with the C function that is to be imported in Python, and `dice6_c.pyf` is the name of the Fortran 90 module file to be generated. Programmers who know Fortran 90 may want to write the `dice6_c.pyf` file by hand.

The next step is to use the information in `dice6_c.pyf` to generate a (C extension) module `_dice6_c1`. Fortunately, `f2py` generates the necessary code, and compiles and links the relevant files, to form a shared library file `_dice6_c1.so`, by a short command:

Terminal

```
Terminal> f2py -c dice6_c.pyf dice6_c.c
```

We can now test the module:

```
>>> import _dice6_c1
>>> print dir(_dice6_c1)   # module contents
['__doc__', '__file__', '__name__', '__package__',
 '__version__', 'dice6']
>>> print _dice6_c1.dice6.__doc__
dice6 - Function signature:
  dice6 = dice6(n,ndice,nsix)
Required arguments:
  n : input int
  ndice : input int
  nsix : input int
Return objects:
  dice6 : float
>>> _dice6_c1.dice6(N=1000, ndice=4, nsix=2)
0.145
```

The method of calling the C function `dice6` via an `f2py` generated module is referred to as *C via f2py*.

G.3.3   Migrating Loops to C Code via Cython

The Cython tool can also be used to call C code, not only generating C code from the Cython language. Our C code is in the file `dice6_c.c`, but for Cython to see this code we need to create a *header file* `dice6_c.h` listing the definition of the function(s) we want to call from Python. The header file takes the form

```
extern double dice6(int N, int ndice, int nsix);
```

The next step is to make a `.pyx` file with a definition of the C function from the header file and a Python function that calls the C function:

```
cdef extern from "dice6_c.h":
    double dice6(int N, int ndice, int nsix)

def dice6_cwrap(int N, int ndice, int nsix):
    return dice6(N, ndice, nsix)
```

Cython must use this file, named `dice6_cwrap.pyx`, to generate C code, which is to be compiled and linked with the `dice6_c.c` code. All this is accomplished in a `setup.py` script:

```
from distutils.core import setup
from distutils.extension import Extension
from Cython.Distutils import build_ext

sources = ['dice6_cwrap.pyx', 'dice6_c.c']

setup(
  name='Monte Carlo simulation',
  ext_modules=[Extension('_dice6_c2', sources)],
  cmdclass={'build_ext': build_ext},
)
```

This `setup.py` script is run as

```
Terminal> python setup.py build_ext --inplace
```

resulting in a shared library file `_dice6_c2.so`, which can be loaded into Python as a module:

```
>>> import _dice6_c2
>>> print dir(_dice6_c2)
['__builtins__', '__doc__', '__file__', '__name__',
 '__package__', '__test__', 'dice6_cwrap']
```

We see that the module contains the function `dice6_cwrap`, which was made to call the underlying C function `dice6`.

G.3.4   Comparing Efficiency

All the files corresponding to the various techniques described above are available in the directory `src/cython`. A file `make.sh` performs all the compilations, while `compare.py` runs all methods and prints out the CPU time required by each method, normalized by the fastest approach. The results for $N = 450,000$ are listed below (MacBook Air running Ubuntu in a VMWare Fusion virtual machine).

Method	Timing
C program	1.0
Cython stdlib.rand	1.2
Cython numpy.random	1.2
C via f2py	1.2
C via Cython	1.2
vectorized Python, version 1	1.9
Cython random.randint	33.6
plain Python	37.7
vectorized Python, version 2	105.0

The CPU time of the plain Python version was 10 s, which is reasonably fast for obtaining a fairly accurate result in this problem. The lesson learned is therefore that a Monte Carlo simulation can be implemented in plain Python first. If more speed is needed, one can just add type information and create a Cython code. Studying the HTML file with what Cython manages to translate to C may give hints about how successful the Cython code is and point to optimizations, like avoiding the call to `random.randint` in the present case. Optimal Cython code runs here at approximately the same speed as calling a handwritten C function with the time-consuming loops. It is to be noticed that the stand-alone C program here ran faster than calling C from Python, probably because the amount of calculations is not large enough to make the overhead of calling C negligible.

Vectorized Python do give a great speed-up compared to plain loops in Python, if done correctly, but the efficiency is not on par with Cython or handwritten C. Even more important is the fact that vectorized code is not at all as readable as the algorithm expressed in plain Python, Cython, or C. Cython therefore provides a very attractive combination of readability, ease of programming, and high speed.

Technical Topics

H.1 Getting Access to Python

A comprehensive eco system for scientific computing with Python used to be quite a challenge to install on a computer, especially for newcomers. This problem is more or less solved today. There are several options for getting easy access to Python and the most important packages for scientific computations, so the biggest issue for a newcomer is to make a proper choice. An overview of the possibilities together with my own recommendations appears next.

H.1.1 Required Software

The strictly required software packages for working with this book are

- Python[1] version 2.7 [23]
- Numerical Python[2] (NumPy) [19, 20] for array computing
- Matplotlib[3] [8, 9] for plotting

Desired add-on packages are

- IPython[4] [21, 22] for interactive computing
- SciTools[5] [14] for add-ons to NumPy
- pytest[6] or nose[7] for testing programs
- pip[8] for installing Python packages
- Cython[9] for compiling Python to C

---

[1] http://python.org
[2] http://www.numpy.org
[3] http://matplotlib.org
[4] http://ipython.org
[5] https://github.com/hplgit/scitools
[6] http://pytest.org/latest/
[7] https://nose.readthedocs.org
[8] http://www.pip-installer.org
[9] http://cython.org

- SymPy[10] [2] for symbolic mathematics
- SciPy[11] [10] for advanced scientific computing

There are different ways to get access to Python with the required packages:

1. Use a computer system at an institution where the software is installed. Such a system can also be used from your local laptop through remote login over a network.
2. Install the software on your own laptop.
3. Use a web service.

A system administrator can take the list of software packages and install the missing ones on a computer system. For the two other options, detailed descriptions are given below.

Using a web service is very straightforward, but has the disadvantage that you are constrained by the packages that are allowed to install on the service. There are services at the time of this writing that suffice for working with most of this book, but if you are going to solve more complicated mathematical problems, you will need more sophisticated mathematical Python packages, more storage and more computer resources, and then you will benefit greatly from having Python installed on your own computer.

This author's experience is that installation of mathematical software on personal computers quickly becomes a technical challenge. Linux Ubuntu (or any Debian-based Linux version) contains the largest repository today of pre-built mathematical software and makes the installation trivial without any need for particular competence. Despite the user-friendliness of the Mac and Windows systems, getting sophisticated mathematical software to work on these platforms requires considerable competence.

H.1.2   Installing Software on Your Laptop: Mac OS X and Windows

There are various possibilities for installing the software on a Mac OS X or Windows platform:

1. Use `.dmg` (Mac) or `.exe` (Windows) files to install individual packages
2. Use Homebrew or MacPorts to install packages (Mac only)
3. Use a pre-built rich environment for scientific computing in Python:
   - Anaconda[12]
   - Enthought Canopy[13]
4. Use a virtual machine running Ubuntu:
   - VMWare Fusion[14]

---

[10] http://sympy.org
[11] http://scipy.org
[12] https://store.continuum.io/cshop/anaconda/
[13] https://www.enthought.com/products/canopy/
[14] http://www.vmware.com/products/fusion

- VirtualBox[15]
- Vagrant[16]

Alternative 1 is the obvious and perhaps simplest approach, but usually requires quite some competence about the operating system as a long-term solution when you need many more Python packages than the basic three. This author is not particularly enthusiastic about Alternative 2. If you anticipate to use Python extensively in your work, I strongly recommend operating Python on an Ubuntu platform and going for Alternative 4 because that is the easiest and most flexible way to build and maintain your own software ecosystem. Alternative 3 is recommended for those who are uncertain about the future needs for Python and think Alternative 4 is too complicated. My preference is then to use Anaconda for the Python installation and Spyder[17] (comes with Anaconda) as a graphical interface with editor, an output area, and flexible ways of running Python programs.

H.1.3 Anaconda and Spyder

Anaconda[18] is a free Python distribution produced by Continuum Analytics and contains over 400 Python packages, as well as Python itself, for doing a wide range of scientific computations. Anaconda can be downloaded from http://continuum.io/downloads. Choose Python version 2.7.

The Integrated Development Environment (IDE) *Spyder* is included with Anaconda and is my recommended tool for writing and running Python programs on Mac and Windows, unless you have preference for a plain text editor for writing programs and a terminal window for running them.

Spyder on Mac Spyder is started by typing `spyder` in a (new) Terminal application. If you get an error message *unknown locale*, you need to type the following line in the Terminal application, or preferably put the line in your $HOME/.bashrc Unix initialization file:

```
export LANG=en_US.UTF-8; export LC_ALL=en_US.UTF-8
```

Installation of additional packages After installing Anaconda you have the `conda` tool at hand for installing additional packages registered on `binstar.org`[19]. For example,

```
                                 ┌─────────┐
─────────────────────────────────│ Terminal │──────────────────────────────
                                 └─────────┘
Terminal> sudo conda install --channel johannr scitools
─────────────────────────────────────────────────────────────────────────
```

(You do not need `sudo` on Windows.)

---

[15] https://www.virtualbox.org/

[16] http://www.vagrantup.com/

[17] https://code.google.com/p/spyderlib/

[18] https://store.continuum.io/cshop/anaconda/

[19] https://binstar.org

Anaconda also installs the `pip` tool that is handy for installing additional packages that are not on `binstar.org`. In a Terminal application on Mac or in a PowerShell terminal on Windows, write

```Terminal
Terminal> sudo pip install --user packagename
```

(Drop sudo on Windows.)

Installing SciTools on Windows If the commands `conda install` or `pip install` do not succeed on Windows, the safest procedure is to download the source of the packages you want and run `setup.py`. To install SciTools, go to https://github.com/hplgit/scitools/ and click on *Download ZIP*. Double-click on the downloaded file in Windows Explorer and perform *Extract all files* to create a new folder with all the SciTools files. Find the location of this folder, open a PowerShell window, and move to the location, e.g.,

```Terminal
Terminal> cd C:\Users\username\Downloads\scitools-2db3cbb5076a
```

Installation is done by

```Terminal
Terminal> python setup.py install
```

H.1.4   VMWare Fusion Virtual Machine

A virtual machine allows you to run another complete computer system in a separate window. For Mac users, I recommend VMWare Fusion over VirtualBox for running a Linux (or Windows) virtual machine. (VMWare Fusion's hardware integration seems superior to that of VirtualBox.) VMWare Fusion is commercial software, but there is a free trial version you can start with. Alternatively, you can use the simpler VMWare Player, which is free for personal use.

Installing Ubuntu The following recipe will install a Ubuntu virtual machine under VMWare Fusion.

1. Download Ubuntu[20]. Choose a version that is compatible with your computer, usually a 64-bit version nowadays.
2. Launch VMWare Fusion (the instructions here are for version 7).
3. Click on *File - New* and choose to *Install from disc or image*.
4. Click on *Use another disc or disc image* and choose your `.iso` file with the Ubuntu image.

---

[20] http://www.ubuntu.com/desktop/get-ubuntu/download

5. Choose *Easy Install*, fill in password, and check the box for sharing files with the host operating system.

6. Choose *Customize Settings* and make the following settings (these settings can be changed later, if desired):

   - *Processors and Memory*: Set a minimum of 2 Gb memory, but not more than half of your computer's total memory. The virtual machine can use all processors.
   - *Hard Disk*: Choose how much disk space you want to use inside the virtual machine (20 Gb is considered a minimum).

7. Choose where you want to store virtual machine files on the hard disk. The default location is usually fine. The directory with the virtual machine files needs to be frequently backed up so make sure you know where it is.

8. Ubuntu will now install itself without further dialog, but it will take some time.

9. You may need to define a higher resolution of the display in the Ubuntu machine. Find the *System settings* icon on the left, go to *Display*, choose some display (you can try several, click *Keep this configuration* when you are satisfied).

10. You can have multiple keyboards on Ubuntu. Launch *System settings*, go to *Keyboard*, click the *Text entry* hyperlink, add keyboard(s) (*Input sources to use*), and choose a shortcut, say `Ctrl+space` or `Ctrl+backslash`, in the *Switch to next source using* field. Then you can use the shortcut to quickly switch keyboard.

11. A terminal window is key for programmers. Click on the Ubuntu icon on the top of the left pane, search for `gnome-terminal`, right-click its new icon in the left pane and choose `Lock to Launcher` such that you always have the terminal easily accessible when you log in. The `gnome-terminal` can have multiple tabs (`Ctrl+shift+t` to make a new tab).

Installing software on Ubuntu   You now have a full Ubuntu machine, but there is not much software on a it for doing scientific computing with Python. Installation is performed through the Ubuntu Software Center (a graphical application) or through Unix commands, typically

---

Terminal

```
Terminal> sudo apt-get install packagename
```

To look up the right package name, run `apt-cache search` followed by typical words of that package. The strength of the `apt-get` way of installing software is that the package *and all packages it depends on* are automatically installed through the `apt-get install` command. This is in a nutshell why Ubuntu (or Debian-based Linux systems) are so user-friendly for installing sophisticated mathematical software.

Python packages are better installed via `pip`:

Terminal

```
Terminal> sudo pip install --user packagename
```

To install a lot of useful packages for scientific work, go to http://goo.gl/RVHixr,
click on `install_minimal_ubuntu.sh`, click on *Raw*, download the file, and run
it:

```
                          Terminal
Terminal> cd ~/Downloads
Terminal> bash install_minimal.sh
```

The program `install_minimal.sh` will run `pip install` and `apt-get install`
commands for quite some time, hopefully without problems. If it stops, set a comment sign # in front of the line where it stopped and rerun.

File sharing The Ubuntu machine can see the files on your host system if you
download *VMWare Tools*. Go to the *Virtual Machine* pull-down menu in VMWare
Fusion and choose *Install VMWare Tools*. A tarfile is downloaded. Click on it and it
will open a folder `vmware-tools-distrib`, normally in your home folder. Move
to the new folder and run `sudo perl vmware-install.pl`. You can go with the
default answers to all the questions.

On a Mac, you must open *Virtual Machine - Settings...* and choose *Sharing* to
bring up a dialog where you can add the folders you want to be visible in Ubuntu.
Just choose your home folder. Then turn on the file sharing button (or turn off and
on again). Go to Ubuntu and check if you can see all your host system's files in
`/mnt/hgfs/`.

If you later detect that `/mnt/hgfs/` folder has become empty, VMWare Tools
must be reinstalled by first turning shared folders off, and then running

```
                          Terminal
Terminal> sudo /usr/bin/vmware-config-tools.pl
```

Occasionally it is necessary to do a full reinstall by `sudo perl vmware-`
`install.pl` as above.

Backup of a VMWare virtual machine on a Mac

The entire Ubuntu machine is a folder on the host computer, typically with
a name like `Documents/Virtual Machines/Ubuntu 64-bit`. Backing up
the Ubuntu machine means backing up this folder. However, if you use tools
like Time Machine and work in Ubuntu during backup, the copy of the state
of the Ubuntu machine is likely to be corrupt. You are therefore strongly recommended to shut down the virtual machine prior to running Time Machine or
simply copying the folder with the virtual machine to some backup disk.

If something happens to your virtual machine, it is usually a straightforward
task to make a new machine and import data and software automatically from
the previous machine.

H.1.5 Dual Boot on Windows

Instead of running Ubuntu in a virtual machine, Windows users also have the option of deciding on the operating system when turning on the machine (so-called dual boot). The Wubi[21] tool makes it very easy to get Ubuntu on a Windows machine this way. There are problems with Wubi on Windows 8, see instructions[22] for how to get around them. It is also relatively straightforward to perform a direct install of Ubuntu by downloading an Ubuntu image, creating a bootable USB stick on Windows[23] or Mac[24], restarting the machine and finally installing Ubuntu[25]. However, with the powerful computers we now have, a virtual machine is more flexible since you can switch between Windows and Ubuntu as easily as going from one window to another.

H.1.6 Vagrant Virtual Machine

A vagrant machine is different from a standard virtual machine in that it is run in a terminal window on a Mac or Windows computer. You will write programs in Mac/Windows, but run them inside a Vagrant Ubuntu machine that can work with your files and folders on Mac/Windows. This is a bit simpler technology than a full VMWare Fusion virtual machine, as described above, and allows you to work in your original operating system. There is need to install VirtualBox and Vagrant, and on Windows also Cygwin. Then you can download a Vagrant machine with Ubuntu and either fill it with software as explained above, or you can download a ready-made machine. A special machine[26] has been made for this book. We also have a larger and richer machine[27]. The username and password are *fenics*.

Pre-i3/5/7 Intel processors and 32 vs. 64 bit

If your computer has a pre-i3/5/7 Intel processor and the processor does not have VT-x enabled, you cannot use the pre-packaged 64-bit virtual machines referred to above. Instead, you have to download a plain 32-bit Ubuntu image and install the necessary software (see Sect. H.1.4). To check if your computer has VT-x (hardware virtualization) enabled, you can use this tool: https://www.grc.com/securable.htm.

---

[21] http://wubi-installer.org
[22] https://www.youtube.com/watch?v=gZqsXAoLBDI
[23] http://www.ubuntu.com/download/desktop/create-a-usb-stick-on-windows
[24] http://www.ubuntu.com/download/desktop/create-a-usb-stick-on-mac-osx
[25] http://www.ubuntu.com/download/desktop/install-ubuntu-desktop
[26] http://goo.gl/hrdhGt
[27] http://goo.gl/uu5Kts

H.2   How to Write and Run a Python Program

You have basically three choices to develop and test a Python program:

1. use the IPython notebook
2. use an Integrated Development Environment (IDE), like Spyder, which offers a window with a text editor and functionality to run programs and observe the output
3. use a text editor and a terminal window

The IPython notebook is briefly descried in Sect. H.4, while the other two options are outlined below.

H.2.1   The Need for a Text Editor

Since programs consist of plain text, we need to write this text with the help of another program that can store the text in a file. You have most likely extensive experience with writing text on a computer, but for writing your own programs you need special programs, called *editors*, which preserve exactly the characters you type. The widespread word processors, Microsoft Word being a primary example, are aimed at producing nice-looking reports. These programs *format* the text and are *not* acceptable tools for writing your own programs, even though they can save the document in a pure text format. Spaces are often important in Python programs, and *editors* for plain text give you complete control of the spaces and all other characters in the program file.

Spyder   Spyder is a graphical application for developing and running Python programs, available on all major platforms. Spyder comes with Anaconda and some other pre-built environments for scientific computing with Python. On Ubuntu it is conveniently installed by `sudo apt-get install spyder`.

The left part of the Spyder window contains a plain text editor. Click in this window and write `print 'Hello!'` and return. Choose *Run* from the *Run* pull-down menu, and observe the output `Hello!` in the lower right window where the output from programs is visible.

You may continue with more advanced statements involving graphics:

```
import matplotlib.pyplot as plt
import numpy as np
x = np.linspace(0, 4, 101)
y = np.exp(-x)*np.sin(np.pi*x)
plt.plot(x,y)
plt.title('First test of Spyder')
plt.savefig('tmp.png')
plt.show()
```

Choosing *Run - Run* now leads to a separate window with a plot of the function $e^{-x} \sin(\pi x)$. Figure H.1 shows how the Spyder application may look like.

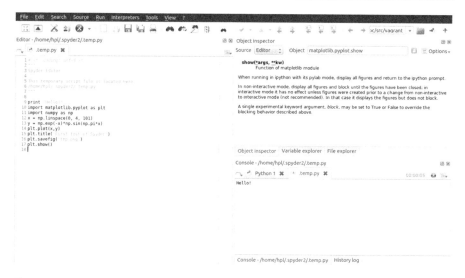

Fig. H.1 The Spyder Integrated Development Environment.

The plot file we generate in the above program, `tmp.png`, is by default found in the Spyder folder listed in the default text in the top of the program. You can choose *Run - Configure . . .* to change this folder as desired. The program you write is written to a file `.temp.py` in the same default folder, but any name and folder can be specified in the standard *File - Save as. . .* menu.

A convenient feature of Spyder is that the upper right window continuously displays documentation of the statements you write in the editor to the left.

Text editors The most widely used editors for writing programs are Atom, Sublime Text, Emacs, and Vim, which are available on all major platforms. Some simpler alternatives for beginners are

- Linux: Gedit
- Mac OS X: TextWrangler
- Windows: Notepad++

We may mention that Python comes with an editor called Idle, which can be used to write programs on all three platforms, but running the program with command-line arguments is a bit complicated for beginners in Idle so Idle is not my favorite recommendation.

Gedit is a standard program on Linux platforms, but all other editors must be installed in your system. This is easy: just google the name, download the file, and follow the standard procedure for installation. All of the mentioned editors come with a graphical user interface that is intuitive to use, but the major popularity of Atom, Sublime Text, Emacs, and Vim is due to their rich set of key commands so that you can avoid using the mouse and consequently edit at higher speed.

H.2.2   Terminal Windows

To run the Python program, you need a *terminal window*. This is a window where you can issue Unix commands in Linux and Mac OS X systems and DOS commands in Windows. On a Linux computer, `gnome-terminal` is my favorite, but other choices work equally well, such as `xterm` and `konsole`. On a Mac computer, launch the application *Utilities - Terminal*. On Windows, launch *PowerShell*.

You must first move to the right folder using the `cd foldername` command. Then running a python program `prog.py` is a matter of writing `python prog.py`. Whatever the program prints can be seen in the terminal window.

Using a plain text editor and a terminal window

1. Create a folder where your Python programs can be located, say with name `mytest` under your home folder. This is most conveniently done in the terminal window since you need to use this window anyway to run the program. The command for creating a new folder is `mkdir mytest`.
2. Move to the new folder: `cd mytest`.
3. Start the editor of your choice.
4. Write a program in the editor, e.g., just the line `print 'Hello!'`. Save the program under the name `myprog1.py` in the `mytest` folder.
5. Move to the terminal window and write `python myprog1.py`. You should see the word `Hello!` being printed in the window.

H.3   The SageMathCloud and Wakari Web Services

You can avoid installing Python on your machine completely by using a web service that allows you to write and run Python programs. Computational science projects will normally require some kind of visualization and associated graphics packages, which is not possible unless the service offers IPython notebooks. There are two excellent web services with notebooks: *SageMathCloud* at https://cloud.sagemath.com/ and *Wakari* at https://www.wakari.io/wakari. At both sites you must create an account before you can write notebooks in the web browser and download them to your own computer.

H.3.1   Basic Intro to SageMathCloud

Sign in, click on *New Project*, give a title to your project and decide whether it should be private or public, click on the project when it appears in the browser, and click on *Create or Import a File, Worksheet, Terminal or Directory....* If your Python program needs graphics, you need to choose *IPython Notebook*, otherwise you can choose *File*. Write the name of the file above the row of buttons. Assuming we do not need any graphics, we create a plain Python file, say with name `py1.py`. By clicking *File* you are brought to a browser window with a text editor where you can write Python code. Write some code and click *Save*. To run the program, click on the plus icon (*New*), choose *Terminal*, and you have a plain Unix terminal window where you can write `python py1.py` to run the program. Tabs over the terminal (or editor) window make it easy to jump between the editor and the terminal. To download the file, click on *Files*, point on the relevant line with the file, and a download icon appears to the very right. The IPython notebook option works much in the same way, see Sect. H.4.

H.3.2   Basic Intro to Wakari

After having logged in at the `wakari.io` site, you automatically enter an IPython notebook with a short introduction to how the notebook can be used. Click on the *New Notebook* button to start a new notebook. Wakari enables creating and editing plain Python files too: click on the *Add file* icon in pane to the left, fill in the program name, and you enter an editor where you can write a program. Pressing *Execute* launches an IPython session in a terminal window, where you can run the program by `run prog.py` if `prog.py` is the name of the program. To download the file, select `test2.py` in the left pane and click on the *Download file* icon.

There is a pull-down menu where you can choose what type of terminal window you want: a plain Unix shell, an IPython shell, or an IPython shell with Matplotlib for plotting. Using the latter, you can run plain Python programs or commands with graphics. Just choose the type of terminal and click on *+Tab* to make a new terminal window of the chosen type.

H.3.3   Installing Your Own Python Packages

Both SageMathCloud and Wakari let you install your own Python packages. To install any package `packagename` available at PyPi[28], run

```
                                            Terminal
pip install --user packagename
```

To install the SciTools package, which is useful when working with this book, run the command

```
                                            Terminal
pip install --user -e \
    git+https://github.com/hplgit/scitools.git#egg=scitools
```

---

[28] https://pypi.python.org/pypi

H.4   Writing IPython Notebooks

The IPython notebook is a splendid interactive tool for doing science, but it can also be used as a platform for developing Python code. You can either run it locally on your computer or in a web service like SageMathCloud or Wakari. Installation on your computer is trivial on Ubuntu, just `sudo apt-get install ipython-notebook`, and also on Windows and Mac[29] by using Anaconda or Enthought Canopy for the Python installation.

The interface to the notebook is a web browser: you write all the code and see all the results in the browser window. There are excellent YouTube videos on how to use the IPython notebook, so here we provide a very quick "step zero" to get anyone started.

H.4.1   A Simple Program in the Notebook

Start the IPython notebook locally by the command `ipython notebook` or go to SageMathCloud or Wakari as described above. The default input area is a *cell* for Python code. Type

```
g = 9.81
v0 = 5
t = 0.6
y = v0*t - 0.5*g*t**2
```

in a cell and *run the cell* by clicking on *Run Selected* (notebook running locally on your machine) or on the "play" button (notebook running in the cloud). This action will execute the Python code and initialize the variables g, v0, t, and y. You can then write `print y` in a new cell, execute that cell, and see the output of this statement in the browser. It is easy to go back to a cell, edit the code, and re-execute it.

To download the notebook to your computer, choose the *File - Download as* menu and select the type of file to be downloaded: the original notebook format (`.ipynb` file extension) or a plain Python program version of the notebook (`.py` file extension).

H.4.2   Mixing Text, Mathematics, Code, and Graphics

The real strength of IPython notebooks arises when you want to write a report to document how a problem can be explored and solved. As a teaser, open a new notebook, click in the first cell, and choose *Markdown* as format (notebook running locally) or switch from *Code* to *Markdown* in the pull-down menu (notebook in the cloud). The cell is now a text field where you can write text with Markdown[30] syntax. Mathematics can be entered as LaTeX code. Try some text with inline mathematics and an equation on a separate line:

---

[29] http://ipython.org/install.html
[30] http://daringfireball.net/projects/markdown/syntax

```
Plot the curve $y=f(x)$, where

$$
f(x) = e^{-x}\sin (2\pi x),\quad x\in [0, 4]
$$
```

Execute the cell and you will see nicely typeset mathematics in the browser. In the new cell, add some code to plot $f(x)$:

```
import numpy as np
import matplotlib.pyplot as plt
%matplotlib inline  # make plots inline in the notebook

x = np.linspace(0, 4, 101)
y = np.exp(-x)*np.sin(2*pi*x)
plt.plot(x, y, 'b-')
plt.xlabel('x'); plt.ylabel('y')
```

Executing these statements results in a plot in the browser, see Fig. H.2. It was popular to start the notebook by `ipython notebook -pylab` to import everything from `numpy` and `matplotlib.pyplot` and make all plots inline, but the `-pylab` option is now officially discouraged[31]. If you want the notebook to behave more as MATLAB and not use the `np` and `plt` prefix, you can instead of the first three lines above write `%pylab`.

Fig. H.2   Example on an IPython notebook.

---

[31] http://carreau.github.io/posts/10-No-PyLab-Thanks.ipynb.html

H.5 Different Ways of Running Python Programs

Python programs are compiled and interpreted by another program called `python`. To run a Python program, you need to tell the operating system that your program is to be interpreted by the `python` program. This section explains various ways of doing this.

H.5.1 Executing Python Programs in iPython

The simplest and most flexible way of executing a Python program is to run it inside IPython. See Sect. 1.5.3 for a quick introduction to IPython. You start IPython either by the command `ipython` in a terminal window, or by double-clicking the IPython program icon (on Windows). Then, inside IPython, you can run a program `prog.py` by

```
In [1]: run prog.py arg1 arg2
```

where `arg1` and `arg2` are command-line arguments.

This method of running Python programs works the same way on all platforms. One additional advantage of running programs under IPython is that you can automatically enter the Python debugger if an exception is raised (see Sect. F.1). Although we advocate running Python programs under IPython in this book, you can also run them directly under specific operating systems. This is explained next for Unix, Windows, and Mac OS X.

H.5.2 Executing Python Programs in Unix

There are two ways of executing a Python program `prog.py` in Unix-based systems. The first explicitly tells which Python interpreter to use:

```
Terminal
Unix> python prog.py arg1 arg2
```

Here, `arg1` and `arg2` are command-line arguments.

There may be many Python interpreters on your computer system, usually corresponding to different versions of Python or different sets of additional packages and modules. The Python interpreter (`python`) used in the command above is the first program with the name `python` appearing in the folders listed in your PATH environment variable. A specific `python` interpreter, say in `/home/hpl/local/bin`, can easily be used as default choice by putting this folder name first in the PATH variable. PATH is normally controlled in the `.bashrc` file. Alternatively, we may specify the interpreter's complete file path when running `prog.py`:

```
Terminal
Unix> /home/hpl/bin/python prog.py arg1 arg2
```

The other way of executing Python programs in Unix consists of just writing the name of the file:

```
Terminal
Unix> ./prog.py arg1 arg2
```

The leading ./ is needed to tell that the program is located in the current folder. You can also just write

```
Terminal
Unix> prog.py arg1 arg2
```

but then you need to have the dot in the PATH variable, which is not recommended for security reasons.

In the two latter commands there is no information on which Python interpreter to use. This information must be provided in the first line of the program, normally as

```
#!/usr/bin/env python
```

This looks like a comment line, and behaves indeed as a comment line when we run the program as python prog.py. However, when we run the program as ./prog.py, the first line beginning with #! tells the operating system to use the program specified in the rest of the first line to interpret the program. In this example, we use the first python program encountered in the folders in your PATH variable. Alternatively, a specific python program can be specified as

```
#!/home/hpl/special/tricks/python
```

H.5.3   Executing Python Programs in Windows

In a DOS or PowerShell window you can always run a Python program by

```
Terminal
PowerShell> python prog.py arg1 arg2
```

if prog.py is the name of the program, and arg1 and arg2 are command-line arguments. The extension .py can be dropped:

```
Terminal
PowerShell> python prog arg1 arg2
```

If there are several Python installations on your system, a particular installation can be specified:

```
                                    ┌──────────┐
──────────────────────────────────┤ Terminal ├──────────────────────────────
                                    └──────────┘
PowerShell> E:\hpl\myprogs\Python2.7.5\python prog arg1 arg2
─────────────────────────────────────────────────────────────────────────────
```

Files with a certain extension can in Windows be associated with a file type, and a file type can be associated with a particular program to handle the file. For example, it is natural to associate the extension .py with Python programs. The corresponding program needed to interpret .py files is then python.exe. When we write just the name of the Python program file, as in

```
                                    ┌──────────┐
──────────────────────────────────┤ Terminal ├──────────────────────────────
                                    └──────────┘
PowerShell> prog arg1 arg2
─────────────────────────────────────────────────────────────────────────────
```

the file is always interpreted by the specified python.exe program. The details of getting .py files to be interpreted by python.exe go as follows:

```
                                    ┌──────────┐
──────────────────────────────────┤ Terminal ├──────────────────────────────
                                    └──────────┘
PowerShell> assoc .py=PyProg
PowerShell> ftype PyProg=python.exe "%1" %*
─────────────────────────────────────────────────────────────────────────────
```

Depending on your Python installation, such file extension bindings may already be done. You can check this with

```
                                    ┌──────────┐
──────────────────────────────────┤ Terminal ├──────────────────────────────
                                    └──────────┘
PowerShell> assoc | find "py"
─────────────────────────────────────────────────────────────────────────────
```

To see the programs associated with a file type, write ftype name where name is the name of the file type as specified by the assoc command. Writing help ftype and help assoc prints out more information about these commands along with examples.

One can also run Python programs by writing just the basename of the program file, i.e., prog.py instead of prog.py, if the file extension is registered in the PATHEXT environment variable.

Double-clicking Python files   The usual way of running programs in Windows is to double click on the file icon. This does not work well with Python programs without a graphical user interface. When you double click on the icon for a file prog.py, a DOS window is opened, prog.py is interpreted by some python.exe program, and when the program terminates, the DOS window is closed. There is usually too little time for the user to observe the output in this short-lived DOS window.

One can always insert a final statement that pauses the program by waiting for input from the user:

```
raw_input('Type CR:')
```

or

```
sys.stdout.write('Type CR:'); sys.stdin.readline()
```

The program will hang until the user presses the Return key. During this pause the DOS window is visible and you can watch the output from previous statements in the program.

The downside of including a final input statement is that you must always hit Return before the program terminates. This is inconvenient if the program is moved to a Unix-type machine. One possibility is to let this final input statement be active only when the program is run in Windows:

```
if sys.platform[:3] == 'win':
    raw_input('Type CR:')
```

Python programs that have a graphical user interface can be double-clicked in the usual way if the file extension is .pyw.

H.5.4   Executing Python Programs in Mac OS X

Since a variant of Unix is used as core in the Mac OS X operating system, you can always launch a Unix terminal and use the techniques from Sect. H.5.2 to run Python programs.

H.5.5   Making a Complete Stand-Alone Executable

Python programs need a Python interpreter and usually a set of modules to be installed on the computer system. Sometimes this is inconvenient, for instance when you want to give your program to somebody who does not necessarily have Python or the required set of modules installed.

Fortunately, there are tools that can create a stand-alone executable program out of a Python program. This stand-alone executable can be run on every computer that has the same type of operating system and the same chip type. Such a stand-alone executable is a bundling of the Python interpreter and the required modules, along with your program, in a single file.

The leading tool for creating a stand-alone executable (or alternatively a folder with all necessary files) is PyInstaller[32]. Say you have program myprog.py that you want to distribute to people without the necessary Python environment on their computer. You run

```
                            | Terminal |
Terminal> pyinstaller --onefile myprog.py
```

---

[32] http://www.pyinstaller.org/

and a folder `dist` is created with the (big) stand-alone executable file `myprog` (or `myprog.exe` in Windows).

H.6   Doing Operating System Tasks in Python

Python has extensive support for operating system tasks, such as file and folder management. The great advantage of doing operating system tasks in Python and not directly in the operating system is that the Python code works uniformly on Unix/Linux, Windows, and Mac (there are exceptions, but they are few). Below we list some useful operations that can be done inside a Python program or in an interactive session.

Make a folder Python applies the term directory instead of folder. The equivalent of the Unix `mkdir mydir` is

```
import os
os.mkdir('mydir')
```

Ordinary files are created by the `open` and `close` functions in Python.

Make intermediate folders Suppose you want to make a subfolder under your home folder:

```
$HOME/python/project1/temp
```

but the intermediate folders `python` and `project1` do not exist. This requires each new folder to be made separately by `os.mkdir`, or you can make all folders at once with `os.makedirs`:

```
foldername = os.path.join(os.environ['HOME'], 'python',
                          'project1', 'temp')
os.makedirs(foldername)
```

With `os.environ[var]` we can get the value of any environment variable `var` as a string. The `os.path.join` function joins folder names and a filename in a platform-independent way.

Move to a folder The `cd` command reads `os.chdir` and `cwd` is `os.getcwd`:

```
origfolder = os.getcwd()    # get name of current folder
os.chdir(foldername)        # move ("change directory")
...
os.chdir(origfolder)        # move back
```

Rename a file or folder The cross-platform `mv` command is

```
os.rename(oldname, newname)
```

List files Unix wildcard notation can be used to list files. The equivalent of `ls *.py` and `ls plot*[1-4]*.dat` reads

```
import glob
filelist1 = glob.glob('*.py')
filelist2 = glob.glob('plot*[1-4]*.dat')
```

List all files and folders in a folder The counterparts to `ls -a mydir` and just `ls -a` are

```
filelist1 = os.listdir('mydir')
filelist1 = os.listdir(os.curdir)   # current folder (directory)
filelist1.sort()                    # sort alphabetically
```

Check if a file or folder exists The widely used constructions in Unix scripts for testing if a file or folder exist are `if [ -f $filename ]; then` and `if [ -d $dirname ]; then`. These have very readable counterparts in Python:

```
if os.path.isfile(filename):
    inputfile = open(filename, 'r')
    ...

if os.path.isdir(dirnamename):
    filelist = os.listdir(dirname)
    ...
```

Remove files Removing a single file is done with `os.rename`, and a loop is required for doing `rm tmp_*.df`:

```
import glob
filelist = glob.glob('tmp_*.pdf')
for filename in filelist:
    os.remove(filename)
```

Remove a folder and all its subfolders The `rm -rf mytree` command removes an entire folder tree. In Python, the cross-platform valid command becomes

```
import shutil
shutil.rmtree(foldername)
```

It goes without saying that this command must be used with great care!

Copy a file to another file or folder The `cp fromfile tofile` construction applies `shutil.copy` in Python:

```
shutil.copy('fromfile', 'tofile')
```

Copy a folder and all its subfolders The recursive copy command `cp -r` for folder trees is in Python expressed by `shell.copytree`:

```
shutil.copytree(sourcefolder, destination)
```

Run any operating system command The simplest way of running another program from Python is to use `os.system`:

```
cmd = 'python myprog.py 21 --mass 4'   # command to be run
failure = os.system(cmd)
if failure:
    print 'Execution of "%s" failed!\n' % cmd
    sys.exit(1)
```

The recommended way to run operating system commands is to use the `subprocess` module. The above command is equivalent to

```
import subprocess
cmd = 'python myprog.py 21 --mass 4'
failure = subprocess.call(cmd, shell=True)

# or
failure = subprocess.call(
          ['python', 'myprog.py', '21', '--mass', '4'])
```

The output of an operating system command can be stored in a string object:

```
try:
    output = subprocess.check_output(cmd, shell=True,
                                     stderr=subprocess.STDOUT)
except subprocess.CalledProcessError as e:
    # Raise a more informative exception
    msg = 'Execution of "%s" failed! (error code: %s)' + \
          '\nOutput: %s' % (cmd, e.returncode, e.output)
    raise subprocess.CalledProcessError(msg)
    # or do sys.exit(1)

# Process output
for line in output.splitlines():
    ...
```

The `stderr` argument ensures that the `output` string contains everything that the command cmd wrote to both standard output and standard error.

The constructions above are mainly used for running stand-alone programs. Any file or folder listing or manipulation should be done by the functionality in the `os` and `shutil` modules.

Split file or folder name Given `data/file1.dat` as a file path relative to the home folder `/users/me` (`$HOME/data/file1.dat` in Unix). Python has tools for extracting the complete folder name `/users/me/data`, the basename `file1.dat`, and the extension `.dat`:

```
>>> path = os.path.join(os.environ['HOME'], 'data', 'file1.dat')
>>> path
'/users/me/data/file1.dat'
>>> foldername, basename = os.path.split(path)
>>> foldername
'/users/me/data'
>>> basename
'file1.dat'
>>> stem, ext = os.path.splitext(basename)
>>> stem
'file1'
>>> ext
'.dat'
>>> outfile = stem + '.out'
>>> outfile
'file1.out'
```

H.7 Variable Number of Function Arguments

Arguments to Python functions are of four types:

- positional arguments, where each argument has a name,
- keyword arguments, where each argument has a name and a default value,
- a variable number of positional arguments, where each argument has no name, but just a location in a list,
- a variable number of keyword arguments, where each argument is a name-value pair in a dictionary.

The corresponding general function definition can be sketched as

```
def f(pos1, pos2, key1=val1, key2=val2, *args, **kwargs):
```

Here, `pos1` and `pos2` are positional arguments, `key1` and `key2` are keyword arguments, `args` is a tuple holding a variable number of positional arguments, and `kwargs` is a dictionary holding a variable number of keyword arguments. This appendix describes how to program with the `args` and `kwargs` variables and why these are handy in many situations.

H.7.1 Variable Number of Positional Arguments

Let us start by making a function that takes an arbitrary number of arguments and computes their sum:

```
>>> def add(*args):
...     print 'args:', args
...     s = 0
...     for arg in args:
...         s = s + arg
```

```
...      return s
...
>>> add(1)
args: (1,)
1
>>> add(1,5,10)
args: (1, 5, 10)
16
```

We observe that `args` is a tuple and that all the arguments we provide in a call to add are stored in `args`.

Combination of ordinary positional arguments and a variable number of arguments is allowed, but the `*args` argument must appear after the ordinary positional arguments, e.g.,

```
def f(pos1, pos2, pos3, *args):
```

In each call to `f` we must provide at least three arguments. If more arguments are supplied in the call, these are collected in the `args` tuple inside the `f` function.

Example Consider a mathematical function with one independent variable t and a parameter v_0, as in $y(t; v_0) = v_0 t - \frac{1}{2} g t^2$. A more general case with n parameters is $f(x; p_1, \ldots, p_n)$. The Python implementation of such functions can take both the independent variable and the parameters as arguments: `y(t, v0)` and `f(x, p1, p2, ...,pn)`. Suppose that we have a general library routine that operates on functions of one variable. The routine can, e.g., perform numerical differentiation, integration, or root finding. A simple example is a numerical differentiation function

```
def diff(f, x, h):
    return (f(x+h) - f(x))/h
```

This `diff` function cannot be used with functions `f` that take more than one argument. For example, passing an `y(t, v0)` function as `f` leads to the exception

```
TypeError: y() takes exactly 2 arguments (1 given)
```

Section 7.1.1 provides a solution to this problem where y becomes a class instance. Here we shall describe an alternative solution that allows our `y(t, v0)` function to be used as is.

The idea is that we pass additional arguments for the parameters in the f function *through* the diff function. That is, we view the f function as `f(x, *f_prms)` in diff. Our `diff` routine can then be written as

```
def diff(f, x, h, *f_prms):
    print 'x:', x, 'h:', h, 'f_prms:', f_prms
    return (f(x+h, *f_prms) - f(x, *f_prms))/h
```

Before explaining this function in detail, we demonstrate that it works in an example:

```
def y(t, v0):
    g = 9.81
     return v0*t - 0.5*g*t**2

dydt = diff(y, 0.1, 1E-9, 3)   # t=0.1, h=1E-9, v0=3
```

The output from the call to `diff` becomes

```
x: 0.1 h: 1e-09 f_prms: (3,)
```

The point is that the `v0` parameter, which we want to pass on to our `y` function, is now stored in `f_prms`. Inside the `diff` function, calling

```
f(x, *f_prms)
```

is the same as if we had written

```
f(x, f_prms[0], f_prms[1], ...)
```

That is, `*f_prms` in a call takes all the values in the tuple `*f_prms` and places them after each other as positional arguments. In the present example with the `y` function, `f(x, *f_prms)` implies `f(x, f_prms[0])`, which for the current set of argument values in our example becomes a call `y(0.1, 3)`.

For a function with many parameters,

```
def G(x, t, A, a, w):
    return A*exp(-a*t)*sin(w*x)
```

the output from

```
dGdx = diff(G, 0.5, 1E-9, 0, 1, 0.6, 100)
```

becomes

```
x: 0.5 h: 1e-09 f_prms: (0, 1, 1.5, 100)
```

We pass here the arguments `t`, `A`, `a`, and `w`, in that sequence, as the last four arguments to `diff`, and all the values are stored in the `f_prms` tuple.

The `diff` function also works for a plain function `f` with one argument:

```
from math import sin
mycos = diff(sin, 0, 1E-9)
```

In this case, `*f_prms` becomes an empty tuple, and a call like `f(x, *f_prms)` is just `f(x)`.

The use of a variable set of arguments for sending problem-specific parameters through a general library function, as we have demonstrated here with the `diff` function, is perhaps the most frequent use of `*args`-type arguments.

H.7.2   Variable Number of Keyword Arguments

A simple test function

```
>>> def test(**kwargs):
...         print kwargs
```

exemplifies that `kwargs` is a dictionary inside the `test` function, and that we can pass any set of keyword arguments to `test`, e.g.,

```
>>> test(a=1, q=9, method='Newton')
{'a': 1, 'q': 9, 'method': 'Newton'}
```

We can combine an arbitrary set of positional and keyword arguments, provided all the keyword arguments appear at the end of the call:

```
>>> def test(*args, **kwargs):
...         print args, kwargs
...
>>> test(1,3,5,4,a=1,b=2)
(1, 3, 5, 4) {'a': 1, 'b': 2}
```

From the output we understand that all the arguments in the call where we provide a name and a value are treated as keyword arguments and hence placed in `kwargs`, while all the remaining arguments are positional and placed in `args`.

Example We may extend the example in Sect. H.7.1 to make use of a variable number of keyword arguments instead of a variable number of positional arguments. Suppose all functions with parameters in addition to an independent variable take the parameters as keyword arguments. For example,

```
def y(t, v0=1):
    g = 9.81
    return v0*t - 0.5*g*t**2
```

In the `diff` function we transfer the parameters in the f function as a set of keyword arguments `**f_prms`:

```
def diff(f, x, h=1E-10, **f_prms):
    print 'x:', x, 'h:', h, 'f_prms:', f_prms
    return (f(x+h, **f_prms) - f(x, **f_prms))/h
```

In general, the `**f_prms` argument in a call

```
f(x, **f_prms)
```

implies that all the key-value pairs in `**f_prms` are provided as keyword arguments:

```
f(x, key1=f_prms[key1], key2=f_prms[key2], ...)
```

In our special case with the y function and the call

```
dydt = diff(y, 0.1, h=1E-9, v0=3)
```

`f(x, **f_prms)` becomes `y(0.1, v0=3)`. The output from `diff` is now

```
x: 0.1 h: 1e-09 f_prms: {'v0': 3}
```

showing explicitly that our v0=3 in the call to `diff` is placed in the `f_prms` dictionary.

The G function from Sect. H.7.1 can also have its parameters as keyword arguments:

```
def G(x, t=0, A=1, a=1, w=1):
    return A*exp(-a*t)*sin(w*x)
```

We can now make the call

```
dGdx = diff(G, 0.5, h=1E-9, t=0, A=1, w=100, a=1.5)
```

and view the output from `diff`,

```
x: 0.5 h: 1e-09 f_prms: {'A': 1, 'a': 1.5, 't': 0, 'w': 100}
```

to see that all the parameters get stored in `f_prms`. The h parameter can be placed anywhere in the collection of keyword arguments, e.g.,

```
dGdx = diff(G, 0.5, t=0, A=1, w=100, a=1.5, h=1E-9)
```

We can allow the f function of one variable and a set of parameters to have the general form `f(x, *f_args, **f_kwargs)`. That is, the parameters can either be positional or keyword arguments. The `diff` function must take the arguments `*f_args` and `**f_kwargs` and transfer these to f:

```
def diff(f, x, h=1E-10, *f_args, **f_kwargs):
    print f_args, f_kwargs
    return (f(x+h, *f_args, **f_kwargs) -
            f(x,   *f_args, **f_kwargs))/h
```

This `diff` function gives the writer of an f function full freedom to choose positional and/or keyword arguments for the parameters. Here is an example of the G function where we let the t parameter be positional and the other parameters be keyword arguments:

```
def G(x, t, A=1, a=1, w=1):
    return A*exp(-a*t)*sin(w*x)
```

A call

```
dGdx = diff(G, 0.5, 1E-9, 0, A=1, w=100, a=1.5)
```

gives the output

```
(0,) {'A': 1, 'a': 1.5, 'w': 100}
```

showing that t is put in f_args and transferred as positional argument to G, while A, a, and w are put in f_kwargs and transferred as keyword arguments. We remark that in the last call to diff, h and t *must* be treated as positional arguments, i.e., we cannot write h=1E-9 and t=0 unless *all* arguments in the call are on the name=value form.

In the case we use both *f_args and **f_kwargs arguments in f and there is no need for these arguments, *f_args becomes an empty tuple and **f_kwargs becomes an empty dictionary. The example

```
mycos = diff(sin, 0)
```

shows that the tuple and dictionary are indeed empty since diff just prints out

```
() {}
```

Therefore, a variable set of positional and keyword arguments can be incorporated in a general library function such as diff without any disadvantage, just the benefit that diff works with different types of f functions: parameters as global variables, parameters as additional positional arguments, parameters as additional keyword arguments, or parameters as instance variables (Sect. 7.1.2).

The program varargs1.py in the src/varargs[33] folder implements the examples in this appendix.

H.8   Evaluating Program Efficiency

H.8.1   Making Time Measurements

The term *time* has multiple meanings on a computer. The *elapsed time* or *wall clock time* is the same time as you can measure on a watch or wall clock, while *CPU time* is the amount of time the program keeps the central processing unit busy. The *system time* is the time spent on operating system tasks like I/O. The concept *user time* is the difference between the CPU and system times. If your computer is occupied by many concurrent processes, the CPU time of your program might be very different from the elapsed time.

---

[33] http://tinyurl.com/pwyasaa/tech

The time module Python has a `time` module with some useful functions for measuring the elapsed time and the CPU time:

```
import time
e0 = time.time()      # elapsed time since the epoch
c0 = time.clock()     # total CPU time spent in the program so far
<do tasks...>
elapsed_time = time.time() - e0
cpu_time = time.clock() - c0
```

The term *epoch* means initial time (`time.time()` would return 0), which is 00:00:00 January 1, 1970. The `time` module also has numerous functions for nice formatting of dates and time, and the newer `datetime` module has more functionality and an improved interface. Although the timing has a finer resolution than seconds, one should construct test cases that last some seconds to obtain reliable results.

Using timeit from IPython To measure the efficiency of a certain set of statements, an expression, or a function call, the code should be run a large number of times so the overall CPU time is of order seconds. Python's `timeit` module has functionality for running a code segment repeatedly. The simplest and most convenient way of using `timeit` is within an IPython shell. Here is a session comparing the efficiency of `sin(1.2)` versus `math.sin(1.2)`:

```
In [1]: import math

In [2]: from math import sin

In [3]: %timeit sin(1.2)
10000000 loops, best of 3: 198 ns per loop

In [4]: %timeit math.sin(1.2)
1000000 loops, best of 3: 258 ns per loop
```

That is, looking up `sin` through the `math` prefix degrades the performance by a factor of $258/198 \approx 1.3$.

Any statement, including function calls, can be timed the same way. Timing of multiple statements is possible by using `%%timeit`. The `timeit` module can be used inside ordinary programs as demonstrated in the file `pow_eff.py`.

Hardware information Along with CPU time measurements it is often convenient to print out information about the hardware on which the experiment was done. Python has a module `platform` with information on the current hardware. The function `scitools.misc.hardware_info` applies the `platform` module and other modules to extract relevant hardware information. A sample call is

```
>>> import scitools.misc, pprint
>>> pprint.pprint(scitools.misc.hardware_info())
{'numpy.distutils.cpuinfo.cpu.info': [
 {'address sizes': '40 bits physical, 48 bits virtual',
```

```
 'bogomips': '4598.87',
 'cache size': '4096 KB',
 'cache_alignment': '64',
 'cpu MHz': '2299.435',
 ...
},
'platform module': {
 'identifier': 'Linux-3.11.0-12-generic-x86_64-with-Ubuntu-13.10',
 'python build': ('default', 'Sep 19 2013 13:48:49'),
 'python version': '2.7.5+',
 'uname': ('Linux', 'hpl-ubuntu2-mac11', '3.11.0-12-generic',
           '#19-Ubuntu SMP Wed Oct 9 16:20:46 UTC 2013',
           'x86_64', 'x86_64')}}
}
```

H.8.2   Profiling Python Programs

A profiler computes the time spent in the various functions of a program. From the timings a ranked list of the most time-consuming functions can be created. This is an indispensable tool for detecting bottlenecks in the code, and you should always perform a profiling before spending time on code optimization. The golden rule is to first write an easy-to-understand program, then verify it, then profile it, and then think about optimization.

▶ *Premature optimization is the root of all evil.*
Donald Knuth, computer scientist, 1938–.

Python 2.7 comes with two recommended profilers, implemented in the modules cProfile and profiles. The section The Python Profilers[34] in the Python Standard Library documentation [3] has a good introduction to the usage of these modules. The results produced by the modules are normally processed by a special statistics utility pstats developed for analyzing profiling results. The usage of the profile, cProfile, and pstats modules is straightforward, but somewhat tedious. The SciTools package therefore comes with a command scitools profiler that allows you to profile any program (say) m.py by just writing

```
Terminal
Terminal> scitools profiler m.py c1 c2 c3
```

Here, c1, c2, and c3 are command-line arguments to m.py.
    A sample output might read

```
    1082 function calls (728 primitive calls) in 17.890 CPU s

Ordered by: internal time
List reduced from 210 to 20 due to restriction <20>
```

---

[34] http://docs.python.org/2/library/profile.html

```
ncalls  tottime  percall  cumtime  percall filename:lineno
     5    5.850    1.170    5.850    1.170 m.py:43(loop1)
     1    2.590    2.590    2.590    2.590 m.py:26(empty)
     5    2.510    0.502    2.510    0.502 m.py:32(myfunc2)
     5    2.490    0.498    2.490    0.498 m.py:37(init)
     1    2.190    2.190    2.190    2.190 m.py:13(run1)
     6    0.050    0.008   17.720    2.953 funcs.py:126(timer)
 ...
```

In this test, `loop1` is the most expensive function, using 5.85 seconds, which is to be compared with 2.59 seconds for the next most time-consuming function, `empty`. The `tottime` entry is the total time spent in a specific function, while `cumtime` reflects the total time spent in the function and all the functions it calls. We refer to the documentation of the profiling tools in the Python Standard Library documentation for detailed information on how to interpret the output.

The CPU time of a Python program typically increases with a factor of about five when run under the administration of the `profile` module. Nevertheless, the relative CPU time among the functions are not much affected by the profiler overhead.

H.9   Software Testing

Unit testing is widely a used technique for verifying software implementation. The idea is to identify small units of code and test each unit, ideally in a way such that one test does not depend on the outcome of other tests. Several tools, often referred to as testing frameworks, exist for automatically running all tests in a software package and report if any test failed. The value of such tools during software development cannot be exaggerated. Below we describe how to write tests that can be used by either the nose[35] or the pytest[36] testing frameworks. Both these have a very low barrier for beginners, so there is no excuse for not using nose or pytest as soon as you have learned about functions in programming.

Model software   We need a piece of software we want to test. Here we choose a function that runs Newton's method for solving algebraic equations $f(x) = 0$. A very simple implementation goes like

```python
def Newton_basic(f, dfdx, x, eps=1E-7):
    n = 0   # iteration counter
    while abs(f(x)) > eps:
        x = x - f(x)/dfdx(x)
        n += 1
    return x, f(x), n
```

---

[35] https://nose.readthedocs.org/
[36] http://pytest.org/latest/

H.9.1   Requirements of the Test Function

The simplest way of using the pytest or nose testing frameworks is to write a set of test functions, scattered around in files, such that pytest or nose can automatically find and run all the test functions. To this end, the test functions need to follow certain conventions.

Test function conventions
1. The name of a test function starts with `test_`.
2. A test function cannot take any arguments.
3. Any test must be formulated as a boolean condition.
4. An `AssertionError` exception is raised if the boolean condition is false (i.e., when the test fails).

There are many ways of raising the `AssertionError` exception:

```
# Formulate a test
tol = 1E-14                    # comparison tolerance for real numbers
success = abs(reference - result) < tol
msg = 'computed_result=%d != %d' % (result, reference)

# Explicit raise
if not success:
    raise AssertionError(msg)

# assert statement
assert success, msg

# nose tools
import nose.tools as nt
nt.assert_true(success, msg)
# or
nt.assert_almost_equal(result, reference, msg=msg, delta=tol)
```

This book contains a lot of test functions that follow the conventions of the pytest and nose testing frameworks, and we almost exclusively use the plain `assert` statement to have full control of what the test condition really is.

H.9.2   Writing the Test Function; Precomputed Data

Newton's method for solving an algebraic equation $f(x) = 0$ results in only an approximate root x_r, making $f(x_r) \neq 0$, but $|f(x_r)| \leq \epsilon$, where ϵ is supposed to be a prescribed number close to zero. The problem is that we do not know beforehand what x_r and $f(x_r)$ will be. However, if we strongly believe the function we want to test is correctly implemented, we can record the output from the function in a test case and use this output as a reference for later testing.

Assume we try to solve $\sin(x) = 0$ with $x = -\pi/3$ as start value. Running `Newton_basic` with a moderate-size eps (ϵ) of 10^{-2} gives $x = 0.000769691024206$, $f(x) = 0.000769690948209$, and $n = 3$. A test function can now compare new computations with these reference results. Since new computations on another

computer may lead to rounding errors, we must compare real numbers with a small
tolerance:

```
def test_Newton_basic_precomputed():
    from math import sin, cos, pi

    def f(x):
        return sin(x)

    def dfdx(x):
        return cos(x)

    x_ref = 0.000769691024206
    f_x_ref = 0.000769690948209
    n_ref = 3

    x, f_x, n = Newton_basic(f, dfdx, x=-pi/3, eps=1E-2)

    tol = 1E-15  # tolerance for comparing real numbers
    assert abs(x_ref - x) < tol       # is x correct?
    assert abs(f_x_ref - f_x) < tol   # is f_x correct?
    assert n == 3                     # is n correct?
```

The `assert` statements involving comparison of real numbers can alternatively be
carried out by `nose.tools` functionality:

```
nose.tools.assert_almost_equal(x_ref, x, delta=tol)
```

For simplicity we dropped the optional messages explaining what went wrong if
tests fail.

H.9.3   Writing the Test Function; Exact Numerical Solution

Approximate numerical methods are sometimes exact in certain special cases. An
exact answer known beforehand is a good starting point for a test since the implementation should reproduce the known answer to machine precision. For Newton's
method we know that it finds the exact root of $f(x) = 0$ in one iteration if $f(x)$ is
a linear function of x. This fact leads us to a test with $f(x) = ax + b$, where we
can choose a and b freely, but it is always wise to choose numbers different from 0
and 1 since these have special arithmetic properties that can hide the consequences
of programming errors.

   The test function contains the problem setup, a call to the function to be verified,
and `assert` tests on the output, this time also with an error message in case tests
fail:

```
def test_Newton_basic_linear():
    """Test that a linear func. is handled in one iteration."""
    f = lambda x: a*x + b
    dfdx = lambda x: a
    a = 0.25; b = -4
```

```
x_exact = 16
eps = 1E-5
x, f_x, n = Newton_basic(f, dfdx, -100, eps)

tol = 1E-15  # tolerance for comparing real numbers
assert abs(x - 16) < tol, 'wrong root x=%g != 16' % x
assert abs(f_x) < eps, '|f(root)|=%g > %g' % (f_x, eps)
assert n == 1, 'n=%d, but linear f should have n=1' % n
```

H.9.4   Testing of Function Robustness

Our `Newton_basic` function is very basic and suffers from several problems:

- for divergent iterations it will iterate forever,
- it can divide by zero in `f(x)/dfdx(x)`,
- it can perform integer division in `f(x)/dfdx(x)`,
- it does not test whether the arguments have acceptable types and values.

A more robust implementation dealing with these potential problems look as follows:

```
def Newton(f, dfdx, x, eps=1E-7, maxit=100):
    if not callable(f):
        raise TypeError(
            'f is %s, should be function or class with __call__'
            % type(f))
    if not callable(dfdx):
        raise TypeError(
            'dfdx is %s, should be function or class with __call__'
            % type(dfdx))
    if not isinstance(maxit, int):
        raise TypeError('maxit is %s, must be int' % type(maxit))
    if maxit <= 0:
        raise ValueError('maxit=%d <= 0, must be > 0' % maxit)

    n = 0  # iteration counter
    while abs(f(x)) > eps and n < maxit:
        try:
            x = x - f(x)/float(dfdx(x))
        except ZeroDivisionError:
            raise ZeroDivisionError(
                'dfdx(%g)=%g - cannot divide by zero' % (x, dfdx(x)))
        n += 1
    return x, f(x), n
```

The numerical functionality can be tested as described in the previous example, but we should include additional tests for testing the additional functionality. One can have different tests in different test functions, or collect several tests in one test function. The preferred strategy depends on the problem. Here it may be natural to have different test functions only when the $f(x)$ formula differs to avoid repeating code.

To test for divergence, we can choose $f(x) = \tanh(x)$, which is known to lead to divergent iterations if not x is sufficiently close to the root $x = 0$. A start value $x = 20$ reveals that the iterations are divergent, so we set `maxit=12` and test that the actual number of iterations reaches this limit. We can also add a test on x, e.g., that x is a big as we know it will be: $x > 10^{50}$ after 12 iterations. The test function becomes

```
def test_Newton_divergence():
    from math import tanh
    f = tanh
    dfdx = lambda x: 10./(1 + x**2)

    x, f_x, n = Newton(f, dfdx, 20, eps=1E-4, maxit=12)
    assert n == 12
    assert x > 1E+50
```

To test for division by zero, we can find an $f(x)$ and an x such that $f'(x) = 0$. One simple example is $x = 0$, $f(x) = \cos(x)$, and $f'(x) = -\sin(x)$. If $x = 0$ is the start value, we know that a division by zero will take place in the first iteration, and this will lead to a `ZeroDivisionError` exception. We can explicitly handle this exception and introduce a boolean variable `success` that is `True` if the exception is raised and otherwise `False`. The corresponding test function reads

```
def test_Newton_div_by_zero1():
    from math import sin, cos
    f = cos
    dfdx = lambda x: -sin(x)
    success = False
    try:
        x, f_x, n = Newton(f, dfdx, 0, eps=1E-4, maxit=1)
    except ZeroDivisionError:
        success = True
    assert success
```

There is a special `nose.tools.assert_raises` helper function that can be used to test if a function raises a certain exception. The arguments to `assert_raises` are the exception type, the name of the function to be called, and all positional and keyword arguments in the function call:

```
import nose.tools as nt

def test_Newton_div_by_zero2():
    from math import sin, cos
    f = cos
    dfdx = lambda x: -sin(x)
    nt.assert_raises(
        ZeroDivisionError, Newton, f, dfdx, 0, eps=1E-4, maxit=1)
```

Let us proceed with testing that wrong input is caught by function `Newton`. Since the same type of exception is raised for different type of errors we shall now also examine (parts of) the exception messages. The first test involves an argument `f` that is not a function:

```
def test_Newton_f_is_not_callable():
    success = False
    try:
        Newton(4.2, 'string', 1.2, eps=1E-7, maxit=100)
    except TypeError as e:
        if "f is <type 'float'>" in e.message:
            success = True
```

As seen, `success = True` demands that the right exception is raised and that its message starts with `f is <type 'float'>`. What text to expect in the message is evident from the source in function `Newton`.

The `nose.tools` module also has a function for testing the exception type and the message content. This is illustrated when `dfdx` is not callable:

```
def test_Newton_dfdx_is_not_callable():
    nt.assert_raises_regexp(
        TypeError, "dfdx is <type 'str'>", Newton,
        lambda x: x**2, 'string', 1.2, eps=1E-7, maxit=100)
```

Checking that `Newton` catches `maxit` of wrong type or with a negative value can be carried out by these test functions:

```
def test_Newton_maxit_is_not_int():
    nt.assert_raises_regexp(
        TypeError, "maxit is <type 'float'>",
        Newton, lambda x: x**2, lambda x: 2*x,
        1.2, eps=1E-7, maxit=1.2)

def test_Newton_maxit_is_neg():
    nt.assert_raises_regexp(
        ValueError, "maxit=-2 <= 0",
        Newton, lambda x: x**2, lambda x: 2*x,
        1.2, eps=1E-7, maxit=-2)
```

The corresponding support for testing exceptions in pytest is

```
import pytest
with pytest.raises(TypeError) as e:
    Newton(lambda x: x**2, lambda x: 2*x, 1.2, eps=1E-7, maxit=-2)
```

H.9.5   Automatic Execution of Tests

Our code for the `Newton_basic` and `Newton` functions is placed in a file `eq_solver.py` together with the tests. To run all test functions with names of the form `test_*()` in this file, use the `nosetests` or `py.test` commands, e.g.:

```
                                    ┌─────────┐
─────────────────────────────────  │Terminal │  ───────────────────────────
                                    └─────────┘
Terminal> nosetests -s eq_solver.py
..........
────────────────────────────────────────────────────────────────────────
Ran 10 tests in 0.004s

OK
```

The -s option causes all output from the called functions in the program eq_
solver.py to appear on the screen (by default, nosetests and py.test suppress
all output). The final OK points to the fact that no test failed. Adding the option
-v prints out the outcome of each individual test function. In case of failure, the
AssertionError exception and the associated message, if existing, are displayed.
Pytest also displays the code that failed.

Warning

Do not use more than one period in the names of files with test functions
(e.g., avoid names like ex7.23.py). Also, do not use hyphens in the name of
(sub)directories. Both constructions confuse nosetests and py.test.

One can also collect test functions in separate files with names starting with
test. A simple command nosetests -s -v will look for all such files in this
folder as well as in all subfolders if the folder names start with test or end with
_test or _tests. By following this naming convention, nosetests can auto-
matically run a potentially large number of tests and give us quick feedback. The
py.test -s -v command will look for and run all test files in the entire tree of
any subfolder.

Remark on classical class-based unit testing

The pytest and nose testing frameworks allow ordinary functions, as explained
above, to perform the testing. The most widespread way of implementing unit
tests, however, is to use class-based frameworks. This is also possible with nose
and with a module unittest that comes with standard Python. The class-based
approach is very accessible for people with experience from JUnit in Java and
similar tools in other languages. Without such a background, plain functions
that follow the pytest/nose conventions are faster and cleaner to write than the
class-based counterparts.

References

1. D. Beazley. *Python Essential Reference*. Addison-Wesley, 4th edition, 2009.

2. O. Certik et al. SymPy: Python library for symbolic mathematics. http://sympy.org/.

3. Python Software Foundation. The Python standard library. http://docs.python.org/2/library/.

4. C. Führer, J. E. Solem, and O. Verdier. *Computing with Python - An Introduction to Python for Science and Engineering*. Pearson, 2014.

5. J. E. Grayson. *Python and Tkinter Programming*. Manning, 2000.

6. Richard Gruet. Python quick reference. http://rgruet.free.fr/.

7. D. Harms and K. McDonald. *The Quick Python Book*. Manning, 1999.

8. J. D. Hunter. Matplotlib: a 2d graphics environment. *Computing in Science & Engineering*, 9, 2007.

9. J. D. Hunter et al. Matplotlib: Software package for 2d graphics. http://matplotlib.org/.

10. E. Jones, T. E. Oliphant, P. Peterson, et al. SciPy scientific computing library for Python. http://scipy.org.

11. D. E. Knuth. Theory and practice. *EATCS Bull.*, 27:14–21, 1985.

12. H. P. Langtangen. Quick intro to version control systems and project hosting sites. http://hplgit.github.io/teamods/bitgit/html/.

13. H. P. Langtangen. *Python Scripting for Computational Science*, volume 3 of *Texts in Computational Science and Engineering*. Springer, 3rd edition, 2009.

14. H. P. Langtangen and J. H. Ring. SciTools: Software tools for scientific computing. http://code.google.com/p/scitools.

15. L. S. Lerner. *Physics for Scientists and Engineers*. Jones and Barlett, 1996.

16. M. Lutz. *Programming Python*. O'Reilly, 4th edition, 2011.

17. M. Lutz. *Learning Python*. O'Reilly, 2013.

18. J. D. Murray. *Mathematical Biology I: an Introduction*. Springer, 3rd edition, 2007.

19. T. E. Oliphant. Python for scientific computing. *Computing in Science & Engineering*, 9, 2007.

20. T. E. Oliphant et al. NumPy array processing package for Python. http://www.numpy.org.

21. F. Perez and B. E. Granger. IPython: a system for interactive scientific computing. *Computing in Science & Engineering*, 9, 2007.

22. F. Perez, B. E. Granger, et al. IPython software package for interactive scientific computing. http://ipython.org/.

23. Python programming language. http://python.org.

24. G. J. E. Rawlins. *Slaves of the Machine: The Quickening of Computer Technology*. MIT Press, 1998.

25. G. Ward and A. Baxter. Distributing Python modules. http://docs.python.org/2/distutils/.

26. F. M. White. *Fluid Mechanics*. McGraw-Hill, 2nd edition, 1986.

Index

Editorial Policy

§1. Textbooks on topics in the field of computational science and engineering will be considered. They should be written for courses in CSE education. Both graduate and undergraduate textbooks will be published in TCSE. Multidisciplinary topics and multidisciplinary teams of authors are especially welcome.

§2. Format: Only works in English will be considered. For evaluation purposes, manuscripts may be submitted in print or electronic form, in the latter case, preferably as pdf- or zipped ps-files. Authors are requested to use the LaTeX style files available from Springer at: https://www.springer.com/gp/authors-editors/book-authors-editors/manuscript-preparation/5636 (Click on $\longrightarrow$ Templates $\longrightarrow$ LaTeX $\longrightarrow$ monographs)
Electronic material can be included if appropriate. Please contact the publisher.

§3. Those considering a book which might be suitable for the series are strongly advised to contact the publisher or the series editors at an early stage.

General Remarks

Careful preparation of manuscripts will help keep production time short and ensure a satisfactory appearance of the finished book.

The following terms and conditions hold:

Regarding free copies and royalties, the standard terms for Springer mathematics textbooks hold. Please write to martin.peters@springer.com for details.

Authors are entitled to purchase further copies of their book and other Springer books for their personal use, at a discount of 33.3% directly from Springer-Verlag.

Texts in Computational Science
and Engineering

For further information on these books please have a look at our mathematics catalogue at the following URL: www.springer.com/series/5151

Monographs in Computational Science
and Engineering

For further information on this book, please have a look at our mathematics catalogue at the following URL: www.springer.com/series/7417

Lecture Notes in Computational Science and Engineering

53. H.-J. Bungartz, M. Schäfer (eds.), *Fluid-Structure Interaction.*

54. J. Behrens, *Adaptive Atmospheric Modeling.*

55. O. Widlund, D. Keyes (eds.), *Domain Decomposition Methods in Science and Engineering XVI.*

56. S. Kassinos, C. Langer, G. Iaccarino, P. Moin (eds.), *Complex Effects in Large Eddy Simulations.*

57. M. Griebel, M.A Schweitzer (eds.), *Meshfree Methods for Partial Differential Equations III.*

58. A.N. Gorban, B. Kégl, D.C. Wunsch, A. Zinovyev (eds.), *Principal Manifolds for Data Visualization and Dimension Reduction.*

59. H. Ammari (ed.), *Modeling and Computations in Electromagnetics: A Volume Dedicated to Jean-Claude Nédélec.*

60. U. Langer, M. Discacciati, D. Keyes, O. Widlund, W. Zulehner (eds.), *Domain Decomposition Methods in Science and Engineering XVII.*

61. T. Mathew, *Domain Decomposition Methods for the Numerical Solution of Partial Differential Equations.*

62. F. Graziani (ed.), *Computational Methods in Transport: Verification and Validation.*

63. M. Bebendorf, *Hierarchical Matrices.* A Means to Efficiently Solve Elliptic Boundary Value Problems.

64. C.H. Bischof, H.M. Bücker, P. Hovland, U. Naumann, J. Utke (eds.), *Advances in Automatic Differentiation.*

65. M. Griebel, M.A. Schweitzer (eds.), *Meshfree Methods for Partial Differential Equations IV.*

66. B. Engquist, P. Lötstedt, O. Runborg (eds.), *Multiscale Modeling and Simulation in Science.*

67. I.H. Tuncer, Ü. Gülcat, D.R. Emerson, K. Matsuno (eds.), *Parallel Computational Fluid Dynamics 2007.*

68. S. Yip, T. Diaz de la Rubia (eds.), *Scientific Modeling and Simulations.*

69. A. Hegarty, N. Kopteva, E. O'Riordan, M. Stynes (eds.), *BAIL* 2008 – *Boundary and Interior Layers.*

70. M. Bercovier, M.J. Gander, R. Kornhuber, O. Widlund (eds.), *Domain Decomposition Methods in Science and Engineering XVIII.*

71. B. Koren, C. Vuik (eds.), *Advanced Computational Methods in Science and Engineering.*

72. M. Peters (ed.), *Computational Fluid Dynamics for Sport Simulation.*

73. H.-J. Bungartz, M. Mehl, M. Schäfer (eds.), *Fluid Structure Interaction II – Modelling, Simulation, Optimization.*

74. D. Tromeur-Dervout, G. Brenner, D.R. Emerson, J. Erhel (eds.), *Parallel Computational Fluid Dynamics 2008.*

75. A.N. Gorban, D. Roose (eds.), *Coping with Complexity: Model Reduction and Data Analysis.*

76. J.S. Hesthaven, E.M. Rønquist (eds.), *Spectral and High Order Methods for Partial Differential Equations.*

77. M. Holtz, *Sparse Grid Quadrature in High Dimensions with Applications in Finance and Insurance.*

78. Y. Huang, R. Kornhuber, O.Widlund, J. Xu (eds.), *Domain Decomposition Methods in Science and Engineering XIX.*

79. M. Griebel, M.A. Schweitzer (eds.), *Meshfree Methods for Partial Differential Equations V.*

80. P.H. Lauritzen, C. Jablonowski, M.A. Taylor, R.D. Nair (eds.), *Numerical Techniques for Global Atmospheric Models.*

81. C. Clavero, J.L. Gracia, F.J. Lisbona (eds.), *BAIL 2010 – Boundary and Interior Layers, Computational and Asymptotic Methods.*

82. B. Engquist, O. Runborg, Y.R. Tsai (eds.), *Numerical Analysis and Multiscale Computations.*

83. I.G. Graham, T.Y. Hou, O. Lakkis, R. Scheichl (eds.), *Numerical Analysis of Multiscale Problems.*

84. A. Logg, K.-A. Mardal, G. Wells (eds.), *Automated Solution of Differential Equations by the Finite Element Method.*

85. J. Blowey, M. Jensen (eds.), *Frontiers in Numerical Analysis – Durham 2010.*

86. O. Kolditz, U.-J. Gorke, H. Shao, W. Wang (eds.), *Thermo-Hydro-Mechanical-Chemical Processes in Fractured Porous Media – Benchmarks and Examples.*

87. S. Forth, P. Hovland, E. Phipps, J. Utke, A. Walther (eds.), *Recent Advances in Algorithmic Differentiation.*

88. J. Garcke, M. Griebel (eds.), *Sparse Grids and Applications.*

89. M. Griebel, M.A. Schweitzer (eds.), *Meshfree Methods for Partial Differential Equations VI.*

90. C. Pechstein, *Finite and Boundary Element Tearing and Interconnecting Solvers for Multiscale Problems.*

91. R. Bank, M. Holst, O. Widlund, J. Xu (eds.), *Domain Decomposition Methods in Science and Engineering XX.*

92. H. Bijl, D. Lucor, S. Mishra, C. Schwab (eds.), *Uncertainty Quantification in Computational Fluid Dynamics.*

93. M. Bader, H.-J. Bungartz, T. Weinzierl (eds.), *Advanced Computing.*

94. M. Ehrhardt, T. Koprucki (eds.), *Advanced Mathematical Models and Numerical Techniques for Multi-Band Effective Mass Approximations.*

95. M. Azaïez, H. El Fekih, J.S. Hesthaven (eds.), *Spectral and High Order Methods for Partial Differential Equations ICOSAHOM 2012.*

96. F. Graziani, M.P. Desjarlais, R. Redmer, S.B. Trickey (eds.), *Frontiers and Challenges in Warm Dense Matter.*

97. J. Garcke, D. Pflüger (eds.), *Sparse Grids and Applications – Munich 2012.*

98. J. Erhel, M. Gander, L. Halpern, G. Pichot, T. Sassi, O. Widlund (eds.), *Domain Decomposition Methods in Science and Engineering XXI.*

99. R. Abgrall, H. Beaugendre, P.M. Congedo, C. Dobrzynski, V. Perrier, M. Ricchiuto (eds.), *High Order Nonlinear Numerical Methods for Evolutionary PDEs – HONOM 2013.*

100. M. Griebel, M.A. Schweitzer (eds.), *Meshfree Methods for Partial Differential Equations VII.*

101. R. Hoppe (ed.), *Optimization with PDE Constraints – OPTPDE 2014.*

102. S. Dahlke, W. Dahmen, M. Griebel, W. Hackbusch, K. Ritter, R. Schneider, C. Schwab, H. Yserentant (eds.), *Extraction of Quantifiable Information from Complex Systems.*

103. A. Abdulle, S. Deparis, D. Kressner, F. Nobile, M. Picasso (eds.), *Numerical Mathematics and Advanced Applications – ENUMATH 2013.*

104. T. Dickopf, M.J. Gander, L. Halpern, R. Krause, L.F. Pavarino (eds.), *Domain Decomposition Methods in Science and Engineering XXII.*

105. M. Mehl, M. Bischoff, M. Schäfer (eds.), *Recent Trends in Computational Engineering – CE2014.* Optimization, Uncertainty, Parallel Algorithms, Coupled and Complex Problems.

106. R.M. Kirby, M. Berzins, J.S. Hesthaven (eds.), *Spectral and High Order Methods for Partial Differential Equations – ICOSAHOM'14.*

For further information on these books please have a look at our mathematics catalogue at the following URL: www.springer.com/series/3527